Advanced Concepts in Photovoltaics

RSC Energy and Environment Series

Series Editors:
Laurence Peter, *University of Bath, UK*
Heinz Frei, *Lawrence Berkeley National Laboratory, USA*
Roberto Rinaldi, *Max Planck Institute for Coal Research, Germany*
Tim S. Zhao, *The Hong Kong University of Science and Technology, Hong Kong*

Titles in the Series:

How to obtain future titles on publication:
A standing order plan is available for this series. A standing order will bring delivery of each new volume immediately on publication.

For further information please contact:
Book Sales Department, Royal Society of Chemistry, Thomas Graham House, Science Park, Milton Road, Cambridge, CB4 0WF, UK
Telephone: +44 (0)1223 420066, Fax: +44 (0)1223 420247
Email: booksales@rsc.org
Visit our website at www.rsc.org/books

Advanced Concepts in Photovoltaics

Edited by

Arthur J Nozik
National Renewable Energy Laboratory, Colorado, USA
Email: anozik@nrel.gov

Gavin Conibeer
University of New South Wales, Sydney, Australia
Email: g.conibeer@unsw.edu.au

Matthew C Beard
National Renewable Energy Laboratory, Colorado, USA
Email: matt.beard@nrel.gov

THE QUEEN'S AWARDS
FOR ENTERPRISE:
INTERNATIONAL TRADE
2013

RSC Energy and Environment Series No. 11

ISBN: 978-1-84973-591-9
PDF eISBN: 978-1-84973-995-5
ISSN: 2044-0774

A catalogue record for this book is available from the British Library

Published by The Royal Society of Chemistry,
Thomas Graham House, Science Park, Milton Road,
Cambridge CB4 0WF, UK

Registered Charity Number 207890

For further information see our web site at www.rsc.org

Printed and bound in Great Britain by CPI Group (UK) Ltd, Croydon, CR0 4YY

Preface

Over the past several years, developments in photovoltaic (PV) solar cells have enabled enormous progress to occur in reducing the cost of PV electrical energy (\$/kWh) such that it is rapidly approaching the cost of electrical energy produced by power plants fuelled by coal, natural gas, and nuclear reactors; this cost goal is generally termed 'achieving grid parity'. In 1977, when U.S. President Carter's administration created the Solar Energy Research Institute in Golden, Colorado, the price of PV electricity was about \$4/kWh, and in 2014 it is about \$0.15/kWh; electricity from the grid is on average about \$0.10/kWh, but varies depending on country region. This dramatic reduction in PV electricity cost has followed an 80% 'learning curve' (which means the module cost drops by 20% for every doubling of cumulative production capacity). Currently, total global PV module capacity is well above 100 GW (expressed as delivered peak power at the surface of the earth when the sun is at high noon with no clouds and has a standard AM1.5 solar spectrum (*i.e.*, the sun is filtered through 1.5 atmospheres). The yearly solar capacity factor (which takes into account the yearly average of cloud cover, the diurnal cycle, and the changing azimuthal angle of the sun) is about 20–30%, depending upon geographical location on the earth's surface. Thus, 100 peak GWs corresponds to about 25 GWs averaged over a year; this is equivalent to the average power capacity of 25 large (1 GW) coal-fired or nuclear power plants. Thus, PV presently represents about 0.5% of yearly total global electrical energy production. Global efforts are under way to increase this fraction by increasing PV power conversion efficiency and further lowering PV system costs.

This book presents a series of chapters that describe various approaches and research directions that have the potential to significantly reduce the energy cost of PV systems, and thus increase the penetration of PV electricity into the global electric power producing market. It covers: (1) PV systems

RSC Energy and Environment Series No. 11
Advanced Concepts in Photovoltaics
Edited by Arthur J Nozik, Gavin Conibeer and Matthew C Beard
© The Royal Society of Chemistry 2014
Published by the Royal Society of Chemistry, www.rsc.org

based on crystalline silicon that dominate the present PV market (termed 1st generation PV); (2) thin film compound semiconductors (for example, thin polycrystalline CdTe films), dye-sensitized nanocrystalline solar cells and organic-based PV (termed 2nd generation PV); and (3) various novel approaches that are in the early stages of research and development (termed future generation PV). These chapters are briefly described below.

In Chapter 1, Stefan Glunz presents an overview of the exciting technological developments in research and production of crystalline Si cells, which since the early days of PV in the 1950s have been the workhorse of the PV industry. Great progress has been achieved in decreasing production costs from hundreds of US dollars per W_p down to values well below $1/$W_p$. A loss analysis is used to categorize the different cell structures and technologies such as selective emitters, dielectric rear passivation, p-type *versus* n-type silicon, back-contact structures, heterojunctions and passivated contacts.

In Chapter 2, Christophe Ballif, Mathieu Boccard, Karin Söderström, Grégory Bugnon, Fanny Meillaud, and Nicolas Wyrsch show how thin-film silicon (TFSi) technology combines the advantages of being based on silicon (abundant, cheap and non-toxic raw material) with the inherent advantages of thin-film technologies (sparse use of raw material, aesthetics, upscalability, flexibility). TFSi opens the road to multiple-junction devices that can be processed in unique equipment, at a low cost, and with high theoretical efficiencies, and can be manufactured on various materials—including flexible foils—with a diversity of transparency, colors and shapes. Multi-junction devices that use hydrogenated amorphous (a-Si:H) as a top cell with a μc-Si:H as the bottom cell are discussed.

Timothy Gessert, Brian McCandless, and Chris Ferekides discuss in Chapter 3 the historical development and present understanding of polycrystalline CdTe-based photovoltaic devices. Recently, laboratory devices with performance approaching about 20% conversion efficiency have been reported with module efficiencies of ∼16%. By presenting the historical development of CdTe technology the authors provide a unique perspective on the current device design and what factors still limit its further development. The authors discuss current limitations and likely advancements in understanding that are needed in order to increase laboratory-device and commercial-module efficiency.

In Chapter 4 Simon P. Philipps and Andreas W. Bett discuss PV cells based on III–V compound semiconductors that consist of elements from the main groups III and V of the periodic table. Through proper mixing of the elements, materials with a wide range of bandgaps are available and can be realized in excellent crystal quality. The highest efficiencies of any photovoltaic technology, so far, have been reached with solar cells consisting of such III–V compound semiconductors. In particular, this is enabled by stacking solar cells of several III–V compound semiconductors, which absorb different parts of the solar spectrum. These III–V multi-junction solar cells have become standard in space and in terrestrial concentrator systems.

Diego Colombara, Phillip Dale, Laurence Peter, Jonathan Scragg and Susanne Siebentritt address in Chapter 5 the issue of the abundance of the elements that constitute the materials in PV cells that could limit their large scale utilization (TW scale). They present a detailed analysis and discussion of various new PV compositions synthesized with earth-abundant elements. They discuss generic fundamental issues and specific properties of these new PV materials and their chemical, thermodynamic, phase equilibria, and optoelectronic properties related to applications for PV cells.

In Chapter 6, Peng Gao and Mohammad Khaja Nazeeruddin introduce the operating principles of dye-sensitized solar cells, and the molecular engineering aspect of sensitizers and redox mediators. The design strategies of ruthenium sensitizers consisting of polypyridyl ligands with and without thiocyanate ligands are demonstrated. Organic sensitizers based on donor–π-spacer–acceptor (D-π-A) architecture, in which electron-rich (donor) and electron-poor (acceptor) are connected through a conjugated (π) bridge and the anchoring group is attached with the acceptor part, donor–chromophore–acceptor family diketopyrrolopyrrole (DPP) and ullazine sensitizers and their photovoltaic properties are discussed. Molecular engineering aspect of a porphyrin core with the bulky donor and strong acceptor groups to obtain panchromatic response is shown. The last section highlights organic–inorganic hybrid perovskites for thin-film photovoltaics, which recently came into the limelight because of their high efficiency, low cost and the ease to make these materials solution processing yielding over 15% efficiency.

A review and update on the new recent remarkable rapid advances in perovskite-based materials for high efficiency and potentially low cost solution-processed thin-film PV cells is presented in Chapter 7 by Nam-Gyu Park. Rapid progress has been made recently. As a result, power conversion efficiencies as high as 15% have been achieved in 2013, and PCEs of 20% are predicted in 2014. Because organolead halide perovskites have high absorption coefficients, long charge diffusion lengths, and balanced electron and hole transporting behavior, they currently show the best PV performance for new PV materials.

Another interesting new class of materials is discussed in Chapter 8 by Sven Rühle and Arie Zaban. These materials are oxides used in 'all-oxide' PV cells (except for the back metal contact). They show great chemical stability, are non-toxic, abundant, have low manufacturing cost at ambient conditions, and could show sufficiently high conversion efficiency. Furthermore, they are inexpensive, stable and environmentally safe. The focus is on Cu_2O based devices, but work on $BiFeO_3$ and Fe_2O_3 are also reviewed, as well as strategies for searching for novel multi-component metal oxide composites.

Tracey M. Clarke, Guanran Zhang, and Attila J Mozer discuss a possible route to 15% PCE from donor–acceptor bulk heterojunction solar cells in Chapter 9. They show through an extensive literature review of more than 30 high performing co-polymer donor/acceptor heterojunctions that achieving a PCE through a high fill factor using relatively thick (>300 nm) active layers is very challenging. Twelve different models aimed at explaining reduced bimolecular recombination in bulk heterojunction solar cells are discussed and compared.

In Chapter 10 Justin C. Johnson and Josef Michl present a description of a relatively new approach to high efficiency in molecular PV cells based on a phenomenon known as singlet fission (SF), in which a molecular chromophore with an excited singlet state shares energy from a single absorbed photon with a close neighboring ground state chromophore to produce two triplet excited chromophores, *i.e.*, one triplet on each of two coupled chromophores. In principle, the process permits the absorption of a single photon to produce two electrons and two holes, leading to a theoretical solar cell efficiency of about 43%, well above the Shockley–Queisser limit of 33%. The process is the molecular analog of multiple exciton generation (MEG) (discussed in Chapter 11), whereby two excitons are formed from a single absorbed photons in semiconductor quantum dots. The prerequisites for efficient singlet fission are considered, both in terms of the properties of individual chromophores and in terms of their mutual electronic coupling. The design rules for efficient chromophores derived from first principles led to the formulation of a model system, 1,3-diphenylisobenzofuran, and this molecule is used to illustrate the singlet fission process and the complications that can arise.

In Chapter 11, Matthew C. Beard, Alexander H. Ip, Joseph Luther, Edward H. Sargent, and Arthur J. Nozik explain the principles and status of the process that creates multiple excitons from single absorbed photons that have energies at least twice the semiconductor bandgap. This process is very efficient in semiconductor nanocrystals (*viz.*, quantum dots and rods) and is termed multiple exciton generation (MEG) in these quantized structures since the photogenerated electron–hole pairs are electronically coupled and exist as excitons; the process is also termed carrier multiplication (CM) when the excitons are dissociated in optoelectronic devices, like PV cells, and free carriers are separated and collected. The progress toward using MEG/CM in quantum-dot PV cells is reviewed.

Hot carriers are electrons and holes that are created when photons larger than the semiconductor bandgap are absorbed. The energy of the electrons and holes created in excess of the bandgap energy exists as kinetic energy, and in all present day commercial PV cells, the excess kinetic energy of hot carriers is converted into heat through rapid electron–phonon scattering (in sub-picosecond to picosecond time scales), and is thus lost for producing electrical free energy. If the energy of hot carriers can be utilized for electric power generation before they cool to the lattice temperature and create heat, the PCE can reach very high levels equivalent to multi-junction PV solar cells; the thermodynamic limit of such a hot carrier solar cell is 66% with one semiconductor photomaterial. In Chapter 12, Gavin Conibeer, Jean-François Guillemoles, Feng Yu and Hugo Levard present the principles of hot carrier solar cells and review progress in their development.

Another strategy to increase the PCE above the Shockley–Queisser limit of 33% is to utilize incident solar photons that are normally not absorbed by the semiconductor photomaterial because the photons have energies less than the semiconductor bandgap. In Chapter 13, Yoshitaka Okada, Tomah Sogabe and Yasushi Shoji describe intermediate band solar cells (IBSCs) that have energy bands created within the bandgap which can absorb the sub-bandgap photons in the solar spectrum and then through a 2nd photo-excitation of the intermediate band, allow the photogenerated sub-bandgap electrons to be excited further into the conduction band. Recent developments and future research opportunities in high-efficiency intermediate band photovoltaics technology based on high-density quantum dot arrays are reviewed.

In Chapter 14, Wilfried van Sark, Jessica de Wild, Zachar Krumer, Celso de Mello Donegá and Ruud Schropp present two examples of photon management for high efficiency PV cells through spectral manipulation that involves up-conversion and down-conversion of solar photons. Up-conversion is applied to thin film silicon solar cells, and efficiency improvements using lanthanides as the up-converter material are presented. Down-conversion is demonstrated in luminescent solar concentrators, and material issues hampering efficiency improvements are discussed. A new class of semiconductor hetero-nanocrystals is shown to be an excellent candidate for surpassing the 10% luminescent solar concentrator efficiency barrier.

Timothy W. Schmidt and Murad J. Y. Tayebjee discuss triplet–triplet annihilation up-conversion to enhance PCE in PV cells in Chapter 15. They introduce the photon–ratchet model of a quantum PV energy convertor and determine the Shockley–Queisser limit; they then extend the model to include generalized photon up-convertors. The photochemistry underpinning the triplet–triplet annihilation up-conversion (TTA-UC) scheme is described and efficiency considerations are discussed. Applications of TTA-UC to thin-film PV devices are described.

In Chapter 16, Feng Yu, Garret Moddel and Richard Corkish discuss quantum rectennas for PV. This PV application is a special case in the rapidly growing field of optical antennas. An optical antenna is 'a device that converts freely propagating optical radiation into localized energy, and *vice versa*'. Quantum antennas for PV are specifically required to couple optical solar radiation to a load, commonly *via* a rectifier, and the combined antenna and rectifier is termed a 'rectenna'. This device converts electro-magnetic energy propagating through space to direct current electricity in a circuit. It has one or more elements, each consisting of an antenna, filter circuits and a rectifying diode or bridge rectifier, either for each antenna element or for the power from several elements combined. There are significant overlaps and common interests with radio astronomy. A solar rectenna is similar to a simple radio telescope or radiometer but differs in that the radio telescope needs to measure the intensity of the radiative power received by the antenna, while the rectenna needs to convert that power to useful DC electricity. In terms of a solar cell analogy, the radiotelescope observes the open circuit voltage while the rectenna extracts power at the maximum power point. Solar rectennas are under active consideration as alternatives to conventional PV solar cells and are reviewed here.

In parallel with progress in the development and improvement of established, commercial PV cells (1st and 2nd generation PV), newer concepts and approaches (next generation PV) are described in Chapters 5–16. Given the impressive improvements in recent years of the power conversion efficiencies of nearly all types of PV cells, Pabitra K. Nayak and David Cahen ask in Chapter 17 if there are other efficiency limits for the potentially low-cost cells, in addition to the Shockley–Queisser (S-Q) limit. This issue is important as it can set practical goals, allow for realistic prognoses for the utility of new PV approaches, and stimulate efforts to circumvent such limits. No material property other than the optical band gap of the absorber is taken into account to calculate the S-Q limit; in real devices other material properties play a critical role in the possible PCE. For a given solar cell design the question then is, how good can we expect it to be? The S-Q efficiency analysis is for an ideal model system. For the real world limit, one has to take into account many factors that are simplified or ignored in the S-Q treatment. This chapter discusses and summarizes these other real-world issues and limits.

The transformational possibilities of PV were recognized immediately after their invention in the 1950s, but it has taken more than 50 years to begin to realize their potential on a large commercial scale. In addition to the great advances in PV technology, energy policy in various countries has played a critical role in the rapid advance of manufacturing and deployment of PV systems. The key challenge now facing policy makers is to appropriately transition from the PV specific policies that have played such a valuable role in driving deployment up and costs down, to addressing the broader energy market policies and regulatory arrangements that will play a key role in PV's future success. Particular challenges include managing the threat that PV poses to some incumbent market participants and current business models, while facilitating its deployment in ways which maximize the value that it brings to the electricity industry. Chapter 18 by Muriel Watt and Iain MacGill follows the growth of PV markets, the policy support strategies employed to stimulate deployment to date, and the new impacts and implications of low-cost PV for electricity systems the world over.

Arthur J. Nozik, Gavin Conibeer and Mathew C. Beard
Boulder, Colorado, USA

Contents

RSC Energy and Environment Series No. 11
Advanced Concepts in Photovoltaics
Edited by Arthur J Nozik, Gavin Conibeer and Matthew C Beard
© The Royal Society of Chemistry 2014
Published by the Royal Society of Chemistry, www.rsc.org

CHAPTER 1

Crystalline Silicon Solar Cells with High Efficiency

STEFAN W. GLUNZ

Fraunhofer Institute for Solar Energy Systems ISE, Heidenhofstr. 2, 79110 Freiburg, Germany
Email: stefan.glunz@ise.fraunhofer.de

1.1 Introduction

Crystalline silicon photovoltaics is the dominant solar cell technology, with a market share of around 85% in 2012. Silicon has several advantages: It is non-toxic and abundantly available in the earth's crust. Crystalline silicon-based photovoltaic (PV) modules have proven their long-term stability over decades in the field and not only in accelerated module tests. The price reduction of silicon PV modules in the last 30 years can be described very well by a learning factor of 20%.[1] Due to strong competition this price decline was even stronger in the last years, resulting in module prices well below $1/W_p$. This is an excellent situation for customers and PV installers, but rather challenging for producers of silicon solar cells and modules. Thus cost reduction is still a major task.

The cost distribution of a crystalline silicon PV module is clearly dominated by material costs, especially by the cost of the silicon wafer and encapsulation materials (see Figure 1.1). Therefore, besides improved production technologies, the efficiency of the cells and modules is the main leverage to bring down the cost even more, especially when considering the full levelized cost of PV electricity.

RSC Energy and Environment Series No. 11
Advanced Concepts in Photovoltaics
Edited by Arthur J Nozik, Gavin Conibeer and Matthew C Beard
© The Royal Society of Chemistry 2014
Published by the Royal Society of Chemistry, www.rsc.org

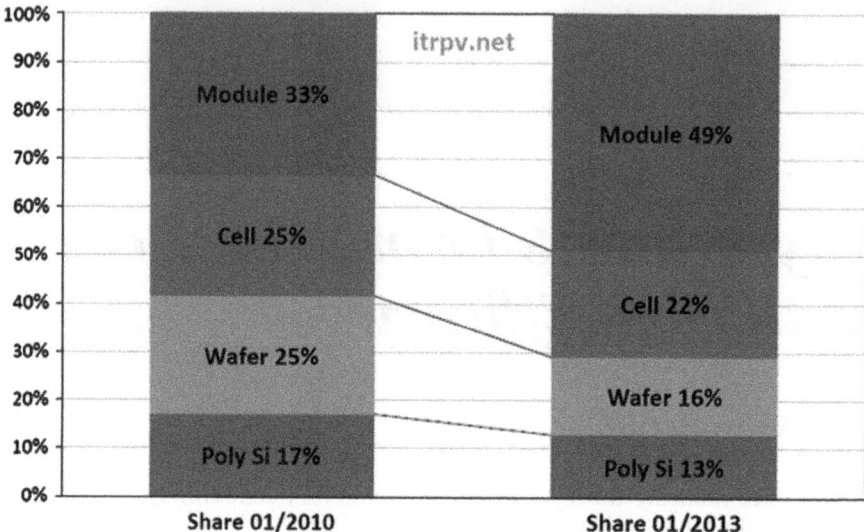

Figure 1.1 Comparison of the proportion of cost attributable to different module elements between 01/2010 and 01/2013 (1.87$ and 0.69$/W). Taken from ITRPV.[1] Permission granted by the SEMI PV Group.

The International Technology Roadmap for Photovoltaic recommends in their latest report:[1]

(i) Continue the cost reduction per piece, along the whole value chain, but especially at the module level, by the efficient use of Si and non-Si materials, and (ii) Improve the module power/cell efficiency without significant increasing processing costs.

This chapter will mainly focus on the second route by using a detailed loss analysis to investigate the current developments and cell architectures in industry and research.

1.2 Efficiency Limitations

1.2.1 Theoretical Limitations: The Auger Limit

Based on a detailed balance calculation, Shockley and Queisser determined already as early as 1961[2] a maximum theoretical efficiency for solar cells. Using the AM1.5 spectrum and assuming no light concentration, the limit for a single junction solar cell is 33%.[3] The main limitation of such cells is the thermalization of hot carriers generated by photons with energies greater than the bandgap energy and the non-absorption of photons with energy smaller than the energy gap (see Figure 1.2).

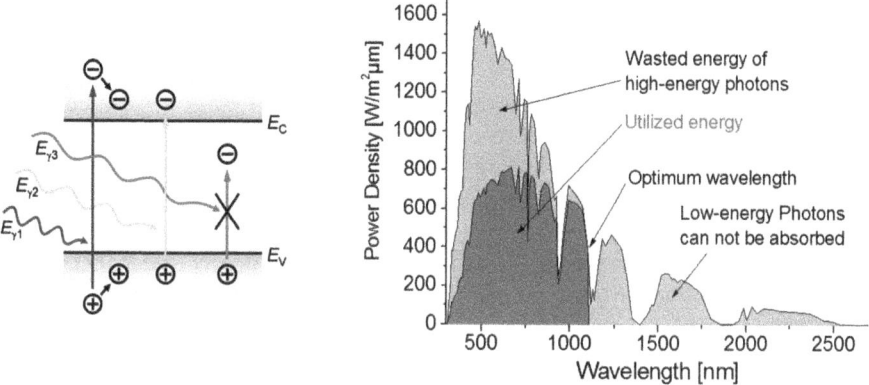

Figure 1.2 Basic losses in a solar cell with one bandgap energy.

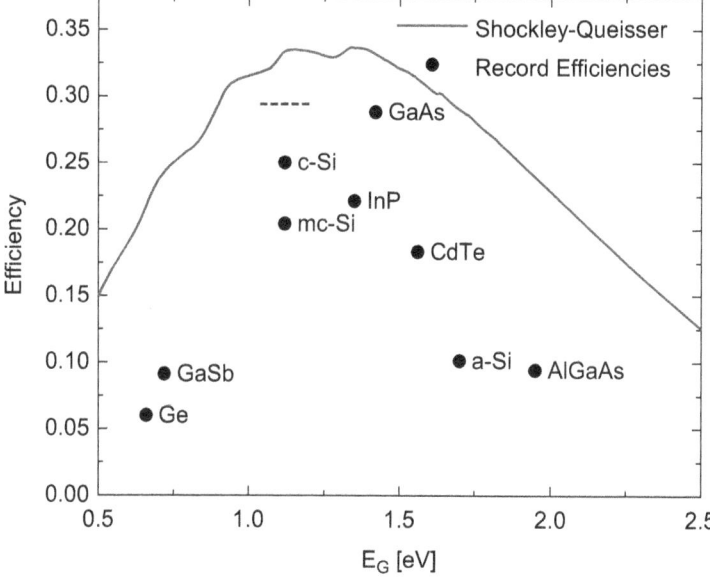

Figure 1.3 Shockley–Queisser limit (solid line) and record efficiencies as a function of bandgap energy, E_G. Dashed horizontal line = Auger limit for crystalline silicon.

Thus, it is important to choose the right bandgap energy in order to balance and minimize these losses. Fortunately, semiconductors like silicon and GaAs are very close to the optimum bandgap energy (see Figure 1.3). However, comparing the best experimental values with the Shockley–Queisser limit (SQL), it is obvious that GaAs gets much closer to this limit.

This is actually not a consequence of a more advanced technological development, but due to the fact that the SQL is not the relevant limitation for solar cells fabricated from the indirect semiconductor crystalline silicon. In their groundbreaking article Shockley and Queisser write:

> It is radiative recombination that determines the detailed balance limit for efficiency. If radiative recombination is only a fraction f_c of all recombination, then the efficiency is substantially reduced below the detailed balance limit.[2]

Other recombination channels can be caused by defect recombination, which could be controlled perfectly in theory. But there also other unavoidable intrinsic recombination channels such as non-radiative Auger recombination, which are given by the physical properties of the semiconductor. As silicon is an indirect semiconductor, radiative recombination also involves a phonon, thereby making the process quite unlikely (and silicon LEDs quite inefficient). Figure 1.4 shows the charge carrier lifetime limitation due to radiative and Auger recombination as a function of photogenerated excess carrier density. It is obvious that Auger recombination is the dominating intrinsic recombination channel for crystalline silicon in the ideal case of defect-free material.

Therefore it is very important to determine and parameterize the Auger recombination as a function of doping concentration and excess carrier

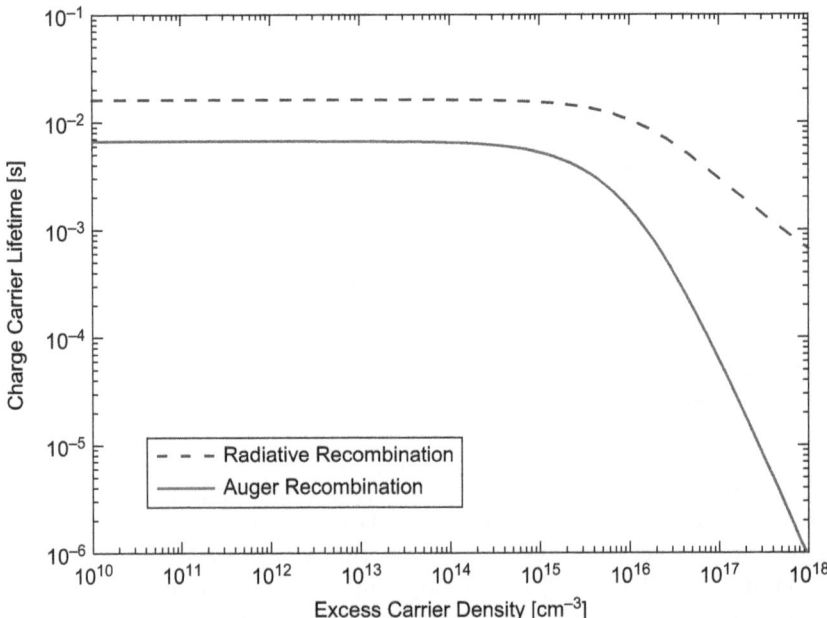

Figure 1.4   Charge carrier lifetime as a function of excess carrier density limited by intrinsic recombination (*i.e.*, radiative or Auger recombination) for an acceptor concentration of 1.5×10^{16} cm^{-3}.

density.[4,5] The most recent parameterization of the intrinsic recombination (including Auger recombination) was published by Richter *et al.*[5]

$$\tau_{\text{intr,adv}} =$$

$$\frac{\Delta n}{\left(np - n_{i,\text{eff}}^2\right)\left[(2.5 \times 10^{-31} g_{\text{eeh}} n_0) + (8.5 \times 10^{-32} g_{\text{ehh}} p_0) + (3 \times 10^{-29} \Delta n^{0.92}) + B_{\text{rel}} B_{\text{low}}\right]}$$

(1.1)

with n and p being the electron and hole density, respectively, n_0 and p_0 the electron and hole thermal equilibrium density, also respectively, Δn the excess carrier density, $n_{i,\text{eff}}$ the effective intrinsic carrier concentration, B_{low} the radiative recombination coefficient for lowly-doped and lowly-injected silicon (4.73×10^{-15} cm^3 s^{-1} at 300 K; see Trupke *et al.*[6]), B_{rel} the relative radiative recombination coefficient and the enhancement factors:[7]

$$g_{\text{eeh}}(n_0) = 1 + 13 \left[1 - \tanh\left(\left(\frac{n_0}{N_{0,\text{eeh}}}\right)^{0.66}\right)\right]$$

(1.2)

and

$$g_{\text{ehh}}(p_0) = 1 + 7.5 \left[1 - \tanh\left(\left(\frac{p_0}{N_{0,\text{ehh}}}\right)^{0.63}\right)\right]$$

(1.3)

with $N_{0,\text{eeh}} = 3.3 \times 10^{17}$ cm^{-3} and $N_{0,\text{ehh}} = 7.0 \times 10^{17}$ cm^{-3}.

Auger recombination included, it is possible to calculate the upper limit for crystalline silicon solar cells.[8,9] With the most recent parameterization [see Richter *et al.*[5] and Equation (1.1)] the maximum value was determined to be 29.4%[8] (see Figure 1.5).

These maximum values are depicted as a dashed horizontal line in Figure 1.3. This makes it clear that the ratio of experimental record values and theoretical maximum values of GaAs and silicon are in both cases in the range of 85%.

1.2.2 Practical Limitations

Such idealized devices considered for the calculation of the efficiency limitations without additional recombination at contacts *etc.* are only of interest in theory and can not be realized. For a realistic, yet optimized silicon solar cell, an efficiency limit of 26% was predicted.[10]

However, most of the cells in the industry and in development are far from this practical efficiency limit. They suffer from a variety of optical, recombination and resistive losses. In order to get a breakdown of the losses of different solar cell architectures, it is advantageous to analyse the loss currents at maximum power point (mpp). There are also very good calculations investigating the power losses at mpp[11] or the free energy loss analysis,[12] but for didactical reasons we have preferred to analyse the loss currents.

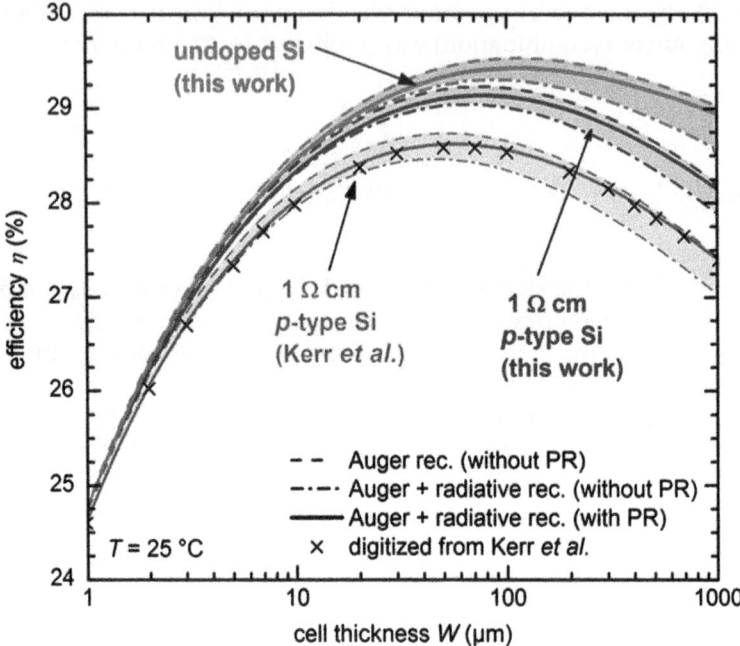

Figure 1.5 Theoretical efficiency limit for crystalline silicon including Auger recombination.
Taken from Richter *et al.*[8] Permission granted by IEEE.

The starting point is j_{max}, *i.e.*, the theoretical maximum photogeneration for a certain thickness. We have kept the cell thickness at 160 μm for all investigated cell structures to keep the different calculations comparable. For thickness of 160 μm and ideal Lambertian light-trapping[13,14] this theoretical maximum is calculated to be 43.6 mA cm^{-2}. The first loss is then the optical losses, $\Delta j_{optical}$, due to the primary surface reflection, parasitic absorption and escape light. $\Delta j_{optical}$ is the difference between j_{max} and j_{gen}, the latter referring to the photocurrent generated in the actual device. j_{gen} was calculated for the individual device structures using the ray-tracing program in Sentaurus Device.[15]

A part of j_{gen} is lost by charge carrier recombination. This recombination loss current Δj_{rec} is calculated with Senaturus Device at maximum power point (see Figure 1.6) In our simulation we have separated the recombination in different parts of the cells by spatial integration of the two- or three-dimensional simulation output:

- Δj_{front}, the recombination loss current in the front emitter, at the front surface and at the front contacts (for interdigitated back-junction cells: front surface and in the front surface field).
- $\Delta j_{Auger,base}$, the intrinsic Auger recombination in the undiffused base region of the cell.

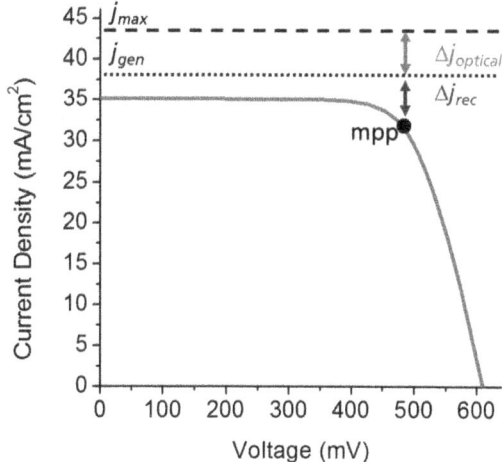

Figure 1.6 Current losses at maximum power point (mpp).

- $\Delta j_{\text{SRH,base}}$, the Shockley–Read–Hall recombination *via* defects and impurities.
- Δj_{rear}, the recombination loss current in the back surface field and at the rear surface (for interdigitated back junction cells: rear emitter, back surface field and rear surface).

These losses are shown in Figure 1.7 for different cell structures on monocrystalline silicon. The black parts of the columns show the obtained current at mpp, j_{mpp}, while the shaded sections denote the optical and re-combination losses. j_{mpp} and the loss currents add up to j_{max}. Additionally, the actual cell parameters V_{OC}, V_{mpp} and efficiency are given above the figure. For our calculation, realistic resistive losses have been assumed. Since we have decided to discuss current losses and not power losses, it is not possible to show such power losses in the figure; however, they have been included in the calculation of the efficiency.

An important characteristic of our study is the fact that we used a consistent set of parameters, such as the bulk lifetime or surface recombination velocities, for example, to allow for a transparent comparison of the different cell structures. Details on the chosen parameters and the simulation procedures are given in Rüdiger *et al.*[16] In the following we will discuss the cell structures shown in Figure 1.7 step by step.

1.3 Screen-printed Al-BSF Solar Cells on p-type Silicon

For decades the working horse of the PV industry has been the screen-printed aluminium back surface field (Al-BSF) cell on p-type silicon (Figure 1.8). Although some new process steps such as the fire-through

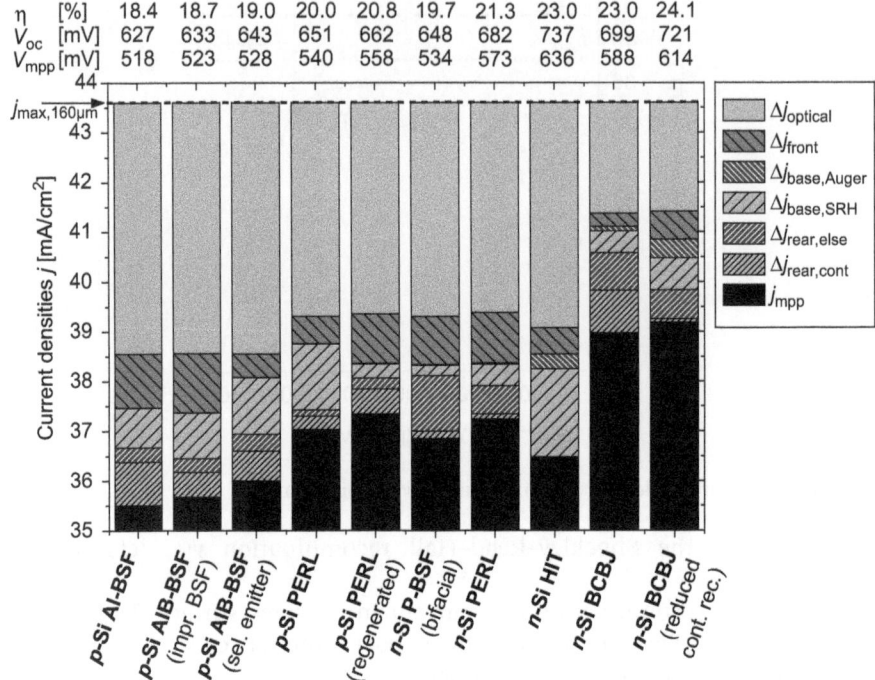

Figure 1.7 Distribution of current losses at mpp and cell parameters for different solar cell architectures on monocrystalline silicon.

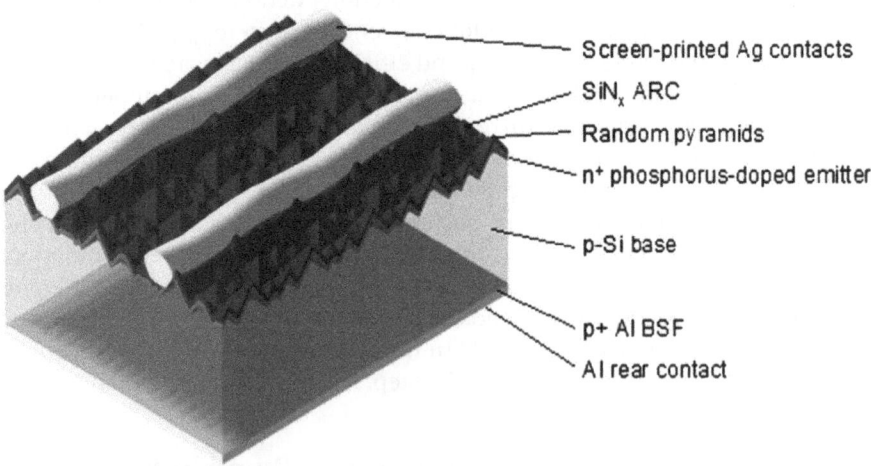

Figure 1.8 Structure of a screen-printed Al-BSF solar cell.

process[17] were introduced, most of the impressive increase in cell efficiency has been due to evolutionary improvements. Improved metal pastes and printing processes, increased emitter sheet resistance and better front surface passivation layers (SiN$_x$ by plasma-enhanced vapour deposition, PECVD[18]) are a few examples of the changes which have been responsible for the increase in average monocrystalline silicon efficiency limits over the years, with values at around 14% in the 1990s to more than 18% today.

1.3.1 Standard Al-BSF Cell

Our calculation results in an efficiency of 18.4% ($V_{OC} = 627$ mV) for standard screen-printed Al-BSF cells. As can be seen in Figure 1.7, the biggest losses are the optical losses due to front reflectance, transmission losses and the recombination at the rear side of the cell. The Al-BSF at the rear side was already described in the 1970s.[19] In today's standard process sequence, it is created by a rather elegant alloying process of screen-printed Al-paste with the base silicon.[20] This alloying process takes place during a very short firing process in an inline belt furnace. During the cool-down phase the silicon that has been dissolved into the molten aluminium, recrystallizes and aluminium is incorporated in the silicon lattice according to the solubility at the actual temperature. Due to this process, the doping profile of Al-BSF cells shows a characteristic shape (black dots in Figure 1.9).

The main limitation of such profiles is the maximum Al doping concentration, which is in the range of 7×10^{18} cm^{-3}, resulting from to the rather low solubility of Al in Si. Since the effectiveness of a BSF, *i.e.*, its capability to reduce the surface recombination velocity, S, or the dark saturation current, J_0, at the rear side, depends strongly on the doping step $N_{acc,BSF}/N_{acc,base}$, a higher doping concentration would be desirable.

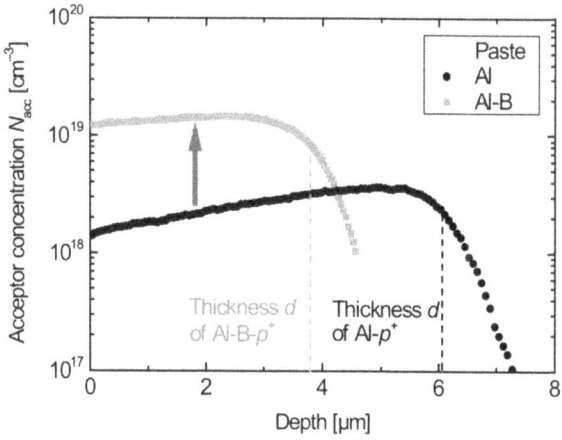

Figure 1.9 Doping profile of aluminium back surface fields. Taken from Rauer *et al.*[22] Permission granted by IEEE.

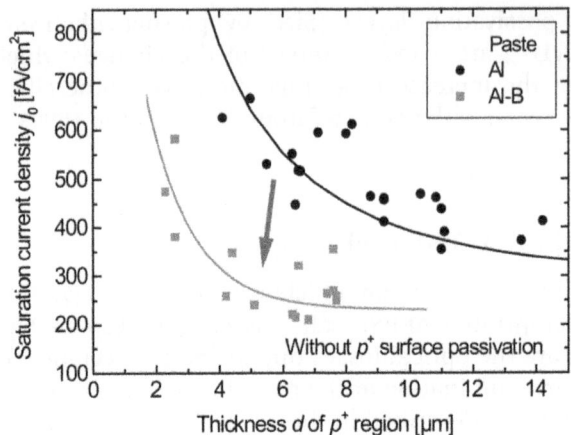

Figure 1.10   Dark saturation currents of Al-BSF with and without boron co-doping. Taken from Rauer *et al.*[22] Permission granted by IEEE.

1.3.2   Improved Al-BSF Formation by Boron Co-doping

The limitation of a maximum doping concentration in the Al-BSF can be overcome by co-doping with boron. This was suggested by Lölgen *et al.*[21] already in 1994 and recently investigated in depth by Rauer *et al.*[22] The pale grey squares in Figure 1.9 show the doping profile a BSF alloyed from an Al paste with B co-doping. Due to the higher solubility of boron, a significantly higher acceptor concentration is achieved. This makes the BSF more effective and the related dark saturation values j_0 are reduced significantly (see Figure 1.10).

This reduction of j_0 corresponds to the reduction of Δj_{rear} (see second column in Figure 1.7) and leads to an improved V_{OC}. The calculated efficiency of the cell is now 18.7%.

1.3.3   Improved Emitter

A second high-recombination region of a standard solar cell is the relatively highly phosphorous-doped emitter. Using a highly doped emitter has two advantages. Firstly, a low sheet resistance is achieved which allows one to increase the pitch between the front grid lines and thus reduces shadowing losses. Secondly, the high surface doping concentration of such emitters, $N_{\text{dop,surface}}$, makes it easy to create a metal contact with low ohmic losses, even when metals are used, which are not optimal for contacting n-type silicon, such as silver, for instance.

However, on the down side, such high doping concentrations N_{dop} will reduce the carrier lifetime due to Auger recombination ($\tau_{\text{Auger}} \sim 1/N_{\text{dop}}^2$), not to mention that the front surface recombination velocity S_{front} increases with $N_{\text{dop,surface}}$ as determined by Cuevas *et al.*[23] Consequently, the dark saturation current increases and the current generated by high-energy photons with short absorption depths (blue response) is reduced. This dilemma can be

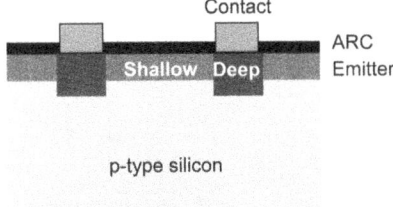

Figure 1.11　Structure of a selective emitter.

solved in two ways. The first strategy is to improve the understanding of the contact formation[24-26] and develop improved metallization pastes, which make it possible to contact lowly doped emitters. There has been a lot of progress in this field and pastes are now commercially available, which meet these demands. A second strategy is known as selective emitter or two-step emitter, whereby the region under the contact is highly doped to facilitate good contact properties and the region in-between the contacts is lowly doped ('shallow') to improve the blue response and reduce the dark saturation current (see Figure 1.11). The highly doped emitter under the contacts ('deep') emitter has a second advantage of 'shielding' the highly recombinative metal/silicon contact and additionally reduces recombination. A very good overview on different technological approaches to fabricate such structures is given by Hahn.[27] By using such a selective emitter structure, Δj_{front} can be reduced significantly with the conversion efficiency reaching a value of 19%.

1.4　Solar Cells with Dielectric Rear Passivation on p-type Silicon

Even with improved rear side Al pastes, as discussed in section 1.3.2, the Al-BSF structure is limited on account of the rear side recombination still being high and due to the fact that the internal reflection for deeply penetrating long-wavelength light reaching the rear side of the cell is only in the range of 65%.[28] The most effective way to overcome this problem is through the introduction of a dielectric rear side passivation with local contact points or lines (partial rear contact, PRC). The PRC structure is the basis for a variety of successful cell architectures like the passivated emitter and rear cell (PERC),[29] the passivated emitter, rear locally-diffused (PERL)[30] or local back surface field structure (LBSF)[31] and the passivated, rear totally diffused cells (PERT).[32] PERL cells have set the actual world record of 25% for crystalline silicon-based solar cells[32,33] (Figure 1.12).

With such a rear structure, low effective surface rear recombination velocities in the range of 60 to 200 cm s^{-1} and internal reflectance of 95% are achievable.[28,34,35] Since the recombination at the rear side is very low resulting in a lower Δj_{rear} and long-wavelength light is not lost due to transmission which reduces $\Delta j_{\text{optical}}$, efficiencies far above 20% are

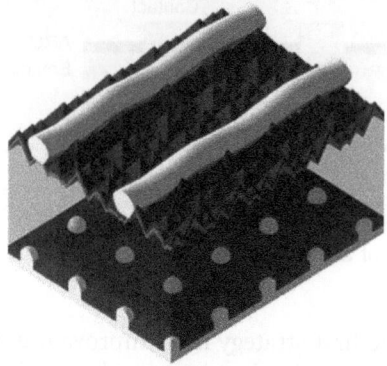

Figure 1.12 Structure of a PRC silicon solar cell with surface passivation (dark grey) at front and rear surface. Only small point-like metal points (pale grey) form the base contact at the rear.

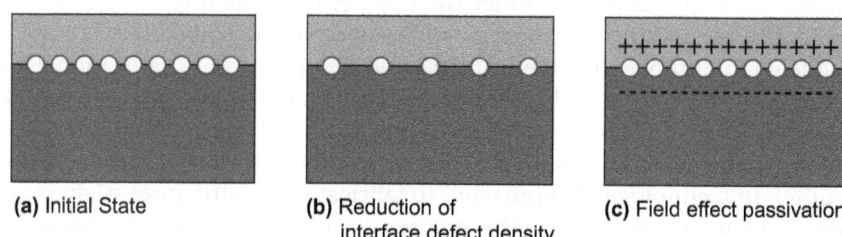

(a) Initial State   **(b)** Reduction of   **(c)** Field effect passivation
                                     interface defect density

Figure 1.13 Strategies for the reduction of surface recombination velocity (dark grey = silicon, pale grey = passivation layer, circles = defect states).

attainable. Although this cell structure achieved excellent efficiencies already as early as 1989,[29] it took 20 years until the transfer into industrial production was performed.[36–39] Efficiencies of 20.2% were recently shown using industrial production equipment.[40]

The PRC structure can be easily combined with the so-called metal-wrap-through (MWT) cell architecture.[41] When this is done, the shadowing loss due to the busbars can be strongly reduced. Efficiencies above 20% have been achieved with such MWT-PERC structures.[42,43]

1.4.1 Rear Passivation Layers

Obviously the surface passivation layer plays a crucial role for this cell structure. Traditionally, the preferred layer type was silicon dioxide (SiO_2), which was thermally grown. This is a technology well known by MOS technology, which was successfully introduced to photovoltaics in the 1980s.[44] The reduction of surface recombination velocity in this case is mainly due to a reduction of interface defect density, D_{it} (Figure 1.13).

Another route to reduce the surface recombination velocity, S, is to incorporate fixed charges in the dielectric layer. This leads to a band bending in the silicon bulk and a strong asymmetry of electron and hole density at the surface. For a high surface recombination, a similar amount of electrons and holes is required. Therefore, the strong inversion (high population of minority carriers) or accumulation (high population of majority carriers) leads to a strong reduction of surface recombination velocity.[45] Although both inversion and accumulation lead to an excellent passivation quality, it was shown that in a finished solar cell, accumulation is indeed the better option. The inversion channel can be shorted by the local point contacts and the effect of field effect passivation is diminished.[46] Therefore, for solar cells on p-type silicon, layers with a high negative charge density are preferred. While SiN_x shows a positive charge, aluminium oxide (AlO_x) layers exhibit a strong negative charge[47] and are now the preferred passivation layers for PRC cells.[48,49] They can be deposited by atomic layer deposition,[48,50,51] by plasma-enhanced chemical vapour deposition[52] or by sputtering.[53] In many cases this AlO_x layer gets a second SiN_x top layer to protect itself from damage by the aggressive printed Al rear metal paste.[39,49] Further effective passivation layers are oxide–nitride stacks, which have led to very good efficiencies in industrial production.[54] A rather new and very interesting passivation layer is sputtered aluminium nitride (AlN_x).[55,56] It also exhibits a very strong negative charge, and low surface recombination velocities ($S = 8$ cm s^{-1}) can thus be achieved.

1.4.2 Contacting Schemes

The local contact points, which cover approximately 1% of the rear surface, are fabricated in the laboratory using photolithography. On account of this process being too complex for industrial production (at least in the case of PV), alternatives had to be found. One possibility is to open the contact points in the passivation layer by laser[49,54,57] or by printing etching pastes.[58] Then an Al paste is printed on the full rear surface and an alloying step is applied. This creates local Al-BSFs which reduce the recombination at the contact points. However, it was shown that it is necessary to adapt the Al paste *e.g.*, by adding silicon powder.[59] Otherwise voids can be observed in the contact areas, which are due to localised alloying process and reduce the performance of the cells.[60]

A second possibility to form local contacts is to leave the passivation layer untouched before the following metallization step. After the Al metal is either screen-printed or evaporated onto the rear, a laser is used to drive the metal through the passivation layer into the silicon. During this so-called laser-fired contact process,[61] an effective local Al-BSF is formed as well.[62,63] A new interesting route is to make use of commercial Al foil as the rear electrode, thereby avoiding complex equipment like screen printers or evaporation systems.[64]

1.4.3 Lifetime Limitations in Boron-doped p-type Silicon

Unfortunately, such cells are normally fabricated on boron-doped Czochralski-grown (Cz) silicon. This material is known to suffer from a metastable defect[65–69] related to boron and oxygen, which is activated by illumination or carrier injection (light-induced degradation).[70] Thus, the bulk lifetime is limited to such an extent that the efficiency is reduced to 20%[36,39,71] (see Figure 1.7, column 'p-Si PERL'). If the defect is deactivated[72] efficiencies of around 21% on large area[49] can be achieved (see Figure 1.7, column 'p-Si PERL regenerated').

There are several options available to reduce light-induced degradation: the use of thinner wafers to improve the ratio diffusion length/cell thickness,[73] decreasing or avoiding the boron doping or oxygen contamination,[67,74] or the application of the regeneration process.[72,75]

A strong reduction of the oxygen concentration to values below 1 ppma results in a perfect suppression of light-induced degradation. Magnetic Czochralski silicon (MCz) has shown a very high efficiency potential.[43,76,77] Another material, PV-FZ,[78] also with a negligible oxygen concentration, was considered a few years ago to be introduced into the large scale PV production. This material type has shown very high and stable carrier lifetimes, but unfortunately is not available for mass production.

A third material type with low oxygen concentration is cast silicon. A very interesting option is the so-called quasi-mono material or mono-like material.[79,80] The quasi-mono crystallization process is based on the standard directional solidification process but utilizes a seed of monocrystalline silicon. The final ingot consists of a rather large fraction of monocrystalline silicon surrounded by multicrystalline silicon. Due to the low oxygen concentration, light-induced degradation is reduced compared to standard Czochralski crystals.[81] Solar cells on quasi-mono p-type material have shown very promising efficiencies.[40,82]

In order to avoid boron doping, alternative acceptors such as gallium can be used. In fact, cells from gallium-doped Cz-silicon show no degradation.[76] The only issue occurring with this material might be the large variation of doping concentration over the ingot due to the low segregation coefficient of gallium. Nevertheless, adapted cell structures show excellent results over a wide doping range.[83]

Boron can also be avoided as a dopant if n-type silicon with phosphorus doping is used. n-type Czochralski-grown material shows no light-induced degradation even in the presence of a significant oxygen concentration.[68,84] Also for multicrystalline silicon, excellent minority carrier lifetimes have been measured.[85] This superior material quality is mainly due to reduced sensitivity to the most relevant impurities such iron.[86,87]

1.5   Solar Cells on n-type Silicon

Due to these excellent material properties, solar cells on n-type silicon are at the centre both of research and of the industry. Obviously, the well-approved phosphorus diffusion[88] cannot be used to create the junction of the cell anymore. One of the following techniques is commonly used to create the p–n junction on n-type substrates:

- Boron diffusion.[89,90]
- Al alloying at the rear side of the cell.[91-94]
- Heterostructure using p-doped amorphous silicon.[95]

1.5.1 n-type BSF Cell Structures

If the boron emitter is used on the front side, a simple cell p^+–nn^+ structure can be created by using phosphorus diffusion at the rear side. This is the analogue to the Al-BSF cell on p-type silicon with inverted polarities (Figure 1.14).

Due to the low complexity of the cell structure, it was used by various companies to enter into n-type cell technology. Efficiencies in the range of 20% with open-circuit voltages of around 650 mV have been achieved.[96,97] Yingli Solar has commercialized this cell structure under the brand name Panda.[98] Since n-type silicon is used, the recombination in the bulk is strongly reduced and the main limitation arises from recombination at the rear side (see Figure 1.7). As Figure 1.7 also shows, the disadvantage of this cell structure is the rather high recombination at the rear side. Analogous to p-type silicon, this problem can be solved by introducing a dielectric rear surface passivation (see next section).

An issue when working with boron-doped emitters is the fact that the well-established SiN_x PECVD passivation layer, which is used for phosphorus-doped emitters, is not well suited anymore.[99-101] This is due to the fact that SiN_x carries a positive built-in charge which results in a majority carrier depletion or inversion in the boron emitter. This was proven by an experiment with phosphorus and boron-doped emitters with SiO_2 passivation layers. When positive charges are applied by corona charging to the SiO_2

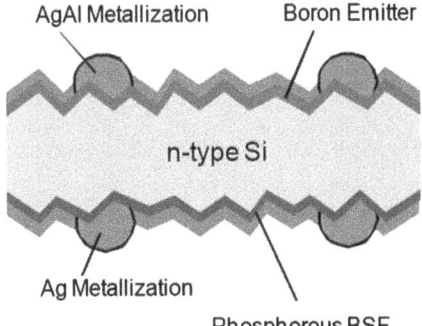

AgAl Metallization   Boron Emitter

n-type Si

Ag Metallization

Phosphorous BSF

Figure 1.14 p^+–nn^+ solar cell structures with boron-diffused front emitter and phosphorus-doped rear emitter.
Taken from Böscke *et al.*[96] Permission granted by IEEE.

layer, the implied open-circuit voltage is severely reduced.[102] The same negative effect can be achieved with negative charges on phosphorus-doped emitters. On the other hand, negative charges have improved the performance of boron-doped emitters. Therefore, a negatively charged dielectric passivation layer would be beneficial. AlO_x was proven to be an ideal layer to improve the passivation and to reduce the dark saturation current of boron emitters.[47] Using such a layer for high-efficiency solar cells on n-type silicon resulted in very high efficiencies of up to 23.9%.[71,102] A very promising alternative for AlO_x layers are sputtered AlN_x layers,[56,103] since they offer a higher refractive index and therefore also serve as an antireflective coating. Another layer which has proven to result in rather low dark saturation currents and increased blue response is an oxide created by a wet-chemical process in a solution of nitric acid.[104]

Another challenge when introducing n-type silicon into production is the rather low segregation coefficient of phosphorus in silicon. This results in a variation of the base resistivity by a factor of around 6 over the whole Czochralski crystal, *i.e.*, the tail of the crystal shows a lower phosphorus concentration than the top. A higher doping concentration or lower resistivity, respectively, is advantageous for the conductivity in the base and a higher fill factor can be achieved in the tail region (see Figure 1.15). On the other hand,

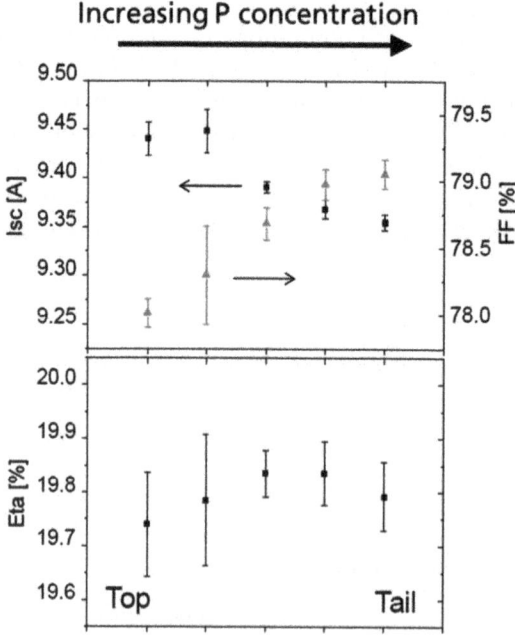

Figure 1.15 Variation of solar cell parameters as a function of position in the crystal.
Slightly supplemented version of Böscke *et al.*[96] Permission granted by IEEE.

a lower doping concentration results in higher lifetimes and longer diffusion lengths, respectively, resulting in a higher short-circuit current, J_{SC}. These two opposite trends compensate and the final cell efficiency is quite constant over the crystal.

A third point of concern revealed by various research teams is the fact that the recombination at the silver/aluminium contact at the front side limits the performance.[96] This leads to losses in open-circuit voltage of around 20 mV.[96,105] Possible solutions to this problem are improved pastes or alternative metallization techniques, such as Ni/Cu contacts.[106]

1.5.2 n-type Cell Structures with Dielectric Rear Passivation

Solar cell structures with dielectric rear passivation on n-type silicon have a huge efficiency potential. On lab-type cells featuring AlO_x passivation of the boron emitter, local phosphorus back-surface fields and SiO_2 rear passivation layer record efficiencies of 23.9% ($V_{OC} = 705$ mV) have been achieved.[71,102] However, process simplifications are mandatory in order to transfer this cell type into industrial production. An elegant process to create the local phosphorus back surface fields (BSF) is the so-called PassDop process.[107] After depositing the phosphorus-containing PassDop layer on the rear side using PECVD, a laser is used to open the contact points in the dielectric passivation layer. The phosphorus contained in the layer is simultaneously driven into the silicon and forms a very effective local BSF ($S_{cont} = 2000$ cm s^{-1}). Efficiencies of 22.4% with excellent open-circuit voltages of 701 mV have been achieved on small areas. This first generation of PassDop layer was limited to low-temperature metallization, such as evaporation or Ni/Cu plating. Recently, a second generation called f-PassDop was developed, exhibiting excellent passivation quality ($S_{pass} = 4$ cm s^{-1}) and an very good doping efficiency was developed as well.[108] This novel layer type is compatible with standard firing processes and can therefore be used with screen-printed metallization.

1.5.3 Heterojunction Solar Cells

Even for advanced cell structures with small contact areas and superior dielectric passivation, a large fraction of charge carrier recombination takes place at the surfaces. For the 23.9% n-type PERL cell described in the previous section, about 66% of the front recombination happens at the very small front contacts.[109] This recombination channel can be reduced by high-low junction as diffused or alloyed back-surface fields.[20,110] But even then, it will always play a major role.

The most effective way to suppress the recombination at the metal semiconductor interface is in the application of heterojunction, *i.e.*, through the deposition of a semiconductor material with wider bandgap on the crystalline silicon base.[111] The most common approach to realizing such a structure is the deposition of doped amorphous silicon.[112–116] An important

Figure 1.16 Structure of heterojunction solar cell and the related band diagram. Taken from De Wolf *et al.*[118] Permission granted by De Gruyter.

feature of the best cell structure is the introduction of an intrinsic amorphous silicon layer between the crystalline bulk and the n- or p-doped a-Si layer.[114] Such cells are well-known under the name HIT and have been optimized by Sanyo/Panasonic to a superior degree. Efficiencies of 24.7% have been achieved on 102 cm² cells with a remarkable open-circuit voltage of 750 mV.[117] Since the surfaces of this structure are perfectly passivated, it is advantageous to reduce the cell thickness. In fact, the best efficiencies were obtained on n-type wafers with a thickness of only 98 μm (Figure 1.16).

Due to the high efficiency potential of this cell structure, strong research and development activities have been launched in various companies and research laboratories.[119–125] Typical efficiencies achieved by these groups are in the range between 21% and 22%. An excellent overview of this topic was recently given by de Wolf.[118]

A major issue of this cell type is photon absorption in the amorphous silicon layer. Although these layers passivate the surface perfectly for carriers which are generated in the crystalline silicon bulk, the diffusion length in the material is too small so that carriers which are photogenerated within these layers recombine with a high probability. This leads to a rather low blue response of these cells[126] and is also reflected in our loss analysis (see Figure 1.7). The recombination losses are strongly increased due to such 'parasitic' absorption in the front region (p-doped and intrinsic a-Si) and also the optical losses are high because light is absorbed in the top TCO layers.

Something else which is rather astonishing when examining Figure 1.7 is the fact that the *absolute* recombination currents of this cell type at maximum power point are quite high, even though the cell parameters like the open-circuit voltage are superior. The high absolute losses are actually a direct consequence of the very good cell parameters, especially the high V_{mpp} (636 mV). A high V_{mpp} is equivalent to a high carrier concentration, *i.e.*, a high number of carriers, which can recombine even when the relative recombination characteristics like dark saturation currents J_0 are low.

Another remarkable fact resulting from the very effective suppression of the other recombination channels like surface and contact recombination is that the intrinsic recombination channel Auger-recombination becomes important. This also shows that we are getting closer to the physical limit of cell performance.

1.5.4   Back-contacted Back-junction Solar Cells

While the main strategy of heterojunction solar cells is the reduction of surface recombination, back-contact solar cells aim to reduce optical losses. There are several types of back-contact cells.

- *Metal wrap through (MWT) solar cells:*[41] In this cell structure only the busbars of the front electrode are placed on the rear side. The rear busbar is interconnected through *via* holes with the front grid.
- *Emitter wrap through (EWT) solar cells:*[127] In this case the contact fingers and the busbars are both located at the rear side while the collecting emitter stays at the front. The front emitter is connected *via* a high number of diffused holes with rear side electrodes.
- *Back-contact back-junction (BCBJ) solar cells:*[128] This cell type features an interdigitated structure contact grid and diffusion of both polarities at the rear side.

Interdigitated back-junction back-contact cells were originally developed for concentrator applications[44] since the grid can be designed for high current densities without increasing shadowing losses. They have been used mostly for special applications such as solar car races[129] up to the beginning of this millennium, when SunPower entered mass production.[130]

Although it is a very elegant and high-potential cell structure, it is also demanding in its production. Since most of the photogenerated carriers have to diffuse through the whole cell thickness, the diffusion length of the used material has to be excellent. Thus, high-quality n-type silicon has to be used.

Another challenge is the structuring of the rear side. For concentrator applications, photo-lithography can be used; however, for 1-sun mass production cells, more cost-effective techniques such as screen-printing or ink-jetting of masks have to be applied. This inevitably results in wider distances between n- and p-contacts in the range of a few millimetres. Therefore, the path of charge carriers in the cell is mostly lateral and not vertical. This can lead to fill factor and current losses.[131,132] By reducing the contact distance at the rear or by decoupling metallization and diffusion geometry,[133] excellent efficiencies of 22.4% have been achieved in production.[131] A very interesting option to achieve the structuring of the rear side is ion implantation. Ion implantation allows the use of *in situ* masking and very low dark saturation currents can be achieved.[134–141]

Figure 1.7 shows that the optical losses of back-contact back-junction cells are strongly reduced. The main recombination takes place at the rear, which includes the rear surface, the back-surface field and the emitter in this case. A closer look shows that 53% of the recombination in the rear region takes place at the metal/semiconductor interface. A reduction of this recombination channel by using 'passivated contacts' would be very beneficial.

1.5.5   Back-contacted Back-junction Solar Cells with Passivated Contacts

The reduction of contact recombination, also known as 'passivated contacts', would help to get even closer to the practical limit of around 26%.[10] In fact, we have already discussed a very effective way to implement passivated contacts, namely heterojunction solar cells (see 1.5.3). Since heterojunction cells also suffer from parasitic light absorption in the front TCO and amorphous silicon emitter, a combination with a back-junction cell is considered as the ideal solar cell. Strong research activities are being carried out in this field.[142–146] The best efficiencies so far were achieved by LG Electronics.[147,148] Their cells show excellent efficiencies of 23.4% and very high open-circuit voltages of 723 mV. Research to further improve the fill factor is being conducted.[149]

Another option to realizing passivated contacts are doped polysilicon layers. Using polysilicon for application in silicon photovoltaics was already suggested some time ago.[150–153] Recently, excellent passivation quality and good transport quality were achieved with doped polysilicon layers on a very thin tunnel oxide.[154] Using these passivated contacts as a full rear contact of both-side contacted solar cells led to efficiencies of 23% with an excellent V_{OC} of 698 mV and high fill factors of 81.1%.[154]

SunPower has combined its successful back contact back-junction solar cell with the passivated contact technology and boosted the open-circuit voltage to values as high as 721 mV. The resulting efficiency of 24.2% is a milestone for a solar cell using commercial technologies. Our loss analysis (Figure 1.7) shows that this cell structure combines the advantages of heterojunction solar cells (high voltage) with those of back-contact back-junction cells (high current) very effectively. The rather high amount of intrinsic Auger recombination shows that this cell structure gets very close to the physical limit.

1.6   Conclusion

This chapter shows that also for a technology such as crystalline silicon solar cells, which has been the leader in the field for more than 50 years, there is still sufficient room for further technological improvement. Thus, the reduction of production costs will be accompanied by a strong push towards high-efficiency cell structures in production and the development of new process technologies. This combined effort will help to further reduce the already very low costs of photovoltaic electricity in the coming decades and allow photovoltaics to be a significant pillar of the worldwide energy supply.

Acknowledgements

The author would like to thank the whole crystalline silicon PV team at Fraunhofer ISE and especially Marc Rüdiger, Heiko Steinkämper and Sebastian Schröer for numerical simulation and Armin Richter, Michael Rauer and Christian Reichel for fruitful discussion and proof reading.

References

1. ITRPV, *International Technology Roadmap for Photovoltaics*, 4th edn, 2013.
2. W. Shockley and H. J. Queisser, *J. Appl. Phys.*, 1961, **32**, 510.
3. P. Würfel, *Physics of Solar Cells – From Principles to New Concepts*, Wiley-Vch Verlag GmbH & Co KgaA, Weinheim, 2005.
4. M. J. Kerr and A. Cuevas, *J. Appl. Phys.*, 2002, **91**, 2473.
5. A. Richter, S. W. Glunz, F. Werner, J. Schmidt and A. Cuevas, *Phys. Rev. B*, 2012, **86**, 165202.
6. T. Trupke, M. A. Green, P. Würfel, P. P. Altermatt, A. Wang, J. Zhao and R. Corkish, *J. Appl. Phys.*, 2003, **94**, 4930.
7. P. P. Altermatt, F. Geelhaar, T. Trupke, X. Dai, A. Neisser and E. Daub, *Appl. Phys. Lett.*, 2006, **88**, 261901.
8. A. Richter, M. Hermle and S. W. Glunz, *IEEE J. Photovoltaics*, 2013, **3**, 1184.
9. M. J. Kerr, A. Cuevas and P. Campbell, *Prog. Photovoltaics Res. Appl.*, 2003, **11**, 97.
10. R. M. Swanson, *Proceedings of the 31st IEEE Photovoltaic Specialists Conference, Orlando, USA*, 2005, 889.
11. P. J. Verlinden, M. Aleman, N. Posthuma, J. Fernandez, B. Pawlak, J. Robbelein, M. Debucquoy, K. Van Wichelen and J. Poortmans, *Sol. Energy Mater. Sol. Cells*, 2012, **106**, 37.
12. R. Brendel, S. Dreissigacker, N. P. Harder and P. P. Altermatt, *Appl. Phys. Lett.*, 2008, **93**, 173503.
13. R. Brendel, *Thin-Film Crystalline Silicon Solar Cells – Physics and Technology*, WILEY-VCH GmbH & Co. KGaA, Weinheim, 2003.
14. E. Yablonovitch and G. D. Cody, *IEEE Trans. Electron Devices*, 1982, **29**, 300.
15. Sentaurus TCAD, *Technology Computer Aided Design* (TCAD), ed. Synopsys, Zürich.
16. M. Rüdiger, H. Steinkämper, M. Hermle and S. W. Glunz, *IEEE J. Photovoltaics*, 2013, submitted for publication.
17. J. Szlufcik, K. De Clercq, P. De Schepper, J. Poortmans, A. Buczkowski, J. Nijs and R. Mertens, *Proceedings of the 12th European Photovoltaic Solar Energy Conference, Amsterdam*, 1994, 1018.
18. A. Aberle, T. Lauinger and R. Hezel, *Proceedings of the 14th European Photovoltaic Solar Energy Conference, Barcelona*, 1997.
19. J. Mandelkorn and J. H. Lamneck Jr., *J. Appl. Phys.*, 1973, **44**, 4785.
20. E. L. Ralph, *Proceedings of the 11th IEEE Photovoltaic Specialists Conference, Scottsdale, Arizona*, 1975, 315.

21. P. Lölgen, W. C. Sinke, C. Leguijt, A. W. Weeber, P. F. A. Alkemade and L. A. Verhoef, *Appl. Phys. Lett.*, 1994, **65**, 2792.
22. M. Rauer, C. Schmiga, M. Glatthaar and S. W. Glunz, *IEEE J. Photovoltaics*, 2012, 1.
23. A. Cuevas, G. Giroult-Matlakowski, P. A. Basore, C. DuBois and R. R. King, *Proceedings of the 1st World Conference on Photovoltaic Energy Conversion – WCPEC, Waikoloa, Hawaii*, 1994, 1446.
24. G. Schubert, F. Huster and P. Fath, *Sol. Energy Mater. Sol. Cells*, 2006, **90**, 3399.
25. M. M. Hilali, B. To and A. Rohatgi, *Proceedings of the 14th Workshop on Crystalline Silicon Solar Cells and Modules NREL, Winter Park, Colorado*, 2004, 109.
26. M. Hörteis, T. Gutberlet, A. Reller and S. W. Glunz, *Adv. Funct. Mater.*, 2010, **40**, 476.
27. G. Hahn, *Proceedings of the 25th European Photovoltaic Solar Energy Conference and Exhibition, Valencia*, 2010, 1091.
28. S. W. Glunz, *Adv. OptoElectron.*, **2007**, 97370.
29. A. W. Blakers, A. Wang, A. M. Milne, J. Zhao and M. A. Green, *Appl. Phys. Lett.*, 1989, **55**, 1363.
30. J. Zhao, A. Wang and M. A. Green, *Proceedings of the 21st IEEE Photovoltaic Specialists Conference, Kissimmee, Florida*, 1990, 333.
31. J. Knobloch, A. Aberle, W. Warta and B. Voss, *Proceedings of the 5th International Photovoltaic Science and Engineering Conference, Kyoto*, 1990.
32. J. Zhao, A. Wang and M. A. Green, *Prog. Photovoltaics Res. Appl.*, 1999, **7**, 471.
33. M. A. Green, *Prog. Photovoltaics*, 2009, **17**, 183.
34. D. Kray, M. Hermle and S. W. Glunz, *Prog. Photovoltaics Res. Appl.*, 2008, **16**, 1.
35. O. Schultz, A. Mette, M. Hermle and S. W. Glunz, *Prog. Photovoltaics Res. Appl.*, 2008, **16**, 317.
36. A. Mohr, P. Engelhart, C. Klenke, S. Wanka, A. A. Stekolnikov, M. Scherff, R. Seguin, S. Tardon, T. Rudolph, M. Hofmann, F. Stenzel, J. Y. Lee, S. Diez, J. Wendt, W. Brendle, S. Schmidt, J. W. Müller, P. Wawer, M. Hofmann, P. Saint-Cast, J. Nekarda, D. Erath, J. Rentsch and R. Preu, *Proceedings of the 26th European Photovoltaic Solar Energy Conference and Exhibition, Hamburg*, 2011, 2150.
37. A. Mohr, S. Wanka, A. Stekolnikov, M. Scherff, R. Seguin, P. Engelhart, C. Klenke, J. Y. Lee, S. Tardon, S. Diez, J. Wendt, B. Hintze, R. Hoyer, S. Schmidt, J. W. Müller and P. Wawer, *Energy Procedia*, 2011, **8**, 390.
38. P. Engelhart, D. Manger, B. Klöter, S. Hermann, A. A. Stekolnikov, S. Peters, H.-C. Ploigt, A. Eifler, C. Klenke, A. Mohr, G. Zimmermann, B. Barkenfelt, K. Suva, J. Wendt, T. Kaden, S. Rupp, D. Rychtarik, M. Fischer and J. W. Müller, *Proceedings of the 26th European Photovoltaic Solar Energy Conference and Exhibition, Hamburg*, 2011, 821.

39. B. Sun, J. Sheng, S. Yuan, C. Zhang, Z. Feng, and Q. Huang, *Proceedings of the 38th IEEE Photovoltaic Specialists Conference, Austin, Texas*, 2012, 1125.
40. Y. Gassenbauer, K. Ramspeck, B. Bethmann, K. Dressler, J. D. Moschner, M. Fiedler, E. Brouwer, R. Drossler, N. Lenck, F. Heyer, M. Feldhaus, A. Seidl, M. Muller and A. Metz, *IEEE J. Photovoltaics*, 2013, **3**, 125.
41. E. van Kerschaver, S. De Wolf and J. Szlufcik, *Proceedings of the 28th IEEE Photovoltaics Specialists Conference, Anchorage, Alaska*, 2000, 209.
42. B. Thaidigsmann, J. Greulich, E. Lohmüller, S. Schmeißer, F. Clement, A. Wolf, D. Biro and R. Preu, *Sol. Energy Mater. Sol. Cells*, 2012, **106**, 89.
43. B. Thaidigsmann, M. Linse, A. Wolf, F. Clement, D. Biro and R. Preu, *Green*, 2012, **2**, 171.
44. R. A. Sinton, Y. Kwark, J. Y. Gan and R. M. Swanson, *IEEE Electron Device Lett.*, 1986, **7**, 567.
45. S. W. Glunz, D. Biro, S. Rein and W. Warta, *J. Appl. Phys.*, 1999, **86**, 683.
46. S. Dauwe, L. Mittelstädt, A. Metz and R. Hezel, *Prog. Photovoltaics Res. Appl.*, 2002, **10**, 271.
47. B. Hoex, J. Schmidt, R. Bock, P. P. Altermatt, M. C. M. van de Sanden and W. M. M. Kessels, *Appl. Phys. Lett.*, 2007, **91**, 112107/1.
48. J. Schmidt, A. Merkle, R. Brendel, B. Hoex, M. C. M. van de Sanden and W. M. M. Kessels, *Prog. Photovoltaics Res. Appl.*, 2008, **16**, 461.
49. A. Metz, D. Adler, S. Bagus, H. Blanke, M. Bothar, E. Brouwer, S. Dauwe, K. Dressler, R. Droessler, T. Droste, M. Fiedler, Y. Gassenbauer, T. Grahl, N. Hermert, W. Kuzminski, A. Lachowicz, T. Lauinger, N. Lenck, M. Manole, M. Martini, R. Messmer, C. Meyer, J. Moschner, K. Ramspeck, P. Roth, R. Schönfelder, B. Schum, J. Sticksel, K. Vaas, M. Volk and K. Wangemann, *Sol. Energy Mater. Sol. Cells*, 2014, **120**, 417.
50. P. Saint-Cast, J. Benick, D. Kania, L. Weiss, M. Hofmann, J. Rentsch, R. Preu and S. W. Glunz, *IEEE Electron Device Lett.*, 2010, **31**, 695.
51. A. Richter, F. M. Souren, D. Schuldis, R. M. Görtzen, J. Benick, M. Hermle and S. W. Glunz, *Proceedings of the 27th European Photovoltaic Solar Energy Conference and Exhibition, Frankfurt*, 2012, 1133.
52. P. Saint-Cast, D. Kania, M. Hofmann, J. Benick, J. Rentsch and R. Preu, *Appl. Phys. Lett.*, 2009, **95**, 151502.
53. T.-T. Li and A. Cuevas, *Phys. Status Solidi RRL*, 2009, **3**, 160.
54. K. A. Münzer, J. Schöne, A. Teppe, M. Hein, R. E. Schlosser, M. Hanke, J. Maier, A. Yodyungyong, J. Isenberg, T. Friess, C. Ehling, K. Varner, S. Keller and P. Fath, *Proceedings of the 26th European Photovoltaic Solar Energy Conference and Exhibition, Hamburg*, 2011, 843.
55. G. Krugel, W. Wolke, F. Wagner, J. Rentsch and R. Preu, *Proceedings of the 27th European Photovoltaic Solar Energy Conference and Exhibition, Frankfurt*, 2012, 1958.
56. G. Krugel, A. Sharma, W. Wolke, J. Rentsch and R. Preu, *Phys. Status Solidi RRL*, 2013, to be published.

57. S. W. Glunz, R. Preu, S. Schaefer, E. Schneiderlöchner, W. Pfleging, R. Lüdemann and G. Willeke, *Proceedings of the 28th IEEE Photovoltaics Specialists Conference, Anchorage, Alaska*, 2000, 168.

58. M. Bähr, G. Heinrich, O. Doll, I. Köhler, C. Maier and A. Lawerenz, *Proceedings of the 26th European Photovoltaic Solar Energy Conference, Hamburg*, 2011, 1206.

59. M. Rauer, R. Woehl, K. Rühle, C. Schmiga, M. Hermle, M. Hörteis and D. Biro, *IEEE Electron Device Lett.*, 2011, **32**, 916.

60. E. Urrejola, K. Peter, H. Plagwitz and C. Schubert, *Appl. Phys. Lett.*, 2011, **98**, 1.

61. E. Schneiderlöchner, R. Preu, R. Lüdemann and S. W. Glunz, *Prog. Photovoltaics Res. Appl.*, 2002, **10**, 29.

62. S. W. Glunz, E. Schneiderlöchner, D. Kray, A. Grohe, H. Kampwerth, R. Preu and G. Willeke, *Proceedings of the 19th European Photovoltaic Solar Energy Conference, Paris*, 2004, 408.

63. D. Kray and S. Glunz, *Prog. Photovoltaics*, 2006, **14**, 195.

64. J.-F. Nekarda, F. Lottspeich, A. Wolf and R. Preu, *Proceedings of the 25th European Photovoltaic Solar Energy Conference and Exhibition, Valencia*, 2010, 2211.

65. H. Fischer and W. Pschunder, *Proceedings of the 10th IEEE Photovoltaic Specialists Conference, Palo Alto, California*, 1973, 404.

66. S. W. Glunz, S. Rein, W. Warta, J. Knobloch and W. Wettling, *Proceedings of the 2nd World Conference on Photovoltaic Energy Conversion, Vienna*, 1998, 1343.

67. S. W. Glunz, S. Rein, J. Y. Lee and W. Warta, *J. Appl. Phys.*, 2001, **90**, 2397.

68. J. Schmidt, A. G. Aberle and R. Hezel, *Proceedings of the 26th IEEE Photovoltaic Specialists Conference, Anaheim, California*, 1997, 13.

69. J. Schmidt and K. Bothe, *Phys. Rev. B Condens. Matter*, 2004, **69**, 0241071.

70. S. W. Glunz, E. Schäffer, S. Rein, K. Bothe and J. Schmidt, *Proceedings of the 3rd World Conference on Photovoltaic Energy Conversion, Osaka*, 2003, 919.

71. S. W. Glunz, J. Benick, D. Biro, M. Bivour, M. Hermle, D. Pysch, M. Rauer, C. Reichel, A. Richter, M. Rüdiger, C. Schmiga, D. Suwito, A. Wolf and R. Preu, *Proceedings of the 35th IEEE Photovoltaic Specialists Conference, Honolulu, Hawaii*, 2010.

72. A. Herguth, G. Schubert, M. Kaes and G. Hahn, *Proceedings of the 4th World Conference on Photovoltaic Energy Conversion, Waikoloa, Hawaii*, 2006, 940.

73. K. A. Münzer, K. T. Holdermann, R. E. Schlosser and S. Sterk, *Proceedings of the 2nd World Conference on Photovoltaic Energy Conversion, Vienna*, 1998, 1214.

74. S. Rein, W. Warta and S. W. Glunz, *Proceedings of the 28th IEEE Photovoltaics Specialists Conference, Anchorage, Alaska*, 2000, 57.

75. B. Lim, V. V. Voronkov, R. Falster, K. Bothe and J. Schmidt, *Appl. Phys. Lett.*, 2011, **98**, 1.
76. S. W. Glunz, S. Rein, J. Knobloch, W. Wettling and T. Abe, *Prog. Photovoltaics Res. Appl.*, 1999, **7**, 463.
77. J. Zhao, A. Wang and M. A. Green, *Sol. Energy Mater. Sol. Cells*, 2001, **66**, 27.
78. J. Vedde, T. Clausen and L. Jensen, *Proceedings of the 3rd World Conference on Photovoltaic Energy Conversion, Osaka*, 2003, 943.
79. N. Stoddard, B. Wu, I. Witting, M. Wagener, Y. Park, G. Rozgonyi and R. Clark, *Solid State Phemomena*, 2008, **131–133**, 1.
80. K. Fujiwara, W. Pan, N. Usami, K. Sawada, M. Tokairin, Y. Nose, A. Nomura, T. Shishido and K. Nakajima, *Acta Mater.*, 2006, **54**, 3191.
81. X. Gu, X. G. Yu, K. X. Guo, L. Chen, D. Wang and D. R. Yang, *Sol. Energy Mater. Sol. Cells*, 2012, **101**, 95.
82. C.-H. Yang, J.-C. Pu, Y.-W. Chang, L.-S. Liao, S.-K. Tzeng, W.-P. Chen and Y.-C. Chen, *Proceedings of the 39th IEEE Photovoltaic Specialists Conference, Tampa, Florida*, 2013, to be published.
83. S. W. Glunz, S. Rein and J. Knobloch, *Proceedings of the 16th European Photovoltaic Solar Energy Conference, Glasgow*, 2000, 1070.
84. T. Yoshida and Y. Kitagawara, *Proceedings of the 4th International Symposium on High Purity Silicon IV, San Antonio, TX*, 1996, 450.
85. A. Cuevas, M. J. Kerr, C. Samundsett, F. Ferrazza and G. Coletti, *Appl. Phys. Lett.*, 2002, **81**, 4952.
86. D. Macdonald and L. J. Geerligs, *Appl. Phys. Lett.*, 2004, **85**, 4061.
87. B. Michl, J. Benick, A. Richter, M. Bivour, J. Yong, R. Steeman, M. C. Schubert and S. W. Glunz, *Energy Procedia*, 2012, **33**, 41.
88. J. Mandelkorn, C. McAfee, J. Kesperis, L. Schwart and W. Pharo, *J. Electrochem. Soc.*, 1962, **109**, 313.
89. R. R. King and R. M. Swanson, *IEEE Trans. Electron Devices*, 1991, **38**, 1399.
90. J. Benick, B. Hoex, O. Schultz and S. W. Glunz, *Proceedings of the 33rd IEEE Photovoltaic Specialists Conference, San Diego*, 2008.
91. D. L. Meier, H. P. Davis, R. A. Garcia, J. Salami, A. Rohatgi, A. Ebong and P. Doshi, *Sol. Energy Mater. Sol. Cells*, 2001, **65**, 621.
92. C. Schmiga, M. Hörteis, M. Rauer, K. Meyer, J. Lossen, H.-J. Krokoszinski, M. Hermle and S. W. Glunz, *Proceedings of the 24th European Photovoltaic Solar Energy Conference, Hamburg*, 2009, 1167.
93. C. Schmiga, M. Rauer, M. Rüdiger, K. Meyer, J. Lossen, H.-J. Krokoszinski, M. Hermle and S. W. Glunz, *Proceedings of the 25th European Photovoltaic Solar Energy Conference and Exhibition, Valencia*, 2010, 1163.
94. R. Bock, S. Mau, J. Schmidt and R. Brendel, *Appl. Phys. Lett.*, 2010, **96**, 1.
95. M. Tanaka, M. Taguchi, T. Takahama, T. Sawada, S. Kuroda, T. Matsuyama, S. Tsuda, A. Takeoka, S. Nakano, H. Hanafusa and Y. Kuwano, *Prog. Photovoltaics Res. Appl.*, 1993, **1**, 85.

96. T. S. Boscke, D. Kania, A. Helbig, C. Schollhorn, M. Dupke, P. Sadler, M. Braun, T. Roth, D. Stichtenoth, T. Wutherich, R. Jesswein, D. Fiedler, R. Carl, J. Lossen, A. Grohe and H. J. Krokoszinski, *IEEE J. Photovoltaics*, 2013, **3**, 674.

97. L. J. Geerligs, I. G. Romijn, A. R. Burgers, N. Guillevin, A. W. Weeber, J. H. Bultman, W. Hongfang, L. Fang, Z. Wenchao, L. Gaofei, H. Zhiyan, X. Jingfeng and A. Vlooswijk, *Proceedings of the 38th IEEE Photovoltaic Specialists Conference, Austin, Texas*, 2012, 1701.

98. A. R. Burgers, R. C. G. Naber, A. J. Carr, P. C. Barton, L. J. Geerligs, X. Jingfeng, L. Gaofei, S. Weipeng, A. Haijiao, H. Zhiyan, P. R. Venema and A. H. G. Vlooswijk, *Proceedings of the 25th European Photovoltaic Solar Energy Conference and Exhibition, Valencia*, 2010, 1106.

99. P. P. Altermatt, H. Plagwitz, R. Bock, J. Schmidt, R. Brendel, M. J. Kerr and A. Cuevas, *Proceedings of the 21st European Photovoltaic Solar Energy Conference, Dresden*, 2006, 647.

100. M. J. Kerr, Surface, emitter and bulk recombination in silicon and development of silicon nitride passivated solar cells. PhD Dissertation, Australian National University, 2002.

101. J. Libal, R. Petres, T. Buck, R. Kopecek, G. Hahn, R. Ferre, M. Vetter, I. Martín, K. Wambach, I. Roever and P. Fath, *Proceedings of the 20th European Photovoltaic Solar Energy Conference, Barcelona*, 2005, 793.

102. J. Benick, B. Hoex, M. C. M. van de Sanden, W. M. M. Kessels, O. Schultz and S. W. Glunz, *Appl. Phys. Lett.*, 2008, **92**, 253504/1.

103. G. Krugel, A. Sharma, A. Moldovan, W. Wolke, J. Rentsch and R. Preu, *Proceedings of the 39th IEEE Photovoltaic Specialists Conference, Tampa, Florida*, 2013, to be published.

104. V. D. Mihailetchi, Y. Komatsu and L. J. Geerligs, *Appl. Phys. Lett.*, 2008, **92**, 1.

105. A. Edler, V. Mihailetchi, R. Kopecek, R. Harney, T. Boscke, D. Stichtenoth, J. Lossen, K. Meyer, R. Hellriegel, T. Aichele and H. J. Krokoszinski, *Energy Procedia (SiliconPV 2011)*, 2011, **8**, 493.

106. J. Bartsch, A. Mondon, M. Kamp, A. Kraft, M. Wendling, N. Wehkamp, M. Jawaid, A. Lorenz, F. Clement, C. Schetter, M. Glatthaar and S. Glunz, *Proceedings of the 27th European Photovoltaic Solar Energy Conference and Exhibition, Frankfurt*, 2012, 604.

107. D. Suwito, U. Jäger, J. Benick, S. Janz, M. Hermle and S. W. Glunz, *IEEE Trans. Electron Devices*, 2010, **57**, 2032.

108. B. Steinhauser, M. bin Masoor, U. Jäger, J. Benick and M. Hermle, *accepted for publication at the European Photovoltaic Solar Energy Conference and Exhibition, Paris*, 2013.

109. J. Benick, B. Hoex, M. C. M. van de Sanden, W. M. M. Kessels, O. Schultz and S. W. Glunz, *Appl. Phys. Lett.*, 2008, **92**.

110. M. P. Godlewski, C. R. Baraona and H. W. Brandhorst, Jr., *Proceedings of the 10th IEEE Photovoltaic Specialists Conference, Palo Alto, California*, 1973, p. 40.

111. E. Yablonovitch, T. Gmitter, R. M. Swanson and Y. H. Kwark, *Appl. Phys. Lett.*, 1985, **47**, 1211.
112. W. Fuhs, K. Niemann and J. Stuke, *Bull. Am. Phys. Soc.*, 1974, **19**, 394.
113. K. Okuda, H. Okamoto and Y. Hamakawa, *Jpn. J. Appl. Phys.*, 1983, **22**, L605.
114. M. Taguchi, M. Tanaka, T. Matsuyama, T. Matsuoka, S. Tsuda, S. Nakano, Y. Kishi and Y. Kuwano, *Proceedings of the 5th International Photovoltaic Science and Engineering Conference, Kyoto*, 1990, p. 689.
115. M. Taguchi, E. Maruyama and M. Tanaka, *Jpn. J. Appl. Phys.*, 2008, **47**, 814.
116. M. Taguchi, Y. Tsunomura, H. Inoue, S. Taira, T. Nakashima, T. Baba, H. Sakata and E. Maruyama, *Proceedings of the 24th European Photovoltaic Solar Energy Conference, Hamburg*, 2009, p. 1690.
117. M. Taguchi, A. Yano, S. Tohoda, K. Matsuyama, Y. Nakamura, T. Nishiwaki, K. Fujita and E. Maruyama, *Proceedings of the 39th IEEE Photovoltaic Specialists Conference, Tampa, Florida, USA*, 2013, to be published.
118. S. De Wolf, A. Descoeudres, Z. C. Holman and C. Ballif, *Green*, 2012, **2**, to be published.
119. J. L. Hernandez, K. Yoshikawa, A. Feltrin, N. Menou, N. Valckx, E. Van Assche, D. Schroos, K. Vandersmissen, H. Philipsen, J. Poortmans, D. Adachi, M. Yoshimi, T. Uto, H. Uzu, T. Kuchiyama, C. Allebe, N. Nakanishi, T. Terashita, T. Fujimoto, G. Koizumi and K. Yamamoto, *Jpn. J. Appl. Phys.*, 2012, **51**.
120. D. L. Bätzner, Y. Andrault, L. Andreetta, A. Büchel, W. Frammelsberger, C. Guerin, N. Holm, D. Lachenal, J. Meixenberger, P. Papet, B. Rau, B. Strahm, G. Wahli and F. Wünsch, *Proceedings of the 1st International Conference on Silicon Photovoltaics, Freiburg*, 2011, p. 153.
121. A. Descoeudres, Z. C. Holman, L. Barraud, S. Morel, S. De Wolf and C. Ballif, *IEEE J. Photovoltaics*, 2013, **3**, 83.
122. D. Muñoz, T. Desrues, A. S. Ozanne, N. Nguyen, S. De Vecchi, F. Souche, S. Martin de Nicholas, C. Denis and P. J. Ribeyron, *Proceedings of the 26th European Photovoltaic Solar Energy Conference and Exhibition, Hamburg*, 2011, p. 861.
123. J.-H. Choi, S.-K. Kim, J.-C. Lee, H. Park, W.-J. Lee and E.-C. Cho, *Proceedings of the 26th European Photovoltaic Solar Energy Conference and Exhibition, Hamburg*, 2011, p. 3302.
124. L. Korte, E. Conrad, H. Angermann, R. Stangl and M. Schmidt, *Sol. Energy Mater. Sol. Cells*, 2009, **93**, 905.
125. M. Bivour, C. Reichel, M. Hermle and S. W. Glunz, *Sol. Energy Mater. Sol. Cells*, 2012, **106**, 11.
126. Z. C. Holman, A. Descoeudres, L. Barraud, F. Z. Fernandez, J. P. Seif, S. De Wolf and C. Ballif, *IEEE J. Photovoltaics*, 2012, **2**, 7.
127. J. M. Gee, W. K. Schubert and P. A. Basore, *Proceedings of the 23rd IEEE Photovoltaic Specialists Conference, Louisville, Kentucky*, 1993, p. 265.

128. M. D. Lammert and R. J. Schwartz, *IEEE Trans. Electron Devices*, 1977, **ED-24**, 337.

129. P. J. Verlinden, R. M. Swanson and R. A. Crane, *Prog. Photovoltaics Res. Appl.*, 1994, **2**, 143.

130. K. R. McIntosh, M. J. Cudzinovic, D. D. Smith, W. P. Mulligan and R. M. Swanson, *Proceedings of the 3rd World Conference on Photovoltaic Energy Conversion, Osaka*, 2003, p. 971.

131. D. De Ceuster, P. Cousins, D. Rose, D. Vicente, P. Tipones and W. Mulligan, *Proceedings of the 22nd European Photovoltaic Solar Energy Conference, Milan*, 2007, p. 816.

132. M. Hermle, F. Granek, O. Schultz-Wittmann and S. W. Glunz, *Proceedings of the 33rd IEEE Photovoltaic Specialists Conference, San Diego*, 2008.

133. C. Reichel, F. Granek, M. Hermle and S. W. Glunz, *Prog. Photovoltaics Res. Appl.*, 2012, 1.

134. N. Bateman, P. Sullivan, C. Reichel, J. Benick and M. Hermle, *Proceedings of the 1st International Conference on Silicon Photovoltaics, Freiburg*, 2011, p. 509.

135. M. Hermle, J. Benick, N. Bateman and S. W. Glunz, *Proceedings of the 26th European Photovoltaic Solar Energy Conference and Exhibition, Hamburg*, 2011, p. 875.

136. R. Müller, J. Benick, N. Bateman, J. Schön, C. Reichel, A. Richter, M. Hermle and S. W. Glunz, *Sol. Energy Mater. Sol. Cells*, 2013, in press.

137. Y. Tao, Y.-W. Ok, F. Zimbardi, A. D. Upadhyaya, J.-H. Lai, S. Ning, V. D. Upadhyaya and A. Rohatgi, *Proceedings of the 39th IEEE Photovoltaic Specialists Conference, Tampa, Florida*, 2013, to be published.

138. A. D. Upadhyaya, Y.-W. Ok, M. Kadish, V. Upadhyaya, K. S. Ryu, M. H. Kang, A. Gupta and A. Rohatgi, *Proceedings of the 39th IEEE Photovoltaic Specialists Conference, Tampa, Florida*, 2013, to be published.

139. J. Benick, R. Müller, N. Bateman and M. Hermle, *Proceedings of the 27th European Photovoltaic Solar Energy Conference and Exhibition, Frankfurt*, 2012, p. 676.

140. C. E. Dubé, B. Tsefrekas, D. Buzby, R. Tavares, W. Zhang, A. Gupta, R. J. Low, W. Skinner and J. B. Mullin, *Proceedings of the 1st International Conference on Silicon Photovoltaics, Freiburg*, 2011, p. 706.

141. J. Wu, Y. Liu, X. Wang and L. Zhang, *Proceedings of the 39th IEEE Photovoltaic Specialists Conference, Tampa, Florida*, 2013, to be published.

142. T. Desrues, S. De Vecchi, F. Souche, D. Djicknoum, D. Munoz, M. Gueunier-Farret, J.-P. Kleider and P.-J. Ribeyron, *Proceedings of the 1st International Conference on Silicon Photovoltaics, Freiburg, Germany*, 2011, p. 294.

143. M. Lu, S. Bowden, U. Das and R. Birkmire, *Appl. Phys. Lett.*, 2007, **91**, 1.

144. M. Tucci, L. Serenelli, E. Salza, S. De Iuliis, L. J. Geerligs, G. De Cesare, D. Caputo and M. Ceccarelli, *Proceedings of the 22nd European Photovoltaic Solar Energy Conference, Milan*, 2007, p. 1600.

145. R. Stangl, M. Bivour, E. Conrad, I. Didschuns, L. Korte, K. Lips and M. Schmidt, *Proceedings of the 22nd European Photovoltaic Solar Energy Conference, Milan*, 2007, p. 870.
146. N. Mingirulli, J. Haschke, R. Gogolin, R. Ferré, T. F. Schulze, J. Düsterhöft, N.-P. Harder, L. Korte, R. Brendel and B. Rech, *Physica Status Solidi RRL*, 2011, **5**, 159.
147. K.-S. Ji, H. Syn, J.-H. Choi, H.-M. Lee and D. Kim, *Jpn. J. Appl. Phys.*, 2012, **51**, 10NA05.
148. J.-H. Choi, *2nd NPV Workshop, Amsterdam*, 2012.
149. S.-Y. Lee, H. Choi, H. Li, K.-s. Ji, S. Nam, J.-H. Choi, S.-W. Ahn, H.-M. Lee and B. Park, *Sol. Energy Mater. Sol. Cells*, 2013, in press.
150. E. Yablonovitch, T. Gmitter, R. M. Swanson and Y. H. Kwark, *Appl. Phys. Lett.*, 1985, **47**, 1211.
151. N. G. Tarr, *IEEE Electron Device Lett.*, 1985, **6**, 655.
152. F. A. Lindholm, A. Neugroschel, M. Arienzo and P. A. Iles, *IEEE Electron Device Lett.*, 1985, **6**, 363.
153. I. R. C. Post, P. Ashburn and G. R. Wolstenholme, *IEEE Trans. Electron Devices*, 1992, **39**, 1717.
154. F. Feldmann, M. Bivour, C. Reichel, M. Hermle and S. W. Glunz, *Sol. Energy Mater. Sol. Cells*, 2014, **120**, 270.

CHAPTER 2

Tandem and Multiple-junction Devices Based on Thin-film Silicon Technology

CHRISTOPHE BALLIF, MATHIEU BOCCARD,
KARIN SÖDERSTRÖM, GRÉGORY BUGNON,
FANNY MEILLAUD AND NICOLAS WYRSCH*

Ecole Polytechnique Fédérale de Lausanne (EPFL), Institute of
Microengineering (IMT), Photovoltaics and Thin film Electronics
Laboratory, Rue Bréguet 2, CH-2000 Neuchâtel, Switzerland
*Email: nicolas.wyrsch@epfl.ch

2.1 Introduction

Thin-film silicon (TFSi) technology combines the advantages of being based on silicon (abundant, cheap and non-toxic raw materials) with the inherent advantages of thin-film technologies (sparse use of raw material, aesthetics, upscalability, flexibility). As TFSi is based on the use of plasma activated deposition, it offers the possibility to make a virtual infinite range of semiconductors with various bandgaps (typically from 1 eV to 2 eV by alloying and phase-change from microcrystalline to completely amorphous). This opens the road to multiple-junction devices that can be processed in unique equipment, at a low cost, and with high theoretical efficiencies and good temperature coefficient ($-0.17\%/^\circ$C to $-0.3\%/^\circ$C). TFSi offers, in principle, a unique route to the tera-watt photovoltaics (PV) society, both in the form of giga-watt-scale solar power plants, but also in the form of building

RSC Energy and Environment Series No. 11
Advanced Concepts in Photovoltaics
Edited by Arthur J Nozik, Gavin Conibeer and Matthew C Beard
Published by the Royal Society of Chemistry, www.rsc.org

integrated PV. It was proven to be a low-cost building element (extremely attractive in terms of price per m^2) with a high reliability, and it can be manufactured on various materials—including flexible foils—with a diversity of transparency, colours and shapes.

In the 1990s the demonstration of efficient hydrogenated microcrystalline (µc-Si:H) silicon solar cells by IMT in Neuchâtel has created a strong renewed interest in the TFSi technology, in particular the possibility to combine the historically used hydrogenated amorphous (a-Si:H) top cell with a µc-Si:H bottom cell. Japanese companies have been the first to demonstrate and produce efficient tandem a-Si:H/µc-Si:H 'micromorph' modules on glass. In parallel to this approach multiple junctions of a-Si:H and amorphous silicon alloyed with germanium (a-Si$_{1-x}$Ge$_x$:H) have been realized by other companies on other types of substrate (*e.g.*, Unisolar in USA on stainless steel substrates). During the years 2000 to 2010, an intense effort has been made by several module producers and equipment makers to upscale the technology. In particular, a synergy with the flat-panel-display industry could be established for the plasma deposition of homogeneous Si layers on large areas (> 1 m^2). Upscaling a technology to production plants over 30–50 MW has taken several hundred million euros in investments and a huge industrialization effort. Triggered by the growing PV market and by the polysilicon shortage of the years 2006–2009, several tens of companies have purchased and/or installed and ramped-up full production lines. Some of these 'second-generation' production lines operate with remarkably high yield (up to 98% reported for thin-film modules on glasses) and illustrate the fantastic industrial developments realized by the technology. Indeed it is now possible to purchase efficient production lines with module efficiency in the range of 10% and with reasonable investment costs.

In parallel and in support to these developments, the years 2000 to 2010 have also seen an intense effort at solving some of the key fundamental challenges of TFSi. On the industry side, this includes the scaling up of processes for transparent conductive oxide (TCO) and plasma coatings of the active absorbers (*e.g.*, standing waves in plasma systems). On the research side, a focus was put on making thinner devices for reducing the light-induced degradation of a-Si:H (Staebler–Wronski effect) and for reducing the cost of the µc-Si:H cell deposition; reducing the parasitic losses in transparent conductive oxides and in doped layers; reaching a better understanding of the strongly interlinked interaction between substrate morphology, growth of the layers, plasma process and light harnessing.

After this period of more fundamental research, the last years have again seen a rise in the certified stabilized device efficiency (now up to 13.4%[1] on small area), after a decade focusing at solving the industrial issues. Micromorph modules from pilot lines are now reaching close to 11% stabilized efficiencies.[2]

The strong drop in standard multi-crystalline modules price has, however, put all thin-film technologies and moderate-volume producers in a difficult situation in the years 2012–2013, which was also the case for several TFSi module suppliers. However, companies with niche product (architecture, flexible products), or companies with downstream integration or captive markets have been able to continue production. The clear challenge for the future of TFSi is two-fold: (1) in terms of research, better exploitation of the potential of all new absorber materials, device structures, and light-management schemes needs to be achieved to raise the device efficiency by several points; and (2) in terms of business plans, lower-cost mounting systems are required, as well as the development of a large specific market for this technology, such as in the low-cost building-integration field.

In this chapter, the materials composing the multiple-junction devices and the means of depositing them are first reviewed in section 2.2. Then, section 2.3 discusses the keys aspects of the designs of multiple-junction devices and their essential requirements for obtaining high efficiencies. After a short discussion of the state of the art in section 2.4, section 2.5 focuses on drawing the current limitations that face the technology, the recent advances made to tackle them and, in addition, discusses the future directions and perspectives to bring further improvements. Finally, section 2.6 will conclude this chapter.

2.2   Material Properties

2.2.1   Hydrogenated Amorphous Silicon (a-Si:H) and its Alloys

The first investigation of a-Si:H was reported by Chittik *et al.* in 1969.[3] It was observed that growth of this material from plasma enhanced chemical vapour deposition (PE-CVD, also referred to as glow discharge) from silane led to a much lower defect density compared to evaporated or sputtered amorphous silicon. Such a low defect density allowed for the n- and p-doping of the material[4] and the demonstration of a first solar cell (with 2% efficiency) based on this material.[5]

a-Si:H without long-range order and a lack of crystalline structure is classified as a disordered material. Short-range order (as characterized by bond lengths and bond angles) is conserved (compared to crystalline silicon c-Si). Incorporation of disorder in the material leads to the creation of dangling bonds (DB or broken Si–Si bonds) by the presence of under-coordinated Si atoms. The presence of DB can be strongly reduced by the incorporation of hydrogen into the material. As a matter of fact, a-Si:H is an alloy of silicon with hydrogen with an atomic content of hydrogen than can vary between *ca.* 2% and 20%. Hydrogen incorporated in the material originates from silane, or from additional hydrogen added to the process gas mixture.

Despite the fact that a-Si:H can also be deposited using a variety of other techniques,[6] PE-CVD remains the most common one for state-of-the-art material properties. Deposition is usually performed from the decomposition of silane gas in a plasma inside a vacuum chamber. The plasma is commonly formed between parallel plates upon application of a DC or an AC excitation current (at 13.56 MHz or higher frequency). Other silicon-containing gas can be used, but hydrogen incorporation is necessary to achieve low enough defect density, by saturating dangling bonds introduced by the disorder in the material. Doping is achieved by adding to silane PH_3 for n-type material, and by adding B_2H_6 or TMB (trimethylboron) for p-type material. Increasing the presence of atomic hydrogen in the plasma (supplied from the decomposition of silane or hydrogen that is added to silane) can lead to a transition regime with the growth of crystalline phases in the material and later to almost fully crystalline films.[7,8] Amorphous material deposited at the onset of the crystalline growth, known as proto-crystalline,[9] exhibiting dense structure, is often sought for high quality devices.[10]

The lowest defect density is usually obtained at low deposition rate; increasing the latter leads to the formation of poly-silicon radicals in the gas phase, which can aggregate in the form of particles or powders and can be incorporated in the growing films. This occurrence is, in most cases, detrimental to the quality of the film. Nevertheless, transitional material of high quality, incorporating nano-crystals embedded in an amorphous Si tissue, so called polymorphous material,[11] is also gaining attention.

Microstructural view of the a-Si:H network has been so far treated as a continuous random network with isolated randomly distributed DB as the defect sites and randomly distributed hydrogen atoms. However, this view is challenged by the observation by IR spectroscopy which reveals the configuration of incorporated hydrogen, like the predominance of hydrogenated di-vacancies in a dense a-Si:H network and the predominance of hydrogenated nano-sized voids in less dense a-Si:H.[12,13]

The absence of long range order in a-Si:H compared to c-Si results in a broadening of the electron-density bands, forming bandtails and a continuous distribution of localized states in the bandgap; these states are a result of the disorder and structural defects such as broken bonds, nano-voids and vacancies.[14] Instead of a strict forbidden energy gap, we observe in these types of amorphous semiconductors, a mobility gap given by the separation between localized states—not contributing to transport—and extended states, in which electrons and holes can move as free carriers.[15] An optical gap is also commonly defined to characterize optical absorption which reflects the distribution of states in the bandgap and band edges. Most reported optical gap values are those of Tauc,[16] and Cody,[17] as well as E_{04} values that is defined as the photon energy for which an optical absorption coefficient of $10\,000$ cm^{-1} is measured. Typical values for the Tauc gap are around 1.7 eV. Note that all (optical) gap definitions suffer from various drawbacks leading to slightly different values. Within the localized states in the bandgap, the ones located in the middle of the gap (or mid-gap states) are mostly attributed to DBs. These DBs take of the three possible charge states (D^+ positive, D^0 neutral or D^- negative) and are located at energy levels that can vary from one a-Si:H to the other. Distributions of these states are usually represented as Gaussians in the bandgap.

Due to the relatively high bandgap, the disordered nature leading to a low mobility, and the transport through extended states, a-Si:H is a semi insulating material with resistivity values higher than 10^{10} Ω cm. This resistivity can be changed by more than seven orders of magnitude by n- or p-doping,[4] by moving the Fermi level. Transport is thermally activated (at room temperature) with an activation energy of 0.8–0.9 eV. DBs play an important role in photoconductivity as they act as recombination centers. However doping is also creating additional defects, through chemical equilibrium process, which rapidly degrades lifetime in this material.[18] This creation of defect (DBs) upon doping puts limits on the doping efficiency and shifts of the Fermi level towards conduction or valence band.

In 1977, Staebler and Wronski observed reduction in photoconductivity and dark conductivity when the material was exposed to light and that these changes could be reversed upon annealing for several hours at temperatures $\geq$150 °C.[19] This metastable effect, known as the Steabler–Wronski effect, is due to an increase in deep defect density (increase of the dangling bond density) created by the breaking of weak bonds. The creation of deep defects (or DBs) is not directly linked to the photon absorption, but to recombination events. A similar defect creation mechanism can be observed when electron-hole pairs recombined following double injection[20] (in the absence of light illumination). Under light-soaking, defect density increases and reaches an equilibrium value, when the rate of defect creation equals the rate of defect annihilation by thermal annealing. This simple 'weak bond model' has attracted a lot of attention and is still the subject of controversy. Several models have been designed to give a microscopic description of the phenomenon but so far with no satisfactory results.[20] However, it is established that SWE and its related defect creation is linked to hydrogen content and hydrogen bonding.[21] Depending on the latter, susceptibility to light-induced degradation can be controlled to some extent: For example, protocrystalline materials[10] as well as a-Si:H materials deposited at relatively high temperature[22,23] exhibit reduced degradation. Degradation rate can also differ significantly depending on their microstructure details[14] and/or deposition methods and conditions.[23,24]

Bandgap of a-Si:H can be changed by varying the incorporation of hydrogen[25] or by alloying with carbon[26] or oxygen[27] (for higher bandgap) or with germanium[28] (for lower bandgap). Germanium has been extensively used for the development of narrow bandgap a-SiGe:H for tandem or triple junction solar cells.[29] However incorporation of large amount of Ge, C, or O usually has a detrimental effect of the material quality (higher defect content) and low material stability (against light soaking).[30]

2.2.2 Hydrogenated Microcrystalline Silicon (μc-Si:H) and its Alloys

The first report on μc-Si:H, which at that time was obtained by chemical transport in low-pressure hydrogen plasma, dates back to 1968.[31] Due to very low growth rate, typically below 1 Å s $^{-1}$, and bad quality, its use was then limited to doped layers.[32,33] The application of µc-Si:H as absorber layer of a photovoltaic device only became possible with the development of PE-CVD.[34,35] A breakthrough in the research and application of µc-Si:H for solar cells application occurred in 1994 at IMT Neuchâtel when Meier *et al.* demonstrated an efficiency of 4.6% for a µc-Si:H solar cell in superstrate configuration.[36] This was achieved due to a slight intentional boron compensation of the absorber layer through micro-doping, as recently investigated by Guha *et al.*[37] Two years later, better control of oxygen contamination was found to be decisive for obtaining a high-quality material with improved transport properties.[38] Since then, substrate morphology and subsequent µc-Si:H growth, as well as plasma conditions, have been demonstrated as crucial to obtain state-of-the art single-junction and multi-junction solar cells. More in-depth details on the work carried out these last 10 years on µc-Si:H can be found in recent overviews.[39-41]

µc-Si:H is typically obtained from PE-CVD using silane diluted in hydrogen (H_2) at excitation frequencies of 13.56 MHz or above. Over the years, other diluent gases such as argon or helium were studied, but no significant advantage was demonstrated over the standard H_2 dilution for high quality µc-Si:H. On the other hand, recent observations have been made on the possibility that fluorinated gases can, under specific deposition conditions, favour the growth of larger crystallites in µc-Si:H.[42]

The bandgap value of µc-Si:H material is similar to that of crystalline silicon, around 1.1 eV, though localized states lie within the bandgap as in the case of a-Si:H. This narrow bandgap enabling infrared absorption makes it an ideal bottom sub-cell in multi-junction solar cell devices.[43,44] However, the indirect bandgap nature of µc-Si:H leads to low light absorption (which also depends on the crystallinity fraction); thick absorber layers (above 1 µm) are thus required in actual solar cells to generate a sufficiently high current density. A key bottleneck for the use of µc-Si:H in thin-film silicon solar cells is therefore its low deposition rate, typically below 5 Å s^{-1}, with a consequent deposition time of typically 45 min to 1 h for a 1-µm-thick absorber layer. Deposition rate can be increased, apart from increasing the excitation frequency, by using *e.g.*, larger plasma excitation power or higher plasma pressures, *i.e.*, the so-called HPD (high-pressure depletion) regimes.[45,46] Various alternative deposition methods have been evaluated for µc-Si:H including microwave plasmas,[47,48] inductively coupled plasmas (ICPs),[49] atmospheric PE-CVD[50,51] modified-cyclotron-resonance (ECR) plasmas,[52,53] expanding thermal (ETPs) plasmas,[54] *etc.* However, capacitively coupled PE-CVD, alone or with HWCVD[55] has been the only one that allows the fabrication of very high-quality µc-Si:H for PV applications. Other promising approaches based on different electrodes—compatible with large-area processing—were also studied and showed some strong potential, such as the use of ladder-shaped electrodes[56] or a multi-hollow cathode.[45,57]

It was already established in early works that the structure of µc-Si:H layers could easily be varied, ranging from completely amorphous to highly crystalline, by appropriately adjusting deposition parameters such as the silane to hydrogen ratio, power, substrate temperature and pressure.[58,59] The evolution of the microstructure as a function of SiH_4/H_2 is schematically presented in Figure 2.1. μc-Si:H is a complex, mixed-phase, material, consisting of small crystalline grains (typically in the range of 3–30 nm) embedded in an amorphous matrix and later arranging in large conglomerates with a typical size of hundreds of nm. The material also comprises disordered regions and voids. The evolution from completely amorphous to highly crystalline material can be monitored by Raman spectroscopy and the assessment of the Raman crystallinity factor.[60] It was demonstrated that for integration as an absorber layer in single-junction or tandem solar cells, the optimum in the Raman crystallinity factor lies around 60%.[29,61]

Three major models have been proposed to explain μc-Si:H growth and formation as a function of plasma species, with a particular focus on hydrogen and deposition conditions.[62] These models are the surface diffusion model, the selective etching model and the chemical-annealing model; the basic requirement for crystallization being that a sufficiently high atomic hydrogen flow impinges upon the growing film. In silane-based plasmas, this can be determined by the silane depletion fraction.[63] Ion bombardment and the growth surface effective temperature can also play a role,[64,65] while both the substrate morphology and chemical nature are also known to play critical roles in the nucleation process.[66]

Once deposited on top of textured substrates, as is typically the case for solar cells, a secondary defective nanoporous phase (sometimes referred to as 'cracks') appears, specifically at the encountering of the growth fronts[67] (dark lines on the right part of Figure 2.2). The material quality can therefore not be assessed only by evaluating the bulk-phase defect density when deposited on flat substrates, nor by using Fourier-transform photocurrent spectroscopy (FTPS),[68] since two distinct μc-Si:H material phases contribute to the solar cell performances.[69] Detailed analysis on how the substrate geometrical parameters influence this secondary phase has been made.[70–73] The PE-CVD process conditions (growth rate, plasma excitation frequency, hydrogen input flow rate, pressure, *etc.*)[74] also have a strong impact on the

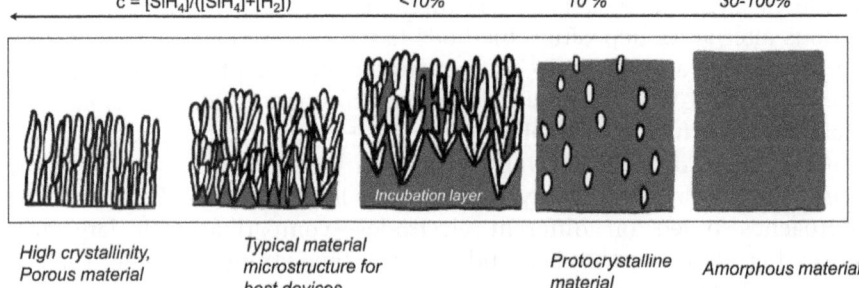

Figure 2.1   Schematics of the evolution of microcrystalline silicon layers microstructure as a function of the silane concentration c in the plasma gas phase. The gray areas represent the amorphous phase.

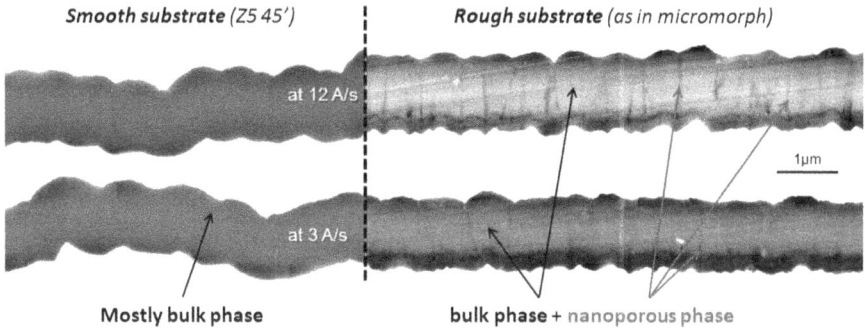

Figure 2.2   Scanning electron microscopy (SEM) images of μc-Si:H layers deposited at 3 and 12 Å s^{-1} on two substrates with different morphologies.

formation of these nanoporous regions. The influence of both the substrate morphology and the process conditions can be seen, for example, in Figure 2.2. This substrate-induced nanoporous zone development also affects the performance stability of cells over time. The nanoporous regions act like bad diodes with low open-circuit potential (V_{OC}) values, thus generating issues similar to shunting in the solar cells: An equivalent-circuit model consisting of parallel-connected diodes with different electronic quality showed that variations of the local saturation current density result in a degradation of the performances, more precisely the open-circuit voltage and fill factor (FF).[75] Hence, the deposition requirements for homogeneously dense μc-Si:H on rough substrates remain to be understood.

Also, when grown under unfavourable conditions (high growth rate, non-adapted power/pressure combinations, *etc.*), the bulk material itself can have a less dense microstructure, being considered more 'porous' within the large conglomerates as well. This degradation of bulk quality can typically be observed by Fourier-transform infrared (FTIR) spectroscopy, with a poor passivation of the crystallites leading to poor solar cell performances associated with post-oxidation and electrical instabilities of the μc-Si:H films over time.[45,76–78]

During these last years a lot of research has been done on μc-Si:H alloys as well. First, alloying with carbon (C) or oxygen (O) allowed for the fabrication of highly transparent doped layers to limit parasitic absorption which impedes the solar cell overall performances.[79–81] Such materials are mixed-phase, as the alloying with silicon will occur within the amorphous phase only, while silicon crystallites will still manage to grow in the film. As a result the optical bandgap of the amorphous phase is enhanced, with a reduced absorption, and the conductivity usually decreases. However, this decrease in conductivity has been shown to be an advantage: indeed, mostly lateral conductivity is affected, which leads to a reduction of shunt path interconnection (*i.e.*, shunt-quenching), while the transverse conductivity is still maintained sufficiently high by the silicon crystallites (appearing as dendrites) so that it does not impact the series resistance of the device.[80,82,83] Based on similar processes, silicon oxide (SiO$_x$) layers with even further lowered refractive indices are also interesting for use as intermediate reflecting layers within multi-junction solar cells.

2.3   Basis of Thin-film Silicon-based Multiple-junction Devices

2.3.1   Solar Cells Based on Thin Films of Silicon

Due to the large defect density in doped layers made out of thin-film silicon, minority carrier lifetime is too low to make a p–n junction as is typically done, *e.g.*, for crystalline silicon. An intrinsic layer is thus inserted, to form a p–i–n device, leading to an electric field in the intrinsic layer. Collection occurs therefore by drift, even though diffusion was shown to contribute as well in the case of μc-Si:H cells. Then, due to the low mobility in thin-film silicon materials, electrodes have to be used to provide lateral transport of collected carriers.[5] As a front electrode, a TCO is used, typically made out of tin oxide (SnO_2), zinc oxide (ZnO) or indium-tin-oxide (ITO).[84] Due to the low absorption coefficient of thin-film silicon-based material, a back reflector is required to reflect light that was not absorbed in a single pass back to the cell. A metallic electrode (typically silver) can be used,[74] or the combination of a TCO layer with a dielectric white reflector.[85]

Figure 2.3 presents the two major configurations that are considered in thin-film silicon technology: In the superstrate configuration light enters the device from the substrate side while light enters from the last layer deposited in the substrate configuration. Due to the poor hole mobility, especially in a-Si:H after light-induced degradation, light enters first the p-doped layer in both cases, making the layers growing sequence p–i–n in the superstrate case and n–i–p in the substrate case. For the substrate configuration, an opaque substrate (like any plastic or metallic foil) can be used, whereas for superstrate configuration, glass is most widely used.

To finally make a module, series interconnections of linear cells can be made by laser scribing, without the need of other metallization than the final wiring of the first and last cell. Interconnection of the front and back electrodes can be made by a combination of three parallel laser ablation steps, firstly after the first electrode deposition, secondly after the Si deposition, and thirdly after the second electrode deposition.

2.3.2   Possible Multiple-junction Devices Based on Thin Films of Silicon

The large versatility of TFSi based materials that can be made with a unique deposition equipment makes many combinations possible. As seen in previous sections, the bandgap-energy range that is accessible ranges from below 1 eV for a germanium-rich μc-SiGe:H alloy[86] to above 2 eV for a-SiC:H and a-SiO:H alloys with almost any intermediate value accessible through

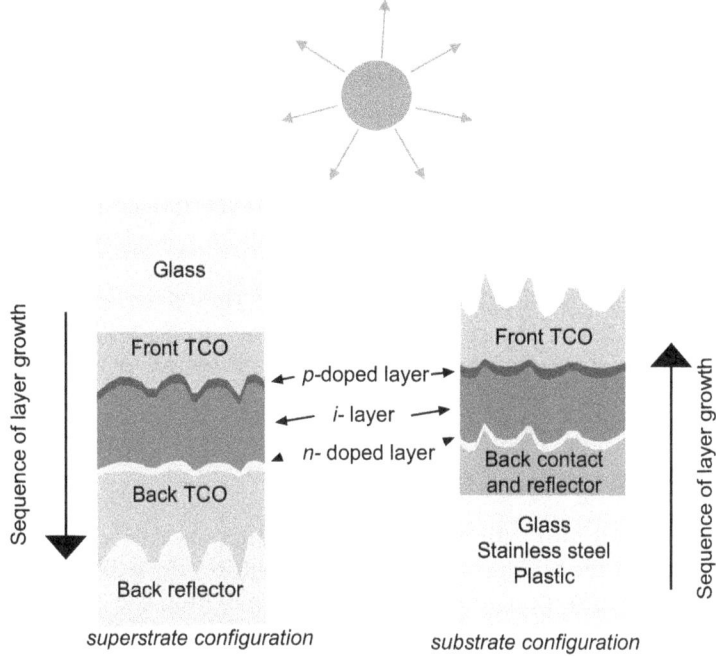

Figure 2.3   Schematics of single-junction thin-film silicon solar cells in the superstrate and in the substrate configurations.

μc-Si:H, a-SiGe:H and a-Si:H. This makes multiple-junction stacks a natural approach to improve the efficiency of TFSi based photovoltaic devices, and different combinations are of interest. For the reference AM1.5G spectrum of light, the theoretically most efficient combination of bandgaps for a tandem device is obtained by combining a top absorber with a bandgap of 1.75 eV with a bottom absorber with a bandgap of 1.1 eV.[87] This corresponds exactly to a combination a-Si:H/μc-Si:H, which is the most standard multiple-junction device based on thin-film silicon. Going further, a triple-junction device combining materials with bandgap values of 1.86 eV, 1.34 eV and 0.93 eV was shown to be optimal, though for an AM1.5D spectrum.[88] Even though an a-SiO:H/a-SiGe:H/μc-SiGe:H combination would come closest to these values, the large light-induced degradation of the a-Si:H-based alloys, and the still limited efficiency of thick μc-SiGe:H devices makes a-Si:H/μc-Si:H/μc-Si:H the present best combination as will be seen in section 2.4. Even though the possibility to make multi-junction devices with up to five junctions was experimentally demonstrated with a-Si:H cells;[89] in practice, going above three junctions has not proven any efficiency gain.

The efficiency gain originating from multiple-junction devices compared to single-junction devices have different causes: Firstly, by stacking different materials together, a better use of the spectrum can be made, minimizing the thermalization losses while ensuring absorption form a large part of the sun spectrum; secondly, stacking two a-Si:H (or alloyed a-Si:H) cells in a tandem device can also be of interest despite the very similar bandgaps of the two subcells (roughly between 1.5 eV and 1.9 eV). Because of the possibility to make at least one very thin i-layer, a lower degradation occurs compared to a single junction of identical total absorption. Higher stable efficiencies are then obtained compared to single-junction devices even though initial efficiencies of a tandem a-Si:H/a-Si:H devices and of a single-junction a-Si:H device can be similar. Finally, in tandem devices, the monolithic interconnection enables the operating voltage of the device to be increased while decreasing its current density.[89] Therefore, as a result of this increased impedance, the resistive losses in the electrodes are decreased, making possible the use of more resistive, thus more transparent, electrodes that lower parasitic absorption losses.

2.3.3   Matching Considerations

For monolithically interconnected multiple-junction devices, the current flowing through the complete device under illumination will be the current generated in the limiting sub-cell. Therefore, matching the current density generated in each sub-cell is required to get the highest efficiency out of a multiple-junction device. To adjust the current density generated in each sub-cell, one can tune the thicknesses of each sub-cell, introduce selective intermediate reflectors between selected sub-cells, or adjust the light-scattering properties of the device to favour light incoupling and light trapping for a specific wavelength range.

It should be noted that due to the different operating voltages and collection efficiency (translated in different short-circuit resistance values) of each sub-cell, the optimal matching conditions (matching at maximum power point, MPP) usually deviate slightly from perfect short-circuit current density matching conditions.[90,91]

In the case of a-Si:H/μc-Si:H tandem devices, the higher operating voltage and poorer collection efficiency of the a-Si:H top-cell compared to the μc-Si:H bottom-cell makes perfect MPP-matching occur for a bottom limitation in short-circuit current density (J_{SC}) conditions by over 1.0 mA cm^{-2} after light-induced degradation. The need to keep the a-Si:H cell as thin as possible to reduce the light-induced degradation therefore makes the use of selective reflectors between the top-cell and the bottom-cell essential to reach high stable efficiency values.

2.3.4   Combining Light Management and High-quality Absorber Layers

For a high efficiency to be reached, coupling light of each wavelength range in its respective individual sub-cell, and providing a template for growing high quality absorber-layers for all sub-cells is needed. A general strategy to improve the coupling of light from a transparent electrode of low refractive index (1.5–2) to the Si layers of higher refractive index (3.5–4) is the use of textured interfaces. Whereas a micrometre-scale roughness improves light coupling in crystalline silicon cells *via* multiple reflection, the nano-roughness features employed in thin-film silicon devices, with typical size lower than the wavelength of considered light, rather acts as an anti-reflective layer with an effective refractive index corresponding to the average between the two materials.

Rough interfaces will also scatter light, and are typically used to elongate the light path in thin-silicon-layers, both in superstrate and substrate configurations. The features composing the rough interface are typically of similar size as the wavelengths for which light would be scattered. Then, because of the higher refractive index of Si compared to the surrounding layers (typically $n = 3.5$–4 compared to 1.5–2 for TCO layers and 1 for the surrounding air), light can be trapped by total internal reflection in the thin Si layers. This can enhance the path of light into the cell quite significantly, with a theoretical maximum enhancement factor of $4n^2$ for a Lambertian light trapping. This leads to an enhancement factor above 50, as anticipated by Yablonovitch and Cody.[92]

Yet, it was observed that the growth of thin-film silicon on a rough surface from a vapour phase was leading to inhomogeneous layers that contain locally porous areas.[67,70,93] These defects limit the electrical transport in the active layer, reducing the V_{OC} and FF that could be obtained on an equivalent device grown on a smooth substrate. Therefore, the adequate surface features of the electrode that are commonly used are made out from a compromise between a texture with a high roughness—leading to high J_{SC} enhancement—and a smooth texture that is suitable for the growth of an absorber layer with high quality to obtain devices with high V_{OC} and FF.

2.4 State of the Art

Tables 2.1 and 2.2 report notable certified efficiencies of devices that were manufactured recently. Table 2.1 reports small area devices and Table 2.2

Table 2.1 Selection of best certified efficiencies reported for small-area multiple-junction devices.

Device structure (remarks)	Area (cm^2)	Stable efficiency (%)	Laboratory
a-Si:H/μc-Si:H (certified by Newport)	1.003	12.63	IMT Neuchâtel EPFL[94]
a-Si:H/μc-Si:H (certified by AIST)	0.962	12.3	Kaneka[95]
a-Si:H/μc-Si:H (certified by NREL)	1.227	11.9	TEL Solar[96]
a-Si:H/μc-Si:H (certified by AIST; customized LS degradation)	1.003	11.9	AIST[97]
a-Si:H/μc-Si:H/μc-Si:H (certified by NREL)	1.006	13.4	LG Electronics[1]
a-Si:H/μc-Si:H/μc-Si:H (certified by NREL)	0.27	12.5	United Solar[98]

Table 2.2 Selection of best certified efficiencies for mini-modules and large-area modules.

Device structure (remarks)	Area (cm^2)	Stable efficiency (%)	Laboratory
a-Si:H/μc-Si:H (certified by AIST)	14.23	11.7	Kaneka[95]
a-Si:H/a-SiGe:H/μc-Si:H (certified by NREL)	399.8	12.0	United Solar[99]
a-Si:H/a-SiGe:H/μc-Si:H (certified by AIST)	14 305	10.9	LG Electronics[94]

large area devices. The results achieved with small-area cells show the actual highest efficiencies that can be achieved for each type of device while the lower values of efficiencies that are reached on large area modules exhibit the current losses when up-scaling the technology. Much progress has been made to close this gap and efforts are continually made to tackle this issue. Certified record efficiencies of 10.1%[85], 10.7%[100] and 10.8%[94] have been reported, respectively, for a-Si:H and μc-Si:H single-junction devices with area above 1 cm^2. However, as discussed previously, it is necessary to include more than one junction in a single device to achieve higher efficiencies because of the light-induced degradation of a-Si:H and of the lower build-in-field in very thick μc-Si:H which decreases the device V_{OC}. The multiple junctions that were intensively investigated these last years both by academic laboratories and by industrial research and development laboratories are: a-Si:H/μc-Si:H, a-Si:H/a-Si$_{1-x}$Ge$_x$:H/a-Si$_{1-x}$Ge$_x$:H and a-Si:H/a-Si$_{1-x}$Ge$_x$:H/μc-Si:H. Due to the recent development of high-quality μc-Si:H materials at higher deposition rate interest has now also grown for a-Si:H/μc-Si:H/μc-Si:H devices which require thick μc-Si:H layers.

There are also remarkable efficiencies which were not certified but are worth mentioning: First, 5.7 m^2 modules with efficiencies over 10% stable efficiency have been reported by Stanovski *et al.*[101] demonstrating high efficiency on very large areas. According to the producer companies of micromorph modules stable efficiencies between 10.3% and 11.7% have been achieved[102–104] on modules with areas above 1 m^2. Finally, LG Electronics reported a module with 1.4 m^2 with a stable efficiency of 11.2% using a triple-junction a-Si:H/a-SiGe:H/μc-Si:H.[105] Although further increases in the efficiencies reported above are necessary, it is observed that this technology is now mature and that the production of large-area modules with efficiencies well above 10% has been achieved.

2.5 Current Limitations and Prospective Concepts

To push forward the device efficiencies shown in section 2.4, the current limitation that faces the technology should be better understood. The following discussions present the recent advances made to tackle the technology limitations and perspectives to bring novel improvements. There are two different but interlinked domains of research which should be addressed to improve the current technology status: First, because of the useful thickness employed in thin-film silicon devices and because silicon is not highly absorbing at long wavelengths, improving light absorption in the active area of the device is extremely important and this will be discussed in section 2.5.1. The second line of research, discussed in section 2.5.2, concentrates on improving the quality of the active materials and on designing novel materials with optical and electrical properties that are better designed for each specific application than standard a-Si:H and μc-Si:H materials.

2.5.1   Increasing Light Absorption in the Absorber

The total reflection of the device and all parasitic absorption (*i.e.*, all absorption that does not take place in the intrinsic layer) has to be lowered in order to increase light absorption in the active absorber layer of a device. The reduction of the total reflectance of the device is achieved through the design of better light-management schemes while the reduction of parasitic absorption has to be achieved by improving the different materials that composes the solar cell.

Currently, high solar-cell efficiencies are obtained by growing devices in superstrate configuration on front electrodes made of ZnO:Al or of ZnO:B and devices in substrate configuration on a silver reflector covered with a buffer layer of thin ZnO:Al. Each one of these films possess a texture with specific features: Smooth sputtered ZnO:Al films develop, after a wet etching step, crater-like features,[106] ZnO:B films naturally develop pyramidal features when grown by CVD for a certain range of temperature[107] and silver films develop grain-like features when sputtered on a substrate hold at high temperature.[108,109] Since these films play both the role of electrode and of light-scattering texture, there is obviously an interlink between the properties required for both functions, and, unfortunately, these requirements often enter in contradiction. However, it is clear that excellent trade-off between conductivity properties and optical properties have already been found to reach the efficiencies discussed in section 2.4. However, to emancipate the technology from the usual necessary compromises, novel techniques such as nano-imprinting,[110–113] and nanomoulding[114] were recently used to decouple the light-management texture from the conductive layer. These novel tools allow implementing in devices specifically designed photonic texture that can be first realized on a template even *via* expensive method such as electron beam lithography.[115] Figure 2.4 presents an example of the use of nano-imprinting in superstrate configuration in which this technique allowed improving the commonly used transparency/conductivity/light-scattering texture trade off. Figure 2.4a shows a scanning electron microscopy (SEM) view of a micromorph device cross-section in which UV nano-imprinting allowed for the implementation of multi-scaled textures which is beneficial for light absorption in such device as discussed by several groups.[116–118,157] Large-scale features that are ideal for scattering light of long wavelength in the bottom μc-Si:H cell were reproduced in a

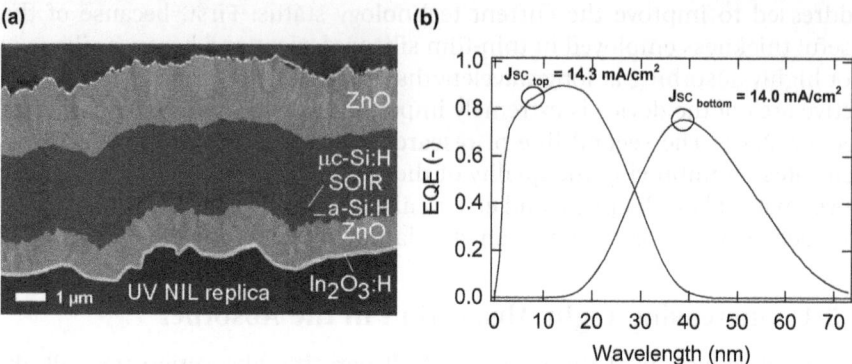

Figure 2.4 (a) SEM view of a micromorph device cross-section in which features with large-scale were imprinted into a UV sensitive resin. The deposition of a thin ZnO:B layer with small-scale features on top of the resin results in a multi-scale texture with excellent transparency and good light scattering properties for micromorph devices. (b) Initial EQE of the micromorph cell.

transparent UV sensitive resin while a thin conductive ZnO:B layer with optimal small-scale features to scatter short-wavelength light in the top cell was added on top of the resin. The high transparency of this electrode resulted in an improved device J_{SC} while the multi-scaled features were seen to be beneficial for the quality of the silicon material grown on top compared to a standard texture. This combination thus allowed reaching a high initial efficiency of 14% in a micromorph device with a surface area of 1 cm²;[117,118] the initial EQE is reproduced in Figure 2.4b.

As these tools made it possible to realize specifically designed texture and implement them directly into working devices, their use also promoted a strong effort on the simulation of the best light-scattering texture. Although random textures are widely used in practice and simulations have been able to successfully reproduce experimental results[119,120] much more work has been dedicated to the simulation of periodic structures[121–126] since periodicity allows using periodic boundary conditions which simplify the simulation and since several groups have discussed that gratings could, at least on a limited wavelength range, provide enhancement over the standard thermodynamic limit of $4n^2$.[127,128] However, several recent research results have shown that, even for a simplified structure such as a one- or two-dimensional grating, it is difficult to determine exactly the absorption in each layer and that it is for instance important to account for the exact shape of each interface within the device.[129–131] Since the conformality of the layers depends on the deposition conditions and on the thickness of each layer, it is therefore difficult to obtain valid predictions for a broad range of devices. Furthermore, the exact determination of the optical constant of each layer is not an easy task and is of foremost importance as observed by Solntsev and Zeman[129] for the silicon material and by Pahud *et al.*[132] for the silver. Finally, it is also not even clear whether the optimization of standard light-scattering texture through simulation or through experiments can really lead to significant improvements in the light absorption in the active part of the device. Indeed, even if the actual light-absorption enhancement is still far from the $4n^2$ limit provided by random textures that scatter light in a Lambertian way, this limit only represents the ideal case where no light is absorbed parasitically and may not be valid in our devices since the discreet nature of the photonic states have to be taken into account within the extremely thin layers that are typically used.[125,133] Furthermore, a simple model recently showed that the main light-management limitation in present thin-film silicon devices rather comes from the parasitic absorption of light—due to our conventional cell designs of doped layers and textured electrode—than from a sub-optimal trapping of light.[158]

Thus, to go beyond the actual light-absorption enhancement obtained, photonic designs that are not based on light scattering *via* rough textured substrates have also been proposed. First, the decoupling of the optical light-path from the electrical path by using nanowires or nanorods has been proposed by several groups[134–136] to combine an electrically thin device to reach high V_{OC} and FF values with an optically thick device with high J_{SC} values. Secondly, it was proposed to texture the absorber layer in itself to create a photonic crystal that modify the density of photonic states within the intrinsic layer.[137–139] Unfortunately, with the conventional techniques of Si deposition from a vapour phase, it has yet been difficult to realize efficient devices using these two types of designs. However, recent progress has been reported using nanowires and an initial efficiency of 8.1% was shown for a single-junction a-Si:H cell by Misra *et al.*[140] Even if this results is still far from the best stable single-junction a-Si:H record efficiency of 10.1% it is a promising advance to bring this method to a next step since several difficulties that arise in such configuration, such as the creation of shunt paths, could to be mitigated. Finally, in a totally different approach, the use of plasmonic effects to enhance light absorption has been heavily studied in recent years[141–144] since the properties of the different types of resonance are extremely attractive. Surface plasmon and localized plasmon resonances offer the possibility to spatially localize the electromagnetic field while light can be scattered at very high angles when interacting with plasmonic resonances occurring within metallic nanoparticles. Furthermore, the resonant energy can be tuned by modifying the shapes and the materials composing the plasmonic design. However, because of the low absorption coefficient of silicon in the wavelength region desired for increased light-absorption it is doubtful whether the absorption enhancement provided by such designs can surpass the one provided by a more standard one since part of the light will be absorbed by the metal.[142,145,146] These plasmonic photonic designs remain however of high interest in other types of solar cells possessing active material with high absorption coefficient.

The use of high substrate-roughness to obtain increased light absorption with traditional light-management schemes relying on textured interfaces conflicts with the growth of high-quality material as shown by several groups both for a-SiH[147] and μc-Si:H.[67,70] Therefore, different researchers have focused on finding novel approaches to benefit from improved surface texture for the cell growth while maintaining a high light-management level. The use of multi-scale textures is one of such approach:[117] It was shown that the number of defective zones was less important when Si is grown on a combination of large features and small features than when it is grown on the features with medium-range scale that are traditionally used. Thereby, the multi-scaled texture provided an improvement in the quality of the silicon layers and thus in V_{OC}. Furthermore, some work has been done on the use of an asymmetric intermediate reflector. While, as discussed in section 2.3.3, an intermediate reflector improves the stability of multiple-junction devices by allowing the use of thinner top-cells,[148,149] they can also be used at the same time to smoothen the interface for the subsequent cell growth to improve the quality of the deposited material and the V_{OC} of the devices as shown by Boccard *et al.* in a superstrate configuration[150] and by Biron *et al.* in a substrate configuration.[151] Finally, the substrate itself can be made flat to guarantee an optimal growth of the silicon material. To ensure sufficient light-absorption with a flat substrate a promising light-trapping design that decouples the optically rough interface form the growth surface has been proposed for devices grown in a substrate configuration: A flat silver reflector is covered with textured doped-ZnO layer which is then covered with n-doped[152] or intrinsic a-Si:H.[153] In a next step, the stack is mechanically polished to obtain a flat surface which is a perfect template for the subsequent growth of defect-free silicon layers. Figure 2.5a presents a scheme of such a substrate onto which devices in a substrate configuration can be grown, and Figure 2.5b shows the EQE of a triple-junction a-Si:H/μc-Si:H/μc-Si:H device grown on such a substrate which exhibited a high stable-efficiency of 13% on a 0.5 cm^2 area.[154]

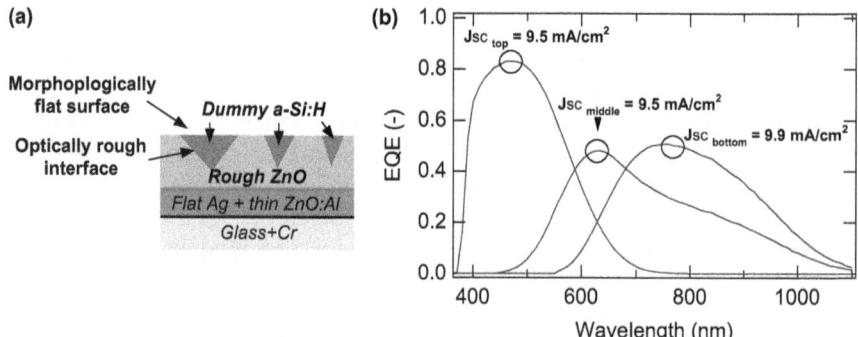

Figure 2.5 (a) Scheme of a flat light-scattering substrate, (b) EQE of a triple-junction a-Si:H/μc-Si:H/μc-Si:H with an area of 0.5 cm^2 grown on such a substrate exhibiting a stable efficiency of 13%.[154]

Since the growth rate and the crystalline fraction of the μc-Si:H layer can be increased with less dramatic effect on V_{OC} and FF on a flat surface than on a rough surface as shown by Bugnon *et al.*,[155] this type of substrate may also allow decreasing drastically the deposition time required for the device growth. The drawback of these substrates arise from parasitic absorption in the dummy Si that is part of the substrate itself but its impact can be limited by using intrinsic a-Si:H material as filler and using optically-thick devices such as triple-junction as demonstrated by Söderström *et al.*[154,156]

The reduction of parasitic absorption is indeed, aside from improving light-management schemes, crucial since they are one of the main J_{SC} limitations as shown by several groups.[157,158] The main sources of parasitic absorptions in standard cell designs are the TCO used as contacts, the doped layers and the back reflector. The current focus for TCO's improvement is on reaching higher carrier-mobility inside the crystal grains that compose the film to increase its conductance without having a high carrier-concentration that absorbs light. Ding *et al.* have shown that the trade-off transparency/conductance can be improved in LP-CVD ZnO:B, by combining bi-layers doped differently[159] or by using post-deposition plasma treatments.[160] Wimmer *et al.*[161] and Hüpkes *et al.*[162] have shown that higher carrier-mobility and stability in humid atmosphere was achieved in sputtered-etched ZnO:Al by using post-deposition thermal treatments which improve as well the transparency/conductance trade-off. Finally, for indium-based TCO, Koida *et al.* have shown that extremely high carrier-mobility beyond $100 \text{ cm}^2 \text{ V}^{-1} \text{ s}^{-1}$ can be reached in hydrogenated indium oxide In_2O_3:H.[163] Compared to the generally used ITO compound, the high mobility of In_2O_3:H leads to layers with low sheet-resistance despite low carrier-concentration. In_2O_3:H layers were successfully used as front contact in TFSi solar cells and have shown excellent transparency compared to ITO[164] or LP-CVD ZnO:B.[165]

Significant reduction of parasitic absorption has been achieved in the doped layers through alloying them with carbon,[166,167] or oxygen[168] thereby obtaining higher bandgap. Recently, Cuony *et al.*[169,170] have revealed in detail the microstructure of $μc$-$Si_{1-x}O_x$ layers bringing a better understanding of the material and showing that a multitude of films with different properties can be made with the same gas precursors. These $μc$-$Si_{1-x}O_x$ materials have now been adopted by many research groups and manufacturers and are used within n- and p-doped layers as well as for growing intermediate reflectors. Finally, important parasitic absorption can also take place in the back reflector especially when made with rough metallic films as several groups have shown.[171–175] The mitigation of absorption in the metal can be made by reducing the light coupling to plasmonic resonances by using either a flat metallic reflector or by engineering the electromagnetic field in such way that its intensity is reduced in the vicinity of the metal. This is generally achieved by adding a buffer layer made of a material with low index of refraction between the metal and the silicon.[176] Finally it was also shown that the reflectance of silver layers grown on rough surfaces could be improved by using thermal annealing at a low temperature of 150 °C.[177]

2.5.2   Improvements in Silicon Materials

The complexity of the disorganized materials used in TFSi solar cells implies that research on the materials is far from being complete. It is first important to understand why the V_{OC} of a-Si:H and μc-Si:H single junctions, which are at best typically 1 V and 600 mV, respectively, are far from their respective bandgap values of 1.7 eV and 1.1 eV. The disorganized silicon-networks which introduce band-tails in the conduction- and valence-bands are often related to this discrepancy between real obtained V_{OC} and the bandgap. It is therefore important to know if it is possible to reduce the band-tails using the actual deposition techniques or if novel deposition tools are required to improve the materials quality. However, aside from the bulk quality, the influence and the importance of the interfaces and of the doped layers on the V_{OC} and FF values have to be better understood. The implementation of alloyed doped-layers with large optical bandgaps were shown to be beneficial not solely for their optical properties as discussed before but for the improvement they brought to the device electrical properties as well. The use of high-gap p-doped layers containing carbon[178] and oxygen[168,179] was seen to lead to a V_{OC} increase that probably results from the band offset that increases the work function, and to a decrease in the recombinations which take place at the i/p interface. It was also observed that by using oxide-doped layers the sensitivity of the device to the substrate morphology could be decreased[180] thereby increasing the V_{OC} and FF values of devices grown on rough substrates which allowed the use of more aggressive texture for light management. This effect was attributed to the low conductivity of oxide layers which introduce a shunt quenching effect by preventing high current densities to flow through the localized shunt paths created by the growth of porous areas above sharp substrate features. Then the use of carbon[181] or oxide[182] alloyed intrinsic-buffer layers at the i/p interface were also seen to improve devices performances. It has to be mentioned that alloying silicon is always an extremely delicate task since it often induces the creation of defects. Thus, by implementing these layers, initial performances are often increased drastically but severe degradation might occur. Still, the multiple successful implementations of alloyed layers in different position in multijunction devices shown by different laboratories proved that by using adequate designs beneficial effects can be obtained after degradation as well. To avoid the use of alloyed buffers, unalloyed intrinsic buffer layers were also inserted at the i/n,[183,184] or at the i/p[185] interfaces with nice results in μc-Si:H devices. Still, the beneficial impact of these intrinsic buffer layers was mostly observed for devices with thin intrinsic-layers for which the limitation due to the interfaces is important compare to the limitations induced by the quality of the bulk material.[183,185] It is thus likely that to further improve the solar cell performance in thicker junctions the quality of the bulk material should be improved. To improve the quality of the bulk material further research is required in the area of plasma and deposition conditions. Indeed recent advances in these domains have been made both for a-Si:H and for μc-Si:H materials: Matsui and Kondo have shown that the use of a remote plasma technique using a triode configuration in PE-CVD can lower to 10% the light-induced degradation of a-Si:H single-junction device with a standard thickness of 250 nm for the intrinsic layer while this degradation is generally at least of 15% for devices grown with standard PE-CVD deposition conditions for the same thickness.[186] They reported at the same time reasonable degradation for larger thicknesses of the intrinsic layer up to 390 nm which is extremely promising for their application in the micromorph configuration. Indeed one of the actual limitations to reach high efficiencies using the micromorph configuration is the current density produced by the top a-Si:H cell: More than 14 mA cm^{-2} is required and the micromorph stability is poor if such high current is obtained using top-cells with a thick intrinsic-layer. This is why many studies have been done on using an intermediate reflector. However, highly stable top cells with thicker intrinsic-layers could be more beneficial than using an intermediate reflector as the latter often results in a loss of total absorption by reflecting part of the light out of the device. Unfortunately, the deposition rate using the triode configuration is almost ten times lower than the deposition rate obtained with conventional techniques but it still demonstrates that it is possible to improve the bulk quality of a-Si:H materials by adjusting the plasma processes in an appropriate fashion.

Improvements of the understanding of plasma deposition conditions has also led recently to improvement in the quality of the μc-Si:H intrinsic material especially for high deposition rates which are important for industrial applications: Chaudhary *et al.*[187] and Matsui and Kondo[186] have shown that the use of higher pressure using very-high-frequency conditions led to improved devices performances for high growth rates. This was achieved by using small inter-electrode gap in the range of 4–5 mm that allows high pressure without the detrimental formation of large powder particles. It is probable that high pressure allows the deposition of dense material despite the high growth-rate, whereas for the lower-pressure-regime an increase in deposition rate was shown to drastically enhance the detrimental formation of defective porous areas during the film growth.[155] For improving light absorption of long wavelength when μc-Si:H layers are used as bottom cells in multiple-junction devices, it is important to increase the crystalline-volume-fraction. Unfortunately, it was observed that an increase in the crystalline-volume-fraction induces losses in the device V_{OC} often attributed to a lower bandgap and increased defects.[188] However, to solve this problem, promising results were obtained using SiF$_4$ as a gas source for the silicon atoms: Zhang *et al.*[189] obtained a high V_{OC} of 523 mV for a single-junction μc-Si:H device with a crystalline volume fraction of 80%. Following this work, Dornstetter *et al.*[190] have shown that it was possible to maintain high V_{OC} in single-junction μc-SiH solar cells even with 3.8-μm-thick intrinsic-layer. This is promising since no V_{OC} and FF dependences on the intrinsic-layer thickness were observed, indicating that, in this case, the bulk material was not limiting the carrier transport. It has also been reported by Abramov and

Roca i Cabarrocas[191] that the addition of a small SiF_4 flow in the generally used SiH_4/H_2 gas precursors helps decrease the oxygen content which improves the electronic transport in μc-Si:H bulk material. All these examples show that the deposition conditions, the plasma properties and the gas precursors have to be better studied and their relations to the material properties should be further understood to reach even better material qualities.

Up to now the discussion focused on how to improve unalloyed a-Si:H and μc-Si:H material but as discussed in section 2.3.2, the choice of materials for the intrinsic layers of the different sub-cells in a multiple-junction device is dependent on its position in the device and on the total number of junctions that composes the device. Yunaz *et al.*[43] found that the optimal top-cell bandgap in triple-junction devices is around 2V. Thus, even if top cells exhibiting 1 V of V_{OC} were demonstrated using a-Si:H by Yan *et al.*[192] and Kim *et al.*,[1] an effort in alloying intrinsic a-Si:H with carbon[193] or oxygen[194,195] to obtain even higher bandgap and possibly higher V_{OC} has been made. However, as indicated before, alloying silicon strongly hampers the film's stability due to an increased density of defects and thus further research on the stability of $a\text{-}Si_{1-x}\text{:}C_x$ and $a\text{-}Si_{1-x}\text{:}O_x$ films is required.

One way or another, high-gap top cells are necessary and with these, triple-junction devices may be good candidates to further push the efficiencies of current devices. It is a complicated enough device composed of three sub-junctions which gaps can be tuned independently to optimize the light absorption in each, and, at the same time, it is a device simple enough to be manufactured. The material chosen as the intrinsic layer for the middle cell is often made of $a\text{-}Si_{1-x}Ge_x\text{:}H$ since bandgaps between 1.7eV and 1.4 eV can be obtained by tuning the Ge content. This material has been highly studied and triple-junction devices with the $a\text{-}Si\text{:}H/a\text{-}Si_{1-x}Ge_x\text{:}H/a\text{-}Si_{1-x}Ge_x\text{:}H$ structure were implemented into production lines by United Solar.[196] Profiling the Ge content in each sub-cell was the main improvement made to decrease the high light-induced degradation.[197] To limit the degradation in triple-junction devices, the actual trend is to replace either the bottom cell or both the middle and the bottom cells by μc-Si:H, both structures having been realized as shown in Tables 2.1 and 2.2. To improve further the absorption of long wavelengths, $\mu c\text{-}Si_{1-x}Ge_x\text{:}H$ has been studied to be used as the bottom cell. Since the first growth of this type of film by Ganguly *et al.* in 1996[86] efficiencies between 3% and 8% have been reported for single junction devices.[198-201] Matsui *et al.* have been able to demonstrate the low degradation of this material, proving therefore that such material is a better candidate than $a\text{-}Si_{1-x}Ge_x\text{:}H$ for being used in multiple-junction devices.[202] Matsui *et al.* have been able to grow tandem devices with this material demonstrating nicely that, in this configuration, the thickness of $\mu c\text{-}Si_{0.83}Ge_{0.17}\text{:}H$ required to produce the desired current is less than half the thickness required with μc-Si:H to produce the same current.[200] The lower use of material is very interesting since it induces a large throughput gain and decreases also the cleaning time of the machines. Initial efficiencies of multiple-junction devices made with $\mu c\text{-Si}_{1-x}\text{Ge}_x\text{:H}$ bottom-cells were already reported: Matsui *et al.* reported initial efficiencies of 11.2%[200] for tandem a-Si:H/$\mu c\text{-Si}_{1-x}\text{Ge}_x\text{:H}$, and of 11.6%[203] in triple-junction a-Si:H/μc-Si:H/$mc\text{-Si}_{1-x}\text{Ge}_x\text{:H}$ while Cao *et al.* reported an initial efficiency of 12.0%[195] in triple-junction a-Si:H/a-Si$_{1-x}$Ge$_x$:H/$mc\text{-Si}_{1-x}\text{Ge}_x\text{:H}$. Further studies are required to improve the $\mu c\text{-Si}_{1-x}\text{Ge}_x\text{:H}$ material since the reported FF values for single-junction solar cells are lower than for single-junction μc-Si:H and drop rapidly for increased Ge content and layer thickness.[200]

The degradation within alloyed a-Si:H single-junction solar-cells and the defect density in $\mu c\text{-Si}_{1-x}\text{Ge}_x\text{:H}$ material are probably the main reasons why the current highest certified efficiency reported for TFSi solar cells was made using an a-Si:H/μc-Si:H/μc-Si:H structure.[1] This latter configuration has several advantages since it allows the use of middle cells with a thin in-trinsic-layer—thus with high V_{OC}—and a low degradation between 3% and 6% was observed by United Solar for this type of device.[95] To conclude this discussion, a-Si:H and μc-Si:H material offers extraordinary possibilities since the means of depositing them provide the opportunity to alloy them and to deposit a whole range of different materials with different bandgaps but the quality of these new materials must still be increased to further push the device efficiencies.

2.6 Conclusions and Perspectives

Demonstrated for the first time close to 40 years ago, TFSi solar cells have been turned into a true technology over the last decades, in one of the largest industrial efforts ever made in photovoltaics at developing mass production solutions for thin films. The technology can be readily deployed, and is fully compatible with TWs of PV. It has potential to reach low production costs (0.35–0.5$ per watt) and efficiencies well over 10% while exhibiting the requested durability. Also, this technology which involves no toxic element has shown amazing examples of how solar could be integrated in virtually all buildings, with a potential to come at the price of the tiles.

In parallel, much progress in the understanding of the TFSi devices has been made in the past decades. The role of substrate morphology, plasma processes, parasitic absorption have now been disentangled, opening the route for new designs to obtain devices with higher efficiencies. A current estimated efficiency potential around 15% is foreseen; going beyond this value will require additional material-improvements (or new alloy/materials) and a reworking of device interfaces to exploit the potential provided by heterojunction or passivated contacts.

References

1. S. Kim, J.-W. Chung, H. Lee, J. Park, Y. Heo and H.-M. Lee, *Sol. Energy Mater. Sol. Cells*, 2013, **119**, 26.

2. A. Shah, E. Moulin and C. Ballif, *Sol. Energy Mater. Sol. Cells*, 2013, **119**, 311.
3. R. C. Chittik, J. H. Alexander and H. E. Sterling, *J. Electrochem. Soc.*, 1969, **116**, 77.
4. W. E. Spear and P. G. LeComber, *Solid State Commun.*, 1975, **17**, 1193.
5. D. E. Carlson and C. R. Wronski, *Appl. Phys. Lett.*, 1976, **28**, 671.
6. W. Luft and Y. S. Tsuo, *Hydrogenated Amorphous Silicon Alloy Deposition Process*, Marcel Dekker, New York, 1993.
7. E. Vallat-Sauvain, U. Kroll, J. Meier, N. Wyrsch and A. Shah, *J. Non-Cryst. Solids*, 2000, **266–269**, 125.
8. E. Vallat-Sauvain, A. Shah and J. Bailat, *Advances in Microcrystalline Silicon Solar Cell Technologies*, John Wiley & Sons, Chichester, 2006, pp. 133–165.
9. J. Koh, H. Fujiwara, Y. Lu, C. R. Wronski and R. W. Collins, *Thin Solid Films*, 1998, **313–314**, 469.
10. Y. Lee, L. Jiao, H. Liu, Z. Lu, R. W. Collins and C. R. Wronski, Proceedings of the 25th IEEE photovoltaic specialists conference, 1996, p. 1165.
11. P. Roca i Cabarrocas, A. Fontcuberta i Morral and Y. Poissant, *Thin Solid Films*, 2002, **403**, 39.
12. A. H. M. Smets, W. M. M. Kessels and M. C. M. van de Sanden, *Appl. Phys. Lett.*, 2003, **82**, 1547.
13. A. H. M. Smets, C. R. Wronski, M. Zeman and M. C. M. van de Sanden, *Mater. Res. Soc. Symp. Proc.*, 2010, **1245**, A.14.02.
14. A. H. M. Smets and M. C. M. van de Sanden, *Phys. Rev. B*, 2007, **76**, 073202.
15. C. R. Wronski, S. Lee, M. Hicks and S. Kumar, *Phys. Rev. Lett.*, 1989, **63**, 1420.
16. R. Grigorovici and A. Vancu, *Physica Status Solidi*, 1966, **15**, 627.
17. G. D. Cody, T. Tiedje, B. G. Brooks and Y. Goldstein, *Phys. Rev. Lett.*, 1982, **47**, 1480.
18. R. A. Street, *Phys. Rev Lett.*, 1982, **49**, 1187.
19. D. L. Staebler and C. W. Wronski, *Appl. Phys. Lett.*, 1977, **31**, 292.
20. A. Kolodziej, *Opto-Electron. Rev.*, 2004, **12**, 21.
21. P. Stardins, Proceedings of the 25th IEEE photovoltaic specialists conference, 2010, p. 142.
22. C. Hof, N. Wyrsch and A. Shah, *J. Non-Cryst. Solids*, 1998, **287–291**, 287.
23. A. Matsuda, M. Takai, T. Nishimoto and M. Kondo, *Sol. Energy Mater. Sol. Cells*, 2003, **78**, 3.
24. M. Stuckelberger, M. Despeisse, G. Bugnon, J.-W. Schüttauf, F.-J. Haug and C. Ballif, *J. Appl. Phys.*, 2013, **114**, 154509.
25. P. Zanzucchi, C. R. Wronski and D. E. Carlson, *J. Appl. Phys.*, 1977, **48**, 5227.
26. A. Morimoto, T. Miura, M. Kumeda and T. Shimizu, *J. Appl. Phys.*, 1982, **53**, 7299.
27. K. Haga and H. Watanabe, *Jpn. J. Appl. Phys.*, 1990, **29**, 636.

28. M. Stutzmann, R. A. Street, C. C. Tsai, J. B. Boyce and S. E. Ready, *J. Appl. Phys.*, 1989, **66**, 569.
29. J. Yang, A. Banerjee and S. Guha, *Appl. Phys. Lett.*, 1997, **70**, 2975.
30. Y. Nakata, A. Yokata, H. Sannomiya, S. Moriuchi, Y. Inoue, K. Nomoto, M. Itoh and K. Tsui, *Jpn. J. Appl. Phys.*, 1992, **31**, 168.
31. S. Veprek and V. Marecek, *Solid-State Electronics*, 1968, **11**, 683.
32. A. Matsuda, *J. Non-Cryst. Solids*, 1983, **59–60**, 767.
33. G. Willeke, in *Amorphous Silicon and Microcrystalline Devices-Materials and Device Physics*, ed. Kanicki J. Artech House, Norwood, MA, ISBN 0-89006-379-6, 1991, p. 55.
34. A. Matsuda, S. Yamasaki, K. Nakagawa, H. Okushi, K. Tanaka, S. Iizima, M. Matsumura and H. Yamamoto, *Jpn. J. Appl. Phys.*, 1980, **19**, L305.
35. C. Tsai, *Amorphous Silicon and Related Materials*, ed. H. Fritzsche, World Scientific, 1988, p. 123.
36. J. Meier, R. Flückiger, H. Keppner and A. Shah, *Appl. Phys. Lett.*, 1994, **65**, 860.
37. G. Yue, B. Yan, L. Sivec, Y. Zhou, J. Yang and S. Guha, *Sol. Energy Mater. Sol. Cells*, 2011, **104**, 109.
38. P. Torres, J. Meier, R. Flückiger, U. Kroll, J. A. Anna Selvan, H. Keppner, A. Shah, S. D. Littelwood, I. E. Kelly and P. Giannoulès, *Appl. Phys. Lett.*, 1996, **69**, 1373.
39. A Shah, *Thin-Film Silicon Solar Cells*, EPFL Press, 2010.
40. J. K. Rath, *Sol. Energy Mater. Sol. Cells*, 2003, **76**, 431.
41. S. Guha, J. Yang and B. Yan, *Sol. Energy Mater. Sol. Cells*, 2013, **119**, 1.
42. Y. Djeridane. PhD thesis, LPICM- Laboratoire de Physique des Interfaces et des Couches Minces, Ecole polytechnique, 2008.
43. I. A. Yunaz, A. Yamada and M. Konagai, *Jpn. J. Appl. Phys.*, 2007, **46**, L1152.
44. J. Meier, S. Dubail, R. Flückiger, D. Fischer, H. Keppner and A. Shah, Proceedings of the 24th IEEE PVSC, Haiwai, 1994, p. 409.
45. A. H. M. Smets, T. Matsui and M. Kondo, *J. Appl. Phys.*, 2008, **104**, 034508.
46. L. Guo, M. Kondo, M. Fukawa, K. Saitoh and A. Matsuda, *Jpn. J. Appl. Phys.*, 1998, **37**, 1116.
47. W. J. Soppe, A. C. W. Biebericher, C. Devilee, H. Donker and H. Schlemm, Proceedings of 3rd World Conference on Photovoltaic Energy Conv., 2003, **2**, 1655.
48. H. Jia, H. Shira and I. M. Kondo, *J. Appl. Phys.*, 2007, **101**, 114912.
49. E. Takahashi, Y. Nishigami, A. Tomyo, M. Fujiwara, H. Kaki, K. Kubota, T. Hayashi, K. Ogata, A. Ebe and Y. Setsuhara, *Jpn. J. Appl. Phys.*, 2007, **46**, 1280.
50. H. Kakiuchi, H. Ohmi, Y. Kuwahara, M. Matsumoto, Y. Ebata, K. Yasutake, K. Yoshii and Y. Mori, *Jpn. J. Appl. Phys.*, 2006, **45**, 3587.
51. H. Kakiuchi, H. Ohmi, K. Ouchi, K. Tabuchi and K. Yasutake, *J. Appl. Phys.*, 2009, **106**, 013521.

52. T. Hai Dao, PhD thesis, Laboratoire de Physique des interfaces et des Couches Minces, 2007.
53. L. Kroely, PhD thesis, Laboratoire de Physique des interfaces et des Couches Minces, 2010.
54. A. C. Bronneberg, PhD thesis, Technische Universiteit Eindhoven, 2012.
55. M. N. van den Donker, B. Rech, F. Finger, W. M. M. Kessels and M. C. M. van de Sanden, *Appl. Phys. Lett.*, 2005, **87**, 263503.
56. Y. Takeuchi, H. Mashima, M. Murata, S. Uchino and Y. Kawai, *Jpn. J. Appl. Phys.*, 2001, **40**, 3405.
57. C. Niikura and M. Kondo, *J. Non-Cryst. Solids*, 2004, **338–340**, 42.
58. O. Vetterl, F. Finger, R. Carius, P. Hapke, L. Houben, O. Kluth, A. Lambertz, A. Mück, B. Rech and H. Wagner, *Sol. Energy Mater. Sol. Cells*, 2000, **62**, 97.
59. E. Vallat-Sauvain, U. Kroll, J. Meier, A. Shah and J. Pohl, *J. Appl. Phys.*, 2000, **87**, 3137.
60. E. Bustarret, M. A. Hachicha and M. Brunel, *Appl. Phys. Lett.*, 1988, **52**, 1675.
61. F. Meillaud, A. Shah, E. Vallat-Sauvain, X. Niquille, M. Dubey and C. Ballif, *Proc. of the 20th EU-PVSEC*, 2005, 1509.
62. A. Matsuda, *Jpn. J. Appl. Phys.*, 2004, **43**, 7909.
63. B. Strahm, A. A. Howling, L. Sansonnens and Ch. Hollenstein, *Plasma Sources Sci. Technol.*, 2007, **16**, 80.
64. M. Kondo, *Sol. Energy Mater. Sol. Cells*, 2005, **78**, 543.
65. B. Kalache, A. I. Kosarev, R. Vanderhaghen and P. Roca i Cabarrocas, *J. Appl. Phys.*, 2003, **93**, 1262.
66. E. Vallat-Sauvain, J. Bailat, J. Meier, X. Niquille, U. Kroll and A. Shah, *Thin Solid Films*, 2005, **485**, 77.
67. Y. Nasuno, M. Kondo and A. Matsuda, *Jpn. J. Appl. Phys.*, 2001, **40**, L303.
68. M. Vanecek and A. Poruba, *Appl. Phys. Lett.*, 2002, **80**, 719.
69. G. Bugnon, PhD thesis, EPFL, Available: http://dx.doi.org/10.5075/epfl-thesis-5991, 2013.
70. M. Python, E. Vallat-Sauvain, J. Bailat, D. Dominé, L. Fesquet, A. Shah and C. Ballif, *J Non-Cryst. Solids*, 2008, **354**, 2258.
71. M. Python, O. Madani, D. Dominé, F. Meillaud, E. Vallat-Sauvain and C. Ballif, *Sol. Energy Mater. Sol. Cells*, 2009, **93**, 1714.
72. H. B. T. Li, R. H. Franken, J. K. Rath and R. E. I. Schropp, *Sol. Energy Mater. Sol. Cells*, 2009, **93**, 338.
73. Y. Naruse, M. Matsumoto, T. Sekimoto, M. Hishida, Y. Aya, W. Shinohara, A. Fukushima, S. Yata, A. Terakawa, M. Iseki and M. Tanaka, Proceedings of the 38th IEEE PV Conference, 2012, 003118.
74. G. Bugnon, PhD thesis, EPFL, 2013.
75. P. O. Grabitz, U. Rau and J. H. Werner, *Physica Status Solidi A*, 2005, **202**, 2920.
76. T. Matsui, M. Kondo and A. Matsuda, *Jpn. J. Appl. Phys.*, 2003, **42**, 901.

77. F. Finger, R. Carius, T. Dylla, S. Klein, S. Okur and M. Gunes, *Circuits, Devices and Systems, IEE Proc.*, 2003, **150**, 300.
78. A. C. Bronneberg, A. H. M. Smets, M. Creatore and M. C. M. van de Sanden, *J. Non-Cryst. Solids*, 2011, **357**, 884.
79. N. Pingate, D. Yotsaksri and P. Sichanugrist, Proceeings of the 21st European Photovoltaic Solar Energy Conference, Dresden, 2006, p. 1601.
80. M. Despeisse, G. Bugnon, A. Feltrin, M. Stueckelberger, P. Cuony, F. Meillaud, A. Billet and C. Ballif, *Appl. Phys. Lett.*, 2010, **96**, 073507.
81. T. Chen, Y. Huang, A. Dasgupta, M. Luysberg, L. Houben, D. Yang, R. Carius and F. Finger, *Sol. Energy Mater. Sol. Cells*, 2012, **98**, 370.
82. P. Cuony, D. T. L. Alexander, I. Perez-Wurfl, M. Despeisse, G. Bugnon, M. Boccard, T. Söderström, A. Hessler-Wyser, C. Hébert and C. Ballif, *Adv. Mater.*, 2012, **24**, 1521.
83. G. Bugnon, G. Parascandolo, T. Söderström, P. Cuony, M. Despeisse, S. Hänni, J. Holovský, F. Meillaud and C. Ballif, *Adv. Funct. Mater.*, 2012, **22**, 1616.
84. W. Beyer, J. Hüpkes and H. Stiebig, *Thin Solid Films*, 2007, **516**, 147.
85. S. Benagli, D. Borrello, E. Vallat-Sauvain, J. Meier, U. Kroll, J. Hötzel, J. Bailat, J. Steinhauser, M. Marmelo, G. Monteduro and L. Castens, Proceedings of 24th EU-PVSEC, 2009, 3B0.9.3, 2293.
86. G. Ganguly, T. Ikeda, T. Nishimiya, K. Saitoh, M. Kond and A. Matsuda, *Appl. Phys. Lett.*, 1996, **69**, 4224.
87. F. Meillaud, A. Shah, C. Droz, E. Vallat-Sauvain and C. Miazza, *Sol. Energy Mater. Sol. Cells*, 2006, **90**, 2952.
88. J. F. Geisz, D. J. Friedman, J. S. Ward, A. Duda, W. J. Olavarria, T. E. Moriarty, J. T. Kiehl, M. J. Romero, A. G. Norman and K. M. Jones, *Appl. Phys. Lett.*, 2008, **93**, 123505.
89. Y. Hamakawa, H. Okamoto and Y. Nitta, *Appl. Phys. Lett.*, 1979, **35**, 187.
90. M. Bonnet-Eymard, M. Boccard, G. Bugnon, F. Sculati-Meillaud, M. Despeisse and C. Ballif, *Sol. Energy Mater. Sol. Cells*, 2013, **117**, 120.
91. C. Ulbrich, C. Zahren, A. Gerber, B. Blank, T. Merdzhanova, A. Gordijn and U. Rau, *Int. J. Photoenergy*, 2013, **2013**, 314097.
92. E. Yablonovitch and G. Cody, *IEEE T. on Electron Dev.*, 1982, **29**, 300.
93. H. B. T. Li, R. H. Franken, J. K. Rath and R. E. I. Schropp, *Sol. Energy Mater. Sol. Cells*, 2009, **93**, 338.
94. M. Boccard, M. Despeisse, J. Escarre, X. Niquille, G. Bugnon, S. Hänni, M. Bonnet-Eymard, F. Meillaud and C. Ballif, *IEEE Journal of Photovoltaics*, to be published.
95. M. A. Green, K. Emery, Y. Hishikawa, W. Warta and D. Dunlop, *Prog. Photovoltaics*, 2013, **21**, 1.
96. J. Bailat, L. Fesquet, J.-B. Orhan, Y. Djeridane, B. Wolf, P. Madliger, J. Steinhauser, S. Benagli, D. Borrello, L. Castens, G. Monteduro, M. Marmelo, B. Dehbozorghi, E. Vallat-Sauvain, X. Multone, D. Romang, J.-F. Boucher, J. Meier, U. Kroll, M. Despeiss, G. Bugnon, C. Ballif, S. Marjanovic, G. Kohnke, N. Borrelli, K. Koch, J. Liu,

R. Modavis, D. Thelen, S. Vallon, A. Zakharian and D. Weidman, *Proc. of 25th EU-PVSEC*, 2010, **3B0.11.5**, 2720.

97. T. Matsui, H. Sai, K. Saito and M. Kondo, *Prog. Photovoltaics*, 2013, **21**, 1363.

98. B. Yan, G. Yue, X. Xu, J. Yang and S. Guha, *Phys. Status Solidi A*, 2010, **207**, 671.

99. S. Guha, J. Yang and B. Yan, *Sol. Energy Mater. Sol. Cells*, 2013, **119**, 1.

100. S. Hänni, G. Bugnon, G. Parascandolo, M. Boccard, J. Escarré, M. Despeisse, F. Meillaud and C. Ballif, *Prog. Photovoltaics*, 2013, **21**, 821.

101. B. Stannowski, O. Gabriel, S. Neubert, S. Kimer, S. Calna, S. Riing, S. Schönau, F. Ruske, M. Zelt, B. Rau, B. Rech, R. Schlatmann, T. Frijnts, H. Zollondz, A. Heidelberg, C. Schultz and B. Szyszka, Presented at the 28th EU-PVSEC, 3BO.5.6, 2013.

102. O. Kluth, J. Kalas, M. Fecioru-Morariu, P. A. Losio and J. Hoetzel, *Proc. of 26th EU-PVSEC*, 2011, 3B0.3.6, 2354.

103. A. Terakawa, M. Hishida, S. Yata, W. Shinohara, A. Kitahara, H. Yoneda, Y. Aya, I. Yoshida, M. Iseki and M. Tanaka, *Proc. of 26th EU-PVSEC*, 2011, 3B0.4.2, 2365.

104. 11.7% efficiency reached by TEL Solar on the module size, private communication.

105. D. J. You, S. H. Kim, H. Lee, J.-W. Chung, S.-T. Hwang, Y. H. Heo, S. Lee and H.-M. Lee, not yet published.

106. O. Kluth, B. Rech, L. Houben, S. Wieder, G. Schöpe, C. Beneking, H. Wagner, A. Löffl and H. W. Schock, *Thin Solid Films*, 1999, **351**, 247.

107. S. Fa, U. Kroll, C. Bücher, E. Vallat-Sauvain and A. Shah, *Sol. Energy Mater. Sol. Cells*, 2005, **86**, 385.

108. J. A. Thornton, *J. Vac. Sci. Technol., A*, 1986, **4**, 3059.

109. R. H. Franken, R. L. Stolk, C. H. M. Li, H. van der Werf, J. K. Rath and R. E. I. Schropp, *J. Appl. Phys.*, 2007, **102**, 014503.

110. J. Bailat, V. Terrazzoni-Daudrix, J. Guillet, F. Freitas, X. Niquille, A. Shah, C. Ballif, T. Scharf, R. Morf, A. Hansen, D. Fischer, Y. Ziegler and A. Closset, *Proc. of the 20th EU-PVSEC*, 2005, **3AO.8.6**, 1529.

111. K. Söderström, J. Escarré, O. Cubero, F.-J. Haug, S. Perregaux and C. Ballif, *Prog. Photovoltaics*, 2011, **19**, 202.

112. A. Bessonov, Y. Cho, S.-J. Jung, E.-A. Park, E.-S. Hwang, J.-W. Lee, M. Shin and S. Lee, *Sol. Energy Mater. Sol. Cells*, 2011, **95**, 2886.

113. U. W. Paetzold, W. Zhang, M. Prömpers, J. Kirchhoff, T. Merdzhanova, S. Michard, R. Carius, A. Gordijn and M. Meier, *J. Mater. Sci. Eng. B*, 2013, **178**, 617.

114. C. Battaglia, J. Escarré, K. Söderström, M. Charrière, M. Despeisse, F.-J. Haug and C. Ballif, *Nat. Photonics*, 2011, **5**, 535.

115. V. E. Ferry, M. A. Verschuuren, H. B. T. Li, R. E. I. Schropp, H. A. Atwater and A. Polman, *Appl. Phys. Lett.*, 2009, **95**, 183503.

116. O. Isabella, J. Krc and M. Zeman, *Appl. Phys. Lett.*, 2010, **97**, 101106.

117. M. Boccard, C. Battaglia, S. Hänni, K. Söderström, J. Escarré, S. Nicolay, F. Meillaud, M. Despeisse and C. Ballif, *Nano Lett.*, 2012, **12**, 1344.

118. A. Tamang, A. Hongsingthong, P. Sichanugrist, V. Jovanov, M. Konagai and D. Knipp, *IEEE J. Photovoltaics*, 2014, **4**, 16.

119. K. Bittkau and T. Beckers, *Physica Status Solidi A*, 2010, **207**, 661.

120. C. Rockstuhl, S. Fahr, K. Bittkau, T. Beckers, R. Carius, F.-J. Haug, T. Söderström, C. Ballif and F. Lederer, *Opt. Express*, 2010, **18**, A335.

121. C. Haase and H. Stiebig, *Prog. Photovoltaics*, 2006, **14**, 629.

122. A. Campa, J. Krc and M. Topic, *J. Appl. Phys.*, 2009, **105**, 083107.

123. K. Söderström, F.-J. Haug, J. Escarré, O. Cubero and C. Ballif, *Appl. Phys. Lett.*, 2010, **96**, 213508.

124. A. Naqavi, K. Söderström, F.-J. Haug, V. Paeder, T. Scharf, H. P. Herzig and C. Ballif, *Opt. Express*, 2011, **19**, 128.

125. F.-J. Haug, K. Söderström, A. Naqavi and C. Ballif, *J. Appl. Phys.*, 2011, **109**, 084516.

126. O. Isabella, S. Solntsev, D. Caratelli and M. Zeman, *Prog. Photovoltaics*, 2013, **21**, 94.

127. P. Sheng, A. N. Bloch and R. S. Stepleman, *Appl. Phys. Lett.*, 1983, **43**, 579.

128. Z. Yu, A. Raman and S. Fan, *Proc. Natl. Acad. Sci. U. S. A.*, 2010, **107**, 17491.

129. S. Solnstev and M. Zeman, *Energy Procedia*, 2011, **10**, 308.

130. M. Sever, B. Lipovšek, J. Krc, A. Èampa, G. Sánchez Plaza, F.-J. Haug, M. Duchamp, W. Soppe and M. Topic, *Sol. Energy Mater. Sol. Cells*, 2013, **119**, 59.

131. V. Jovanov, U. Palanchoke, P. Magnus, H. Stiebig, J. Hüpkes, P. Sichanugrist, M. Konagai, S. Wiesendanger, C. Rockstuhl and D. Knipp, *Opt. Express*, 2013, **21**, A595.

132. C. Pahud, O. Isabella, A. Naqavi, F.-J. Haug, M. Zeman, H. P. Herzig and C. Ballif, *Opt. Express*, 2013, **21**, A786.

133. H. R. Stuart and D. G. Hall, *J. Opt. Soc. Am. A*, 1997, **14**, 3001.

134. J. Zhu, Z. Yu, G.-F. Burkhard, C.-M. Hsu, S.-T. Connor, Y. Xu, Q. Wang, M. McGehee, S. Fan and Y. Cui, *Nano Lett.*, 2009, **9**, 279.

135. M. J. Naughton, K. Kempa, Z. F. Ren, Y. Gao, J. Rybczynski, N. Argenti, W. Gao, Y. Wang, Y. Peng, J. R. Naughton, G. McMahon, T. Paudel, Y. C. Lan, M. J. Burns, A. Shepard, M. Clary, C. Ballif, F.-J. Haug, T. Söderström, O. Cubero and C. Eminian, *Physica Status Solidi (RRL)*, 2010, **4**, 181.

136. M. Vanecek, O. Babchenko, A. Purkrt, J. Holovsky, N. Neykova, A. Poruba, Z. Remes, J. Meier and U. Kroll, *Appl. Phys. Lett.*, 2011, **98**, 163503.

137. D. Duché, L. Escoubas, J.-J. Simon, P. Torchio, W. Vervisch and F. Flory, *Appl. Phys. Lett.*, 2008, **92**, 193310.

138. A. Chutinan, N. P. Kherani and S. Zukotynski, *Opt. Express*, 2009, **17**, 8871.

139. X. Meng, G. Gomard, O. El Daif, E. Drouard, R. Orobtchouk, A. Kaminski, A. Fave, M. Lemiti, A. Abramov, P. Roca i Cabarrocas and C. Seassal, *Sol. Energy Mater. Sol. Cells*, 2011, **95**, S32.

140. S. Misra, L. Yu, M. Foldynaa and P. Roca I Cabarrocas, *Sol. Energy Mater. Sol. Cells*, 2013, **90**, 118.

141. S. Pillai, K. R. Catchpole, T. Trupke and M. A. Green, *J. App. Phys.*, 2007, **101**, 093105.

142. H. A. Atwater and A. Polman, *Nat. Mater.*, 2010, **9**, 205.

143. C. Eminian, F.-J. Haug, O. Cubero, X. Niquille and C. Ballif, *Prog. Photovoltaics*, 2011, **19**, 260.

144. N. N. Lal, H. Zhou, M. Hawkeye, J. K. Sinha, P. N. Bartlett, G. A. J. Amaratunga and J. J. Baumberg, *Phys. Rev. B*, 2012, **85**, 245318.

145. F.-J. Haug, K. Söderström, A. Naqvi, C. Battaglia and C. Ballif, *Mater. Res. Soc. Symp. Proc.*, 2012, 1391.

146. E. A. Schiff, *J. App. Phys.*, 2011, **110**, 104501.

147. H. Sakai, T. Yoshida, T. Hama and Y. Ichikawa, *Jpn. J. Appl. Phys.*, 1990, **29**, 630.

148. P. Buehlmann, J. Bailat, D. Dominé, A. Billet, F. Meillaud, A. Feltrin and C. Ballif, *Appl. Phys. Lett.*, 2007, **91**, 143505.

149. T. Söderström, F.-J. Haug, X. Niquille, V. Terrazzoni and C. Ballif, *Appl. Phys. Lett.*, 2009, **94**, 063501.

150. M. Boccard, C. Battaglia, N. Blondiaux, R. Pugin, M. Despeisse and C. Ballif, *Sol. Energy Mater. Sol. Cells*, 2013, **119**, 12.

151. R. Biron, S. Hänni, M. Boccard, C. Pahud, K. Söderström, M. Duchamp, R. Dunin-Borkowski, G. Bugnon, L. Ding, S. Nicolay, G. Parascandolo, F. Meillaud, M. Despeisse, F.-J. Haug and C. Ballif, *Sol. Energy Mater. Sol. Cells*, 2013, **114**, 147.

152. H. Sai, Y. Kanamori and M. Kondo, *Appl. Phys. Lett.*, 2011, **98**, 113502.

153. K. Söderström, G. Bugnon, F.-J. Haug, S. Nicolay and C. Ballif, *Sol. Energy Mater. Sol. Cells*, 2012, **101**, 193.

154. K. Söderström, PhD thesis EPFL No 56714, 2013.

155. G. Bugnon, G. Parascandolo, T. Söderström, P. Cuony, M. Despeisse, S. Hänni, J. Holovsky, F. Meillaud and C. Ballif, *Adv. Funct. Mater.*, 2012, **22**, 3665.

156. K. Söderström, G. Bugnon, R. Biron, C. Pahud, F. Meillaud, F.-J. Haug and C. Ballif, *J. Appl. Phys.*, 2012, **112**, 114503.

157. R. Dewan, I. Vasilev, V. Jovanov and D. Knipp, *J. Appl. Phys.*, 2011, **110**, 013101.

158. M. Boccard, C. Battaglia, F.-J. Haug, M. Despeisse and C. Ballif, *Appl. Phys. Lett.*, 2012, **101**, 151105.

159. L. Ding, M. Boccard, G. Bugnon, M. Benkhaira, S. Nicolay, M. Despeisse, F. Meillaud and C. Ballif, *Sol. Energy Mater. Sol. Cells*, 2012, **98**, 331.

160. L. Ding, S. Nicolay, J. Steinhauser, U. Kroll and C. Ballif, *Adv. Funct. Mater.*, 2013, **23**, 5177.

161. M. Wimmer, F. Ruske, S. Scherf and B. Rech, *Thin Solid Films*, 2012, **520**, 4203.
162. J. Hüpkes, J. I. Owen, M. Wimmer, F. Ruske, D. Greiner, R. Klenk, U. Zastrow and J. Hotovy, *Thin Solid Films*, 2014, **555**, 48.
163. T. Koida, H. Fujiwara and M. Kondo, *Jpn. J. Appl. Phys.*, 2007, **46**, L685.
164. T. Koida, H. Sai and M. Kondo, *Thin Solid Films*, 2010, **518**, 2930.
165. C. Battaglia, L. Erni, M. Boccard, L. Barraud, J. Escarré, K. Söderström, G. Bugnon, A. Billet, L. Ding, M. Despeisse, F.-J. Haug, S. De Wolf and C. Ballif, *J. Appl. Phys.*, 2011, **109**, 114501.
166. A. Banerjee and S. Guha, *Mater. Res. Soc. Symp. Proc.*, 1990, **192**, 57.
167. Y. Hattori, D. Kruangam, K. Katoh, Y. Nitta, H. Okamoto and Y. Hamakawa, Proc. 19th IEEE Photovoltaic Conf., 1987, 689.
168. P. Sichanugrist, T. Sasaki, A. Asano, Y. Ichikawa and H. Sakai, *Sol. Energy Mater. Sol. Cells*, 1994, **34**, 42.
169. P. Cuony, M. Marending, D. T. L. Alexander, M. Boccard, G. Bugnon, M. Despeisse and C. Ballif, *Appl. Phys. Lett.*, 2010, **97**, 213502.
170. P. Cuony, D. T. L. Alexander, I. Perez-Wurfl, M. Despeisse, G. Bugnon, M. Boccard, T. Söderström, A. Hessler-Wyser, C. Hébert and C. Ballif, *Adv. Mater.*, 2012, **24**, 1182.
171. J. Springer, B. Rech, W. Reetz, J. Müller and M. Vanecek, *Sol. Energy Mater. Sol. Cells*, 2005, **85**, 1.
172. D. Sainju, P. J. van den Oever, N. J. Podraza, M. Syed, J. A. Stoke, J. Chen, X. Yang, X. Deng and R. W. Collins, 4th IEEE World Conference on Photovoltaic Energy Conversion, 2006, **2**, 1732.
173. F.-J. Haug, T. Söderström, O. Cubero, V. Terrazzoni-Daudrix and C. Ballif, *J. Appl. Phys.*, 2008, **104**, 064509.
174. B. Yan, G. Yue, L. Sivec, J. Owens-Mawson, J. Yang and S. Guha, *Sol. Energy Mater. Sol. Cells*, 2012, **104**, 13.
175. U. Palanchoke, V. Jovanov, H. Kurz, P. Obermeyer, H. Stiebig and D. Knipp, *Opt. Express*, 2012, **20**, 6340.
176. F.-J. Haug, T. Söderström, O. Cubero, V. Terrazzoni-Daudrix and C. Ballif, *J. Appl. Phys.*, 2009, **106**, 044502.
177. K. Söderström, F.-J. Haug, J. Escarré, C. Pahud, R. Biron and C. Ballif, *Sol. Energy Mater. Sol. Cells*, 2011, **95**, 3585.
178. Y. Tawada, H. Okamoto and Y. Hamakawa, *Appl. Phys. Lett.*, 1981, **39**, 237.
179. R. Biron, C. Pahud, F.-J. Haug, J. Escarré, K. Söderström and C. Ballif, *J. Appl. Phys.*, 2011, **110**, 125411.
180. M. Despeisse, C. Battaglia, M. Boccard, G. Bugnon, M. Charrière, P. Cuony, S. Hänni, L. Löfgren, F. Meillaud, G. Parascandolo, T. Söderström and C. Ballif, *Physica Status Solidi A*, 2011, **208**, 1863.
181. C. Beneking, B. Rech, J. Fölsch and H. Wagner, *Physica Status Solidi B*, 1996, **194**, 41.
182. G. Bugnon, G. Parascandolo, S. Hänni, M. Stuckelberger, M. Charrière, M. Despeisse, F. Meillaud and C. Ballif, *Sol. Energy Mater. Sol. Cells*, 2013, **120**, 143.

183. S. Hänni, G. Bugnon, M. Boccard, M. Despeisse, J.-W. Schütauf, F.-J. Haug, F. Maillaud and C. Ballif, in preparation.

184. T. Söderström, F.-J. Haug, V. Terrazzoni-Daudrix, X. Niquille, M. Python and C. Ballif, *J. Appl. Phys.*, 2008, **104**, 104505.

185. G. Yue, B. Yan, C. Teplin, J. Yang and S. Guha, *J. Non-Cryst. Solids*, 2008, **354**, 2440.

186. T. Matsui and M. Kondo, *Sol. Energy Mater. Sol. Cells*, 2013, **119**, 156.

187. D. Chaudhary, C. Goury, J. Hötzel, S. Jost, M. Klindworth, E. Kügler, G.-F. Leu, A. Salabas and A. Sublet, Proc. of the 27th EU-PVSEC, 2012, 3AO.4.1, 2094.

188. C. Droz, E. Vallat-Sauvain, J. Bailat, L. Feitknecht, J. Meier and A. Shah, *Sol. Energy Mater. Sol. Cells*, 2004, **81**, 61.

189. Q. Zhang, E. V. Johnson, Y. Djeridane, A. Abramov and P. Roca i Cabarrocas, *Physica Status Solidi (RRL)*, 2008, **2**, 154.

190. J.-C. Dornstetter, S. Kasouit and P. Roca i Cabarrocas, *IEEE J. Photovoltaics*, 2013, **3**, 581.

191. A. Abramov and P. Roca i Cabarrocas, *Physica Status Solidi C*, 2010, **7**, 529.

192. B. Yan, J. Yang and S. Guha, *Appl. Phys. Lett.*, 2003, **83**, 782.

193. I. A. Yunaz, H. Nagashima, D. Hamashita, S. Miyajima and M. Konagai, *Sol. Energy Mater. Sol. Cells*, 2011, **95**, 107.

194. S. Inthisang, T. Krajangsang, I. A. Yunaz, A. Yamada, M. Konagai and C. R. Wronski, *Physica Status Solidi C*, 2011, **8**, 2990.

195. D. Y. Kim, R. A. C. M. M. van Swaaij and M. Zeman, *IEEE J. Photovoltaics*, 2014, **4**, 22.

196. J. Yang, A. Banerjee and S. Guha, *Sol. Energy Mater. Sol. Cells*, 2003, **78**, 597.

197. S. Guha, J. Yang, A. Pawlikiewicz, T. Glatfelter, R. Ross and S. R. Ovshinski, *Appl. Phys. Lett*, 1989, **54**, 2330.

198. R. Carius, J. Fölsch, D. Lundszien, L. Houben and F. Finger, *Mater. Res. Soc. Symp. Proc.*, 1998, **507**, 813.

199. M. Isomura, K. Nakahata, M. Shima, S. Taira, K. Wakisaka, M. Tanaka and S. Kiyama, *Sol. Energy Mater. Sol. Cells*, 2002, **74**, 519.

200. T. Matsui, H. Jia and M. Kondo, *Prog. Photovoltaics*, 2010, **18**, 48.

201. Y. Cao, J. Zhang, C. Li, T. Li, Z. Huang, J. Ni, Z. Hu, X. Geng and Y. Zhao, *Sol. Energy Mater. Sol. Cells*, 2013, **114**, 161.

202. T. Matsui, C.-W. Chang, T. Takada, M. Isomura, H. Fujiwara and M. Kondo, *Appl. Phys. Express*, 2008, **1**, 031501.

203. T. Matsui, H. Jia, M. Kondo, K. Mizuno, S. Tsuruga, S. Sakai and Y. Takeuchi, Proc. of the 35th IEEE-PVSEC, 2010, 000311.

CHAPTER 3

Thin-film CdTe Photovoltaic Solar Cell Devices

TIMOTHY GESSERT,*[a] BRIAN MCCANDLESS[b] AND CHRIS FEREKIDES[c]

[a] National Renewable Energy Laboratory (NREL), Golden, Colorado 80401, USA; [b] Institute of Energy Conversion (IEC), Newark, Delaware 19716, USA; [c] University of South Florida (USF), Tampa, Florida 33620, USA
*Email: tim.gessert@nrel.gov

3.1 Introduction

Thin-film photovoltaic (PV) modules based on the polycrystalline absorber material cadmium telluride (CdTe) are the second most widely deployed form of PV technology in the world, being surpassed only by modules based on large-grained polycrystalline silicon. Further, and perhaps more noteworthy, many believe CdTe-based PV could become a *dominant* technology in future renewable-energy production because of combined material and production advantages. In part, this presumption is based on the following: both the amount of energy consumed and the chemical by-product produced during CdTe PV module production have the potential to be significantly lower than competing technologies.[1–3] The basic technologies and material functionality used to produce the present generation of commercial CdTe PV modules are generally available.[4–9] However, few reports adequately detail the evolutionary nature of certain process steps that have now become central to commercial module production, nor do they detail where significant uncertainty remains in critical scientific understanding. The following chapter reviews the major processes used in present polycrystalline

RSC Energy and Environment Series No. 11
Advanced Concepts in Photovoltaics
Edited by Arthur J Nozik, Gavin Conibeer and Matthew C Beard
© The Royal Society of Chemistry 2014
Published by the Royal Society of Chemistry, www.rsc.org

CdTe PV module manufacture; but more importantly, it identifies and discusses several processes where improved understanding would likely lead to significantly improved device performance, reduced manufacturing cost, or both.

3.1.1   History of CdTe Photovoltaic Devices

In the 1950s, CdTe was identified as having a nearly ideal bandgap match to the solar spectrum with high optical absorptivity, and CdTe thin-film solar cell research and development (R&D) escalated during the 1970s under academic, corporate, and government sponsorship. Today, laboratory-scale cell conversion efficiency has reached almost 20%,[10] modules with efficiencies exceeding 16%[10] have been demonstrated, and the technology has transitioned from the laboratory to being a market commodity. Despite this progress, critical challenges remain for future development and deployment of CdTe photovoltaics: overcoming the open-circuit voltage deficit between theoretical expectations and obtained performance; reducing CdTe film thickness to address future tellurium supply and demand cost issues; and developing alternative device architectures that enhance light capture and improve form factor for integrated PV applications.

The pedigree of high-performance thin-film CdTe cells can be traced to the development of superstrate front-wall thin-film cells by Adirovich *et al.*[11] in the late 1960s, in the former Soviet Union. In their superstrate cell configuration, light enters through the glass as in today's CdTe cells. Thin films of CdS, CdTe, and copper or chrome were sequentially evaporated onto a glass superstrate coated with a transparent conductive oxide (TCO) made of tin oxide. The work reported several aspects of the device that remain as active research areas today including parasitic absorption in the CdS, nonideal diode behavior with ideality factor $A = 2$, formation of $CdTe_{1-x}S_x$ alloy, and the importance of the back-contact metal work function. Research and development of similar thin-film cell structures, with the n-type CdS deposited onto the TCO, was widely explored as a platform to evaluate different CdTe deposition methods, post-deposition treatments, and back-contact schemes. By the mid-1970s, several groups had demonstrated thin-film CdTe cells with conversion efficiencies approaching 10%, with the CdTe film deposited by screen printing,[12] electrodeposition,[13,14] metal–organic chemical vapor deposition,[15] and evaporation.[16] Cell efficiencies continued to suffer from low photocurrent due to parasitic optical loss in the CdS film, which produced little verifiable contribution to collected photoelectrons.

The device design adopted by Adirovich and subsequent researchers owed much to the state of understanding of CdTe and CdS fundamental properties, much of which had been discovered in the 1950s and early 1960s. In particular, the physical chemistry of defects in II–VI materials had been codified by the preeminent Dutch chemist Ferdinand A. Kroeger, while working at Philips Electrical in the Netherlands. His graduate student Dirk de Nobel performed the first systematic and definitive experiments on the relationship between Cd-Te phase equilibria and CdTe semiconducting properties, establishing the existence region of the single-crystal phase and the role of stoichiometric deviation in controlling conductivity. This work was published in 1959 as a two-part paper in the Philips Research Reports, Eindhoven. de Nobel's 1958 patent (1955 application date) for a method of making a CdTe semiconducting device, set the stage for processing routes that could control the electronic properties of CdTe. This work, coupled with emerging analytic models of minority-carrier transport by Hovel[17] and others,[18] enabled different device architectures and contact schemes to be aggressively pursued through the 1960s by groups at GE[19] and RCA.[20] The CdS heteropartner had been even more intensively studied during this same period, having found applications as both a photoresistor[21] and photocell,[22] in conjunction with copper sulfide.[23] Added to this, the role of halogens and compounds thereof—in crystal growth, as flux agents, and as activators for dopants and luminescence centers—had been under intense investigation from the 1950s and well into the 1970s.

The CdTe thin-film superstrate cell story was advanced in the early 1980s by Monosolar's electrodeposited n-CdS/p-CdTe cells, showing enhanced photocurrent for cells deposited with chlorine in the CdTe deposition bath followed by post-deposition air anneals of the CdS/CdTe stack.[24] This technology was sold to Sohio, later BP, who continued to refine the CdTe deposition bath, post-deposition treatments, and back contacts containing Cu.[25] In the mid-1980s, Ametek's electrodeposited cells focused initially on n-CdTe with Au grid contacts and later shifted to the CdS/CdTe/ZnTe:Cu superstrate cell, adding a post-deposition thermal anneal step of the semi-conductor stack after coating with a layer of cadmium chloride ($CdCl_2$).[26] Ametek reached greater than 10% efficiency before abandoning the technology. Meanwhile, Eastman Kodak was depositing CdTe by close-spaced sublimation (CSS) at substantially elevated temperatures compared to electrodeposited films and demonstrated enhanced cell performance when small amounts of oxygen were added to the growth ambient.[27] Kodak achieved single-cell efficiency greater than 10% and also made the first CdTe mini-module, with 30 cm^2 area comprised of 12 cells and an efficiency of 8%. Kodak abandoned CdTe work before the ideas of high-temperature growth, chloride and air treatment, and copper addition would be combined. This occurred in 1991 by researchers at the University of South Florida (USF), who ultimately achieved greater than 15% conversion efficiency using CSS-deposited CdTe followed by a $CdCl_2$ air treatment, nitric acid-based etch, and Cu-doped graphite + HgTe contact.[28] Their cell performance benefited from thinner CdS and correspondingly higher photocurrent than obtained in earlier cells. At this period, cell performance and stability was sufficiently robust to permit detailed analysis of cell operation, performance-limiting issues, and cell-level durability issues, and the enhancing role of interlayers on each side of the device was identified. The addition of a high-resistance transparent (HRT) interlayer between the TCO and CdS allowed thinner CdS to be used, yielding higher photocurrent, especially when the cell was fabricated on high-transparency research-grade glass (see also section 3.2.1). On the back side of the cell, reactive modification of the CdTe surface created a layer that enabled alternative contact materials to form low-resistance contacts, yielding higher fill factors. In 2001, researchers at the National Renewable Energy Laboratory (NREL) succeeded in replacing the historic SnO_2-based TCO(s) with a bi-layer TCO structure based on low-resistance Cd_2SnO_4 and a higher-resistance Zn_2SnO_4 buffer layer. The improved optical and electrical performance of this front-contact structure ultimately led to devices that attained greater than 16% efficiency.[29] With the demonstration of these levels of record efficiencies, the U.S. Department of Energy established and supported a CdTe National Team under the Thin-Film Photovoltaic Partnership Program from about 1994 until about 2004. During this time, the mission of the National Teams included fostering collaboration and education among researchers and industries involved in thin-film CdTe PV research. Under this structure, several important joint investigations were undertaken including studies of the function of the TCO buffer layers, evolution of the junction during typical processes such as the $CdCl_2$ process and contacting with Cu, and reliability studies (see also section 3.2).

Parallel industrial efforts sought to develop CdTe modules, with emphasis on area uniformity and product throughput. Photon Energy, founded in 1975 by John Jordan in El Paso, Texas, used high-temperature spray deposition for all layers, including the tin oxide.[30] CdTe was deposited from $CdCl_2$ and Te oxide precursors. After acquisition by Coors and becoming Golden Photon in 1992, they achieved about 8% module efficiency and greater than 12% cell efficiency. The spray technology enabled deposition of a robust TCO/HRT stack that, in the best cells, ensured high photocurrent since the CdTe spray step completely consumed the CdS film. Photocurrent benefited as well from enhanced ultraviolet photogeneration and extended near-infrared response due to $CdTe_{1-x}S_x$ alloy formation. It appeared that difficulties in controlling the CdTe film density and lateral cell uniformity eventually led to abandoning the technology.

British Petroleum, having acquired the Sohio/Monosolar electrodeposited technology, formed BP Solar, which developed the Apollo module in the UK and then built a factory in Fairfield, California, in 2000. The electrodeposition batch line was capable of simultaneous deposition of CdTe onto eighty $0.75\,m^2$ modules. Although modules with 12% efficiency were demonstrated in 2002, concern over cadmium toxicity was cited as one reason behind the BP corporate decision to abandon CdTe development.

Glasstech Solar, founded in 1984 by the inventor Harold McMaster, envisioned a solar cell plant located at the exit end of a sheet-glass refinery. Original work aimed at amorphous silicon technology as a low-cost competitor to crystalline silicon; however, alternative work began in the late 1980s to develop a CdTe process, combining high-speed growth with high film-growth temperature. The new venture, called Solar Cells, Inc., focused on a vapor transport (VT) deposition method developed by Rick Powell and a small team that included Peter Meyers, who had led the earlier Ametek effort. In 1999, the company was sold to True North Partners, LLC, and the name changed to First Solar, Inc. Over the next decade, and leading to the present, First Solar became the thin-film solar industry leader in manufacturing and installed capacity, by combining robust fabrication processes with a cradle-to-grave sales policy, taking advantage of German and Malaysian production incentives. In 2011, First Solar had a production capacity of about 2.3 GW/year.[31]

After a several-decade hiatus from active CdTe R&D, General Electric (GE) re-entered the field from 2007 until 2013, during which time their group performed in-house development of CSS devices and acquired PrimeStar Solar, Inc., in Golden, Colorado. At this writing, the highest verified AM1.5 thin-film CdTe efficiencies are about 20% for cells and about 16% for modules.[10] The present $\sim 20\%$ cell conversion efficiency represents 59% of the Shockley–Queisser (SQ) efficiency upper limit, $\eta_{SQ} \sim 32\%$, expected for a material having an optical bandgap $E_G = 1.5$ eV.[32] The efficiency advancements by GE and First Solar in 2013, attributable to optimization of short-circuit current density (J_{SC}) and fill factor (FF), highlight open-circuit voltage (V_{OC}) as the critical challenge for developing cells with efficiency approaching the SQ efficiency limit. Based on SQ limit estimates, an ultimate $V_{OC} = 1.2$ V is expected for $E_G = 1.5$ eV assuming the following: entropy of loss in directivity of spontaneous emission, no light-trapping losses, and no quantum efficiency losses. The present maximum reported room-temperature V_{OC} for a record-efficiency polycrystalline thin-film CdTe/CdS cells is 0.857 V[10] (note that the present maximum confirmed V_{OC} for a *non-record* CdTe/CdS polycrystalline cell is 0.903 V[33]), while that for a bulk single-crystal CdTe/ITO cell is 0.91 V, with an acceptor concentration of 10^{15} cm^{-3}.[34] The V_{OC} deficit (which is the SQ V_{OC} minus the measured V_{OC}) in CdTe thin-film cells is thus ~ 0.34 V, or rather, the 0.857 V value is about 72% of the ultimate V_{OC}. The highest V_{OC} single-crystal CdTe cell has a V_{OC} deficit of 0.29 V, corresponding to 76% of the ultimate V_{OC}. In contrast, single-crystal GaAs ($E_G = 1.43$ eV) and Si ($E_G = 1.15$ eV) solar cells, have reached 99% and 83% of their SQ V_{OC} limits, respectively. Resolving the CdTe V_{OC} problem, where both single-crystal and thin-film cells fall far short of expectation, will require discovery of the predominant limiting mechanism(s) and development of processing routes capable of mitigating those mechanisms. Some pathways toward understanding these issues are suggested in section 3.2.2.

3.1.2 Layer-specific Process Description for Superstrate CdTe Devices

The historic development of CdS/CdTe polycrystalline PV devices described above has resulted in a series of fabrication process steps that are surprisingly similar throughout the academic research and industrial CdTe PV communities. The following section presents typical process steps that are used to produce CdS/CdTe devices where vacuum processing is assumed to be the primary deposition method. When properly optimized, these process steps can be expected to produce devices with efficiencies greater than 10%, and possibly as high as about 17%. Regarding recent world-record efficiencies of nearly 20% for thin-film CdTe PV, it should be noted that this higher level of performance has been achieved primarily through processes improvements that have increased J_{SC} (greater than about 26 mA cm^{-2}) while maintaining other critical parameters. Although pathways to increase J_{SC} will be suggested in the following sections, at the time of this writing, specific processes associated with the recent J_{SC} improvements have not yet been described in the literature.

3.1.2.1   Superstrate Materials

Presently, the majority of published research on CdTe thin-film PV devices has related to the 'superstrate design' (described above), where light enters the device through a transparent glass superstrate. This design results in a device consistent with that shown in Figure 3.1. Typical superstrate materials include historic barium silicate glass (*e.g.*, Corning 7059), borosilicate glass, alumino-borosilicate glass, or soda-lime float glass (for commercial-module applications). Presently, there are no definitive studies that compare material and/or performance differences resulting from these different glass choices. However, it is known that important parameters of the glass

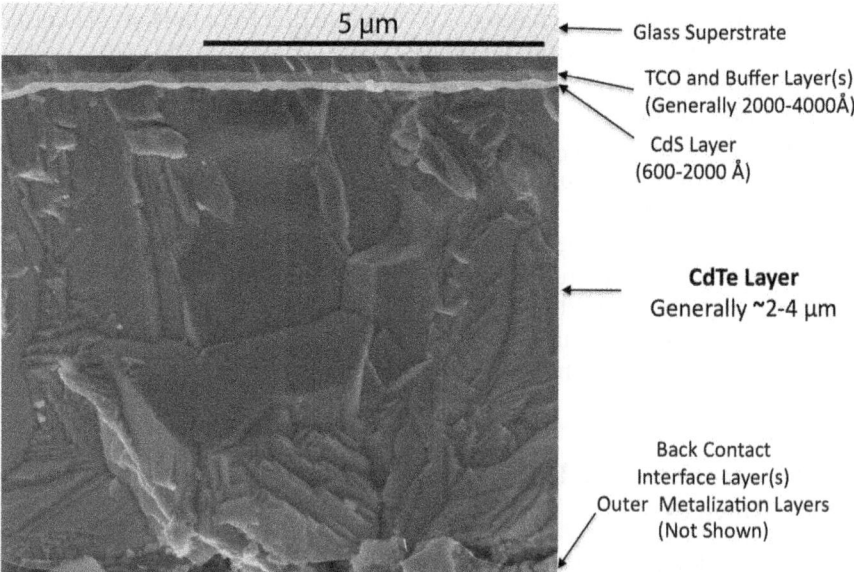

Figure 3.1   SEM micrograph showing cross section of CdS/CdTe superstrate device produced at NREL. Certain layers have been color-enhanced for clarity.

superstrate include mechanical stability as a function of temperature (*e.g.*, softening point); coefficient of thermal expansion (CTE) relative to the CTE of important active layers (*e.g.*, the CdTe); and glass composition as it relates to potential for detrimental impurity diffusion and the need for alkali diffusion barrier layer(s).

3.1.2.2   The TCO Layer

The TCO layer provides a lateral conduction pathway for electrons from the n-type side of the device, so it must be highly conductive. However, because incident light must pass through the TCO layer for subsequent absorption in the junction, it must also be highly transparent. Finally, because subsequent processing steps are often at high temperature (*e.g.*, 400 °C to 600 °C), the TCO must be electrically and compositionally stable (or aligned) with these thermal conditions. Partly due to its extensive use in low-emissivity (low-e) glass for residential and appliance applications, fluorine-doped SnO_2 (SnO_2:F) has been the historical choice for the TCO layer of CdS/CdTe PV devices. SnO_2:F is chemically and thermally stable at the CdTe deposition temperature, and is not greatly affected by environmental moisture or diffusion from other materials in the PV device. It may further provide a degree of protection from Na diffusion from soda-lime glass. In addition to SnO_2:F, many research groups have investigated alternative TCOs for CdTe devices.[35] For example, one former world-record thin-film CdTe device used a Cd_2SnO_4 TCO layer.[36–38] An alternative avenue to improve the transmission of TCOs for PV applications involves forming TCO alloys with higher dielectric permittivity.[39]

3.1.2.3   Buffer Layer

Most present high-performance CdTe PV devices place a high-resistance oxide layer between the TCO and the CdS layers (see also section 3.2.1). This layer is often called a buffer layer (BL) but has also been called a high-resistivity transparent (HRT) layer. Many attributes have been ascribed to this layer, and it is not presently clear if all attributes manifest in all CdTe device processes, or if certain BL attributes assist only certain types of device designs and/or fabrication processes. Suggested benefits of the BL include the following: the BL may allow a thinner CdS layer to be used; the BL may provide tolerance to shunt (or short) paths from the back contact to the front contact; the BL may allow higher temperatures or longer times to be used during the $CdCl_2$ treatment; the BL may consume some of the CdS layer, reducing the amount of (optically absorbing) CdS, and producing instead a wider bandgap CdZnS material.[37]

3.1.2.4   CdS Layer

The CdS layer is the n-type 'heteroface' partner of the CdS/CdTe device, but for most device processes, the resulting electrical junction results between the CdTe absorber layer and an intermixed CdSTe layer that forms during CdTe deposition (see also section 3.1.2.6). Because of this intermixing, the device is not generally considered a true heterojunction, but instead, a quasi-homojunction.[40] Although materials other than CdS have been tested for the heteroface partner layer, CdS has been found to yield the highest device performance. Moreover—and unlike the situation in CuInSe$_2$-alloy devices—because a much thicker Cd-containing layer still exists in the device (the CdTe layer), there has been little motivation to eliminate the Cd from the heteroface layer based solely on perceived toxicity concerns.

3.1.2.5 CdTe Layer

A primary production advantage of CdTe, relative to other known thin-film absorber layers, is that it can be deposited very quickly using vacuum techniques (rates greater than 20 µm min^{-1}). This means that the physical length of the CdTe deposition zone (and related deposition hardware costs) can be much smaller compared with other thin-film PV technologies that require slower deposition rates. A variety of deposition techniques have been used to deposit CdTe including: various forms of vacuum sublimation (*e.g.*, evaporation, close-spaced sublimation, vapor transport deposition),[41–44] sputtering,[45,46] electrodeposition,[13,47] screen printing,[48] spray techniques,[49,50] and printing of nanoparticles.[51] Following proper post-deposition processing, nearly all these techniques have yielded CdS/CdTe PV devices with efficiency greater than 10%.

3.1.2.6 CdSTe Layer

CdTe substrate temperatures higher than about 425 °C yield progressively more interdiffusion between the CdTe and the underlying CdS layer, forming what is often called the CdSTe-alloy region (or CdSTe layer). For devices deposited at temperatures higher than about 425 °C, a CdSTe-alloy region forms primarily during CdTe growth, and is altered further during the CdCl$_2$ process.[41,52] If a lower-temperature CdTe deposition process is used, a CdSTe alloy region can also be established during the CdCl$_2$ treatment.[52] Although it is known that most high-performance CdTe devices contain a CdSTe-alloy region, the function and preferred attributes of this region remain areas of active debate.[40,41,53,54]

3.1.2.7 CdCl$_2$ Activation

Since about 1985, nearly all thin-film CdTe PV devices have incorporated a CdCl$_2$ 'treatment' or 'activation step'. The incorporation of this step is one of the main reasons why so many different groups, using different CdTe deposition processes and temperatures, have been able to achieve conversion efficiencies greater than 10%. Many call the CdCl$_2$ process an 'equalizer' step because it results in CdTe polycrystalline thin-film material that demonstrates minority-carrier lifetimes that are similar and sufficiently high to enable good device performance. The CdCl$_2$ process usually involves dipping or spraying a CdCl$_2$-saturated methanol or water solution onto the CdTe surface after deposition, followed by drying and annealing in an oxygen-containing ambient at about 400 °C for about 10 min. The CdCl$_2$ treatment can also be performed by combining the application and thermal steps into a CdCl$_2$ vapor process at about 400 °C. For solution-based CdTe deposition processes, chlorine can be included in the CdTe precursor with results similar to a post-deposition CdCl$_2$ treatment.

For small-grained CdTe (*e.g.*, solution or low-temperature CdTe growth), the CdCl$_2$ process typically leads to complete recrystallization that produces larger polycrystalline grains.[52] Although for larger-grained (higher-temperature) CdTe growth the CdCl$_2$ process generally does not lead to significant re-crystallization; however, both small- and large-grained material demonstrate higher minority-carrier lifetime after the CdCl$_2$ treatment. For large-grained material, the improvement appears to be due primarily to passivation at grain boundaries.[55] As with most of the process steps discussed, the particular defect modifications associated with the CdCl$_2$ treatment continues to be debated.[56–59]

One artifact of the CdCl$_2$ process that is often not discussed relates to residuals left behind on the CdTe surface. It has been shown that if the CdCl$_2$ process is wet (application of a saturated methanol or aqueous solution), then residuals are either cadmium oxychlorides or oxytellurides.[55] Removing these residuals will often occur during pre-contact chemical etching step(s), and they are therefore not a major concern for wet-contact processing. However, if a dry-contacting process is desired, sufficient removal of CdCl$_2$ residuals requires the use of a dry removal process (*e.g.*, ion-beam or ion-etch processes).[55] If a vapor process is used for the CdCl$_2$ process, the resulting residual is CdCl$_2$, which can be removed using a thermal pre-contact processing.[60]

3.1.2.8 Back Contact

For superstrate CdTe PV devices, the last step in device formation is fabricating the back contact. As described in section 3.2.2, a major role of the contacting process for most CdS/CdTe device processes is to enable Cu to diffuse into the CdTe and create an appropriate concentration and extent of acceptor defects that form the junction. As described in section 3.1.2.7, developing a back-contacting process generally includes determining if/how any residuals from the CdCl$_2$ process should be removed.[55] Removal process options include thermal, chemical, and physical (*e.g.*, ion-beam treatment). Historically, a chemical option has been preferred because it often produces a Te-rich layer that assists certain types of contact formation.[61,62] It has also been suggested that gettering of excess Cu by etch-formed Te inclusions can enhance device stability.[63] One also needs to consider if the process used to remove CdCl$_2$ residuals should produce a stoichiometric surface.

For example, a p-ZnTe contact interface requires a stoichiometric CdTe surface to facilitate valence-band alignment.

3.2   Important and Under-reported Processes

The early development of thin-film CdTe solar cells was guided by the known properties of CdTe, which constrained the architecture and processing sequence necessary to extract modest photovoltage and photocurrent. Over time, empirical studies, based on feedback from experiments using either alternative device components or controlled quantities of chemical additives, led to the present level of conversion efficiencies needed for commercial competitiveness. Although laboratory efficiencies of about 20% have now been achieved, the photovoltage of all these cells falls about 340 mV below (or about 72% of) the Shockley–Quiesser estimates for the 1.5 eV bandgap. Given the wide array of effort already expended, many CdTe technologists believe a new approach is required to overcome what appear to be limiting characteristics associated with CdTe doping and electron lifetime. To develop a new approach, it is necessary to understand where other approaches have succeeded and failed. The following section reviews some of these new directions for CdTe polycrystalline thin-film devices.

3.2.1   Buffer Layers

As indicated in section 3.1.1, one of the key developments toward reaching today's about 20% record cell efficiencies has been the use of a HRT layer—also referred to as a *buffer layer*—inserted between the low-resistance front TCO contact and the n-type CdS window layer, as shown in Figure 3.1. One example of this is the stack composed of SnO_2:F/HRT/CdS. The first CdTe technology to benefit from the use of the HRT layer is most likely Golden Photon's process, where reduction of the CdS layer thickness during processing was quite extensive relative to other technologies at the time. The leading factor for efficiency advancements since the early 1990s has been improvements in the front optics of the device (glass and/or TCO), or the buffer layer, including the USF and NREL cells in 1993 and 2000, respectively.[28,37] The details of First Solar's and GE's recent high-performance device results are not known at this time; however, based on the available quantum efficiency data, it appears that the CdS absorption has been significantly reduced either through complete consumption of the CdS layer during the cell fabrication process or its replacement with a wider-gap material.[10,33] Although purely speculative, we believe that this accomplishment could be attributed to an improved *buffer layer*.

  The development of the buffer layer came about at a time (during the NREL-sponsored Thin-Film Photovoltaic Partnership Program) when making efficiency advancements focused on minimizing losses due to absorption in the CdS window. Thinning the CdS layer would yield higher currents, but below a certain thickness (typically in the range of 50–80 nm, depending on the process and processing conditions), the current gains were accompanied by losses in the V_{OC} and *ff*, with the overall effect being *lower efficiency*.[64,65] The losses were attributed to the formation of SnO_2/CdTe *microdiodes*;[66] the combined effects of the polycrystalline nature of the films (the *rough* surfaces and interfaces); the use of very-thin CdS, which increased the probability of pinhole formation; and the consumption of part of the CdS during subsequent processing. This CdS consumption was suspected to lead to microscopic regions where the CdTe may come in direct contact with the SnO_2, forming an inferior SnO_2/CdTe junction. In general, when compared to the CdS/CdTe junction, SnO_2/CdTe produces lower V_{OC} and *ff*. One study reports that the SnO_2/CdTe junction outperformed CdS/CdTe: however, the level of performance in this instance was around 8–9%, with the voltages only being about 700 mV, and it appears that the CdS/CdTe junction was not fully optimized.[67] The weak SnO_2/CdTe microdiodes act as shunts to the CdS/CdTe junction, lowering its voltage. The use of the HRT buffer increases the lateral resistivity and therefore minimizes the shunting effect of the microdiodes.

To date, several materials have been used as effective buffers. These include undoped SnO_2, In_2O_3, Ga_2O_3, and $ZnSnO_x$.[37,64,68,69] The resistivity of the buffer is reported to be in the rather wide range of 0.1–2 Ω m.[69] These values represent the as-deposited resistivity of the films; it is possible that in the finished device the resistivity could be higher due to exposure of the buffers to high temperatures and/or oxygen ambient. Depending on the material, effective buffers with thicknesses down to 50 nm have been used, although more typical thicknesses are in the range of 0.1–0.2 μm.

With regards to structure, all reports indicate that the buffer layers are polycrystalline. If the as-deposited films are amorphous (typical of room-temperature depositions), then a crystallization thermal annealing step is necessary for achieving high performance.[37,68,69] The use of an amorphous buffer layer (Zn–Sn–O) has been shown to enhance chemical reactions between the CdS and buffer layers; however, these cells consistently exhibited higher dark currents, presumably due to increased recombination and lower voltages than their counterparts fabricated with an annealed polycrystalline Zn–Sn–O buffer.[69] The work at NREL on $ZnSnO_x$ buffers has shown that this particular buffer compound contributes to the consumption of the CdS.[37,70] The $ZnSnO_x$ buffer also appeared to improve adhesion and reproducibility.[70]

A key challenge associated with the CdTe cell is the interdependence among the various cell components and processes. The binary compounds (SnO_2, In_2O_3, and Ga_2O_3) to first order appear to be robust; they do not undergo significant changes during the cell fabrication process and do not appear to interact with the CdS window layer. Incorporating these buffer layers in the device structure does not present significant challenges. On the other hand, although much is known about the most successful reported buffer $ZnSnO_x$ (note that the buffer details for the First Solar and GE record cells are currently unknown), it appears that its incorporation in the cell may not be trivial judging from the limited reports on this buffer following the success reported by NREL.

3.2.2   Incorporation of Cu

Incorporation of Cu into the polycrystalline CdTe layer from the back contact of a superstrate CdS/CdTe device has been used to improve device performance since the early days of the technology. Until about 1995, many reports suggested that the primary purpose of Cu was to form a contact interface layer of 'highly doped' CdTe to facilitate ohmic contact formation. This was often suggested even though Cu as an acceptor even in crystalline CdTe had not been shown to achieve degenerated doping levels.[71-76] Additionally, implied in these early reports was an assumption that a junction of reasonable quality had formed between the n-CdS and p-CdTe prior to contacting, and the primary purpose of the contact was to enable low-resistance current flow from this junction. As described below, for most polycrystalline CdTe PV devices, this historic assumption appears to have been largely incorrect.

Most contacts used for CdS/CdTe devices during these early years involved chemical treatments that yielded a reproducible CdTe surface stoichiometry (often Te rich). Although generally not stated, removal of residues from the preceding $CdCl_2$ process was also an important attribute of the chemical pre-contact treatment.[55] The next step was to apply a thin contact-interface layer (CIFL) that contained a small amount of Cu, and sometimes other elements that were found to be beneficial (*e.g.*, Te, Hg). On top of the CIFL was usually placed thicker metal layer(s) that would provide for lateral conduction, as well as mechanical and chemical robustness needed for probing and/or solder contacting. The general belief was that Cu from the CIFL diffused into the near-surface region of the CdTe layer and doped it p-type to reduce the width of the contact barrier formed between the CdTe and CIFL. This would enable partial tunneling through the barrier *via* the process of thermionic-field emission transport.[76,77] Because most CIFLs were either metals or semi-metals, resistance between the CIFL and the outer metallization was expected to be low.

Although many of these historic contact processes yielded devices with reasonable performance and stability, relationships between barrier formation and other contacting parameters (*e.g.*, chemical etching, contact metallization, annealing conditions) did not easily allow analysis of how Cu diffusion might be affecting the junction. During this time, it was believed that a 'good' contact process produced a contact barrier that was sufficiently small to allow the *pre-existing* junction parameters to be observed. In contrast, a 'bad' contact process yielded a contact barrier that dominated the device performance. For these types of contacts it was also difficult to control the amount of Cu in the CIFL that diffused into the CdTe layer, and device performance and stability would depend on the extent of this Cu diffusion. Performance reduction was typically ascribed to Cu forming conduction pathways along either grain boundaries and/or pinholes, thereby producing progressively lower device shunt resistance. Because of these beliefs, much research at this time was directed toward developing 'Cu-free' contacts, because it was believed that Cu diffusion into the CdTe was something that could and should be avoided.

In contrast to CIFLs, which were believed to function by reducing contact-barrier width, certain tellurides (most notably ZnTe) were also identified that should demonstrate a high degree of valence-band continuity with CdTe (a manifestation of the Common Anion Rule).[78,79] If true, hole conduction at the CdTe/ZnTe interface could proceed by low-resistance valence-band transport, while a narrow barrier could be formed in the ZnTe—because it was known that ZnTe could more easily be doped to degenerate p-type levels than CdTe. Moreover, the necessary doping concentration in ZnTe could be enabled by a dopant species shallower and/or possibly more stable than Cu. With proper design, these attributes could be expected to improve both performance and stability. With this use in mind, ZnTe CIFLs using various dopants (*e.g.*, P, Sb, N) were developed and shown to demonstrate both high doping and yield ohmic contacts between the p-ZnTe and several outer metallization.[80] However, and very surprisingly, when these low-Cu or Cu-free CIFLs were applied in CdS/CdTe devices, junction performance was found to be very poor, with PV devices generally demonstrating voltages of 400 mV or lower. Exceptions to this were generally found to use process steps that likely introduced Cu as a significant, but uncontrolled, impurity (*e.g.*, Cu impurity in $CdCl_2$ that diffused into CdTe during the about 400 °C $CdCl_2$ treatment). One compelling explanation for many of these observations was that Cu diffusion from the contact was not only assisting back-contact formation, but also was significantly altering the junction functionality.

Once it was realized that Cu incorporation from the contact could be significantly altering the CdS/CdTe junction evolution and subsequent device optimization, investigations began to study how different contact processes sourced Cu as a function of various contact-process parameters. Parameters investigated included the amount of Cu in the CIFL, thermal sequence of the contacting process, and stoichiometry of the surface onto which the contact was applied.[81–84] These studies suggested that an *optimum* device resulted when an optimum amount of Cu diffusion had occurred during the contacting process. Moreover, the optimum amount of Cu was a delicate balance between the amount of Cu diffusing from the CIFL and the temperature sequence during diffusion.[83] The main resulting sequence of CdS/CdTe junction evolution can be viewed as follows: Cu diffusion from a Cu-containing CIFL increases the net acceptor concentration in the CdTe layer $(N_a - N_d)$, thereby reducing the depletion width (W_D) of the junction. Optimum PV performance is attained when W_D is narrow enough to produce a drift field in the CdTe absorber that is sufficiently strong to overcome the relatively poor lifetime of the minority carriers (*i.e.*, τ_n, electrons in CdTe), but still wide enough to limit effects of voltage-dependent collection (*i.e.*, photocarriers should be generated primarily within, or very near to, the depletion region when the device is biased near the maximum power point [MPP]). Cu diffusion from the contact can also increase τ_n in the CdTe, if the diffusion is performed at a temperature of about 250 °C to 320 °C.[85]

Therefore, 'optimum Cu diffusion' produces a W_D of about 0.5 μm when the device is biased near the MPP, and a τ_n in the space-charge region of about greater than 0.5–1 ns. Cu incorporation from the contact beyond its optimum concentration causes $N_a - N_d$ to increase further. Although this can produce lower values of reverse saturation current density (J_o), and can potentially increase V_{OC}, this high level of Cu incorporation has been found to reduce τ_n. Furthermore, excessive Cu diffusion can lead to Cu incorporation into CdS. This will reduce the net *donor* density ($N_d - N_a$) in the CdS, reduce W_D in the CdTe, and yield a manifestation of photoconductivity, as confirmed by red-light-bias quantum efficiency (QE) measurements.[86] These complications of excess Cu diffusion combine to limit the benefits of a lower J_o, often reducing both V_{OC} and FF of the device. Similar (but longer-term) redistribution of Cu into the CdS layer has also been linked to device instability. This insight has helped establish several benchmarks to guide the development of CdS/CdTe junctions that use Cu contacts. One of the more useful involves monitoring the amount of Cu in the CdS layer so that it remains less than that needed to produce significant 'apparent' QE artifacts under red-light bias conditions.[86]

3.2.3   Defects and Defect Modeling

The above discussion suggests that typical polycrystalline CdTe films are produced at conditions farther from equilibrium than for bulk single-crystal or epitaxial-film growth. Although theoretical calculations generally assume conditions much nearer to equilibrium, these calculations have been found to be useful in understanding the likelihood of formation and properties of defects in polycrystalline CdTe. Until recently, most of these calculations used first-principles band-structure calculations within the density-functional theory (DFT) and local-density approximation (LDA).[87,88] For CdS/CdTe PV devices, DFT + LDA calculations have provided valuable insight into why oxygen, chlorine, and copper have historically been found to improve device performance. The calculations further suggested that the most likely mid-gap defects would include V_{Te}, the Te interstitial (Te_i), and the Te antisite (Te_{Cd}). As expected, V_{Te} is most likely to form under Cd-rich conditions, whereas Te_i and Te_{Cd} are more likely under Te-rich conditions. It is also possible for V_{Te} and V_{Cd} defects to co-exist and form Schottky pairs[87,89] Although the extent that different deposition processes and/or source types may influence chemical potential is not well understood, it remains noteworthy that most reported CdTe deposition processes do not actively control chemical potential.

A schematic diagram of the ionization energies of intrinsic defects from the DFT + LDA calculations is shown in Figure 3.2. This historic DFT + LDA guidance also indicated that, although the formation energy of Te_i would be relatively high for p-type materials, the formation energy of V_{Te} and Te_{Cd} was expected to be much lower.[90] Therefore, the V_{Te} and Te_{Cd} mid-gap recombination centers were believed to be the most concerning mid-gap

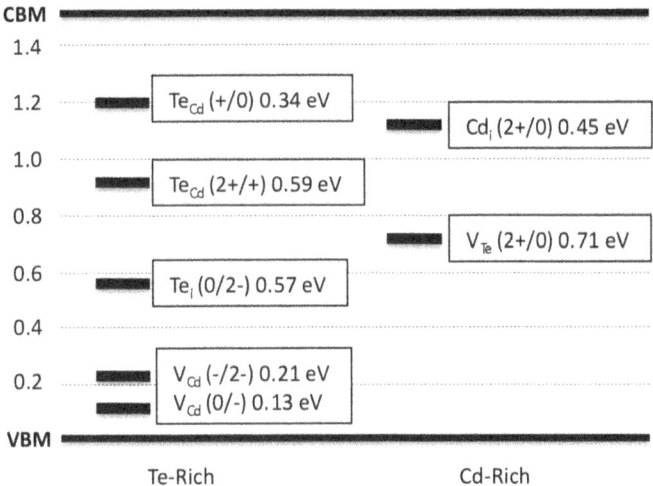

Figure 3.2 Position of intrinsic defects in CdTe calculated using DFT + LDA. Defects are grouped assuming Te-rich and Cd-rich chemical potentials. Box indicates ionization energy measured from VBM for acceptors, and from CBM for donors.
(From Wei and Zhang.[87])

defects, with V_{Te} being more problematic due to its predicted location very close to mid-gap and its lower formation energy. However, it is well known that DFT + LDA calculations underestimate the semiconductor bandgap: thus, it is acknowledged that the predicted defect level may contain large uncertainty due to the empirical treatment of the required bandgap corrections,[87] especially for deep levels such as V_{Te}.

With the historic DFT + LDA insight, Figure 3.3 suggests how the traditional O-, Cl-, and Cu-based enhancement processes used to make high-performance CdTe may influence the effect of the V_{Te} mid-gap defect. The evolution begins by noting that oxygen is typically incorporated intentionally or unintentionally during high-rate CdTe deposition. Oxygen is thermodynamically favored to substitute onto a V_{Te}, forming an O_{Te} isoelectronic defect that may form a defect pair with a V_{Cd} ($V_{Cd} + O_{Te}$).[27,56,91] During the CdCl$_2$ treatment, Cl is predicted to substitute onto V_{Te}, forming a different defect pair with V_{Cd} ($V_{Cd} + Cl_{Te}$), often described as A-Center.[87,92] These defect pairs should have lower ionization energy than the V_{Cd} defect, and thus, their benefit to the device could be twofold.[87] Finally, when both O and Cu are present, a defect pair composed of a Cu interstitial (Cu_i) and O_{Te} ($Cu_i + O_{Te}$) has been reported that may be a deep donor (about 150 meV below the conduction band).[56] Although the effect on $N_A - N_D$ of all these defect formations remains uncertain, with proper concentration, they would all be expected to reduce mid-gap V_{Te}. This could benefit the CdTe device, by reducing Shockley–Read–Hall recombination more than producing detrimental impacts on $N_A - N_D$.

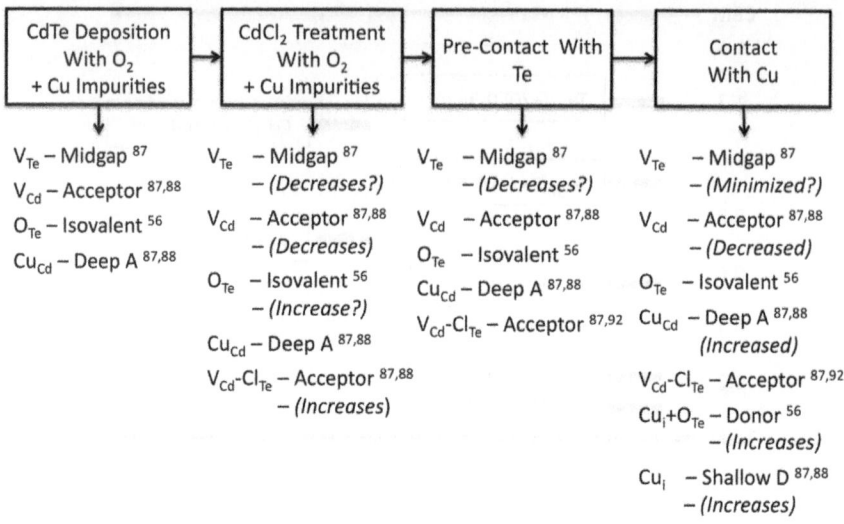

Figure 3.3   Possible defect-evolution sequence for intrinsic, O-, Cl-, and Cu-related
defects in CdTe superstrate polycrystalline devices based on observations and historic DFT + LDA guidance (see references). Te_i and Te_{Cd}
defects are not indicated based on calculated ionization and formation
energies. Question marks indicate likely functionality but where trend
has not been studied or reported.

The question marks indicated in Figure 3.3 indicate where few (if any)
studies exist that directly probe changes in V_{Te} concentration as a function of
process step. Therefore, the extent to which V_{Te} may limit minority-carrier
lifetime in polycrystalline CdTe films remains uncertain. One reason is that
nearly all groups capable of producing polycrystalline CdTe devices use
processes and equipment that implicitly assume a nearly stoichiometric
chemical potential during CdTe growth. Efforts to controllably alter chemical potential during growth or post-growth require insightful consideration
of how these changes may impact beneficial defect formation, as suggested
in Figure 3.3. One recent attempt to alter the intrinsic defects involved exposing a CdCl$_2$-treated CdTe back surface of a superstrate device to a small
amount of Te vapor at relatively high temperature. Exposure occurred during
an ion-beam milling process, but prior to application of a ZnTe:Cu/Ti back-
contacting layer.[93] Devices exposed to the Te vapor demonstrated performance reduction when compared to devices without Te-vapor exposure. Although this result may at first seem surprising within the context of
Figures 3.2 and 3.3 (*i.e.*, the figures suggest that processes that promote the
elimination of V_{Te} should be beneficial), the result supports the idea that
efforts to beneficially reduce V_{Te} must consider how O, Cl, and Cu processing
may also impact the concentration of V_{Te}. Another recent study probing the
effect of chemical potential has used large-grained (about 1 mm) CdTe
polycrystals fabricated to yield Cd-rich or Te-rich stoichiometries to the
extent allowed by equilibrium phase boundaries.[90,94] In this study, bulk

recombination rates were significantly greater in Te-rich CdTe (Cd/Te = 0.99) relative to Cd-rich CdTe (Cd/Te = 1.01).[90] Finally, a third study was conducted where the CdTe film was deposited at lower temperature to encourage a Cd-rich chemical potential (480 °C *versus* 600 °C). Although this study indicates that lower-temperature growth requires re-optimization of all other process steps to achieve optimum performance, initial device results again suggest higher τ for devices produced at growth conditions that should favor a Cd-rich chemical potential.[90] In all three of these cases, observations suggest that a Te-rich chemical potential does not produce improvement in the optoelectronic quality of the CdTe layer. This observation appears regardless of which stage of the process the chemical potential is varied, suggesting Cd-rich growth or post-growth conditions may be worth considering.

The observations noted above suggest that improved theoretical guidance related to intrinsic defects and mid-gap defect formation could enable the CdTe community to make significant improvements in device performance. The primary question would be: Would a Te-rich or a Cd-rich chemical potential result in a higher minority-carrier lifetime? Progress toward answering this question has involved both a more thorough treatment of the effect of chemical potential, and replacing LDA with more advanced (but also more computationally intensive) density functionals such as the hybrid functional developed by Heyd, Scuseria and Ernzerhof (HSE).[95] The HSE approximation combines parts of both LDA and Hartree–Fock approximations such that the resulting bandgap energy is more accurate than for either approximation alone. Compared to the historic LDA calculations, these 'hybrid' approximations are believed to provide improved predictions of defect formation and ionization energies, especially for intrinsic semiconductor defects when the states are fully occupied or fully empty. Figure 3.4 shows some of the main results of these recent DFT + HSE calculations on CdTe.[90] The most important ramifications of these studies pertain to mid-gap intrinsic defects. Unlike in LDA calculations, the HSE calculation finds the V_{Te} defect to be a relatively shallow donor with an $(2 + /0)$ ionization energy about 70 meV below the conduction-band minimum (CBM).[90] Therefore, in the HSE calculations, the dominant mid-gap deep levels arise from Te_i and Te_{Cd}, which also have low defect formation energies, especially for p-type material under Te-rich chemical potentials.[90] Another ramification of the HSE calculation is that V_{Cd} is now predicted to be an even deeper acceptor with a $(0/2-)$ transition energy level at about 0.4 eV above the valence-band maximum (VBM).[90]

Figure 3.5 shows a modification of the defect-evolution map shown in Figure 3.4, but now considering the predictions from the LDA + HSE calculations.[90] Although V_{Te} is now assigned as a shallow donor rather than a mid-gap acceptor, Figure 3.5 shows that the traditional polycrystalline CdTe process sequence would still be expected to reduce the concentration of this defect. This would benefit the device by reducing compensation (*i.e.*, increasing on $N_A - N_D$), which is consistent with observed device performance. Further, an increase in the concentration of mid-gap Te_i and Te_{Cd} defects that

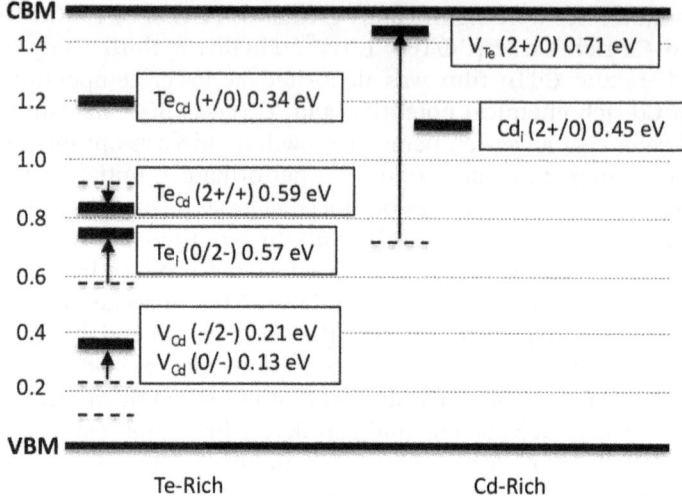

Figure 3.4   Position of ionization energy of intrinsic defects in CdTe calculated using DFT + HSE (dark lines) with some positions from LDA calculations shown for comparison (dashed lines). Defects are grouped assuming Te-rich and Cd-rich chemical potentials. Box indicates ionization energy measured from VBM for acceptors, and from CBM for donors. (From Ma *et al.*[90])

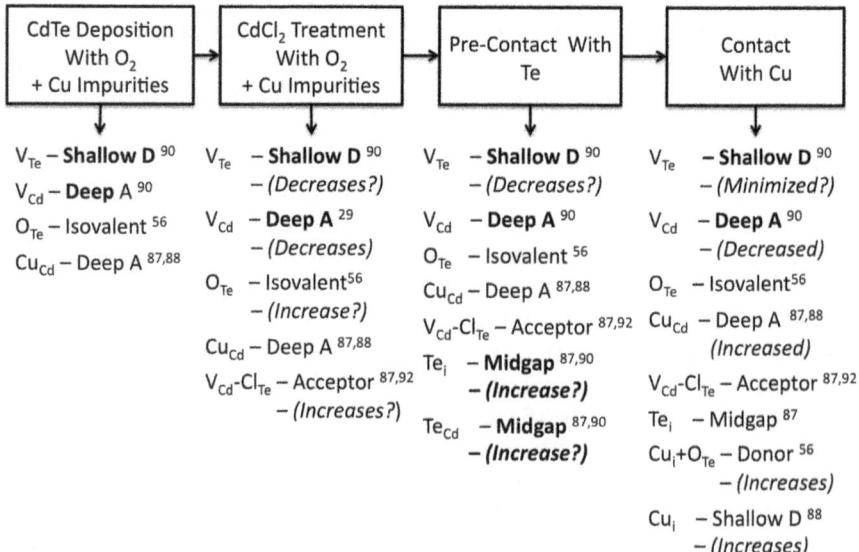

Figure 3.5   Possible defect-evolution sequence for intrinsic, O-, Cl-, and Cu-related defects in CdTe superstrate polycrystalline devices based on observations and recent DFT + HSE guidance (see references). Question marks indicate likely functionality but where trend has not been well reported. Main defect changes due to applying the HSE approximation are shown in bold.

may occur during a pre-contact step at high temperature is more consistent with recent observations of decreased performance after Te exposure.[93]

One might expect that the assignment of V_{Te} as a donor should lead to reported changes in conductivity type (from n- to p-type) as the device processing proceeds from left to right in Figure 3.5. However, reliable determination of the electrical properties of as-deposited CdTe has been very difficult to assess in polycrystalline devices prior to device completion.[96] Even if sequential electrical measurements were possible at each step, processing with O, Cl, and Cu also affects the electrical properties at grain boundaries.[97,98] Therefore, isolating effects of the V_{Te} point defect within the intragrain regions from effects at grain boundaries remains a challenging task.

3.2.4 Junction Formation and Location

A final question remains regarding where the semiconductor junction is actually located. We now know that the specific parameters of the CdS and CdTe deposition combine with the parameters of the CdCl$_2$ treatment to affect the formation and extent of the CdSTe layer. We also know that diffusion from the CIFL additionally alters the electrical properties of the CdTe, thereby altering the junction. However, we still have not discussed where the junction is.

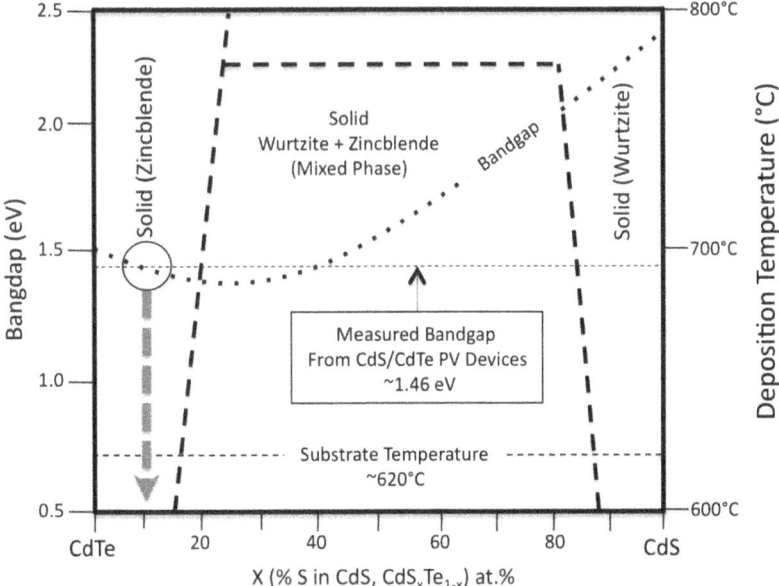

Figure 3.6 Schematic illustration of the effect of band bowing and phase separation in the CdS$_x$Te$_{1-x}$ layer. The circle at $x = \sim 0.1$ and 620 °C indicates the likely sulfur concentration, based on the measured bandgap from quantum-efficiency measurements, the effective bandgap altered by band bowing, and the available single-phase region.

To understand this better, one also must consider that when sulfur is diffused into the CdTe layer (forming a CdSTe interdiffused layer), the bandgap of the CdSTe alloy decreases relative to CdTe or CdS. This is called band bowing, and is shown schematically in Figure 3.6. At a deposition temperature of about 620 °C, depending on the stoichiometry of the CdSTe alloy, it can exist as a single-phase, hexagonal material (wurtzite, about 85–100% sulfur), a single-phase cubic material (zincblende, 0 to about 15% sulfur), or a mixed phase CdSTe material (between about 15 and about 85% sulfur).[99] During intermixing of the CdS and CdTe layers at the CdTe deposition temperature, it is believed that all three of these phases will be present in what has so far been described simply as the CdSTe layer. Many technologists believe that it is reasonable to assume that the CdSTe layer will be primarily in the wurtzite (*i.e.*, S-rich) phase nearest the CdS layer, primarily in the zincblende (Te-rich) phase nearest the CdTe layer, and in the mixed-phase between these two (primarily) single-phase endpoints. However, only a few studies of this region have been reported.[100]

For the type of devices produced at NREL, we believe the electrical junction is formed between the non-alloyed CdTe and a diffusion-formed zincblende CdSTe.[40] Analysis of the infrared absorption characteristics of QE

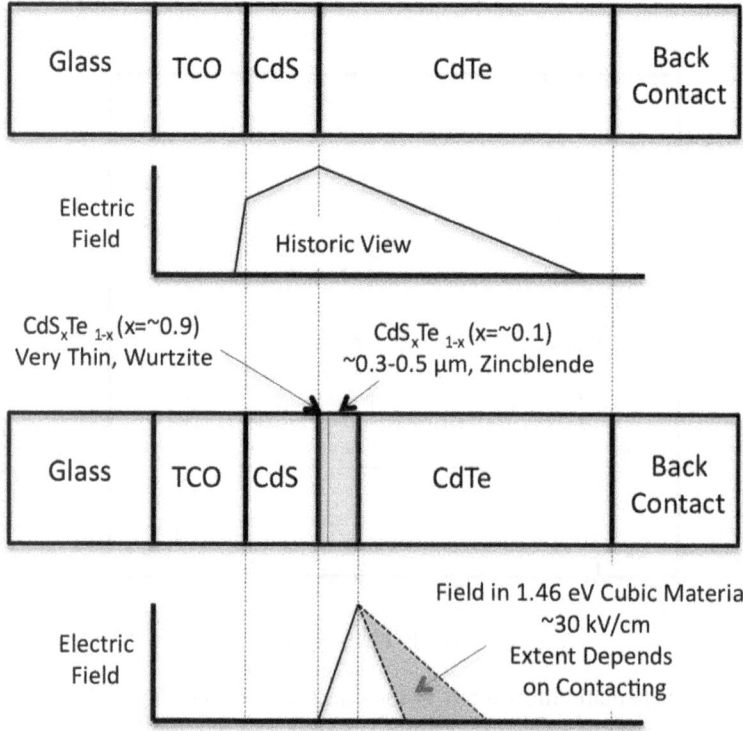

Figure 3.7   Schematic illustration of the location of the junction in a CdS/CdTe PV device from a historic (top) and more recent (bottom) perspective.

data indicates the effective sulfur concentration in the junction region is about 10%. This is estimated by calculating the effective bandgap from device QE (about 1.46 eV), while accounting for the known band bowing.[101] Figure 3.6 shows this estimation graphically. Figure 3.7 shows two schematic diagrams illustrating both the historic and more recent understanding of the layer structure, as well as the associated electric fields.

3.3 Conclusions

This chapter has presented a technological description of components and process consistent with the present generation of CdTe thin-film PV devices. The chapter also suggests why thin-film CdTe PV technology may embody considerable opportunity for becoming a significant part of future large-scale electricity production. These opportunities include improved understanding of the effect(s) of the buffer layer to reduce absorption in the CdS, development of back contacts that yield high reproducibility in junction parameters and stability, and ultimately, the ability to produce cost-effective polycrystalline CdTe absorber layers with sufficient acceptor doping and minority-carrier lifetimes in a way that does not have to rely on a back-contact processing.

Acknowledgements

The authors wish to thank the Group of SuHuai Wei of NREL for assistance with material modeling. Portions of this work were supported under U.S. Department of Energy Contract No. DE-AC36-08-GO28308 to NREL and DE-EE0005401 to USF.

References

1. M. Raugei, S. Bargigli and S. Ulgiati, *Energy*, 2007, **32**, 1310.
2. V. M. Fthenakis, *Renewable Sustainable Energy Rev.*, 2004, **8**, 303.
3. M. Held and R. Ilg, 'Life cycle assessment (LCA) of CdTe thin film PV modules and material flow analysis (MFA) of cadmium within EU27,' Proc. 23rd Eur. Photovoltaic Solar Energy Conf., Sept. 1–5, Valencia, (2008).
4. D. Bonnet, in *Clean Energy from Photovoltaics*, ed. M. Archer and R. Hill, Imperial College Press, London, 1st edn, 2001, Series on Photoconversion of Solar Energy, vol. 1, ch 6.
5. B. E. McCandless and J. R. Sites, in *Handbook of Photovoltaics Science and Engineering*, ed. A. Luque and S. Hegedus, Wiley, West Sussex, 2003, ch. 14.
6. M. Burgelman, in *Thin Film Solar Cells: Fabrication, Characterization, and Applications*, ed. J. Poortmans and V. Arkhipov, Wiley, West Sussex, 2006, ch. 7.

7. R. Birkmire, in *Solar Cells and Their Applications*, ed. Lewis Fraas and Larry Partain, Wiley, New Jersey, 2nd edn, 2010, ch. 6.

8. T. A. Gessert, in *Comprehensive Renewable Energy*, ed. W. van Sark, Elsevier, 2012, vol. 1, ch. 1.19.

9. T. A. Gessert and D. Bonnet, in *Clean Energy from Photovoltaics*, ed. M. Archer and M. Green, World Series Publishing, London, 2nd edn, (expected 2014). Series on Photoconversion of Solar Energy, vol. 1, ch. 5.

10. M. A. Green, K. Emery, Y. Hishikawa, W. Warta and E. D. Dunlop, *Prog. Photovolt Res. Appl.*, 2013, **21**, 827.

11. Z. N. Adirovich, F. M. Yuabov and G. R. Yagudaev, *Fiz. Tekh. Poluprovodn.*, 1969, **4**, 270.

12. N. Suyama, T. Arita, Y. Nishiyama, N. Ueno, S. Kitamura and M. Murosono, Proc. 21st IEEE Photovoltaic Specialists Conf., Kissimmee, FL, pp. 498–503 (1990).

13. B. M. Basol, *J. Appl. Phys.*, 1984, **55**, 601.

14. P. V. Meyers, 'Polycrystalline Cadmium Telluride n-i-p Solar Cells,' SERI Subcontract Final Report No. ZL-7-06031-2, Available through National Renewable Energy Laboratory (NREL) Library, Golden, Colorado.

15. H. Ohyama, T. Aramoto, S. Kumazawa, H. Higuchi, T. Arita, S. Shibutani, T. Nishio, J. Nakajima, M. Tsuji, A. Hanafusa, T. Hibino, K. Omura and M. Murozono, '16.0% Efficient Thin-Film CdS/CdTe Solar Cells,' Proc. 26th IEEE Photovoltaic Specialists Conf., Anaheim, CA, 1997, pp. 343–346.

16. D. Bonnet and H. Rabenhorst, 'New Results on the Development of Thin Film p-CdTe/n-CdS Heterojunction Solar Cell,' Proc. 9th IEEE Photovoltaic Specialists Conf., Silver Springs, MD, 1972, pp. 129–132.

17. H. J. Hovel, *Solar Energy Mater.*, 1980, **2**, 277.

18. D. L. Feucht, *J. Vac. Sci. Technol.*, 1977, **14**, 57.

19. D. Cusano, General Electric Res. Lab. Report, No. 4582, 1963.

20. P. Rappaport, *RCA Rev.*, 1959, **20**, 373.

21. P. K. Weimar, The TFT a new thin-film transistor, *Proc. IRE*, 1962, **50**, 1462.

22. Y. S. Deev, *Soviet J. Atomic Energy*, 1960, **6**, 321.

23. A. L. Fahrenbruch and R. H. Bube, in *Fundamentals of Solar Cells*, Academic Press, San Diego, CA, 1983, ch. 10, pp. 417–463.

24. B. Basol, Proc. 21st IEEE Photovoltaic Specialist Conf., pp. 588–594 (1990).

25. L. F. Szabo and W. J. Biter, U.S. Patent 4,735,662 (1988).

26. P. Meyers, X. Liu and T. Frey, U.S. Patent 4,710,589 (1987).

27. Y.-S. Tyan, F. Vazan and T. S. Barge, Effect of Oxygen on Thin-Film CdS/CdTe Solar Cells, Proc. 17th IEEE Photovoltaic Specialists Conf., IEEE, Piscataway, NJ, 1982, pp. 840–844.

28. J. Britt and C. Ferekides, *Appl. Phys. Lett.*, 1993, **62**, 2851.

29. X. Wu, J. C. Keane, R. G. Dhere, C. DeHart, D. A. Albin, A. Duda, T. A. Gessert, S. A. Asher, D. H. Levi and P. Sheldon, 16.5% Efficient

CdS/CdTe Polycrystalline Thin-Film Solar Cell, Proc. 17th European Photovoltaic Solar Energy Conf., October 22–26, 2001, WIP-Renewable Energies, Munich, pp. 995–1000 (2002).

30. J. F. Jordan, Photovoltaic Cell and Method, U.S. Patent No. 5261968 (1993).
31. http://www.firstsolar.com/Projects/ ∼ /media/ 4DF9E6D853334F6BA37D27B3E13F4157.ashx.
32. W. Shockley and H. J. Queisser, *J. Appl. Phys.*, 1961, **32**, 510.
33. M. Gloeckler, I. Sankin and Z. Zhao, CdTe solar cells at the threshold to 20% efficiency, *IEEE J. Photovoltaics*, 2013, **3**, 1389.
34. T. Nakazawa, K. Takamizawa and K. Ito, *Appl. Phys. Lett.*, 1987, **50**, 279.
35. R. G. Dhere, M. Bonnet-Eymard, E. Charlet, E. Peter, J. N. Duenow, J. V. Li, D. Kuciauskas and T. A. Gessert, *Thin Solid Films*, 2011, **519**, 7142.
36. T. J. Coutts, D. L. Young and X. Li, *MRS Bull.*, August 2000, **58**.
37. X. Wu, *Solar Energy*, 2004, **77**, 803.
38. X. Wu, W. P. Mulligan and T. J. Coutts, *Thin Solid Films*, 1996, **286**, 274.
39. T. A. Gessert, J. Burst, X. Li, M. Scott and T. J. Coutts, *Thin Solid Films*, 2011, **519**, 7146.
40. R. G. Dhere, Y. Zhang, M. J. Romero, S. E. Asher, M. Young, B. To, R. Noufi and T. A. Gessert, Investigation of junction properties of CdS/CdTe solar cells and their correlation to device properties, Proc. 33rd IEEE PVSC, Manuscript No. 279 (2008).
41. B. McCandless, I. Youm and R. Birkmire, *Prog. Photovolt.*, 1999, **7**, 21.
42. T. L. Chu, S. S. Chu, J. Britt, C. Ferekides, C. Wang, C. Q. Wu and H. S. Ullal, *IEEE Elect. Dev. Lett.*, 1992, **13**, 303.
43. D. H. Rose, F. S. Hasoon, R. G. Dhere, D. S. Albin, R. M. Ribelin, X. S. Li, Y. Mahathongdy, T. A. Gessert and P. Sheldon, *Prog. Photovolt. Res. Appl.*, 1999, **7**, 331.
44. R. A. Sasala, R. C. Powell, G. L. Dorer and N. Reiter, *AIP Conf. Proc Ser.*, 1996, **394**, 171.
45. X. Li, T. A. Gessert, R. J. Matson, J. F. Hall and T. J. Coutts, *J. Vac. Sci. Technol. A.*, 1993, **12**, 1608.
46. R. Wendt, A. Fischer, D. Grecu and A. D. Compaan, *J. Appl. Phys.*, 1998, **84**, 2920.
47. G. Fulop, M. Doty, P. Meyers, J. Betz and C. H. Liu, *Appl. Phys. Lett.*, 1982, **40**, 327.
48. N. Suyama, T. Arita, Y. Nishiyama, N. Ueno, S. Kitamura and M. Murosono, CdS/CdTe Solar Cells by the Screen-Printing_Sintering Technique, Proc. 21th IEEE Photovolt. Spec. Conf. IEEE, Piscataway, New Jersey, 1990, pp. 498–503.
49. J. Jordan, International Patent App. WO93/14524 (1993).
50. J. L. Kester, S. Albright, V. Kaydanov, R. Ribelin, L. M. Wods and J. A. Phillips, *AIP Conf. Proc Ser.*, 1996, **394**, 162.
51. I. Gur, N. A. Fromer, M. L. Geier and A. P. Alivisatos, *Science*, 2005, **310**, 462.

52. H. R. Moutinho, R. G. Dhere, M. M. Al-Jassim, D. H. Levi and L. L. Kazmerski, *J. Vac. Sci. Technol. A*, 1999, **17**, 1793.

53. D. S. Albin, Y. Yan and M. M. Al-Jassim, *Prog. Photovolt. Res. Appl.*, 2002, **10**, 309.

54. R. G. Dhere, D. S. Albin, D. H. Rose, S. E. Asher, K. M. Jones, M. M. Al-Jassim, H. R. Moutinho and P. Sheldon, Intermixing at the CdS/CdTe interface and its effect on device performance, *Mater. Res. Soc. Symp. Proc.* vol. 426, MRS, Warrendale, PA, 1996, pp. 361–366.

55. T. A. Gessert, M. J. Romero, C. L. Perkins, S. E. Asher, R. Matson, H. Mountinho and D. Rose, Microscopic analysis of residuals on polycrystalline CdTe following wet CdCl2 treatment, *Mat. Res. Soc. Symp. Proc.*, vol. 669, MRS, Warrendale, PA, 2001, pp. H1.10.1–H1.10.6.

56. C. Corwine, J. R. Sites, T. A. Gessert, W. K. Metzger, J. Li, A. Duda and G. Teeter, *Appl. Phys. Lett.*, 2005, **86**, 221901.

57. K. Zanio, in *Semiconductors and Semimetals*, Academic Press, New York, 1978, vol. 13, pp. 164–186.

58. D. P. Halliday, M. D. G. Potter, D. S. Boyle and K. Durose, Photo-luminescence characterization of ion implanted CdTe, *Mater. Res. Soc. Symp. Proc.* vol. 668, MRS, Warrendale, PA, 2001, p. H1.8.1.

59. V. Valdna, J. Hiie and A. Gavrilov, *Solid State Phenom.*, 2001, **80**, 155.

60. D. M. Waters, D. Niles, T. Gessert, D. Albin, D. Rose and P. Sheldon, Surface analysis of CdTe after various pre-contact treatment, Proc. 2nd World Conf. on Photovoltaic Solar Energy Conversion, Vienna, (European Commission, Luxembourg, 1988) p. 1031.

61. X. Li, D. W. Niles, F. S. Hasoon, R. J. Matson and P. Sheldon, *J. Vac. Sci. Technol. A*, 1999, **17**, 805.

62. D. Levi, D. Albin and D. King, *Prog. Photovolt Res. Appl.*, 2000, **8**, 591.

63. D. S. Albin, S. H. Demtsu and T. J. McMahon, *Thin Solid Films*, 2006, **515**, 2659.

64. B. E. McCandless and K. D. Dobson, *Sol. Energy*, 2004, **77**, 839.

65. C. S. Ferekides, D. Marinskiy, V. Viswanathan, B. Tetali, V. Palekis, P. Selvaraj and D. L. Morel, *Thin Solid Films*, 2000, **361–362**, 520.

66. V. G. Karpov, A. D. Compaan and D. Shvydka, Micrononuniformity Effects in Thin-film Photovoltaics, Proc. 29th IEEE Photovolt. Spec. Conf., 19–24 May, pp. 708–711 (2002).

67. S. K. Das and G. C. Morris, *J. Appl. Phys.*, 1993, **73**, 782.

68. R. Mamazza, Ternary Spinel Cd2SnO4, CdIn2O4, and Zn2SnO4 and Binary SnO2 and In2O3 Transparent Conducting Oxides as Front Contact materials for CdS/CdTe PhotoVoltaic Devices, Ph.D. Dissertation, University of South Florida, Fl, (2003).

69. C. S. Ferekides, R. Mamazza, U. Balasubramanian and D. L. Morel, *Thin Solid Films*, 2005, **480–481**, 224.

70. X. Wu, R. G. Dhere, D. S. Albin, T. A. Gessert, C. DeHart, J. C. Keane, A. Duda, T. J. Coutts, S. Asher, D. H. Levi, H. R. Moutinho, Y. Yan, T. Moriarty, S. Johnston, K. Emery and P. Sheldon, High-Efficiency CTO/ZTO/CdS/CdTe Polycrystalline Thin-Film Solar Cells, Report No.

NREL/CP-520-3102, Proc. NCPV Program Review Meeting, Lakewood, Colorado, October 2001.

71. K. Zanio, in *Semiconductors and Semimetals*, Academic Press, New York, 1978, vol. 13, pp. 115–163.
72. H. H. Woodbury and M. Aven, *J. Appl. Phys.*, 1968, **39**, 5485.
73. R. B. Hall and H. H. Woodbury, *J. Appl. Phys.*, 1968, **39**, 5361.
74. J. L. Pautrat, J. M. Francou, N. Magnea, E. Molva and K. Saminadayar, *J. Crystal Growth*, 1985, **72**, 194.
75. D. Kim, A. L. Fahrenbruch, A. Lopez-Otero and R. H. Bube, *J. Appl. Phys.*, 1994, **75**, 2673.
76. J. Ponpon, *Solid-State Electron*, 1985, **28**, 689.
77. E. H. Rhoderick and R. H. Williams, in *Monographs in Electrical and Electronic Engineering*, ed. P. Hammond and R. L. Grimsdale, Clarendon Press, Oxford, 1988, No. 19, Chapter 2.
78. J. Tersoff, *Phys. Rev. Lett.*, 1986, **56**, 2755.
79. D. Rioux, D. W. Niles and H. Hochst, *J. Appl. Phys.*, 1993, **73**, 8381.
80. S. O. Ferreira, H. Sitter, W. Faschinger and G. Brunthaler, *J. Crystal Growth*, 1994, **140**, 282.
81. T. A. Gessert, M. J. Romero, R. G. Dhere and S. E. Asher, Analysis of the ZnTe:Cu Contact on CdS/CdTe Solar Cells, *Mater. Res. Soc. Symp. Proc.* vol. 763 (MRS, Warrendale, PA, 2003), pp. 133–138.
82. T. A. Gessert, S. Asher, S. Johnston, A. Duda and M. R. Young, Formation of ZnTe:Cu/Ti Contacts at High Temperature for CdS/CdTe Devices, Proc. 4th WCPVEC, IEEE, Piscataway, NJ, 2006, pp. 432–435.
83. T. A. Gessert, S. Asher, S. Johnston, M. Young, P. Dippo and C. Corwine, *Thin Solid Films*, 2007, **515**, 6103.
84. D. S. Albin, Accelerated Stress Testing and Diagnostic Analysis of Degradation in CdTe Solar Cells, 2008 SPIE Optophotonics Meeting, Reliability of Photovoltaic Cells, Modules, Components, and Systems, San Diego, CA, Aug. 10–14, 2009.
85. T. A. Gessert, W. K. Metzger, P. Dippo, S. E. Asher, R. G. Dhere and M. R. Young, Dependence of carrier lifetime on Cu-contacting temperature and ZnTe:Cu thickness in CdS/CdTe thin film solar cells, *Thin Solid Films*, 2009, **517**, 2370–2373.
86. T. A. Gessert, W. K. Metzger, S. E. Asher, M. R. Young, S. Johnston, R. G. Dhere and T. Moriarty, Effect of Cu Diffusion from ZnTe:Cu/Ti Contacts on Carrier Lifetime of CdS/CdTe Thin Film Solar Cells, Proc. 33rd IEEE PVSC, 2008, Manuscript No. 14.
87. S.-H. Wei and S. B. Zhang, *Phys. Rev. B*, 2002, **66**, 155221.
88. J. Ma, S-H. Wei, T. A. Gessert and K. K. Chin, *Phys. Rev. B*, 2011, **83**, 245201.
89. M.-H. Du, H. Takenaka and D. J. Singh, *J. Appl. Phys.*, 2008, **104**, 093521.
90. J. Ma, D. Kuciauskas, D. Albin, R. Bhattacharya, M. Reese, T. Barnes, T. Gessert and S.-H. Wei, *Phys. Rev. Lett.*, 2013, **111**, 067402.
91. S. A. Awadalla, A. W. Hunt, K. G. Lynn, H. Glass, C. Szeles and S.-H. Wei, *Phys. Rev. B*, 2004, **6**, 075210.

92. A. Castaldini, A. Cavallini, B. Fraboni, L. Polenta, P. Fernandez and J. Piqueras, *Mater. Sci. Eng.*, 1996, **B42**, 302.
93. T. A. Gessert, J. M. Burst, S.-H. Wei, J. Ma, D. Kuciauskas, W. L. Rance, T. M. Barnes, J. N. Duenow, M. O. Reese, J. V. Li, M. R. Young and P. Dippo, *Thin Solid Films*, 2013, **535**, 237.
94. V. Lyahovitskaya, L. Chernyak, J. Greenberg, L. Laplan and D. Cahen, *J. Appl. Phys.*, 2000, **88**, 3976.
95. A. V. Krukau, O. A. Vydrov, A. F. Izmaylov and G. E. Scuseria, *J. Chem. Phys.*, 2006, **125**, 224106.
96. B. McCandless, Effects of Treatments on CdTe Film Conductivity, NREL Subcontract Report #ADJ-1-30630-12, Dec. 23, 2003.
97. H. R. Moutinho, R. G. Dhere, M. J. Romero, C.-S. Jiang, B. To and M. M. Al-Jassim, *J. Vac. Sci. Technol. A*, 2008, **26**, 1068.
98. P. R. Edwards, S. A. Galloway and K. Durose, *Thin Solid Films*, 2000, **361–362**, 364.
99. S.-Y. Nunoue, T. Hemmi and E. Kato, *J. Electrochem. Soc.*, 1990, **137**, 1248.
100. R. G. Dhere, Y. Zhang, M. J. Romero, S. E. Asher, M. Young, B. To, R. Noufi and T. A. Gessert, Investigation of Junction Properties of CdS/CdTe Solar Cells and Their Correlation to Device Properties, Proc. 33rd IEEE PVSC, Manuscript No. 279 (2008).
101. D. Compaan, Z. Feng, G. Contreas-Puente, C. Narayanswamy and A. Fisher, Properties of Pulsed Laser Deposited CdS_xTe_{1-x} Films on Glass, *Mater. Res. Soc. Symp. Proc.*, vol. 426, MRS, Warrendale, PA, 1996, pp. 367–372.

CHAPTER 4

III–V Multi-junction Solar Cells

SIMON P. PHILIPPS* AND ANDREAS W. BETT

Fraunhofer Institute for Solar Energy Systems ISE, Heidenhofstr. 2, 79110 Freiburg, Germany
*Email: simon.philipps@ise.fraunhofer.de; andreas.bett@ise.fraunhofer.de

4.1 Introduction

The highest efficiencies of any photovoltaic technology, so far, have been reached with solar cells made of III–V compound semiconductors. These compound semiconductors consist of elements from the main groups III and V of the Periodic Table and offer a wide range of bandgaps which can be realized in excellent crystal quality. Consequently, they are particularly suitable for multi-junction solar cells, in which solar cells with different bandgaps are used to absorb distinct parts of the solar spectrum. Monolithically stacked III–V multi-junction solar cells have reached efficiencies of 38.8% under the global reference spectrum AM1.5g, 35.1% under AM0 for space applications and 44.7% under the concentrated direct reference spectrum AM1.5d.[1–3]

Such high efficiencies are not of pure scientific interest but III–V multi-junction solar cells have become the standard cell in space applications and in terrestrial high concentration systems. In space, the benchmark cost is based on power per weight, motivating the use of cells with highest efficiency. This is due to higher costs to launch satellites into space orbits.[4] Thus, higher production cost at cell level is feasible. Looking at terrestrial applications, the cost per generated kWh is the benchmark. The levelized

RSC Energy and Environment Series No. 11
Advanced Concepts in Photovoltaics
Edited by Arthur J Nozik, Gavin Conibeer and Matthew C Beard
© The Royal Society of Chemistry 2014
Published by the Royal Society of Chemistry, www.rsc.org

cost of energy (LCOE), which is defined as the total cost of a system divided by its lifetime energy production, is often used to rate the cost competiveness of different photovoltaic technologies. In general, high efficiencies are essential for lowering the levelized cost of energy (LCOE) as more energy can be produced from the same installation area.[5] With respect to concentrating photovoltaics (CPV) it is necessary that the higher systems costs, due to the usage of a tracking system, do not outweigh the benefit of higher energy production. Moreover, III–V multi-junction solar cells are essentially more expensive than conventional, *e.g.*, silicon-based solar cell technologies of the same cell area. The main cost driver is the substrate used: the price of typical GaAs or Ge wafers can be up to a hundred-fold higher than for a silicon wafer. Thus, even if their efficiency is about twice as high, substrate-based III–V multi-junction solar cells are currently too expensive for the use in flat-plate modules on Earth. Hence the development of high concentrating photovoltaic (HCPV) systems with concentration factors above 300 was essential to introduce III–V multi-junction solar cells to the terrestrial market.[6]

It should be noted here for completeness that one possibility to reduce the impact of high substrate costs could be to re-use the expensive semiconductor substrate after carrying out a lift-off of the thin solar cell structure.[7,8] The resulting thin-film solar cells have high efficiencies and can be used in flexible modules. Lately this approach has been followed commercially and is seen as a way to enable the use of III–V solar cells in flat-plate modules in building-integrated as well as rooftop applications in the near future.[9,10] However, the proof of cost competitiveness with conventional flat-plate modules is still pending. In addition, not all designs of III–V multi-junction solar cells allow lift-off techniques.

Yet, large-area III–V multi-junction solar cells are nowadays standard in space applications. Besides cost aspects III–V multi-junction solar cells are also particularly suitable for the need in space due to their radiation hardness, *i.e.*, the high end-of-life efficiency,[11] small temperature coefficients, high reliability and the combination of high voltage and low currents.[12,13] Due to these advantages, the early development of III–V multi-junction solar cells had been driven by the space demands. However, the space market (see Figure 4.1, left) is comparably stable and has not grown for a long time. Therefore, it was of interest to introduce III–V multi-junction solar cells into the strongly growing terrestrial market. Eventually this was successful, not in flat-plate modules but in HCPV systems. Here the use of cost-efficient concentrating optics reduces the need for expensive cell area, which allows competitive LCOE. The right graph in Figure 4.1 shows the development of the new HCPV installations per year and the cell technology used. Since 2009 III–V multi-junction solar cells have been standard in HCPV systems. However, this application necessitates adaption in the multi-junction solar cell structure. Today, the on-going development of III–V multi-junction solar cells is driven by the needs of space and terrestrial market and both benefit from each other.

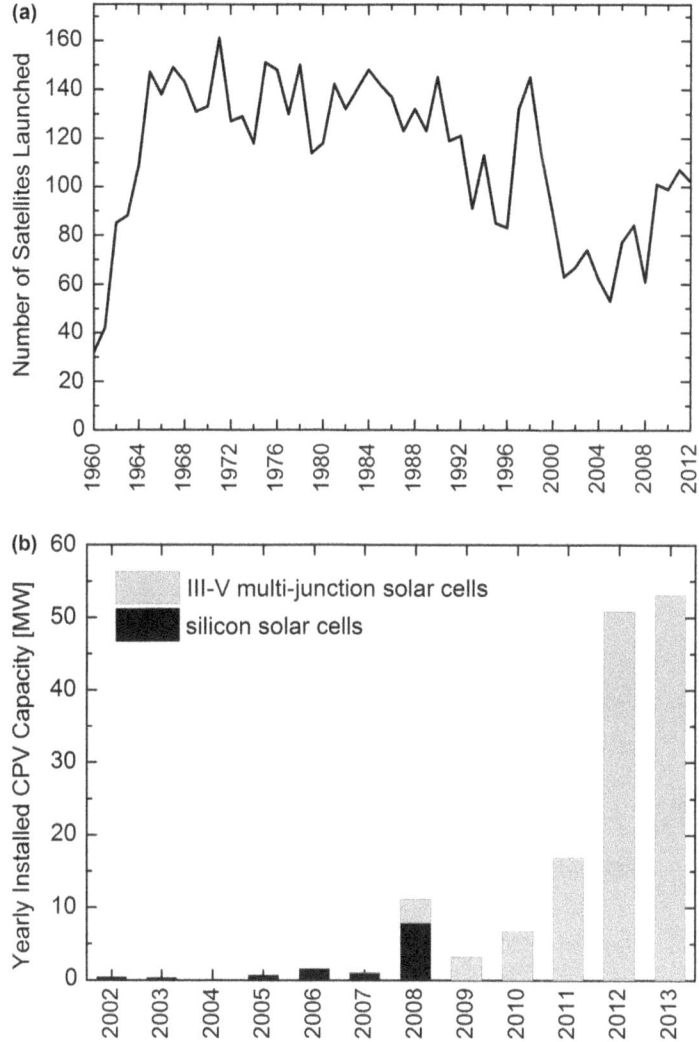

Figure 4.1 (a) Number of yearly launched satellites (Source of data: Hiriart and Saleh[21] for 1960–2008; www.satelliteonthenet.co.uk for 2009–2012). (b) Yearly installed CPV capacity with indication of the solar cells used. III–V multi-junction solar cells have become standard in CPV systems.

It should be noted that III–V solar cells are also used in several niche applications such as laser power converters and thermo-photovoltaics (TPV).[14–20] Their central benefit for these applications is again the wide range of possible bandgaps, which allows adapting the absorber profile to the emission spectrum of the light source, *e.g.*, a laser or a thermal emitter.

The current landmark is the MOVPE (metal–organic vapour phase epitaxy) grown $Ga_{0.50}In_{0.50}P/Ga_{0.99}In_{0.01}As/Ge$ triple-junction, which only contains

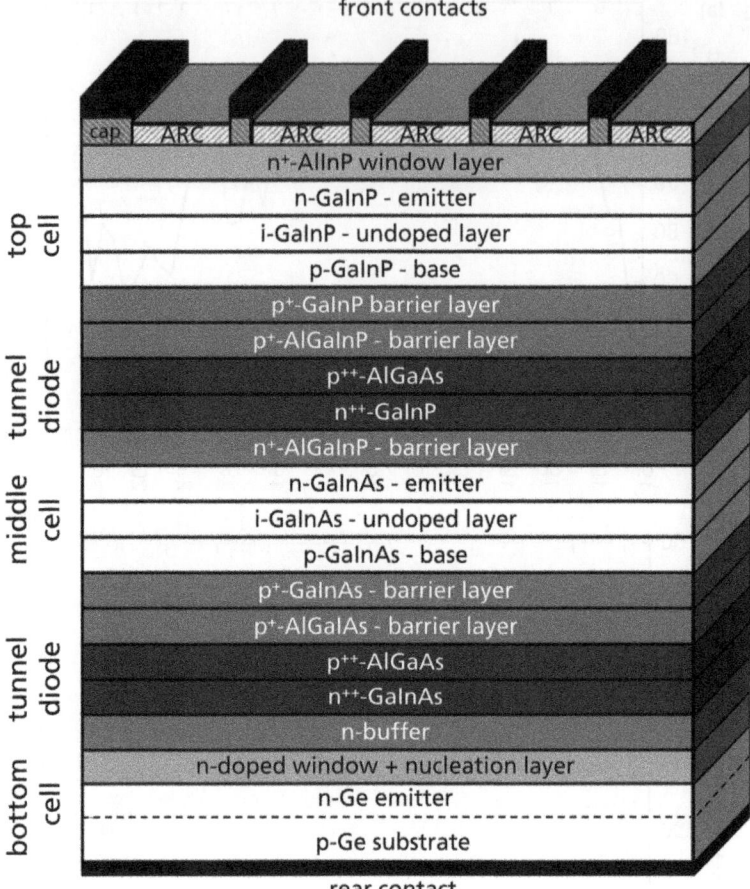

front contacts

Figure 4.2   Simplified structure of a lattice-matched $Ga_{0.50}In_{0.50}P/Ga_{0.99}In_{0.01}As/Ge$ triple-junction solar cell.

lattice-matched layers. The internal architecture of this monolithically grown solar cell is sophisticated since tunnel diodes, barrier and passivation layers as well as differently doped layers are needed. Figure 4.2 shows an internal structure (not to scale) in order to demonstrate the complexity. Anyhow, this device has achieved a record efficiency of 41.6% (AM1.5d, 364 suns) and is commercially available with efficiencies around 40% under concentrated sunlight.[22–25] Due to the mature development status of this device new concepts are being investigated to increase the efficiencies further. A variety of materials as well as an extensive technological toolbox is available. Consequently, diverse solar cell architectures have been realized, which allow a more efficient use of the solar spectrum. Several approaches have already reached efficiencies higher than the lattice-matched triple-junction solar cell.

4.2 On the Efficiency of III–V Multi-junction Solar Cells

The efficiency potential of a photovoltaic cell is strongly influenced by i) the spectral distribution of the incident photons and ii) by the bandgap of the materials used in its pn-junction. For example, photons from the sun's spectrum with energies below the bandgap cannot be absorbed and are transmitted (transmission losses). But also a part of the energy of absorbed high-energy photons is lost due to collisions of the created charge carriers with the crystal-lattice (thermalisation losses). Due to these losses, which are inevitable in conventional solar cells, it is instructive to first investigate the optimal bandgap(s) for a solar cell without considering possible restrictions due to other material properties such as material quality, technical feasibility and availability. Several models exist for this optimization (see Kurtz *et al.*[26] for an overview). A prominent method is Shockley and Queissers' *detailed balance approach*, which results in an upper theoretical limit for the conversion efficiency of a solar cell under a certain spectrum.[27] The central assumption is that radiative recombination is the only inevitable recombination mechanism. Other recombination channels are hence not considered in Shockley–Queissers' approach. Solar cells are further assumed to be ideal in the sense that they have an external quantum efficiency of unity for all photons with energies above the bandgap and behave according to the one-diode model. The calculated efficiencies are to be seen as upper limits, but give valuable guidance on the optimal bandgap. A comparison with realized solar cells indicates that—as rule of thumb—between 70% and 80% of the theoretical efficiencies can be achieved in reality.

The detailed balance approach can easily be extended to multi-junction solar cells. It was, for example, implemented in the programme *etaOpt*.[28] In order to evaluate multi-junction devices *etaOpt* assumes that current from upper sub-cells can be balanced with lower ones to improve current-matching, which is necessary for maximum conversion efficiency. In real solar cells this is achieved by thinning the absorbing layers. In the following *etaOpt* will be used to calculate upper limits for the efficiencies of different solar cell concepts under various spectral conditions.

4.2.1 Photovoltaic Cells and Monochromatic Light: A Perfect Match

A PV cell exhibits its highest efficiencies if it is illuminated with monochromatic light near the energy of its bandgap.[29,30] Figure 4.3 shows the theoretical efficiency limit for this scenario as the dependence on the wavelength and the intensity of the monochromatic illumination. The calculation was carried out with *etaOpt*. High efficiencies are particularly possible for semiconductors with high bandgaps (*i.e.*, for photons with short wavelengths). This is due to the fact that radiative recombination, which is

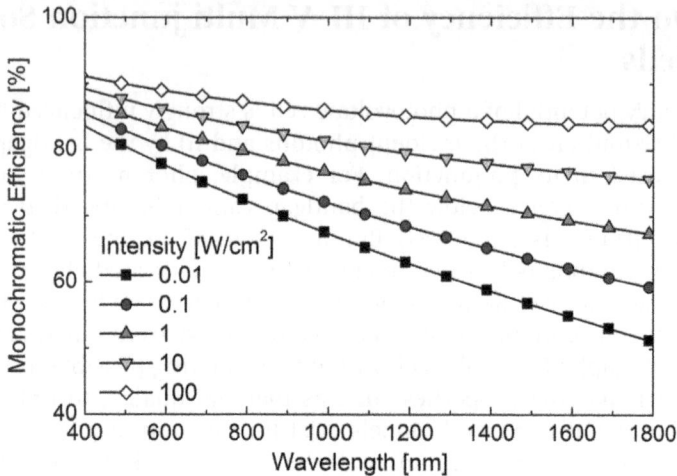

Figure 4.3 Theoretical limits for the monochromatic efficiency of ideal PV cells, whose bandgap equals the energy of the incident photons $(E_g = h\nu_{mono})$.

the only recombination channel in the Shockley–Queisser approach, decreases with increasing bandgap. Moreover and important to note, the efficiency increases with higher illumination intensity as the open-circuit voltage increases logarithmically with the number of generated charge carriers.

Photovoltaic cells for laser power conversion have already been realized with different semiconductor materials. GaInP can be used for energy conversion in the visible wavelength range, *e.g.*, 500 to 650 nm.[29] GaAs is suitable for the wavelength range from 800 to 860 nm. Corresponding laser power converters have already been realized with high efficiencies, *e.g.*, 55% under an illumination of 37 W cm^{-2} and 56% at 50 W cm^{-2}.[16,31] An efficiency value around 45% has been reported for a Silicon photovoltaic cell, which was illuminated with monochromatic light of 1020 nm and an intensity of 1 W cm^{-2}.[32] The range of the near-infrared (1-2 µm) could be covered with ternary and quaternary III–V compound semiconductors. Monochromatic efficiencies of such materials have been comparably low, *e.g.*, around 34% with In$_{0.53}$Ga$_{0.47}$As at 1.55 µm and 22% with In$_{0.72}$Ga$_{0.28}$As at 2.10 µm.[14] However, an InGaAsP device with an efficiency of 44.6% under monochromatic illumination at 1.55 µm and a power density of 1 kW m^{-2} was recently presented.[33] Significant optimization potential still remains. Another option is GaSb, which has already achieved efficiencies up to 49% under high irradiances of 75 W cm^{-2} with a 1680 nm laser.[16]

Photovoltaic cells for monochromatic illumination are not only of scientific interest. In recent years there is growing activity to use such cells as laser power converters.[17,18,34–36] Thereby, typically laser or LED light is converted to electrical energy through a PV cell with an aligned bandgap. The electrical energy can for example be used to power microelectronic sensors. This approach is often called *power-over-fibre* as the laser light is usually transmitted through optical fibres. The central benefit is that the electronic components, which are powered by the PV cell, are galvanically isolated from other electronics, *e.g.*, the control unit. This facilitates operation in high-voltage, hazardous or explosive environments. Possible fibre-less applications are the powering of aerial vehicles, or the beaming of energy by light from space to Earth.[37,38] Such systems, which would require very high-power lasers, are still in the early prototype and conceptual stage, respectively.

One central challenge of laser power converters with one pn-junction is that the voltage is comparably low for the operation of microelectronic devices and circuits. Therefore, innovative approaches are being investigated to increase the voltage of the photovoltaic cells. One promising option is to series connect several single-junction elements, *e.g.*, by stacking them on top of each other or through a segmentation of the devices.[18,35,39,40] Taking into consideration that very limited resources have been devoted to the development of laser power converters up to now, high optimization potential remains.

4.2.2 Towards a Match with the Solar Spectrum: Stacking Photovoltaic Cells

Thermalisation and transmission losses are inevitable for solar cells due to the wide range of photon energies in the solar spectrum. Figure 4.4a indicates the comparably small part of the AM1.5g spectrum that can be transferred into electrical energy by a single-junction solar cell with the optimal bandgap for this spectrum. A better match of the solar cell to the solar spectrum can be achieved if the solar spectrum is split. This can be achieved with advanced optical elements, for example holograms, beam splitters, prisms *etc.*[41–43] The alternative approach is stacking PV cells with decreasing bandgaps from top to bottom, either by the so-called mechanical or by monolithically stacking. This approach is illustrated in Figure 4.4b for a six-junction device with optimal bandgaps for each sub-cell. Thermalisation losses are reduced as photons are mostly absorbed in layers with a bandgap close to the photon's energy. Moreover, transmission losses are reduced as the absorption range of the multi-junction solar cell is usually wider than for single-junction devices.

It is evident that a better use of the solar spectrum through the stacking of solar cells allows significantly higher efficiencies. Figure 4.5 shows this increase for ideal solar cells with 1 to 6 pn-junctions under different spectral conditions. The calculation was carried out with *etaOpt*. An ideal single-junction under the extra-terrestrial reference spectrum AM0 could theoretically achieve an efficiency of 30.3%, whereas a six-junction boosts this limit to 56.7%. Slightly higher efficiencies are possible under the global, terrestrial reference spectrum AM1.5g. This spectrum applies to solar cells on Earth under non-concentrated sunlight, *e.g.*, flat-plate modules, as it includes photons, which come directly from the sun as well as from scattered light.

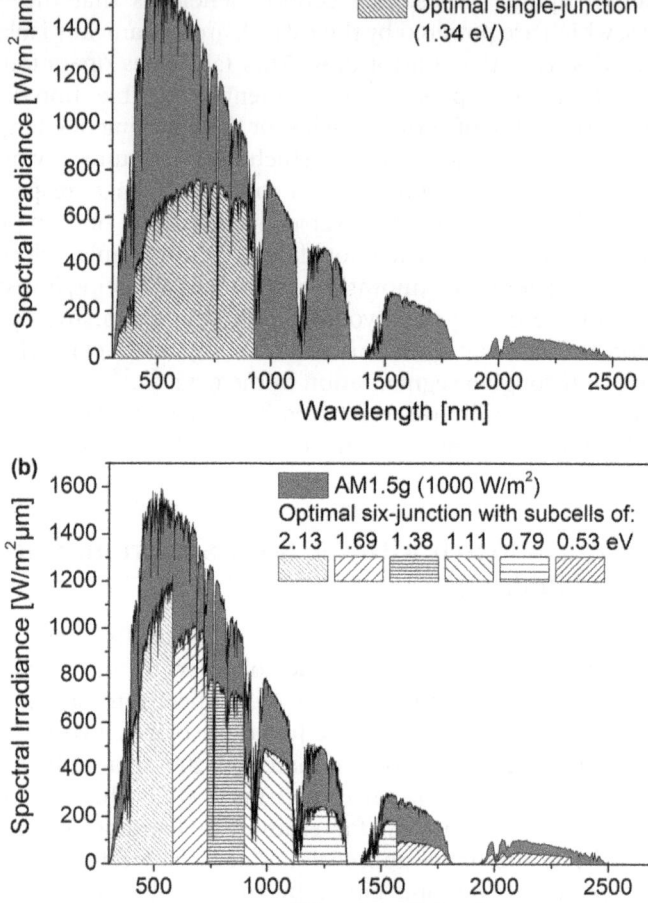

Figure 4.4 Parts of the energy content of the AM1.5g reference spectrum (1000 W m^{-2}; IEC 60904-3, ed. 2) that can be converted by an ideal single-junction solar cell (a) and a six-junction solar cell with an optimal bandgap combination (b).

It might seem contradictory that higher efficiencies are possible for AM1.5g compared to AM0 although the latter has a higher irradiance of 1367 W m^{-2} *versus* 1000 W m^{-2} for AM1.5g. The origin lies in a different relative distribution of the light in the spectra. In particular the higher share of high energy photons in the AM0 leads to stronger thermalisation losses compared to AM1.5g. An additional increase of efficiency is possible by concentrating the sunlight as the solar cell's voltage increases logarithmically with the number of generated charge carriers.[44] This effect is exemplified in Figure 4.5 for a concentration of 500 suns of the direct, terrestrial reference spectrum AM1.5d, which only accounts for the direct sunlight and is hence applicable for concentrator applications on Earth.

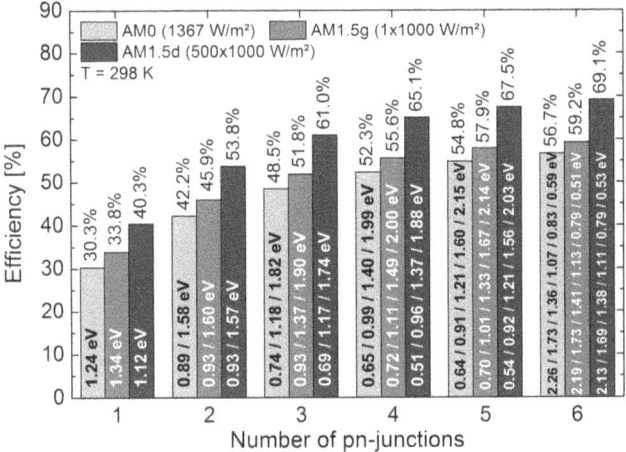

Figure 4.5 Dependence of the theoretical efficiency limits on the number of p–n junctions for the reference spectra AM0 (extraterrestrial; 1367 W m^{-2}; ISO 15387, ed. 1), AM1.5g (terrestrial, global; 1000 W m^{-2}; IEC 60904-3, ed. 2) as well as AM1.5d (terrestrial, direct; 500×1000 W m^{-2}; ASTM G-173-03 direct).

Figure 4.5 also indicates the ideal bandgap combinations for each reference spectrum. In order to harvest the efficiency potential, solar cell stacks with appropriate bandgaps need to be realized in very good material quality. III–V semiconductors have become standard for such multi-junction approaches due to the wide range of available bandgaps. In addition, the use of direct III–V semiconductors facilitates a high absorption of light even in comparably thin layers. These are the main reasons why III–V multi-junction solar cells reach the highest efficiencies of any photovoltaic technology.[1] Figure 4.6 shows the development of efficiencies of III–V multi-junction solar cells under concentrated sunlight (x*AM1.5d) and under AM0 for space applications. Record efficiencies of international laboratory cells show an impressive increase from 32.6% in 1993 to the current record value of 44.7%.[3,45] Note that the record in 1993 was achieved with a dual-junction solar cell, whereas a four-junction solar cell reached the current record of 44.7%. All efficiency records in between have been achieved with triple-junction solar cells. Different structural and technological concepts have been used as will be discussed in the following. As also shown in the graph average commercial cell efficiencies for concentrator cells are only about 10%$_{relative}$ lower than the record values.

4.3 The Technological Toolbox to Fabricate III–V Multi-junction Solar Cells

Various technological methods and concepts can be used for the realization of III–V multi-junction solar cells. The lattice-matched Ga$_{0.50}$In$_{0.50}$P/ Ga$_{0.99}$In$_{0.01}$As/Ge triple-junction solar cell can serve as a reference as this

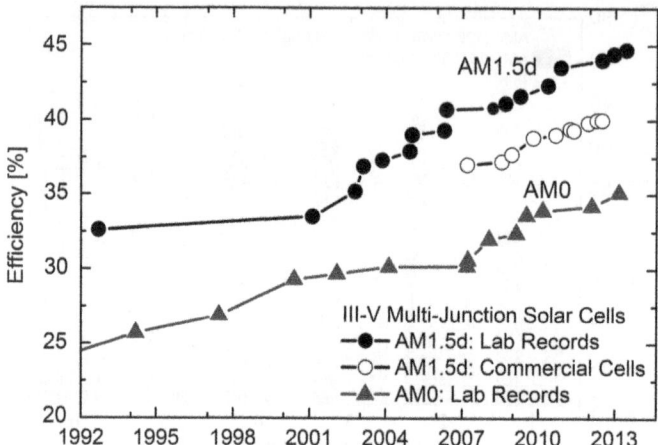

Figure 4.6   Development of III-V multi-junction solar cell record efficiencies under concentrated light (x*AM1.5d) and AM0. Examples for typical commercial concentrator cell efficiencies (different concentration levels) are also indicated. (AM1.5d lab records according to Green *et al.*, Solar Cell Efficiency Tables from 1993[45] to 2014;[1] AM0 lab records based on various individual publications, and McClure and Gaddy[46] and Wilt[47]; AM1.5d commercial efficiencies averaged from company product sheets).

device is the state-of-the-art approach in space and terrestrial concentrator systems. The device is grown with high throughput in commercial metal–organic vapour phase epitaxy (MOVPE) reactors. As the essential feature all semiconductors in this structure have the same lattice constant as the Ge substrate, which facilitates crystal growth in high material quality. However, its bandgap combination is neither optimal for space nor for terrestrial applications, as the bottom cell generates significantly more current than the upper two cells. Nevertheless, a record efficiency of 41.6% (AM1.5d, 364 suns) has been achieved and average production efficiencies are approaching the 40%-mark under concentrated light and 30% for space solar cells.[22–25,48] Various new concepts and technologies are under investigation to increase the efficiencies further. Figure 4.7 shows elements of the toolbox to make advanced III–V multi-junction solar cells.

4.3.1   Epitaxial Growth Methods

Starting on a substrate the thin crystalline semiconductor layers in III–V solar cells are formed layer upon layer in a process called epitaxy. In this process liquid or gaseous materials or even single molecule/atom beams are brought onto the substrate (for details on the epitaxial process see, for example, Pohl[49]). Under adequate growth conditions semiconductor layers are being deposited on the substrate surface. For high efficiencies it is essential that the growth process results in well-defined crystal layers with a related orientation. Several epitaxial methods are used for the growth of III–V solar

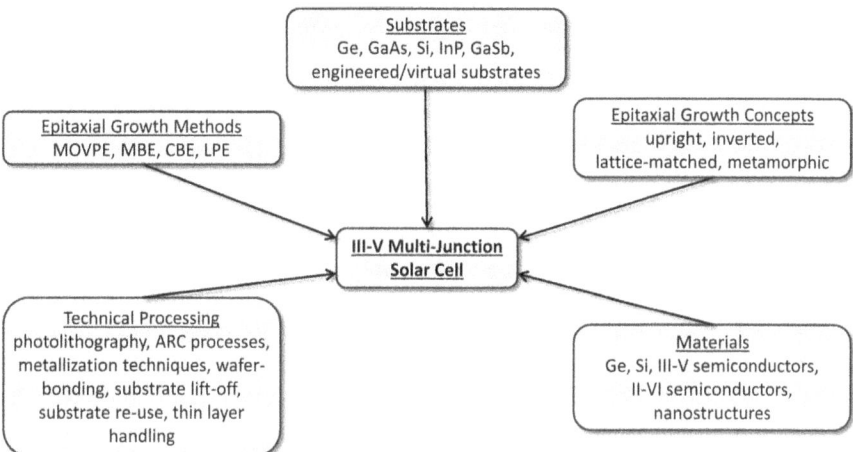

Figure 4.7   Non-exclusive toolbox for the realization of advanced III–V multi-junction solar cells.

cells. The oldest method is liquid phase epitaxy (LPE), in which epitaxial layers are deposited on a substrate from materials in the liquid phase.[50] The required semiconductor materials are dissolved in a melt until the melt is saturated for a certain temperature. The temperature is then selectively reduced in order to exceed the solubility limit, which results in the deposition of epitaxial layers. LPE was used in the early days of III–V solar cells. However, it turned out that the process was very difficult to control if the structure consisted of different materials, as is the case for monolithic III–V multi-junction solar cells. The more sophisticated structures can be grown more easily and with high throughput using metal–organic vapour phase epitaxy (MOVPE), in which gaseous materials are delivered to the substrate inside a reaction chamber.[51] The name originates from the metal–organic precursors, which are used for bringing some of the central elements to grow the compound semiconductor into the reactor. Trimethylgallium (TMGa), trimethylaluminium (TMAl) or trimethylindium (TMIn) are for example often used for the group-III elements, whereas dimethylzinc (DMZn) could be used for doping. In addition, often gases such as arsine or phosphine are used to provide a source for the group-V elements. By applying the correct temperature and pressure to the right mixture of gases a chemical reaction takes place resulting in the deposition of crystalline layers on top of the substrate. MOVPE enables a high precision with respect to atomic composition and growth rates and delivers highly reproducible semiconductor structures. The epitaxial process is relatively fast with growth rates typically in the range up to 10 μm h^{-1}. Challenges are impurities for example oxygen and carbon which are easily incorporated into the semiconductor material, thus cleanliness of the reactor and the precursors are essential. Currently, some semiconductors, *e.g.*, dilute nitride-alloys, cannot be grown with sufficiently high quality using MOVPE technology.[22,52,53]

Very high purity of these and other epitaxially grown materials has been demonstrated with molecular beam epitaxy (MBE).[54–57] The technique generates beams of molecules/atoms by heating typically solid sources.[58] The molecules/atoms react on the crystalline surface of a heated substrate to form semiconductor layers. The process takes place in ultra-high vacuum to enable high material quality and to avoid collisions of the molecule beams with impurities in the reactor. As the beams can be controlled quickly, very abrupt hetero-structures can be obtained with MBE. A drawback for large-scale production is the relatively slow growth rate of typically $1\ \mu m\ h^{-1}$ and the lower throughput. Another method for epitaxy is chemical beam epitaxy (CBE), which can be seen as a hybrid of MBE and MOVPE.[59] The technical setting is similar to MBE as molecular beams are directed onto a heated substrate in an ultra-high vacuum. However, CBE uses metal–organic and hydride precursors similar to MOVPE, whereas single atom beams are mainly used in MBE.

Anyway, MOVPE is the standard method for epitaxy of III–V multi-junction solar cell structures. Large-area commercial MOVPE reactors are available from several suppliers. However, activities with other epitaxial methods are on-going. These aim in particular at materials, which are difficult to realize with MOVPE in high crystal quality, *e.g.*, (In)GaNAs alloys with CBE and MBE.[54–57,60,61] MBE is also used to realize nanostructures in III–V multi-junction solar cells.[62]

4.3.2   Substrates

Different substrates are available for the growth of III–V multi-junction solar cells. These differ in technical characteristics like bandgap, lattice constant, off-cuts and available doping of the substrate material as well as in economic aspects such as price and availability. In many architectures for multi-junction solar cells the substrate not only functions as a seed layer for epitaxial growth, but also becomes the bottom cell of the sub-cell stack, *e.g.*, through diffusion of doping material into the upper part of the substrate. Hence for these cell architectures the bandgap of the substrate should be in accordance with corresponding calculations of the optimal bandgap combination (see Figure 4.5). On the other hand the lattice constant of the substrate material should match that of the upper solar cell layers since this facilitates epitaxial growth with high material quality. Figure 4.8 exemplarily shows various binary and ternary III–V semiconductors as well as common substrates. It is evident that a wide range of bandgaps can, in principle, be covered with III–V semiconductors. However, the limited number of available substrates puts a restriction on the technically feasible material combinations. Currently, germanium is the standard substrate for commercial III–V triple-junction solar cells due to its relatively low bandgap, the lattice-match with suitable III–V semiconductors and its commercial availability. Alternatives are still being investigated as Ge is relatively expensive. One low cost alternative is silicon, which has again found increasing attention in the last years. Silicon could enable high efficiencies in dual- and triple-junction solar cells. However, for multi-junctions with more

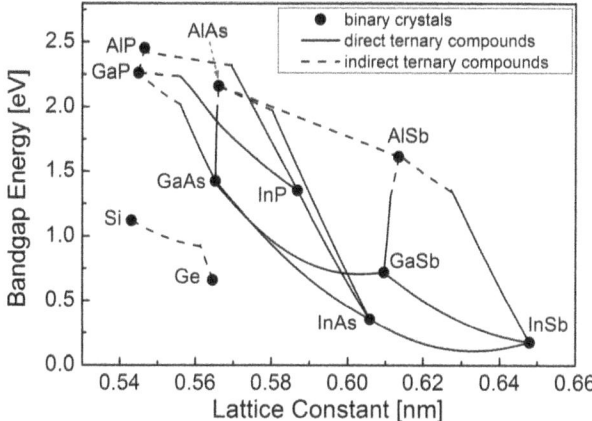

Figure 4.8 Relation between band gap energy and lattice constant for exemplary semiconductors (based on Bett *et al.*[11]). Lines between binary crystals represent direct (solid lines) or indirect (broken lines) ternary compounds. An extension to quaternary and quintenary alloys allows for even more options in band gap designs.

than three p–n junctions its bandgap is too high. The main technical challenge lies in the epitaxial growth of suitable III–V semiconductors on silicon due to its smaller lattice constant. Various strategies such as metamorphic growth or wafer bonding are being investigated to overcome the lattice-mismatch (for details see section 4.4.3). GaAs has been used as a substrate material for single- and dual-junction solar cells for a long time.[8,63–66] As a direct semiconductor high absorption is facilitated and several common III–V compounds with similar lattice-constant are available to build solar cell stacks in good material quality. However, its high bandgap prohibits multi-junction architectures with more than two junctions with a GaAs bottom cell.

Several research groups are also investigating epitaxial growth of III–V multi-junction solar cells on InP substrates as this allows lattice-matched growth of materials for high efficiency bandgap combinations.[67–70] Although scientific results are promising high substrate costs currently pose a challenge to wide commercial adoption.

Another substrate material is GaSb, which has mostly been investigated with a focus on thermo-photovoltaics (TPV), and mechanically stacked multi-junction devices.[71–75] GaSb has fallen out of focus in research in recent years, but might be revisited as an infrared absorber in the future due to its low bandgap. The substrate is currently even more expensive than InP. However, significant cost reductions are possible as antimony is, in principle, available in large quantities.

4.3.3 Epitaxial Growth Concepts

Depending on the substrate used and the intended materials needed to make a specific solar cell architecture different epitaxial growth concepts can be applied. The straight forward approach is to grow III–V multi-junction solar cells, which only contain lattice-matched materials in reference to the substrate, in an upright direction. This means that the solar cell structure is grown from bottom cell to top cell on the substrate. However, only a restricted range of materials and thus bandgaps can be grown lattice-matched to the available substrates. Using the upright lattice-matched concept on Ge substrates, as is the standard today, leads to twice the current in the Ge subcell. In other words, the efficiency for this structure is not maximized. Therefore, the concept of metamorphic growth has been developed. Here, materials with different lattice constant are grown on top of each other. This allows the adaption of the bandgaps of the top and middle cell on a Ge-substrate to maximize the efficiency. However, metamorphic growth is challenging since the change in the lattice constant introduces defects and strain. Therefore, very specially designed buffer structures are usually implemented.[76–79] They gradually transition the lattice constant and eventually provide a substrate surface for growth with a new lattice constant. It is very important that all dislocations and other crystal defects, which result from the difference in the lattice constants, are confined to within the buffer so that the subsequent solar cell layers can be grown strain-relaxed and have good material quality (see Figure 4.9). In general it is extremely challenging to realize buffer structures which absorb all defects. Some dislocations might propagate into the growing layers. Therefore, it is particularly

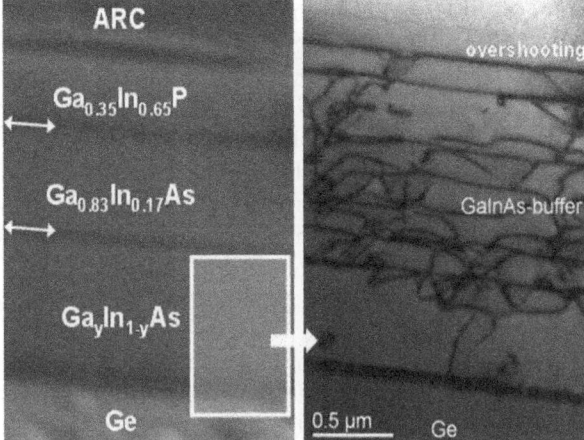

Figure 4.9 REM image of the epitaxial structure of a metamorphic $Ga_{0.35}In_{0.65}P$/$Ga_{0.83}In_{0.17}As$/Ge triple-junction solar cell (left). In addition to the sub cells the structure contains tunnel diodes (marked with arrows), an anti-reflective coating (ARC) and a $Ga_{1-y}In_yAs$ buffer. A 220 bright field TEM cross section shows that dislocations are confined in the step-graded buffer (right). This allows epitaxial growth of the upper sub-cells in high material quality.
(Reprinted with permission from Guter *et al.*[78] Copyright 2009, AIP Publishing LLC).

challenging if a buffer is implemented early during the growth phase. In such a situation inverted metamorphic growth is preferred.[25,80–86] In this approach the multi-junction solar cell is grown inversely, *i.e.*, with the top and middle cell being grown first on a lattice-matched substrate followed by another sub-cell, which is grown upon a metamorphic buffer. After growth lift-off techniques are used to remove the substrate from the solar cell and the grown structure is flipped around. From a technical point of view the inverted metamorphic growth concept has mainly two advantages compared to upright growth. First, the growth of the buffer is postponed to the latest possible growth phase, while the major part of the multi-junction solar cell can be grown lattice matched to the substrate. Thus, threading dislocations due to the transition of the lattice constant do not affect the upper cells. Second, the bandgap of the bottom cell can be chosen with fewer restrictions as the cell is grown epitaxially and not created in the Ge substrate. Economically a cost benefit in production could arise if the same substrate is used for several epitaxial runs. Yet this might be counter-balanced by higher production costs and lower yield due to the complexity of cell processing. For space applications the possible lower power to mass ratio and the possibility for flexible modules could be additional benefits.

4.3.4 Materials

The wide choice of III–V materials allows for various designs of III–V multi-junction solar cells and is the key to high efficiencies. However, the prerequisite for high efficiencies is excellent material quality. Realizing theoretically optimal multi-junction solar cell stacks with sufficient material quality might be the main challenge in the R&D of III–V solar cells. The central parameter to evaluate material quality is the charge carrier lifetime. It determines in particular the diffusion length of minority carriers. If the lifetime and thus the diffusion length in the emitter or base layer are too short, minority carriers will recombine before reaching the p–n junction and the cell current will be low. The quality of the bulk material is obviously affected by the purity of the epitaxial sources and by residues from previous growth in the epitaxial reactor. But even in a pure environment material quality is strongly influenced by the parameters of the epitaxy, *e.g.*, pressure, temperature, and III/V ratio in MOVPE growth. Moreover, lattice mismatch and thermal stress within the layer stack need to be dealt with adequately in order to limit the number of defects within the structure.

Particular challenges arise from the high number of (hetero)-interfaces within the layer stack.[87–89] Defects at the interface can lead to high recombination velocities. Moreover, depending on the band alignment at the interface charge carriers might flow in unintended directions, *e.g.*, away from the p–n junction. In order to reduce these effects emitter and base layer are enclosed by passivation layers. For these front-surface-field (window) and back-surface-field layers III–V semiconductors are chosen, which shall allow for a sufficient interface quality and for a beneficial band alignment (Figure 4.2).

One example of a material system, which would be highly valuable for III–V multi-junction solar cells are dilute nitrides (GaIn)(NAsP). It offers a wide range of bandgaps from 1-2 eV for different lattice-constants. However, in MOVPE-growth material quality has been limited leading to short minority carrier diffusion lengths.[52,90,91] Currently, dilute nitrides can only be grown with MBE in sufficiently high material quality.[92,93]

A more complex class of materials are nanostructures. Research efforts are on-going to increase the efficiency of III–V solar cells through the integration of nanostructures (for an overview see Tsakalakos[94] and Cristóbal[95]). Specific nanostructures can be integrated into the absorbing layers or placed outside the solar cell. In some approaches the complete device is structured on the nanoscale.

Multiple quantum wells (MQW) or quantum dots (QD) were proposed to be incorporated into the intrinsic region of a sub-cell in a multi-junction solar cell.[62,96–102] This gives a new technological option to achieve current matching conditions and thus to increase the efficiency of the standard $Ga_{0.50}In_{0.50}P/Ga_{0.99}In_{0.01}As/Ge$ triple-junction solar cell. Multiple quantum wells can be realized by thin alternating layers of semiconductors with higher and lower bandgap compared to the host material, which leads to a confinement of charge carriers in one dimension and hence to discrete energy values. By placing such a stack into the intrinsic region of the p- and n-doped layers of a solar cell, the absorption can be extended to longer wavelengths. In the standard triple-junction solar cell this leads to higher current densities. An opposing trend is the decrease of open-circuit voltage, which might balance a possible efficiency increase.[97] However, it was found that strained-balanced MQWs minimize the voltage loss.[100,103] In addition, to higher current densities MQW solar cells might also be beneficial as they can lead to higher radiation hardness in space.[104]

Instead of MQWs, QDs can also be used to achieve the current matching.[105,106] Also investigated was whether QD can be used to form an intermediate absorption band within the bandgap of a semiconductor.[107] Corresponding intermediate band solar cells could increase the current, while preserving the voltage. Although experimental proofs of concept were provided, structures with high efficiencies have not yet been realized.[108] An overview of other applications of quantum dots in solar cells can be found in Wu and Wang.[109]

Another recent approach is the growth of a forest of nanowires from III–V materials, which avoids dislocations in metamorphic material combinations and reduces overall material consumption.[110–112] Moreover, as growth on Si substrates is possible, the high costs of III–V substrates can be avoided.[113] Although still in the early prototype phase, nanowire solar cells are promising. Recently an efficiency of 13.8% has been achieved with single-junction InP nanowires.[110] An overview about the experimental status and the theoretical background of III–V nanowire solar cells can be found in LaPierre *et al.*[114]

4.3.5 Post-growth Technological Processing

After epitaxial growth the stack of semiconductor layers is processed into solar cell devices. Various post-growth technologies are used. These include photolithography and etching for defining the structures, deposition of the ARC as well as metallization techniques for top and rear contacts. As discussed above monolithic growth of optimal multi-junction solar cell stacks can be challenging, in particular for devices with more than three junctions. Therefore, advanced post-growth technologies which allow realizing monolithic multi-junction solar cells without the requirement of monolithic growth of the complete structure are interesting.

One post-growth technology wafer bonding, which combines independently grown (multi-junction) solar cells, has recently been under increasing attention.[115–121] It can help to overcome challenges arising from different lattice constants or thermal expansion coefficients in monolithic growth. This gives a higher degree in flexibility for substrate and hence lattice constant and bandgap choice. One stack is usually grown inversely. During the bonding process the cell stacks are, in principle, just pressed together. However, as the bonding interface lies between sub-cells within the solar cell stack it needs to fulfil several requirements.[119,122] In particular, high optical transparency is important in order to avoid parasitic absorption. This is why direct wafer bonds are favourable in contrast to metal bonding. Moreover, the wafer bond needs to have high mechanical strength and a low electrical resistance. One option is to bond two wafers at room temperature after deoxidizing the surfaces with wet-chemical etching and use thermal annealing at high temperatures to ensure a good bond.[122] However, the induced thermal stress might result in cracks. Another option is to activate the bond surfaces with highly energetic argon atoms followed by a bond at room temperature (see Figure 4.10).[119,123] Through rapid thermal annealing as well as high doping levels at the interface, low resistances have already been demonstrated with fast atom beam activated direct wafer bonding.[121]

After the bonding step the substrate of the upper stack needs to be removed, which can for example be achieved with laser, stress induced, ion implantation or wet chemical lift-off processes.[124–127] An aim for the future is to re-use the substrate several times in order to reduce costs. Although the total process is technologically challenging, promising results have already been achieved.[115–118,128]

Another option for combining independently grown solar cells, which was mainly used before MOVPE-growth was well established, is to use mechanical stacking.[73–75] In this approach solar cells are arranged on top of each other without direct contact and without removing the substrate(s). However, due to the precise positioning required, the use of several substrates and the necessity of duplicative processing costs, it does not seem to be competitive approaches with monolithically grown or probably wafer bonded devices. Moreover optical losses in the substrate and at the interfaces decrease the efficiency of mechanical stacks.

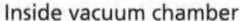

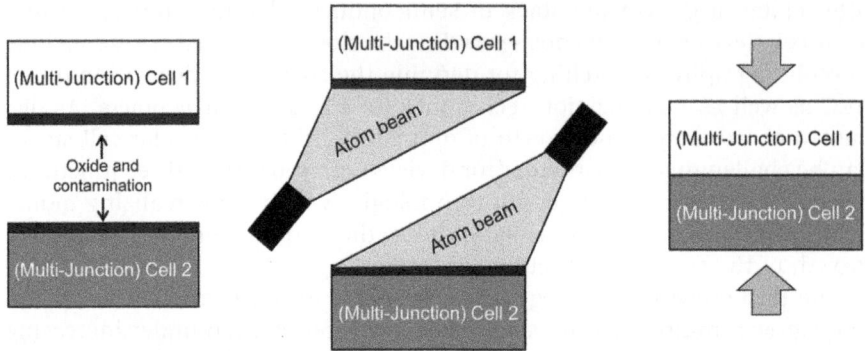

Figure 4.10 Schematic illustration of fast atom beam activated wafer bonding. Before the actual bonding of the two (multi-junction) solar cells their surfaces are cleaned with atom beams in order to allow for a bond interface with low resistance and high mechanical stability.

Another approach is based on spectrum-splitting architectures, which split the solar spectrum and direct the light toward single- or dual-junction solar cells with adequate absorption ranges.[41,42,129,130] Thereby the challenge of lattice-match is circumvented. Several devices in prototype stage have already been realized. However, as for mechanical stacks competitive costs will be difficult to achieve due to greater assembly effort, more extensive processing and the use of several substrates.

4.4 Some Members of the III–V Multi-junction Solar Cell Family

The wide variety of III–V semiconductors available in conjunction with the extensive technical toolbox has resulted in the heterogeneity of different III–V multi-junction solar cells. In the remainder of this chapter we will discuss some concepts already investigated, which use and combine elements of this toolbox for III–V multi-junction solar cells, illustrated with examples. Figure 4.11 presents different cell architectures which are discussed in this chapter. Note that only an incomplete overview can be given. For other reviews see, for example, several other studies.[12,131–137]

4.4.1 Upright Metamorphic Devices on Ge Substrates

The large excess current in the bottom cell of the lattice-matched triple-junction results from the high bandgap difference between the Ge bottom cell (0.66 eV) and the $Ga_{0.99}In_{0.01}As$ middle cell (1.41 eV). Thus, lower bandgaps for the upper two cells could increase the overall current, but would also lower the voltage. Calculations show that higher theoretical efficiencies and higher energy yields can be achieved on a Ge bottom cell if

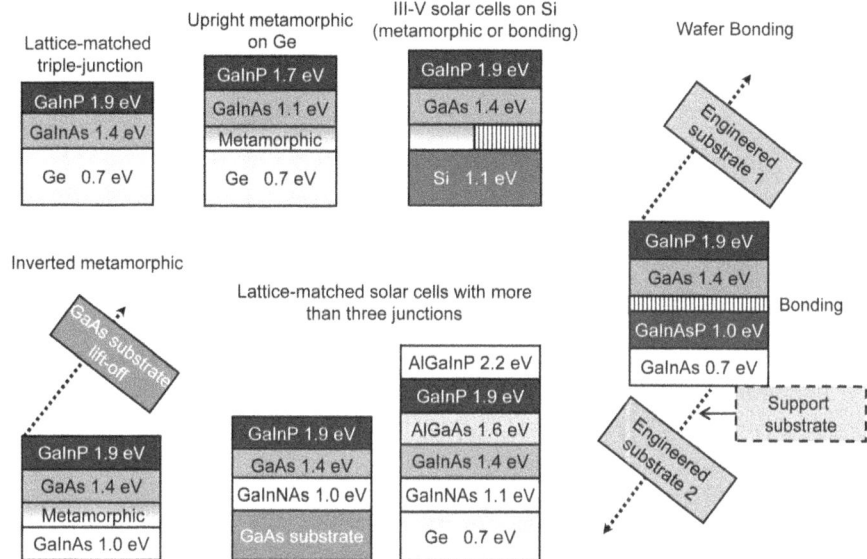

Figure 4.11 Schematic illustration of some currently investigated III–V multi-junction solar cell concepts. Some concepts have also been realized with different materials and band gap combinations.

the bandgap of the middle and top cells are lowered.[78,138] A suitable bandgap combination can be obtained by increasing the Indium content in $Ga_xIn_{1-x}As$ and $Ga_yIn_{1-y}P$. However, as the lattice constant also increases, the metamorphic growth concept is necessary. Direct growth of the materials on top of the Ge bottom cell would cause misfit dislocations and poor material quality. The effect of dislocation can be reduced through the implementation of buffer structures between the Ge and the GaInAs sub-cell, which increase the lattice constant gradually (Figure 4.9).[76,78] Corresponding upright metamorphic triple-junction solar cells have already been realized with efficiencies above 41% under concentrated sunlight.[78] Theoretical calculations underline that there is still room for significantly higher efficiencies.[138] It is expected that efficiencies up to 45% will be realized with this concept.

4.4.2 Inverted Metamorphic Multi-junction Solar Cells

High efficiencies for both terrestrial reference spectra and for the AM0 spectrum have been realized with inverted metamorphic (IMM) multi-junction solar cells.[1,25,82,84,85,139] Triple-junction solar cells achieved a value of 37.9% under AM1.5g, whereas 44.4% was realized under AM1.5d and a concentration of 302 suns.[139] It should be noted here that this efficiency value is close to what can be expected for a triple-junction device, *i.e.*, 45 to 46%. Figure 4.12 illustrates the concept of this MOVPE-grown IMM device. The solar cell layers are grown upside down on a GaAs substrate, meaning

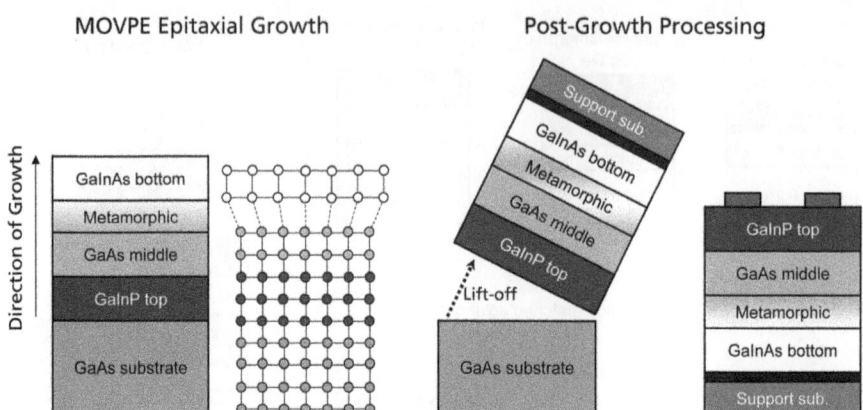

Figure 4.12   Schematic illustration of the production of an inverted metamorphic triple-junction solar cell.

that the $Ga_{0.51}In_{0.49}P$ top cell is grown first followed by a GaAs middle cell. Both cells are grown lattice-matched to the substrate. Then a buffer layer is implemented, which gradually increases the lattice constant in order to grow the final $Ga_{0.69}In_{0.31}As$ (1.0 eV) bottom cell in high crystal quality. During cell fabrication the GaAs substrate is etched off and the triple-junction solar cell is flipped and mounted on a substrate which is often silicon.

In recent years efficiencies above 40% (AM1.5d) have been reached with different internal designs of IMM triple-junction solar cells, which under-lines the high flexibility of this approach.[25,82,85,139] Yet, the approach is not limited to three junctions and IMM cells with four, five and six sub-cells have also been investigated.[22,25,86] A four-junction solar cell reached an efficiency of 34.2% under AM0.[86] A combination of upright and inverted growth—called bifacial growth—is also noteworthy. Sub-cells are grown on both sides of a GaAs wafer with a metamorphic GaInAs bottom cell being grown first on the wafers' backside. After flipping the wafer in the MOVPE reactor a GaAs middle and a GaInP top cell are grown lattice-matched. One technological advantage is that the substrate within the structure protects the upper two cells from dislocations originating from the metamorphic growth on the other side of the wafer. The approach has recently led to an efficiency of 42.3% (AM1.5d, 406 suns).[140]

The achievement of high efficiencies underlines the technical feasibility of the IMM approach. In order to make the devices cost-efficient for mass production the post-growth technical processing needs to be optimized and simplified. Moreover, the reproducibility has to be improved and re-use of the substrate should be enabled.

4.4.3   III–V on Si

The expensive Ge substrate in state-of-the-art III–V multi-junction solar cells makes up for a high share of the production costs.[141] Therefore, research efforts are ongoing to grow III–V multi-junction solar cells on lower-cost Silicon substrates. As the Ge bottom cell in the lattice-matched triple-junction solar cell has a large excess current, its replacement with a higher bandgap Silicon bottom cell would not decrease the overall current significantly, but could enable higher voltages. Thus, a triple-junction with upper cells of $Ga_{0.51}In_{0.49}P$ and GaAs on a Silicon bottom cell is promising. However, a technical challenge arises from the 4.1% difference in lattice constant as well as from the different thermal expansion coefficients between Si and GaAs of 2.6×10^{-6} °C^{-1} and 5.7×10^{-6} °C^{-1}, respectively.[142] Two different approaches are being investigated to overcome this difference: direct growth on the Si substrate and wafer bonding.

It was found that direct epitaxial growth of GaAs on Silicon substrates with MOVPE leads to a prohibitive number of defects in the GaAs layers.[143,144] It was already reported in 1998 that the best GaAs single-junction solar cell grown on an inactive Silicon substrate achieved an efficiency of 21.3% under AM0.[145] Since metamorphic growth has recently been introduced, this concept is currently under investigation to develop adequate buffer layers to gently transition the lattice constant. Different strategies are examined (for an overview see Roesener *et al.*[146]). One option is the creation of a Ge layer either directly or through the use of SiGe compounds.[147–149] Another option is to realize a GaP nucleation which is nearly lattice-matched to silicon (see Figure 4.8), followed by a buffer of $Ga_{1-x}In_xP$ or $GaAs_xP_{1-x}$.[146,150] However, the achieved efficiencies are still far below the corresponding structures on GaAs or Ge reference substrates: $Ga_{0.51}In_{0.49}P$/GaAs dual-junction solar cells, which were grown with buffers on Si substrates, have achieved efficiencies between 16 and 17% under AM1.5g.[149,151] An alternative strategy for direct growth on silicon substrates without the necessity for buffer layers is to use dilute nitrides (GaIn)(NAsP), which allows choosing lattice-matched materials to silicon. However, the short diffusion length of minority carriers currently limits this approach.[52]

The challenges of direct growth of III–V semiconductors on Si can be circumvented by using the wafer bonding technology.[115,116] Therein, III–V solar cells are grown upright or inverted onto a suitable (*e.g.*, GaAs) substrate. After removal of the substrate the III–V structure is wafer bonded onto an independently produced silicon bottom cell. Only few results of this concept have been published so far. An $Al_{0.1}Ga_{0.9}As$/Si dual-junction solar cell achieved an non-calibrated efficiency of 25.2% under non-concentrated sunlight.[122] Recently, a wafer-bonded GaInP/GaAs//Si triple-junction solar cell was realized with an efficiency of 26.0% under AM1.5g and 27.9% under AM1.5d (48 suns).[137,151] Further research is necessary to improve the technological processing, the quality of the bonding interface and the solar cell layer structure. For cost-competiveness a process, which ensures the re-use of the III–V substrate, is also mandatory. Yet, the results achieved show the high promise of wafer bonding of III–V solar cells on silicon.

4.4.4   Wafer-bonded Multi-junction Solar Cells

Wafer-bonding is not only used to realize III–V solar cells on comparatively cheap silicon substrates, as discussed in the previous section. It has, for example, also been proposed to create multi-junction solar cells with more than three junctions using wafer bonding.[152] Recently, a five-junction solar cell with 35.1% under AM0 and 38.8% under AM1.5g was presented.[1,2] Three upper sub-cells of AlGaInP (2.2 eV), AlGa(In)As (1.7 eV) and Ga(In)As (1.4 eV) were grown inverted on a GaAs or Ge substrate and then combined *via* wafer bonding with sub-cells of GaInPAs (1.1 eV) and GaIn(P)As (0.7 eV). The latter were grown upright on an InP substrate. All sub-cells are grown lattice-matched to their respective substrate. An even higher efficiency of 33.5% under AM0 is achieved with four-junction solar cell using a similar approach.[153] Another semiconductor-bonded four-junction solar cell with different bandgaps (1.9/1.4/1.0/0.73 eV) achieved an efficiency around 33.5% under AM0.[118,153] Recently, a four-junction solar cell consisting of a GaInP/GaAs dual-junction wafer bonded to a GaInAsP/GaInAs dual-junction for terrestrial concentrator applications was developed. The challenge here is to obtain a very low ohmic resistance at the wafer bonding interface.[121] This device achieved a new record efficiency of 44.7% (AM1.5d, 297 suns).[3]

4.4.5   Lattice-matched Growth of more than Three Junctions

The straightforward approach for III–V multi-junction solar cells with more than three junctions is to grow lattice-matched on Ge substrates. Theoretical calculations show that a 1.0 eV sub-cell placed in between the $Ga_{0.99}In_{0.01}As$ middle cell and the Ge bottom cell of the standard lattice-matched triple-junction solar cell would lead to a nearly optimal four-junction device. Yet, the realization of such a 1.0 eV material in a lattice-matched configuration is challenging. The promising candidate (GaIn)(NAs), *e.g.*, with 7% In and 2% N, suffers from a low minority carrier diffusion length if the material is grown in MOVPE reactors.[22,52,53,154] However, a former world record triple-junction solar cell with an efficiency of 44.0% (AM1.5d, 942 suns) is composed of GaInP/GaAs/GaInNAs grown on a GaAs substrate.[57,92] This device was grown by MBE, which might also be an option for future four-junction solar cells with dilute nitrides. Obviously for industrial scale production the production costs must be competitive with MOVPE grown devices.

For MOVPE growth an option to work around the limited GaInNAs sub-cell is to move to five- or six-junction solar cells, which require lower currents for each sub-cell. Several such devices have already been realized with promising efficiencies.[22,53,155]

4.5   Conclusion

The use of III–V semiconductors for solar cells enables the highest efficiencies due to the wide range of available bandgaps and the possibility to fabricate complex solar cell stacks with high quality material. An extensive toolbox of technological methods facilitates creativity and new ideas. Consequently, various novel and heterogeneous designs for III–V multi-junction solar cells are being investigated to exceed the high efficiencies already reached by the lattice-matched triple-junction solar cell. Many concepts have already been realized with efficiencies beyond 40% under concentrated sunlight. Several others are still in prototype status, but show great promise concerning efficiency and cost reduction. It can be expected that efficiencies towards 50% will be reached within the next decade.

Acknowledgements

The authors wish to thank all members of the department 'III–V – Epitaxy and Solar Cells' at Fraunhofer ISE for their valuable input. A special thanks to Tom Tibbits for proofreading.

References

1. M. A. Green, K. Emery, Y. Hishikawa, W. Warta and E. D. Dunlop, *Prog. Photovoltaics*, 2014, **22**, 1–9.
2. P. T. Chiu, D. C. Law, R. L. Woo, S. B. Singer, D. Bhusari, W. D. Hong, A. Zakaria, J. Boisvert, S. Mesropian, R. R. King and N. H. Karam, *IEEE J. Photovoltaics*, 2014, **4**, 493–497.
3. F. Dimroth, M. Grave, P. Beutel, U. Fiedeler, C. Karcher, T. N. D. Tibbits, E. Oliva, G. Siefer, M. Schachtner, A. Wekkeli, A. W. Bett, R. KrauseM. Piccin, N. Blanc, C. Drazek, E. Guiot, B. Ghyselen, T. Salveatat, A. Tauzin, T. Signamarcheix, A. Dobrich, K. Schwarzburg, T. Hannappel, *Prog. Photovoltaics*, 2014, **22**, 277–282.
4. S. G. Bailey, R. Raffaelle and K. Emery, *Prog. Photovoltaics*, 2002, **10**, 399–406.
5. X. T. Wang, L. Kurdgelashvili, J. Byrne and A. Barnett, *Renewable Sustainable Energy Rev.*, 2011, **15**, 4248–4254.
6. M. Wiesenfarth, H. Helmers, S. P. Philipps, M. Steiner and A. W. Bett, Proceedings of the 27th European Photovoltaic Solar Energy Conference and Exhibition, Frankfurt, 2012.
7. G. J. Bauhuis, P. Mulder, E. J. Haverkamp, J. C. C. M. Huijben and J. J. Schermer, *Sol. Energy Mater. Sol. Cells*, 2009, **93**, 1488–1491.
8. B. M. Kayes, H. Nie, R. Twist, S. G. Spruytte, F. Reinhardt, I. G. Kizilyalli and G. S. Higashi, Proceedings of the 37th IEEE Photovoltaic Specialists Conference, Seattle, Washington, 2011.
9. H. A. Atwater and T. J. Watson, Proceedings of the 37th IEEE Photovoltaic Specialists Conference, Seattle, Washington, 2011.
10. L. S. Mattos, S. R. Scully, M. Syfu, E. Olson, L. L. Yang, C. Ling, B. M. Kayes and G. He, Proceedings of the 38th IEEE Photovoltaic Specialists Conference, Austin, Texas, 2012.

11. A. W. Bett, F. Dimroth, G. Stollwerck and O. V. Sulima, *Appl. Phys.*, 1999, **69**, 119–129.

12. D. J. Friedman, J. M. Olson and S. Kurtz, in *Handbook of Photovoltaic Science and Engineering*, ed. A. Luque and S. Hegedus. John Wiley & Sons, Ltd., Chichester, 2011, pp. 314–364.

13. S. Bailey and R. Raffaelle, in *Handbook of Photovoltaic Science and Engineering*, ed. A. Luque and S. Hegedus, John Wiley & Sons, Ltd., Chichester, 2011, pp. 365–399.

14. S. J. Wojtczuk, Proceedings of the 26th IEEE Photovoltaic Specialists Conference, Anaheim, California, 1997.

15. C. Algora and V. Díaz, *IEEE Trans. Electron Devices*, 1998, **45**, 2047–2054.

16. V. Andreev, V. Khvostikov, V. Kalinovsky, V. Lantratov, V. Grilikhes, V. Rumyantsev, M. Shvarts, V. Fokanov and A. Pavlov, Proceedings of the 3rd World Conference on Photovoltaic Energy Conversion, Osaka, 2003.

17. G. Boettger, M. Dreschmann, C. Klamouris, M. Hubner, M. Roger, A. W. Bett, T. Kueng, J. Becker, W. Freude and J. Leuthold, *IEEE Photonics Technol. Lett.*, 2008, **20**, 39–41.

18. J. Schubert, E. Oliva, F. Dimroth, W. Guter, R. Löckenhoff and A. W. Bett, *IEEE Trans. Electron Devices*, 2009, **56**, 170–175.

19. V. Andreev, V. Khvostikov and A. S. Vlasov, in *Concentrator Photovoltaics*, ed. A. Luque and V. Andreev. Springer, Heidelberg, 2007, pp. 175–198.

20. T. Bauer, *Thermophotovoltaics: Basic Principles and Critical Aspects of System Design*, Springer, Berlin, 2011.

21. T. Hiriart and J. H. Saleh, *Space Policy*, 2010, **26**, 53–60.

22. R. King, A. Boca, W. Hong, D. Larrabee, K. M. Edmondson, D. C. Law, C. Fetzer, S. Mesropian and N. H. Karam, Proceedings of the 24th European Photovoltaic Solar Energy Conference and Exibition, Hamburg, 2009.

23. W. Guter, R. Kern, W. Köstler, T. Kubera, R. Löckenhoff, M. Meusel, M. Shirnow and G. Strobl, Proceedings of the 7th International Conference on Concentrating Photovoltaic Systems, Las Vegas, Nevada, 2011.

24. J. H. Ermer, R. K. Jones, P. Hebert, P. Pien, R. R. King, D. Bhusari, R. Brandt, O. Al-Taher, C. Fetzer, G. S. Kinsey and N. Karam, *IEEE J. Photovoltaics*, 2012, **2**, 209–213.

25. D. Aiken, E. Dons, S.-S. Je, N. Miller, F. Newman, P. Patel and J. Spann, *IEEE J. Photovoltaics*, 2013, **3**, 542–547.

26. S. R. Kurtz, D. Myers, W. E. McMahon, J. Geisz and M. Steiner, *Prog. Photovoltaics*, 2008, **16**, 537–546.

27. W. Shockley and H. J. Queisser, *J. Appl. Phys.*, 1961, **32**, 510–519.

28. G. Létay and A. W. Bett, Proceedings of the 17th European Photovoltaic Solar Energy Conference, Munich, 2001.

29. A. W. Bett, F. Dimroth, R. Löckenhoff, E. Oliva and J. Schubert, Proceedings of the 33rd IEEE Photovoltaic Specialists Conference, San Diego, California, 2008.

30. R. K. Jain, *IEEE Trans. Electron Devices*, 1993, **40**, 1893–1895.
31. E. Oliva, F. Dimroth and A. W. Bett, *Prog. Photovoltaics*, 2008, **4**, 289–295.
32. M. A. Green, J. Zhao, A. Wang and S. R. Wenham, *IEEE Electron Device Lett.*, 1992, **13**, 317–318.
33. J. Mukherjee, S. Jarvis, M. Perren and S. J. Sweeney, *J. Phys. D Appl. Phys.*, 2013, **46**, 264006.
34. Y. Tanaka, M. Kinoshita, A. Takahashi and T. Kurokawa, *Jpn. J. Appl. Phys.*, 2011, **50**, 112501.
35. S. Park, D. A. Borton, M. Y. Kang, A. V. Nurmikko and Y. K. Song, *Sensors*, 2013, **13**, 6014–6031.
36. S. Sohr, R. Rieske, K. Nieweglowski and K.-J. Wolter, Proceedings of the IEEE Electronic Components & Technology Conference, Las Vegas, Nevada, 2013.
37. M. C. Achtelik, J. Stumpf, D. Gurdan and K. M. Doth, Proceedings of the IEEE/RSJ International Conference on Intelligent Robots and Systems, San Francisco, CA, 2011.
38. D. M. Flournoy, *Solar Power Satellites*, Springer, New York, 2012.
39. M. B. Spitzer, R. W. McClelland, B. D. Dingle, J. E. Dingle, D. S. Hill and B. H. Rose, Proceedings of the 22nd IEEE Photovoltaic Specialists Conference, Las Vegas, Nevada, 1991.
40. J. G. Werthen, Proceedings of the Conference on Optical Fiber Communication, San Diego, California, 2008.
41. A. Barnett, D. Kirkpatrick, C. Honsberg, D. Moore, M. W. Wanlass, K. Emery, R. J. Schwartz, D. Carlson, S. Bowden, D. Aiken, A. Gray, S. Kurtz, L. L. Kazmerski, M. Steiner, J. L. Gray, T. Davenport, R. Buelow, L. Takacs, S. Shatz, J. Bortz, O. Jani, K. Goossen, F. Kiamilev, A. Doolittle, I. Ferguson, B. Unger, G. Schmidt, E. Christensen and D. Salzman, *Prog. Photovoltaics*, 2009, **17**, 75–83.
42. B. Mitchell, G. Peharz, G. Siefer, M. Peters, T. Gandy, J. C. Goldschmidt, J. Benick, S. W. Glunz, A. W. Bett and F. Dimroth, *Prog. Photovoltaics*, 2011, **19**, 61–72.
43. H. A. Atwater, M. D. Escarra, C. N. Eisler, E. D. Kosten, E. C. Warmann, S. Darbe, J. Lloyd and C. Flowers, Proceedings of the 39th IEEE Photovoltaics Specialists Conference, Tampa, Florida, 2013.
44. H. Helmers, M. Schachtner and A. W. Bett, *Sol. Energy Mater. Sol. Cells*, 2013, **116**, 144–152.
45. M. A. Green, K. Emery, D. L. King, S. Igari and W. Warta, *Prog. Photovoltaics*, 1993, **12**, 55–62.
46. E. McClure and E. Gaddy, Proceedings of the 35th IEEE Photovoltaic Specialists Conference, Honolulu, Hawaii, 2010.
47. D. M. Wilt, Keynote Presentation at SPIE 8256, San Francisco, California, 2012.
48. D. C. Law, X. Q. Liu, J. C. Boisvert, E. M. Redher, C. M. Fetzer, S. Mesropian, R. R. King, K. M. Edmondson, B. Jun, R. L. Woo, D. D. Krut, P. T. Chiu, D. M. Bhusari, S. K. Sharma and N. H. Karam,

Proceedings of the 38th IEEE Photovoltaic Specialists Conference, Austin, Texas, 2012.

49. U. W. Pohl, *Epitaxy of Semiconductors: Introduction to Physical Principles*, Springer, Heidelberg, 2013.

50. E. Kuphal, *Appl. Phys. A Mater. Sci. Process.*, 1991, **52**, 380–409.

51. G. B. Stringfellow, *Organometallic Vapor-Phase Epitaxy: Theory and Practice*, 2 edn., Academic Press, San Diego, 1999.

52. K. Volz, W. Stolz, J. Teubert, P. J. Klar, W. Heimbrodt, F. Dimroth, C. Baur and A. W. Bett, in *Dilute III-V Nitride Semiconductors and Material Systems*, ed. E. Ayse, Springer, Berlin, 2008, vol. 15, pp. 369–404.

53. S. Essig, E. Stämmler, S. Rönsch, E. Oliva, M. Schachtner, G. Siefer, A. W. Bett and F. Dimroth, Proceedings of the 9th European Space Power Conference, Saint-Raphael, 2011.

54. A. J. Ptak, D. J. Friedman, S. Kurtz and R. C. Reedy, *J. Appl. Phys.*, 2005, **98**, 094501.

55. D. B. Jackrel, S. R. Bank, H. B. Yuen, M. A. Wistey and J. S. Harris, *J. Appl. Phys.*, 2007, **101**, 114916.

56. A. Aho, A. Tukiainen, V. Polojärvi, J. Salmi and M. Guina, Proceedings of SPIE 8620, Physics, Simulation, and Photonic Engineering of Photovoltaic Devices, San Francisco, California, 2013.

57. V. Sabnis, H. Yuen and M. Wiemer, Proceedings of the 8th International Conference on Concentrating Photovoltaic Systems, Toledo, 2012.

58. S. Sze and M.-K. Lee, *Semiconductor Devices - Physics and Technology, International Student Version*, 3rd edn, John Wiley & Sons, Singapore, 2012.

59. W. T. Tsang, *Appl. Phys. Lett.*, 1984, **45**, 1234–1236.

60. Y. Ohshita, H. Suzuki, N. Kojima, T. Tanaka, T. Honda, M. Inagaki and M. Yamaguchi, *J. Cryst. Growth*, 2011, **318**, 328–331.

61. G. Kolhatkar, A. Boucherif, C. E. Valdivia, S. G. Wallace, S. Fafard, V. Aimez and R. Arès, *J. Cryst. Growth*, 2013, **380**, 256–260.

62. T. Sugaya, A. Takeda, R. Oshima, K. Matsubara, S. Niki and Y. Okano, *Appl. Phys. Lett.*, 2012, **101**, 133110.

63. A. W. Bett, F. Lutz, T. Louis and K. Leo, Proceedings of the 8th European Photovoltaic Solar Energy Conference, Florence, 1988.

64. S. R. Kurtz, J. M. Olson and A. Kibbler, Proceedings of the 21st IEEE Photovoltaic Specialists Conference, Kissimimee, Florida, 1990.

65. J. M. Olson, S. R. Kurtz, A. E. Kibbler and P. Faine, *Appl. Phys. Lett.*, 1990, **56**, 623–625.

66. I. Garcia, I. Rey-Stolle, B. Galiana and C. Algora, *Appl. Phys. Lett.*, 2009, **94**, 053509.

67. M. W. Wanlass, T. J. Coutts, J. S. Ward, K. A. Emery, T. A. Gessert and C. R. Osterwald, Proceedings of the 22nd IEEE Photovoltaic Specialists Conference, Las Vegas, Nevada, 1991.

68. M. P. Lumb, M. K. Yakes, M. Gonzalez, R. HoheiseI, C. G. Baile, W. Yoon and R. J. Walters, Proceedings of the 37th IEEE Photovoltaic Specialists Conference, Seattle, Washington, 2011.

69. C. T. Lin, W. E. McMahon, J. S. Ward, J. F. Geisz, M. W. Wanlass, J. J. Carapella, W. Olavarria, M. Young, M. A. Steiner, R. M. Frances, A. E. Kibbler, A. Duda, J. M. Olson, E. E. Perl, D. J. Friedman and J. E. Bowers, Proceedings of the 38th IEEE Photovoltaic Specialists Conference, Austin, Texas, 2012.
70. M. S. Leite, R. L. Woo, J. N. Munday, W. D. Hong, S. Mesropian, D. C. Law and H. A. Atwater, *Appl. Phys. Lett.*, 2013, **102**, 033901.
71. A. W. Bett and O. V. Sulima, *Semicond. Sci. Technol.*, 2003, **18**, 184–190.
72. M. G. Mauk and V. M. Andreev, *Semicond. Sci. Technol.*, 2003, **18**, S191–S201.
73. L. M. Fraas, G. R. Girard, J. E. Avery, B. A. Arau, V. S. Sundaram, A. G. Thompson and J. M. Gee, *J. Appl. Phys.*, 1989, **66**, 3866–3870.
74. A. W. Bett, S. Keser, G. Stollwerck, O. V. Sulima and W. Wettling, Proceedings of the 26th IEEE Photovoltaic Specialists Conference, Anaheim, California, 1997.
75. V. M. Andreev, S. V. Sorokina, N. K. Timoshina, V. P. Khvostikov and M. Z. Shvarts, *Semiconductors*, 2009, **43**, 668–671.
76. A. W. Bett, C. Baur, F. Dimroth and J. Schöne, *Mater. Res. Soc. Symp. Proc.*, 2005, **836**, 223–234.
77. J. Schöne, E. Spiecker, F. Dimroth, A. W. Bett and W. Jäger, *Appl. Phys. Lett.*, 2008, **92**, 081905.
78. W. Guter, J. Schöne, S. P. Philipps, M. Steiner, G. Siefer, A. Wekkeli, E. Welser, E. Oliva, A. W. Bett and F. Dimroth, *Appl. Phys. Lett.*, 2009, **94**, 223504.
79. R. R. King, D. C. Law, K. M. Edmondson, C. M. Fetzer, G. S. Kinsey, H. Yoon, R. A. Sherif and N. H. Karam, *Appl. Phys. Lett.*, 2007, **90**, 183516.
80. M. W. Wanlass, R. K. Ahrenkiel, D. S. Albin, J. Carapella, A. Duda, K. Emery, D. Friedman, J. F. Geisz, K. M. Jones, A. E. Kibbler, J. Kiel, S. Kurtz, W. E. McMahon, T. Moriarty, J. M. Olson and A. J. Ptak, Proceedings of the 4th World Conference on Photovoltaic Energy Conversion, Waikoloa, Hawaii, 2006.
81. A. B. Cornfeld, M. Stan, T. Varghese, J. Diaz, A. V. Ley, B. Cho, A. Korostyshevsky, D. J. Aiken and P. R. Sharps, Proceedings of the 33rd IEEE Photovoltaic Specialists Conference, San Diego, California, 2008.
82. J. F. Geisz, D. J. Friedman, J. S. Ward, A. Duda, W. J. Olavarria, T. E. Moriarty, J. T. Kiehl, M. J. Romero, A. G. Norman and K. M. Jones, *Appl. Phys. Lett.*, 2008, **93**, 123505.
83. H. Yoon, M. Haddad, S. Mesropian, J. Yen, K. Edmondson, D. Law, R. R. King, D. Bhusari, A. Boca and N. H. Karam, Proceedings of the 33rd IEEE Photovoltaic Specialists Conference, San Diego, California, 2008.
84. A. Yoshida, T. Agui, N. Katsuya, K. Murasawa, H. Juso, K. Sasaki and T. Takamoto, Proceedings of the 21st International Photovoltaic Science and Engineering Conference, Fukuoka, 2011.

85. R. M. France, J. F. Geisz, M. A. Steiner, D. J. Friedman, J. S. Ward, J. M. Olson, W. Olavarria, M. Young and A. Duda, *IEEE J. Photovoltaics*, 2013, **3**, 893–898.

86. P. Patel, D. Aiken, A. Boca, B. Cho, D. Chumney, M. B. Clevenger, A. Cornfeld, N. Fatemi, Y. Lin, J. McCarty, F. Newman, P. Sharps, J. Spann, M. Stan, J. Steinfeldt, C. Strautin and T. Varghese, *IEEE J. Photovoltaics*, 2012, **2**, 377–381.

87. B. Galiana, I. Rey Stolle, M. Baudrit, I. Garcia and C. Algora, *Semicond. Sci. Technol.*, 2006, **21**, 1387–1392.

88. A. S. Gudovskikh, N. A. Kaluzhniy, V. M. Lantratov, S. A. Mintairov, M. Z. Shvarts and V. M. Andreev, *Thin Solid Films*, 2008, **516**, 6739–6743.

89. R. Hoheisel and A. W. Bett, *IEEE J. Photovoltaics*, 2012, **2**, 398–402.

90. M. Kondow, K. Uomi, A. Niwa, T. Kitatani, S. Watahiki and Y. Yazawa, *Jpn. J. Appl. Phys.*, 1996, **35**, 1273–1275.

91. J. F. Geisz and D. J. Friedman, *Semicond. Sci. Technol.*, 2002, **17**, 769–777.

92. M. A. Green, K. Emery, Y. Hishikawa, W. Warta and E. D. Dunlop, *Prog. Photovoltaics*, 2013, **21**, 1–11.

93. T. J. Garrod, J. Kirch, P. Dudley, S. Kim, L. J. Mawst and T. F. Kuech, *J. Cryst. Growth*, 2011, **315**, 68–73.

94. L. Tsakalakos, (ed.), *Nanotechnology for Photovoltaics*, CRC Press Taylor & Francis Group, Boca Raton, 2010.

95. A. B. Cristóbal, A. Martí and A. Luque, (ed.), *Next Generation of Photovoltaics*, Springer, Berlin, 2012.

96. A. Freundlich and I. Serdiukova, Proceedings of the 2nd World Conference on Photovoltaic Energy Conversion, Vienna, 1998.

97. N. J. Ekins-Daukes, K. W. J. Barnham, J. P. Connolly, J. S. Roberts, J. C. Clark, G. Hill and M. Mazzer, *Appl. Phys. Lett.*, 1999, **75**, 4195–4197.

98. R. Kellenbenz, R. Hoheisel, P. Kailuweit, W. Guter, F. Dimroth and A. W. Bett, Proceedings of the 35th IEEE Photovoltaic Specialists Conference, Honolulu, 2010.

99. K. W. J. Barnham, I. M. Ballard, B. C. Browne, D. B. Bushnell, J. P. Connolly, N. J. Ekins-Daukes, M. Fuhrer, R. Ginige, G. Hill, A. Ioannides, D. C. Johnson, M. C. Lynch, M. Mazzer, J. S. Roberts, C. Rohr and T. N. D. Tibbits, in *Nanotechnology for Photovoltaics*, ed. L. Tsakalakos. CRC Press Taylor & Francis Group, Boca Raton, 2010, pp. 187–210.

100. J. G. J. Adams, B. C. Browne, I. M. Ballard, J. P. Connolly, N. L. A. Chan, A. Ioannides, W. Elder, P. N. Stavrinou, K. W. J. Barnham and N. J. Ekins-Daukes, *Prog. Photovoltaics*, 2011, **19**, 865–877.

101. K.-H. Lee, K. W. J. Barnham, J. P. Connolly, B. C. Browne, R. J. Airey, J. S. Roberts, M. Führer, T. N. D. Tibbits and N. J. Ekins-Daukes, *IEEE J. Photovoltaics*, 2012, **2**, 68–74.

102. C. G. Bailey, D. V. Forbes, R. P. Raffaelle and S. M. Hubbard, *Appl. Phys. Lett.*, 2011, **98**, 163105.

103. R. Kellenbenz, W. Guter, P. Kailuweit, E. Oliva and F. Dimroth, Proceedings of the 8th European Space Power Conference, Konstanz, 2008.

104. R. J. Walters, G. P. Summers, S. R. Messenger, A. Freundlich, C. Monier and F. Newman, *Prog. Photovoltaics*, 2000, **8**, 349–354.
105. R. P. Raffaelle, S. Sinharoy, J. Andersen, D. M. Wilt and S. G. Bailey, Proceedings of the 4th World Conference on Photovoltaic Energy Conversion, Waikoloa, Hawaii, 2006.
106. C. G. Bailey, D. V. Forbes, S. J. Polly, Z. S. Bittner, Y. S. Dai, C. Mackos, R. P. Raffaelle and S. M. Hubbard, *IEEE J. Photovoltaics*, 2012, **2**, 269–275.
107. A. Marti, E. Antolin, P. G. Linares, I. Ramiro, I. Artacho, E. Lopez, E. Hernandez, M. J. Mendes, A. Mellor, I. Tobias, D. F. Marron, C. Tablero, A. B. Cristobal, C. G. Bailey, M. Gonzalez, M. Yakes, M. P. Lumb, R. Walters and A. Luque, *J. Photonics Energy*, 2013, **3**, 031299.
108. A. Luque and A. Martí, *Adv. Mater.*, 2009, **22**, 160–174.
109. J. Wu and Z. M. Wang, (eds.), *Quantum Dot Solar Cells*, Springer, New York, 2013.
110. J. Wallentin, N. Anttu, D. Asoli, M. Huffman, I. Åberg, M. H. Magnusson, G. Siefer, P. Fuss-Kailuweit, F. Dimroth, B. Witzigmann, H. Q. Xu, L. Samuelson, K. Deppert and M. T. Borgström, *Science*, 2013, **339**, 1057–1060.
111. H. Goto, K. Nosaki, K. Tomioka, S. Hara, K. Hiruma, J. Motohisa and T. Fukui, *Appl. Phys. Express*, 2009, **2**, 035004.
112. P. Kailuweit, M. Peters, J. Leene, K. Mergenthaler, F. Dimroth and A. W. Bett, *Prog. Photovoltaics*, 2011, **20**, 945–953.
113. T. Martensson, C. P. T. Svensson, B. A. Wacaser, M. W. Larsson, W. Seifert, K. Deppert, A. Gustafsson, L. R. Wallenberg and L. Samuelson, *Nano Lett.*, 2004, **4**, 1987–1990.
114. R. R. LaPierre, A. C. E. Chia, S. J. Gibson, C. M. Haapamaki, J. Boulanger, R. Yee, P. Kuyanov, J. Zhang, N. Tajik, N. Jewell and K. M. A. Rahman, *Physica Status Solidi RRL*, 2013, **7**, 815–830.
115. J. M. Zahler, K. Tanabe, C. Ladous, T. Pinnington, F. D. Newman and H. A. Atwater, *Appl. Phys. Lett.*, 2007, **91**, 012108.
116. M. J. Archer, D. C. Law, S. Mesropian, M. Haddad, C. M. Fetzer, A. C. Ackerman, C. Ladous, R. King and H. A. Atwater, *Appl. Phys. Lett.*, 2008, **95**, 103503.
117. D. C. Law, D. M. Bhusari, S. Mesropian, J. C. Boisvert, W. D. Hong, A. Boca, D. C. Larrabee, C. M. Fetzer, R. R. King and N. H. Karam, Proceedings of the 34th IEEE Photovoltaic Specialists Conference, Philadelphia, Pennsylvania, 2009.
118. J. Boisvert, D. Law, R. King, D. Bhusari, X. Liu, A. Zakaria, W. Hong, S. Mesropian, D. Larrabee, R. Woo, A. Boca, K. Edmondson, D. Krut, D. Peterson, K. Rouhani, B. Benedikt and N. H. Karam, Proceedings of the 35th IEEE Photovoltaic Specialists Conference, Honolulu, Hawaii, 2010.
119. K. Derendorf, S. Essig, E. Oliva, V. Klinger, T. Roesener, S. P. Philipps, J. Benick, M. Hermle, M. Schachtner, G. Siefer, W. Jäger and F. Dimroth, *IEEE J. Photovoltaics*, 2013, **3**, 1423–1428.

120. S. Essig, O. Moutanabbir, A. Wekkeli, H. Nahme, E. Oliva, A. W. Bett and F. Dimroth, *J. Appl. Phys.*, 2013, **113**, 203512.
121. S. Essig and F. Dimroth, *ECS J. Solid State Sci. Technol.*, 2013, **2**, Q178–Q181.
122. K. Tanabe, K. Watanabe and Y. Arakawa, *Sci. Rep.*, 2012, **2**, 1–6.
123. T. Yu, M. M. R. Howlader, F. Zhang and M. Bakr, *ECS Trans.*, 2011, **35**, 3–10.
124. M. K. Kelly, O. Ambacher, R. Dimitrov, R. Handschuh and Stutzmann, *Physica Status Solidi A*, 1997, **159**, R3–R4.
125. F. Dross, J. Robbelein, B. Vandevelde, E. Van Kerschaver, I. Gordon, G. Beaucarne and J. Poortmans, *Appl. Phys. A Mater. Sci. Process.*, 2007, **89**, 149–152.
126. J. W. Mayer, *1973 International Electron Devices Meeting Technical Digest*, 1973, 3–55.
127. P. Demeester, I. Pollentier, P. De Dobbelaere, C. Brys and P. Van Daele, *Semicond. Sci. Technol.*, 1993, **8**, 1124–1135.
128. K. Dreyer, E. Fehrenbacher, E. Oliva, S. Essig, V. Klinger, T. Roesener, A. Leimenstoll, F. Schätzle, M. Hermle, A. Bett and F. Dimroth, DPG Spring Meeting, Dresden, 2011.
129. R. L. Moon, L. W. James, H. A. Vander Plas, T. O. Yep, G. A. Antypas and Y. Chai, Proceedings of the 13th IEEE Photovoltaic Specialists Conference, Washington, D.C., 1978.
130. D. Vincenzi, A. Busato, M. Stefancich and G. Martinelli, *Physica Status Solidi A*, 2009, **206**, 375–378.
131. D. C. Law, R. R. King, H. Yoon, M. J. Archer, A. Boca, C. M. Fetzer, S. Mesropian, T. Isshiki, M. Haddad, K. M. Edmondson, D. Bhusari, J. Yen, R. A. Sherif, H. A. Atwater and N. H. Karam, *Sol. Energy Mater. Sol. Cells*, 2008, **94**, 1314–1318.
132. D. J. Friedman, *Curr. Opin. Solid State Mater. Sci.*, 2010, **14**, 131–138.
133. A. Luque, *J. Appl. Phys.*, 2011, **110**, 031301.
134. S. P. Philipps, W. Guter, E. Welser, J. Schöne, M. Steiner, F. Dimroth and A. W. Bett, in *Next Generation of Photovoltaics*, ed. A. B. Cristóbal López, A. Martí Vega and A. Luque López. Springer Verlag, Berlin, 2012, pp. 1–21.
135. S. P. Philipps, F. Dimroth and A. W. Bett, in *Practical Handbook of Photovoltaics*, Academic Press, Boston, 2nd edn, 2012, vol. pp. 417–448.
136. R. R. King, D. Bhusari, D. Larrabee, X. Q. Liu, E. Rehder, K. Edmondson, H. Cotal, R. K. Jones, J. H. Ermer, C. M. Fetzer, D. C. Law and N. H. Karam, *Prog. Photovoltaics*, 2012, **20**, 801–815.
137. A. W. Bett, S. P. Philipps, S. Essig, S. Heckelmann, R. Kellenbenz, V. Klinger, M. Niemeyer, D. Lackner and F. Dimroth, Proceedings of the 28th European Photovoltaic Solar Energy Conference and Exhibition, Paris, 2013.
138. S. P. Philipps, G. Peharz, R. Hoheisel, T. Hornung, N. M. Al-Abbadi, F. Dimroth and A. W. Bett, *Sol. Energy Mater. Sol. Cells*, 2010, **94**, 869–877.

139. K. Sasaki, T. Agui, K. Nakaido, N. Takahashi, R. Onitsuka and T. Takamoto, Proceedings of the 9th International Conference on Concentrator Photovoltaic Systems, Miyazaki, 2013.

140. S. Wojtczuk, P. Chiu, X. Zhang, D. Pulver, C. Harris and M. Timmons, *IEEE J. Photovoltaics*, 2012, **2**, 371–376.

141. S. D'Souza, J. Haysom, H. Anis and K. Hinzer, Proceedings of the IEEE Electrical Power and Energy Conference, Winnipeg, Manitoba, 2011.

142. M. Levinshtein, S. Rumyantsev and M. Shur, (ed.), *Si, Ge, C (Diamond), GaAs, GaP, GaSb, InAs, InP, InSb*, World Scientific Publishing, Singapore, 1996.

143. S. F. Fang, K. Adomi, S. Iyer, H. Morkoç, H. Zabel, C. Choi and N. Otsuka, *J. Appl. Phys.*, 1990, **68**, R31–R58.

144. R. K. Ahrenkiel, M. M. Al-Jassim, B. Keyes, D. Dunlavy, K. M. Jones, S. M. Vernon and T. M. Dixon, *J. Electrochem. Soc.*, 1990, **137**, 996–1000.

145. M. Umeno, T. Soga, K. Baskar and T. Jimbo, *Sol. Energy Mater. Sol. Cells*, 1998, **50**, 203–212.

146. T. Roesener, H. Döscher, A. Beyer, S. Brückner, V. Klinger, A. Wekkeli, P. Kleinschmidt, C. Jurecka, J. Ohlmann, K. Volz, W. Stolz, T. Hannappel, A. W. Bett and F. Dimroth, Proceedings of the 25th European Photovoltaic Solar Energy Conference and Exhibition, Valencia, 2010.

147. S. A. Ringel, J. A. Carlin, C. L. Andre, M. K. Hudait, M. Gonzalez, D. M. Wilt, E. B. Clark, P. Jenkins, D. Scheiman, A. Allerman, E. A. Fitzgerald and C. W. Leitz, *Prog. Photovoltaics*, 2002, **10**, 417–426.

148. V. K. Yang, M. Groenert, C. W. Leitz, A. J. Pitera, M. T. Currie and E. A. Fitzgerald, *J. Appl. Phys.*, 2003, **93**, 3859–3865.

149. M. R. Lueck, C. L. Andre, A. J. Pitera, M. L. Lee, E. A. Fitzgerald and S. A. Ringel, *IEEE Electron Device Lett.*, 2006, **27**, 142–144.

150. T. J. Grassman, M. R. Brenner, M. Gonzalez, A. M. Carlin, R. R. Unocic, R. R. Dehoff, M. J. Mills and S. A. Ringel, *IEEE Trans. Electron Devices*, 2010, **57**, 3361–3369.

151. F. Dimroth, T. Roesener, S. Essig, C. Weuffen, A. Wekkeli, E. Oliva, G. Siefer, K. Volz, T. Hannappel, D. Häussler, W. Jäger and A. W. Bett, *IEEE J. Photovoltaics*, 2014, **4**, 620–625.

152. J. M. Zahler, A. Fontcuberta i Morral, C. G. Ahn, H. A. Atwater, M. W. Wanlass, C. Chu and P. A. Iles, Proceedings of the 29th IEEE Photovoltaic Specialists Conference, New Orleans, Louisiana, 2002.

153. D. Bhusari, D. Law, R. Woo, J. Boisvert, S. Mesropian, D. Larrabee, H. F. Hong and N. H. Karam, Proceedings of the 37th IEEE Photovoltaic Specialists Conference, Seattle, Washington, 2011.

154. K. Volz, J. Koch, B. Kunert and W. Stolz, *J. Cryst. Growth*, 2003, **248**, 451–456.

155. F. Dimroth, M. Meusel, C. Baur, A. W. Bett and G. Strobl, Proceedings of the 31st IEEE Photovoltaic Specialists Conference, Orlando, Florida, 2005.

CHAPTER 5

Thin-film Photovoltaics Based on Earth-abundant Materials

DIEGO COLOMBARA,[a] PHILLIP DALE,[a] LAURENCE PETER,*[b] JONATHAN SCRAGG[c] AND SUSANNE SIEBENTRITT[d]

[a] Laboratory for Energy Materials, University of Luxembourg, 41, rue du Brill, L-4422 Belvaux, Luxembourg; [b] Department of Chemistry, University of Bath, Bath BA2 7AY, UK; [c] Ångström Solar Centre, Solid State Electronics, Uppsala University, Box 534, 751 21 Uppsala, Sweden; [d] Laboratory for Photovoltaics, University of Luxembourg, 41, rue du Brill, L-4422 Belvaux, Luxembourg
*Email: l.m.peter@bath.ac.uk

5.1   Introduction

5.1.1   Future Requirements for Photovoltaics: 2050 Scenarios

Mankind's current dependence on carbon-based fuels to provide increasing amounts of energy for an expanding world economy represents a threat to long term sustainability. The pressing need to develop alternative renewable energy sources has led to a sustained increase in the deployment of solar photovoltaics (PV). A key question in the context of this chapter is whether PV can make a major contribution to the world's energy requirements over the coming decades. According to BP's 2013 statistical review of world energy,[1] global primary energy consumption in 2012 was equivalent to ca. 12.5 gigatonnes (Gtoe) of oil. This rate of primary energy consumption

RSC Energy and Environment Series No. 11
Advanced Concepts in Photovoltaics
Edited by Arthur J Nozik, Gavin Conibeer and Matthew C Beard
© The Royal Society of Chemistry 2014
Published by the Royal Society of Chemistry, www.rsc.org

corresponds to a thermal output power of around 16 TW. 87% of energy consumption was associated with burning carbon-based fuels (oil, gas and coal), ca. 7% was generated as hydroelectricity, 4% came from nuclear power stations, and only 2% came from renewables (wind, geothermal, solar, biomass and biofuels).[1] With the rapid expansion of Asian economies and with the world's population set to reach 8–11 billion by 2050, most estimates agree that energy consumption will at least double from its current value, posing enormous problems in terms of climate change and sustainability.

A huge expansion of installed PV will be necessary if solar PV is to make a significant contribution to the world's energy needs. A recent report from the Intergovernmental Panel on Climate Change (IPCC)[2] assesses the possible contribution of PV to global electricity supply for different CO_2 amelioration scenarios. The most optimistic scenario predicts that up to 10^{20} J per year could be generated by PV, which corresponds roughly to a power of 3 TW or 10% of the total power requirement in 2050. A more conservative scenario considered by the International Energy Agency (IEA)[3] predicts a smaller figure of around 0.5 TW. Since the average output of PV is generally less than 20% of the installed peak capacity, we are clearly dealing in either case with TWp (terawatt peak) figures in terms of installed capacity. The IEA estimated the total installed PV capacity at the end of 2012 to be around 100 GWp,[4] with Germany and Italy having the capacity to generate nearly 6% of their electricity by PV. Achieving any of the 2050 targets will require at least a 100-fold expansion of the current installed capacity, potentially posing severe problems associated with materials availability.

5.1.2 Resource Implications for Thin-film Photovoltaics

The majority of currently installed PV is based on crystalline silicon. The high efficiency of silicon PV has to be set against the high energy input required in manufacture. Key metrics for any PV technology are the energy pay-back time (EPT) and the CO_2 emission rate. EPT values for thin film technologies are currently under half of the values for crystalline silicon as a consequence of the use of much thinner absorber films and the lower embodied energy of the absorber.[5] In fact, the contribution of the absorber to the EPT is almost negligible for thin film PV. For this reason, interest in thin film PV has increased as efficiencies have risen, and it is anticipated that it will gain an increasing market share in the future.

Current commercial (non-silicon) thin film technologies are based on either CdTe or Cu(In,Ga)Se$_2$ (CIGSe) absorbers. A recent European Commission report[6] and a report from the US Department of Energy (DOE)[7] both list gallium, indium and tellurium as critical in terms of economic importance and supply risk. Green[8] has discussed likely trends in In and Te production and prices based on an analysis of direct mining of known ores and the increased demand arising from deployment of thin film CdTe and CIGSe PV. This analysis identifies the supply of Te and In as critical issues for PV deployment. Ga, In and Te are all by-products. Gallium is a by-product of production of aluminium and to a lesser extent zinc; world production in 2011 was estimated to be 216 tonnes.[9] The main source of indium is slag and dust produced during zinc refining, and around 50% of the annual production is currently used for LCD displays; the annual production in 2011 was 640 tonnes (excluding recycling).[10] The main source of tellurium is the anode sludge formed during copper refining. Reliable figures for the annual production rate of Te are not available, but a figure of around 110 tonnes is given by the US Geological Survey.[11] The production rates of In, Ga and Te are therefore largely determined by the demand for aluminium, zinc and copper respectively, and the only way that annual production could be increased significantly in the mid-term is by improvements in processing. Accurate prediction of supply and demand for Ga, In and Te is difficult of course, but Forbes and Peter[12] have used a simple logistical growth model to illustrate the fact that the market share of CIGSe and CdTe thin film PV is likely to become limited by the supply of indium and tellurium. The need for supply security therefore provides a rationale for the search for alternative absorber materials containing earth-abundant elements in order to ensure that supply limitations do not impact on the exponential expansion of PV that is foreseen in the scenarios mentioned above. On the downside, Anctil and Fthenakis[13] have pointed out that increases in materials complexity combined with low end of life value are likely to discourage recycling of PV modules based on earth-abundant elements.

5.1.3 Earth-abundant Absorbers

The requirements for an absorber material to replace CIGSe or CdTe are well understood. The material needs to have a bandgap close to the Shockley and Queisser[14] optimum range (1.15–1.35 eV for AM 1.5 illumination). It needs to have suitable optoelectronic properties (direct bandgap, preferably p-type conductivity, appropriate doping, and low defect density) and it should be stable under the processing conditions used to fabricate solar cell structures. The earth-abundant materials that have excited the most interest as suitable replacements for CIGSe are the quaternary kesterite compounds Cu_2ZnSnS_4 (CZTS) and $Cu_2ZnSnSe_4$ (CZTSe) and their alloys. The name kesterite comes from the sulfide mineral $Cu_2(Zn,Fe)SnS_4$, in which zinc and iron atoms occupy the same sites in the crystal structure. Although these compounds have been known for some time,[15] systematic exploration of their potential as earth-abundant replacements for CIGSe absorbers in thin film solar cells dates back to papers in the late 1990s by Nakayama and Ito.[16,17] These were followed by work by Katagiri et al.[18,19] and by Friedlmeier et al.[20] By 2001, Katagiri et al.[21] were able to report fabrication of a CZTS cell with an efficiency of 2.67%, and two years later the same group had already reached an efficiency of 5.45%.[22] However CZTS remained a minority interest until 2010, when the IBM group of Mitzi, Todorov et al. made a remarkable breakthrough to efficiencies close to 10% with hydrazine-based liquid processing of CZTS and CZTSSe cells.[23–26] These papers sparked an upsurge in interest in CZTSSe as an absorber material, and the last 3 years have seen an exponential growth of the number of publications dealing with kesterite solar cells. The current status of the practical development of kesterite cells has been reviewed in a number of recent publications,[27–31] and Walsh *et al.*[32] have outlined the considerable progress that has been made in the theoretical modelling of the materials. The current situation is that hydrazine-processed kesterite cells still hold the record for efficiency, but the majority of the international research effort is focussed on developing alternative safe and environmentally friendly routes to fabrication. The methods being pursued include reactive sputtering,[33] selenization/sulfurization of co-evaporated precursors,[34,35] co-evaporation,[36,37] sulfurization/selenization of electrodeposited metal precursors,[38–40] direct electrodeposition,[41] direct liquid coating or dip coating,[42–44] sol–gel processing,[45] spray pyrolysis[46] and atmospheric CVD.[47] Published efficiencies are rapidly catching up with the IBM champion cells, and with over 300 papers published in 2013, it is clear that the kesterites have moved centre stage in the development of alternative absorber materials.

The substitution of In by Zn and Sn in going from CIGSe to CZTSSe leads to an increase in material complexity that may have a negative impact on processing and recycling routes.[13] For this reason, interest is beginning to be shown in ternary Cu-Sn chalcogenides. For example, the preparation of the ternary compounds Cu_2SnS_3, $Cu_5Sn_2S_7$ and Cu_3SnS_4 by successive ionic layer adsorption and reaction (SILAR) has been reported by Su *et al.*[48] With bandgaps of 1.0 eV, 1.45 eV and 1.47 eV respectively, these materials clearly deserve further study. Avallaneda *et al.*[49] have used chemical bath deposition (CBD) to prepare Cu_2SnS_3 and Cu_4SnS_4 thin films, which they suggest may be suitable for photovoltaic applications, whereas Bouaziz *et al.*[50] have prepared Cu_2SnS_3 films by sulfurization of metal precursors. Several attempts to fabricate thin layer cells from Cu_2SnS_3 (CTS) have been reported recently[43,51,52] with a best efficiency to date of 2.84%.[52] It remains to be seen whether this simpler ternary compound will be able to compete with CZTSSe, but recent work demonstrates that we should be looking at a wider range of earth-abundant materials as replacements for materials containing indium gallium or tellurium.

Another class of ternary minerals that has been investigated for possible application in PV are the sulfosalts.[53] Naturally occurring examples of the ABX_2 type include emplectite ($CuBiS_2$) and chalcostibite ($CuSbS_2$).The structure and electronic properties of these two materials have been modelled recently by Dufton *et al.*[54] The A_3BX_3 type is exemplified by Cu_3BiS_3, which has been proposed as an absorber material.[55–58] Although the bandgap of Cu_3BiS_3 (1.5–1.7 eV) is rather too high for optimum light harvesting, the material appears to absorb more strongly than CIGSe or CZTSSe.[58] Colombara *et al.*[59] have prepared $CuSbS_2$ and $CuSbSe_2$ thin films *via* chalcogenization of Sb-Cu metal precursors and have reported bandgaps of 1.5 and 1.2 eV respectively.

Finally, binary sulfide materials have a long history as absorber layers in thin film photovoltaics. Cu_2S/CdS solar cells[60,61] were well on the way to commercialization in the 1980s, but problems with the solid state chemistry of the devices limited their long-term stability. Other binary materials that could be interesting include the tin sulfides SnS[62] and Sn_2S_3,[63] as well as Bi_2S_3, which (although n-type) has an optimum bandgap.[64] SnS solar cells have been fabricated, but they have low efficiencies. This is probably due to poor band alignment with the back contact and buffer layers,[65] so further work taking this into account could be productive. Other binary sulfide absorber layers include Sb_2S_3, which has been used in extremely thin absorber (ETA) layer solar cells,[66] and FeS_2, which has been investigated for many years[67,68] but appears to be problematic due to its complex defect chemistry. Clearly it will only be possible to exploit the cost/availability advantages of any of these materials if cells can be fabricated with efficiencies that rival those of CIGSe.

5.1.4   The Scope of the Chapter

The rapid expansion of work on earth-abundant photovoltaics—in particular on kesterite-based devices—means that any review of progress is bound to be considerably out of date by the time it is published. For this reason, the authors have chosen to focus attention on fundamental properties, emerging issues and generic methodologies rather than on cell performance. For example, this chapter highlights the importance of the thermodynamics of phase equilibria for the preparation and thermal processing of absorber layers. In this context, the differences between the thermal stability of the kesterites CZTS/CZTSe and the chalcopyrite CIGSe are very relevant to an understanding of the need to control the processing atmosphere for kesterite absorbers. Of course thermodynamics is only useful as a guide; kinetic factors are equally important, and therefore these are also considered. Since current attention is strongly focussed on the kesterite systems, we have used kesterite as a case study to illustrate structural, electronic and thermodynamic properties. However, there is a much wider range of earth-abundant materials that have the potential for application in devices. Therefore some other candidate materials are reviewed, again with emphasis on phase equilibria and thermodynamic aspects. The authors hope that the usefulness of the present chapter will be enhanced by its consideration of earth-abundant PV from a more generic point of view, with emphasis on fundamental physiochemical aspects.

5.2   Kesterite: a Case Study

5.2.1   Iso-electronic Substitution: An Introduction to $Cu_2ZnSnS(Se)_4$

This section begins by outlining a systematic approach for the derivation of inorganic semiconductor compounds (see also Walsh *et al.*[32]). This approach, which is based on *iso-electronic* or *cross-substitution*, was first devised by Goodman and Pamplin in the 1950s–1960s as a way to rationalise super-cell structures.[69–71] For explanatory purposes, the older notation is employed to identify the groups of the periodic table (see, for example, Deming[72]) rather than the more recent IUPAC designation.[73] The older commonly used notation is based on the assignment of roman numerals, which in most cases correspond to the number of valence shell electrons for each group. We start with the elemental semiconductors Si and Ge, which belong to group IV of the periodic table and have four outer shell electrons. Their crystalline structure can be described by the well-known cubic lattice of diamond or by the lesser-known hexagonal lattice of the rare allotrope lonsdaleite. In both of these crystalline structures, all atoms are tetrahedrally coordinated and their positions are defined by two interpenetrating face-centred cubic sub-lattices (diamond) and two interpenetrated hexagonal close-packed arrays (lonsdaleite). By replacing the group IV atoms of the two sub-lattices with, for example, atoms from groups III and V or II and VI re-spectively, the average number of valence electrons is unaltered. Such a ra-tional derivation can be extended further, involving the replacement of the atoms within the sub-lattice and so forth, as long as the overall atomic substitution satisfies Equation (5.1):[69]

$$\frac{\sum_i n_i \nu_i}{\sum_i n_i} = 4 \tag{5.1}$$

where n_i is the number of atoms of the i-th kind with ν_i valence shell electrons in the newly formed compound. The genealogical derivation based on iso-electronic substitution of the group IV elements gives rise to the so-called *adamantine family* of compounds.[74] The substitution process is demonstrated by the arrows on the portion of the periodic table depicted in Figure 5.1.

Adamantine compounds exhibit crystalline structures that are very similar to those of diamond and lonsdaleite. Understandably, the high symmetry of the parent unit cell is lost progressively as the group IV atoms are replaced by two or more sets of different atoms. When the difference in electronegativity of the sets of replacing atoms is significant, the crystal structure is better described by the two substructures populated by anions and cations respectively. Tet-rahedral coordination minimizes the free energy, since each cation is sur-rounded solely by anions and vice versa. Focusing on the progeny based on the diamond structure, Table 5.1 lists a number of binary (III–V, II–VI), ternary (I–II–VI$_2$, I–III–VI$_2$) and quaternary (I$_2$–II–IV–VI$_4$, I–II$_2$–III–VI$_4$) compounds together with some of their structural and electronic properties.

The electronic properties exhibited by the compounds listed in Table 5.1 are a consequence of the modified nature of the chemical bonds between the constituent atoms. Referring back to Figure 5.1, one can note as a simple guide that arrows pointing sideways in the periodic table lead to compounds with increased ionicity due to the increased difference of electronegativity. An increased ionic character of the chemical bond is generally associated with higher energy bandgaps. On the other hand, arrows, pointing down in

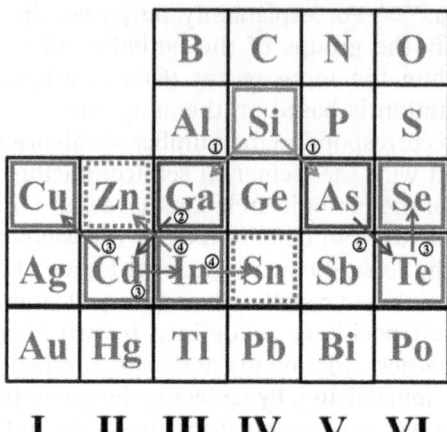

Figure 5.1 Portion of the Periodic Table illustrating the derivation of adamantine compounds by isoelectronic substitution. In step 1, (①) atoms of the parent group IV element, Si, in the diamond structure are exchanged with an equal number of groups III and V atoms to form GaAs. Subsequent steps (②→③→④) show the derivation of compounds such as CdTe, CuInSe$_2$ and Cu$_2$ZnSnSe$_4$.

the periodic table lead to compounds with decreased covalency and increased metallic character, *i.e.*, decreased energy bandgaps. For example, AlP is expected to have stronger ionicity than Si (arrows pointing sideways), while GaAs is expected to be more metallic than AlP (arrows pointing down). This is consistent with the measured bandgaps of the compounds listed in Table 5.1: E_g(AlP) > E_g(Si), and E_g(GaAs) < E_g(AlP). An intrinsic feature of iso-electronic substitution is that every substitutional step results in an increase of the formal oxidation state of one or more of the constituents. This has important consequences for the thermodynamic stability of the compounds, as discussed later in this chapter.

In terms of the crystal properties of the compounds listed in Table 5.1, the symmetry of the structure decreases at each level of substitution, as can be inferred from the decreased space group number.[115] The sequence is as follows: no. 227 (*Fd3̄m*) for IV→no. 216 (*F4̄3m*) for binary III–V and II–VI → no. 122 (*I4̄2d*) for ternary I–II–VI$_2$ → no. 121 (*I4̄2m*) or no. 82 (*I4̄*) for quaternary I$_2$–II–IV–VI$_4$. The last level of substitution deserves a more thorough description, and is now analysed using the example of two relevant compounds: the quaternaries Cu$_2$FeSnS$_4$ (*I-42m*) and Cu$_2$ZnSnS$_4$ (*I4̄*) known respectively as the minerals stannite and kesterite.

Stannite may occur in nature as the mineral Cu$_2$Fe(Zn)SnS$_4$ with Fe partially substituted by Zn. Similarly, kesterite can show some substitution of Zn by Fe, leading to Cu$_2$Zn(Fe)SnS$_4$. Due to the formula similarity and the reported natural intergrowth between these two minerals, the pseudo-binary system Cu$_2$FeSnS$_4$-Cu$_2$ZnSnS$_4$ has been investigated synthetically by Springer[116] in order to establish whether it displays complete miscibility.

Table 5.1 Non-exhaustive list of binary (III–V, II–VI), ternary (I–II–VI$_2$, I$_2$–IV–VI$_3$) and quaternary (I$_2$–II–IV–VI$_4$, I–II$_2$–III–VI$_4$) adamantine compounds derived from the cubic structure of diamond.

Type and formula	Mineral	Space group	E_g (eV)	E_g type	Ref.
IV		227 (Fd$\bar{3}$m)			
C	Diamond		5.5	Indirect	75
Si			1.12	Indirect	76
Ge			0.66	Indirect	77
α-Sn	Grey tin		0.08	Indirect	78
Si$_{1-x}$Ge$_x$		–		Direct	79,80
Ge$_{1-x}$Sn$_x$		–		Indirect	79–82
III–V		216 (F$\bar{4}$3m)			
β-BN			6.27	Indirect	83
BP			2.0	Indirect	83
Bas			1.5	Indirect	84
BSb			0.59	Indirect	85
AlP			2.45	Indirect	86,87
AlAs			2.163	Indirect	86,87
AlSb			1.58	Indirect	86,87
GaP			2.261	Indirect	86,87
GaAs			1.424	Direct	86,87
GaSb			0.726	Direct	86,87
InP			1.351	Direct	86,87
InAs			0.360	Direct	86,87
InSb			0.172	Direct	86,87
II–VI		216 (F$\bar{4}$3m)			
ZnS	Zincblende		3.67–3.80	Direct	88,89
ZnSe	Stilleite		2.67–2.82	Direct	90,91
ZnTe			2.39	Direct	87
CdS	Hawleyite		2.42–2.45	Direct	92
CdSe			–	–	87
CdTe			1.47	Direct	86,87
I–II–VI$_2$		122 (I$\bar{4}$2d)			
CuAlSe$_2$			2.67	Direct	93
CuAlS$_2$			2.4–3.5	Direct	94,95
CuGaSe$_2$			1.68	Direct	96
CuGaS$_2$	Gallite		2.15	Direct	97
CuInSe$_2$			1.0	Direct	98,99
CuInS$_2$	Roquesite		1.55	Direct	100
CuFeS$_2$	Chalcopyrite		0.6	Direct	95
AgGaS$_2$			2.73	Direct	101
AgGaSe$_2$			1.83	Direct	101
I$_2$–IV–VI$_3$					
Cu$_2$GeS$_3$		9 (Cc)	0.53	–	102,103
Cu$_2$GeSe$_3$		44 (Imm2)	0.78	Direct	104
Cu$_2$SnS$_3$		121 (I$\bar{4}$2m)	0.93–1.51	–	51,105
Cu$_2$SnSe$_3$		9 (Cc)	0.4–1.7	Direct	106–108
I$_2$–II–IV–VI$_4$					
Cu$_2$ZnGeS$_4$		121 (I$\bar{4}$2m)	2.28–2.13	Direct	102,109,110
Cu$_2$ZnGeSe$_4$		–	1.63	Direct	111
Cu$_2$ZnSnS$_4$	Kesterite	82 (I$\bar{4}$)	1.5	Direct	112
Cu$_2$ZnSnSe$_4$		82 (I$\bar{4}$)	1.0	Direct	112
Cu$_2$FeSnS$_4$	Stannite	121 (I$\bar{4}$2m)	1.1	–	113
I–II$_2$–III–VI$_4$					
CuZn$_2$AlSe$_4$		–	–	–	74,114
CuCd$_2$InSe$_4$		–	–	–	74,114

Data corresponding to lonsdaleite (wurtzite)-based structures are not reported.

Continuity of the cell parameters with composition was observed at high temperature,[117] while the discontinuity observed at low temperature suggests that the system is biphasic.[118] At temperatures higher than ~ 680 °C, the Fe-rich member undergoes a transition into a phase with increased X-ray reflections, pointing to an unusual cation reordering from a higher to a lower symmetry unit cell with the increase of the temperature. For the Zn-rich member, a transition was observed at much higher temperature (~ 876 °C),[119] but this is attributed to a phase change from the tetragonal to the cubic (zincblende) structure (*i.e.*, from lower to higher symmetry with increasing temperature) that is commonly observed also for chalcopyrite compounds such as $CuInS_2$.

The structural distinction between stannite and kesterite has been questioned,[120–122] but neutron diffraction experiments—required to easily discriminate the contributions of Cu^+ and Zn^{2+} [115,123]—have confirmed that the two compounds exhibit subtle but meaningful differences of cation ordering. The diverse ordering in the sub-lattices is ultimately responsible for the different space groups displayed by Cu_2FeSnS_4 ($I\bar{4}2m$) and Cu_2ZnSnS_4 ($I\bar{4}$), as shown in Figure 5.2. For the Cu_2ZnSnS_4 phase, assignment of the $I\bar{4}$ space group was based on refinement of both powder[122] and single-crystal[124,125] neutron scattering measurements.

In pure stannite, the sequence of cation planes as shown in Figure 5.2 is Fe-Sn/Cu-Cu, while in pure kesterite it is Cu-Sn/Cu-Zn. Such an arrangement means that the two mirror planes at (110) and (−110), the two-fold rotation axes, the double-glide planes and the screw axes that are present in the unit cell of stannite (Figure 5.2a) are, by contrast, missing in the unit cell of kesterite (Figure 5.2b). As a consequence, the symmetry of kesterite drops down to the space group no. 82 ($I\bar{4}$). Recent findings[119] indicate that Cu and Zn occupancy on the planes parallel to (001) intersecting the unit cell at $\frac{1}{4} \cdot c$ and $\frac{3}{4} \cdot c$ in CZTS samples quenched at 750 °C is completely random (raising objection to the formal attribution of the $I\bar{4}$ space group in this case), while a 60% kesterite ordering was found for the same samples cooled slowly (1°C h^{-1}) to room temperature. These results are very interesting because they suggest that the heating and cooling rates employed during the synthesis of CZTS affect greatly the structural properties of the material.

In addition to the stannite and kesterite structures, it is possible to describe CZTS with three more cation ordering configurations, all of which display coordination of the S anions by two Cu, one Zn and one Sn cations,[127] and two more configurations where the octet rule is not followed.[119,128] Density functional theory (DFT) calculations[127] of the energetic stability of the five CZTS structures with the octet rule suggest that domains of all these structures may coexist in actual CZTS samples (*cf.* Table 5.2). In section 5.2.3 we speculate whether these aspects of ordering can be related to the experimentally assessed optoelectronic properties of CZTS(Se) thin films.

(a) Stannite (I-42m)   **(b)** Kesterite (I-4)

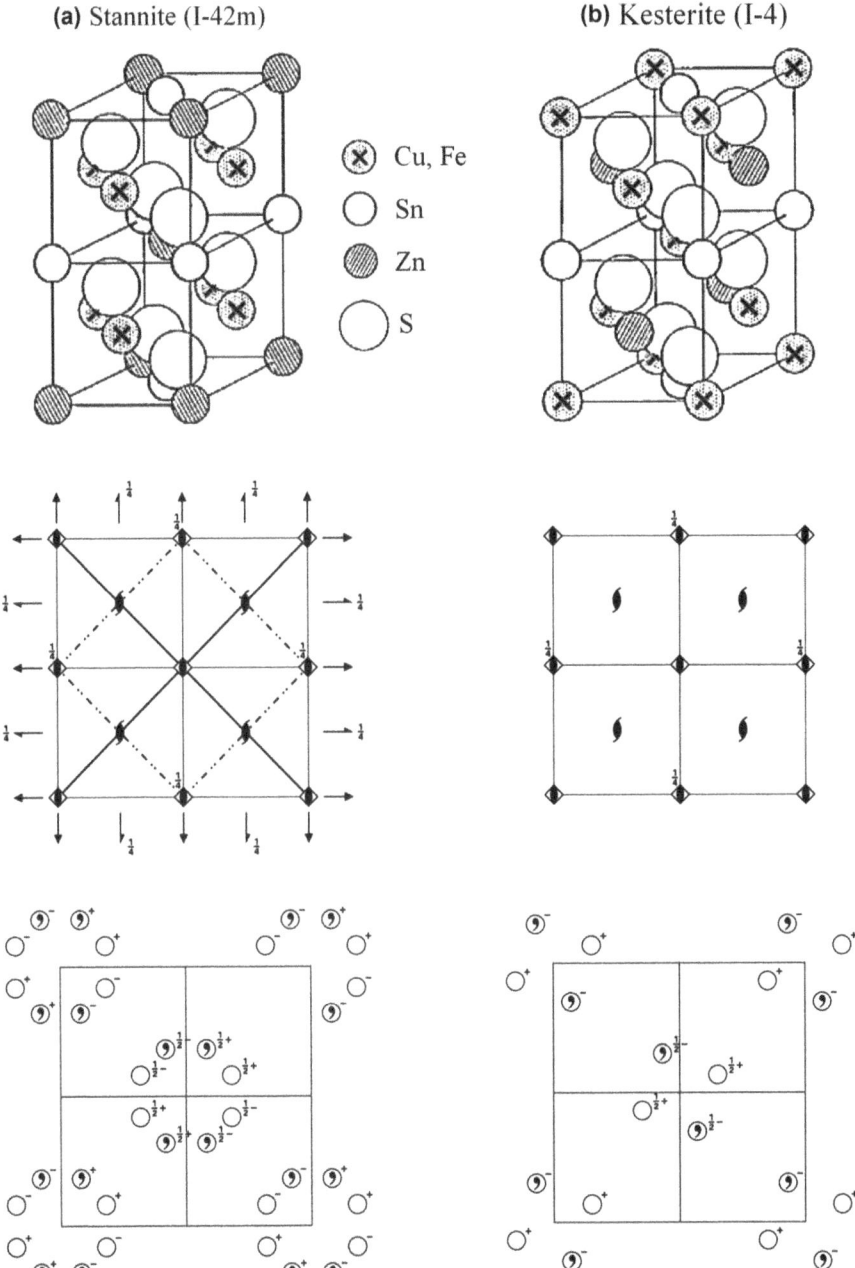

⊗ Cu, Fe
○ Sn
◍ Zn
○ S

Figure 5.2   Representation of the unit cell and cation ordering in (a) stannite and (b) kesterite and corresponding *I$\bar{4}$2m* and *I$\bar{4}$* four space group diagrams.[126]
Adapted with permission from Hall *et al.*[121]

Table 5.2 Stabilities of seven conceivable crystalline structure configurations of CZTS as computed by DFT and expressed in kJ mol^{-1} relative to the most stable structure (kesterite).

Structure (sp. gr. no.)	PBE	HSE	GGA	GGA	Octet rule
Kesterite I$\bar{4}$ (82)	0	0	0	0	Followed
Stannite I$\bar{4}$2m (121)	4.4	5.2	5.4	4.7	Followed
P$\bar{4}$2c (112)	0.48	11.6	15.5	—	Followed
Structure 4[127]	—	—	3.5	—	Followed
Structure 2[127]	—	—	7.2	—	Followed
PMCA P$\bar{4}$2$_1$m (113)	18.8	26.2	—	5.2	Not followed
P2 (3)	26.1	37.6	—	—	Not followed
Reference	119,128		127	129	—

5.2.2 A Comparison of Phase Equilibria in the Cu-In-Se and Cu-Zn-Sn-Se Systems

Reference is made to the selenide systems throughout this section, but the concepts outlined are largely transferrable to the corresponding sulfide systems. The reader should refer to Binsma et al.[130] for the Cu$_2$S-In$_2$S$_3$ pseudo-binary phase diagram and to Olekseyuk et al.[131] as well as Lafond et al.[132] for the Cu$_2$S-ZnS-SnS$_2$ pseudo-ternary phase diagram.

In spite of increasing interest in Cu$_2$ZnSnSe$_4$ (CZTSe), phase equilibria in the Cu-Zn-Sn-Se system are still less studied compared with those for Cu-In-Se. It is known that the solubility of Cu$_2$Se in CuInSe$_2$ is very limited,[133] and it is commonly acknowledged that the extension of the homogeneity region of CuInSe$_2$ towards In-rich compositions is wider. Indeed, apart from the first report,[134] early works on the phase relations in the Cu-In-Se system showed a solubility of In$_2$Se$_3$ in CuInSe$_2$ between 5.3 and 7 mol.%,[135-138] whereas the latest studies have narrowed this range to about 3 mol.%.[139,140]

Due to the high vapour pressure of Se, the quaternary nature of the Cu-Zn-Sn-Se system should be described by a five-dimensional graph: a three-dimensional plot only suffices to describe the isothermal phase relations of the quaternary system at fixed Se partial pressure.[141] Current knowledge of the equilibrium phase relations in the Cu-Zn-Sn-Se system is confined to the work of Dudchak and Piskach,[142] who reduced it to the pseudo-ternary Cu$_2$Se-ZnSe-SnSe$_2$ system. Due to the imposed stoichiometric constraints, this approach does not take into account the equilibrium relations with other potential intermediate phases such as CuSe, Cu$_{2-x}$Se, Sn$_2$Se$_3$ and SnSe that may form during the selenization of Cu-Zn-Sn precursors or the processing of other precursor configurations. For the Cu$_2$Se-ZnSe-SnSe$_2$ system, the isothermal section constructed at ∼400 °C displays a homogeneity region smaller than 3 mol.% for CZTSe.[142] Generally, the solubility range of substitutional solid solutions increases with the temperature, but it seems unlikely that it would deviate substantially from 3 mol.% at the normal temperatures employed for the synthesis of CZTSe thin films (500–600 °C). This information is very important, as it allows us to predict the type and extent of secondary phase segregation during the synthesis of CZTSe and only partially justifies the historical belief that the difficulty in growing single-phase CZTSe films is associated with its narrow single-phase field.[143]

For readers who are not fully familiar with phase diagrams, the concept of phase separation at equilibrium can be understood in terms of the excess free energy that would be required to maintain a single-phase system subject to increasing concentrations of substitutional defects. As the substitution process is performed, the crystalline structure of the base material becomes increasingly distorted as a result of accommodating the atomic exchange. If the chemistries of the exchanging atoms are too different, this process does not proceed indefinitely, and phase separation will occur when a critical extent of substitution is reached, *i.e.*, when the phase separation process becomes thermodynamically favourable (and kinetically possible).[144] A convenient way to represent the composition of synthesized films and to predict the extent and type of co-existing secondary phases is by the use of triangular phase plots. An alternative (and more intuitive) way of displaying the phase composition of a CZTSe film is shown in Figure 5.3. In this graph, the binary phases as well as Cu_2SnSe_3 are marked with different colours and their mole fractions in the ternary fields are indicated by the correspondingly coloured lines and scales. The graph was constructed by explicit application of the lever rule to the phases in the system.[145]

If the composition of the film falls within one of the triangular fields or on one of the tie lines shown in Figure 5.3, the film is expected to be triphasic or biphasic respectively, provided that it can be assumed that synthesis occurs under equilibrium conditions. If the samples are not fully converted or the growth conditions are too far from thermodynamic equilibrium, co-existence of secondary phases may violate the ternary plot in Figure 5.3. Furthermore, $SnSe_2$ is often not detected because this phase decomposes into Se and SnSe under the normal conditions employed for the synthesis of CZTSe thin films. Therefore, SnSe is more likely to be encountered, despite its relatively high vapour pressure at the normal annealing temperatures. $SnSe_2$ has reportedly been observed on samples grown in the condensed state.[146] The best CZTSe-based devices are commonly obtained with absorbers for which the composition is reported to be non-stoichiometric. The corresponding compositions are generally situated in region (d) of Figure 5.3 in proximity to CZTSe, *i.e.*, they are slightly Cu deficient and Zn rich.[147]

Table 5.3 lists the secondary phases that may be encountered in synthesized CZTSe samples, their bandgaps, their reported or expected impact on the optoelectronic properties of the devices and the etching procedures employed for their removal. Given the similar symmetries and lattice parameters of ZnSe, Cu_2SnSe_3 and CZTSe, these three phases display very similar XRD patterns.[148] Therefore, Table 5.3 also includes alternative characterization techniques that can be used to identify secondary phases.

It has been inferred that secondary phases with bandgaps lower than CZTSe can cause a reduction of the open circuit voltage of the devices due to complete shunting.[30] This is extremely detrimental for the devices, and it has been

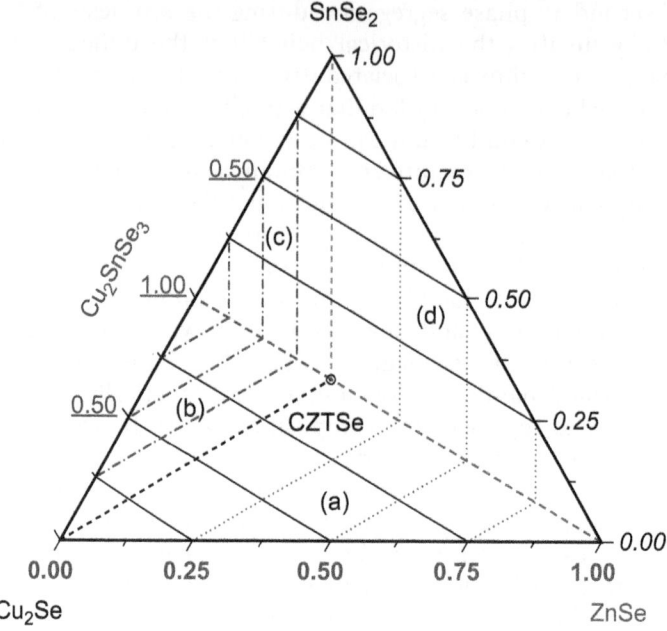

Figure 5.3   Triangular plot showing an alternative way of displaying the phase
composition in the Cu_2Se-ZnSe-$SnSe_2$ pseudo-ternary system, based on
the explicit application of the lever rule to CZTSe, Cu_2Se (black), $SnSe_2$
(italic type), ZnSe (bold type) and Cu_2SnSe_3 (underlined). The dashed
lines represent the pseudobinary joints (tie lines) Cu_2Se-CZTSe (black),
$SnSe_2$-CZTSe (grey), ZnSe-CZTSe (dotted lines) and Cu_2SnSe_3-CZTSe (dot–
dash lines). Samples with composition falling on these joints are
biphasic with CZTSe + one secondary phase, and the mole fraction can
be read on the correspondingly coloured scale (solid lines are a guide
to the eye). Samples with composition falling in each of the four
triangles defined by the dashed lines are triphasic. These are respectively: CZTSe + Cu_2Se + ZnSe (a), CZTSe + Cu_2Se + Cu_2SnSe_3 (b), CZTSe +
Cu_2SnSe_3 + $SnSe_2$ (c), and CZTSe + $SnSe_2$ + ZnSe (d). Note that the occurrence of $SnSe_2$ secondary phase depends strongly on the partial pressure
of Se present during the synthesis and the cooling stages (see text).
(Graph adapted with kind permission from Rabie Djemour.) [A colour
version of this figure is available as supplementary information on the
RSC website.]

reported for absorbers that most likely contain Cu_2SnSe_3[158] and Cu_2Se[149]
respectively. On the other hand, the presence of ZnSe, a secondary phase with
a bandgap larger than CZTSe, has been shown to cause a reduction of the
short circuit current density[161] due to a current blocking phenomenon[155]
when it is located at the interface between CZTSe and CdS.[156] This current
reduction was shown to be linearly dependent on the amount of ZnSe phase
and inversely proportional to the microstructural size of the ZnSe crystals.[161]

In summary, it may be argued that the best-performing CZTSe devices
contain absorber layers with under-stoichiometric Cu and over-stoichio-
metric Zn content because this compositional region favours the formation

Table 5.3 Potential secondary phases encountered in synthesized CZTSe films, listed with their corresponding band gaps, their reported or expected impact on the optoelectronic properties of the devices, procedures employed for their removal and possible characterization techniques.

Phase	Band gap	Effect on CZTSe devices	Etchant	Detection technique
$Cu_{2-x}Se$	Semi-metallic	Shunting[149]	KCN[149,150]	XRD,[151,152] Raman[149]
CuSe	2.0, 2.8 eV[153]	—	KCN	XRD, Raman
ZnSe	2.7 eV[154]	Conditional current blocking[155-157]	$Br_2/MeOH$[158] or sequential HCl/ KCN/HCl	EBIC,[155] SEM,[159] Raman,[152,160] PL,[160] spectral response[161]
$SnSe_2$	1.6 eV[161], 1.5 eV[162]	—	$Br_2/MeOH$	XRD,[163] Raman[164]
SnSe	0.95 eV[163]	Reduced V_{OC}[30]	$Br_2/MeOH$	XRD[163]
Cu_2SnSe_3	0.84 eV[108]	Reduced V_{OC}	$Br_2/MeOH$[158]	PL,[164] Raman[107]

of the least detrimental secondary phase for CZTSe devices, namely ZnSe. If this hypothesis is correct it would imply that the best-performing CZTSe devices do not contain single-phase absorbers, but display traces of ZnSe and/or SnSe. Alternatively, the actual homogeneity region of CZTSe could be larger than currently thought,[131] as recently suggested for the related sulfide system.[132] More work to clarify this issue would be beneficial for the CZTSe community. The homogeneity range of CZTSe thin films may also be larger due to the presence of impurities, that may have been absent in the bulk investigation.[131] It has been proposed that this kind of behaviour in the case of $CuInSe_2$ is due to Na incorporation.[133]

5.2.3 Electronic Properties

The electronic properties of any material are decisive for its use in a solar cell. Figure 5.4 shows a typical band diagram of a thin film solar cell made with a kesterite absorber.

The kesterite solar cell structure is copied directly from $Cu(In,Ga)Se_2$ solar cells, where the p–n junction is formed between the p-type absorber and the n-type TCO, with a CdS layer and a thin undoped ZnO film as buffers in between. Important features of this band diagram are (1) the bandgap of the absorber, which determines the absorption edge, (2) the band bending in the absorber, which determines the extension of the space charge region and (3) the band alignment between the buffer and the absorber, which has a decisive influence on the dominant recombination channel.

The room temperature bandgap of kesterite material can be varied between approximately 1.0 eV for $Cu_2ZnSnSe_4$ and 1.5 eV for Cu_2ZnSnS_4.[112] The bowing of the bandgap with alloying has been found to be small, *i.e.*, there is an almost linear dependence of the bandgap on the composition.[165,166] The low-temperature bandgap of Cu_2ZnSnS_4 was found to be 1.64 eV, with a rather high temperature dependence of almost 1 meV K^{-1} above 200 K.[167] The room temperature bandgaps cover the same range as $CuIn(S,Se)_2$; however, in the field of chalcopyrites it has proven easier to vary the bandgap

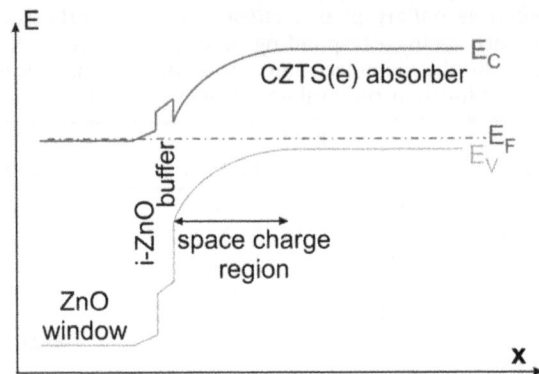

Figure 5.4   Typical band diagram of a kesterite thin film cell (E_C: conduction band edge, E_V: valence band edge, E_F: Fermi level).

by cation alloying than by anion alloying. These bandgaps range from 1 eV for CuInSe$_2$ to 1.65 eV for CuGaSe$_2$.[168] A similar approach has been followed in the field of kesterites by alloying with Ge,[169–172] which allows access to bandgaps up to 1.6 eV for Cu$_2$ZnGeSe$_4$[173] and 2.0eV for Cu$_2$ZnGeS$_4$.[174] Whether the addition of Ge still qualifies the material as earth-abundant is open to question. Certainly, the semiconductors of the family Cu$_2$Zn(Sn,Ge)(S,Se)$_4$ cover the bandgap range of interest for solar cells. The measured absorption spectra of the kesterites[175–178] indicate a direct bandgap, which is also predicted by DFT, hybrid DFT and GW calculations.[129,179,180] A direct bandgap is essential for thin film solar cell applications in order to avoid having to use light trapping methods.

A problem with kesterite materials (which is much less important in chalcopyrites) is the occurrence of different crystal modifications, as discussed in section 5.2.1 (*cf.* Table 5.2). Cu$_2$ZnSn(S,Se)$_4$ can occur as kesterite or as stannite or as a structure derived from the Cu-Au ordering of chalcopyrite.[112] These structures differ in the distribution of the Cu and Zn atoms on the available sites. The stability[128] and the bandgaps[128,129,179,180] of the different structures have been predicted theoretically by a range of different methods, ranging from LDA to GW. The general finding[112] is that the kesterite structure (space group $I\bar{4}$) is the most stable one, in agreement with experimental results.[119] However, the other crystal modifications are only slightly less stable, by a few meV per atom. This is not enough to reliably stabilise the kesterite structure at the typical growth or annealing temperatures of around 500 °C. Different crystal modifications have also been found in the case of CuInS$_2$,[181,182] where different modifications are as likely to occur as in Cu$_2$ZnSn(S,Se)$_4$ because the difference in binding energy is also only a few meV per atom.[183] In contrast, the chalcopyrite structure is stabilised in Cu(In,Ga)Se$_2$ because of the larger difference in binding energy, particularly for Ga-containing compounds.[183] The existence of different crystal modifications of Cu$_2$ZnSn(S,Se)$_4$ poses a serious problem for solar cells since they are expected to have bandgaps differing by around 100 meV.

Bandgap variations on the order of 100 meV critically reduce the efficiency of solar cells even in the radiative recombination limit.[184] The presence of different modifications in kesterite solar cell absorbers is likely to be responsible for the often uncontrolled variations of the bandgaps determined by quantum efficiency (QE) measurements[185] as well as for different emission peak energies observed in room-temperature photoluminescence (PL).[186]

An example of controlled variation of the bandgap of sulfide absorbers is shown in Figure 5.5.[185] The absorbers were prepared from metallic precursors, deposited by electrodeposition and annealed at 560 °C in S atmosphere. The bandgap shows a clear variation with the duration of sulfurization. A likely explanation of the observed behaviour is that stannite with its lower bandgap[128,129,179,180] may form at the early stages of Cu_2ZnSnS_4 synthesis and that a slow conversion into the thermodynamically more stable kesterite is achieved after longer annealing times. The minimum could be explained by a strong bowing behaviour of the bandgap of the solid solution between kesterite and stannite.[185]

Several room temperature PL spectra of $Cu_2ZnSnSe_4$ absorbers that very clearly exhibit different peaks are shown in Figure 5.6.[186] These absorbers resulted in solar cell efficiencies between 4% and 6%. Intensity-dependent studies show that the luminescent transitions between 0.8 and 1.05 eV are in fact due to band–band transitions. This implies that, in general, four different bandgaps are present in the absorbers. The higher energy emission around 1.3 eV is due to a ZnSe secondary phase.[160] Even in the PL spectra of record devices, several peaks or shoulders (or at least significant broadening) are visible, indicating the presence of several bandgaps.[28,185,187] The presence of different bandgap materials in the absorber will certainly limit the efficiency of the ensuing solar cells, even in the most ideal case of a Shockley–Queisser device, where the only loss mechanism is radiative recombination. Basically, what happens is that the highest bandgap

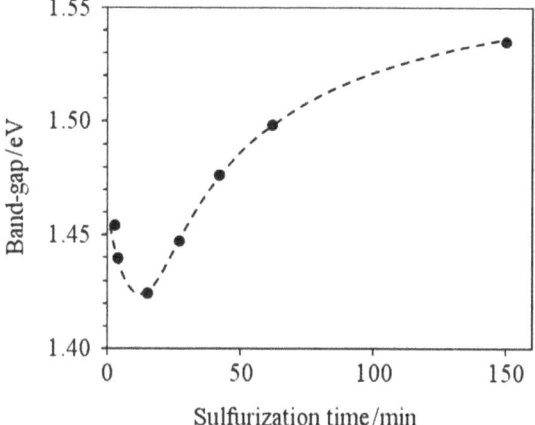

Figure 5.5 Band-gap variation in CZTS as determined from QE spectra. Taken from Scragg[185] with permission.

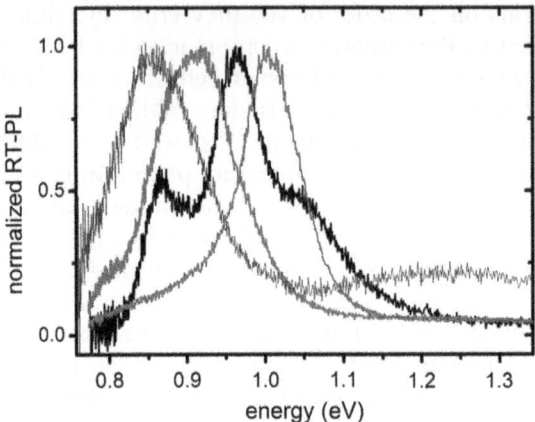

Figure 5.6 Room temperature PL spectra of various $Cu_2ZnSnSe_4$ absorbers from solar cells with efficiencies between 4 and 6%.
Taken from Djemour *et al.*[186] with permission.

determines the available current, whereas the lowest bandgap determines the voltage.[186] In real devices further complications arise from the presence of different materials, since at their interfaces band bending and band offsets and thus transport barriers will occur. It will be crucial for the development of kesterite solar cells to significantly reduce the presence of different crystal modifications.

All reports on the kesterite conduction type in the literature indicate p-type doping (as suggested by various studies[16,19,20,173,178]). There are no reports on n-type doping. The doping level is critical for the band bending.[168] If the doping is too low (typically $< 10^{15}$ cm^{-3}), the built-in potential will be low and the space charge region wide; both effects will reduce the open circuit voltage. The first effect is due to the fact that in most p–n junction devices the built-in voltage limits the open-circuit voltage. For the second effect it needs to be taken into account that in the best case current kesterite devices are limited by recombination in the space charge region, thus a wider space charge region increases the recombination current and thus reduces the open-circuit voltage. If the doping is too high (typically $> 2 \times 10^{16}$ cm^{-3}), recombination in the space charge region will be enhanced by tunnelling effects, again reducing the open circuit voltage. Early reports[17,21,176–178,189,190] on doping in $Cu_2ZnSnS(e)_4$ indicated levels that are far too high for a solar cell absorber. However, state-of-the-art devices have more suitable doping levels in the range of 10^{15}–10^{16} cm^{-3}.[191] It is likely that doping levels in kesterites depend on the composition, as is the case for chalcopyrite materials,[192] but no detailed and systematic study has been made up to now. It was found that Ge alloying increases the doping level,[193] and for $Cu_2ZnSnSe_4$, it was shown that a Cu-rich step during absorber preparation improves transport properties.[194]

Generally, doping is determined by shallow defects, whereas deep defects act as recombination centres for photogenerated carriers. So far, most of the information available on the shallow and deep defects in CZTSe has been provided by theoretical rather than experimental approaches. Formation energies of intrinsic point defects have been mostly calculated by DFT methods using various GGA functionals;[195-198] only recently have hybrid functionals been used.[199] It is generally agreed that the most abundant acceptor-like defects are the Cu_{Zn} antisite and the V_{Cu} vacancy. Whereas the formation energy of Cu_{Zn} comes out as negative in GGA calculations,[195,196] it becomes positive in hybrid functional calculations[199] and in GGA calculation with a larger supercell,[198] indicating that Cu_2ZnSnS_4 is thermodynamically stable. The Cu vacancy has higher formation energies than in Cu(In,-Ga)Se$_2$,[197] and is therefore expected to be less abundant in $Cu_2ZnSn(S,Se)_4$ and also less abundant than the Cu_{Zn} antisite.[195,196,199] Nevertheless, the Cu vacancy is considered to be the main acceptor responsible for the p-type doping in Cu-poor and Zn-rich kesterite absorbers, which are used in solar cells,[200] because of its lower energy level.[199] Deep defect levels near the gap centre, which would act as recombination centres, have been attributed to the antisites Sn_{Cu},[198] Sn_{Zn},[198,199,201] to the cluster defect $(Cu_3)_{Sn}$[199] and to the defect cluster $Cu_{Zn} + Sn_{Zn}$.[202] The neutral defect complex of the antisites $Cu_{Zn} + Zn_{Cu}$ is found to be very stable,[196,199] in agreement with observations from neutron scattering.[119] The observation that Sn_{Zn} is found by various calculation methods to act as a deep trap and that solar cells of reasonable efficiency are only prepared in the Zn-rich, Sn-poor composition range provides a hint that Sn_{Zn} is a serious recombination centre which is avoided in the suitable composition range.

Admittance spectroscopy (AS) or photoluminescence (PL) provide experimental access to defects. However, PL studies can only be used for defect spectroscopy if the emission energies are not dominated by fluctuating potentials due to a high degree of compensation, which causes broad and asymmetric emissions.[203,204] Almost all PL measurements of the kesterites in the literature show these broad emissions.[205-208] The thermal quenching behaviour of such emissions can be used to extract information about defect ionisation energies. In this way, a defect at 140 meV was detected in the sulfide kesterite[209] and a 70 meV defect in the selenide kesterite.[164] Ionisation energies between 25 and 45 meV[210] are usually attributed to Cu-Sn-Se compounds.[164] Only two reports exist in the literature which show narrow emissions (which allow direct defect spectroscopy), one for sulfide[211] and one for selenide.[209] Both studies find rather low defect energies of about 30 meV for acceptors and about 5 meV for donors. Interestingly, the compositions of both samples were in the Cu-rich, slightly Sn-poor region.

The interpretation of admittance spectra is never straightforward because not only defects contribute to the temperature and frequency dependent capacitance steps but also interfaces,[212] transport effects,[213] and the circuit response of the device itself.[214] The latter is particularly serious in kesterite cells, since they generally have a high series resistance, at least at low temperatures. A careful investigation of the activation behaviour of the series resistance and the AS behaviour of selenide solar cells reveals that the main capacitance step is not a direct defect response, but due to the series resistance with an activation energy of 130 meV, which represents either a contact barrier or the ionisation energy of the main acceptor dopant.[215] This value was found to increase continually to 200 meV for the sulfide kesterite.[191] Another study found an activation energy of 60 meV for a sulfoselenide with low sulfur content, although this is not necessarily associated with a direct defect response.[216] An AS study of selenide kesterite taking a potential backside contact barrier into account indicates activation energies around 45 meV and 110 meV.[217] A deep defect 800 meV above the valence band was found by photocapacitance measurements independent of the S/Se ratio.[218] At the moment, no conclusion is possible concerning the experimental observation of defects in these materials, other than that shallow defects are present and that great care has to be taken in the interpretation of capacitance measurements.

The dominating recombination path in the solar cells depends critically on the conduction band alignment at the absorber/buffer interface,[168,212] and also to a certain degree at the buffer/window interface.[219,220] If the conduction band edge of the buffer is lower in energy than the conduction band edge of the absorber, a so-called 'cliff', the interface bandgap becomes lower than the absorber bandgap, enhancing interface recombination. Preferable is a 'spike', where the conduction band edge of the buffer is higher than that of the absorber, as long as the spike is smaller than about 0.4 eV. The band offsets at the interface of Cu_2ZnSnS_4 and $Cu_2ZnSnSe_4$ with CdS have been studied experimentally[221,222] and theoretically.[166,223] These studies indicate a cliff for Cu_2ZnSnS_4 and a spike for $Cu_2ZnSnSe_4$. It should be noted though, that there is also a study based on UPS spectra under additional illumination, which concludes a spike for Cu_2ZnSnS_4 as well.[224] However, this result depends on the assumption that the bandgap at the surface of the absorber is the same as in the bulk and that the illumination achieves flatband conditions, which might not be the case. One theoretical study also predicts a spike for Cu_2ZnSnS_4.[225] In a study of solar cells with different S/Se ratio,[226] it was shown that pure $Cu_2ZnSnSe_4$ solar cells do not exhibit interface recombination, whereas solar cells based on $Cu_2ZnSn(S,Se)_4$ are dominated by interface recombination. This result supports the presence of a spike for $Cu_2ZnSnSe_4$ changing to a cliff as soon as a significant amount of sulfur is added to the absorber.

5.3   Preparative Routes to Earth-abundant Absorber Films

A 2 μm film of a typical semiconductor contains approximately 10^{18} atoms in every square centimetre. Of all the possible combinations of so many atoms, only a few are useful from a technological perspective: we generally require an extended crystalline lattice with a defect density of perhaps a few parts per million, and we may be able to tolerate a grain boundary every few hundred nanometres or so. Assembling such a perfect film is not a simple task because the extent of human control over the process is essentially limited to providing the right types and proportions of atoms and to setting up basic variables such as temperature, pressure and time. If we choose these variables well, the crystalline film emerges spontaneously, according to the laws of chemistry. Determining the right conditions for a particular material, however, may not be easy, and there is no one-size-fits-all synthesis method. Fortunately, we do not need to rely on trial and error alone. This section discusses how some relevant aspects of thermodynamic and kinetic theory can help in choosing variables for a successful synthesis process, even for materials about which comparatively little is known. We direct the interested reader to dedicated textbooks for more detail.[227–229]

5.3.1 Thermodynamic Considerations

Thermodynamics is concerned with the total energy of systems. For our purposes, the 'system' includes the substrate upon which the film is being deposited (which may or may not participate in the process), the film itself and its constituent elements, with atomic fractions $X_1, X_2, \ldots X_n$, and any gases present at partial pressures $p_1, p_2, \ldots p_n$. The entire system is assumed to share a uniform temperature, T. Although many synthesis methods involve steps performed at or near room temperature, formation of high-quality semiconductor films invariably involves higher temperatures that are in the range 450–600 °C for the materials of interest here. Thus, the following discussion is concerned with processes occurring at such elevated temperatures. From a thermodynamic standpoint, the total energy of different arrangements of the system—corresponding to different chemical reaction products or groups of products—can be determined as a function of the parameters X_i, p_i and T. For example, one possibility is that the solid and gas elements combine together to form a crystalline film on the substrate. A second is that the film decomposes to release material into the gas phase. A third is that the film phase-separates into several different solid phases, and yet another is that the film reacts with the substrate material to form a new compound. There are many possibilities, which multiply as the number of elements concerned is increased. At given values of X_i, p_i and T, the particular arrangement that minimises the free energy of the system can be termed the 'thermodynamic product': it is this arrangement that is predicted to form *and to be chemically stable* under the given synthesis conditions. Conversely, all other possible arrangements of the system under the same conditions are chemically *unstable*, and will tend to convert spontaneously to the thermodynamic product.

Although kinetic effects can hinder the rate of conversion to the thermodynamic product (see below), the energy-minimization argument is fundamental. Thus, the very first consideration when designing a synthesis should be to determine the ranges of X_i, p_i and T in which the desired product is the thermodynamic product. If this is not done, difficulties in preparing single-phase, high-quality material are likely to be encountered. Conveniently, the thermodynamic products have already been determined for many materials as a function of, for example, composition, pressure, and temperature, and are represented in phase diagrams. As an example, Figure 5.7 shows two phase diagrams for mixtures of Sn and S. Using these two diagrams together, we now show how reasonable conditions for synthesising α-SnS can be determined, choosing an arbitrary synthesis temperature of 500 °C. Figure 5.7a shows that α-SnS exists in the temperature range 0–602 °C, and that Sn and S must be mixed in an exact 1:1 ratio to avoid impurities of either a Sn-rich phase (if the ratio is greater than 1) or Sn_2S_3 (ratio less than 1). This diagram on its own, however, does not provide all the information that is needed for thin-film synthesis. In construction of composition–temperature phase diagrams like Figure 5.7a, the elements are usually confined in a closed system, such as a quartz ampoule, during heating. The pressure of any volatile substances, such as S, can easily reach saturation in such a system, which explains how liquid S can be present in the phase diagram at temperatures in excess of its normal boiling point. In the context of a film synthesis, where larger, open, or flowing reactors or vacuum chambers are used, such pressures are not readily obtained, and the phase diagram of Figure 5.7a may not be directly relevant. More insight can be gained from Figure 5.7b, which shows how the pressure of S_2 in the vapour phase affects the formation of the various Sn-S phases. A clear feature of the diagram is that higher pressures of S_2 yield compounds containing more sulfur atoms. The same pattern occurs for all compounds that have a volatile component, *e.g.*, sulfides, selenides, oxides and phosphides.

Figure 5.7b shows that SnS will be the sole thermodynamic product at 500 °C if the S_2 pressure is in the range from 10^{-11} to 10^{-2} mbar: impurities of Sn and Sn_2S_3 can therefore be avoided by keeping the S_2 pressure between these limits. It is also important to consider the vapour pressure of the desired product, because a stable solid can only be obtained in the presence of its own saturated vapour. For many solids of interest, this pressure is insignificant, but the vapour pressure of SnS is unusually high (it evaporates as a molecule), and it is indicated by the dashed line in Figure 5.7b. To deposit a solid film of SnS at 500 °C, we must supply SnS vapour at a pressure of at least 10^{-3} mbar. In summary, a sensible starting point for synthesising pure SnS films at 500 °C is to heat or deposit Sn in an atmosphere of *at most* 10^{-2} mbar of S_2 and *at least* 10^{-3} mbar of SnS. Since the *total* pressure of this synthesis must always be at least 10^{-3} mbar, SnS will be much more difficult to produce in a PVD system than, say ZnS. SnS_2 would be even more difficult, requiring an S_2 pressure in excess of 0.4 mbar at the same temperature. Note however that Figure 5.7b shows that a reduction in synthesis temperature reduces the pressure limits, extending the range of possible synthesis conditions.

This example highlights that the choice of synthesis method ought to reflect the ranges of conditions (in terms of temperature, pressure and composition) in which the material of interest is the thermodynamic product: not all methods are suited to all materials. If an unsuitable method is

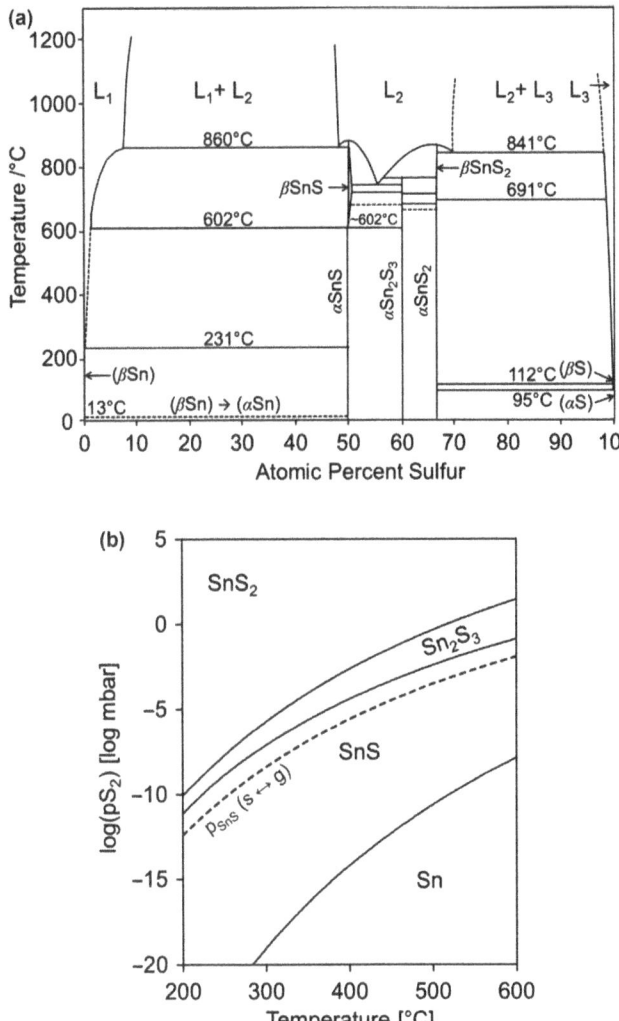

Figure 5.7 Phase diagrams showing thermodynamic products in the Sn–S system. (a) Composition–temperature diagram, after Sharma and Chang;[230] (b) solid lines: S_2 pressure–temperature diagram, derived using data from Vaughan[227] and Mills.[231] The dashed line indicates the saturated vapour pressure of SnS (see text for explanation).

employed, then the material will be hard to make, but other methods may work well. For SnS_2, a higher pressure technique such as closed-space sublimation may be more suitable.[232] Some materials, for example Sn_2S_3, have extremely narrow ranges of thermodynamic stability; this can also be seen in Figure 5.7a and b. Such materials will require very tight control of conditions, and thus may be difficult to synthesise regardless of the method employed.

If phase diagrams for a material do not exist, thermodynamic calculations can help us to estimate stability ranges (although this is no substitute for experimental data) by notionally deconstructing the material into simpler components about which data *is* available. This is illustrated below for $CuInSe_2$ and CZTSe:

$$Cu_2ZnSnSe_4 \rightleftharpoons Cu_2Se + ZnSe + SnSe_2 \tag{5.2}$$

$$2CuInSe_2 \rightleftharpoons Cu_2Se + In_2Se_3 \tag{5.3}$$

For all the compounds on the right hand side, thermodynamic properties and phase diagrams are well established. We can, for example, consider their thermal stability to get an insight into the stability of the parent compounds, making a small correction for the energy difference between the multinary compounds $CuInSe_2$ or CZTSe and their constituent units. This exercise is described in detail for the above compounds in the study by Scragg *et al.*,[233] where it was predicted that the range of Se pressures for which CZTSe is stable is many orders of magnitude smaller than for $CuInSe_2$. This lower stability range is connected to the relative weakness of Sn-Se bonds, making it an inherent feature of CZTSe (and also of the sulfide variant). Experimentally, the effects of this instability are manifested as the following thermal decomposition reaction, which occurs when the S(e) pressure is too low at a given temperature.[234]

$$Cu_2ZnSnS(e)_4 \rightleftharpoons Cu_2S(e) + ZnS(e) + SnS(e) + \tfrac{1}{2}S(e)_2 \ (g) \tag{5.4}$$

For sulfide CZTS, the minimum pressure of S required to avoid this reaction was presented as a line on the *p–T* phase diagram in the paper by Scragg *et al.*,[235] and it is shown in Figure 5.5. CZTS is only the thermodynamic product above this line. Figure 5.8 shows that in a temperature window of 500 to 600 °C, CZTS requires an absolute minimum S pressure of 10^{-5} to 10^{-3} mbar of S_2 to be stable. The exact position of the boundary also depends on the available SnS partial pressure, again due to the volatility of SnS. If the SnS partial pressure is much lower than its equilibrium value above CZTS, then the required minimum pressure of S_2 increases rapidly.[235] This limited stability window of CZTS may not coincide with the normal conditions used in certain vacuum-based processing methods. In contrast, no such limitations are encountered for $CuInSe_2$: the greater strength of In-Se bonding pushes the *p–T* boundary to much lower pressures, making it perfectly compatible with vacuum-based synthesis.

Aside from thermal decomposition, the weaker Sn-S(e) bonding in CZTS(e) results in a reaction at the interface with the layer of Mo that is usually used as a back contact for chalcopyrite absorber films, and is thus in contact with them during synthesis. When evaluating the likely thermodynamic product in a given system, one should consider all the elements in the system, which includes this Mo layer. One finds that the thermodynamic product in the Cu-Zn-Sn-Mo-S system is not 'CZTS + Mo', but rather

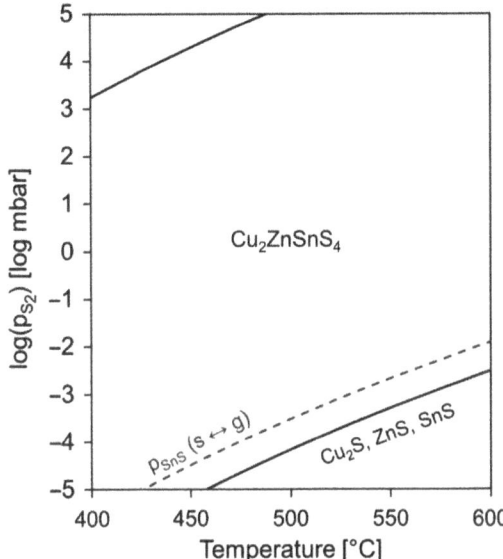

Figure 5.8 S_2 pressure–temperature diagram for Cu_2ZnSnS_4, showing limits of stability at lower S_2 pressures. The saturation pressure of SnS is also indicated by the dashed line.
Based on data from Scragg *et al.*[235]

'$Cu_2S + ZnS + SnS + MoS_2$', implying that sulfur atoms can be preferentially removed from CZTS in order to form MoS_2, resulting also in the formation of secondary phases at the CZTS/Mo interface.[233] This has indeed been observed experimentally.[236] On the other hand, the $CuInSe_2$/Mo interface is inert, because the thermodynamic product is '$CuInSe_2 + Mo$'. This example should serve as a lesson: for new materials, it is important to check chemical compatibility with the established device structures—an example of this analysis for $CuInSe_2$ can be found in the paper by Guillemoles.[237] For CZTSe at least, alternative, more inert back contact materials such as transition metal nitrides may be beneficial.[188]

The reason that Sn-S(e) bonds in CZTS(e) are relatively weak is that Sn cations, which have an oxidation state of +IV in CZTS(e), are easily converted to their +II oxidation state, as in SnS, whereupon the kesterite structure can no longer exist. In general, reactions in which the oxidation state is decreased become easier for higher oxidation states like Sn(IV). For this reason, the weakest bonds in a given compound will usually be those involving the cation with the highest oxidation state, and compounds containing high oxidation state elements will only be the thermodynamic product over smaller ranges of processing conditions.[233] As pointed out in section 5.2.1, the derivation of higher multinary compounds by iso-electronic substitution leads inevitably to an increase in maximum oxidation state. Such substitutions are likely to lead to less and less thermodynamically stable compounds, and give rise to more challenges in synthesis.

5.3.2   Kinetic Considerations

Thermodynamic arguments predict the ultimate outcome of chemical reactions, given long enough reaction times and high enough temperatures
that all the necessary chemical rearrangements required to reach the thermodynamic product(s) are able to occur. Kinetic theories relate instead to
the *rates* of reactions and other processes, such as diffusion, that occur in
the system, which can also be decisive for the outcome of synthesis. Some
key aspects are discussed below.

5.3.2.1   Reaction Rate

Once elements or compounds are physically close enough to react there are
two factors that together determine the reaction rate. The first of these is the
activation energy, E_a, which is the energy input required to initiate the necessary chemical rearrangement—*e.g.*, bond breaking—that precedes formation of new products. The reaction rate exponentially increases with
temperature according to a Boltzmann distribution with a characteristic
energy of E_a. Fast reactions (with low activation energies) will dominate over
slow ones, even if their products are less thermodynamically stable. This
leads to the concept of the 'kinetic product': if there is a large difference in
activation energy between reaction pathways, the reaction with the lowest
activation energy will dominate, especially at lower temperatures and short
reaction times. Thus, if the thermodynamic product has a high activation
energy relative to the process temperature, it may not form on a relevant
timescale. Conversely, a substance that is thermodynamically unstable
under certain conditions may persist if the activation energy for its decomposition is high. These kinetic effects can strongly modify predictions
based only on thermodynamic arguments. Unfortunately, activation energies and the kinetic product cannot be reliably determined just from theory,
meaning that experimental input is always of great value.

Reaction rates are also proportional to the concentrations of the substances involved. For our purposes, the elemental concentration—in atoms
per unit volume—is constant for most solids, unless a large homogeneity
range exists (for example in $Cu_2ZnSn(S_xSe_{1-x})_4$ in which the concentration of
S and Se is variable, with x in the range 0–1). Conversely, the concentration
of gaseous species, expressed as their partial pressure, can easily vary over
many orders of magnitude, and is of great importance in determining the
rate of solid–gas reactions. The possibility for the concentration of a component to vary gives rise to reversible reaction processes, *i.e.*, reactions that
can operate in either direction depending on the concentration of particular
species, as in Reaction (5.5):

$$Cu_2ZnSnSe_4 + 2S_2(g) \rightleftharpoons Cu_2ZnSnS_4 + 2Se_2(g) \qquad (5.5)$$

According to Equation (5.5), one prepares sulfoselenides $Cu_2ZnSn(S_xSe_{1-x})_4$
with different values of x in the range 0 to 1 by altering the balance of S and

Se partial pressures.[238] By contrast, all substances in Reaction (5.6) are solids with fixed compositions, and thus the reaction can only proceed in one direction:

$$2Cu_2ZnSnS_4 + Mo \rightarrow 2Cu_2S + 2ZnS + 2SnS + MoS_2 \tag{5.6}$$

After the initial moments of reaction, however, there is another extremely important kinetic effect that must be considered when evaluating the continuation of solid state reactions, which is the rate at which fresh material is able to reach the reacting zone. This is determined by mass transport processes.

5.3.2.2   Mass Transport Processes

Mass transport processes are especially important in solid state reactions, because the motion of elements within a solid matrix is so restricted compared to gas and liquid phase reactions. This makes mass transport slower, and the slowest step in a reaction is what determines the overall rate. Consider a sequentially deposited bilayer system, *e.g.*, Cu_2S/SnS_2, that is then heated to react the layers. According to the phase diagram, Cu_2SnS_3 is the thermodynamic product.[131] However, once Cu_2SnS_3 starts forming at the interface of the two layers, the area of contact between them starts to diminish, and the overall reaction rate decreases in proportion. Once the entire interface is blocked by Cu_2SnS_3, a barrier to further reaction is formed. Diffusion of S, Cu or Sn across this barrier is now required for the reaction to continue. It is therefore quite normal for diffusion to become the rate-determining step in the reaction as the interlayer grows. As an example, the previously mentioned reaction between CZTS and the Mo back contact follows the time dependence expected for a diffusion controlled process.[236] In a related case, formation of $MoSe_2$ due to reaction between Mo and Se vapour has been heavily suppressed by coating the Mo with a diffusion barrier of TiN—a material with a dense crystal structure that strongly inhibits diffusion of Se atoms to the Mo layer.[239] Whilst diffusion is possible through most materials (often *via* their grain boundaries), and true (bulk) diffusion barriers are special cases, the rate of diffusion can often be the controlling factor in the rate of synthesis of the desired product. Slow diffusion in the solid state can be one reason why the thermodynamic product may not be reached, at least at shorter annealing times, and can also be responsible for the formation of layered structures or gradients in composition.[240] However, longer annealing times and higher temperatures ought eventually to lead to the thermodynamic product.

Mass transport in the gas phase can also be an important issue. In annealing of CZTS(e), an inert background pressure (argon or nitrogen) is often supplied, as well as a reactive gas (*e.g.*, S, Se, or SnS(e)). The transport of the reactive gas from its source to the sample is determined by the source-to-sample distance but is also affected strongly by the total pressure—diffusive transport in the gas phase is slowed at higher total pressures. It is the *local* concentration of reactive gases at the sample surface that determines both the rate of the reaction and the thermodynamic product, and it is important to note that the local concentration can vary with time and be significantly different from the overall concentration, due to mass transport effects. A background pressure of inert gas will enhance the effective concentration of gases that are produced at the sample surface, but reduce the effective concentration of gases that originate from other sources.

5.3.3   Preparative Methods

This section summarises some of the methods that have been used to prepare CZTS(e) films. Where possible, attention is drawn to connections between the practical aspects of synthesis and the more theoretical considerations discussed in the previous section.

The two essential aspects of film synthesis are (1) *deposition*, *i.e.*, getting the right combination of elements to the substrate, and (2) *reaction*, *i.e.*, inducing them to combine to form the desired crystalline compound. In so-called 'single-stage' methods, deposition and reaction are performed simultaneously, and no further processing is required. Single-stage methods are attractive for their simplicity, but are inherently less flexible because the process conditions need to be co-optimised for both deposition and reaction. 'Two-stage' methods separate the deposition and reaction steps, allowing for individual optimisation. In the following, we pick out some instructive examples of both these process types, using the example of CZTS(e).

5.3.3.1   Single-stage Processes

For reasons presumably related to the lower stability range of CZTS shown in Figure 5.5, examples of true single-stage processes for CZTS(e) are rare. Two examples are co-evaporation and sputtering.

5.3.3.1.1   Co-evaporation. Despite the excellent results obtained for $Cu(In,Ga)Se_2$ (CIGSe) using co-evaporation, it became apparent early on that the same technique was problematic when employed for CZTS, mainly with respect to the difficulty of incorporation of Sn at deposition temperatures above 450 °C.[20] The reason for this problem is most likely the lower thermal stability of CZTS which was already discussed: when deposited at high temperature in the absence of sufficient S(e) and SnS(e) pressures, CZTS is not the thermodynamic product and decomposes *via* Reaction (5.4). These 'sufficient pressures' appear to be somewhat higher than the typical background pressure used for co-evaporation. However, some success has been achieved for CZTSe by developing a modified co-evaporation process in which the temperature is lowered and an *in situ* anneal in a Sn and Se rich environment is performed after deposition, to ensure incorporation of enough Sn and Se, and a ZnSe cap is deposited before cooling.[187] The 9.15% efficient devices from this modified method perhaps indicate that CZTSe can become the thermodynamic product even in co-evaporation, albeit with a smaller process window. Although strictly speaking this is no longer a single-stage method, carrying out all steps in one system obviously retains the main advantage of single-stage processing.

5.3.3.1.2 (Reactive) Sputtering. Sputtering is another PVD method that could have some advantages for CZTS single-stage processing, although to date the device efficiency from single-stage reactive sputtering is only 1.35%.[241] Sputtering is carried out at higher background pressures than co-evaporation, which (according to Figure 5.8) ought to allow higher temperatures to be used while retaining stability of the compound. Furthermore, since the incoming atoms are already ionised, the activation energy for chemical reaction is reduced. A possible drawback is damage induced in the film due to ion bombardment.[242]

5.3.3.2 Two-stage Processes

Processing CZTSe at higher temperatures is made easier in a two-stage preparation route, where the ability to employ a truly optimised annealing step without the constraints imposed by single-stage processing is a distinct advantage. There are two main types of two-stage methods, depending on the chalcogen content in the precursor. This can range from zero (*i.e.*, a metallic precursor) up to the stoichiometric amount (around 50 at.%) and in some cases *excess* chalcogen may be incorporated. If the precursor already contains sufficient chalcogen to form CZTS(e), the second step (heat treatment) only needs to provide stabilizing conditions while the precursor crystallizes. If less than the stoichiometric amount of chalcogen is included, then the heat treatment step is used to introduce the rest by supplying excess sulfur or selenium (in elemental form or as $H_2S(e)$) to react with the precursor. This process is referred to as 'sulfurization' or 'selenization'. For the sake of brevity, not all of the multitude of two-stage methods used for CZTS(e) are listed here; instead we consider some illustrative examples.

5.3.3.2.1 Chalcogen-containing Precursors. We can divide the chalcogen-containing precursor routes into two categories: those that use solution-based techniques and those that use physical vapour deposition (PVD). The first category of precursor deposition methods relies on dissolving or suspending chemical compounds containing the required elements in some kind of solvent, possibly with other additives, and subsequently coating this solution onto a substrate. In some cases, the films are dried to remove excess solvent, and the coating may be repeated several times to build up a thick enough layer. The resulting film can then be annealed, whereupon CZTS(e) crystallizes and the remaining solvent and other additives are lost to the gas phase. The main approaches are given in Table 5.4. Some advantages of these routes are that the required

Table 5.4 A summary of notable two-stage synthesis routes for CZTS(e) involving the deposition of chalcogen-containing precursors by solution based methods.

Solvent	Solute	Coating method	Anneal conditions	Best device efficiency	Ref.
Hydrazine	Cu_2S, $Zn(S)$ (suspended), SnSe, S(e)	Spin coating 5 layers	$500 + °C$, chalcogen vapour, several minutes	11.1%	28
Dimethyl sulfoxide	Cu acetate, $ZnCl_2$, $SnCl_2$, thiourea	Doctor blading + drying at 300 °C, 8 repeats	540 °C, Se vapour, 6 min	7.5%	44
Hexanethiol	CZTS nanoparticles (from hot-injection synthesis, suspended)	Knife coating + drying at 300 °C, 2 repeats	500–550 °C, Se, 15–40 min	9.15%	244
Alkanethiol	Cu_2SnS_3, ZnS, SnS nanoparticles	Bar coating + drying at 350 °C, 6 repeats	560 °C, Se, atmospheric pressure	8.5%	245

equipment and raw materials are lower in cost, and the material utilization can be very high compared to PVD methods. It is simple to control the composition of the precursor and it can be relatively easy to add different elements, *e.g.*, Ge.[243] It remains to be seen whether the inclusion of impurities from the solvent or additives will be a problem for higher efficiency devices, although as can be seen from Table 5.4 the results to date are very promising. Producing uniform and pinhole-free yet thin films over larger areas is a technological challenge with these methods.

In contrast to the solution based methods, PVD processes are carried out in vacuum. As discussed, this makes them inherently less suitable for high temperature processing of CZTS(e), but they still have the advantages of extremely high purity and uniformity. For this reason, a number of 'cold' PVD approaches have been developed to prepare CZTS(e) precursors with full chalcogen content, deposited at no more than a few hundred degrees. Table 5.5 gives some examples of cold PVD for CZTS(e) precursors. Some further advantages of using PVD are that the films are typically dense and mixed at the atomic scale. Annealing, carried out in a separate system, is less complicated compared with solution-based precursors, since no additional substances (solvents, counter-ions, ligands *etc.*) need to be removed. Disadvantages of PVD methods include the fact that control of composition is much more complicated, requiring robust process monitoring. Materials utilisation can be lower, and the required equipment is more complex and expensive, and it is less easy to include many different elements in one deposition system compared with the solution processes. On the other hand, large scale film preparation by PVD is commonplace in industrial processes.

After precursor deposition by either PVD or solution processing, the anneal step is used to crystallize the CZTS(e) film, and, in the latter case to remove any solvents or additives. In a well-mixed precursor with full chalcogen content, CZTS(e) forms very rapidly, because the reaction rate is not limited by diffusion (see, for example, Todorov *et al.*[28] and Scragg *et al.*[33]). CZTS(e) films are almost always prepared with excess Zn and Sn content, and if outside the homogeneity range, the excess material will segregate outside the growing grains of CZTS(e) as ZnS(e) and/or $SnS(e)_{(2)}$, ending up at the top surface, in between CZTS(e) grains or at the back contact. $SnS(e)_{(2)}$ formed near the surface may evaporate, but if formed deeper in the film it is

Table 5.5 A summary of notable two-stage synthesis routes for CZTS(e) involving the deposition of chalcogen-containing precursors by physical vapour deposition.

Deposition method	Material sources	Anneal conditions	Best device efficiency	Ref.
Reactive sputtering	CuSn alloy, Zn, Sn + H$_2$S	560 °C, excess S, 10 min	7.9%	246
Co-sputtering	CuSe$_2$, ZnS, SnS	580 °C, SnS + S$_2$	9.3%	247
Co-evaporation	Cu, Zn, Sn, S or Se	5 min, 570 °C, S or Se	8.4% (S), 8.9% (Se)	188

trapped.[236] Regardless of the method used to prepare the precursor, there is a basic requirement, alluded to already, which is to maintain CZTS(e) as the thermodynamic product throughout the annealing. This requires supply of a sufficient pressure of S(e) and SnS(e). S(e) is almost always deliberately added to the annealing atmosphere, whereas SnS(e) may be added purposefully,[33] or it may be generated from excess Sn/S(e) content in the precursor itself (this is likely in most cases due to the preference for Cu-poor material). A background pressure of inert gas and a small reaction volume will hinder the loss of S(e) and SnS(e) from the reaction zone, and thus help to keep the reaction conditions in the desired range.

5.3.3.2.2 Metallic Precursors. In contrast to the deposition of chalcogen-containing precursors, deposition of high-quality metallic films is relatively uncomplicated. Table 5.6 lists some of the methods used for metallic CZTS(e) precursors. Metal layers are most often deposited by sputtering, but can also be prepared from solution using electrodeposition or spin coating from metal salt solutions. The elements of the precursor can either be deposited sequentially or simultaneously, and in some cases, a sub-stoichiometric amount of S is also included in the precursor.[240,248]

While the precursor deposition may be simpler, the annealing process required to convert metallic precursors into CZTS(e) is far more complicated, because S or Se must be added uniformly to the precursor, which results in volume expansion and the formation of various intermediate phases. The process proceeds in three (not necessarily sequential) stages: alloying, sulfur/selenization and crystallization. Alloying of the metallic elements

Table 5.6 A summary of notable two-stage synthesis routes for CZTS(e) involving the deposition of metallic precursors.

Deposition method	Deposit structure	Anneal conditions	Best device efficiency	Ref.
Electrodeposition	Layered: Cu/Zn/ Sn or Cu/Sn/Zn	i. 210–350 °C in N_2 (alloying) ii. 585 °C, S, 12 min	7.3%	249
Spin coating from salt solution	$Cu(NO_3)_2 \cdot xH_2O$, $Zn(NO_3)_2$, $SnCl_4$ dissolved in ethanol/ propane-diol + ethyl cellulose	i. drying at 230 °C in air ii. up to 660 °C, Se, 26 min	6.2%	250
Sputtering	Layered Cu, Zn and Sn	450–520 °C, H_2Se, 5–30 min	9.2%	251, 252
Sputtering	Unreported	Sulfurization + selenization	10.8%	253
Sputtering	~100 layers of Cu/Zn/Sn	i. 1700 nm Se layer evaporated onto sample ii. RTP in S ambient, 550 °C	6.6%	254
Sputtering	ZnS/Sn/Cu/ZnS	580 °C, 20% H_2S–N_2, 10–80 min	7.6%	248

yields binary Cu-Zn and Cu-Sn phases as the thermodynamic product; no ternary phases with a composition near Cu$_2$ZnSn exist.[255] As a result, co-deposited metallic films will phase separate, whilst the mixing of sequentially deposited layers is only partial.[249] During alloying, care must be taken to avoid melting of Sn or Zn-rich phases, which could lead to de-wetting of the film from the substrate. At the same time, the high vapour pressure of Zn can cause a loss of this element during heating. Upon arrival of the S or Se source (from elemental precursors or from H$_2$S(e)), sulfur/selenization can begin. The three different metals have rather different reactivity towards chalcogen. Cu reacts vigorously, in part as a result of the rapid diffusion of Cu vacancies through Cu$_{2-x}$S.[256] ZnS(e), which is the most thermodynamically stable binary, also forms relatively quickly by reaction of Zn with S(e). However, these phases do not combine because there are no Cu-Zn-S(e) ternary phases.[131] Only upon sulfur/selenization of the Sn, which is rather sluggish, can the emergence of a ternary phase (Cu-Sn-S(e)) occur, followed by the inclusion of ZnS(e) that finally yields CZTS(e).[257,258] The stacking order in layered precursors can obviously affect the sequence in which the metals are sulfurized or selenized, effectively altering the reaction pathway.[259] Throughout the complex route from metallic precursor to CZTS(e) film there are many kinetic barriers, and diffusion operates to both cause phase separation and recombination at different points in the process. Although CZTS(e) is the thermodynamic product (assuming sufficient sulfur is provided), obtaining uniform crystallization by this route is not as straightforward as it is in the case of chalcogen-containing precursors. On top of this, the same issues that were discussed above apply to annealing of metallic precursors, namely the stabilisation of surface and back contacts during crystallization and grain growth in the CZTS(e) phase. Despite all these complications, some efficient devices have been prepared based on metallic precursors, as shown in Table 5.6.

5.3.4 Summary

Thermodynamics and kinetics are large branches of chemical theory that can be usefully applied in choosing suitable approaches and processing conditions for preparation of new materials. In this section, some basics of thermodynamic and kinetic theory were described, and their relevance to the choice of synthesis method was discussed, using CZTS(e) as a case study. The thermal decomposition of CZTS(e)—which leads to the prevalence of two-stage processing methods—and the reaction with the Mo back contact were given as examples of the power of thermodynamics in explaining and predicting phenomena occurring in synthesis. While kinetic theory is less of a predictive tool, some of the basic concepts covered here can still help to rationalise observations from synthesis and also to design appropriate annealing conditions. As far as possible, we have tried to summarise the many preparative routes used for CZTS(e). The diversity of routes and the fact that good device efficiencies are being achieved by a variety of techniques, holds promise that the 'optimal' method for this material will soon emerge. Whether by accident or design, this method will allow formation of material with the best possible quality by most effectively complementing the particular chemical properties of the CZTS(e) system. Whilst the focus of this section has been on CZTS(e), the same considerations can of course be applied to any given material, and the approach outlined here would certainly be a valuable exercise prior to embarking on any new synthesis.

5.4 Device Fabrication and Characterization

Solar cells[260] based on $Cu_2ZnSn(S,Se)_4$ absorbers are currently made with the same structure as $Cu(In,Ga)Se_2$ solar cells, *i.e.*, a substrate structure with a Mo back contact, onto which the absorber is grown by various methods as discussed in section 5.3. A CdS buffer is normally used, but there are some excellent device results with alternative Zn or In-based buffers,[261–263] reaching efficiencies of 7.6%. The device is completed with a sputtered layer of undoped ZnO, followed by a TCO layer (either Al-doped ZnO or ITO), and finally a metal grid.

At the time of writing, the best solar cell efficiency (11.1%) had been achieved with a mixed S-Se absorber with a low S/(S + Se) ratio of 5%.[28] Cells utilizing pure sulfide[264] and pure selenide[187] absorbers have both reached efficiencies of 9.2%. It should be noted, however, that the sulfo-selenide and the selenide record devices are small laboratory cells with areas somewhat smaller than 0.5 cm², whereas the record sulfide device is a 7 cell, 14 cm² sub-module. Table 5.7 compares the bandgaps determined from the linear extrapolation of the low energy edge of the quantum efficiency spectra of some of the recent record devices together with the open circuit voltages and the loss in open circuit voltage compared to an ideal Shockley–Queisser device.[14]

The V_{OC} loss in the low gap devices is around 300 mV, while it is 450 mV for the wide gap devices. A similar trend is observed in chalcopyrites, independent of the type of alloying, *i.e.*, whether the bandgap is increased by adding Ga or by adding S.[268] While there are many differences between the wide gap chalcopyrites and the low gap chalcopyrites that can contribute to losses in the solar cells, one of the main problems of the wide gap chalcopyrites is the unfavourable cliff like band alignment to the CdS buffer. As discussed in

Table 5.7 Open-circuit voltages of a number of kesterite record devices, together with their loss in V_{OC} compared to the ideal Shockley–Queisser cell.

Absorber	E_g (eV)	V_{OC} (mV)	$\Delta V_{OC,SQ}$ (mV)	Ref.
$Cu_2ZnSnSe_4$	0.90	377	281	187
$Cu_2ZnSnSe_4$	0.98	423	309	265
$Cu_2ZnSn(S,Se)_4$	1.02	460	293	28
$Cu_2ZnSn(S,Se)_4$	1.08	517	308	266
Cu_2ZnSnS_4	1.39	661	452	188
Cu_2ZnSnS_4	1.44	711	448	267

The band gaps do not necessarily agree with the band gaps given in the references, since different methods of band gap determination are used.

section 5.2.3, the situation appears to be similar in kesterite solar cells, which is confirmed by the observed trend of V_{OC} losses. Although the trend is comparable to $Cu(In,Ga)Se_2$, the absolute values of the V_{OC} losses are considerably larger in kesterites: V_{OC} losses in chalcopyrite record cells are below 200 meV.[269,270] Low V_{OC} has been identified as the main loss mechanism in several investigations (see, for example, Choudhury *et al.*[216] and Mitzi *et al.*[260]) *i.e.*, there are recombination channels in the solar cells that need to be eliminated to improve solar cell efficiency.[271] The V_{OC} loss in sulfur-containing films is mostly attributable to interface recombination, probably due to unfavourable band alignment to the CdS buffer. By contrast, selenide solar cells tend to be dominated by recombination inside the absorber. A major recombination path is likely to arise from the presence of secondary phases. These will form interfaces inside the absorber and they might create unwanted doping profiles. The major secondary phase for the compositions used for solar cells is $ZnS(e)$.[30] It has been found, at least for certain processes, to form a nanometre-sized network inside the kesterite absorber,[143] which creates a large interface and is likely to be responsible for recombination losses.[194]

The immediate contact materials of the absorber—the CdS buffer and the Mo back contact—have simply been adopted from $Cu(In,Ga)Se_2$ solar cells. The CdS buffer appears suitable for selenide kesterites, but a different buffer with a higher conduction band energy would be more suitable for sulfur-containing kesterites. However, choosing another buffer may not solve the problem because the conduction band of the window layer must also be high enough in energy to avoid interface recombination.[219] The Mo back contact has been identified as problematic in two aspects: it has been discussed as responsible for the high series resistance[260] observed in kesterite solar cells, particularly those based on sulfur containing absorbers, and it has been related to the instability of kesterite absorbers due to decomposition into binary secondary phases[233,272] (see section 5.3.1). However it seems more likely that the series resistance is linked to the presence of a ZnSe secondary phase in the absorber.[155] Thus, in terms of series resistance, Mo is a suitable back contact. The problem of decomposition is currently under discussion, and it may turn out that Mo is not a suitable back contact and a less reactive material would be more suitable.

5.5 Other Earth-abundant Materials

The previous sections have focussed on the kesterites, but in this section we extend consideration to a wider range of potentially useful earth-abundant absorber materials. Discussion is restricted to p-type compound semiconductors; it will be assumed that a suitable earth-abundant n-type partner can be found for the p-type absorber layer. This section is organized into two main parts. The first part provides a brief overview of the essential thermodynamic and optoelectronic criteria for a candidate absorber layer. The second part identifies suitable absorber layers and reviews the current state of the art in terms of the criteria outlined in the first part.

5.5.1   Phase Equilibria Considerations

Phase purity is a very important aspect of thin-film PV technology. This naturally becomes a crucial property of a good quality semiconductor absorber layer, since the p–n junction is the core of the solar cells. The search for alternative inorganic light-harvesting materials for thin-film PV has prompted investigation of the properties of multinary compounds, including (but not limited to) those conceptually derived by iso-electronic substitution, such as Cu_2ZnSnS_4 (*cf.* section 5.2.1). As a consequence of this approach, the materials systems under study are of increased complexity.

The successful synthesis of single-phase compound thin films requires a detailed understanding of the phase equilibria involved in the corresponding material system. Whether the phase diagram displays a wide or a narrow homogeneity region for the phase of interest has a profound impact on the feasibility and ease of synthesis. Any compositional deviation from the equilibrium single-phase field will lead to formation of secondary phases that may degrade the optoelectronic properties of the devices.

A typical example is that of $CuInSe_2$ (CIS). CIS can be described as an intermediary phase in the $Cu_2Se-In_2Se_3$ pseudo-binary system. At room temperature, its single-phase field as a bulk material is limited to $CuInSe_2$ stoichiometry on the Cu-rich side, while the In-rich side accepts up to 2.5 mol% of In_2Se_3.[136] The best grain sizes of CIS thin films are attained with Cu-rich precursors,[273,274] and photoluminescence measurements suggest that a higher open-circuit voltage may be achieved with CIS grown under Cu excess.[275] Even so, the presence of a $Cu_{2-x}Se$ secondary phase in films synthesized under Cu excess is detrimental for device performance since it acts as parasitic shunt path,[276] and its removal by etching in aqueous KCN solutions is a common practice to improve the solar cell parameters.[150]

Secondary phase formation is particularly relevant if the synthetic procedure consists of a two-stage route (*cf.* section 5.3) where elemental precursor films are first deposited (typically metals) and secondly react with a gas (typically a non-metal, such as a chalcogen or a chalcogen-bearing reactant) to form the compound film. During such a route, the formation of secondary phases (as intermediate lower order compounds) is largely unavoidable,[277] but the key question is whether these secondary phases react together to form a compact single-phase film, or whether they segregate in such a way as to impair the quality of the absorber film. Factors such as the heterogeneous reaction rates between each of the elements that constitute the system and the reacting gas,[55,59,278] as well as the rates of interdiffusion in the solid state dictate the crystallite size of the intermediate phases and the extent of their segregation, ultimately affecting morphology, crystallinity and phase purity of the resulting compound film.

Section 5.2.2 considers the example of Cu-Zn-Sn-S(Se) systems, with the focus on the target compositions $Cu_2ZnSnS(Se)_4$ (CTZS(Se)). The difficulty of growing single-phase CZTS(Se) layers is commonly attributed to the fact that the phase equilibria in these systems exhibit a narrow compositional homogeneity range for CTZS(Se).[143] Apart from thermodynamic stability issues (*cf.* section 5.3.1, or Scragg *et al.*[233]), the synthesis of a phase-pure CZTS(Se) is challenging simply as a consequence of the very different chemical nature of Cu, Zn and Sn. This becomes apparent when comparing the early results from synthesis routes based on solution processing[25]— where the four/five components are molecularly intermixed[279]—with the results of routes involving chalcogen incorporation into a metallic precursor film.[259] The intrinsically different reactivity of the metals towards chalcogens has recently prompted the search for alternative metal precursor microstructures, such as columnar,[249] dendritic[280] or nanoporous[55] in attempts to obtain phase-pure absorber films by compensating for the different bulk solid–gas kinetics.

To conclude, knowledge of the equilibrium phase diagram of the systems under investigation is of paramount importance because it allows realistic prediction of the phase composition of the film after processing. However, the morphology of compound films obtained *via* the two-stage route is largely driven by kinetics, which dictates the reaction trajectory and associated phase development.[57] Unfortunately, knowledge of the phase diagram is sometimes insufficient to predict whether a compound of interest is thermodynamically stable under the processing conditions at each of its interfaces. Ideally, candidate absorber materials should display a good thermodynamic stability in order to permit a large processing window. The following section analyses and provides suggestions for the case when thermodynamic stability of the candidate compounds is an issue.

5.5.2 Phase Stability Considerations

As seen in section 5.3, in processes where all elements are present at the precursor stage, compound formation is driven mostly by the entropic gain, while in the two-stage process, where metals are oxidized at the expense of the reactive gas, the enthalpy term dominates the free energy change of the process. Regardless of the preparative route employed to synthesize the absorber film, suitably designed heat treatments are employed to form the compound, ensuring at the same time improved homogeneity and enhanced crystal quality. With processing temperatures typically ranging between 400 and 600 °C, activation barriers to diffusion are overcome and, given enough time, phase equilibrium is established with emergence of the most stable compounds from the mixture of possible intermediate phases.

In section 5.2.1 it was shown that the iso-electronic substitution method implies an increase of the formal oxidation state of the elements involved at each step of substitution. Although a free energy gain is associated with formation of an increased number of bonds, elements with high oxidation states are generally less stable in comparison to those with lower oxidation states.[281] It follows that decomposition of a compound *via* reduction of cations from higher to lower oxidation states may be thermodynamically allowed.[233] In general, high-temperature processing can result in chemical decomposition processes that are not thermodynamically or kinetically favoured at low temperature. This may represent an issue for the synthesis of phase pure absorber films and may upset the interfacial phase compatibility of the device layers, especially between back contact and absorber film.[236,282] In order to prevent such detrimental processes from taking place or to minimise their effects, any preparative procedure that involves a high temperature treatment should not only consider the choice of materials sequence in the device but should also be designed to take into account the thermodynamic stability of the absorber phase and of phases in direct contact with it. A well-documented example is 'SnS(Se) loss' occurring in CZTS(Se).[234] Lesser known cases are those concerning In in CIS(Se),[283,284] Sb in CuSbS$_2$[59,285] and Bi in Cu$_3$BiS$_3$.[286]

A generic compound decomposition reaction may be written in the form:

$$A_{(s)} = \sum_i c_i C_{i(s,l)} + \sum_j g_j G_{j(g)} \tag{5.7}$$

where $A_{(s)}$ is the generic compound subject to decomposition reaction, C_i are the condensed phases (solid or liquid) and G_j are the gaseous products of the reaction. For such a reaction at equilibrium at a specific temperature, the equilibrium (pressure) constant K_p is given by:

$$K_p = \prod_j p(G_j)^{g_j} = \exp\left(-\frac{\Delta G^\circ}{RT}\right) \tag{5.8}$$

where $p(G_j)$ is the equilibrium gas pressure of the component G_j and ΔG° is the free energy change associated with the reaction. The product in Equation (5.8) is constant at constant temperature, which means that the system will adapt to any induced deviations from the equilibrium (Le Chatelier principle). More of the compound $A_{(s)}$ will decompose into the products $C_{i(s,l)}$ and $G_{j(g)}$ if the atmosphere above the compound is displaced, for example by a flux of inert gas. The opposite behaviour also applies: since the pressures of the gaseous species in equilibrium with the decomposing compound are related by Equation (5.8), compound decomposition can be minimized by deliberate introduction into the processing chamber of one or more of the gaseous species. Decomposition of inorganic compounds may release the gaseous non-metal in elemental form, associated with simultaneous reduction of one or more of the cation species,[235] and the product of the cation rearrangement may itself be volatile at the processing temperature.[235]

In principle, complete suppression of decomposition phenomenon should be achieved if the composition of the annealing atmosphere is such that the pressures of all the gaseous species exceed their equilibrium pressures at the temperature of interest. It has become clear that control of the atmospheric composition during synthesis is the key to controlling potential bulk decomposition due to thermodynamic instability of compounds. The most accurate way to experimentally determine pressure and composition of all the evolving species is by mass spectrometry. Measurements should be performed with the compound of interest placed in a closed system at the temperature pertinent for the process. The information obtained will allow definition of the appropriate atmospheric composition so that the compound will remain stable during processing. Suppression of the decomposition by strict control of the annealing atmosphere is also possible if the thermodynamic properties of the chemical species involved in the reaction are known, preferably as a function of the temperature. This allows calculation of the value of K_p and the pressures of the gaseous species, if the stoichiometry of the reaction is known. Valuable sources of thermochemical data for inorganic compounds that can assist in this exercise include Knacke and Kubashewski,[287] Barin,[288] Mills[231] and the National Institute of Standards and Technology database.[289] Unfortunately, thermochemical data may not be available for certain compounds. In this case, experimental procedures such as those based on calorimetry,[290,291] the torsion–weighing effusion vapour pressure method,[292,293] the Knudsen mass spectrometric or weight loss method,[290,294] or the electromotive force (emf) method[295] can be used to determine thermochemical properties, although phase pure materials are often required for accurate measurements.

In the case where thermochemical data is missing for chalcogenides, a first approximation for the chemical potential was inferred by Craig and Barton[296] and Vaughan and Craig[297] based on their extensive research on sulfosalts, and this approach may be extended to other multinary compounds. In their 'sulfide sum' procedure, complex sulfides are assumed to behave as if they consist of ideal mixtures of their simple end-members, assuming a zero enthalpy change of mixing, hence a purely entropic free energy change. An example of this approach is shown below

$$Cu_2S + SnS_2 \rightarrow Cu_2SnS_3 \qquad (5.9)$$

Density functional theory (DFT) calculations suggest that the assumption of zero enthalpy change made in the 'sulfide sum' method is too strong, and may lead to large errors in the estimation of K_p. For example the enthalpy change for the reaction shown in Equation (5.9) has been estimated to be ca. -69.8 kJ mol^{-1}.[197] Due to the fact that multinary compounds display a negative free energy change when formed from the binaries, K_p values calculated based on the 'sulfide sum' or on the thermochemical data of the binary compounds are expected to be over-estimates, simply based on consideration of Hess's law[277] (*cf.* experimental sequence of SnS loss rates: $Cu_2ZnSnS_4 > Cu_2SnS_3 > SnS$[234]). Such over-estimation may not be an issue for some industrial applications, and no further information may be needed to design the conditions for compound synthesis and processing; conditions that in any case would be conservative and would certainly prevent the compound from decomposing. On the other hand, a range of thermodynamic data of inorganic compounds estimated from DFT is available.[32,54,63,166] Although the chemical potentials of condensed phases are usually computed by DFT at 0 K,[298] it would be possible to compute and include the appropriate entropic terms for more sound estimations of K_p; this may be a fertile field for computational chemistry.

Other approaches can be considered for the design of processing conditions capable of preserving compound stability or minimising the extent of decomposition. These may be less effective, in so far as thermodynamic suppression of the decomposition reaction is not ensured, but they may provide viable alternatives—especially if wisely applied in conjunction with each other—when the options listed above are subject to physical constraints or when there are technological barriers for industrial upscaling. These aspects may also help the design of effective procedures for compound film synthesis on the laboratory scale. These points and related parameters can be divided into the two following categories: (1) *kinetic*: (a) time, (b) static background pressure of inert gas; and (2) *thermodynamic*: (a) temperature, (b) volume, (c) pressure of one of the volatile components.

5.5.2.1   Kinetics: Time

It is worth remembering that Equation (5.8) is only valid when the system has reached thermodynamic equilibrium. It may be obvious, but time plays an important role in defining the extent of compound decomposition, especially if the processing is performed in non-closed environments. The processing time should be minimised as far as possible, and tests should be carried out to investigate the minimum time required to achieve the desired compound properties; surprisingly, in certain cases the relevant processes can occur in the timeframe of seconds or minutes.

5.5.2.2   Kinetics: Background Pressure of Inert Gas

Applying a static background pressure of inert gas may reduce the extent of compound decomposition simply due to a kinetic effect. Because increased background pressures decrease the rate of diffusion of the evolving species from the compound surface, the partial pressure in the proximity of the sample surface is increased. Of course this phenomenon is time-dependent and it does not prevent the compound decomposition. Nevertheless, in certain cases it may suffice, especially if used wisely in conjunction with the information in section 5.5.2.1. This method may not be suitable when compound formation occurs by incorporation of elements with a reacting gas, because the presence of background gas may prevent a uniform reactant distribution from being achieved.

5.5.2.3   Thermodynamics: Temperature

The extent of compound decomposition is strongly dependent on the processing temperature. When it comes to decomposition with evolution of gaseous species, the higher the temperature, the greater the magnitude of $\Delta G°$. As for the information in section 5.5.2.1, it is obvious that the processing temperature should be as low as possible to ensure the formation of the compound film with the desired properties.

5.5.2.4   Thermodynamics: Volume

Although often neglected, this aspect is at least as important as section 5.5.2.3. Maximization of the ratio between volume of processed compound film and volume of the processing environment brings about a linear reduction of the extent of compound decomposition.[277] The experimental set-up or processing line should not be oversized, especially if it is not a closed system and operates at or close to atmospheric pressure.

5.5.2.5   Thermodynamics: Pressure of the Volatile Components

On an industrial scale, it may not be feasible to process the compound film under an atmosphere containing a complex mixture of reactants, especially if the process is performed under static conditions and one or more of the reactants have low diffusion coefficients. An option in such circumstances is to supply an excess pressure of only one or more of the evolving species, for example the non-metal gas or its bearing molecule.[277] This option is very convenient in the case of the two-stage process, as this reactant is the same used for incorporation during the second stage. The stoichiometry of the relevant reaction will determine the effect of the non-metal excess on the extent of reaction suppression. In certain cases, this solution may be enough to minimize the extent of reaction decomposition to industrially acceptable values. Of course, employing a very high partial pressure of the non-metal creates a highly oxidizing environment. Care should be taken that this does not impact the properties of the substrate material or the compound itself due to over-reaction.[299,300]

5.5.3   Opto-electronic Considerations

The aim of this subsection is to summarize the important optoelectronic properties of absorber layers and their minimum/maximum values necessary to give a device with efficient power conversion. The following brief considerations are for a perfect p-type semiconductor absorber layer in a single junction device. It will be assumed that an earth-abundant n-type partner semiconductor with correct band alignment can be found to complete the device. For a more rigorous approach, the reader is referred to reference.[301] The optical properties are considered first as they dictate the necessary electrical properties of the material, which are considered second.

A p-type semiconductor absorber layer should have a direct bandgap transition and a high absorption coefficient (α). A direct optical transition combined with a high absorption coefficient enables the absorber layer to be just a few microns thick, meaning little material is required to make a photovoltaic module. From consideration of the Beer–Lambert law, 99% of the incident light is absorbed within 1 μm if α is 5×10^4 cm^{-1}. Photon absorption in the absorber layer generates electron hole pairs. For maximum device efficiency, all the photo-generated minority carriers (electrons for a p-type absorber) should pass through to the n-type layer, and the excess holes should be transported to the back contact. The Shockley–Queisser calculation of the theoretical limits of device efficiency[271] predicts that an efficiency of 30% or more under AM 1.5 G irradiation is possible for bandgaps of between 0.95 and 1.60 eV.

The most important electrical properties are those related to the transport of the minority carriers. Minority carriers are collected if they are generated within the space charge region (w) or within one diffusion length (L_e) of w. The space charge width is inversely proportional to the square root of the net acceptor concentration (N_A) assuming the n-type layer has a much greater net donor concentration. L_e is given by the square root of the product of minority carrier diffusion coefficient (D_e) and lifetime (τ_e). Typical values for w, L_e, and N_A in high-performance CIGSe devices are 0.45 μm,[302] 3 μm[303] and 10^{16} cm^{-3},[303] respectively. The value of L_e takes on extra significance when considered relative to α. The lower α is, the larger L_e must become to collect all the minority carriers.

In the ideal case where recombination occurs only in the neutral region, the device open circuit voltage is inversely proportional to the reverse saturation current (j_0) of the device, independent of the material system, j_0 is given by:

$$j_0 = e\left(\frac{n_i^2}{N_A} \sqrt{\frac{D_e}{\tau_e}} + \frac{n_i^2}{N_D} \sqrt{\frac{D_h}{\tau_h}} \right) \tag{5.10}$$

where n_i is the intrinsic doping density. The second term on the right hand side of Equation (5.9) relates to the n-type semiconductor and can be neglected due to the normally large value of N_D. Therefore j_0 (and thus device efficiency) depends on N_A and τ_e in the p-type semiconductor. τ_e in turn depends on the dominant recombination mechanism, which is usually due to the concentration of deep defects in the middle of the bandgap. τ_e in good CIGSe devices is on the order of 100–200 ns.[304]

The voltage drop due to transport resistance of the majority hole carriers through the absorber layer to the back contact is a source of power loss in a solar cell. If a 10 mV drop is considered as the threshold for an acceptable loss in a 2 μm thick layer, then the conductivity (σ_h) of the absorber layer should be greater than 10^{-3} Ω^{-1} cm^{-1}. Therefore assuming a N_A of 10^{16} cm^{-3} a hole mobility (μ_h) of 0.5 cm^2 V^{-1} s^{-1} is required (most polycrystalline inorganic semiconductors exceed this value). In summary then the important parameters of an absorber layer are: bandgap, absorption coefficient above the bandgap, net acceptor doping density, electron life time, hole conductivity and mobility.

5.5.4 Application of Criteria of Earth Abundance, Thermodynamics, and Opto-electronic Properties to Other Potential Absorber Materials

In this section we discuss potential candidate earth-abundant p-type semiconductors besides CZTS(Se) for use as absorber layers in single heterojunction photovoltaic devices. The selection criterion for the candidate materials is that they have a bandgap within the Shockley–Queisser limits that would allow theoretical device power conversion efficiency up to 30%. Candidate materials are SnS, Cu_2S, FeS_2, Cu_2SnS_3, $CuSbS_2$ and Cu_3BiS_3, which are discussed in terms of their thermodynamic and optoelectronic properties, following the framework of the previous section. The optoelectronic properties of these materials are tabulated in Table 5.8 and are compared to typical values for $Cu(In,Ga)Se_2$ as well as to reasonable threshold values for an efficient device.

5.5.4.1 FeS₂ (Pyrite)

FeS_2 (pyrite) has a bandgap of 0.95 eV and a high absorption coefficient reaching 5×10^4 cm^{-1} 300 meV above the bandgap. In the early 1990s, photoelectrochemical devices based on FeS_2 n-type single crystals gave a 2.8% power conversion efficiency with external quantum efficiencies above 90% 300 meV above the bandgap.[67] A short-circuit current of 30 mA cm^{-2} could also be achieved with 500 mW cm^{-2} white light illumination and 75 mV open-circuit voltage using the same type of crystals in a Schottky junction device.[67] Antonucci *et al.*[316] used 100 μm thick polycrystalline pyrite photo-anodes, thermally treated with hydrogen, in a photoelectrochemical set-up to achieve a power conversion efficiency of 5.5% under AM 1 illumination. Although this early progress appeared promising, no higher conversion efficiencies were reported, despite numerous efforts. The main properties and problems of the material are outlined below as well as the areas which need further effort if FeS_2 is to become a serious candidate absorber.

When pyrite is heated above 300 °C, it starts to decompose releasing elemental sulfur into the gas phase. This observation underlines the importance of annealing samples in a sufficient partial pressure of sulfur. FeS_2 has two polymorphs, pyrite and marcasite. Pyrite is considered to be the desirable phase, and it can be favoured by using high activities of sulfur in the vapour phase and by the presence of sodium.[68] FeS_2 is found to have the Fe in oxidation state (II) and the S_2 dianion as (II$^-$). Early research was conducted on naturally occurring crystals and synthetic crystals.[67] Synthetic crystals grown by chemical vapour phase transport are n-type, as are most naturally occurring single crystals. The conductivity of synthetic crystals was found to be 1 Ω^{-1} cm^{-1}. The Hall mobility ranged from 50 to 366 cm^2 V^{-1} s^{-1}, and the carrier concentration was 10^{16} cm^{-3}.[67] Although the crystals were found to be n-type, photoelectron spectra gave evidence that the surface Fermi level is 100 meV from the valence band edge, indicating strong band bending and type

Table 5.8 Selected opto-electronic properties of earth-abundant absorber materials compared to Cu(In,Ga)Se$_2$ (CIGSe).

Material	E_g (eV)	$E - E_g$, where α reaches 5×10^4 cm^{-1} (eV)	σ (Ω^{-1} cm^{-1})	μ (cm^2 V^{-1} s^{-1})	N_A (cm^{-3})	Record eff. (%)	Annual production (tonnes)[305,306]
CIGSe	1.00 to 1.65[307]	0.3[308]	10^{-3} to 10^{-1} [309]	20[303]	10^{16} [303]	20.3[269]	In 5.74×10^2; Ga $<10^2$
SnS	1.1 to 1.2 (i)[310,311]	0.4[310,311]	10^{-3} to 10^{-2} [312]	1 to 15[312]	10^{15} to 10^{18} [310]	2.0[310]	Sn, 10^5
Cu$_2$S	1.16 (i,d)[313]	0.70[313]	10 to 200[314]	2 to 4[314]	10^{16} [314]	10.1[60]	Cu, 10^7
FeS$_2$	0.97(i)[68]	0.28[68]	1, 5[68,315]	<2[68]	10^{18} to 10^{19} [68]	5.5[316]	S, 7×10^7
Cu$_2$SnS$_3$	0.93, 0.99[51]	~0.6[317]; ~1.0[49]	10^{49}	80[49]	10^{18} [49]	6.0 (Ge alloy)[318]	Sn, 10^5
CuSbS$_2$	1.38 to 1.89[59,319–323]	0.2[319]; 0.40[324]	3.33×10^{-4} [322]	2.07×10^2 [322]	4.9×10^{13} [322]	3.1[325]	Sb. 10^5
CuBiS$_2$	1.65[326] to 1.8[327]	0.35[324]	—	5 to 141[328]	6 to 14×10^{17} [328]	—	Bi, 7.6×10^3
Cu$_3$BiS$_3$	1.3 to 1.4[55,56,329,330]	0.27[56]	4×10^{-3} to 1×10^{-1} [56,329]	—	3×10^{17} [55]	—	B, 7.6×10^3

inversion, presumed to be due to surface states. Attempts to passivate the surface defects with hydrogen made no difference to the band bending, and it was assumed that the surface states are due to the broken lattice periodicity at the surface, *i.e.*, an intrinsic property of pyrite.[67]

The synthesis of nanocrystals and thin films of FeS_2 has been reported extensively since 2010 and is reviewed by the Law group.[68,331] All thin films of pyrite are reported to be p-type, have a high conductance of $1\ \Omega^{-1}\ cm^{-1}$, high carrier concentrations of between 10^{18} and $10^{20}\ cm^{-3}$, and low mobility $< 2\ cm^2\ V^{-1}\ s^{-1}$, regardless of preparation method. Currently, the doping density is too high for use in p–n$^+$ junctions. The high carrier density found in all polycrystalline thin films is attributed either to a common dopant such as oxygen or to a high concentration of surface states, similar to those found in the single crystal studies.[68,331] Recently, several theoretical studies have been undertaken to try and understand the doping properties of pyrite, and conflicting results were obtained.[332,333] Interestingly, Aravind *et al.*[334] showed that native defects are energetically expensive to make both in the bulk and at the interface and cannot account for the high measured doping densities. However, they observed that surface states reduce the bandgap at the surface, which could explain the low open circuit voltages obtained so far. In summary, although FeS_2 is extremely abundant with a very high absorption coefficient, current growth methods are unable to control the doping density, so that future work must focus on resolving this issue.

5.5.4.2 Cu₂S (Chalcocite)

Cu_xS exists in a wide range of compositions where $x = 1$ to 2; chalcocite (Cu_2S), djurleite ($Cu_{1.94}S$), digenite ($Cu_{1.8}S$), anilite ($Cu_{1.75}S$), and covellite (CuS). All forms of Cu_xS are p-type according to reference[335] except for covellite, which has been found by Mazin to have superconductive properties.[336] Chalcocite also exists in three polymorphs, low, high, and cubic.[335] The synthesis of any film will be sensitive to the sulfur source activity and growth conditions. The opto-electronic properties also depend heavily on the exact composition and synthesis conditions. Before discussing the reported properties of Cu_xS given in Table 5.8, a brief account of Cu_2S in photovoltaics is given.

Cu_2S was one of the first heavily researched thin film photovoltaic absorbers, and when coupled with a thick n-type $Cd_{1-x}Zn_xS$ layer, completed devices reached a maximum power conversion efficiency of 10.2% in 1981.[60] The device consisted of a brass (Cu-Zn alloy) back contact coated with a 20–30 μm thick n-type $Cd_{1-x}Zn_xS$ layer, of which the top 300 nm was subsequently converted to a Cu_2S layer by electrochemical displacement of the Cd and Zn ions. After annealing the p–n structure in hydrogen at 170 °C for several hours, a transparent gold layer was added to complete the device.[60] Burgelman and Pauwels[337] identified several problems with the Cu_2S and with the device structure. The Cu_2S acceptor density of $\sim 10^{19}$ meant that the space charge region in the Cu_2S was very narrow so that collection of minority carriers occurred by diffusion only. A further problem is that the absorption coefficient of Cu_2S only reaches 5×10^4 cm^{-1} at wavelengths around 650 nm,[313] resulting in a considerable current loss due to lack of photon absorption, as the absorber layer was only 300 nm thick. It is not clear whether the measured 1.2 eV bandgap is indirect, direct, or mixed.[313] The thinness of the absorber layer was chosen to minimize the hole transport resistance to the front contact.[338] Interestingly though, the junction appears to have been excellent as the measured reverse saturation current for a cell similar to the record device was of the order of 10^{-11} A cm^{-2},[338] which is similar to the best current CIGSe devices displaying efficiencies above 20%.[269]

Enthusiasm for Cu_2S-based absorber layer devices declined in the middle of the 1980s, presumably due to the inability to increase device efficiency and the inability to stop long term degradation of the absorber layer, which impacted on device performance. The chalcocite phase (Cu_2S) converts to djurleite ($Cu_{1.96}S$) due to out-diffusion of Cu to the CdS at the junction interface and to the front interface to form Cu_2O.[339] Conversion of chalcocite to djurleite reduces the minority carrier diffusion length, increases the bandgap, and reduces the absorption coefficient, degrading device performance.[313]

Clearly, any new attempt to work on the material must overcome the mobility and reactivity of copper. Since 2010, interest in the material has been rekindled as indicated by the large number of papers published, but examination of the recent literature reveals that little attention has been paid to stabilization of the chalcocite phase to prevent copper diffusion. A recent theoretical study by Xu et al.[335] proposes that the $Cu_{1.75}S$ anilite phase may be more suitable than chalcocite as an absorber layer as there is an enthalpic energy barrier to removing Cu from the lattice. Anilite is predicted to be p-type and have a bandgap of around 1.4 eV, but its opto-electronic properties are little studied. Martinson et al.[314] have shown that thin films of chalcocite can be grown on quartz substrates by atomic layer deposition with an acceptor doping density of 10^{16} cm^{-3} when capped with a thin layer of Al_2O_3 to act as an oxygen barrier. This prevents Cu_2O formation and thus avoids depleting Cu from the Cu_2S. A doping density of 10^{16} cm^{-3} also means that the absorber can be used in a p–n$^+$ configuration, such as those used in CIGSe technologies.[337] The challenge remains to see whether an oxygen barrier layer can be incorporated into device designs. Furthermore an n-type partner with suitable band alignment needs to be found that is thermodynamically stable in contact with Cu_2S. The experimental evidence of the instability of chalcocite needs to be borne in mind when considering the electronic properties given in Table 5.8, which were reported by Martinson et al.[314] for oxygen free conditions.

5.5.4.3   SnS

Tin sulfide has been investigated for use as an absorber layer since the late 1980s.[340] Three distinct chemical stoichiometries exist, namely SnS, Sn_2S_3, and SnS_2. The phase formed depends on the activity of sulfur available during the synthesis of the film, and care must be taken to avoid forming a mixture of phases. Only SnS is intrinsically p-type. There are conflicting reports on its optical gap, although a recent theoretical study by Vidal *et al.*[310] and several experimental studies[311,341] reach a consensus that the material has an indirect bandgap between 1.1 and 1.2 eV, and an effective absorption onset at around 1.5 to 1.6 eV.[311] Sinsermsuksakul *et al.*[312] reported that the electrical resistivity of SnS thin films is anisotropic, which they related to the presence of grain boundaries and a preferred crystal orientation in the film. They found the vertical resistivity to be 60 Ω cm, whilst the lateral resistivity was found to be 172 Ω cm for a sample prepared at 120 °C. Moreover they found a doping density in the plane of 10^{16} cm^{-3} which is a suitable value for an absorber layer in a standard p–n$^+$ device configuration.[312]

One of the earliest reports of a SnS solar cell is that of Noguchi *et al.* in 1994.[342] They reported a 0.29% power conversion efficiency with a SnS/CdS junction using a silver back contact and an ITO window layer. The SnS layer was grown by vacuum evaporation, probably leading to a film consisting of a mixture of SnS and Sn. Since then, the highest reported total area power conversion efficiency is 2.04% achieved in 2013 with a Mo/SnS/Zn(O,S)/ZnO/ITO structure.[312] The SnS was deposited by a pulsed chemical vapour deposition method. The external quantum efficiency spectra of the device appears to confirm a band edge around 1.2 eV, but the internal quantum efficiency only rises above 40% at 1.55 eV. However, the main problem appears to be the low open circuit voltage of 244 mV, pointing to high levels of recombination in the device both due to point defects and interface recombination.[312] Burton and Walsh[65] also point out that the valence band and conduction band of SnS lie considerably above other common p-type absorber layers, and therefore perhaps different contact layers are required. In summary, p-type SnS appears to present severe challenges that need to be overcome before it can be considered realistically as an absorber layer. The most obvious deficit is the indirect bandgap at 1.2 eV which leads to a low current collection.

5.5.4.4 Cu$_2$SnS$_3$

p-type Cu$_2$SnS$_3$ is one of 18 ternary phases observed within the Cu-Sn-S phase diagram.[343] Kuku and Fakolujo[317] reported the first Cu$_2$SnS$_3$/indium Schottky junction device with a 0.11% power conversion efficiency in 1987. Since then, the first p–n heterojunction device with 0.54% power conversion efficiency was reported by Berg *et al.*,[51] and most recently the group of Katagiri published a 2.84% efficient device.[52] A spread of bandgap values are reported for the Cu$_2$SnS$_3$, ranging from 0.93 to 1.35 eV.[105,343] Several authors report absorption coefficients greater than 10^4 cm^{-1}.[51,105,344] The origin of the differences in bandgap can be traced back to two different causes. Firstly, Cu$_2$SnS$_3$ is reported to have three different polymorphs which are monoclinic, cubic, and tetragonal. Secondly, as observed by Weber *et al.*,[234] Cu$_2$SnS$_3$ evaporates SnS into the gas phase if no partial pressure of SnS is already present. This leads to a change in stoichiometry and perhaps to formation of other ternary Cu-Sn-S phases such as Cu_4SnS_4. The different values of measured bandgap therefore reflect the different growth conditions used. Interestingly, Berg *et al.*[51] reported a monoclinic polymorph for their device, whilst the Katagiri group reported a cubic or monoclinic polymorph.[52,345] Evidently, the principal challenge with this material is to control the growth conditions to permit the properties of a particular phase pure polymorph to be studied.

Replacing some of the Sn in Cu_2SnS_3 with Ge increases the bandgap. Umehara *et al.*[318] showed that a Ge/(Sn + Ge) ratio of zero yielded a bandgap of 0.93 eV, whilst a ratio of 0.17 yielded a bandgap of 1.01 eV. Moreover, the addition of Ge into the absorber layer increased the device efficiency from 2.1 to 6.0%. All device parameters showed improvement with the addition of Ge; in particular there was a 100 mV gain in open circuit voltage.

In summary, Cu_2SnS_3 is relatively unstudied in comparison to the other alternative earth-abundant absorber layers described in this chapter, although a 6% efficient device has already been reported. Cu_2SnS_3 has a relatively low melting point of around 840 °C,[346] allowing higher quality crystal growth compared to other materials when limited to the same annealing temperature. On the other hand, the material has a very narrow phase space, many alternative Cu-Sn-S phases, and a volatile binary component, making control of the growth of the material challenging.

5.5.4.5 $CuSbS_2$

$CuSbS_2$, also known as the mineral chalcostibite, belongs to the space group no. 62 (Pnma) and exhibits an orthorhombic unit cell containing four formula units.[347] Despite the formula analogy, the structural parallelism of chalcopyrite with chalcostibite is not as strong as with kesterite. In fact, although Sb and In have almost the same ionic radius,[348] the coordination polyhedra of $CuSbS_2$ and $CuInS_2$ are very different.[349] Although Cu ions show the classical tetrahedral coordination, Sb is penta-coordinated by S ions forming square pyramidal SbS_5 units dictated by the accommodation of the Sb lone pairs.[54]

The bandgap of $CuSbS_2$ is reported to lie in the range 1.38–1.89 eV and is generally believed to be direct.[59,319–323] However, very recent DFT work[324] has suggested that its fundamental bandgap of 1.5 eV is indirect, but the band structure displays a direct gap just 0.1 eV higher, which explains why many experimental measurements indicate a direct transition. The absorption coefficient of $CuSbS_2$ exceeds 5×10^4 cm^{-1} between 0.2 eV[319] and 0.4 eV[324] above the bandgap, implying that a substantial amount of radiation can be absorbed within 1 μm thick films (*cf.* section 5.5.1.2). The high absorption coefficient has prompted the investigation of $CuSbS_2$ as a potential extremely thin absorber layer to be employed in solid state dye-sensitized solar cells and three-dimensional devices, and a measurable photovoltaic behaviour has been achieved.[350–352] A working device has also been reported for a p–i–n structure where $CuSbS_2$ acts as the p-layer,[353] while the first

$CuSbS_2$-based device with traditional p–n structure has been reported recently with a promising 3.1% power conversion efficiency.[325]

The Cu-Sb-S system shown in Figure 5.9 displays around 9 compositions, only three of which are ternary: $Cu_{12}Sb_4S_{13}$, Cu_3SbS_3 and $CuSbS_2$, and only the last two lie on the Cu_2S-Sb_2S_3 pseudo-binary joint. $CuSbS_2$ melts congruently at 551 °C.[354] Thin films of $CuSbS_2$ have been produced with a range of deposition techniques such as spray pyrolysis,[352] direct evaporation,[319,355] chemical bath deposition followed by interdiffusion annealing treatment,[320,353] as well as with the two-stage process, namely by electrodeposition/sulfurization.[59,325] The synthesis of phase-pure $CuSbS_2$ films seems to be straightforward even at temperature as low as 350 °C,[59] and this seems consistent with the relatively low melting temperature and associated likely high ion mobility. The defect chemistry of $CuSbS_2$ has been explored using lattice reactions combined with the electroneutrality condition,[356] and it is suggested that mixed electronic–ionic conduction is observed if $CuSbS_2$ deviates from stoichiometry.

The relatively low melting temperature of $CuSbS_2$ is also an indication that the compound may suffer from decomposition if treated at too high temperature. Evidence of Sb depletion from Sb_2S_3 films has been reported in the literature,[59] but—in contrast with other systems—excess chalcogen partial pressure seems to shift the equilibrium towards increased decomposition, due to a chemical vapour transport phenomenon[357] possibly associated with the formation of gaseous Sb_2S_4.[358] On the other hand, evaporated films of $CuSbS_2$[319] have been reported to undergo substantial morphology degradation and Sb depletion to the composition Cu_3SbS_3, when annealed in the absence of chalcogen up to 500 °C.[124] The thermochemistry of Sb_2S_3 has been investigated in order to gain insights into the decomposition phenomenon,[277] but this study was not comprehensive because it did not take into account all the 22 gaseous species that seem to be involved in the volatilization of Sb_2S_3.[124] It is clear that these aspects deserve further investigation. The relatively low melting temperature of $CuSbS_2$ implies that care should be taken during the synthesis so that compound decomposition is prevented, but it may also be an inherent advantage, since large grain size can be achieved with lower processing temperatures.

There is only one report on resistivity, mobility and doping density for $CuSbS_2$,[322] and the sub-optimal values highlight the fact that this is a little-studied material that requires further work. A recent computational investigation infers an alternative way to identify materials with suitable solar absorption properties.[359] The method defines a spectroscopic limited maximum efficiency (SLME), according to which materials such as $CuSbS_2$ and especially Cu_3SbS_4 (iso-electronic with CZTS[71]) are worth considering.

5.5.4.6 Cu_3BiS_3

Cu_3BiS_3 occurs naturally as the sulfosalt mineral wittichenite. The reported value of its bandgap ranges between 1.3 and 1.4 eV,[55,56,329,330] and it is

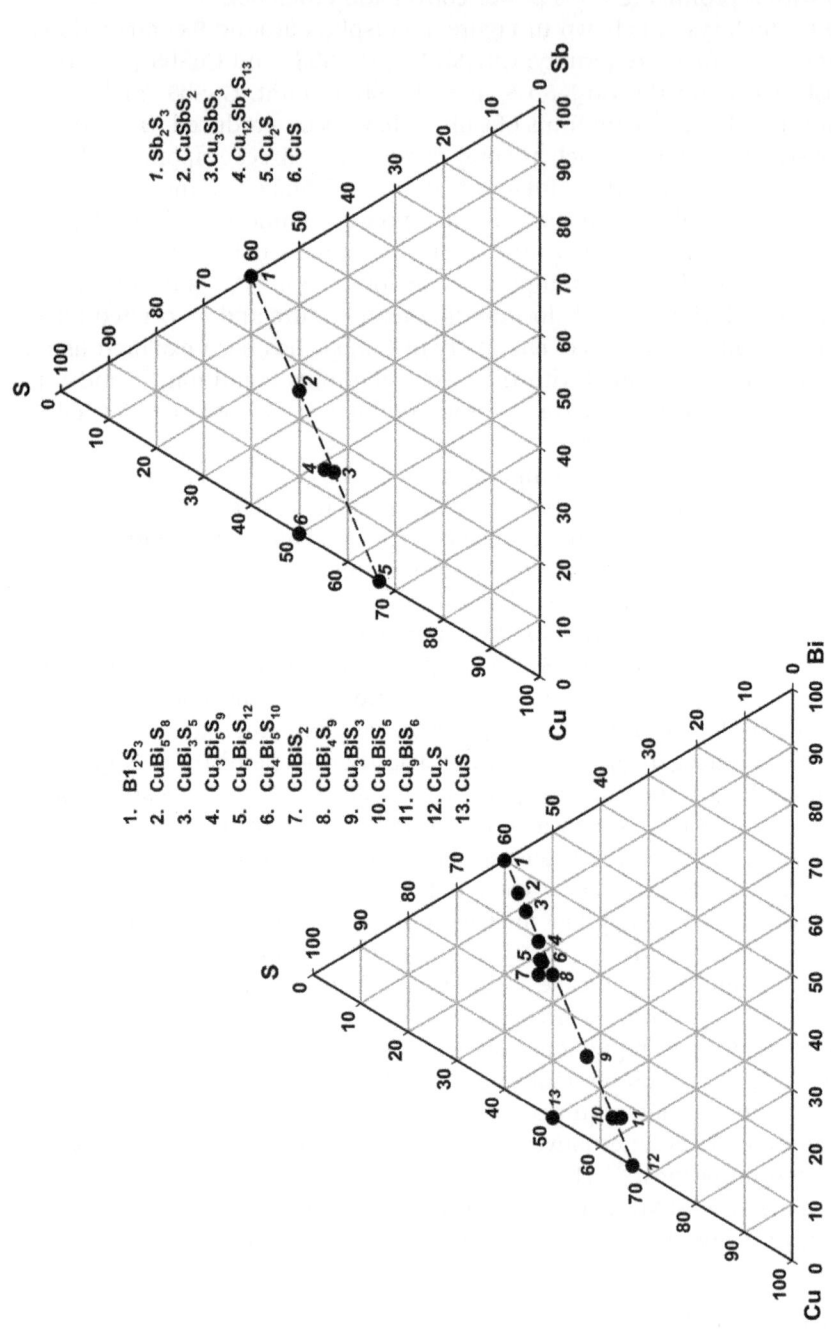

Figure 5.9 Cu-Sb-S and Cu-Bi-S phase diagrams showing the composition of the known binary and ternary sulfide phases.[124,145] The dashed lines are the pseudo-binary joints Sb_2S_3-Cu_2S and Bi_2S_3-Cu_2S.

therefore suitable for application in single heterojunction devices. Although its fundamental bandgap is reported to be direct forbidden,[56] its absorption coefficient exceeds 5×10^4 cm^{-1} less than 300 meV above the bandgap;[56] therefore light absorption does not seem to be an issue for this material. The potential application of Cu_3BiS_3 as a p-type absorber film was first considered by Nair *et al.*[360] Recently Mesa *et al.*[329] have synthesized Cu_3BiS_3 films with suitable electrical conductivity for thin film PV applications.

The crystalline structure of Cu_3BiS_3 cannot be derived by iso-electronic substitution from Si (*cf.* section 5.2.1); its low temperature polymorph belongs to the space group no. 19 (P2$_1$2$_1$2$_1$).[361] Its unit cell is orthorhombic ($a = 7.723$ Å, $b = 10.395$ Å, $c = 6.715$ Å)[362] and contains four formula units. From charge balance considerations within the formula, the oxidation states of Cu and Bi in Cu_3BiS_3 may be assumed to be (I) and (III), respectively. XPS analysis is consistent with the presence of Cu$^+$ ions,[363] while Bi^{3+} is the form that better fits its ionic radius, by comparison with Bi_2S_3.[364,365]

According to Sugaki and Shima,[366] Cu_3BiS_3 is an incongruent melting compound, like CZTS, but its decomposition temperature is much lower (527 °C). Furthermore, from the reported Cu_2S-Bi_2S_3 phase diagram, there is evidence of a limited stoichiometric deviation with Cu_2S loss in the high temperature range, which may hinder the synthesis of stoichiometric films by high-temperature processing. It was reported that Cu_3BiS_3 shows a series of phase transitions at temperatures lower than those generally employed for the annealing of CISe and CZTSe. However, the lattice parameters are only slightly affected by the phase transition and the unit cell volume remains nearly constant.[367] The Cu-Bi-S system is rather complex (see Figure 5.5.) with at least 16 possible compositions, six of which are ternary phases lying on the pseudo-binary joint Cu_2S-Bi_2S_3. Despite this complexity, producing a reasonably phase-pure Cu_3BiS_3 film is surprisingly simple,[56,330,360] even if the synthesis is performed with a two-stage process.[55,277] However, the relatively low decomposition temperature of Cu_3BiS_3 (527 °C) implies that synthesis or processing at high temperature may require control of the atmospheric composition to prevent or minimise detrimental effects. Indeed, early works on the sulfurization of Cu-Bi metal precursors with H_2S by Gerein and Haber[286] reported Bi and S depletion during thermal treatment, leaving only Cu_2S if the treatment was performed in vacuum at 600 °C. As a result of this observation, the processing temperature was restricted to 270 °C, which required very long times (>16 h) for the phase to be formed completely. Nevertheless, recent studies of the solid–gas decomposition equilibria of Cu_3BiS_3 suggest that Bi and S depletion can be suppressed by sulfur excess in the atmosphere.[277] These studies were also supported by thermochemical computation of the extent of the possible decomposition reactions as a function of temperature and sulfur partial pressure, based on the thermodynamic properties of Bi_2S_3:[287]

$$Bi_2S_{3(s)} \rightarrow 2BiS_{(g)} + \tfrac{1}{2}S_{2(g)} \qquad (5.11)$$

$$Bi_2S_{3(s)} \rightarrow 2Bi_{(g)} + \tfrac{3}{2}S_{2(g)} \tag{5.12}$$

$$Bi_2S_{3(s)} \rightarrow Bi_{2(g)} + \tfrac{3}{2}S_{2(g)} \tag{5.13}$$

The stoichiometry of each of the above reactions determines the effect of the chalcogen excess on the decomposition suppression. The $p(S_{2(g)})$–T field was analysed and five regions were identified where all the above decomposition reactions provide a relatively different contribution to the Bi and S losses while occurring simultaneously. It was concluded that macroscopic losses of Bi are likely due to the release of $Bi_{2(g)}$ if processing is performed under vacuum at 500 °C. On the other hand, it was found that Le Chatelier mass action of $S_{2(g)}$ alone is sufficient to minimise Bi losses. No appreciable Bi depletion could be detected on films of Cu_3BiS_3 processed at 550 °C, even up to 16 hours, if a $S_{2(g)}$ partial pressure of 23 mbar was maintained.[55]

In summary, synthesis of Cu_3BiS_3 is relatively facile, but the potential of this compound as an alternative p-type layer still needs to be demonstrated. More research needs to be directed towards the optimization of optoelectronic properties such as conductivity, mobility and doping density.

5.5.4.7 Outlook for Other Earth-abundant Absorber Layers

Clearly the CZTS(Se) system is currently the most high profile earth-abundant absorber material for thin film solar cells. The properties of six other chalcogen containing potential absorber layer materials have been reviewed above. The materials reviewed were chosen for having an optimal bandgap for a single p–n heterojunction solar cell. Cu-S materials have so far reached the highest power conversion efficiency (10.1%) and even pilot line production. The process was abandoned due to the instability of Cu-S, and any future research into the material must overcome the weak bonding of Cu in the compound and its propensity to diffuse. FeS_2 has achieved a 5.5% power conversion efficiency in a liquid electrolyte device in the 1980s. However, in spite of renewed interest in FeS_2 in recent years, there are no reports of better functioning devices. Challenges with this material include the need to reduce doping density and to understand an apparent type inversion at the surface. SnS is similarly well studied, and recently a 2% device was reported. The material suffers from an indirect bandgap and low open circuit voltages. Cu_2SnS_3 is relatively unstudied with around 20 publications at the time this chapter was written, but a device with 6% power conversion efficiency has been demonstrated when the material is alloyed with germanium. The obstacle with this material appears to be controlling the phase purity. Despite their suitable bandgaps and very high absorption coefficients, $CuSbS_2$ and Cu_3BiS_3 are also relatively unstudied, the only reported power conversion efficiency is a recent 3.1% for a $CuSbS_2$-based device.[325] Their relatively low melting/decomposition temperatures make the synthesis of phase pure $CuSbS_2$ and Cu_3BiS_3 films reasonably facile if the annealing is not performed under strong vacuum or at too high temperature.

For these two materials more research needs to be directed towards the optimization of properties such as conductivity, mobility and doping density. Each material has its own set of challenges, and at this stage it is not easy to predict which materials will be successful. Besides optimization of the absorber layers themselves, finding suitable back contact and/or window layers will still be necessary in most cases. It remains to be seen whether any of the materials reviewed will eventually be able to compete with CZTS(Se).

5.6   Summary and Outlook

This chapter has attempted to provide deeper insights into a range of topics that are relevant to the development of earth-abundant PV. CZTS(Se) is certainly the most promising of the materials reviewed here, but several problems remain to be solved before cells can be fabricated with efficiencies approaching those of CIGSe. Key issues include identification and elimination of the losses that lead to reduced open circuit voltage (including the effects of secondary phases), the need for compatible buffer layers to replace CdS, and effective ways of stabilizing the back contact. The learning curve for CZTS(Se) is remarkably similar to that for CIGSe, and we may therefore reasonably expect rapid progress as the international research effort intensifies. The situation with the other chalcogenide materials reviewed is less dynamic, but it is possible that some of them may move centre stage as research continues.

This chapter has been written from an academic perspective, and it is therefore reasonable to ask how soon the PV industry might take up new materials such as CZTS(Se). While the academic PV community is increasingly turning its attention to new materials as part of a forward-looking strategy, the PV industry has been experiencing difficulties in the last few years that have impacted on the development of new technologies. However, there can be no doubt that in the longer term the kind of work described here will provide a platform for technological development. The authors hope that this chapter will inspire researchers to tackle the remaining problems so that earth-abundant photovoltaics can begin to compete with current thin film PV based on rare elements or indeed with highly developed (but more energy intensive) silicon PV.

Acknowledgements

D.C. and P.D. acknowledge funding from the EU commission through the grants No. EP/F029624/1 (EPSRC) and No. 284486. P.D. acknowledges the Fonds National de la Recherche du Luxembourg (ATTRACT Fellowship Grant 07/06). J.S. acknowledges the Swedish Energy Agency (Energimyndigheten) and Research Council (Vetenskapsrådet) for funding his contribution to this work. Su.S. acknowledges the Fonds National de la Recherche du Luxembourg for several projects, enabling the work on kesterite solar cells.

References

1. *BP Statistical Review of World Energy June 2013*, BP plc, 2013.
2. P. B. D. Arvizu, L. Cabeza, T. Hollands, A. Jäger-Waldau, M. Kondo, C. Konseibo, V. Meleshko, W. Stein, Y. Tamura, H. Xu and R. Zilles, in *IPCC Special Report on Renewable Energy Sources and Climate Change Mitigation*, ed. R. P.-M. O. Edenhofer, Y. Sokona, K. Seyboth, P. Matschoss, S. Kadner, T. Zwickel, P. Eickemeier, G. Hansen, S. Schlömer and C. von Stechow, Cambridge University Press, Cambridge, 2011.
3. S. N. P. Frankl, M. Gutschner, S. Gnos and T. Rinke, *Technology Roadmap Solar Photovoltaic Energy*, International Energy Agency, 2010.
4. *PVPS Report. A Snapshot of Global PV 1992-2012*, International Energy Agency, 2013.
5. V. M. Fthenakis and H. C. Kim, *Solar Energy*, 2011, **85**, 1609–1625.
6. *Critical Raw Materials for the EU*, European Commission, 2010.
7. *US Department of Energy Critical Materials Strategy*, US Department of Energy, 2011.
8. M. A. Green, *Prog. Photovoltaics Res. Appl.*, 2009, **17**, 347–359.
9. U.S. Geological Survey, *Mineral Commodity Summaries*, January 2012; Gallium, 2012.
10. U.S. Geological Survey, *Mineral Commodity Summaries*, January 2012; Indium, 2012.
11. U.S. Geological Survey, *Mineral Commodity Summaries*, January 2012; Tellurium, 2012.
12. I. Forbes and L. M. Peter, in *Materials for a Sustainable Future*, ed. T. Letcher and J. Scott, Royal Society of Chemistry, Cambridge, 2012, pp. 558–591.
13. A. Anctil and V. Fthenakis, *Prog. Photovoltaics Res. Appl.*, 2013, **21**, 1253–1259.
14. W. Shockley and H. J. Queisser, *J. Appl. Phys.*, 1961, **32**, 510–519.
15. W. Schafer and R. Nitsche, *Mater. Res. Bull.*, 1974, **9**, 645–654.
16. N. Nakayama and K. Ito, *Appl. Surf. Sci.*, 1996, **92**, 171–175.
17. K. Ito and T. Nakazawa, *Jpn. J. Appl. Phys.*, 1988, **27**(Part 1), 2094–2097.
18. H. Katagiri, M. Nishimura, T. Onozawa, S. Maruyama, M. Fujita, T. Sega and T. Watanabe, Proceedings of the Power Conversion Conference - Nagaoka 1997, Vols I and II, 1997, pp. 1003–1006.
19. H. Katagiri, N. Sasaguchi, S. Hando, S. Hoshino, J. Ohashi and T. Yokota, *Sol. Energy Mater. Sol. Cells*, 1997, **49**, 407–414.
20. T. M. Friedlmeier, H. Dittrich and H. W. Schock, in *Ternary and Multinary Compounds*, ed R. D. Tomlinson, A. E. Hill and R. D. Pilkington. Institute of Physics, Bristol, 1998, vol. 152, pp. 345–345.
21. H. Katagiri, K. Saitoh, T. Washio, H. Shinohara, T. Kurumadani and S. Miyajima, *Sol. Energy Mater. Sol. Cells*, 2001, **65**, 141–145.
22. H. Katagiri, K. Jimbo, K. Moriya and K. Tsuchida, Proceedings of 3rd World Conference on Photovoltaic Energy Conversion, Vols A-C, 2003, 2874–2879.

23. D. B. Mitzi, T. K. Todorov, O. Gunawan, M. Yuan, Q. Cao, W. Liu, K. B. Reuter, M. Kuwahara, K. Misumi, A. J. Kellock, S. J. Chey, T. G. de Monsabert, A. Prabhakar, V. Deline, K. E. Fogel, in *35th IEEE Photovoltaic Specialists Conference*, 2010, pp. 640–645.
24. T. Todorov and D. B. Mitzi, *Eur. J. Inorg. Chem.*, 2010, 17–25.
25. T. K. Todorov, K. B. Reuter and D. B. Mitzi, *Adv. Mater.*, 2010, **22**, E156–E159.
26. O. Gunawan, T. K. Todorov and D. B. Mitzi, *Appl. Phys. Lett.*, 2010, **97**, 233506.
27. D. B. Mitzi, O. Gunawan, T. K. Todorov and D. A. R. Barkhouse, *Philos Trans. R. Soc. London Ser A*, 2013, **371**, 20110432.
28. T. K. Todorov, J. Tang, S. Bag, O. Gunawan, T. Gokmen, Y. Zhu and D. B. Mitzi, *Adv. Energy Mater.*, 2013, **3**, 34–35.
29. I. L. Repins, M. J. Romero, J. V. Li, S.-H. Wei, D. Kuciauskas, C.-S. Jiang, C. Beall, C. DeHart, J. Mann, W.-C. Hsu, G. Teeter, A. Goodrich and R. Noufi, *IEEE J. Photovoltaics*, 2013, **3**, 439–445.
30. S. Siebentritt, *Thin Solid Films*, 2013, **535**, 1–4.
31. H. Katagiri and Ieee, 3rd International Conference on Photonics 2012, 2012, pp. 345–349.
32. A. Walsh, S. Chen, S.-H. Wei and X.-G. Gong, *Adv. Energy Mater*, 2012, **2**, 400–409.
33. J. J. Scragg, T. Ericson, X. Fontané, V. Izquierdo-Roca, A. Pérez-Rodríguez, T. Kubart, M. Edoff and C. Platzer-Björkman, *Prog. Photovoltaics Res. Appl.*, 2012, DOI: 10.1002/pip.2265.
34. G. Zoppi, I. Forbes, R. W. Miles, P. J. Dale, J. J. Scragg and L. M. Peter, *Prog. Photovoltaics*, 2009, **17**, 315–319.
35. A. Redinger, M. Mousel, R. Djemour, L. Gütay, N. Valle and S. Siebentritt, *Prog. Photovoltaics Res. Appl.*, 2013, DOI: 10.1002/pip.2324.
36. H. Katagiri, K. Jimbo, W. S. Maw, K. Oishi, M. Yamazaki, H. Araki and A. Takeuchi, *Thin Solid Films*, 2009, **517**, 2455–2460.
37. W.-C. Hsu, I. Repins, C. Beall, C. DeHart, B. To, W. Yang, Y. Yang and R. Noufi, *Prog. Photovoltaics Res. Appl.*, 2012, DOI: 10.1002/pip.2296.
38. J. J. Scragg, P. J. Dale and L. M. Peter, *Electrochem. Commun.*, 2008, **10**, 639–642.
39. J. J. Scragg, P. J. Dale and L. M. Peter, *Thin Solid Films*, 2009, **517**, 2481–2484.
40. J. J. Scragg, D. M. Berg and P. J. Dale, *J. Electroanal. Chem.*, 2010, **646**, 52–59.
41. W. Septina, S. Ikeda, A. Kyoraiseki, T. Harada and M. Matsumura, *Electrochim. Acta*, 2013, **88**, 436–442.
42. Y. Sun, Y. Zhang, H. Wang, M. Xie, K. Zong, H. Zheng, Y. Shu, J. Liu, H. Yan, M. Zhu and W. Lau, *J. Mater. Chem. A*, 2013, **1**, 6880–6887.
43. D. Tiwari, T. K. Chaudhuri, T. Shripathi, U. Deshpande and R. Rawat, *Sol. Energy Mater. Sol. Cells*, 2013, **113**, 165–170.
44. T. Schnabel, M. Löw and E. Ahlswede, *Sol. Energy Mater. Sol. Cells*, 2013, **117**, 324–325.

45. H. Park, Y. H. Hwang and B.-S. Bae, *J. Sol-Gel Sci. Technol.*, 2013, **65**, 23–27.
46. O. Vigil-Galán, M. Espíndola-Rodríguez, M. Courel, X. Fontané, D. Sylla, V. Izquierdo-Roca, A. Fairbrother, E. Saucedo and A. Pérez-Rodríguez, *Sol. Energy Mater. Sol. Cells*, 2013, **117**, 246–250.
47. T. Washio, T. Shinji, S. Tajima, T. Fukano, T. Motohiro, K. Jimbo and H. Katagiri, *J. Mater. Chem.*, 2012, **22**, 4021–4024.
48. Z. Su, K. Sun, Z. Han, F. Liu, Y. Lai, J. Li and Y. Liu, *J. Mater. Chem.*, 2012, **22**, 16346–16352.
49. D. Avellaneda, M. T. S. Nair and P. K. Nair, *J. Electrochem. Soc.*, 2010, **157**, D346–D352.
50. M. Bouaziz, J. Ouerfelli, S. K. Srivastava, J. C. Bernede and M. Amlouk, *Vacuum*, 2011, **85**, 783–786.
51. D. M. Berg, R. Djemour, L. Guetay, G. Zoppi, S. Siebentritt and P. J. Dale, *Thin Solid Films*, 2012, **520**, 6291–6294.
52. J. Koike, K. Chino, N. Aihara, H. Araki, R. Nakamura, K. Jimbo and H. Katagiri, *Jpn. J. Appl. Phys.*, 2012, **51**, 10NC34.
53. H. Dittrich, A. Stadler, D. Topa, H.-J. Schimper and A. Basch, *Physica Status Solidi (A)*, 2009, **206**, 1034–1041.
54. J. T. R. Dufton, A. Walsh, P. M. Panchmatia, L. M. Peter, D. Colombara and M. S. Islam, *Phys. Chem. Chem. Phys.*, 2012, **14**, 7229–7233.
55. D. Colombara, L. M. Peter, K. Hutchings, K. D. Rogers, S. Schaefer, J. T. R. Dufton and M. S. Islam, *Thin Solid Films*, 2012, **520**, 5165–5171.
56. N. J. Gerein and J. A. Haber, *Chem. Mater.*, 2006, **18**, 6297–6302.
57. N. J. Gerein and J. A. Haber, *Chem. Mater.*, 2006, **18**, 6289–6296.
58. M. Kumar and C. Persson, *Appl. Phys. Lett.*, 2013, **102**, 062109.
59. D. Colombara, L. M. Peter, K. D. Rogers, J. D. Painter and S. Roncallo, *Thin Solid Films*, 2011, **519**, 7438–7443.
60. R. B. Hall, R. W. Birkmire, J. E. Phillips and J. D. Meakin, *Appl. Phys. Lett.*, 1981, **38**, 925–926.
61. K. H. Norian and R. B. Hall, *Thin Solid Films*, 1982, **88**, 55–66.
62. M. T. S. Nair, E. Barrios-Salgado, A. Rosa Garcia, M. Rebeca Aragon-Silva, J. Campos and P. K. Nair, in *Photovoltaics for the 21st Century 7*, ed. M. Tao, L. Deligianni, K. Rajeshwar, C. Claeys, J. G. Park and M. Sunkara, 2011, vol. 41, pp. 177–183.
63. L. A. Burton and A. Walsh, *J. Phys. Chem. C*, 2012, **116**, 24262–24267.
64. L. M. Peter, *J. Electroanal. Chem.*, 1979, **98**, 49–55.
65. L. A. Burton and A. Walsh, *Appl. Phys. Lett.*, 2013, **102**, 132111.
66. T. Muto, G. Larramona and G. Dennler, *Appl. Phys. Express*, 2013, **6**, 072301.
67. A. Ennaoui, S. Fiechter, C. Pettenkofer, N. Alonsovante, K. Buker, M. Bronold, C. Hopfner and H. Tributsch, *Sol. Energy Mater. Sol. Cells*, 1993, **29**, 289–370.
68. N. Berry, M. Cheng, C. L. Perkins, M. Limpinsel, J. C. Hemminger and M. Law, *Adv. Energy Mater.*, 2012, **2**, 1124–1135.
69. B. R. Pamplin, *Nature*, 1960, **188**, 136–137.

70. B. R. Pamplin, *J. Phys. Chem. Solids*, 1964, **25**, 675–684.
71. C. H. L. Goodman, *J. Phys. Chem. Solids*, 1958, **6**, 305–314.
72. H. G. Deming, *General Chemistry*, Wiley, New York, 1952.
73. E. Fluck, *Pure Appl. Chem.*, 1988, **60**, 431–436.
74. B. R. Pamplin, *Prog. Crystal Growth Characterization*, 1981, **3**, 179.
75. J. Walker, *Rep. Prog. Phys.*, 1979, **42**, 1605.
76. M. A. Green, *J. Appl. Phys.*, 1990, **67**, 2944–2954.
77. *Group IV Elements, IV-IV and III-V Compounds. Part b - Electronic, Transport, Optical and Other Properties - Group III Condensed Matter* Volume 41A1b, Landolt-Börnstein, 2002.
78. *Handbook of Chemistry and Physics*, 89th edn., CRC/Taylor and Francis, Boca Raton, 2009.
79. R. A. Soref and C. H. Perry, *J. Appl. Phys.*, 1991, **69**, 539–541.
80. M. Pairot, I. Zoran and H. Paul, *Semicond. Sci. Technol.*, 2007, **22**, 742.
81. O. Gurdal, M.-A. Hasan, J. M. R. Sardela, J. E. Greene, H. H. Radamson, J. E. Sundgren and G. V. Hansson, *Appl. Phys. Lett.*, 1995, **67**, 956–955.
82. W. Wegscheider, K. Eberl, U. Menczigar and G. Abstreiter, *Appl. Phys. Lett.*, 1990, **57**, 875–877.
83. P. Rodríguez-Hernández, M. González-Diaz and A. Muñoz, *Phys. Rev. B*, 1995, **51**, 14705–14705.
84. G. L. W. Hart and A. Zunger, *Phys. Rev. B*, 2000, **62**, 13522–13537.
85. Y. Yao, D. König and M. Green, *Sol. Energy Mater. Sol. Cells*, 2013, **111**, 123–126.
86. J. Singh, *Electronic and Optoelectronic Properties of Semiconductor Structures*, Cambridge University Press, Cambridge, 2003.
87. D. W. Palmer, Available: www.semiconductors.co.uk, 2005.
88. O. Kanehisa, M. Shiiki, M. Migita and H. Yamamoto, *J. Cryst. Growth*, 1990, **86**, 367–371.
89. J. Gutowski, I. Broser and G. Kudlek, *Phys. Rev. B*, 1989, **39**, 3670–3676.
90. M. Cardona, *J. Appl. Phys.*, 1961, **32**, 2151–2155.
91. H. Venghaus, *Phys. Rev. B*, 1979, **19**, 3071–3082.
92. M. Ichimura, F. Goto and E. Arai, *J. Appl. Phys.*, 1999, **85**, 7411–7417.
93. S. Chichibu, A. Iwai, S. Matsumoto and H. Higuchi, *J. Cryst. Growth*, 1993, **126**, 635–642.
94. T. J. Alwan and M. A. Jabbar, *Turk. J. Phys.*, 2010, **34**, 107–116.
95. L. Barkat, N. Hamdadou, M. Morsli, A. Khelil and J. C. Bernède, *J. Cryst. Growth*, 2006, **297**, 426–431.
96. M. Steichen, J. Larsen, L. Gütay, S. Siebentritt and P. J. Dale, *Thin Solid Films*, 2011, **519**, 7254–7255.
97. A. Soni, V. Gupta, C. M. Arora, A. Dashora and B. L. Ahuja, *Sol. Energy*, 2010, **84**, 1481–1489.
98. A. Rockett and R. W. Birkmire, *J. Appl. Phys.*, 1991, **70**, R81–R97.
99. T. Tinoco, C. Rincón, M. Quintero and G. S. Pérez, *Physica Status Solidi (A)*, 1991, **124**, 427–434.
100. B. Tell, J. L. Shay and H. M. Kasper, *Phys. Rev. B*, 1971, **4**, 2463–2471.

101. B. Tell and H. M. Kasper, *Phys. Rev. B*, 1971, **4**, 4455–4459.

102. O. V. Parasyuk, L. V. Piskach, Y. E. Romanyuk, I. D. Olekseyuk, V. I. Zaremba and V. I. Pekhnyo, *J. Alloys Compd.*, 2005, **397**, 85–94.

103. *Semiconductors - Group III Condensed Matter Volume 41E*, Landolt-Börnstein, 2000.

104. B. K. Sarkar, A. S. Verma and P. S. Deviprasad, *Physica B-Condensed Matter*, 2011, **406**, 2847–2850.

105. P. A. Fernandes, P. M. P. Salome and A. F. da Cunha, *J. Phys. D-Applied Physics*, 2010, **43**, 215403.

106. K. M. Kim, H. Tampo, H. Shibata and S. Niki, *Thin Solid Films*, 2013, **536**, 111–114.

107. N. B. M. Amiri and A. Postnikov, *J. Appl. Phys.*, 2012, **112**, 033719–033712.

108. G. Marcano, C. Rincon, L. M. de Chalbaud, D. B. Bracho and G. S. Perez, *J. Appl. Phys.*, 2001, **90**, 1847–1853.

109. M. Leon, S. Levcenko, R. Serna, G. Gurieva, A. Nateprov, J. M. Merino, E. J. Friedrich, U. Fillat, S. Schorr and E. Arushanov, *J. Appl. Phys.*, 2010, **108**, 093502–093505.

110. S. Levcenco, D. Dumcenco, Y. S. Huang, K. K. Tiong and C. H. Du, *Opt. Mater.*, 2011, **34**, 183–185.

111. H. Matsushita, T. Ochiai and A. Katsui, *J. Cryst. Growth*, 2005, **275**, e995–e999.

112. S. Siebentritt and S. Schorr, *Prog. Photovoltaics*, 2012, **20**, 512–519.

113. X. Liang, XianhuaWei and D. Pan, *J. Nanomater.*, 2012, **2012**, 708645.

114. F. W. Ohrendorf, *Präparative Arbeiten und Gitterdynamische Rechnungen an Verbindungen mit Tetraederstrukturen*, Herbert Utz, München.

115. S. Schorr, *Thin Solid Films*, 2007, **515**, 5985–5991.

116. G. Springer, *Can. Mineral.*, 1972, **11**, 535–541.

117. G. P. Bernardini, P. Bonazzi, M. Corazza, F. Corsini, G. Mazzetti, L. Poggi and G. Tanelli, *Eur. J. Mineral.*, 1990, **2**, 219–225.

118. S. A. Kissin and D. R. Owens, *Can. Mineral.*, 1979, **17**, 125–135.

119. S. Schorr, *Sol. Energy Mater. Sol. Cells*, 2011, **95**, 1482–1485.

120. S. R. Hall, S. A. Kissin and J. M. Stewart, *Acta Crystallogr. Sect. A*, 1975, **31**, S67–S67.

121. S. R. Hall, J. T. Szymanski and J. M. Stewart, *Can. Mineral.*, 1978, **16**, 131–132.

122. S. Schorr, H.-J. Hoebler and M. Tovar, *Eur. J. Mineral.*, 2007, **19**, 65–73.

123. P. J. Brown, A. G. Fox, E. N. Maslen, M. A. O'Keefe and B. T. M. Willis, in *International Tables for Crystallography*, ed. E. Prince, International Union of Crystallography, 2006, vol. C, pp. 554–595.

124. D. Colombara, PhD thesis, University of Bath, 2012.

125. D. Colombara, S. Delsante, G. Borzone, J. M. Mitchels, K. C. Molloy, L. H. Thomas, B. G. Mendis, C. Y. Cummings, F. Marken and L. M. Peter, *J. Cryst. Growth*, 2013, **364**, 101–110.

126. in *International Tables for Crystallography - Volume A, Space-group symmetryed* T. Hahn, International Union of Crystallography, 2006, vol. A.

127. M. Ichimura and Y. Nakashima, *Jpn. J. Appl. Phys.*, 2009, **48**, 090202.
128. J. Paier, R. Asahi, A. Nagoya and G. Kresse, *Phys. Rev. B*, 2009, **79**, 115126.
129. S. Chen, X. G. Gong, A. Walsh and S.-H. Wei, *Appl. Phys. Lett.*, 2009, **94**, 041903–041903.
130. J. J. M. Binsma, L. J. Giling and J. Bloem, *J. Cryst. Growth*, 1980, **50**, 429–436.
131. I. D. Olekseyuk, I. V. Dudchak and L. V. Piskach, *J. Alloys Compd.*, 2004, **368**, 135–143.
132. A. Lafond, L. Choubrac, C. Guillot-Deudon, P. Deniard and S. Jobic, *Zeitschrift Fur Anorganische Und Allgemeine Chemie*, 2012, **638**, 2571–2577.
133. T. Haalboom, T. Gödecke, F. Ernst, M. Rühle, R. Herberholz, H. W. Schock, C. Beilharz and K. W. Benz, in *Proceedings of the 11th Conference on Ternary and Multinary Compounds*, Institute of Physics, Salford, 1998, pp. 249–252.
134. L. S. Palatnik and E. I. Rogacheva, *Sov. Phys. Dokl., (Engl. Transl.)*, 1967, **12**, 503.
135. C. W. Folmer, J. A. Turner, R. Noufi and D. Cohen, *J. The Electrochem. Soc*, 1985, **132**, 1319–1327.
136. M. L. Fearheiley, *Sol. Cells*, 1986, **16**, 91–100.
137. U. C. Boehnke and G. Kühn, *J. Mater. Sci.*, 1987, **22**, 1635–1641.
138. K. J. Bachmann, H. Goslowsky and S. Fiechter, *J. Cryst. Growth*, 1988, **89**, 160–164.
139. T. Gödecke, T. Haalboom and F. Ernst, *Z. Metallkd.*, 2000, **91**, 622–634.
140. S. Jianyun, W. K. Kim, S. Shunli, C. Maoyou, C. Song and T. J. Anderson, *Rare Metals*, 2006, **25**, 481–487.
141. E. M. Kartzmark, *J. Chem. Educ.*, 1980, **57**, 125–126.
142. I. V. Dudchak and L. V. Piskach, *J. Alloys Compd.*, 2003, **351**, 145–150.
143. T. Schwarz, O. Cojocaru-Miredin, P. Choi, M. Mousel, A. Redinger, S. Siebentritt and D. Raabe, *Appl. Phys. Lett.*, 2013, **102**, 042101.
144. S. Chen, X. G. Gong, A. Walsh and S.-H. Wei, *Appl. Phys. Lett.*, 2010, **96**, 021902–021903.
145. P. Villars, A. Prince and H. Okamoto, *Handbook of Ternary Alloys Phase Diagrams*, ASM International, 1995.
146. M. Grossberg, J. Krustok, J. Raudoja, K. Timmo, M. Altosaar and T. Raadik, *Thin Solid Films*, 2011, **517**, 2489–2492.
147. P. J. Dale, M. Arasimowicz, D. Colombara, A. Crossay, E. Robert and A. A. Taylor, in *MRS Spring Meeting - Invited contribution*, 2013.
148. A. Shavel, J. Arbiol and A. Cabot, *J. Am. Chem. Soc.*, 2010, **132**, 4514–4515.
149. T. Tanaka, T. Sueishi, K. Saito, Q. Guo, M. Nishio, K. M. Yu and W. Walukiewicz, *J. Appl. Phys.*, 2012, **111**, 053522–053524.
150. Y. Hashimoto, N. Kohara, T. Negami, M. Nishitani and T. Wada, *Jpn. J. Appl. Phys.*, 1996, **35**, 4760–4764.
151. G. Juška, V. Gulbina and A. Jagminas, *Lithuanian J. Phys.*, 2010, **50**, 233–239.

152. A. Redinger, K. Hones, X. Fontane, V. Izquierdo-Roca, E. Saucedo, N. Valle, A. Perez-Rodriguez and S. Siebentritt, *Appl. Phys. Lett.*, 2011, **98**, 101907.

153. B. Pejova and I. Grozdanov, *J. Solid State Chem.*, 2001, **158**, 49–54.

154. S. Venkatachalam, S. Agilan, D. Mangalaraj and S. K. Narayandass, *Mater. Sci. Semiconduct. Proc*, 2007, **10**, 128–132.

155. J. T. Watjen, J. Engman, M. Edoff and C. Platzer-Bjorkman, *Appl. Phys. Lett.*, 2012, **100**, 173510.

156. W.-C. Hsu, I. Repins, C. Beall, C. DeHart, G. Teeter, B. To, Y. Yang and R. Noufi, *Sol. Energy Mater. Sol. Cells*, 2013, **113**, 160–164.

157. D. Colombara, E. V. C. Robert, A. Crossay, A. Taylor, M. Guennou, M. Arasimowicz, J. Malaquias, R. Djemour and P. J. Dale, *Sol. Energy Mater. Sol. Cells*, 2013, **23**, 220–227.

158. M. Mousel, A. Redinger, R. Djemour, M. Arasimowicz, N. Valle, P. Dale and S. Siebentritt, *Thin Solid Films*, 2013, **535**, 83–87.

159. N. Vora, J. Blackburn, I. Repins, C. Beall, B. To, J. Pankow, G. Teeter, M. Young and R. Noufi, *J. Vac. Sci. Technol. A Vacuum, Surfaces, Films*, 2012, **30**, 051201.

160. R. Djemour, M. Mousel, A. Redinger, L. Guetay, A. Crossay, D. Colombara, P. J. Dale and S. Siebentritt, *Appl. Phys. Lett.*, 2013, **102**, 222108.

161. See reference 157.

162. L. Amalraj, M. Jayachandran and C. Sanjeeviraja, *Mater. Res. Bull.*, 2004, **39**, 2193–2201.

163. D. Martínez-Escobar, M. Ramachandran, A. Sánchez-Juárez and J. S. Narro Rios, *Thin Solid Films*, 2013, **535**, 390–393.

164. M. Grossberg, J. Krustok, K. Timmo and M. Altosaar, *Thin Solid Films*, 2009, **517**, 2489–2492.

165. S. Levcenco, D. Dumcenco, Y. P. Wang, Y. S. Huang, C. H. Ho, E. Arushanov, V. Tezlevan and K. K. Tiong, *Opt. Mater.*, 2012, **34**, 1362–1365.

166. S. Chen, A. Walsh, J.-H. Yang, X. G. Gong, L. Sun, P.-X. Yang, J.-H. Chu and S.-H. Wei, *Phys. Rev. B*, 2011, **83**, 125201.

167. P. K. Sarswat and M. L. Free, *Physica B: Condensed Matter*, 2012, **407**, 108–111.

168. R. Scheer and H. W. Schock, *Chalcogenide Photovoltaics. Physics, Technologies and Thin Film Devices*, Wiley-VCH, Weinheim, 2011.

169. C. J. Hages, J. Moore, S. Dongaonkar, M. Alam, M. Lundstrom and R. Agrawal, 38th IEEE Photovoltaic Specialist Conference, Austin, Tx. 2012.

170. Q. J. Guo, G. M. Ford, W. C. Yang, C. J. Hages, H. W. Hillhouse and R. Agrawal, *Sol. Energy Mater. Sol. Cells*, 2012, **105**, 132–136.

171. S. Bag, O. Gunawan, T. Gokmen, Y. Zhu and D. B. Mitzi, *Chem. Mater.*, 2012, **24**, 4588–4593.

172. Q. Shu, J.-H. Yang, S. Chen, B. Huang, H. Xiang, X.-G. Gong and S.-H. Wei, *Phys. Rev. B*, 2013, **87**, 115205.

173. H. Matsushita, T. Maeda, A. Katsui and T. Takizawa, *J. Cryst. Growth*, 2000, **208**, 416–422.

174. G.-Q. Yao, H.-S. Shen, E. D. Honig, K. R. K. Dwight and A. Wold, *Solid State Ionics*, 1987, **24**, 249–252.

175. S. Ahn, S. Jung, J. Gwak, A. Cho, K. Shin, K. Yoon, D. Park, H. Cheong and J. H. Yun, *Appl. Phys. Lett.*, 2010, **97**, 021905.

176. T. Tanaka, T. Nagatomo, D. Kawasaki, M. Nishio, Q. X. Guo, A. Wakahara, A. Yoshida and H. Ogawa, *J. Phys. Chem. Solid.*, 2005, **66**, 1978–1981.

177. J. Zhang, L. X. Shao, Y. J. Fu and E. Q. Xie, *Rare Metals*, 2006, **25**, 315–319.

178. R. A. Wibowo, W. S. Kim, E. S. Lee, B. Munir and K. H. Kim, *J. Phys. Chem. Solid.*, 2007, **68**, 1908–1913.

179. C. Persson, *J. Appl. Phys.*, 2010, **107**, 053710.

180. S. Botti, D. Kammerlander and M. A. L. Marques, *Appl. Phys. Lett.*, 2011, **98**, 241915.

181. J. Alvarez-Garcia, A. Perez-Rodriguez, B. Barcones, A. Romano-Rodriguez, J. R. Morante, A. Janotti, S.-H. Wei and R. Scheer, *Appl. Phys. Lett.*, 2002, **80**, 562.

182. D. S. Su, W. Neumann, R. Hunger, P. Schubert-Bischoff, M. Giersig, H. J. Lewerenz, R. Scheer and E. Zeitler, *Appl. Phys. Lett.*, 1998, **73**, 785.

183. S. H. Wei, S. B. Zhang and A. Zunger, *Phys. Rev. B*, 1999, **59**, 2478–2481.

184. U. Rau and J. Werner, *Appl. Phys. Lett.*, 2004, **84**, 3735.

185. J. J. Scragg, PhD thesis, University of Bath, 2010.

186. R. Djemour, Ph.D thesis, 2014, University of Luxembourg.

187. I. Repins, C. Beall, N. Vora, C. DeHart, D. Kuciauskas, P. Dippo, B. To, J. Mann, W. C. Hsu, A. Goodrich and R. Noufi, *Sol. Energy Mater. Sol. Cells*, 2012, **101**, 154–159.

188. B. Shin, O. Gunawan, Y. Zhu, N. A. Bojarczuk, S. J. Chey and S. Guha, *Prog. Photovoltaics Res. Appl.*, 2013, **21**, 72–76.

189. T. M. Friedlmeier, N. Wieser, T. Walter, H. Dittrich and H. W. Schock, 14th European Photovoltaic Solar Energy Conference, Barcelona, 1997.

190. R. A. Wibowo, E. S. Lee, B. Munir and K. H. Kim, *Physica Status Solidi (A)*, 2007, **204**, 3373–3379.

191. O. Gunawan, T. Gokmen, C. W. Warren, J. D. Cohen, T. K. Todorov, D. A. R. Barkhouse, S. Bag, J. Tang, B. Shin and D. B. Mitzi, *Appl. Phys. Lett.*, 2012, **100**, 253905.

192. S. Siebentritt, L. Gütay, D. Regesch, Y. Aida and V. Deprédurand, *Sol. Energy Mater. Sol. Cells*, 2013, DOI: 10.1016/j.solmat.2013.1004.1014.

193. J. Moore, C. Hages, M. Lundstrom and R. Agrawal, *38th IEEE Photovoltaic Specialist Conference*, 2012.

194. M. Mousel, T. Schwarz, R. Djemour, T. P. Weiss, J. Sendler, J. Malaquias, A. Redinger, O. Cojocaru-Miredin, P. P. Choi and S. Siebentritt, *Adv. Energy Mater.*, 2013, DOI: 10.1002/aenm.201300543.

195. A. Nagoya, R. Asahi, R. Wahl and G. Kresse, *Phys. Rev. B*, 2010, **81**, 113202.

196. S. Chen, J.-H. Yang, X. G. Gong, A. Walsh and S.-H. Wei, *Phys. Rev. B*, 2010, **81**, 245204.
197. T. Maeda, S. Nakamura and T. Wada, *Jpn. J. Appl. Phys.*, 2011, **50**, 04DP07.
198. S. Y. Chen, A. Walsh, X. G. Gong and S. H. Wei, *Adv. Mater.*, 2013, **25**, 1522–1539.
199. D. Han, Y. Y. Sun, J. Bang, Y. Y. Zhang, H.-B. Sun, X.-B. Li and S. B. Zhang, *Phys. Rev. B*, 2013, **87**, 155206.
200. H. Katagiri, K. Jimbo, M. Tahara, H. Araki and K. Oishi, *Mater. Res. Soc. Symp. Proc.*, 2009, **1165**, 125–136.
201. K. Biswas, S. Lany and A. Zunger, *Appl. Phys. Lett.*, 2010, **96**, 201902.
202. S. Y. Chen, L. W. Wang, A. Walsh, X. G. Gong and S. H. Wei, *Appl. Phys. Lett.*, 2012, **101**, 225901.
203. S. Siebentritt, in *Wide Gap Chalcopyrites*, ed. S. Siebentritt and U. Rau, Springer, Berlin, 2006, pp. 113–156.
204. B. I. Shklovskii and A. L. Efros, *Electronic Properties of Doped Semiconductors*, Springer-Verlag, Berlin, 1984.
205. M. J. Romero, I. Repins, G. Teeter, M. A. Contreras, M. Al-Jassim and R. Noufi, *38th IEEE Photovoltaic Specialist Conference*, 2012.
206. M. J. Romero, H. Du, G. Teeter, Y. Yan and M. M. Al-Jassim, *Phys. Rev. B*, 2011, **84**, 165324.
207. M. Grossberg, J. Krustok, J. Raudoja and T. Raadik, *Appl. Phys. Lett.*, 2012, **101**, 102102–102104.
208. K. Tanaka, Y. Miyamoto, H. Uchiki, K. Nakazawa and H. Araki, *Physica Status Solidi (A) Appl. Mater. Sci.*, 2006, **203**, 2891–2896.
209. F. Luckert, D. I. Hamilton, M. V. Yakushev, N. S. Beattie, G. Zoppi, M. Moynihan, I. Forbes, A. V. Karotki, A. V. Mudryi, M. Grossberg, J. Krustok and R. W. Martin, *Appl. Phys. Lett.*, 2011, **99**, 062104.
210. J. P. Leitao, N. M. Santos, P. A. Fernandes, P. M. P. Salome, A. F. da Cunha, J. C. Gonzalez, G. M. Ribeiro and F. M. Matinaga, *Phys. Rev. B*, 2011, **84**, 024120.
211. K. Hoenes, E. Zscherpel, J. J. Scragg and S. Siebentritt, *Physica B*, 2009, **404**, 4949–4952.
212. T. Eisenbarth, T. Unold, R. Caballero, C. A. Kaufmann and H.-W. Schock, *J. Appl. Phys.*, 2010, 107.
213. U. Reislöhner, H. Metzner and C. Ronning, *Phys. Rev. Lett.*, 2010, **104**, 226403.
214. T. P. Weiss, A. Redinger, J. Luckas, M. Mousel and S. Siebentritt, *Appl. Phys. Lett.*, 2013, **102**, 202105.
215. T. P. Weiss, A. Redinger, J. Luckas, M. Mousel and S. Siebentritt, in *39th IEEE Photovoltaic Specialist Conference*, Tampa, 2013.
216. K. R. Choudhury, Y. Cao, J. V. Caspar, W. E. Farneth, Q. Guo, A. S. Ionkin, L. K. Johnson, M. Lu, I. Malajovich, D. Radu, H. D. Rosenfeld and W. Wu, 38th IEEE Photovoltaic Specialist Conference, Austin, Tx. 2012.
217. P. A. Fernandes, A. F. Sartori, P. M. P. Salome, J. Malaquias, A. F. d. Cunha, M. P. F. Graca and J. C. Gonzalez, *Appl. Phys. Lett.*, 2012, **100**, 233504.

218. D. W. Miller, C. W. Warren, O. Gunawan, T. Gokmen, D. B. Mitzi and J. D. Cohen, *Appl. Phys. Lett.*, 2012, **101**, 142106–142104.
219. H. Wilhelm, H.-W. Schock and R. Scheer, *J. Appl. Phys.*, 2011, **109**, 084514.
220. Q. Nguyen, K. Orgassa, I. Koetschau, U. Rau and H. W. Schock, *Thin Solid Films*, 2003, **431–432**, 330–334.
221. M. Bär, B.-A. Schubert, B. Marsen, R. G. Wilks, S. Pookpanratana, M. Blum, S. Krause, T. Unold, W. Yang, L. Weinhardt, C. Heske and H.-W. Schock, *Appl. Phys. Lett.*, 2012, **99**, 222105.
222. A. Santoni, F. Biccari, C. Malerba, M. Valentini, R. Chierchia and A. Mittiga, *J. Phys. D Appl. Phys.*, 2013, **46**, 175101.
223. W. Bao and M. Ichimura, *Jpn. J. Appl. Phys.*, 2012, **51**, 4.
224. R. Haight, A. Barkhouse, O. Gunawan, B. Shin, M. Copel, M. Hopstaken and D. B. Mitzi, *Appl. Phys. Lett.*, 2011, **98**, 253502.
225. A. Nagoya, R. Asahi and G. Kresse, *J. Phys. Condensed Matter*, 2011, **23**, 404203.
226. A. Redinger, M. Mousel, M. H. Wolter, N. Valle and S. Siebentritt, *Thin Solid Films*, 2013, **535**, 291–295.
227. J. R. C. D. J. Vaughan, *Mineral Chemistry of Metal Sulfides*, Cambridge University Press, Cambridge, 1975.
228. R. A. Swalin, *Thermodynamics of Solids*, Wiley-Interscience, 1972.
229. R. C. Ropp, *Solid State Chemistry*, Elsevier, 2003.
230. R. C. Sharma and Y. A. Chang, *Bull. Alloy Phase Diagram.*, 1986, **7**, 269–273.
231. K. C. Mills, *Thermodynamic Data for Inorganic Sulphides, Selenides and Tellurides*, Butterworths, 1974.
232. C. Shi, P. Yang, M. Yao, X. Dai and Z. Chen, *Thin Solid Films*, 2013, **534**, 28–31.
233. J. J. Scragg, P. J. Dale, D. Colombara and L. M. Peter, *ChemPhysChem*, 2012, **13**, 3035–3046.
234. A. Weber, R. Mainz and H. W. Schock, *J. Appl. Phys.*, 2010, **107**, 013516.
235. J. J. Scragg, T. Ericson, T. Kubart, M. Edoff and C. Platzer-Björkman, *Chem. Mater*, 2011, **23**, 4625–4633.
236. J. J. Scragg, J. T. Watjen, M. Edoff, T. Ericson, T. Kubart and C. Platzer-Bjorkman, *J. Am. Chem. Soc.*, 2012, **134**, 19330–19333.
237. J. F. Guillemoles, *Thin Solid Films*, 2000, **361–362**, 338–345.
238. A. Fairbrother, X. Fontane, V. Izquierdo-Roca, M. Espindola-Rodriguez, S. Lopez-Marino, M. Placidi, J. Lopez-Garcia, A. Perez-Rodriguez and E. Saucedo, *ChemPhysChem*, 2013, **14**, 1836–1843.
239. B. Shin, Y. Zhu, N. A. Bojarczuk, S. J. Chey and S. Guha, *Appl. Phys. Lett.*, 2012, **101**, 053903.
240. C. Platzer-Bjorkman, J. Scragg, H. Flammersberger, T. Kubart and M. Edoff, *Sol. Energy Mater. Sol. Cells*, 2012, **98**, 110–117.
241. V. Chawla and B. Clemens, *35th IEEE Photovoltaic Specialists Conference*, 2010, 1902–1905.
242. S. Seeger and K. Ellmer, *Thin Solid Films*, 2009, **517**, 3143–3147.

243. G. M. Ford, Q. Guo, R. Agrawal and H. W. Hillhouse, *Chem. Mater.*, 2011, **23**, 2626–2629.
244. C. K. Miskin, N. J. Carter, W.-C. Yang, C. J. Hages, E. Stach and R. Agrawal, in 39th IEEE Photovoltaic Specialists Conference Tampa, Fl., 2013.
245. Y. Cao, M. S. Denny, Jr., J. V. Caspar, W. E. Farneth, Q. Guo, A. S. Ionkin, L. K. Johnson, M. Lu, I. Malajovich, D. Radu, H. D. Rosenfeld, K. R. Choudhury and W. Wu, *J. Am. Chem. Soc.*, 2012, **134**, 15644–15647.
246. J. J. Scragg, T. Kubart, J. T. Wätjen, T. Ericson, M. K. Linnarsson and C. Platzer-Björkman, *Chem. Mater.*, 2013, **25**, 3162–3171.
247. V. Chawla and B. Clemens, in *2012 38th IEEE Photovoltaic Specialists Conference*, 2012, pp. 2990–2992.
248. T. Fukano, S. Tajima and T. Ito, *Appl. Phys. Express*, 2013, **6**, 062301.
249. S. Ahmed, K. B. Reuter, O. Gunawan, L. Guo, L. T. Romankiw and H. Deligianni, *Adv. Energy Mater.*, 2012, **2**, 253–259.
250. C. M. Fella, A. R. Uhl, Y. E. Romanyuk and A. N. Tiwari, *Physica Status Solidi (A) Appl. Mater. Sci.*, 2012, **209**, 1043–1045.
251. G. Brammertz, M. Buffiere, Y. Mevel, Y. Ren, A. E. Zaghi, N. Lenaers, Y. Mols, C. Koeble, J. Vleugels, M. Meuris and J. Poortmans, *Appl. Phys. Lett.*, 2013, **102**, 013902–013903.
252. M. Buffière, G. Brammertz, N. Lenaers, Y. Ren, C. Koeble, A. E. Zaghi, J. Vleugels, M. Meuris and J. Poortmans, in *Poster presented at the 39th IEEE Photovoltaic Specialists Conference* Tampa, Fl., 2013.
253. H. Hiroi, N. Sakai, T. Kato and H. Sugimoto, in *39th IEEE Photovoltaic Specialists Conference* Tampa, Fl., 2013.
254. R. Lechner, S. Jost, J. Palm, M. Gowtham, F. Sorin, B. Louis, H. Yoo, R. A. Wibowo and R. Hock, *Thin Solid Films*, 2013, **535**, 5–9.
255. N. Bochvar, V. Ivanchenko, E. Lysova and L. Rokhlin, Non-Ferrous Metal Ternary Systems. Selected Soldering and Brazing Systems: Phase Diagrams, Crystallographic and Thermodynamic Data in *Landolt Börnstein Group IV Physical Chemistry*, ed. G. Effenberg and S. Ilyenko, Springer, 2007, vol. 11C3.
256. R. Blachnik and A. Müller, *Thermochim. Acta*, 2001, **366**, 47–59.
257. R. Schurr, A. Hölzing, S. Jost, R. Hock, T. Voå, J. Schulze, A. Kirbs, A. Ennaoui, M. Lux-Steiner, A. Weber, I. Kötschau and H. W. Schock, *Thin Solid Films*, 2009, **517**, 2465–2465.
258. R. A. Wibowo, W. H. Jung, M. H. Al-Faruqi, I. Amal and K. H. Kim, *Mater. Chem. Phys.*, 2010, **124**, 1006–1010.
259. H. Araki, A. Mikaduki, Y. Kubo, T. Sato, K. Jimbo, W. S. Maw, H. Katagiri, M. Yamazaki, K. Oishi and A. Takeuchi, *Thin Solid Films*, 2008, **517**, 1457–1460.
260. D. Mitzi, O. Gunawan, T. Todorov, K. Wang and S. Guha, *Sol. Energy Mater. Sol. Cells*, 2011, **95**, 1421–1436.
261. D. A. R. Barkhouse, R. Haight, N. Sakai, H. Hiroi, H. Sugimoto and D. B. Mitzi, *Appl. Phys. Lett.*, 2012, **100**, 193904.

262. H. Hiroi, N. Sakai, S. Muraoka, T. Katou and H. Sugimoto, 38th IEEE Photovoltaic Specialist Conference, Austin, Tx. 2012.
263. M. T. Htay, Y. Hashimoto, N. Momose, K. Sasaki, H. Ishiguchi, S. Igarashi, K. Sakurai and K. Ito, *Jpn. J. Appl. Phys.*, 2011, 50.
264. T. Kato, H. Hiroi, N. Sakai, S. Muraoka and H. Sugimoto, 27th European Photovoltaic Solar Energy Conference, Frankfurt, 2012.
265. S. Bag, O. Gunawan, T. Gokmen, Y. Zhu, T. K. Todorov and D. B. Mitzi, *Energy Environ. Sci.*, 2012, **5**, 7060–7065.
266. D. A. R. Barkhouse, O. Gunawan, T. Gokmen, T. K. Todorov and D. B. Mitzi, *Prog. Photovoltaics Res. Appl.*, 2012, **20**, 6–11.
267. H. Sugimoto, H. Hiroi, N. Sakai, S. Muraoka and T. Katou, *38th IEEE Photovoltaic Specialist Conference*, 2012.
268. R. Herberholz, V. Nadenau, U. Rühle, C. Köble, H. W. Schock and B. Dimmler, *Sol. Energy Mater. Sol. Cells*, 1997, **49**, 227–237.
269. P. Jackson, D. Hariskos, E. Lotter, S. Paetel, R. Wuerz, R. Menner, W. Wischmann and M. Powalla, *Prog. Photovoltaics*, 2011, **19**, 894–897.
270. I. Repins, M. A. Contreras, B. Egaas, B. DeHart, J. Scharf, C. L. Perkins, B. To and R. Noufi, *Prog. Photovoltaics*, 2008, **16**, 235–239.
271. S. Siebentritt, *Sol. Energy Mater. Sol. Cells*, 2011, **95**, 1471–1476.
272. S. Lopez-Marino, M. Placidi, A. Perez-Tomas, J. Llobet, V. Izquierdo-Roca, X. Fontane, A. Fairbrother, M. Espindola-Rodriguez, D. Sylla, A. Perez-Rodriguez and E. Saucedo, *J. Mater. Chem. A*, 2013, **1**, 8338–8343.
273. R. Klenk, T. Walter, H. W. Schock and D. Cahen, *Adv. Mater.*, 1993, **5**, 114–119.
274. V. Depredurand, Y. Aida, J. Larsen, T. Eisenbarth, A. Majerus and S. Siebentritt, Photovoltaic Specialists Conference (PVSC), 2011 37th IEEE, 2011.
275. L. Gutay, D. Regesch, J. K. Larsen, Y. Aida, V. Depredurand and S. Siebentritt, *Appl. Phys. Lett.*, 2011, **99**, 151912.
276. H. C. Hsieh, *J. Appl. Phys.*, 1982, **53**, 1727–1733.
277. D. Colombara, L. M. Peter, K. D. Rogers and K. Hutchings, *J. Solid State Chem.*, 2012, **186**, 36–46.
278. C. von Klopmann, J. Djordjevic, E. Rudigier and R. Scheer, *J. Cryst. Growth*, 2006, **289**, 121–133.
279. D. B. Mitzi, *Adv. Mater.*, 2009, **21**, 3141–3155.
280. L. Ribeaucourt, G. Savidand, D. Lincot and E. Chassaing, *Electrochim Acta*, 2011, **56**, 6628–6637.
281. A. A. Frost, *J. Am. Chem. Soc.*, 1951, **73**, 2680–2682.
282. B. Shin, Y. Zhu, N. A. Bojarczuk, S. J. Chey and S. Guha, *Appl. Phys. Lett.*, 2012, **101**, 053903.
283. K. T. Ramakrishna Reddy, I. Forbes and R. W. Miles, *Appl. Surf. Sci.*, 2001, **169–170**, 387–391.
284. M. E. Beck and M. Cocivera, *Thin Solid Films*, 1996, **272**, 71–82.
285. F. Al-Saab, B. Gholipour, C. C. Huang and D. W. Hewak, PVSAT-9, Swansea University, Swansea, Wales, 2013.

286. N. J. Gerein and J. A. Haber, Photovoltaic Specialists Conference, 2005. Conference Record of the Thirty-first IEEE, 2005.

287. O. Knacke and O. Kubaschewski, *Thermochemical Properties of Inorganic Substances*, 2nd edn., Springer-Verlag, Berlin, 1991.

288. I. Barin, *Thermochemical Data of Pure Substances*, 3rd edn. Wiley-VCH, 2005.

289. M. W. Chase, *NIST - JANAF Thermochemical Tables*, 4th. edn., 1995.

290. N. Parodi, G. Borzone, G. Balducci, S. Brutti, A. Ciccioli and G. Gigli, *Intermetallics*, 2003, **11**, 1175–1181.

291. R. Ferro, G. Borzone, G. Cacciamani and R. Raggio, *Thermochim. Acta*, 2000, **347**, 103–122.

292. V. Piacente, P. Scardala, D. Ferro and R. Gigli, *J. Chem. Eng. Data*, 1985, **30**, 372–376.

293. B. Brunetti and V. Piacente, *J. Mater. Sci. Lett.*, 1993, **12**, 1738–1740.

294. S. Brutti, G. Balducci, G. Gigli, A. Ciccioli, P. Manfrinetti and A. Palenzona, *J. Cryst. Growth*, 2006, **289**, 578–586.

295. H. Ipser, A. Mikula and I. Katayama, *Calphad*, 2010, **34**, 271–275.

296. J. R. Craig and P. B. Barton, *Econ. Geol.*, 1973, **68**, 493–506.

297. D. J. Vaughan and J. R. Craig, *Mineral Chemistry of Metal Sulfides*, Cambridge University Press, Cambridge, 1975.

298. G.-X. Qian, R. M. Martin and D. J. Chadi, *Phys. Rev. B*, 1988, **38**, 7649.

299. B. Shin, N. A. Bojarczuk and S. Guha, *Appl. Phys. Lett.*, 2013, **102**, 091907–091904.

300. P. A. Fernandes, P. M. P. Salomé, A. F. da Cunha and B.-A. Schubert, *Thin Solid Films*, 2011, **519**, 7382–7385.

301. P. Würfel, *Physics of Solar Cells - From Principles to New Concepts*, Wiley-VCH Verlag GmbH & Co., Weinheim, 2005.

302. I. Repins, M. Contreras, M. Romero, Y. F. Yan, W. Metzger, J. Li, S. Johnston, B. Egaas, C. DeHart, J. Scharf, B. E. McCandless, R. Noufi and Ieee, in Pvsc: 2008 33rd IEEE Photovoltaic Specialists Conference, vols. 1–4, 2008, pp. 1127–1132.

303. R. Scheer, *Thin Solid Films*, 2011, **519**, 7472–7475.

304. W. K. Metzger, I. L. Repins and M. A. Contreras, *Appl. Phys. Lett.*, 2008, **93**, 022110–022110-3.

305. L. M. Peter, *Philos. Trans. R. Soc. London Ser. A Math., Phys. Eng. Sci.*, 2011, **369**, 1840–1856.

306. J. F. Carlin, U.S. Geological Survey, Reston, Virginia, 2011, p. 195.

307. B. J. Stanbery, *Crit. Rev. Solid State Mater. Sci.*, 2002, **27**, 73–117.

308. M. I. Alonso, M. Garriga, C. A. D. Rincon and M. Leon, *J. Appl. Phys.*, 2000, **88**, 5796–5801.

309. A. Virtuani, E. Lotter, M. Powalla, U. Rau, J. H. Werner and M. Acciarri, *J. Appl. Phys.*, 2006, **99**, 14906.

310. J. Vidal, S. Lany, M. d'Avezac, A. Zunger, A. Zakutayev, J. Francis and J. Tate, *Appl. Phys. Lett.*, 2012, **100**, 032104.

311. M. Steichen, R. Djemour, L. Guetay, J. Guillot, S. Siebentritt and P. J. Dale, *J. Phys. Chem. C*, 2013, **117**, 4383–4393.

312. P. Sinsermsuksakul, J. Heo, W. Noh, A. S. Hock and R. G. Gordon, *Adv. Energy Mater.*, 2011, **1**, 1116–1125.
313. L. D. Partain, P. S. McLeod, J. A. Duisman, T. M. Peterson, D. E. Sawyer and C. S. Dean, *J. Appl. Phys.*, 1983, **54**, 6708–6720.
314. A. B. F. Martinson, S. C. Riha, E. Thimsen, J. W. Elam and M. J. Pellin, *Energy Environ. Sci.*, 2013, **6**, 1868–1875.
315. A. Baruth, M. Manno, D. Narasimhan, A. Shankar, X. Zhang, M. Johnson, E. S. Aydil and C. Leighton, *J. Appl. Phys.*, 2012, 112.
316. V. Antonucci, A. S. Arico, N. Giordano, P. L. Antonucci, U. Russo, D. L. Cocke and F. Crea, *Sol. Cells*, 1991, **31**, 119–141.
317. T. A. Kuku and O. A. Fakolujo, *Sol. Energy Mater.*, 1987, **16**, 199–204.
318. M. Umehara, Y. Takeda, T. Motohiro, T. Sakai, H. Awano and R. Maekawa, *Appl. Phys. Express*, 2013, **6**, 044501.
319. M. Kanzari, A. Rabhi and B. Rezig, *Thin Solid Films*, 2009, **517**, 2477–2480.
320. Y. Rodríguez-Lazcano, M. T. S. Nair and P. K. Nair, *J. Cryst. Growth*, 2001, **223**, 399–406.
321. J. Zhou, G.-Q. Bian, Q.-Y. Zhu, Y. Zhang, C.-Y. Li and J. Dai, *J. Solid State Chem.*, 2009, **182**, 259–264.
322. C. Yan, Z. Su, E. Gu, T. Cao, J. Yang, J. Liu, F. Liu, Y. Lai, J. Li and Y. Liu, *RSC Adv.*, 2012, **2**, 10481–10484.
323. C. Garza, S. Shaji, A. Arato, E. Perez Tijerina, G. Alan Castillo, T. K. Das Roy and B. Krishnan, *Sol. Energy Mater. Sol. Cells*, 2011, **95**, 2001–2005.
324. M. Kumar and C. Persson, *J. Renew. Sustain. Energy*, 2013, **5**, 031616.
325. S. Ikeda, Y. Iga, W. Septina, T. Harada and M. Matsumura, 38th IEEE Photovoltaic Specialists Conference, Austin, Tx., 2012.
326. S. Pawar, A. Pawar and P. Bhosale, *Bull. Mater. Sci.*, 1986, **8**, 423–426.
327. P. S. Sonawane, P. A. Wani, L. A. Patil and T. Seth, *Mater. Chem. Phys.*, 2004, **84**, 221–227.
328. V. Balasubramanian, N. Suriyanarayanan, S. Prabahar, S. Srikanth and P. Ravi, *Optoelectron. Adv. Mater.*, 2012, **6**, 104–106.
329. F. Mesa, A. Dussan and G. Gordillo, *Physica Status Solidi (C)*, 2010, **7**, 917–920.
330. V. Estrella, M. T. S. Nair and P. K. Nair, *Semicond. Sci. Technol.*, 2003, **18**, 190–194.
331. S. Seefeld, M. Limpinsel, Y. Liu, N. Farhi, A. Weber, Y. Zhang, N. Berry, Y. J. Kwon, C. L. Perkins, J. C. Hemminger, R. Wu and M. Law, *J. Am. Chem. Soc.*, 2013, **135**, 4412–4424.
332. R. Sun, M. K. Y. Chan, S. Kang and G. Ceder, *Phys. Rev. B*, 2011, **84**, 035212.
333. J. Hu, Y. Zhang, M. Law and R. Wu, *Phys. Rev. B*, 2012, **85**, 085203.
334. K. Aravind, F. W. Herbert, Y. Sidney, J. V. V. Krystyn and Y. Bilge, *J. Phys. Condens. Matter*, 2013, **25**, 045004.
335. Q. Xu, B. Huang, Y. Zhao, Y. Yan, R. Noufi and S.-H. Wei, *Appl. Phys. Lett.*, 2012, 100.

336. I. I. Mazin, *Phys. Rev. B*, 2012, **85**, 115133.
337. M. Burgelman and H. J. Pauwels, *Electron. Lett.*, 1981, **17**, 224–226.
338. J. A. Bragagnolo, A. M. Barnett, J. E. Phillips, R. B. Hall, A. Rothwarf and J. D. Meakin, *IEEE Trans. Electron Devices*, 1980, **27**, 645–651.
339. A. M. Al-Dhafiri, G. J. Russell and J. Woods, *Semicond. Sci. Technol.*, 1992, **7**, 1052.
340. R. D. Engelken, H. E. McCloud, C. Lee, M. Slayton and H. Ghoreishi, *J. Electrochem. Soc.*, 1987, **134**, 2696–2707.
341. K. Hartman, J. L. Johnson, M. I. Bertoni, D. Recht, M. J. Aziz, M. A. Scarpulla and T. Buonassisi, *Thin Solid Films*, 2011, **519**, 7421–7424.
342. H. Noguchi, A. Setiyadi, H. Tanamura, T. Nagatomo and O. Omoto, *Sol. Energy Mater. Sol. Cells*, 1994, **35**, 325–331.
343. S. Fiechter, M. Martinez, G. Schmidt, W. Henrion and Y. Tomm, *J. Phys. Chem. Solid*, 2003, **64**, 1859–1862.
344. M. Bouaziz, M. Amlouk and S. Belgacem, *Thin Solid Films*, 2009, **517**, 2527–2530.
345. K. Chino, J. Koike, S. Eguchi, H. Araki, R. Nakamura, K. Jimbo and H. Katagiri, *Jpn. J. Appl. Phys.*, 2012, **51**, 10C35.
346. E. Belandria, R. Avila and B. J. Fernandez, *Jpn. J. Appl. Phys. Suppl.*, 2000, **39**, 132–133.
347. A. N. Wachtel, *J. Electron. Mater.*, 1980, **9**, 281–297.
348. A. Rabhi, M. Kanzari and B. Rezig, *Mater. Lett.*, 2008, **62**, 3576–3575.
349. A. Kyono and M. Kimata, *Am. Mineral.*, 2005, **90**, 162–165.
350. A. Duta, D. Perniu, L. Isac and A. Enesca, 9th edition of the National Seminar of nanoscience and nanotechnology, Bucharest, 2010.
351. S. Bourdais, C. Chone and Y. Cuccaro, E-MRS Spring meeting, Strasbourg, 2011.
352. S. Manolache, A. Duta, L. Isac, M. Nanu, A. Goossens and J. Schoonman, *Thin Solid Films*, 2007, **515**, 5957–5960.
353. Y. Rodríguez-Lazcano, M. T. S. Nair and P. K. Nair, *J. Electrochem. Soc.*, 2005, **152**, G635–G635.
354. A. Sugaki, H. Shima and A. Kitakaze, Technology Reports of Yamaguchi University, 1973.
355. L. I. Soliman, A. M. A. E. Soad, H. A. Zayed and S. A. E. Ghfar, *Fizika A*, 2003, **11**, 139–152.
356. S. Perniu, A. Duta and J. Schoonman, *IEEE International Semiconductor Conference*, 2006, pp. 245–248.
357. J. Yang, Y.-C. Liu, H.-M. Lin and C.-C. Chen, *Adv. Mater.*, 2004, **16**, 713–716.
358. A. V. Steblevskii, V. V. Zharov, A. S. Alikhanyan, V. I. Gorgoraki and A. S. Pashinkin, *Russ. J. Inorg. Chem.*, 1989, **34**, 891–894.
359. L. Yu, R. S. Kokenyesi, D. A. Keszler and A. Zunger, *Adv. Energy Mater.*, 2013, **3**, 43–45.
360. P. K. Nair, L. Huang, M. T. S. Nair, H. L. Hu, E. A. Meyers and R. A. Zingaro, *J. Mater. Res.*, 1997, **12**, 651–656.

361. E. W. Nuffield, *Econ. Geol.*, 1947, **42**, 147–160.
362. V. Kocman and E. W. Nuffield, *Acta Crystallogr.*, 1973, **B29**, 2525.
363. J. Zhong, W. Xiang, Q. Cai and X. Liang, *Mater. Lett.*, 2012, **70**, 63–66.
364. L. H. Ahrens, *Geochim. Cosmochim. Acta*, 1952, **2**, 155–169.
365. L. F. Lundegaard, E. Makovicky, T. Boffa-Ballaran and T. Balic-Zunic, *Phys. Chem. Miner.*, 2005, **32**, 578–584.
366. A. Sugaki and H. Shima, *Technology Report of Yamaguchi University*, 1972, **1**, 45–70.
367. E. Makovicky, *J. Solid State Chem.*, 1983, **49**, 85–92.

CHAPTER 6

Chemistry of Sensitizers for Dye-sensitized Solar Cells

PENG GAO, MICHAEL GRÄTZEL AND M. D. K. NAZEERUDDIN*

Laboratory for Photonics and Interfaces, Institution of Chemical Sciences and Engineering, School of Basic Sciences, Swiss Federal Institute of Technology, CH-1015 Lausanne, Switzerland
*Email: mdkhaja.nazeeruddin@epfl.ch

6.1   Introduction

Among the several approaches for harnessing solar energy and converting it into electricity, the dye-sensitized solar cells (DSSCs) represents one of the most promising methods for future large-scale power production from renewable energy sources. In these cells the sensitizer is one of the key components, harvesting the solar radiation and converting it into electric current. The electrochemical, photophysical, ground and the excited state properties of the sensitizer play an important role for charge transfer dynamics at the semiconductor interface. Over the last 20 years, ruthenium complexes, and donor–bridge–acceptor organic sensitizers endowed with anchoring groups have maintained a clear lead in generating power conversation efficiencies over 10%. Their validated efficiency record under standard air mass 1.5 reporting conditions stands presently at $12.4 \pm 0.3\%$.[1]

In this chapter, we discuss various sensitizers of metal complexes, organic, porphyrin and perovskite and their design strategies. For metal complexes, the choice of ruthenium is of special interest for a number of reasons: (1) because of its octahedral geometry one can introduce specific ligands in a controlled manner; (2) the photophysical, photochemical and the

RSC Energy and Environment Series No. 11
Advanced Concepts in Photovoltaics
Edited by Arthur J Nozik, Gavin Conibeer and Matthew C Beard
© The Royal Society of Chemistry 2014
Published by the Royal Society of Chemistry, www.rsc.org

electrochemical properties of these complexes can be tuned in a predictable way; and (3) the ruthenium metal possesses stable and accessible oxidation states from ɪ to ɪv. Similar to ruthenium sensitizers, the organic sensitizers should have both good light harvesting and carriers transporting properties. The organic sensitizers based photovoltaic devices use a donor and an acceptor type of dyes, which form a heterojunction favouring the separation of the exciton into charge carriers.[2] Those formed carriers are then transported to the electrodes by charge transporting media. In the last section we introduce organic–inorganic hybrid perovskites thin film photovoltaics, which came to the limelight because of their high efficiency, low cost and the ease to make these materials solution processable. The methyl ammonium lead iodide $(CH_3NH_3PbI_3)$ perovskite material by virtue has a direct bandgap, a large absorption coefficient $(1.5 \times 10^4 \text{ cm}^{-1}$ at 550 nm), very high charge carrier mobility and to low non-radiative recombination rates. These materials have been used in solid state solar cells where it was found to act not only as a sensitizer,[3] but an excellent electron,[4] and hole conductor.[5] Using perovskite as an absorber layer and 2,2′,7,7′-tetrakis-(*N,N*-di-*p*-methoxyphenylamine)-9,9′-spirobifluorene (spiro-OMeTAD) as a hole transport material (HTM), power conversion efficiencies (PCE) of over 15% were obtained.

The first sensitization of a photoelectrode was reported in 1887.[6] However, the operating mechanism of injection of electrons from photo-excited dye molecules into the conduction band of the n-type semiconductor substrates dates only from the 1960s.[7] The concept developed in the following years with first the chemisorption of the dye on the surface of the semiconductor[8,9] and then the use of dispersed particles to provide a sufficient interface area.[10,11] In 1991, O'Regan and Grätzel demonstrated that a film of titania (TiO_2) nanoparticles could act as a mesoporous n-type photo-anode and thereby increase the effective surface area for dye deposition by a factor of more than one thousand.[12] This approach improved light harvesting dramatically and brought power-conversion efficiencies into a range that allowed the DSSC to be viewed as a serious competitor to other solar cell technologies.[13] Unlike the silicon solar cells, the dye-sensitized solar cell (DSSC) technology separates the two main requirements of charge generation and charge transport, which were done by the sensitizer, and TiO_2 and electrolyte, respectively.[14,15] The spectral properties of the sensitizers can be optimized by engineering at the molecular level, while carrier transport properties can be improved by optimizing the semiconductor and the electrolyte composition.

Figure 6.1 shows, a schematic representation of liquid (a) and solid-state (b) dye-sensitized solar cell configuration. The liquid dye-sensitized solar cell contains broadly five components: (1) a mechanical support coated with transparent conductive oxides (TCOs); (2) the semiconductor film, usually TiO_2, ZnO, or SnO_2; (3) a sensitizer adsorbed onto the surface of the semiconductor; (4) an electrolyte containing a redox mediator or hole transporting materials; and (5) a counter-electrode capable of regenerating

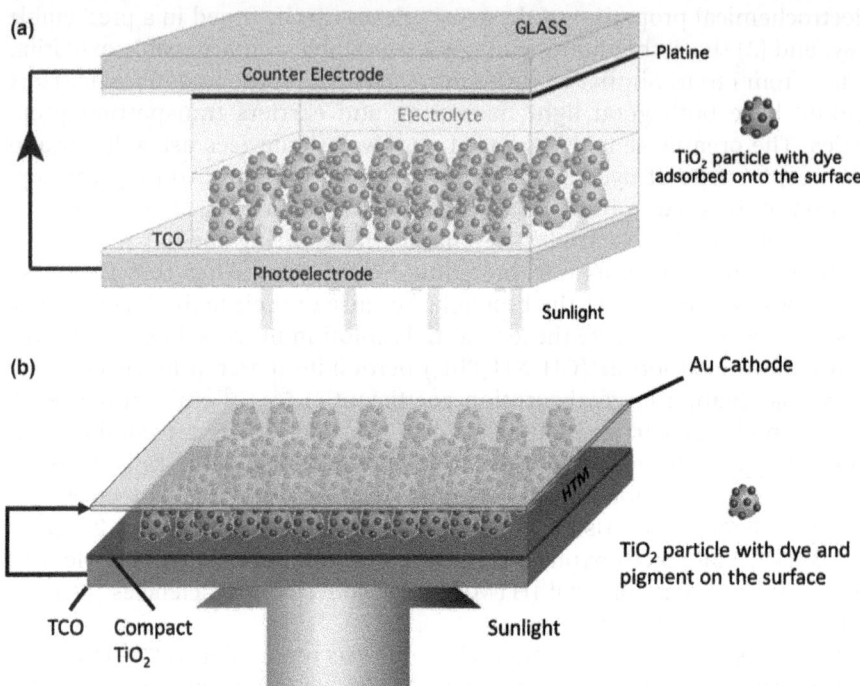

Figure 6.1   A schematic representation of liquid (a) and solid-state (b) dye-sensitized solar cell configuration.

the redox mediator. The solid-state version of a dye-sensitized solar cell
has same components except the liquid electrolyte is replaced with an
organic or inorganic hole transporting materials. Also, in general, the TiO$_2$
film thickness in liquid DSSCs is typically between 4 to 14 μm thick where
as in the solid state-state configuration it is around 2 μm. In this chapter
we will focus on engineering of sensitizers and their photovoltaic
properties.

The details of the operating principles of the dye-sensitized solar cell are
given in Figure 6.2. One of the essential parts of the device is the dye
adsorbed mesoporous TiO$_2$ oxide layer. The photo-excitation of the metal to
ligand charge transfer (MLCT) of the adsorbed dye leads to injection of
electrons into the conduction band of the oxide. The original state of the dye
is subsequently restored by electron donation from an electrolyte, containing the redox system (iodide/triiodide). The injected electron flows
through the semiconductor network to arrive at the back contact and then
through the external load to the counter-electrode. At the counter-electrode,
reduction of triiodide in turn regenerates iodide, which completes the circuit. The voltage generated under illumination corresponds to the difference
between the Fermi level of the electron in the solid and the redox potential of
the electrolyte. The overall conversion efficiency (η) of the dye-senstized

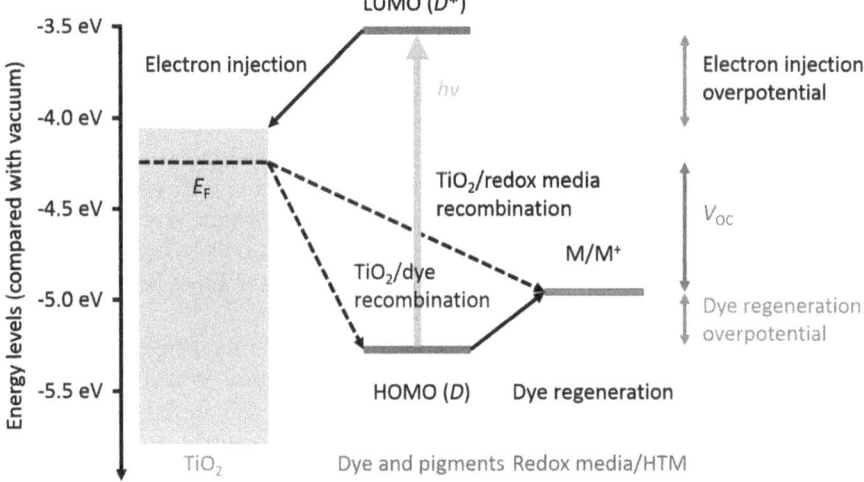

Figure 6.2 Energy diagram of and operating principles of the dye-sensitized solar cell.

solar cell is determined by the photocurrent density (i_{ph}) measured at short circuit, the open-circuit potential (V_{OC}), the fill factor (FF) of the cell and the intensity of the incident light (I_s) as shown in Equation (6.1).

$$\eta_{global} = i_{ph} \times V_{OC} \times FF/I_s \tag{6.1}$$

During the 1990s and the early 2000s, we and others found that the ruthenium(II) polypyridyl complexes provided the highest power-conversion efficiencies using iodide/triiodide redox mediator.[16–22] The power-conversion efficiency of the champion cells rapidly climbed to 10% in the late 1990s and then slowly settled to 11.9% by the beginning of 2012.[23–28] Although the I^-/I_3^- redox couple has been established as a standard for many years and good stability data[29,30] have been reported in combination with ruthenium bipyridine dyes,[31] this system has some inherent disadvantages, *e.g.*, chemically very aggressive, corrodes silver fingers, significant absorption in the visible. Moreover, the I^-/I_3^- redox couple suffers from a low redox potential, necessitating an excessive thermodynamic driving force for the dye-regeneration reaction. This limits the open-circuit potential of current DSSCs to 0.7–0.8 V. Therefore, more recently, much research has focused on the optimization of highly soluble cobalt bi-pyridine and related redox complexes with the aim to overcome all the disadvantages of iodine based electrolyte and stabilize the oxidation potential close to the sensitizer redox couple. This has led to DSSC with photovoltages up to 1 V which is 250 mV higher than in the case of the I^-/I_3^- redox couple.[32] However, up to now the cobalt-based electrolytes are not compatible with ruthenium bi-pyridine dyes as a slow ligand exchange can be expected over time, but they are suitable in combination with organic dyes.

A variety of organic dyes for DSSCs has been explored beyond the initially utilized Ru–polypyridyl dyes, which are inexpensive.[33–35] These organic dyes are promising because they bypass the need of precious metals, exhibit large π–π* excitation cross-sections and absorb strongly in the visible region of the spectrum.[36–39] Generally, organic dyes work well with the I^-/I_3^- redox couple. For example, its combination with MK-type dyes has led to long-term stable cells.[40] Upon the introduction of cobalt bi-pyridine redox complexes, organic DSSCs with efficiencies up to 11% have been reported recently.[37]

Zn-porphyrin dyes, as another example of noble-metal-free sensitizers, which have been incorporated in some of the most efficient DSSCs to date,[41,42] are particularly attractive since they have high molar absorptivity and very attractive green colour. In addition, the spectral and electronic properties can be tuned by the peripheral substituents.[43–45] Champion cells based on Zn-porphyrin sensitizer and cobalt bi-pyridine redox complexes showed efficiency of 13%.[46]

Solid-state DSSCs (ss-DSSCs), which use solid hole conductors instead of a liquid electrolyte, are like a cousin of liquid DSSCs.[47] The hole conductor is typically made from either wide-bandgap small molecules (such as spiro-OMeTAD) or semiconducting polymers (such as PEDOT or P3HT). These DSSCs are, in principle, more industrially compatible than liquid-type DSSCs because they do not contain a corrosive and volatile liquid electrolyte, which requires careful packaging. The highest values of V_{OC} (>1 V) achieved so far have been demonstrated in devices that exploit a small-molecule hole conductor, but the best efficiency (7.1%) is much lower compared to that of liquid DSSCs.[48,49]

The recent discovery that the sensitizer in DSSC can be replaced efficiently with a organo-lead halide pigment gave ss-DSSCs a renaissance.[50,51] The perovskite materials have very interesting properties for the realization of low-cost solar cells. A certified cell efficiency of 14.1% was reached in the middle of 2013[52] and by the end of 2013, 16.2% efficiency was certified as well.[53] The fast development has triggered a large research interest and started discussions about the underlying principles and the potential for improvement of efficiency.[54–60] An important finding is that, as the perovskite currently under study is a strong light absorber with an energy gap of 1.5 eV, the electrode spacing can be smaller than 0.5 μm, and the p-selective contact at the back side can be achieved with an even thinner layer of < 50 nm. More importantly, $CH_3NH_3PbX_3$ (where X = I, Cl, Br) showed balanced charge carrier mobility (ambipolarity) with diffusion length of excitons (LD) from 100 nm to exceeding 1 μm.[61,62] This high value reinforces hope for the future of hybrid perovskite solar cells, because it makes possible the fabrication of devices with thicker active layers, where the absorption of light can be increased without affecting the collection efficiency of the generated charges (Table 6.1).

Table 6.1 Summary of a state-of-the-art champion results in dye-sensitized solar cells.

Dye	Couple/conductor	PCE (%)	J_{SC} (mA cm^{-2})	V_{OC} (mV)	ff (%)	Colour appearance	Type	Laboratory	Reference
N719	I$^-$/I$_3^-$	11.18	17.73	846	74.5	Red to dark brown	Ru metal organic	EPFL	Nazeeruddin et al.[28]
CYC-B11	I$^-$/I$_3^-$	11.5	20.1	743	77	Green to black	Ru metal organic	EPFL	Chen et al.[63]
N/A	I$^-$/I$_3^-$	11.9	22.58	744	71.2	Green to black	Ru metal organic	FujifilmSharp	Komiya et al.[27]
C259 + C239	Co(bpy)$_3$	11.5	17.85	891	72.2	Green to black	Organic	CAS	Zhang et al.[37]
SM315	Co(bpy)$_3$	13	18.1	910	78	Green to black	Organic zinc porphyrin	EPFL	Mathew et al.[46]
CH$_3$NH$_3$PbI$_3$	Spiro-OMeTAD	14.1	21.34	1.007	65.7	Dark brown to black	Lead halide perovskite	EPFL	Burschka et al.[52]
CH$_3$NH$_3$PbI$_3$	PTAA	17.9	N/A	N/A	N/A	Dark brown to black	Lead halide perovskite	KRICT	The NREL[53]

The table shows the power-conversion efficiency (PCE), short-circuit current density (J_{SC}), open-circuit potential (V_{OC}), fill factor (ff), colour appearance, type of dye and laboratory for best-in-class dye-sensitized solar cells (DSSCs). NREL, National Renewable Energy Laboratory.

6.2   Ruthenium Sensitizers

The optimal sensitizer for the dye-senstitized solar cell should be pan-chromatic, *i.e.*, absorb visible light of all colours. Ideally, all photons below a threshold wavelength of about 920 nm should be harvested and converted to electric current. This limit is derived from thermodynamic considerations showing that the conversion efficiency of any single-junction photovoltaic solar converter peaks at approximately 33% near threshold energy of 1.4 eV. In addition, the sensitizer should fulfill several demanding conditions:

- It must be firmly grafted onto the semiconductor oxide surface and inject electrons into the conduction band with a quantum yield of unity.
- The excited state oxidation potential of the dye must be energetically lie above the conduction band edge of the semiconductor in order to inject electrons quantitatively.
- Its ground state redox potential should be sufficiently high that it can be regenerated rapidly *via* electron donation from the electrolyte or a hole conductor.
- The extinction coefficient of the dye should be high over the whole absorption spectrum to absorb most of the light.
- The dye should be soluble in some solvent for adsorption on TiO_2 surface and should not be desorbed by the electrolyte solution.
- It should be stable enough to sustain at least 10^8 redox turnovers under illumination corresponding to about 20 years of exposure to natural sunlight.

Molecular engineering of ruthenium complexes that can act as pan-chromatic charge transfer sensitizers for TiO_2-based solar cells presents a challenging task as several requirements have to be fulfilled by the dye which are very difficult to be met simultaneously. The lowest unoccupied molecular orbitals (LUMO) and the highest occupied molecular orbitals (HOMO) have to be maintained at levels where photo-induced electron transfer into the TiO_2 conduction band and regeneration of the dye by iodide can take place at practically 100% yield. This restricts greatly the options available to accomplish the desired red-shift of the metal to ligand charge transfer transitions (MLCT) to about 900 nm.

The spectral and redox properties of ruthenium polypyridyl complexes can be tuned in two ways: firstly, by introducing a ligand with a low lying p* molecular orbital, and secondly by destabilization of the metal t_{2g} orbitals through the introduction of a strong donor ligand. Meyer *et al.* have used these strategies to tune considerably the MLCT transitions in ruthenium complexes.[64] Heteroleptic complexes containing bidentate ligands with low lying p* orbitals together with others having strong sigma donating properties show indeed impressive panchromatic absorption properties.[64] However, the extension of the spectral response into the near IR was gained at the expense of shifting the LUMO orbital to lower levels from where charge injection into the TiO_2 conduction band can no longer occur.[65,66]

Near infrared response can also be gained by upward shifting of the Ru t_{2g} (HOMO) levels. However, it turns out that the mere introduction of strong sigma donor ligands into the complex often does not lead to the desired spectral result as both the HOMO and LUMO are displaced in the same direction. Furthermore, the HOMO position can not be varied freely as the redox potential of the dye must be maintained sufficiently positive to ascertain rapid regeneration of the dye by electron donation from iodide following charge injection into the TiO$_2$.

Based on extensive screening of hundreds of ruthenium complexes, we discovered that the sensitizer excited state oxidation potential should be negative of at least –0.9 V *versus* SCE, in order to inject electrons efficiently onto TiO$_2$ conduction band. The ground state oxidation potential should be about 0.5 V *versus* SCE, in order to be regenerated rapidly *via* electron donation from the electrolyte (iodide/triiodide redox system) or a hole conductor. A significant decrease in electron injection efficiencies will occur if the excited and ground state redox potentials are lower than these values.

The introduction of the so-called **N3**-sensitizer (Figure 6.3), where two thiocyanate ligands are introduced, set a benchmark for all following dyes.[67] The purpose of thiocyanate ligands was to tune the HOMO level of the metal t_{2g} orbitals, and carboxylate groups are anchoring of the sensitizer to TiO$_2$ surface and as well lowering of LUMO levels.

Using the N3-dye as a paradigm, many research groups have started to tune the electronic and optical properties by exchanging one or more of the ligands. Another noticeable strategy from this basic matrix is a terpyridyl-Ru complex (**N749**) introduced in 2001. Its extended absorption in the near infrared significantly enhances the overall incident photon to current conversion efficiency. Thus, DSSCs sensitized with this so-called 'black dye', have a higher short circuit current. The conversion efficiency of 10.4% (1 cm^2)[68] and 11.1% (0.26 cm^2),[26] are the highest certified efficiencies so far, which makes this dye as the second reference sensitizer. In the N749 sensitizer, the HOMO levels are destabilized by introducing three thiocyanate ligands, and the LUMO levels were lowered by the presence of 4,4′,4″-tricarboxylic acid 2,2′,6,2″-terpyridine ligand reducing the

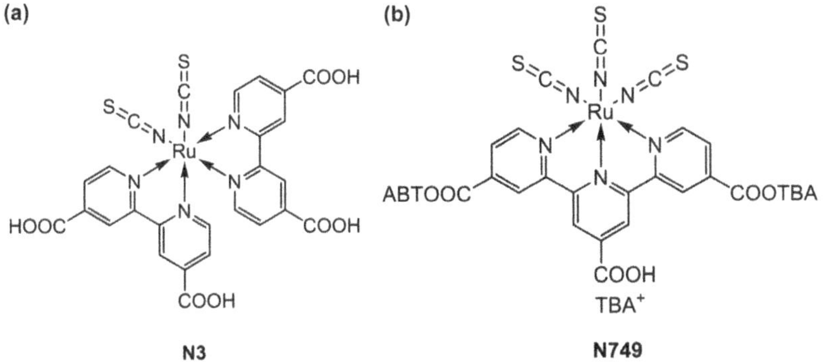

Figure 6.3 Structures of two paradigm dyes **N3** and **N749**.

HOMO-LUMO gap. The resulting sensitizer shows absorption extending in the near IR regions.

When the sensitizer is chemically bound to the TiO_2 surface, the protons of the anchoring group (carboxylic acid and/or phosphonic acid) are partially transferred to the surface of TiO_2. In a comparative study,[23,69] upon gradually exchanging the protons by bulky tetra-*n*-butylammonium (TBA) groups, the photovoltaic performance was correlated with the number of protons left on the **N3**-sensitizer, ranging from zero protons (**N712**) to four protons (**N3**). With increasing number of protons the TiO_2 surface is positively charged and the Fermi level of the TiO_2 is shifted down. The electric field associated with this surface dipole enhances the dye absorption and leads to a higher dye coverage. Another consequence is a higher driving force for charge injection, which increases the short circuit current (**N712**: 13 mA cm^{-2}; **N3**: 19 mA cm^{-2}).[69] At the same time, the energy gap between the quasi-Fermi-level of the TiO_2 and the redox electrolyte, which determines the open-circuit potential, decreases (**N712**: 900 mV; **N3**: 600 mV).[69] The maximum conversion efficiency was found for the doubly deprotonated sensitizer **N719** (11.18%) (Figure 6.4), which had a central role in advancing significantly the DSSC technology. The photovoltaic performance of **N719** is superior to that of compound N3 due to a higher V_{OC} and a comparable J_{SC}.

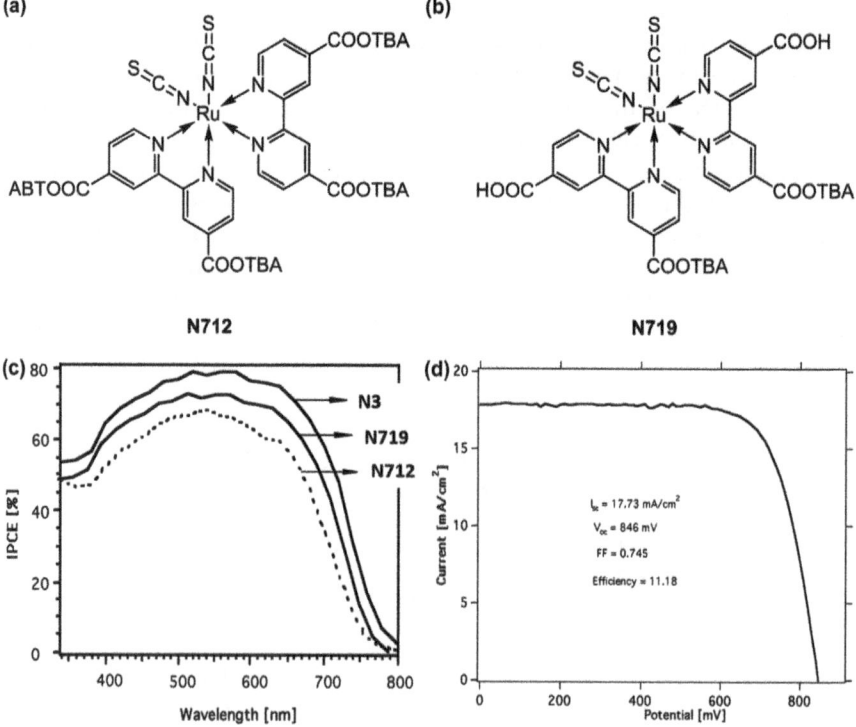

Figure 6.4 Structures of **N3** dye derivatives with different degrees of deprotonation.

6.2.1 High Molar Extinction Coefficient Sensitizers

However, the main drawback of **N719** is the hydrophilicity and relatively low molar extinction coefficients. For example, ε value of **N719** at λ_{max} of ^{1}MLCT is approximately 14 700 M^{-1} cm^{-1}, (for organic dyes, 15 000–60 000 M^{-1} cm^{-1}). Therefore, in order to sufficiently absorb the photons in the NIR region, a thick TiO$_2$ film is needed. However, increase of the TiO$_2$ thickness decreases the V_{OC} value; therefore, increase of the ε of the Ru polypyridyl dye is an important way to improve the cell performances. In order to enhance the extinction coefficient of the Ru dyes, introduction of the π-conjugated molecules into the polypyridyl ligand was carried out.

Following this thread, sensitizers exhibiting hydrophobicity and high molar extinction coefficient appeared (Figure 6.5). It is based on hetero-leptic sensitizers with a ligand incorporating thiophene or furan moieties to increase significantly the spectral properties of the complex and alkyl chains to shield the TiO$_2$ surface from the redox mediator. For example, heteroleptic Ru dyes, **C101** and **C106** possessing thiophene derivatives with hexyl chains on the bpy ligand achieved not only high conversion efficiency of over 11% but also high device durability.[70,71] In these cases, the HOMO energy level was elevated with increasing the electron-donating property of the π-conjugates. Additionally, such long alkyl chains can suppress dye aggregations, and improve the V_{OC} and injection efficiency. The **C101** and **C106** sensitizers exhibit performances similar or superior to **N719**. How-ever, the main drawback of **N719** and its derivatives is the lack of ab-sorption in the red region (band-edge around 780 nm) of the visible spectrum which makes these dyes appear a reddish colour and restricts NIR light harvesting.[72]

6.2.2 Panchromatic Ruthenium Sensitizers

To fulfill the requirement of panchromatic ruthenium complexes, **N749** ('black dye') triisothiocyanato-(2,2′:6′,6″-terpyridyl-4,4′,4″-tricarboxylato) Ru(II) tris(tetra-butylammonium) (Figure 6.3b) has been synthesized in which the ruthenium centre is coordinated to a monoprotonated tri-carboxyterpyridine ligand and three thiocyanate ligands.[73,74]

Figure 6.6 shows the photocurrent action spectrum of a cell containing **N719** and **N749**, where the incident photon to current conversion efficiency is plotted as a function of wavelength. It is evident that the response of the **N749** extends 100 nm further into the infrared compared to the **N719** sen-sitizer. The observed red shift in the black dye is due to combination of lowering the lowest unoccupied molecular orbitals and destabilization of highest occupied molecular orbitals caused by the carboxy-terpyridine lig-and and thiocyanate ligands, respectively. The photocurrent onset is close to 920 nm, *i.e.*, near the optimal threshold for single-junction converters. From that point, the IPCE rises gradually until at 700 nm and it reaches a plateau of over 80%. From the overlap integral of the IPCE curves (Figure 6.6) with the AM 1.5 solar emission, one predicts the short circuit photocurrents (J_{SC})

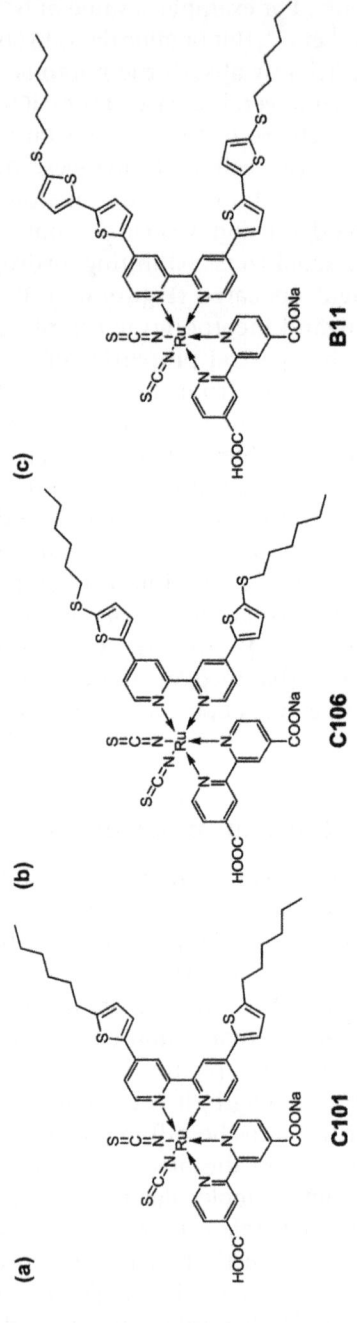

Figure 6.5   Structures of high molar extinction coefficient dyes.

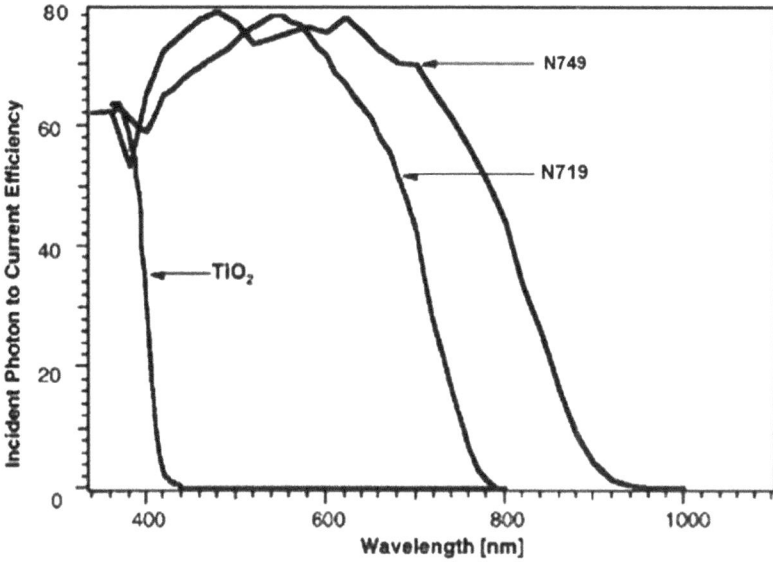

Figure 6.6 IPCE obtained with the **N749** attached to nanocrystalline TiO$_2$ films. The incident photon-to-current conversion efficiency is plotted as a function of the wavelength of the exciting light. IPCE for bare TiO$_2$ and TiO$_2$ sensitized with **N719** have been included for comparison.

of **N719** and **N749**-sensitized cells to be 16.5 and 20.5 mA cm^{-2}.[75] Routinely, the experimental photocurrents obtained with **N749** are in the range of 18–21 mA cm^{-2}.[74] The open-circuit potential (V_{OC}) is 720 mV, and the fill factor (*ff*) is 0.7, yielding for the overall solar (global AM 1.5 solar irradiance 1000 W m^{-2}) to electricity conversion efficiency (η) a value of 10.4%.[74] With the **N749** dye, conversion efficiency of 11.1% has been achieved using high-haze TiO$_2$ electrodes by Han and colleagues.[76]

Sugihara and colleagues have developed another type of panchromatic sensitizers having β-diketonato ancillary ligand in place of monodentate NCS ligands (see Ru(tctpy)(tfac)(NCS) in Figure 6.7a).[77] Ru(tctpy)(tfac)(NCS) exhibits an intense MLCT band at 610 nm with a distinct sholder at 720 nm. Under similar photovoltaic measuring conditions, Ru(tctpy)(tfac)(NCS) showed higher IPCE values between 720 and 900 nm than the N749. However, a low V_{OC} limited the power conversion efficiency. To overcome the drawback, Han and colleagues have synthesized a substituted β-diketonato Ru(II) sensitizer, Ru(tctpy)(tffpbd)(NCS)(TBA)$_2$ (tffpbd = 4,4,4-trifluoro-1-(4-fluorophenyl)butane-1,3-dione), which yielded a conversion efficiency, η, of ∼9.0% and photocurrent density J_{SC} of ∼20.0 mA cm^{-2}.[76,78] However, the substitution of two thiocyanato ligands by a fluorine-substituted tffpbd chelating ligand stabilizes the ground state oxidation potential by withdrawing electron density from the ruthenium centre. This stabilizes the ruthenium t$_{2g}$ orbitals, in turn blue-shifting the lowest energy MLCT band. Funaki *et al.* have shown Ru(II) tricarboxyterpyridyl with a pyridinecarboxylate ligand instead of two NCS ligands (see Ru(tctpy)(pc)(NCS)

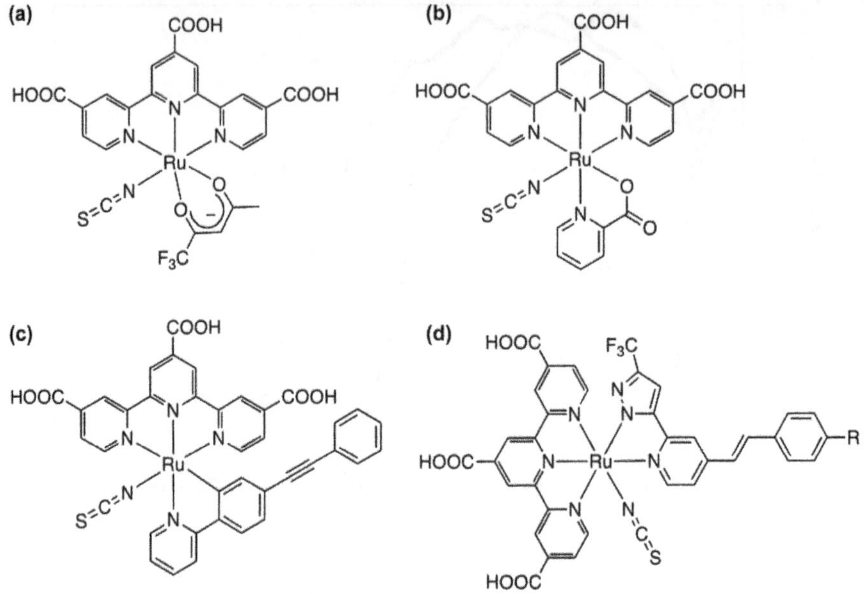

Figure 6.7 Molecular structures of (a) Ru(tctpy)(tfac)(NCS), (b) Ru(tctpy)(pc)(NCS), (c) Ru(tctpy)(C^N)(NCS), and (d) Ru(tctpy)(pypz)(NCS).

in Figure 6.7b).[79] The lowest energy band was blue-shifted compared to that of the **N749** due to the replacement of two NCS ligands with a 2-pyridinecarboxylate ligand. The data show that the electron-donating ability of a single 2-pyridinecarboxylate ligand is inferior to that of two NCS ligands. In spite of the blue-shifted MLCT peak, this sensitizer exhibits broad absorption over the visible region and a panchromatic IPCE similar to the **N749** dye. Both sensitizers showed comparable DSSC performance yielding in the case of Ru(tctpy)(pc)(NCS) J_{SC} of 19.8 mA cm^{-2} and η of 9.66% while J_{SC} of 19.0 mA cm^{-2} and η of 9.58% were obtained for **N749**. The same group has also reported a panchromatic cyclometallated Ru(ii) complex, Ru(tctpy)(C^N)(NCS), where C^N is a bidentate cyclometallating ligand, 2-(4-(2-phenylethynyl) phenyl)pyridinato (see the structure in Figure 6.7c).[80] They observed strong π–π* absorptions for the coordinated ligand in the UV region and broad MLCT absorption in a region of lower energy wavelengths than the UV region. The most notable feature in the absorption spectra is an absorption band above 700 nm with a distinct shoulder around 800 nm. These bands are attributed to a spin-forbidden MLCT absorption.[81] The dye showed absorption maxima at 749 nm ($\varepsilon = 2,700$ M^{-1} cm^{-1}) and 733 nm ($\varepsilon = 4,000$ M^{-1} cm^{-1}), respectively, due to the introduction of a C^N ligand. DSSCs with this dye showed an IPCE value of 10% at 900 nm and an onset IPCE at 1000 nm. Ru(ii) terpyridine bearing pyridine pyrazolate (pypz) (see molecular structure in Figure 6.7d) has been recently reported by Chou and co-workers.[82] The substitution drastically increased the molar extinction coefficient in the wavelength range 400–550 nm even though the MLCT band at 520 is blue shifted when compared to the **N749** dye. Substituting H, OMe, OC_8H_{17}, or *tert*-butyl groups onto the pypz ligand of the complex produced a very comparable J_{SC} (1.0–1.1 times as high as in device with the **N749**). Moreover, a long hydrophobic alkoxy chain or *tert*-butyl group resulted in ~ 30 mV gain in V_{OC} which led to over 10% power conversion efficiency.[82]

The molar extinction coefficient (ε) value of the black dye at λ_{max} of ^{1}MLCT is approximately 7000 M^{-1} cm^{-1}, therefore, the π-expansion concept is also needed to improve the molar extinction coefficients of the Ru tpy complexes. Exchanges of all the three NCS of the black dye for the electron-donating π-conjugated ligand were carried out for Ru bpy dyes.[82–84]

Chou and co-workers also reported TF series (Figure 6.8) featuring the substituting of three NCS groups by 2,6-bis(3-(trifluoromethyl)-5-pyrazolyl)pyridine

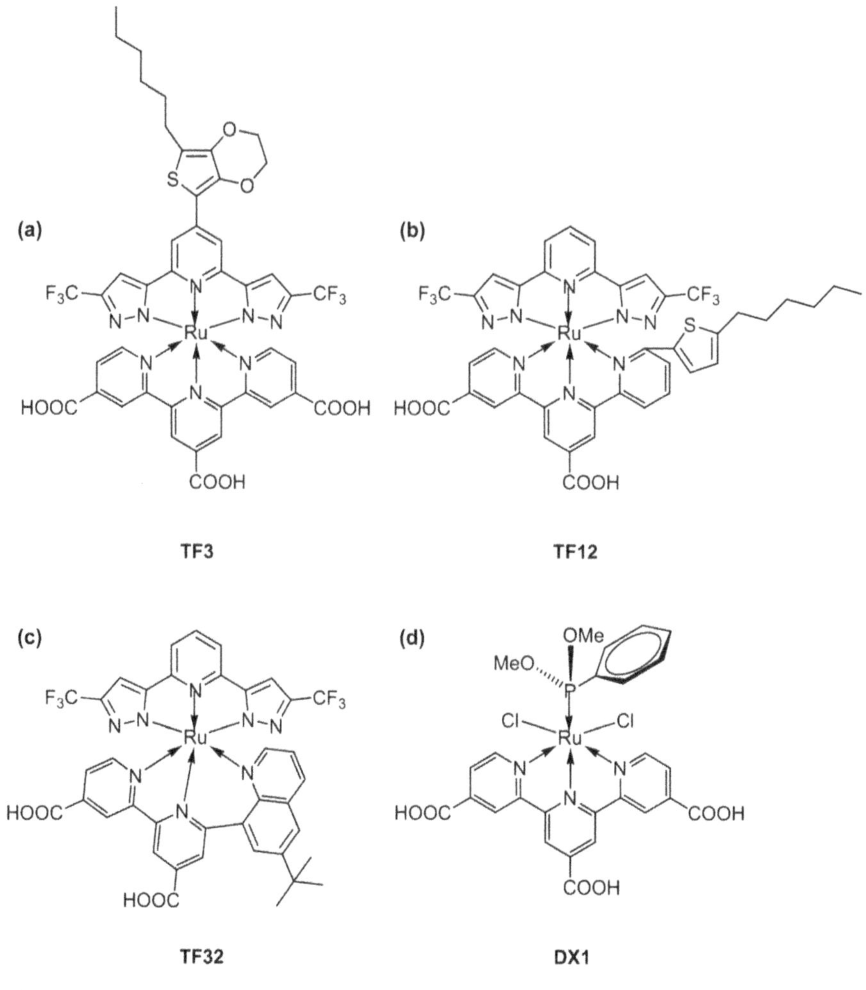

Figure 6.8   Structures of **TF3**, **TF12**, **TF32** and **DX1**.

derivatives.[85] All the TF dyes showed more intense and broader visible absorption bands compared to that of **N749** and therefore gave similar or higher current densities (19–21.4 mA cm^{-2}) than **N749** (20 mA cm^{-2}), reflecting the increased absorption and light-harvesting efficiency upon introduction of the π-conjugated pendant groups. Especially, **TF32** displayed the combined advantages of both black dye and **TF-3**, that is, much superior optical responses in the regime spanning the far-visible and NIR regions; such optical responses have never been documented for Ru(II) sensitizers with the tctpy class of anchors.[85–89] A conversion efficiencies of $\eta = 10.19$ is reported for **TF32**.[90]

DX1 is a phosphine-coordinated Ru(II) sensitizer with near-infrared, spin-forbidden singlet-to-triplet direct transitions reported by Segawa *et al.*[91] (Figure 6.8d). A DSSC using **DX1** generated a photocurrent density of 26.8 mA cm^{-2}, the highest value for an organic photovoltaic cell reported to date. A tandem-type DSSC employing both **DX1** and the traditional sensitizer **N719** is shown to have a power conversion efficiency of >12% under 35.5 mW cm^{-2} simulated sunlight. It is found that **DX1** in DMF solution (at a temperature of 298 K) has an intense ultraviolet absorption band at a wavelength of ∼320 nm, which is assigned to the ligand-centred π–π* transition of the terpyridyl ligand. In addition, broad absorption bands in the 500–900 nm region are assigned to MLCT transitions, which are characteristic of typical Ru(II) polypyridyl complexes. Notably, the MLCT band of **DX1** at ∼620 nm is weaker in intensity than the MLCT bands of **N749**, but the **DX1** exhibits higher-intensity and broad electronic absorption peaks at longer wavelengths centred at 792 nm. This type of ligand design featuring a spin-forbidden transition in the near infrared could be applied to other sensitizers and may be useful for the panchromatic sensitization.

6.2.3 Cyclometallated NCS-free Ruthenium Sensitizers

The thiocyanate ligands are usually considered as the most fragile part of the ruthenium dyes. First, because it is a monodentate ligand, therefore it is easier to de-coordinate than a bidentate ligand like bipyridine. Second, it is an ambidentate ligand, which can coordinate through either the sulfur atom or the nitrogen atom. Efforts have been made to replace thiocyanate ligands without great success as the efficiencies obtained for the devices remain well below 10%.

A promising result was obtained recently by replacing the thiocyanate by a cyclometallated 2,4-difluorophenyl-pyridine, yielding the complex **YE05** (Figure 6.9b).[92] The 2,4-difluorophenyl-pyridine ligand is widely used in iridium emitters for organic light emitting devices (OLEDs).[93] The spectral response is significantly red shifted when compared to **N719** as can be seen in the IPCE spectrum which reach a maximum over 80% at 600 nm extending to 800 nm. It originates from the absorption spectrum where three absorption bands can be observed in the visible instead of the two usually obtained with thiocyanate-based ruthenium complexes. In addition, the lowest energy

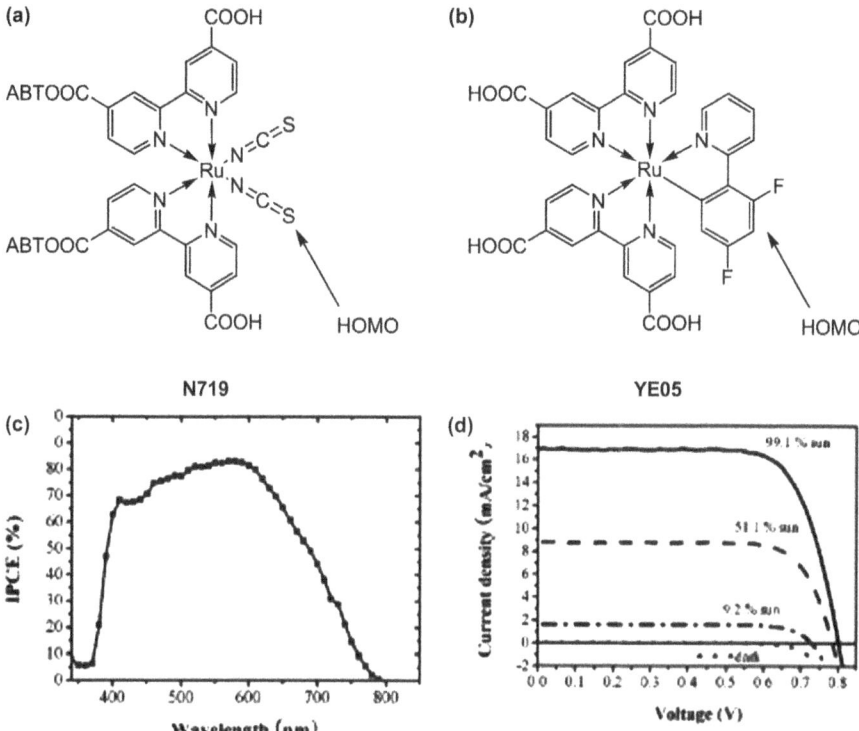

Figure 6.9 Chemical structure of **YE05** (b) and IPCE spectrum (c) and photocurrent voltage curves (d) under various light intensities of AM 1.5 sunlight.

MLCT band in **YE05** is red shifted by 25 nm when compared to **N719**, with overall remarkable high molar extinction coefficient exceeding substantially that of **N719** over the whole visible domain. This is due to the cyclometallated ligand, which is a stronger donor than the two thiocyanate groups. It results in the stronger destabilization of the highest occupied molecular orbital (HOMO) than the lowest unoccupied molecular orbital (LUMO) (Figure 6.9). The presence of the two fluorine atoms allows the fine tuning of the redox potential of the sensitizer. Overall, **YE05** produces a short-circuit photocurrent of 17 mA cm^{-2}, a V_{OC} of 800 mV, and a fill factor of 0.74, corresponding to a conversion efficiency of 10.1% under AM 1.5 standard sunlight. Thus, **YE05** emerges as a prototype for thiocyanate-free cyclometallated ruthenium complexes, exhibiting remarkable spectral and stability properties.

6.2.4 Cyclometallated NCS-free Ruthenium Dyes with a Co^{III}/Co^{II} Redox Shuttle

As has been mentioned in the introduction part, electrolytes that rely on the Co^{III}/Co^{II} redox shuttle can produce high PCE for organic dyes;[94–96] however, among the few examples reported in the literature[97–102] only limited performances have been achieved with ruthenium(II) sensitizers. The highest performing dye (**Z907**) yields efficiencies of up to 6.5% with cobalt[100] *versus* 8.5% with iodine,[103] unless the co-adsorbent is specifically engineered.[104] Molecular dynamic simulations suggest close contact interactions between the cobalt(III) species and the anchored sensitizer(s), which causes undesired recombination of the electrons injected into the TiO$_2$ conduction band to the redox mediator.[105] This concept has been successfully proved for organic dyes through the addition of peripheral bulky substituents, which prolong the electron lifetime in the semiconductor by preventing the electrolyte from accessing the surface[94] (*vide infra*). Implementation of a similar design principle could improve compatibility between ruthenium(II) sensitizers and cobalt electrolytes. Cyclometallated tris-heteroleptic complexes[106] as good examples of NCS-free sensitizers will have chance to yield high-efficiency iodine-free DSSCs.

For example, Berlinguette *et al.* based on the structure of **YE05** synthesized a series of trisheteroleptic cyclometallated Ru^{II} sensitizers with three 'functionalized' ligands[101] (Figure 6.10). In this type of complex, each of the three bipyridines are given specific functions: one ligand is responsible for anchoring onto the TiO$_2$ surface; the second ligand is substituted with a π-delocalized system and hydrophobic chains to enhance light harvesting and protect the TiO$_2$ surface; the third cyclometallating ligand provides electron-donating character. Besides the moderate performance of **B3** in an I$^-$/I$_3$$^-$-based electrolyte (7.3%), they also demonstrated that **B3** give the highest efficiency obtained for Ru-based dye with a Co-based electrolyte at 1 sun before 2011 ($\eta = 5.5\%$). Before that, the highest efficiency for a Ru-based dye using [Co(bpy)3]$^{2+/3+}$ was 1.3%.[99]

Motivated by these encouraging results, Polander *et al.* further designed and synthesized two multifunctional cyclometallated NCS-free ruthenium dyes (**LP1** and **LP2**) (Figure 6.11) aiming at further improving the compatibility of ruthenium dyes with cobalt electrolyte.[107]

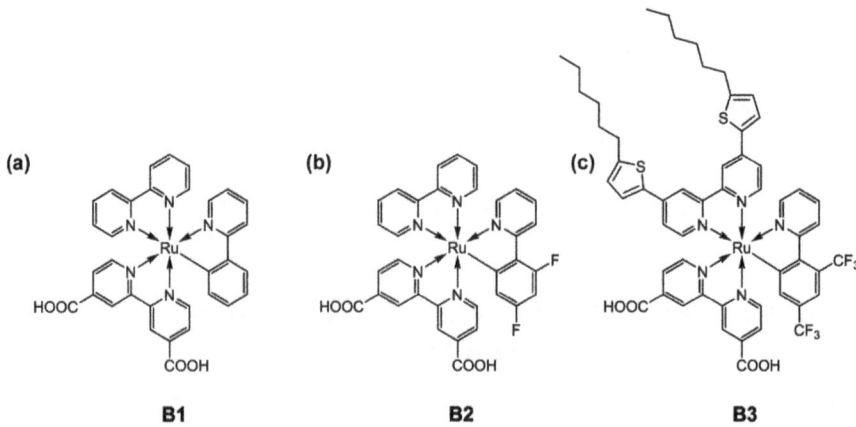

Figure 6.10   Typical cyclometallated NCS-free ruthenium dyes **B1**, **B2** and **B3**.

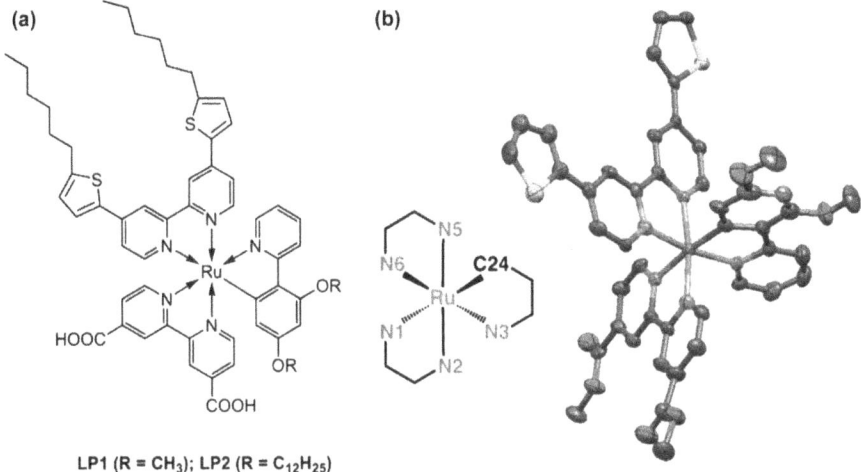

(a)

(b)

LP1 (R = CH₃); LP2 (R = C₁₂H₂₅)

Figure 6.11   Structures of **LP1** and **LP2**. Crystal structure of **LP1**·2CH₂Cl₂. Thermal ellipsoids are drawn at the 50% probability level. The counter-ion, hexyl chains, hydrogen atoms, and solvent molecules are omitted for clarity.

They found 2′,6′-dimethoxy-2,3′-bipyridine to be convenient for tuning the HOMO energy level. The alkoxy substituents are also advantageous as a substitution point to insulate the TiO_2 surface by elongation of the chains. To validate this concept, both methoxy and dodecyloxy derivatives were synthesized. The other two ligands used to complete the coordination sphere of the ruthenium centre have two functions: 2,2′-bis(5-hexylthiophen-2-yl)-2,2′-bipyridine enhances the absorptivity of the complex, compared to an unsubstituted 2,2′-bipyridine ligand,[108] whereas 2,2′-bipyridine-4,4′-dicarboxylic acid serves as an electron-accepting ligand and anchor to the TiO_2. Owing to the rational design, DSSCs exhibiting comparable efficiencies with both cobalt and iodine redox mediators were obtained.

The incident photon-to-current efficiency (IPCE) spectra using a $[Co^{II/III}(phen)_3]^{2+/3+}$ electrolyte are shown in Figure 6.12a. The complex **LP2** was also investigated using I^-/I_3^- for comparison purposes. The three IPCEs have a similar shape in the range of 380–770 nm, but with significantly different intensities. Similar to other ruthenium(II) sensitizers,[18,19,22] the photocurrent action spectra is limited to *ca.* 50% with **LP1**. The presence of C12 alkoxy chains on the cyclometallated ligand results in an increase of up to 70%, as exemplified by **LP2**. Notably, this is lower than what can be obtained when iodine is used as redox mediator. The solar-to-electricity conversion efficiencies were evaluated by recording the current–voltage (*J–V*) characteristics under simulated AM1.5G illumination (100 mW cm^{-2}) and the results are depicted in Figure 6.12b. As expected, the J_{SC} values follow the trend observed in IPCE intensities. The measured V_{OC} values show large differences between **LP1** and **LP2**. The lower V_{OC} obtained with iodine in the

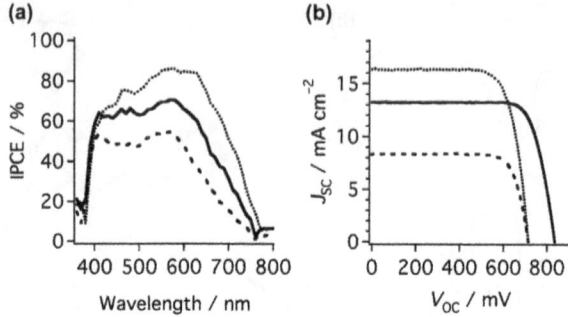

Figure 6.12 (a) Photocurrent action spectra in mesoscopic solar cells, and (b) *J–V* characteristics measured under simulated AM 1.5 G full sun illumination (100 mW cm^{-2}) for devices employing **LP1** and **LP2**. **LP1** + cobalt (- - - - -), **LP2** + cobalt (· · · · · · ·), **LP2** + iodine (· · · · · ·).

case of **LP2** highlights the benefit of using a cobalt electrolyte. The recombination of electrons injected into the semiconductor can occur by reductive electron transfer to the electrolyte and/or to the oxidized form of the sensitizer; however, the presence of C12 alkoxy chains can insulate the TiO$_2$ surface by preventing close proximity with the redox mediator. This effect has a major impact on both the J_{SC} and V_{OC} photovoltaic parameters. Overall, the J_{SC} and V_{OC} obtained for **LP1** (8.3 mA cm^{-2} and 714 mV, respectively) contribute to a power conversion efficiency of 4.7%. As a result of the increased electron lifetime, the J_{SC} and V_{OC} values measured for **LP2** rise to 13.2 mA cm^{-2} and 837 mV, respectively; DSSCs with this sensitizer reach efficiencies of up to 8.6% in the presence of a cobalt electrolyte. Notably, this result closely matches that obtained with iodine (8.7%), which has not been thus far reported for a Ru(ɪɪ) sensitizer.

6.3   Metal-free Organic Sensitizers

Although ruthenium-based complexes work well and have been the most widely used dyes over the past two decades, metal-free pure organic dyes are also promising candidates as sensitizers in DSSCs due to the high molecular extinction coefficient (50 000–200 000 M^{-1} cm^{-1}), low cost, and high design flexibility of the molecules. Although organic sensitizers typically have narrower spectral bandwidths ($\Delta\lambda \approx$ 100–250 nm), great strides have been made in designing and synthesizing new organic dyes for use in DSSCs over the past few years.[109–112]

Almost all of the organic dyes are designed based on donor–π-spacer–acceptor (D-π-A) architecture, in which electron-rich (donor) and electron-poor (acceptor) sections are connected through a conjugated (π) bridge and the anchoring group is attached with the acceptor part as shown in Figure 6.13. The electron-poor section is functionalized with an acidic binding group that couples the molecule to the oxide surface. Photoexcitation causes a

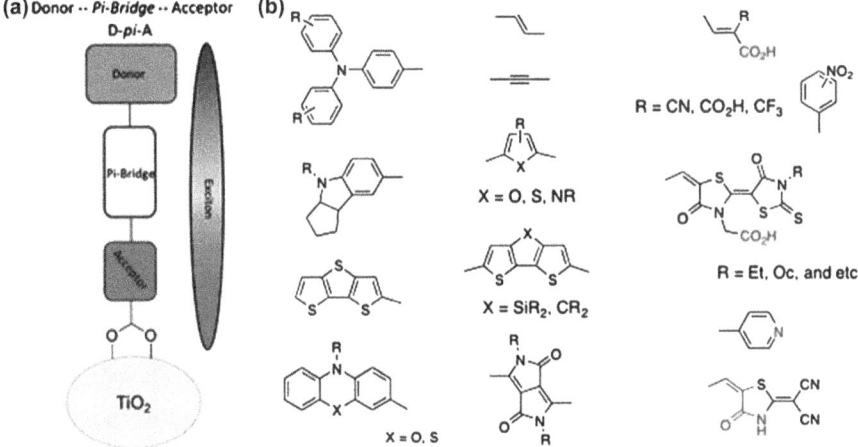

Figure 6.13 Schematic drawing of D-π-A type organic dye and some samples of the donor, π-spacer and acceptor with anchoring components.

net electron transfer from the donor to acceptor sections such that the electron wavefunction couples to the titania conduction band states, while the hole wavefunction resides mostly away from the oxide surface where it is well-positioned to interact with the redox couple.[35,42,113] Alkyl chains are also often attached to the side of the dye to create a barrier between holes in the redox couple and electrons in the titania, thereby inhibiting recombination.

Normally, as donor parts, electron-rich moieties such as triarylamines,[114] carbazoles,[115] indolines,[109,116–118] phenoxazines,[119–121] and pyran[122] are often employed. Corresponding acceptor parts are composed of an electron-withdrawing unit and an anchoring group. Practically, cyanoacrylic acid and its analogues,[123] rhodanines,[109,124–128] and pyridines[129,130] are frequently used owing to their electron-withdrawing properties and coordinating features to the TiO$_2$ electrode. As π-spacer units, π-conjugated systems such as polyenes,[131–134] polyynes,[135,136] thiophenes,[35,137–139] furanes,[140–142] pyrroles,[143] fused thiophenes,[144–147] diketopyrrolopyrrole,[148–150] benzothiadiazole,[118,151–153] and benzotriazole[154,155] are incorporated into D-π-A dyes.

Like ruthenium-based complexes, different types of D-π-A sensitizers have been working well with the I$^-$/I$_3$$^-$ redox couple, and a **C219**-based cell has been reported to have one of the highest performance among the organic dyes ($V_{OC} = 0.77$ V; $J_{SC} = 17.94$ mA cm^{-2}; $\eta = 10.3\%$; FF $= 0.73$).[156] However, due to the eagerness of the community to replace iodine-containing electrolytes, for which the reason has been elaborated in the introduction part, it is now a fashion that, when designing a new organic dye, one should think about the compatibility of it with an iodine-free electrolyte, *e.g.*, cobalt bi-pyridine complexes. In this section we will discuss one successful story of molecular engineering in making cobalt electrolyte compatible organic sensitizers.

6.3.1 Organic Sensitizers and their Cobalt Electrolyte Compatibility

Divalent/trivalent cobalt bipyridine complexes ($[Co(bpy)_3]^{3+/2+}$) are the most studied iodine-free electrolytes. The advantages of using a cobalt-based redox shuttle as an electrolyte include variability of the chemical structure, tunability of the redox potential and transparency to visible light. Despite these merits, however, after the first report in 2001 by Nusbaumer *et al.*[19] few papers have been published on cobalt complex redox shuttles in DSSCs,[32,97,105,157,158] mainly due to the lack of proper sensitizer compatibility with the new electrolyte: it has been shown that when the cobalt-based electrolytes were tested with sensitizers that performed well with I^-/I_3^-, the efficiencies achieved were rather poor in comparison,[98–100,159] because Co^{2+}/Co^{3+} electrolytes suffered from recombination rates that were at least an order of magnitude faster than iodide-based systems.[19,97] Only a few examples have been shown where a sensitizer gave improved efficiency in a cobalt-based electrolyte compared to that of an iodide/triiodide electrolyte.

In 2010, Sun and Hagfeldt demonstrated a significant improvement in the power-conversion efficiency of cobalt-based systems by modification of the base triphenylamine (TPA) donor with two bis(2,4-dibutoxyl)benzene substitution which provide a bulkier and slightly less strongly donating group than the base TPA.[94,99] The 'as synthesized' dye **D35** (Figure 6.14) performed very well with cobalt-based redox shuttles because of the efficient suppression of recombination by the extra bulky bis(butyloxy)phenyl on the TPA moiety with overall conversion efficiencies of 6.7% compared to 6.0%.[94,160]

To further explore the validity of the so-called BPTPA donor (shown as a bold structure in Figure 6.14), Yi *et al.* further synthesized the dye **Y123** with much extended spectral response compared to **D35**. Therefore the dye afforded an improved power conversion efficiency (PCE) of 8.8% with the cobalt-based redox shuttle,[161] and further improvement to 10% by optimizing the TiO$_2$ mesoporous film[162] and the use of a new cobalt bipyridine pyrazole electrolyte[32] (Figure 6.15). Later on, **Y123** was applied as a co-sensitizer with

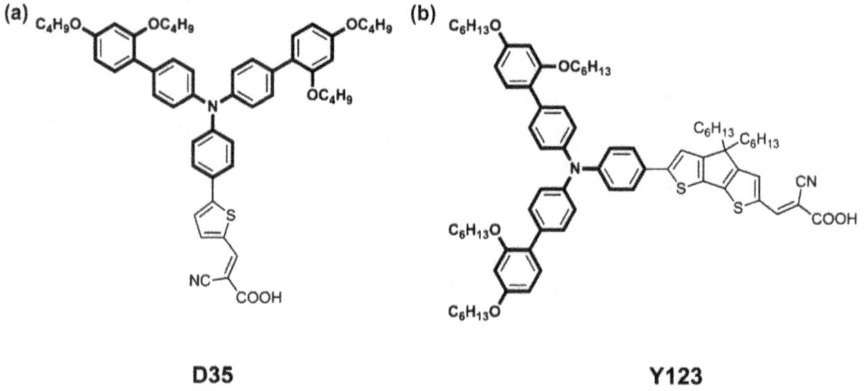

Figure 6.14 Structures of **D35** and **Y123** (BPTPA donor in bold).

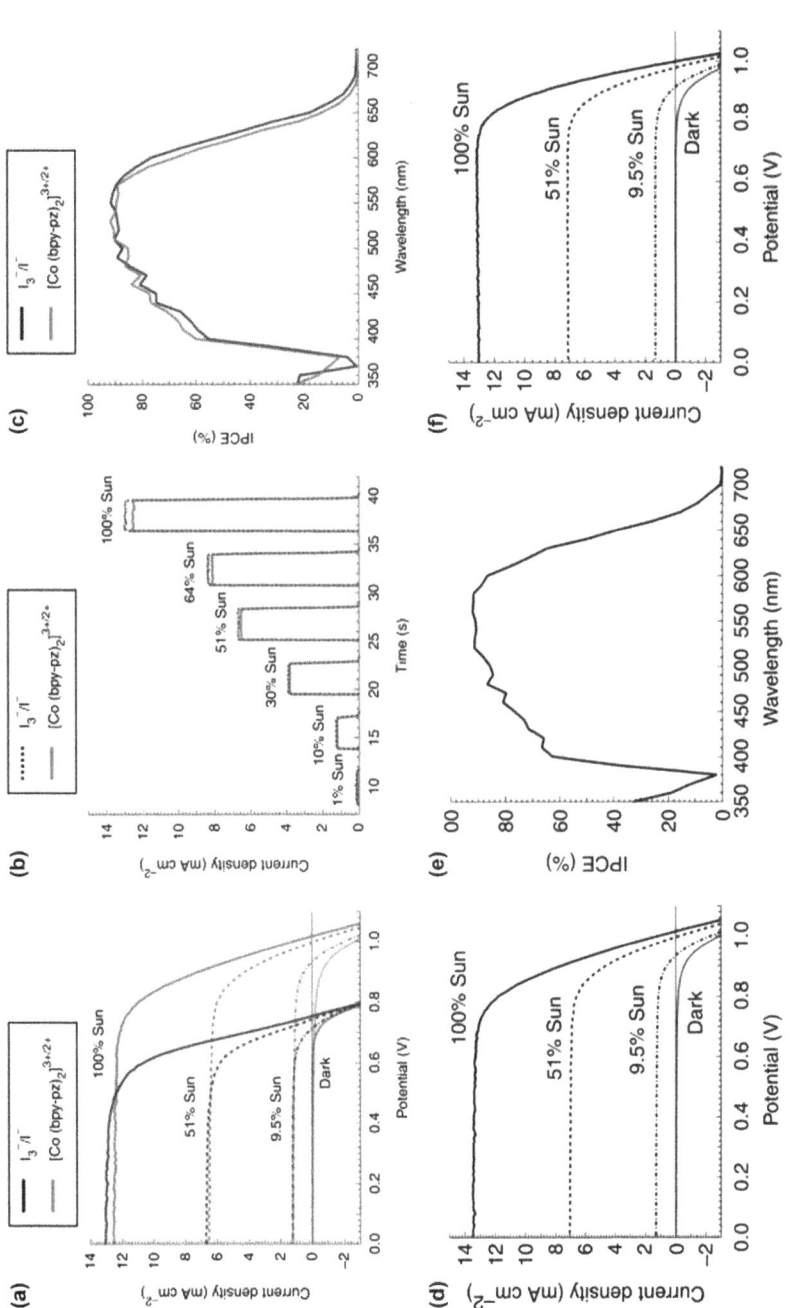

Figure 6.15 Photovoltaic characteristics of **Y123** using cobalt and iodide redox systems. (a) Comparison of *J–V* curve for the [Co(bpy-pz)$_2$]$^{3+/2+}$ (pale gray lines) and I$^-$/I$_3^-$ (black lines). (b) Photocurrent dynamics as a function of light intensity for the [Co(bpy-pz)$_2$]$^{3+/2+}$ (pale gray lines) and I$^-$/I$_3^-$ (black lines). (c) Comparison of IPCE spectra using TiO$_2$ film thickness is 5.6 µm and [Co(bpy-pz)$_2$]$^{3+/2+}$ (pale gray lines) and I$^-$/I$_3^-$ (black lines) redox systems. (d) *J–V* characterization and (e) IPCE of the DSSC employing the double-layered TiO$_2$ (5.6 + 5 µm) and Pt counter-electrode. (f) *J–V* characterization of the DSSC employing the double-layered TiO$_2$ (4.0 + 4.5 µm) and the PProDOT cathode instead of Pt.

a porphyrin chromophore to yield the record efficiency of 12.3%, used in conjunction with the $[Co(bpy)_3]^{3+/2+}$ redox couple.[163] All these examples indicated that the D-π-A dyes with the BPTPA donor are compatible with cobalt-based redox mediators, which are promising alternatives to the I^-/I_3^- redox couple for large-scale manufacturing of DSSCs, for the reasons mentioned above.[164]

6.3.2 Size Effect of the Donor Groups in the Cobalt Electrolyte Compatibility of Dyes

For the purpose of a convenient synthesis and scale up of the BPTPA donor, Gao *et al.* reported a newly designed synthetic route towards the donor moiety before it is applied to construct new sensitizers **NT35** and **G220** (Figure 6.16). Then they evaluated the effectiveness of the BPTPA donor towards hampering the recombination rate in both electrolytes (a loss mechanism) and DSSCs with these two redox shuttles were fabricated for each dye with the same double-layer TiO_2 as the photo-anode and Pt-coated glass as the counter electrode.[165]

Figure 6.17a and b shows the current–voltage characteristics of the DSSCs fabricated with **NT35**, **G220** and **G221** as sensitizers under standard global

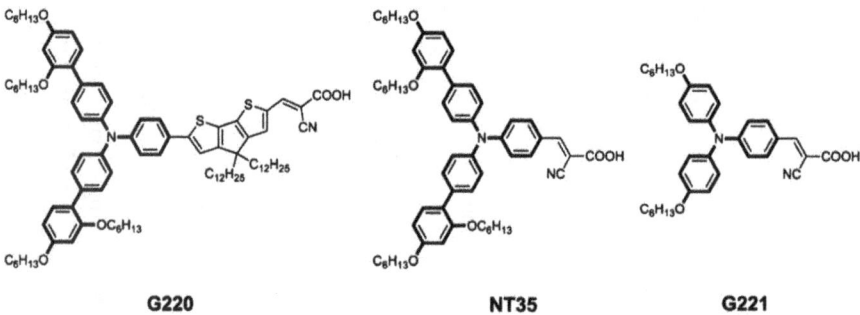

Figure 6.16 Structures of **G220**, **NT35** and **G221**.

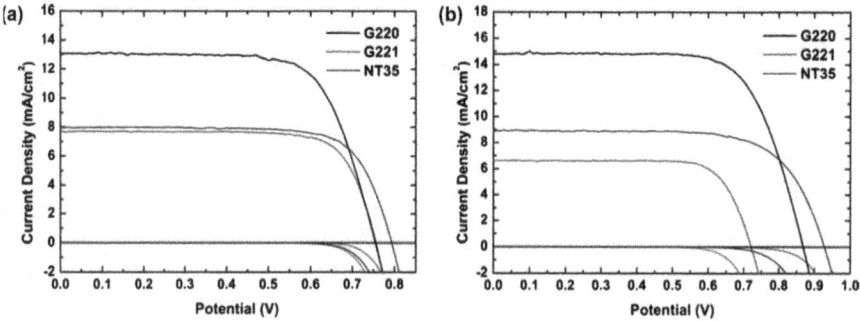

Figure 6.17 *J–V* curves of DSSCs sensitized with the dyes with (a) iodine and (b) cobalt as a redox couple.

AM 1.5 solar light conditions in iodide- and cobalt based electrolytes respectively. **NT35** has higher V_{OC} than that of **G221** in both redox systems because the introduction of the bulky BPTPA donor can act as a barrier, preventing electrolyte from approaching the TiO$_2$ surface, and thus inhibiting charge recombination, which is more dramatic with cobalt-based electrolytes. For **NT35** and **G220**, the J_{SC} values of the cells with the cobalt electrolyte were around 12% higher than that of the iodide/triiodide-based DSSCs. This increase was also reflected in the IPCE curves and could be ascribed to the lack of competitive light absorption by iodine electrolyte. It is noteworthy that the two dyes with the BPTPA donor (**NT35** and **G220**) both have higher open-circuit voltages (V_{OC}) (928 mV and 868 mV) in the cobalt electrolyte than in the iodide/triiodide electrolyte (793 mV and 755 mV) measured under the same conditions and the increase is ascribed mainly to a 210 mV higher oxidation potential of the redox potential of the cobalt complex compared to an iodide system.[161] In contrast, both the J_{SC} and V_{OC} of the **G221** cell with the cobalt-based redox shuttle are lower than those of the same cell with an iodide-based electrolyte system (Figure 6.17). This result is interpreted by rapid recombination of electrons in the TiO$_2$ conduction band with the Co^{3+} redox species, which can be indicated from the changes in the dark current with applied bias (Figure 6.17). The onset of recombination current occurs at the lowest potential for **G221** compared to the other two dyes in the cobalt electrolyte (Figure 6.17). The observed increase in the dark current density of **G221** is essentially due to the inadequate suppression of Co^{3+} reduction to Co^{2+} at the dye-sensitized TiO$_2$ electrode, which is a result of inadequate surface protection by the sensitizing dye. Notably, the dyes with the BPTPA donor all exhibited relatively lower dark current density.

The BPTPA donor showed tangible effects towards solving this recombination problem. That is to say, by introducing the BPTPA donor, we can fully make use of the advantageously tunable redox potential of the cobalt-based electrolytes without worrying about the adverse charge recombination. Consequently, the cell based on the **G220** dye showed the best photovoltaic performance in both iodide-based (6.97%) and cobalt-based electrolytes (9.06%).

6.3.3 Towards Cobalt Electrolyte Compatible Panchromatic Organic Dyes

In donor–bridge–acceptor (D-π-A) organic sensitizers, the highest occupied molecular orbital (HOMO) and lowest unoccupied molecular orbital (LUMO) are spatially differentiated. The HOMO resides preferentially across the donor and the bridge, whereas the LUMO extends predominantly across the bridge and the acceptor. Because of intrinsic electron-donating properties, triphenylamines are frequently used as donors. The choice of donor affects the intramolecular charge transfer (ICT) within the dye, both in wavelength and intensity of the absorption band. Stronger electron donation is expected to reduce the optical gap and produce a red shift in absorption by raising the HOMO energy level more than that measured in dichloromethane and acetonitrile solutions.

Frey *et al.* synthesized a new sensitizer (**JF419**) that emulates the strong electronic push of **C218** and the favourable back-recombination features of **Y123**, all with a single donor (Figure 6.18).[166] The desired effect is obtained using a fluorene moiety without further modification at the π-bridge or the acceptor. Because of this careful design, DSSCs based on **JF419** achieve efficiencies up to 11.1% at 50% sun, and 10.3% overall efficiency at AM 1.5 G simulated sunlight (100 mW cm^{-2}, *i.e.*, full sun) in the presence of a standard cobalt electrolyte.

Figure 6.19 shows the incident photon-to-current conversion efficiency (IPCE), as a function of the light excitation wavelength for DSSCs with this optimized electrolyte. Films coated with **JF419** harvest significantly more photons than films sensitized with **Y123** and **C218**, with the IPCE reaching *ca.* 85% from 400 nm to 620 nm. Therefore, in this region, the internal quantum efficiency for current generation is close to unity when taking into account light reflection by the glass. This is significantly better than **Y123** and **C218** films for which <75% of the incident light is converted from 400 nm to 620 nm. Furthermore, the IPCE spectrum of **JF419** is drastically higher at longer wavelength, with a slow decline until 740 nm. Therefore, through careful design, the authors merged two established concepts, strong electron donation and surface insulation, while taking into the account the intrinsic limitations of the device. The resulting **JF419** sensitizer shows superior incident photon-to-current conversion efficiency (IPCE) response and efficiencies than previous organic dyes, demonstrating the relevance of the design.

Besides increasing the donating strength of donor groups, another strategy to extend the dye absorption further into the red is the extension of the conjugation length of the π-bridges.[167] In this regard, π-bridges with different conjugation lengths are employed to connect bisphenyl triphenyl-amine (BPTPA) and cyanoacrylic acid and four different sensitizers (Figure 6.20) are synthesized by Gao *et al.* to investigate the general influences of the length of the π-bridges (EDOT, bisEDOT, EDOT-CPDT, and bisEDOT-CPDT) on the optoelectronic and photovoltaic properties of DSSCs using a tris(bipyridine)cobalt(II/III) redox shuttle.[38]

The spectroscopic properties showed clearly the effect of extending the conjugation of the bridging moieties on of these dyes. Based on **G188**, the addition of one more EDOT extended the π conjugation and gave **G234**, leading to a bathochromic shift of 15 nm in the absorption spectra compared to that of **G188**. On the other hand, the substitution of one EDOT linker in **G234** by CPDT, giving **G268**, caused a dramatic red shift to 581 nm. A significant red shift of **G268** relative to **G188** and **G234** is derived from the much larger π system and co-planar structure of CPDT, which extends the conjugation length. From **G268** to **G270**, due to a greater length of

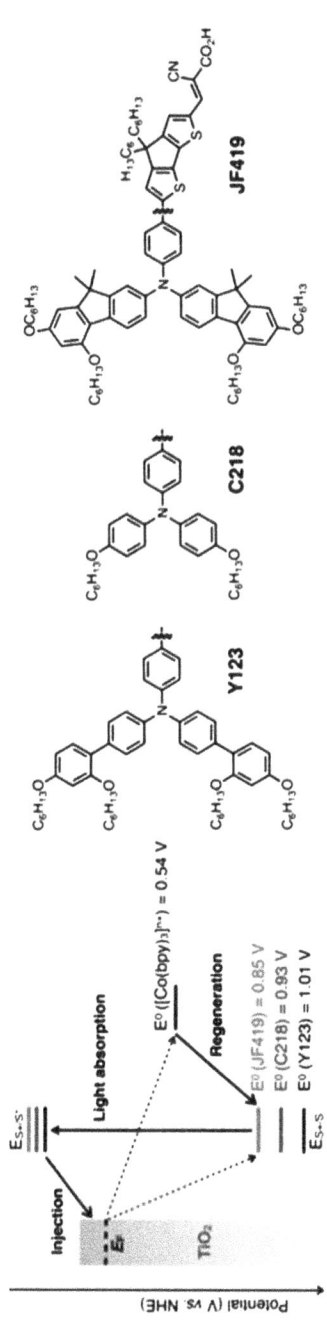

Figure 6.18 Schematic energy diagram of TiO$_2$ photoanodes sensitized with **Y123**, **C218**, and **JF419** with a [Co(bpy)$_3$]$^{2+/3+}$ electrolyte. The electron-donating strength of the donor increases from left to right. Recombination processes are indicated with dashed arrows.

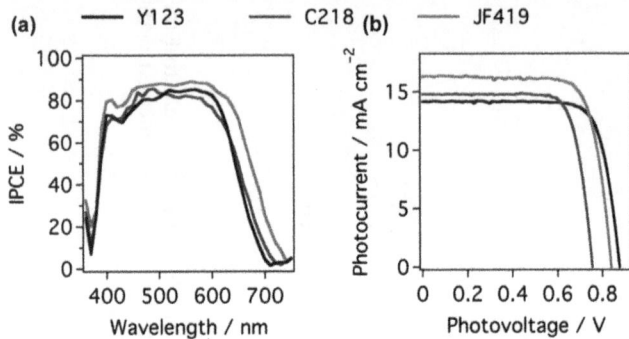

Figure 6.19 Incident photon-to-current conversion efficiency (IPCE) spectra (a) and current–voltage characteristics of **Y123**, **C218**, and **JF419** sensitizers with a $[Co(bpy)_3]^{2+/3+}$ electrolyte.

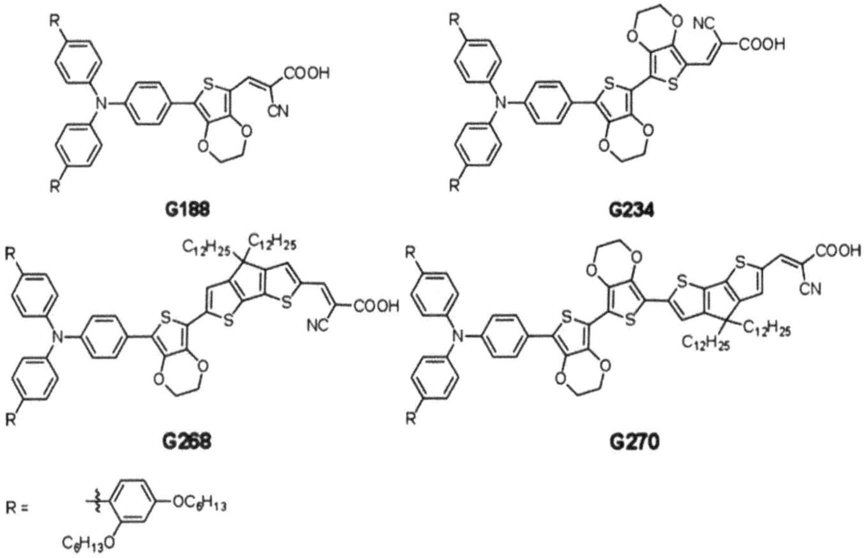

Figure 6.20 Synthesized organic sensitizers with extending conjugation length.

conjugation, a bathochromic shift of 31 nm resembling the trend from **G188** to **G234** was observed. Among the four dyes, **G270** exhibited the highest absorption coefficient and the longest-wavelength charge-transfer absorption band.

The energy-offsets of the dye molecules with respect to the nanocrystalline TiO_2 and redox electrolytes are depicted in Figure 6.21. With the extension of conjugation length, the values of first oxidation potential (E_{ox}) and E_{0-0} of these dyes are decreasing gradually. The low E_{ox} of **G270** could lead to slow dye regeneration, inducing the geminate charge recombination between oxidized dye molecules and photoinjected electrons in the nanocrystalline TiO_2 film.

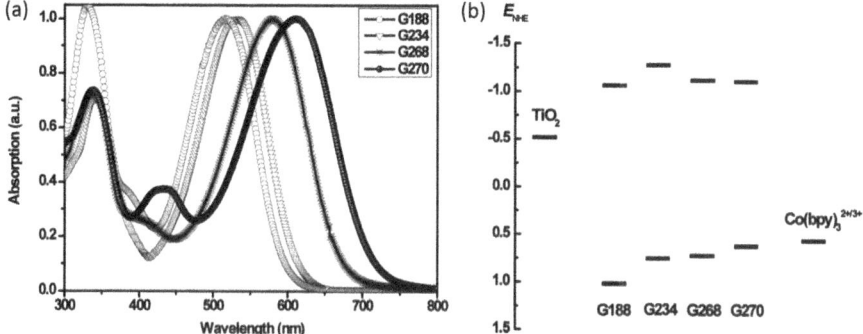

Figure 6.21 UV–visible absorption spectra of **G188**, **G234**, **G268** and **G270** in CH$_2$Cl$_2$ and energy diagram of the sensitizers with [Co(bpy)$_3$]$^{2+/3+}$ electrolyte based on the results of cyclic voltammetry measurements.

With the elongation of the π-bridges, the series of sensitizers showed gradually red-shifted electronic absorption spectra and a persistent decrease in oxidation potential observed *via* cyclic voltammetry measurements, which verified the merit of extending the bridge length to enhance light-harvesting properties. Moreover, the same elongation is also found to affect the charge displacement and dipole of dyes grafted on titania, which have a dramatic impact on the physicochemical properties of the titania/dye interface and thereby alter the TiO$_2$ conduction band.[168–170] In this series of dyes, photocurrent action spectra show that the extension of π-conjugated bridges decreases the open-circuit photovoltage through lowering the TiO$_2$ conduction band. The best performance is shown in G268 with a short-circuit photocurrent density (J_{SC}) of 16.27 mA cm^2, an open-circuit photovoltage (V_{OC}) of 0.83 V, and a fill factor (ff) of 0.67, corresponding to an overall conversion efficiency of 9.24%. Unexpectedly, **G270**, which has the longest π-bridge, showed the lowest J_{SC}, V_{OC}, and efficiency. The authors believe that this structure–property relationship study will shed light on a better molecular design and synthesis of highly efficient organic sensitizers (Figure 6.22).

6.3.4 Donor–Chromophore–Acceptor-based Asymmetric Diketopyrrolopyrrole Sensitizers

A major development in DSSC technology occurred with the advent of high-performance 'metal'-free sensitizers (*i.e.*, without the precious metal ruthenium), which typically rely on the donor–π-bridge–acceptor (D-π-A) motif,[17] where the bridge serves primarily to extend π-conjugation, while efficiently relaying electron density from the donor to the acceptor. Holcombe *et al.* described a dye configuration in which a chromophore core can serve as an inherently coloured π-bridge, leading to the conceptualization of the donor–chromophore–anchor (D-C-A) sensitizer (Figure 6.23).[171] This class of dye, although not called as such, is already under development in

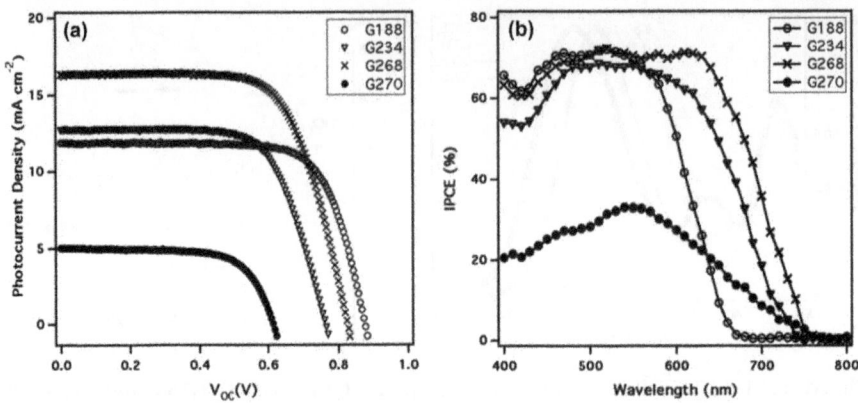

Figure 6.22 (a) Photovoltaic performance obtained at 100 mW cm^{-2} G1.5 A.M. illumination and (b) IPCE spectra of the sensitizers.

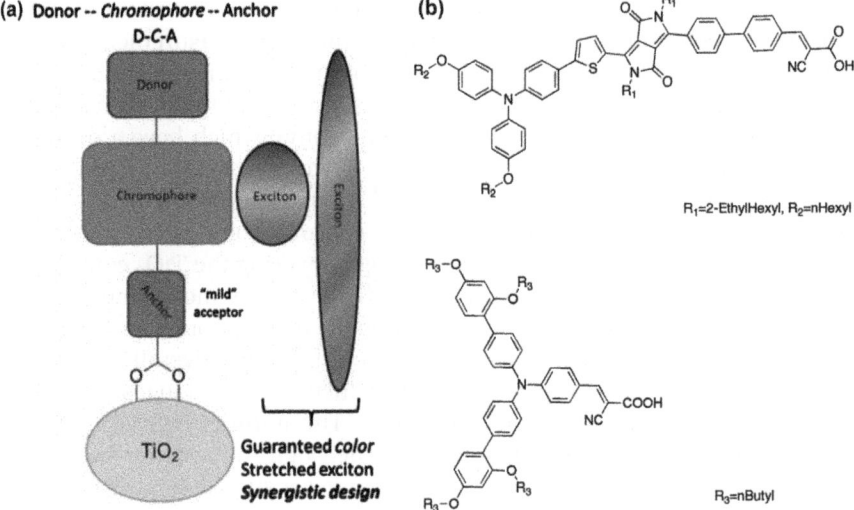

Figure 6.23 Donor–chromophore–acceptor concept (left) and structures of **DPP07** (top right), and **NT35** (down right).

the field of DSSC sensitizers. This approach has achieved light harvesting of 1000 nm with squaraine[172] and holds the world-record PCE with the porphyrin chromophore (*vide infra*).[163]

The DPP core is at the heart of many chemical structures for application in materials technology, from car-paint pigments[173] to small molecule and polymeric organic photovoltaics.[174] However, DPP-based sensitizers have thus far exhibited limited use in DSSCs.[148,175–179] Understanding how to molecularly engineer the industrial pigment DPP into a high-performance D-C-A sensitizer represents under-explored potential within the field of DSSC sensitizer research.

The seminal work on DPP sensitizers was published in 2010,[148] comparing the phenyl-DPP (PhDPP) and thienyl-DPP (ThDPP) bridge. This original study prompted Holcombe *et al.* to investigate DPP, which led to the development of higher efficiency ThDPP-based sensitizers,[180] as well as the discovery of an unexpectedly high performance asymmetric DPP bridge. A mixed thienyl-DPP and phenyl-DPP unit (Th-DPP-Ph) provided enhanced incident photon-to-electron conversion efficiency (IPCE), and the highest PCE of all DPP-based sensitizers (J_{SC}: 15.6, V_{OC}: 0.68, *ff*: 0.73, PCE: 7.67%), at that time (DPP07, Figure 6.23).[149]

Meanwhile, Qu *et al.* showed an explicit red shift in the absorption band owing to replacement of a strong donor such as indoline instead of triphenylamine (TPA) and achieved 7.4%.[181] Another approach combining the DPP and porphyrin chromophores yielded panchromatic sensitizers,[178,179,182] and achieved 7.7% very recently. Furthermore, there exist very few sensitizers with a blue colour,[183] largely because achieving absorption beyond 700 nm with organic sensitizers is a design and synthesis challenge that has not been effectively surmounted, while maintaining high performance.

Very recently, Holcombe *et al.* demonstrated the design and synthesis of high-performing blue-coloured DPP-based sensitizers widening the colour palette of this promising, aesthetically pleasing solar cell technology (Figure 6.24).[150] They increased the red light response of asymmetric DPP dyes by further engineering the bridge conjugation and modulating the donor group. Replacing the phenylcyanoacrylic acid of **DPP07** with the furanylcyanoacrylic acid anchor of **DPP13** is confirmed to enhance absorption properties and the incident photon-to-electron conversion efficiency. By employing strongly electron donating indoline groups in **DPP14**, **DPP15**,

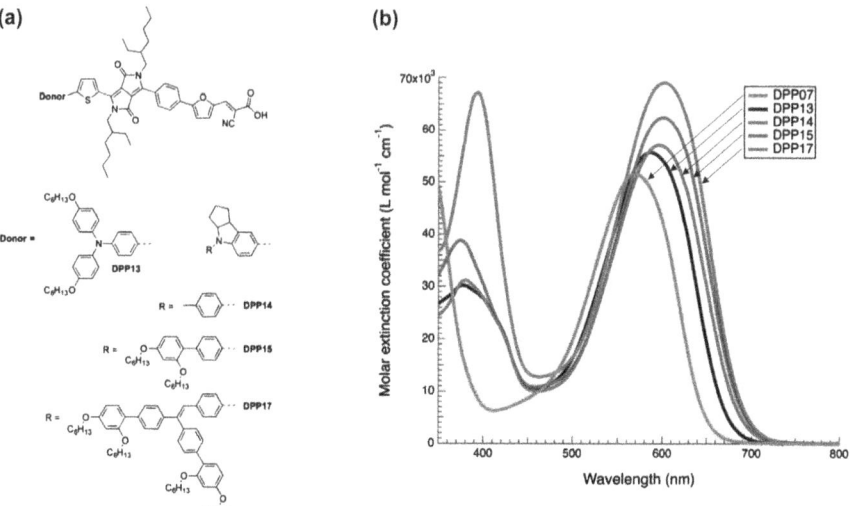

Figure 6.24 (a) Molecular structures of the investigated sensitizers. (b) All UV–visible spectra.

and **DPP17** a further ~10 nm red shift in the low-energy excitation compared to **DPP13** was achieved and in particular the slightly extended conjugation of the indoline donor on **DPP17** drove the high-energy absorption into the visible region. These sensitizers provided good compatibility with cobalt electrolyte and the highest performance for DPP-based sensitizers, to date. Maximum power conversion efficiencies of 8–10% were achieved with the cobalt redox system, in particular over 9% for **DPP15** and greater than 10% for **DPP17**. Incorporating 2,4-bis(hexyloxybiphenyl) steric bulk on the donors of **DPP15** and **DPP17** provided enhanced V_{OC} and PCE for these sensitizers compared to **DPP14**. In fact, **DPP15** and **DPP17** firmly establish DPP-based dyes as one of the highest classes of organic dyes known in DSSCs. Moreover, all dyes showed an aesthetically pleasing blue colour both in solution and on the TiO_2 film. Future rational design for the DPP dye should be towards increasing output voltage, leading to further increased power conversion efficiency.

6.3.5 Ullazine-based Sensitizers

Ullazine, **1**, is a 16 π-electron nitrogen-containing heterocyclic system which is isoelectronic with pyrene and was first reportedly synthesized in 1983 to study radical cation and anion persistence (Figure 6.25).[184,185] An aromatic, 14 π-electron annulene resonance structure consisting of a centralized positive charge upon donation of electron density from the in-plane nitrogen, illustrates the potential donating strength as well as electron stabilizing properties of this nitrogen containing heterocycle (Figure 6.25, centre). Initial anion/cation stability studies[185] and a computational comparison study to pyrene suggest that ullazine may be a good candidate for incorporation into π-conjugated materials with optoelectronic applications. Specifically, ullazine possesses (1) a planar π-system to promote strong ICT; (2) both strong donating (push) and surprising electron accepting (pull) properties; and (3) multiple substitution sites for molecular engineering.

Delcamp *et al.* have developed a rapid, efficient synthesis of ullazine for potential large-scale organic electronic applications. The ullazine core is available from simple commercial starting materials in four scalable steps

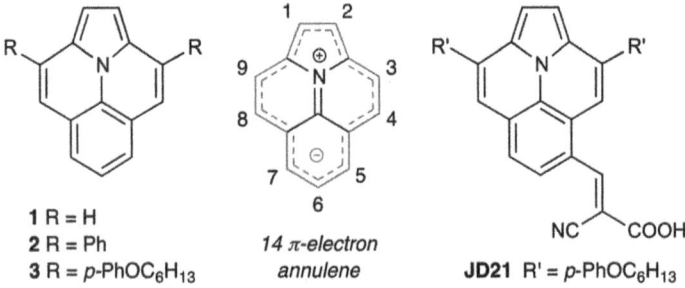

1 R = H
2 R = Ph
3 R = *p*-PhOC$_6$H$_{13}$

14 π-electron annulene

JD21 R' = *p*-PhOC$_6$H$_{13}$

Figure 6.25 Structures of ullazine and **JD21**.

via a double Fürstner cyclization as the key bond-forming reaction.[186] The ullazine core also offers an attractive number of π-conjugated sites for the molecular engineering of ullazine-based functional materials (Figure 6.26). The unique resonance structure contributes to the generation of an electron-accepting centre, and an electron-donating aromatic periphery in part contributes to the remarkable performance exhibited in DSSCs with this heterocycle. DSSC devices sensitized with the metal-free, low molecular weight ullazine-based dye **JD21** demonstrate an excellent visible absorption spectra resulting in a power conversion efficiency of 8.4%. Due to the concise, scale-friendly synthesis and high performance in DSSCs, we expect the ullazine core to find widespread use in organic electronic applications with possible industrial applications for the high-efficiency DSSC organic dye **JD21**.

6.4 Porphyrin Sensitizers

For the porphyrin sensitizers to be useful for an DSSC device, at least one anchoring group appended to the porphyrin ring is required to allow the attachment of the dye to the TiO_2 metal oxide. Inevitably, the anchoring moiety (or bridged-anchor) is also an inherent acceptor and simultaneously acts as an electron-withdrawing group (EWG). A porphyrin features eight β-positions and four *meso*-positions that are potentially available for functionalization with one or more anchoring group, and other substituents. This extensive number of functionalizable positions of the macrocycle (twelve), offers a large panel of possible molecular designs for porphyrin dyes (Figure 6.27). Historically, the first use of porphyrinoid sensitizers in TiO_2-DSSCs was reported in 1993 for β-substituted chlorophyll derivatives and related natural porphyrins, reaching a maximum efficiency of 2.6%. Since then, the performance of DSSCs based on β-substituted porphyrins had not progressed for more than 11 years,[187] until Officer, Grätzel and co-workers reported between 2004 and 2005 β-porphyrin dyes reaching efficiencies of 4.8–5.6%[188,189] and 7.1% in 2007.[190] In 2011, Tan and co-workers reported one of the best performing β-linked porphyrin dye to date (**tda-2b-bd-Zn**) with an efficiency of 7.5%.[191,192] More recently, dimeric-porphyrins,[187,193] fused-porphyrins,[194–197] or dimeric *meso*-fused-porphyrins[193] were reported as promising dyes, featuring extended absorption in the NIR.

Since 2009 and until now, *meso*-porphyrin dyes have been the most efficient porphyrin sensitizers reported in DSSCs.[198] In 2012, Grätzel and co-workers reported the best efficient sensitizer ever reported to date in TiO_2-DSSCs, the *meso*-porphyrin **YD2-o-C8**, reaching 11.9% of efficiency in conjunction with a Co^{III}/Co^{II} electrolyte, and up to 12.3% when co-sensitized with another organic dye.[41] Although the absorption of Ru complexes is limited beyond the NIR region, porphyrins show intense absorption in the region of 400–500 nm (Soret band) and 500–700 nm (Q-band) along with good stability. Thus, porphyrins and their derivatives can be used as panchromatic photosensitizers. This section aims to highlight the recent progress of cobalt electrolyte compatible *meso*-porphyrin sensitizers.

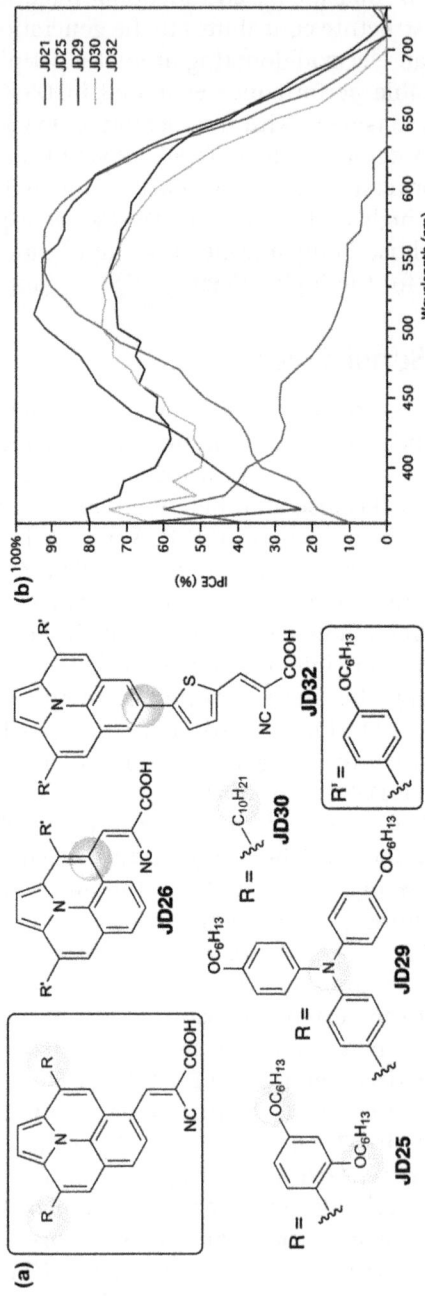

Figure 6.26 Family of ullazine-based dyes and their IPCE spectra.

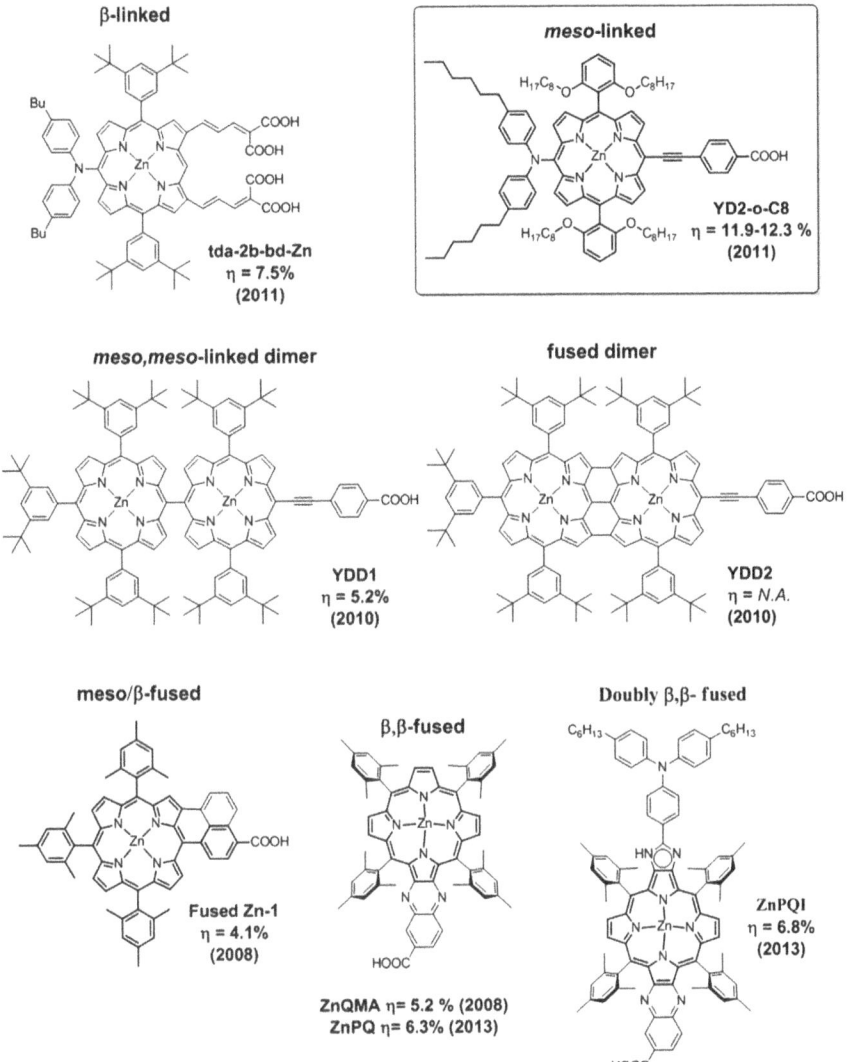

Figure 6.27 Some performing porphyrin dyes in DSSCs by family: **Tda-2B-db-Zn**,[194] **YD2-o-C8**,[166] **Fused Zn-1**,[197] **YDD1**,[196] **YDD2**,[196] **ZnQMA/ZnPQ**[198–200] and **ZnPQI**.[200]

6.4.1 Towards High Efficiency and Cobalt Compatible *meso*-Porphyrin Sensitizers

In 2009, Diau, Yeh and co-workers reported a diarylamino substituted porphyrin 'green dye' **YD1** reaching remarkable high efficiencies in TiO$_2$-DSSCs ($\eta = 5.4\%$ and 6.0% without and with CHENO as co-adsorbent, respectively) that outperformed the best porphyrin sensitizer at the time, and displayed comparable performances to those of the **N3** dye under the same

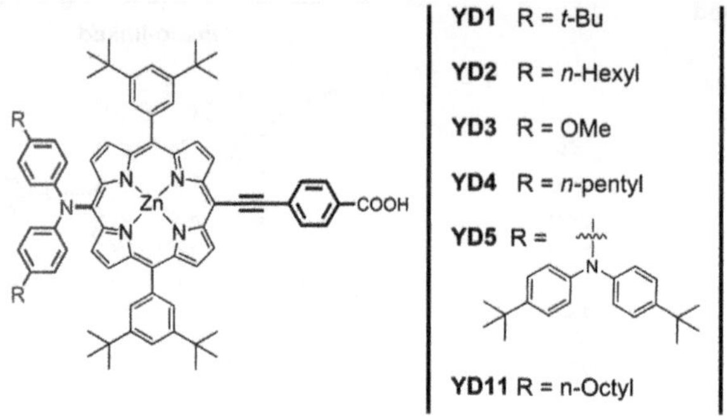

Figure 6.28   Amino-linked donor, alkynylbenzoic acid anchored, di-terbutyl 'push–pull' porphyrin dyes **YD1-5/11**.

conditions[199] (Figure 6.28). Diaryl amino groups have been revealed to be one of the most efficient donors to date in push–pull porphyrin dyes, allowing strong conjugation with the π-conjugated porphyrin macrocycle and a strong electron-'pushing' effect. It was shown that diarylamino groups, directly N-bonded at the *meso*-position of the porphyrin, allow a delocalization of the positive charge of the oxidized dye with the porphyrin ring. Moreover, the extension of π-conjugation of the porphyrin ring over the diaryl amino- and *meso*-alkynylbenzoic acid moieties, both strongly coupled to the macrocycle, significantly widens and red shifts the absorption bands of the dye.

The best performances in TiO_2 DSSCs were obtained for the well-known **YD2** dye reaching an efficiency of 6.56%, which is higher than the optimized TiO_2/**YD1** cell ($\eta = 6.15\%$) under the same conditions. The performance of **YD2** dye in TiO_2 DSSCs was successively improved between 2010 and 2011 ($\eta = 7.1\%$[200] and 7.41%[201]) and an achievement of 10.9% solar-to-electric power conversion efficiency was achieved under optimized and standard conditions (11 μm film and an additional 5 μm scattering particles layer under AM 1.5 G, 100 mW cm^{-2} intensity).[202]

Based on the excellent performances obtained for the **YD2** dye in TiO_2 DSSCs, Diau, Yeh, Grätzel and co-workers reported in 2011 a structurally modified analogue, **YD2-o-C8**, in which the 3,5-di-*tert*-butylphenyl groups have been replaced by 2,6-di-octyloxy-phenyls[163] The two sets of long alkoxy chains located at the *ortho*-positions of the *meso*-phenyl groups are favourably *ortho*-directed atop and above of both faces of the porphyrin, enveloping efficiently the macrocycle. Diau and co-workers later introduced this structural design under the general concept of 'alkoxy-wrapped' push–pull porphyrin dye (Figure 6.29)[200,203] In conjunction with a novel iodine-free electrolyte (coded **AY1**) using a Co$^{III/II}$(trisbipyridyl) complex as redox couple, **YD2-o-C8** sensitized TiO_2 cell strikingly outperformed **YD2** under the same conditions ($\eta = 11.9\%$ *versus* 8.4%, respectively), which is the best performance ever obtained for a porphyrin sensitizer to date. The use of Co$^{III/II}$ complex as redox shuttle offers a main advantage in terms of attainable open-circuit potential due to the flexibility it offers in tuning the redox

(a)   (b)

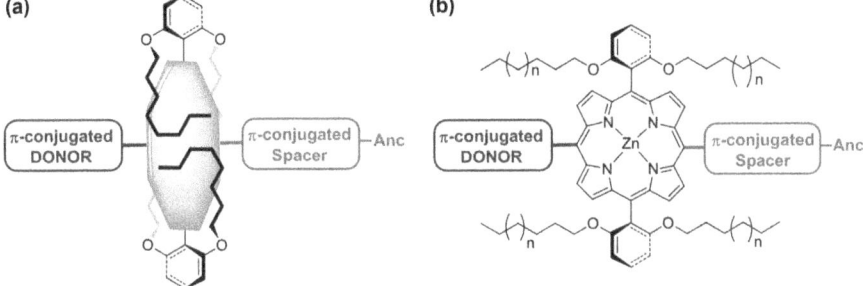

Figure 6.29 Schematic view (left) and general structure (right) of 'alkoxy-wrapped push–pull' porphyrins.

(a)   (b)

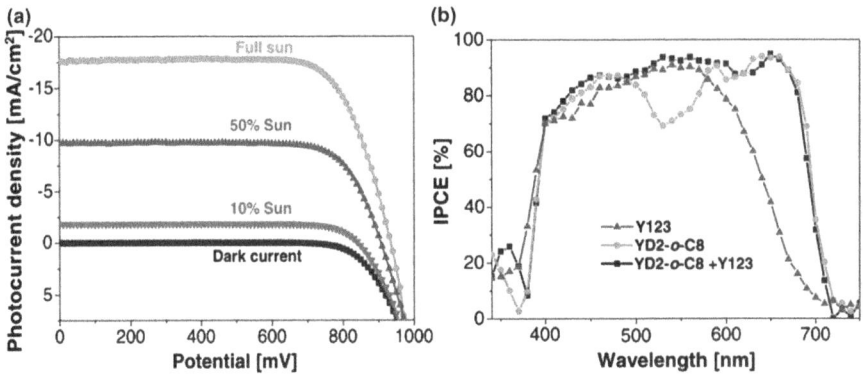

Figure 6.30 (a) *J–V* characteristics of a **YD2-o-C8/Y123** co-sensitized DSSC, measured under standard AM 1.5 global sunlight under various light intensities. The molar **YD2-o-C8/Y123** ratio in the dye solution was 7. (b) Spectral response of the IPCE for **YD2-o-C8** (pale grey dots), **Y123** (dark grey triangles), and **YD2-o-C8/Y123** co-sensitized nanocrystalline TiO$_2$ films (black squares).

potential by varying the ligands.[204] In addition, the over-potential needed for dye regeneration is considerable less compared to the standard iodide/tri-iodide redox mediators.[205–207] Co-sensitization of **YD2-o-C8** with another organic dye (**Y123**) and using the Co^{III}/Co^{II} redox couple in the **AY1** electrolyte, further enhanced the efficiency of the cell up to 12.3%, which exceed those obtained for the best ruthenium sensitizers (Figure 6.30).

6.4.2 Towards Panchromatic, High Efficiency and Cobalt Compatible *meso*-Porphyrin Sensitizers

The development of single-molecule panchromatic D-π-A sensitizers remains a molecular engineering challenge towards improving the overall PCE of the DSSC device. Until now, achieving a panchromatic light-harvesting response in DSSCs relied on co-sensitization,[163] energy-relay strategies, or tandem device configurations.[187,208–210] Although gains in PCE are often realized through improved light harvesting utilizing these strategies, the fabrication and optimization of these devices can be laborious and technically challenging. The development of a single D-π-A sensitizer, possessing panchromatic light response in the DSSC, remains a prominent objective towards realizing maximum PCE with standard device fabrication protocols.[211] Studies unrelated to the DSSC have demonstrated that integration of proquinoidal units into the porphyrin structure causes strong perturbations to the electronic structure of the macrocycle.[212–215] These perturbations in BTD-porphyrin analogues result in improved light harvesting by broadening and red-shifting absorbance of the Soret and Q-bands.

In this regard, Mathew *et al.* re-engineered the prototypical structure of D-π-A porphyrins to simultaneously maximize cobalt-electrolyte compatibility and improve light-harvesting properties.[216] Functionalization of the porphyrin core with the bulky bis(2′,4′-bis(hexyloxy)-[1,1′-biphenyl]-4-yl)amine[145] donor and an 4-ethynylbenzoic acid yielded the green dye **SM371**, which exhibited a slightly improved PCE of 12% compared to the previously reported **YD2-o-C8** (11.9%). Incorporation of the proquinoidal BTD unit into this structure afforded the dye **SM315**, a panchromatic porphyrin sensitizer (Figure 6.31). **SM315** exhibited significant broadening of Soret and Q-band absorbance features compared to **SM371**, yielding improved light harvesting in both the green (500–600 nm) and red (up to 800 nm) region of the spectrum. **SM315** demonstrated an enhancement in green-light absorption resulting in an improved J_{SC} (18.1 *versus* 15.9 mA cm^{-2}, for **SM315** and **SM371** respectively) when utilized in the DSSC.

The dyes **SM371** and **SM315** were utilized in DSSCs using thin (7 μm) mesoporous TiO$_2$ films to enable compatibility with the $[Co(bpy)_3]^{2+/3+}$ redox couple, essential in obtaining DSSCs exhibiting a high V_{OC}. Figure 6.32 shows the *J–V* curve for the two devices measured under AM 1.5 G

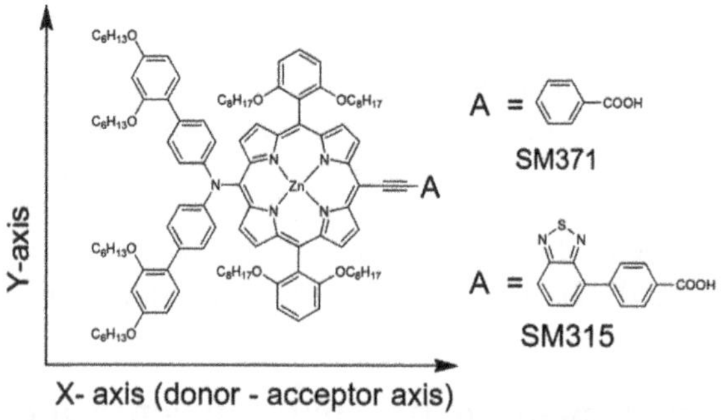

Figure 6.31   Structures of the two dyes synthesized for comparision. The structures have been coded **SM371** and **SM315**. They both feature a porphyrin core and a bulky bis(2′,4′-bis(hexyloxy)-[1,1′-biphenyl]-4-yl)amine donor. Their acceptor groups differs, with **SM315** featuring a benzothiadiazole group.

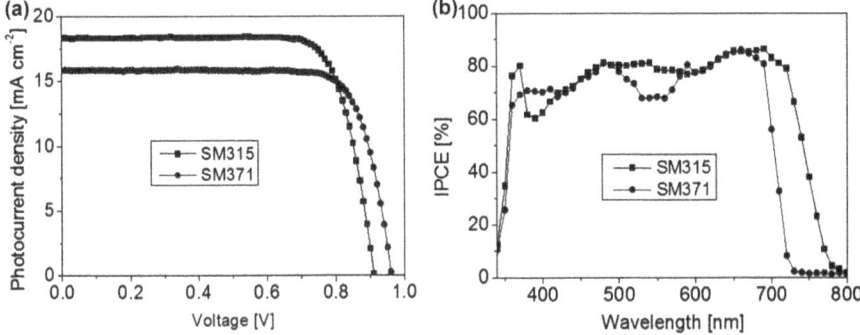

Figure 6.32 Comparison of (a) *J–V* curve under AM 1.5 G illumination (100 mW cm^{-2}) and (b) photocurrent action spectrum of **SM371** and **SM315**.

illumination (1000 W m^{-2} at 298 K). DSSCs fabricated using **SM371** gave a high V_{OC} (0.96 mV) and J_{SC} (15.9 mA cm^{-2}) achieving an overall PCE of 12.0%, a slight improvement compared to the porphyrin sensitizer **YD2-o-C8** possessing a similar structure.[6] Despite exhibiting a slightly lower V_{OC} of 0.91 V, **SM315** attained a higher J_{SC} (18.1 mA cm^{-2}) and an overall PCE of 13.0%, outperforming **SM371** as a result of the improvement in visible and near-infrared light harvesting.

Therefore, judicious molecular engineering of push–pull porphyrins has allowed the realization of two high-performance sensitizers, **SM371** and **SM315**, exhibiting unprecedented solar-to-electric power conversion efficiencies under standard AM 1.5 G illumination.

6.5 Perovskite Sensitizers for Solid-state Solar Cells

A major breakthrough in ss-DSSC was achieved when hybrid inorganic–organic perovskite semiconductor was revisited for the fabrication of mesoscopic solar cells.[3,4] They have been known for over a century, but remain unexplored in solar cells until recently. The surge of interest in hybrid inorganic–organic perovskite semiconductors as light harvesters in mesoscopic solar cells has encouraged new interest in the development of cost effective, highly efficient solar cells and, recently, the first certified record efficiency of 14.1% was achieved by Michael Gratzel's group,[218] with ample room to further optimize the system to harvest more light (Figure 6.33).[219]

Metal-halide-based perovskite compounds, with the general formula $(RNH_3)MX_3$ (R: C_nH_{2n+1}; X: halogen I, Br, Cl; and M: Pb, Cd, Sn, *etc.*), which are used as light harvesters, are typical self-assembled materials. Figure 6.34 shows the X-ray diffraction (XRD) pattern and thermogravimetric analysis (TGA) spectrum of the synthesised $CH_3NH_3PbI_3$ crystal that was prepared by drying 40 wt% of the $CH_3NH_3PbI_3/\gamma$-butyrolactone solution at 100 °C. The strong peaks at 14.08°, 14.19°, 28.24°, and 28.50°, corresponding to the (002), (110), (004), and (220) planes, confirm the formation of a tetragonal

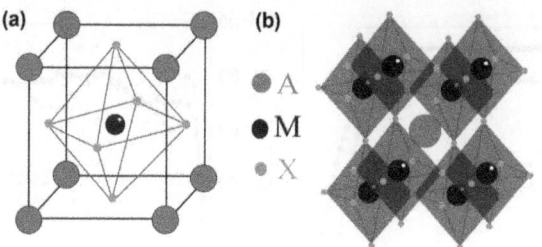

Figure 6.33 (a) Ball and stick model of the basic perovskite structure and (b) their extended network structure connected by the corner-shared octahedra.[217]

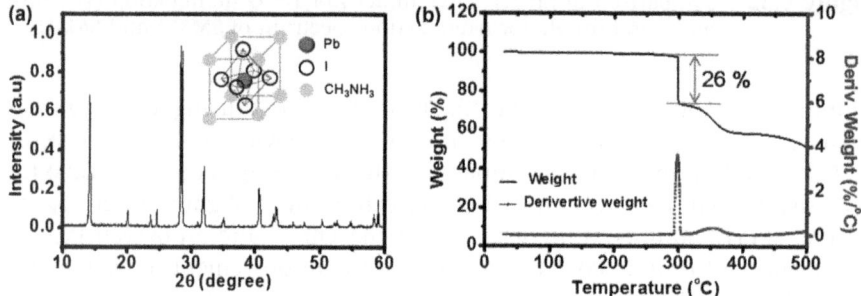

Figure 6.34 (a) XRD pattern and (b) TGA spectrum of $CH_3NH_3PbI_3$ perovskite crystal.

perovskite structure with lattice parameters of $a = b = 8.88$ Å and $c = 12.68$ Å (the inset is the schematic crystal structure of perovskite). The TGA spectrum shows that the perovskite crystal is thermally stable up to 300 °C and quickly decomposes above this temperature, owing to the decomposition of the CH_3NH_3I component, the weight fraction of which is 25.6% in $CH_3NH_3PbI_3$ (Figure 6.34).

It has been reported that organic lead halide perovskite $CH_3NH_3PbX_3$, though simple as it is, can act not only as a light absorber but also as excellent electron transporter and hole transporter.[4,220] In either case, the title compound is deposited and formed *in situ* on top of a substrate *via* a film deposition technique. So far there have been four different methods reported:

- One-step precursor solution deposition[50]
- Two-step sequential deposition[218]
- Dual-source vapour deposition[221]
- Vapour-assisted solution procession[222]

All the first three methods have given efficiencies as high as >15%. The versatility in the chemistry of forming perovskite materials will make this compound more and more popular in the near future.

6.5.1 One-step Precursor Solution Deposition

The precursor solution of perovskite was prepared by mixing the powder of CH_3NH_3I and PbI_2 at a 1:1 mole ratio in GBL at 60 °C for 12 h. This was then used for the *in situ* formation of $CH_3NH_3PbI_3$ by spin casting on a thin layer of TiO_2. The TiO_2 photo-anodes were prepared on an etched fluorine-doped tin oxide (FTO) conductive glass (TEC 7, Pilkington) by depositing a 10 nm compact TiO_2 layer by atomic layer deposition (ALD, TDMAT, 150 °C). The mesoporous film was prepared by spin-coating the TiO_2 paste at 2000 rpm for 30 s, followed by sintering at 500 °C for 30 min in air. The prepared perovskite precursor solution was dropped on the semiconductor surface and spin-coated at 1500 rpm for 30 s in the dry air box. The film was then annealed at 100 °C for 10 min. The hole transport material (HTM), consisting of 0.06 M spiro-OMeTAD, 0.03 M lithium bis(trifluoromethyl sulfonyl)imide (LiTFSI), 0.2 M 4-*tert*-butylpyridine (TBP), and 1% of tris(2-(1*H*-pyrazol-1-yl)-4-*tert*-butylpyridine) cobalt(III) tris (bis(trifluoromethylsulfonyl)imide)) in chlorobenzene, was spin-coated on the top of the perovskite layer with the spin speed of 4000 rpm. Finally, 70 nm of Au was deposited by thermal evaporation under a pressure of 5×10^{-6} Torr on the top of the HTM, as the counter-electrode. The best solar cell device fabricated with the 0.5%Y doped TiO_2 resulted in a short-circuit current density (J_{SC}), open-circuit voltage (V_{OC}) and fill factor (*ff*), respectively, of 18.1 mA cm^{-2}, 942 mV and 0.66, leading to an efficiency of 11.2% (Figure 6.35).

The incident photo-to-current conversion efficiency (IPCE) spectra show a current response from 400 to 800 nm, with a maximum of more than 75% in the wavelength range of 450–600 nm for 0.5%Y-TiO_2, and more than 65% for TiO_2 (Figure 6.36). The inset in Figure 6.36 represents the UV–visible absorbance spectra of the photo-anode, *i.e.*, 0.5%Y-TiO_2/$CH_3NH_3PbI_3$ and TiO_2/$CH_3NH_3PbI_3$ films. Despite employing the same thickness of the mesoporous films and the same $CH_3NH_3PbI_3$ deposition method, the 0.5%Y-TiO_2 film exhibits higher absorbance than TiO_2 from the visible to the NIR. This stronger absorbance indicates that more $CH_3NH_3PbI_3$ is supported by the 0.5%Y-TiO_2. Both 0.5%Y-TiO_2 and TiO_2 pastes have a similar particle size, pore size, and surface area. Thus the presence of merely 0.5%Y substitution not only shows a crucial effect on crystal growth on the surface of the film, but also a significant improvement of perovskite absorber loading, which is consistent with the higher J_{SC} and IPCE.

6.5.2 Two-step Sequential Deposition Method

A two-step deposition method was developed by Burschka *et al.*, where the PbI_2 is first introduced from solution into the nanoporous titania film and subsequently transformed into the perovskite by exposing it to a solution of CH_3NH_3I.[218] The perovskite formation is instantaneous within the nanoporous host upon contacting the two components. The two sequential procedures allow much better control over the perovskite morphology compared

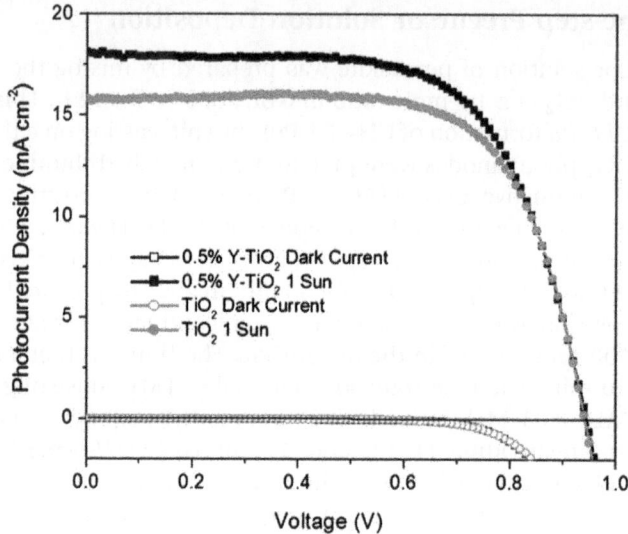

Figure 6.35 Photovoltaic characteristics. Current–voltage characteristics of the heterojunction solar cells based on 0.5%Y-TiO$_2$ (black) and TiO$_2$ (pale grey) measured under dark, and under 100 mW cm^{-2} photon flux (1 sun).

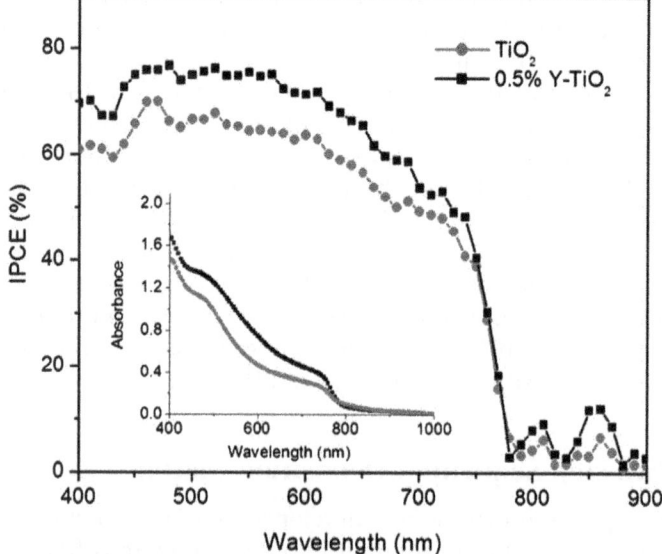

Figure 6.36 IPCE and UV–visible absorbance spectra. The incident photo-to-electron conversion efficiency spectra of the devices based on 0.5%Y-TiO$_2$ (black) and TiO$_2$ (pale gray). Inset shows the UV–visible absorbance spectra of 0.5%Y-TiO$_2$/CH$_3$NH$_3$PbI$_3$ and TiO$_2$/CH$_3$NH$_3$PbI$_3$ electrodes.

to the one-step deposition method. Employing this technique for the fabrication of solid-state mesoscopic solar cells greatly improves the reproducibility of their performance, and allows achieving a new record power conversion efficiency (PCE) of 15%, as measured under standard AM1.5G test conditions (see Figure 6.37). This finding opens up completely new opportunities for the fabrication of solution-processed photovoltaics with high power conversion efficiencies and stability that matches or even exceeds those of today's best thin film photovoltaic devices.

6.5.3 Dual-source Vapour Deposition

The fabricated device is an inverted thin film solar cell consisting of a sublimated methylammonium lead iodide ($CH_3NH_3PbI_3$) perovskite layer that is sandwiched between two very thin electron and hole blocking layers consisting of organic molecules. The organic materials were deposited using solution-based processes, whereas the $CH_3NH_3PbI_3$ perovskite and the metal contact were deposited using thermal evaporation under vacuum. The uniform $CH_3NH_3PbI_3$ perovskite thin film was prepared by co-evaporation of the two starting compounds, CH_3NH_3I and PbI_2. In this structure, which is typical for organic-photovoltaic and light-emitting devices, a transparent conductor was used as the positive charge collecting contact. The device structure consists of a 70 nm poly(3,4-ethylenedioxythiophene):poly(styrene-sulfonic acid) (PEDOT:PSS) layer and a thin layer (<10 nm) of poly[N,N'-bis(4-butylphenyl)-N,N'-bis(phenyl) benzidine] (polyTPD), as the electron blocking

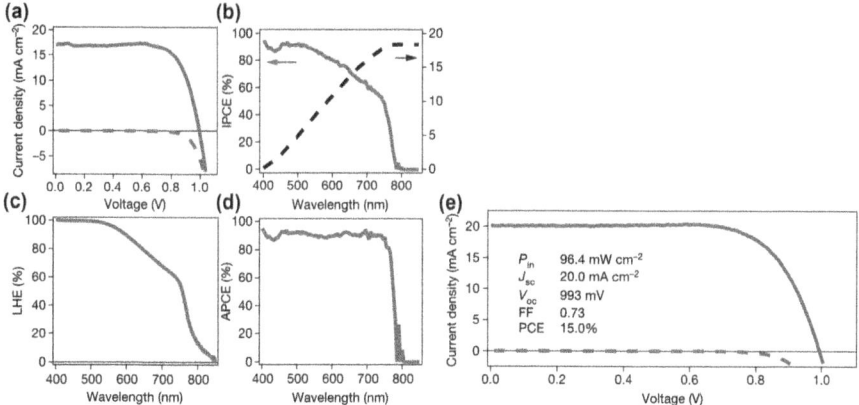

Figure 6.37 (a) *J–V* curves for a photovoltaic device measured at a simulated AM1.5G solar irradiation of 95.6 mW cm^{-2} (solid line) and in the dark (dashed line). (b) Incident photon-to-current conversion efficiency (IPCE) spectrum. (c) Light harvesting efficiency (LHE) spectrum. (d) (Absorbed photon-to-current conversion efficiency) APCE spectrum derived from the IPCE and LHE. (e) *J–V* curves for a best-performing cell measured at a simulated AM1.5G solar irradiation of 96.4 mW cm^{-2} (solid line) and in the dark (dashed line).

layer. On top of this the $CH_3NH_3PbI_3$ was thermally evaporated to a thickness of around 300 nm followed by a thin layer (<10 nm) of [6,6]-phenyl C_{61}-butyric acid methylester (PCBM) as the hole blocking layer. The device is completed by the evaporation of an Au top electrode (100 nm). Both the polyTPD and the PCBM layers were deposited using a meniscus coating process to ensure high-quality films. The thickness of the layers will be established using absorbance measurements.

Figure 6.38a shows the current–voltage (*J–V*) characteristics of a typical small area (0.09 cm^2) perovskite solar cell measured in the dark, and under light intensities of 100, 50 and 10 mW cm^{-2}. The reproducibility from device to device was excellent with less than 10% deviation. The short-circuit current density (J_{SC}), open-circuit voltage (V_{OC}) and fill factor (*ff*), respectively, are 16.12 mA cm^{-2}, 1.05 V and 0.67, leading to a power conversion efficiency of 12.04% measured at 100 mW cm^{-2}. The device at 50 and 10 mW cm^{-2} illumination exhibited very similar efficiencies, 12.04 and 12.00%, respectively. The high open-circuit potential indicates that there are negligible surface and sub-band-gap states in the perovskite film. The device performance under 100 mW cm^{-2} is remarkable in view of the very thin perovskite film (only 285 nm thick). A device with larger area cells (0.98 cm^2) was also prepared and evaluated at 100 mW cm^{-2}.

The incident photon-to-current conversion efficiency (IPCE) spectra reach 74% (Figure 6.38b) wherein the generation of photocurrent is seen to start at 790 nm in agreement with the bandgap of the $CH_3NH_3PbI_3$. It is interesting to note that the IPCE spectra show a very steep onset, contrary to the IPCE spectra reported for TiO_2 and Al_2O_3 mesoscopic-based perovskite cells. A very similar IPCE spectrum to that observed for our solar cells was reported for a solar cell based on a perovskite layer sandwiched in between a C_{60} modified TiO_2 layer and a layer of poly(3-hexylthiophene). As such, the steep onset may be related to the specific interaction of the perovskite with the fullerene

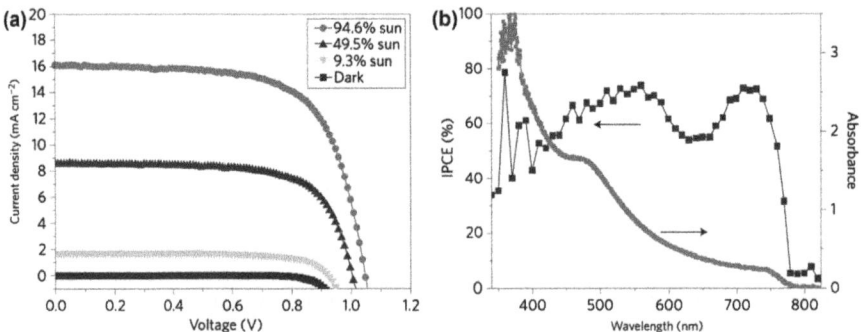

Figure 6.38 Typical *J–V* and IPCE characteristics of the perovskite solar cell. (a) Photocurrent density versus voltage at 100, 50 and 10 mW cm^{-2} and in the dark. (b) IPCE spectrum and absorbance of a 285-nm-thick perovskite layer.

molecules. The IPCE spectrum is almost flat except for a dip at 630 nm, which could be due to the oxidized polyTPD acting as a filter, leading to reduced absorption in the perovskite. Integrating the overlap of the IPCE spectrum with the AM1.5G solar photon flux yields a current density of 16.40 mA cm^{-2}, which is in excellent agreement with the measured photocurrent density of 16.12 mA cm^{-2} measured at the standard solar AM 1.5 intensity of 100 mW cm^{-2}, confirming that the mismatch between the simulated sunlight and the AM1.5G standard is negligible.

6.6 Conclusion

We have shown an overview of molecular engineering aspect of dyes and pigments for dye-sensitized solar cells and their photovoltaic performance: ruthenium complexes-, organic-, porphyrin- and perovskite-based sensitizers. For the ruthenium complexes, we briefly reviewed a series of Ru(II) complexes showing broad light-harvesting characteristics with colour ranging from red to green–black. The photocurrent onset of classic Ru(II) sensitizers is already around 900 nm but the IPCE in the longer wavelength range is still low because of the low extinction coefficient. Judicious molecular engineering of the ruthenium dye structure through gradually deprotonation of the anchor groups, substitution of the NCS donor groups by cyclometallating ligands and the introduction of conjugated π molecules allow for not only further increasing the light harvesting in the 700–900 nm region, but also a beneficial compatability to the Co^{III}/Co^{II} redox shuttle. A nearly vertical rise of the photocurrent close to the 920 nm absorption threshold would increase the short circuit photocurrent from currently 20.5 to about 28 mA cm^{-2} raising the overall efficiency to above 15% assuming the other parameters, V_{OC} and *ff*, are maintained.

The elimination of Ru(II) metal brings the shortening of the absorption onset wavelength, but at the same time the advantage of extremely high molar extinction coefficients from organic dyes. Therefore, thinner TiO$_2$ film can be used. The skeleton of metal-free sensitizers normally consists of three components: electron donor, electron acceptor, and conjugated spacer. A tremendous number of organic metal-free dyes has been synthesized due to the open source of all the components. Arylamine is commonly employed as the electron donor. Acid ligands, such as carboxylic acid, phosphonic acid, and sulfonic acid, may serve as the electron acceptor and the anchoring group between dye molecules and semiconductor surface as well. The conjugated spacer plays an important role in electronic communication between the donor and the acceptor, in enhancing light harvesting, and in dark current suppression. Chemically modifide donor groups have been successfully provided to reduce recombination of the injected electrons with cobalt electrolyte while keeping strong donating capability. Systematic tuning of the conjugation length of the π spacer has been done as well and indicated that a balanced and optimal length should be reached to give a dye with efficient light-to-electricity conversion efficiency. The coining of the concept 'donor–chromophore–acceptor' profiled a type of organic dyes with coloured chromophore as the π bridge, *e.g.*, DPP. sensitizers that are deliberately designed with this method provided much red-shifted absorption and vivid colour for industrial interest.

Donor–π-acceptor porphyrin sensitizers can be seen as another type of noble-metal-free sensitizer. Important work has been done in molecular engineering of different types of porphyrin sensitizers and has achieved higher and higher photovoltaic performance. To date the great majority of the best performing porphyrin sensitizers are push–pull porphyrin sensitizers anchored through an (ethynyl)carboxyphenyl group, with strong donor group. The two *meso* positions consisting of an alkoxy-substituted phenyl ring reduces aggregation and recombination. To make use of the $[Co(bpy)_3]^{2+/3+}$ redox couple, BPTPA donor groups are incorporated into porphyrin to reduce recombination and increase the open circuit. Further molecular engineering of porphyrin comes from the acceptor part by introducing a benzothiadiazole which reduces the gap between HOMO–LUMO further to get the panchromatic response leading to over 18 mA cm^{-2} short circuit current. If one could realize 20 mA cm^{-2} short circuit current, 1 V open-circuit potential and a fill factor of 75%, a power conversion efficiency (PEC) beyond 15% is very promising at full sun (AM 1.5 G, 1000 W m^{-2}) illumination.

In the last section, we have revealed the emerging 'old and new' organic–inorganic hydride perovskite material could be made *via* a number of fabrication methods from solution procession to vacuum deposition. A surge in the research interest and power conversion efficiencies has already been witnessed by the whole photovoltaics community. Perovskite materials as a 'game changer' will bring us not only the doubled power conversion efficiencies but also a revolution in the geometry of the photovoltaic devices. It will not be an exaggeration that, one day, perovskite-based solar cells will match the capability and capacity of their arch-rival, silicon, and power our planet.

Acknowledgements

The authors acknowledge financial contribution from the European Community's Seventh Framework Programme (FP7/2007-2013) under grant agreement. n° 246124 of the SANS project, 'ORION' grant agreement No. NMP-229036, under grant agreement 'ENERGY-261920, ESCORT', and SSSTC (Sino-Swiss Science and Technology Cooperation) and NANO-MATCELL. M.K.N. thanks the Global Research Laboratory (GRL) Program, Korea, and World Class University programs (Photovoltaic Materials, Department of Material Chemistry, Korea University) funded by the Ministry of Education, Science and Technology through the National Research Foundation of Korea (No. R31-2008-000-10035-0).

References

1. M. A. Green, K. Emery, Y. Hishikawa and W. Warta, *Prog. Photovoltaics Res. Appl.*, 2009, **17**, 85–94.
2. C. W. Tang, *Appl. Phys. Lett.*, 1986, **48**, 183–185.
3. A. Kojima, K. Teshima, Y. Shirai and T. Miyasaka, *J. Am. Chem. Soc.*, 2009, **131**, 6050–6051.
4. M. M. Lee, J. Teuscher, T. Miyasaka, T. N. Murakami and H. J. Snaith, *Science*, 2012, **338**, 643–647.
5. W. A. Laban and L. Etgar, *Energy Environ. Sci.*, 2013, **6**, 3249–3253.
6. J. Moser, *Monatsh. Chemie*, 1887, **8**, 373.
7. H. Gerischer and H. Tributsch, *Phys. Chem.*, 1968, **72**, 437.
8. M. P. Dare-Edwards, J. B. Goodenough, A. Hamnett, K. R. Seddon and R. D. Wright, *Faraday Discuss. Chem. Soc.*, 1981, **70**, 285.
9. H. Tsuborama, M. Matsumura, Y. Nomura and T. Amamiya, *Nature*, 1976, **261**, 402–403.
10. D. Duonghong, N. Serpone and M. Grätzel, *Helv. Chim. Acta*, 1984, **67**, 1012–1018.
11. J. Desilvestro, M. Grätzel, L. Kavan, J. Moser and J. Augustynski, *J. Am. Chem. Soc.*, 1985, **107**, 2988–2990.
12. B. O'Regan and M. Gratzel, *Nature*, 1991, **353**, 737–740.
13. M. A. Green, K. Emery, Y. Hishikawa, W. Warta and E. D. Dunlop, *Prog. Photovoltaics Re. Appl.*, 2014, **22**, 1–9.
14. M. Gratzel, *Nature*, 2001, **414**, 338–344.
15. A. Hagfeldt and M. Gratzel, *Acc. Chem. Res.*, 2000, **33**, 269–277.
16. N. Robertson, *Angew. Chem. Int. Ed. Engl.*, 2006, **45**, 2338–2345.
17. A. Mishra, M. K. R. Fischer and P. Bauerle, *Angew. Chem. Int. Ed.*, 2009, **48**, 2474–2499.
18. G. Oskam, B. V. Bergeron, G. J. Meyer and P. C. Searson, *J. Phys. Chem. B*, 2001, **105**, 6867–6873.
19. H. Nusbaumer, J. E. Moser, S. M. Zakeeruddin, M. K. Nazeeruddin and M. Gratzel, *J. Phys. Chem. B*, 2001, **105**, 10461–10464.
20. Z. Zhang, P. Chen, T. N. Murakami, S. M. Zakeeruddin and M. Gratzel, *Adv. Funct. Mater.*, 2008, **18**, 341–346.
21. P. Wang, S. M. Zakeeruddin, J. E. Moser, R. Humphry-Baker and M. Gratzel, *J. Am. Chem. Soc.*, 2004, **126**, 7164–7165.
22. S. Hattori, Y. Wada, S. Yanagida and S. Fukuzumi, *Jpn. J. Appl. Phys.*, 2005, **127**, 9648–9654.
23. M. K. Nazeeruddin, F. De Angelis, S. Fantacci, A. Selloni, G. Viscardi, P. Liska, S. Ito, B. Takeru and M. Gratzel, *J. Am. Chem. Soc.*, 2005, **127**, 16835–16847.
24. F. Gao, *J. Am. Chem. Soc.*, 2008, **130**, 10720–10728.
25. C. Y. Chen, *ACS Nano*, 2009, **3**, 3103–3109.
26. Y. Chiba, *Jpn. J. Appl. Phys.*, 2006, **45**, 638–640.

27. A. Fukui, R. Komiya, N. Murofushi, N. Koide, R. Yamanaka, H. Katayama, in *Technical Digest, 21st International Photovoltaic Science and Engineering Conference, Fukuoka*, Editon edn., November, 2011.

28. M. K. Nazeeruddin, F. De Angelis, S. Fantacci, A. Selloni, G. Viscardi, P. Liska, S. Ito, T. Bessho and M. Graetzel, *J. Am. Chem. Soc.*, 2005, **127**, 16835–16847.

29. N. Jiang, T. Sumitomo, T. Lee, A. Pellaroque, O. Bellon, D. Milliken and H. Desilvestro, *Sol. Energy Mater. Sol. Cells*, 2013, **119**, 36–50.

30. A. Hinsch, J. M. Kroon, R. Kern, I. Uhlendorf, J. Holzbock, A. Meyer and J. Ferber, *Prog. Photovoltaics Res. Appl.*, 2001, **9**, 425–438.

31. J. M. Kroon, N. J. Bakker, H. J. P. Smit, P. Liska, K. R. Thampi, P. Wang, S. M. Zakeeruddin, M. Grätzel, A. Hinsch, S. Hore, U. Würfel, R. Sastrawan, J. R. Durrant, E. Palomares, H. Pettersson, T. Gruszecki, J. Walter, K. Skupien and G. E. Tulloch, *Prog. Photovoltaics Res. Appl.*, 2007, **15**, 1–18.

32. J.-H. Yum, E. Baranoff, F. Kessler, T. Moehl, S. Ahmad, T. Bessho, A. Marchioro, E. Ghadiri, J.-E. Moser, C. Yi, M. K. Nazeeruddin and M. Grätzel, *Nat. Commun.*, 2012, **3**, 631.

33. T. Bessho, E. Yoneda, J. H. Yum, M. Guglielmi, I. Tavernelli, H. Imai, U. Rothlisberger, M. K. Nazeeruddin and M. Gratzel, *J. Am. Chem. Soc.*, 2009, **131**, 5930–5934.

34. M. Grätzel, *Acc. Chem. Res.*, 2009, **42**, 1788–1798.

35. A. Mishra, M. K. R. Fischer and P. Bäuerle, *Angew. Chem. Int. Ed. Engl.*, 2009, **48**, 2474–2499.

36. N. Cai, R. Li, Y. Wang, M. Zhang and P. Wang, *Energy Environ. Sci.*, 2013, **6**, 139–147.

37. M. Zhang, Y. Wang, M. Xu, W. Ma, R. Li and P. Wang, *Energy Environ. Sci.*, 2013, **6**, 2944–2949.

38. P. Gao, H. N. Tsao, C. Yi, M. Grätzel and M. K. Nazeeruddin, *Adv. Energy Mater.*, 2013, DOI: 10.1002/aenm.201301485.

39. P. Gao, H. N. Tsao, M. Gratzel and M. K. Nazeeruddin, *Org. Lett.*, 2012, **14**, 4330–4333.

40. K. Hara, Z.-S. Wang, Y. Cui, A. Furube and N. Koumura, *Energy Environ. Sci.*, 2009, **2**, 1109–1114.

41. A. Yella, H.-W. Lee, H. N. Tsao, C. Yi, A. K. Chandiran, M. K. Nazeeruddin, E. W.-G. Diau, C.-Y. Yeh, S. M. Zakeeruddin and M. Grätzel, *Science*, 2011, **334**, 629–634.

42. T. Bessho, S. M. Zakeeruddin, C. Y. Yeh, E. W. G. Diau and M. Gratzel, *Angew. Chem. Int. Ed.*, 2010, **49**, 6646–6649.

43. H. Imahori, T. Umeyama and S. Ito, *Acc. Chem. Res.*, 2009, **42**, 1809–1818.

44. C. Y. Lin, Y. C. Wang, S. J. Hsu, C. F. Lo and E. W. G. Diau, *J. Phys. Chem. C*, 2010, **114**, 687–693.

45. W. M. Campbell, A. K. Burrell, D. L. Officer and K. W. Jolley, *Coord. Chem. Rev.*, 2004, **248**, 1363–1379.

46. S. Mathew, A. Yella, P. Gao, R. Humphry-Baker, B. F. Curchod, N. Ashari-Astani, I. Tavernelli, U. Röthlisberger, M. K. Nazeeruddin and M. Grätzel, *Nat. Chem.*, 2014, **6**, 242–247.

47. U. Bach, D. Lupo, P. Comte, J. E. Moser, F. Weissortel, J. Salbeck, H. Spreitzer and M. Gratzel, *Nature*, 1998, **395**, 583–585.

48. P. Chen, *Nano Lett.*, 2009, **9**, 2487–2492.

49. J. Burschka, *J. Am. Chem. Soc.*, 2011, **133**, 18042–18045.

50. H.-S. Kim, C.-R. Lee, J.-H. Im, K.-B. Lee, T. Moehl, A. Marchioro, S.-J. Moon, R. Humphry-Baker, J.-H. Yum, J. E. Moser, M. Grätzel and N.-G. Park, *Sci. Rep.*, 2012, **2**, 591.

51. N.-G. Park, *J. Phys. Chem. Lett.*, 2013, **4**, 2423–2429.

52. J. Burschka, N. Pellet, S.-J. Moon, R. Humphry-Baker, P. Gao, M. K. Nazeeruddin and M. Grätzel, *Nature*, 2013, **499**, 316–319.

53. The National Renewable Energy Laboratory (NREL), Editon edn., 2013.

54. M. He, D. Zheng, M. Wang, C. Lin and Z. Lin, *J. Mater. Chem. A*, 2014, **2**, 5994–6003.

55. S. Dharani, H. K. Mulmudi, N. Yantara, P. T. Thu Trang, N. G. Park, M. Graetzel, S. Mhaisalkar, N. Mathews and P. P. Boix, *Nanoscale*, 2014, **6**, 1675–1679.

56. E. Edri, S. Kirmayer, D. Cahen and G. Hodes, *J. Phys. Chem. Lett.*, 2013, **4**, 897–902.

57. J. Qiu, Y. Qiu, K. Yan, M. Zhong, C. Mu, H. Yan and S. Yang, *Nanoscale*, 2013, 3–6.

58. D. Liu and T. L. Kelly, *Nat. Photonics*, 2013, **8**, 133–138.

59. J. M. Ball, M. M. Lee, A. Hey and H. Snaith, *Energy Environ. Sci.*, 2013, **6**, 1739–1743.

60. L. Etgar, P. Gao, Z. Xue, Q. Peng, A. K. Chandiran, B. Liu, M. K. Nazeeruddin and M. Gratzel, *J. Am. Chem. Soc.*, 2012, **134**, 17396–17399.

61. S. D. Stranks, G. E. Eperon, G. Grancini, C. Menelaou, M. J. P. Alcocer, T. Leijtens, L. M. Herz, A. Petrozza and H. J. Snaith, *Science*, 2013, **342**, 341–344.

62. G. Xing, N. Mathews, S. Sun, S. S. Lim, Y. M. Lam, G. M. S. Mhaisalkar and T. C. Sum, *Science*, 2013, **342**, 344–347.

63. C. Y. Chen, M. K. Wang, J. Y. Li, N. Pootrakulchote, L. Alibabaei, C. H. Ngoc-le, J. D. Decoppet, J. H. Tsai, C. Gratzel, C. G. Wu, S. M. Zakeeruddin and M. Gratzel, *ACS Nano*, 2009, **3**, 3103–3109.

64. P. A. Anderson, G. F. Strouse, J. A. Treadway, F. R. Keene and T. J. Meyer, *Inorg. Chem.*, 1994, **33**, 3863–3864.

65. J. A. Treadway, J. A. Moss and T. J. Meyer, *Inorg. Chem.*, 1999, **38**, 4386–4387.

66. M. Alebbi, C. A. Bignozzi, T. A. Heimer, G. M. Hasselmann and G. J. Meyer, *J. Phys. Chem. B*, 1998, **102**, 7577–7581.

67. M. K. Nazeeruddin, *J. Am. Chem. Soc.*, 1993, **115**, 6382–6390.

68. M. K. Nazeeruddin, *J. Am. Chem. Soc.*, 2001, **123**, 1613–1624.

69. M. K. Nazeeruddin, R. Humphry-Baker, P. Liska and M. Gratzel, *J. Phys. Chem. B*, 2003, **107**, 8981–8987.

70. M. K. Nazeeruddin, Q. Wang, L. Cevey, V. Aranyos, P. Liska, E. Figgemeier, C. Klein, N. Hirata, S. Koops, S. A. Haque, J. R. Durrant, A. Hagfeldt, A. B. P. Lever and M. Gratzel, *Inorg. Chem.*, 2006, **45**, 787–797.

71. Y. M. Cao, Y. Bai, Q. J. Yu, Y. M. Cheng, S. Liu, D. Shi, F. F. Gao and P. Wang, *J. Phys. Chem. C*, 2009, **113**, 6290–6297.

72. H. J. Snaith, *Advanced Functional Materials*, 2010, **20**, 13–19.

73. M. K. Nazeeruddin, P. Péchy and M. Grätzel, *Chem. Commun.*, 1997, 1705–1706.

74. M. K. Nazeeruddin, P. Péchy, T. Renouard, S. M. Zakeeruddin, R. Humphry-Baker, P. Comte, P. Liska, L. Cevey, E. Costa, V. Shklover, L. Spiccia, G. B. Deacon, C. A. Bignozzi and M. Grätzel, *J. Am. Chem. Soc.*, 2001, **123**, 1613–1624.

75. M. Grätzel, *J. Photochem. Photobiol. A*, 2004, **168**, 235–235.

76. Y. Chiba, A. Islam, Y. Watanabe, R. Komiya, N. Koide and L. Y. Han, *Jpn. J. Appl. Phys. Part 2*, 2006, **45**, L638–L640.

77. A. Islam, H. Sugihara and H. Arakawa, *J. Photochem. Photobiol. A*, 2003, **158**, 131–138.

78. A. Islam, F. A. Chowdhury, Y. Chiba, R. Komiya, N. Fuke, N. Ikeda and L. Y. Han, *Chem. Lett.*, 2005, **34**, 344–345.

79. T. Funaki, M. Yanagida, N. Onozawa-Komatsuzaki, K. Kasuga, Y. Kawanishi and H. Sugihara, *Chem. Lett.*, 2009, **38**, 62–63.

80. T. Funaki, M. Yanagida, N. Onozawa-Komatsuzaki, K. Kasuga, Y. Kawanishi, M. Kurashige, K. Sayama and H. Sugihara, *Inorg. Chem. Commun.*, 2009, **12**, 842–845.

81. S. Altobello, R. Argazzi, S. Caramori, C. Contado, S. Da Fre, P. Rubino, C. Chone, G. Larramona and C. A. Bignozzi, *J. Am. Chem. Soc.*, 2005, **127**, 15342–15343.

82. B. S. Chen, K. Chen, Y. H. Hong, W. H. Liu, T. H. Li, C. H. Lai, P. T. Chou, Y. Chi and G. H. Lee, *Chem. Commun.*, 2009, 5844–5846.

83. C.-C. Chou, K.-L. Wu, Y. Chi, W.-P. Hu, S. J. Yu, G.-H. Lee, C.-L. Lin and P.-T. Chou, *Angew. Chem. Int. Ed. Engl.*, 2011, **50**, 2054–2058.

84. C. W. Hsu, S. T. Ho, K. L. Wu, Y. Chi, S. H. Liu and P. T. Chou, *Energy Environ. Sci.*, 2012, **5**, 7549–7554.

85. K. L. Wu, C. H. Li, Y. Chi, J. N. Clifford, L. Cabau, E. Palomares, Y. M. Cheng, H. A. Pan and P. T. Chou, *J. Am. Chem. Soc.*, 2012, **134**, 7488–7496.

86. M. Kimura, H. Nomoto, N. Masaki and S. Mori, *Angew. Chem. Int Ed.*, 2012, **51**, 4371–4374.

87. M. Kimura, J. Masuo, Y. Tohata, K. Obuchi, N. Masaki, T. N. Murakami, N. Koumura, K. Hara, A. Fukui, R. Yamanaka and S. Mori, *Chem. A Eur. J.*, 2013, **19**, 1028–1034.

88. Y. Numata, S. P. Singh, A. Islam, M. Iwamura, A. Imai, K. Nozaki and L. Y. Han, *Adv. Funct. Mater.*, 2013, **23**, 1817–1823.

89. T. Funaki, H. Funakoshi, O. Kitao, N. Onozawa-Komatsuzaki, K. Kasuga, K. Sayama and H. Sugihara, *Angew. Chem. Int. Ed.*, 2012, **51**, 7528–7531.

90. C.-C. Chou, F.-C. Hu, H.-H. Yeh, H.-P. Wu, Y. Chi, J. N. Clifford, E. Palomares, S.-H. Liu, P.-T. Chou and G.-H. Lee, *Angew. Chem. Int. Ed.*, 2014, **53**, 178–183.

91. T. Kinoshita, J. Dy, S. Uchida, T. Kubo and H. Segawa, *Nat. Photonics*, 2013, **7**, 535–539.

92. T. Bessho, E. Yoneda, J.-H. Yum, M. Guglielmi, I. Tavernelli, H. Imai, U. Rothlisberger, M. K. Nazeeruddin and M. Graetzel, *J. Am. Chem. Soc.*, 2009, **131**, 5930–5934.

93. E. Baranoff, J.-H. Yum, M. Graetzel and M. K. Nazeeruddin, *J. Organomet. Chem.*, 2009, **694**, 2661–2670.

94. S. M. Feldt, E. A. Gibson, E. Gabrielsson, L. Sun, G. Boschloo and A. Hagfeldt, *J. Am. Chem. Soc.*, 2010, **132**, 16714–16724.

95. D. Zhou, Q. Yu, N. Cai, Y. Bai, Y. Wang and P. Wang, *Energy Environ. Sci.*, 2011, **4**, 2030–2034.

96. H. N. Tsao, C. Yi, T. Moehl, J.-H. Yum, S. M. Zakeeruddin, M. K. Nazeeruddin and M. Grätzel, *ChemSusChem*, 2011, **4**, 591–594.

97. S. A. Sapp, C. M. Elliott, C. Contado, S. Caramori and C. A. Bignozzi, *J. Am. Chem. Soc.*, 2002, **124**, 11215–11222.

98. H. Nusbaumer, S. M. Zakeeruddin, J.-E. Moser and M. Grätzel, *Chem. A Eur. J.*, 2003, **9**, 3756–3763.

99. B. M. Klahr and T. W. Hamann, *J. Phys. Chem. C*, 2009, **113**, 14040–14045.

100. Y. Liu, J. R. Jennings, Y. Huang, Q. Wang, S. M. Zakeeruddin and M. Grätzel, *J. Phys. Chem. C*, 2011, **115**, 18847–18855.

101. P. G. Bomben, T. J. Gordon, E. Schott and C. P. Berlinguette, *Angew. Chem. Int. Ed. Engl.*, 2011, **50**, 10682–10685.

102. Y. Xie and T. W. Hamann, *J. Phys. Chem. Lett*, 2013, **4**, 328–332.

103. S. M. Zakeeruddin, M. K. Nazeeruddin, R. Humphry-Baker, P. Péchy, P. Quagliotto, C. Barolo, G. Viscardi and M. Grätzel, *Langmuir*, 2002, **18**, 952–954.

104. Y. Liu, J. R. Jennings, X. Wang and Q. Wang, *Phys. Chem. Chem. Phys.*, 2013, **15**, 6170–6174.

105. E. Mosconi, J.-H. Yum, F. Kessler, C. J. Gomez-Garcia, C. Zuccaccia, A. Cinti, M. K. Nazeeruddin, M. Grätzel and F. De Angelis, *J. Am. Chem. Soc.*, 2012, **134**, 19438–19453.

106. P. G. Bomben, K. C. D. Robson, P. A. Sedach and C. P. Berlinguette, *Inorg. Chem.*, 2009, **48**, 9631–9643.

107. L. E. Polander, A. Yella, B. F. E. Curchod, A. N. Ashari, J. J. Teuscher, R. Scopelliti, P. Gao, S. Mathew, J.-E. Moser, I. Tavernelli, U. Rothlisberger, M. Gratzel, M. K. Nazeeruddin, J. Frey, N. Ashari Astani and M. Grätzel, *Angew. Chem. Int. Ed.*, 2013, **52**, 4386–4387.

108. F. Gao, Y. Wang, D. Shi, J. Zhang, M. Wang, X. Jing, R. Humphry-Baker, P. Wang, S. M. Zakeeruddin and M. Grätzel, *J. Am. Chem. Soc.*, 2008, **130**, 10720–10728.

109. T. Horiuchi, H. Miura, K. Sumioka and S. Uchida, *J. Am. Chem. Soc.*, 2004, **126**, 12218–12219.

110. J. H. Yum, P. Walter, S. Huber, D. Rentsch, T. Geiger, F. Nüesch, F. De Angelis, M. Grätzel and M. K. Nazeeruddin, *J. Am. Chem. Soc.*, 2007, **129**, 10320–10321.

111. W. M. Campbell, *J. Phys. Chem. C*, 2007, **111**, 11760–11762.

112. J. He, *J. Am. Chem. Soc.*, 2002, **124**, 4922–4932.

113. C. W. Lee, *Chem. Eur. J.*, 2009, **15**, 1403–1412.

114. Z. J. Ning and H. Tian, *Chem. Commun.*, 2009, 5483–5495.

115. Z. S. Wang, N. Koumura, Y. Cui, M. Takahashi, H. Sekiguchi, A. Mori, T. Kubo, A. Furube and K. Hara, *Chem. Mater.*, 2008, **20**, 3993–4003.

116. S. Ito, H. Miura, S. Uchida, M. Takata, K. Sumioka, P. Liska, P. Comte, P. Péchy and M. Grätzel, *Chem. Commun.*, 2008, 5194–5196.

117. Y. Wu, X. Zhang, W. Li, Z.-S. Wang, H. Tian and W. Zhu, *Adv. Energy Mater.*, 2012, **2**, 149–156.

118. K. Pei, Y. Z. Wu, W. J. Wu, Q. Zhang, B. Q. Chen, H. Tian and W. H. Zhu, *Chem. A Eur. J.*, 2012, **18**, 8190–8200.

119. H. N. Tian, X. C. Yang, R. K. Chen, Y. Z. Pan, L. Li, A. Hagfeldt and L. C. Sun, *Chem. Commun.*, 2007, 3741–3743.

120. H. N. Tian, X. C. Yang, R. K. Chen, A. Hagfeldt and L. C. Sun, *Energy Environ. Sci.*, 2009, **2**, 674–677.

121. K. M. Karlsson, X. Jiang, S. K. Eriksson, E. Gabrielsson, H. Rensmo, A. Hagfeldt and L. C. Sun, *Chem. A Eur. J.*, 2011, **17**, 6415–6424.

122. S. Franco, J. Garín, N. Martínez de Baroja, R. Pérez-Tejada, J. Orduna, Y. Yu, M. Lira-Cantú and N. M. de Baroja, *Org. Lett.*, 2012, **14**, 752–755.

123. Y. Numata, I. Ashraful, Y. Shirai and L. Han, *Chem. Commun.*, 2011, **47**, 6159–6161.

124. S. Ito, H. Miura, S. Uchida, M. Takata, K. Sumioka, P. Liska, P. Comte, P. Péchy and M. Grätzel, *Chem. Commun.*, 2008, 5194–5196.

125. H. Tian, X. Yang, R. Chen, Y. Pan, L. Li, A. Hagfeldt and L. Sun, *Chem. Commun.*, 2007, 3741–3743.

126. S. L. Li, K. J. Jiang, K. F. Shao and L. M. Yang, *Chem. Commun.*, 2006, 2792–2794.

127. S. Ito, S. M. Zakeeruddin, R. Humphry-Baker, P. Liska, R. Charvet, P. Comte, M. K. Nazeeruddin, P. Pechy, M. Takata, H. Miura, S. Uchida and M. Gratzel, *Adv. Mater.*, 2006, **18**, 1202–1205.

128. D. Kuang, S. Uchida, R. Humphry-Baker, S. M. Zakeeruddin and M. Gratzel, *Angew. Chem. Int. Ed.*, 2008, **47**, 1923–1927.

129. Y. Ooyama, S. Inoue, T. Nagano, K. Kushimoto, J. Ohshita, I. Imae, K. Komaguchi and Y. Harima, *Angew. Chem. Int. Ed. Engl.*, 2011, **50**, 7429–7433.

130. Y. Ooyama, T. Nagano, S. Inoue, I. Imae, K. Komaguchi, J. Ohshita and Y. Harima, *Chem A Eur. J.*, 2011, **17**, 14837–14843.

131. K. Hara, M. Kurashige, S. Ito, A. Shinpo, S. Suga, K. Sayama and H. Arakawa, *Chem. Commun.*, 2003, 252–253.

132. K. Hara, T. Sato, R. Katoh, A. Furube, Y. Ohga, A. Shinpo, S. Suga, K. Sayama, H. Sugihara and H. Arakawa, *J. Phys. Chem. B*, 2003, **107**, 597–606.

133. T. Kitamura, M. Ikeda, K. Shigaki, T. Inoue, N. A. Anderson, X. Ai, T. Q. Lian and S. Yanagida, *Chem. Mater.*, 2004, **16**, 1806–1812.

134. X. M. Ma, J. L. Hua, W. J. Wu, Y. H. Jin, F. S. Meng, W. H. Zhan and H. Tian, *Tetrahedron*, 2008, **64**, 345–350.

135. J. S. Song, F. Zhang, C. H. Li, W. L. Liu, B. S. Li, Y. Huang and Z. S. Bo, *J. Phys. Chem. C*, 2009, **113**, 13391–13397.

136. C. Teng, X. C. Yang, C. Yang, H. N. Tian, S. F. Li, X. N. Wang, A. Hagfeldt and L. C. Sun, *J. Phys. Chem. C*, 2010, **114**, 11305–11313.

137. Y.-S. Yen, H.-H. Chou, Y.-C. Chen, C.-Y. Hsu and J. T. Lin, *J. Mater. Chem.*, 2012, **22**, 8734–8747.

138. D. P. Hagberg, T. Marinado, K. M. Karlsson, K. Nonomura, P. Qin, G. Boschloo, T. Brinck, A. Hagfeldt and L. Sun, *J Org. Chem.*, 2007, **72**, 9550–9556.

139. N. Koumura, Z.-S. Wang, S. Mori, M. Miyashita, E. Suzuki and K. Hara, *J. Am. Chem. Soc.*, 2006, **128**, 14256–14257.

140. I. Jung, J. K. Lee, K. H. Song, K. Song, S. O. Kang and J. Ko, *J Org. Chem.*, 2007, **72**, 3652–3658.

141. J. T. Lin, P. C. Chen, Y. S. Yen, Y. C. Hsu, H. H. Chou and M. C. P. Yeh, *Org. Lett.*, 2009, **11**, 97–100.

142. S. Y. Qu, C. J. Qin, A. Islam, J. L. Hua, H. Chen, H. Tian and L. Y. Han, *Chem-Asian J.*, 2012, **7**, 2895–2903.

143. S. Paek, H. Choi, C. Kim, N. Cho, S. So, K. Song, M. K. Nazeeruddin and J. Ko, *Chem. Commun.*, 2011, **47**, 2874–2876.

144. P. Gao, H. N. Tsao, M. Gratzel and M. K. Nazeeruddin, *Org. Lett.*, 2012, **14**, 4330–4333.

145. P. Gao, Y. J. Kim, J. H. Yum, T. W. Holcombe, M. K. Nazeeruddin and M. Gratzel, *J. Mater. Chem. A*, 2013, **1**, 5535–5544.

146. W. D. Zeng, Y. M. Cao, Y. Bai, Y. H. Wang, Y. S. Shi, M. Zhang, F. F. Wang, C. Y. Pan and P. Wang, *Chem. Mater.*, 2010, **22**, 1915–1925.

147. L. Y. Lin, C. H. Tsai, K. T. Wong, T. W. Huang, L. Hsieh, S. H. Liu, H. W. Lin, C. C. Wu, S. H. Chou, S. H. Chen and A. I. Tsai, *J. Org. Chem.*, 2010, **75**, 4778–4785.

148. S. Y. Qu, W. J. Wu, J. L. Hua, C. Kong, Y. T. Long and H. Tian, *J. Phys. Chem. C*, 2010, **114**, 1343–1349.

149. J. H. Yum, T. W. Holcombe, Y. Kim, J. Yoon, K. Rakstys, M. K. Nazeeruddin and M. Gratzel, *Chem. Commun.*, 2012, **48**, 10727–10729.

150. J. H. Yum, T. W. Holcombe, Y. Kim, K. Rakstys, T. Moehl, J. Teuscher, J. H. Delcamp, M. K. Nazeeruddin and M. Gratzel, *Sci. Report.*, 2013, **3**, 2446(1–8).

151. Y. Z. Wu, X. Zhang, W. Q. Li, Z. S. Wang, H. Tian and W. H. Zhu, *Adv. Energy Mater.*, 2012, **2**, 149–156.

152. S. Haid, M. Marszalek, A. Mishra, M. Wielopolski, J. Teuscher, J. E. Moser, R. Humphry-Baker, S. M. Zakeeruddin, M. Gratzel and P. Bauerle, *Adv. Funct. Mater.*, 2012, **22**, 1291–1302.
153. Y. Z. Wu, M. Marszalek, S. M. Zakeeruddin, Q. Zhang, H. Tian, M. Gratzel and W. H. Zhu, *Energy Environ. Sci.*, 2012, **5**, 8261–8272.
154. Y. Cui, Y. Z. Wu, X. F. Lu, X. Zhang, G. Zhou, F. B. Miapeh, W. H. Zhu and Z. S. Wang, *Chem. Mater.*, 2011, **23**, 4394–4401.
155. J. Y. Mao, F. L. Guo, W. J. Ying, W. J. Wu, J. Li and J. L. Hua, *Chem-Asian J.*, 2012, **7**, 982–991.
156. W. Zeng, Y. Cao, Y. Bai, Y. Wang, Y. Shi, M. Zhang, F. Wang, C. Pan and P. Wang, *Chem. Mater.*, 2010, **22**, 1915–1925.
157. Y. Bai, J. Zhang, D. Zhou, Y. Wang, M. Zhang and P. Wang, *J. Am. Chem. Soc.*, 2011, **133**, 11442–11445.
158. C. Qin, A. Islam and L. Han, *Dyes Pigm.*, 2012, **94**, 553–560.
159. J. J. Nelson, T. J. Amick and C. M. Elliott, *J. Phys. Chem. C*, 2008, **112**, 18255–18263.
160. D. P. Hagberg, X. Jiang, E. Gabrielsson, M. Linder, T. Marinado, T. Brinck, A. Hagfeldt and L. Sun, *J. Mater. Chem.*, 2009, **19**, 7232–7238.
161. H. N. Tsao, C. Yi, T. Moehl, J.-H. Yum, S. M. Zakeeruddin, M. K. Nazeeruddin and M. Grätzel, *ChemSusChem*, 2011, **4**, 591–594.
162. H. N. Tsao, P. Comte, C. Yi and M. Grätzel, *ChemPhysChem*, 2012, **13**, 2976–2981.
163. A. Yella, *Science*, 2011, **334**, 629–634.
164. T. W. Hamann, *Dalton Trans.*, 2012, **41**, 3111–3115.
165. P. Gao, Y. J. Kim, J.-H. Yum, T. W. Holcombe, M. K. Nazeeruddin, M. Grätzel and M. Graetzel, *J. Mater. Chem. A*, 2013, **1**, 5535–5544.
166. A. Yella, R. Humphry-Baker, B. F. E. Curchod, N. Ashari Astani, J. Teuscher, L. E. Polander, S. Mathew, J.-E. Moser, I. Tavernelli, U. Rothlisberger, M. Grätzel, M. K. Nazeeruddin and J. Frey, *Chem. Mater.*, 2013, **25**, 2733–2739.
167. N. Cai, R. Z. Li, Y. L. Wang, M. Zhang and P. Wang, *Energy Environ. Sci.*, 2013, **6**, 139–147.
168. J. Y. Liu, R. Z. Li, X. Y. Si, D. F. Zhou, Y. S. Shi, Y. H. Wang, X. Y. Jing and P. Wang, *Energy Environ. Sci.*, 2010, **3**, 1924–1928.
169. M. Lu, M. Liang, H. Y. Han, Z. Sun and S. Xue, *J. Phys. Chem. C*, 2011, **115**, 274–281.
170. E. Ronca, M. Pastore, L. Belpassi, F. Tarantelli and F. De Angelis, *Energy Environ. Sci.*, 2013, **6**, 183–193.
171. T. W. Holcombe, J.-H. Yum, Y. Kim, K. Rakstys and M. Gratzel, *J. Mater. Chem. A*, 2013, **1**, 13978–13983.
172. J. Y. Li, C. Y. Chen, W. C. Ho, S. H. Chen and C. G. Wu, *Org. Lett.*, 2012, **14**, 5420–5423.
173. Z. M. Hao and A. Iqbal, *Chem. Soc. Rev.*, 1997, **26**, 203–213.
174. P. M. Beaujuge and J. M. J. Frechet, *J. Am. Chem. Soc.*, 2011, **133**, 20009–20029.

175. C. Kanimozhi, P. Balraju, G. D. Sharma and S. Patil, *J. Phys. Chem. C*, 2010, **114**, 3287–3291.
176. J. Warnan, L. Favereau, Y. Pellegrin, E. Blart, D. Jacquemin and F. Odobel, *J. Photochem. Photobiol. A*, 2011, **226**, 9–15.
177. S. Y. Qu, B. Wang, F. L. Guo, J. Li, W. J. Wu, C. Kong, Y. T. Long and J. L. Hua, *Dyes Pigm.*, 2012, **92**, 1384–1393.
178. L. Favereau, J. Warnan, F. B. Anne, Y. Pellegrin, E. Blart, D. Jacquemin and F. Odobel, *J. Mater. Chem. A*, 2013, **1**, 7572–7575.
179. J. Warnan, L. Favereau, F. Meslin, M. Severac, E. Blart, Y. Pellegrin, D. Jacquemin and F. Odobel, *ChemSusChem*, 2012, **5**, 1568–1577.
180. T. W. Holcombe, J. H. Yum, J. Yoon, P. Gao, M. Marszalek, D. Di Censo, K. Rakstys, M. K. Nazeeruddin and M. Graetzel, *Chem. Commun.*, 2012, **48**, 10724–10726.
181. S. Y. Qu, C. J. Qin, A. Islam, Y. Z. Wu, W. H. Zhu, J. L. Hua, H. Tian and L. Y. Han, *Chem. Commun.*, 2012, **48**, 6972–6974.
182. A. Kay and M. Gratzel, *J. Phys. Chem. US*, 1993, **97**, 6272–6277.
183. A. Burke, L. Schmidt-Mende, S. Ito and M. Gratzel, *Chem. Commun.*, 2007, 234–236.
184. H. Balli and M. Zeller, *Helvet. Chim. Acta*, 1983, **66**, 2135–2139.
185. F. Gerson and A. Metzger, *Helvet. Chim. Acta*, 1983, **66**, 2031–2043.
186. J. H. Delcamp, A. Yella, T. W. Holcombe, M. K. Nazeeruddin and M. Grätzel, *Angew. Chem. Int. Ed. Engl.*, 2013, **52**, 376–380.
187. H. P. Wu, Z. W. Ou, T. Y. Pan, C. M. Lan, W. K. Huang, H. W. Lee, N. M. Reddy, C. T. Chen, W. S. Chao, C. Y. Yeh and E. W. G. Diau, *Energy Environ. Sci.*, 2012, **5**, 9843–9848.
188. M. K. Nazeeruddin, R. Humphry-Baker, D. L. Officer, W. M. Campbell, A. K. Burrell and M. Gratzel, *Langmuir*, 2004, **20**, 6514–6517.
189. Q. Wang, W. M. Carnpbell, E. E. Bonfantani, K. W. Jolley, D. L. Officer, P. J. Walsh, K. Gordon, R. Humphry-Baker, M. K. Nazeeruddin and M. Gratzel, *J. Phys. Chem. B*, 2005, **109**, 15397–15409.
190. W. M. Campbell, K. W. Jolley, P. Wagner, K. Wagner, P. J. Walsh, K. C. Gordon, L. Schmidt-Mende, M. K. Nazeeruddin, Q. Wang, M. Gratzel and D. L. Officer, *J. Phys. Chem. C*, 2007, **111**, 11760–11762.
191. M. Ishida, S. W. Park, D. Hwang, Y. B. Koo, J. L. Sessler, D. Y. Kim and D. Kim, *J. Phys. Chem. C*, 2011, **115**, 19343–19354.
192. J. K. Park, H. R. Lee, J. P. Chen, H. Shinokubo, A. Osuka and D. Kim, *J. Phys. Chem. C*, 2008, **112**, 16691–16699.
193. C. L. Mai, W. K. Huang, H. P. Lu, C. W. Lee, C. L. Chiu, Y. R. Liang, E. W. G. Diau and C. Y. Yeh, *Chem. Commun.*, 2010, **46**, 809–811.
194. S. Hayashi, M. Tanaka, H. Hayashi, S. Eu, T. Umeyama, Y. Matano, Y. Araki and H. Imahori, *J. Phys. Chem. C*, 2008, **112**, 15576–15585.
195. S. Eu, S. Hayashi, T. Urneyama, Y. Matano, Y. Araki and H. Imahori, *J. Phys. Chem. C*, 2008, **112**, 4396–4405.
196. A. Kira, Y. Matsubara, H. Iijima, T. Umeyama, Y. Matano, S. Ito, M. Niemi, N. V. Tkachenko, H. Lemmetyinen and H. Imahori, *J. Phys. Chem. C*, 2010, **114**, 11293–11304.

197. H. Hayashi, A. S. Touchy, Y. Kinjo, K. Kurotobi, Y. Toude, S. Ito, H. Saarenpaa, N. V. Tkachenko, H. Lemmetyinen and H. Imahori, *ChemSusChem*, 2013, **6**, 508–517.

198. T. Bessho, S. M. Zakeeruddin, C. Y. Yeh, E. W. Diau and M. Gratzel, *Angew. Chem. Int. Ed. Engl*, 2010, **49**, 6646–6649.

199. C. Y. Lin, C. F. Lo, L. Luo, H. P. Lu, C. S. Hung and E. W. G. Diau, *J. Phys. Chem. C*, 2009, **113**, 755–764.

200. S. L. Wu, H. P. Lu, H. T. Yu, S. H. Chuang, C. L. Chiu, C. W. Lee, E. W. G. Diau and C. Y. Yeh, *Energy Environ. Sci.*, 2010, **3**, 949–955.

201. E. M. Barea, V. Gonzalez-Pedro, T. Ripolles-Sanchis, H. P. Wu, L. L. Li, C. Y. Yeh, E. W. G. Diau and J. Bisquert, *J. Phys. Chem. C*, 2011, **115**, 10898–10902.

202. T. Bessho, S. M. Zakeeruddin, C. Y. Yeh, E. W. G. Diau and M. Gratzel, *Angew. Chem. Int. Ed.*, 2010, **49**, 6646–6649.

203. T. Ripolles-Sanchis, B. C. Guo, H. P. Wu, T. Y. Pan, H. W. Lee, S. R. Raga, F. Fabregat-Santiago, J. Bisquert, C. Y. Yeh and E. W. G. Diau, *Chem. Commun.*, 2012, **48**, 4368–4370.

204. Y. L. Xie and T. W. Hamann, *J. Phys. Chem. Lett.*, 2013, **4**, 328–332.

205. R. Jono, M. Sumita, Y. Tateyama and K. Yamashita, *J. Phys. Chem. Lett.*, 2012, **3**, 3581–3584.

206. G. Boschloo, E. A. Gibson and A. Hagfeldt, *J. Phys. Chem. Lett.*, 2011, **2**, 3016–3020.

207. J. L. Lan, T. C. Wei, S. P. Peng, C. C. Wan and G. Z. Cao, *J. Phys. Chem. C*, 2012, **116**, 25727–25733.

208. R. Y. Ogura, S. Nakane, M. Morooka, M. Orihashi, Y. Suzuki and K. Noda, *Appl. Phys. Lett.*, 2009, **94**, 073308(1–3).

209. B. E. Hardin, E. T. Hoke, P. B. Armstrong, J. H. Yum, P. Comte, T. Torres, J. M. J. Frechet, M. K. Nazeeruddin, M. Gratzel and M. D. McGehee, *Nat. Photonics*, 2009, **3**, 667–667.

210. J. H. Yum, E. Baranoff, S. Wenger, M. K. Nazeeruddin and M. Gratzel, *Energy Environ. Sci.*, 2011, **4**, 842–857.

211. S. Ito, T. N. Murakami, P. Comte, P. Liska, C. Gratzel, M. K. Nazeeruddin and M. Gratzel, *Thin Solid Films*, 2008, **516**, 4613–4619.

212. K. Susumu, T. V. Duncan and M. J. Therien, *J. Am. Chem. Soc.*, 2005, **127**, 5186–5195.

213. H. J. Song, D. H. Kim, T. H. Lee and D. K. Moon, *Eur. Polym. J.*, 2012, **48**, 1485–1494.

214. T. D. Lash, P. Chandrasekar, A. T. Osuma, S. T. Chaney and J. D. Spence, *J. Org. Chem.*, 1998, **63**, 8455–8469.

215. Y. Y. Huang, L. S. Li, X. B. Peng, J. B. Peng and Y. Cao, *J. Mater. Chem.*, 2012, **22**, 21841–21844.

216. F. Jones, J. B. Farrow and W. van Bronswijk, *Langmuir*, 1998, **14**, 6512–6517.

217. Z. Y. Cheng and J. Lin, *CrystEngComm*, 2010, **12**, 2646–2662.

218. J. Burschka, N. Pellet, S. J. Moon, R. Humphry-Baker, P. Gao, M. K. Nazeeruddin and M. Gratzel, *Nature*, 2013, **499**, 316–319.

219. N. G. Park, *J. Phys. Chem. Lett.*, 2013, **4**, 2423–2429.
220. L. Etgar, P. Gao, Z. S. Xue, Q. Peng, A. K. Chandiran, B. Liu, M. K. Nazeeruddin and M. Gratzel, *J. Am. Chem. Soc.*, 2012, **134**, 17396–17399.
221. M. Liu, M. B. Johnston and H. J. Snaith, *Nature*, 2013, **501**, 395–398.
222. Q. Chen, H. Zhou, Z. Hong, S. Luo, H. S. Duan, H. H. Wang, Y. Liu, G. Li and Y. Yang, *J. Am. Chem. Soc.*, 2014, **136**, 622–625.

CHAPTER 7

Perovskite Solar Cells

NAM-GYU PARK

School of Chemical Engineering and Department of Energy Science,
Sungkyunkwan University, Suwon 440-746, Republic of Korea
Email: npark@skku.edu

7.1 Introduction

Perovskite, named after the Russian mineralogist L.A. Perovski, is a crystal
structure with the formula ABX_3 (X = halogen or oxygen), where the cation A
is stabilized in a cubo-octahedral cage formed by the 12 nearest X anions,
and the cation B coordinates with the six nearest X anions to form octahedral geometry. Oxide perovskite started to receive attention because of its
unique physical properties such as ferroelectricity[1–3] and superconductivity.[4,5] Compared to oxide perovskite, halide perovskite has received relatively less attention because of the lack of unique physical properties.
Replacement of an inorganic element in the A site with organic molecules
was found to lead to a perovskite structure. Organotin halide compounds
with layered perovskite structure gained some attention because of semiconductor-to-metal transition behavior and superconductivity.[6] The three-
dimensional organolead halide $CH_3NH_3PbX_3$ (X = Cl, Br and/or I) was found
to crystallize with a perovskite structure and showed a special feature of
molecular motion of methylammonium ion.[7,8] However, methylammonium
lead halide perovskite attracted little attention because interesting electronic
properties were not discovered. Moreover, no special attention regarding the
opto-electronic property has been paid to organolead halide perovskite. In
2009, Miyasaka *et al.* were the first researchers to find that $CH_3NH_3PbX_3$
possessed light harvesting properties when they were used it as the sensitizer

RSC Energy and Environment Series No. 11
Advanced Concepts in Photovoltaics
Edited by Arthur J Nozik, Gavin Conibeer and Matthew C Beard
© The Royal Society of Chemistry 2014
Published by the Royal Society of Chemistry, www.rsc.org

in a dye-sensitized electrochemical junction-type solar cell.[9] $CH_3NH_3PbBr_3$-deposited nanocrystalline TiO_2 showed a power conversion efficiency (PCE) of 3.1% and $CH_3NH_3PbI_3$ showed a little higher PCE of 3.8%. The better efficiency from iodide perovskite was due to a narrower bandgap energy.[9] Two years later, a higher PCE (6.5%) was reported based on a $CH_3NH_3PbI_3$ light harvester, in which the TiO_2 thin film whose surface was covered sparsely with $CH_3NH_3PbI_3$ nanocrystals was confirmed to show one order of magnitude higher absorption coefficient than the same thick TiO_2 film with a ruthenium bipyridyl-based organometallic dye molecule.[10] Such a superior light absorption property has been considered to be suitable for thin film solar cell technology. Nevertheless, $CH_3NH_3PbX_3$ was not considered as a promising sensitizer because of its chemical instability in polar electrolyte solution. In 2012, a long-term stable and high efficiency all-solid-state mesoscopic solar cell based on $CH_3NH_3PbI_3$ was first developed, where a PCE of 9.7% was achieved from submicrometer-thick TiO_2 film (0.6 μm) whose surface was deposited with $CH_3NH_3PbI_3$ nanodots.[11] The non-sensitization concept using the mixed halide perovskite $CH_3NH_3PbI_{3-x}Cl_x$ coated on an Al_2O_3 surface was proposed, which demonstrated a PCE of 10.9%.[12] Rapid progress was made soon after discovery of the so-called 'perovskite solar cell'. In 2013, the perovskite solar cell achieved much higher PCEs, exceeding 15%, by modification of the deposition method or junction structure.[13,14] Figure 7.1 depicts the progress in perovskite solar cell technology. In this chapter, the synthesis and opto-electronic properties of oganometal halide perovskite and photovoltaic performance of the perovskite solar cell are described.

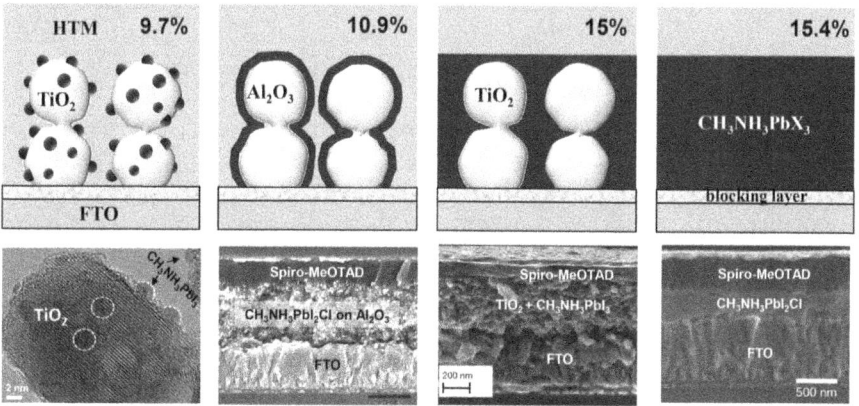

Figure 7.1 Progress in perovksite solar cell showing structural modification from the $CH_3NH_3PbI_3$ perovskite dot deposited on TiO_2 surface to the $CH_3NH_3PbI_2Cl$ perovskite thin layer on scaffold Al_2O_3 surface, the $CH_3NH_3PbI_3$ perovskite-infiltrated nancomposite and the planar pn junction with mesoporous oxide layer.
SEM images were reprinted from Im *et al.*,[10] Lee *et al.*,[12] Burschka *et al.*,[13] and Liu *et al.*,[14] from left to right, respectively.

7.2   Synthesis of Organolead Halide Perovskite

$CH_3NH_3PbX_3$ (X = Cl, Br, I or mixed halides) can be synthesized by reacting CH_3NH_3X with PbX_2. For the case of $CH_3NH_3PbI_3$, CH_3NH_3I is first prepared by reacting methylamine (CH_3NH_2) with hydroiodic acid (HI) at 0 °C. The as-prepared CH_3NH_3I should be washed with diethyl ether in order to remove unreacted chemical species and finally the washed powder should be re-covered by drying at 60 °C overnight. Methylammonium bromide and chloride can be also prepared by the similar way. To prepare $CH_3NH_3PbI_3$, the equimolar of CH_3NH_3I and PbI_2 is dissolved in an aprotic polar solvent such as γ-butyrolactone (GBL) or dimethylformamide (DMF). The solvent evaporation from this solution leads to $CH_3NH_3PbI_3$ formation. To make the film type, the solution is spin-coated on the substrate (bare substrate or oxide film deposited substrate), followed by drying to evaporate the solvent. The detailed synthetic process can be found elsewhere.[10] $CH_3NH_3PbI_3$ film can be prepared by a two-step dipping method. PbI_2 forms first on the substrate and this stage is followed by dipping in CH_3NH_3I solution to form $CH_3NH_3PbI_3$. The mechanism of the two-step method is based on an interaction reaction. Since PbI_2 is known to be a layered structure, CH_3NH_3I can be intercalated into the interlayer of two-dimensional PbI_2. This two-step method is suitable for making thicker films and filling pores of mesoporous oxide films compared to the one-step deposition method. The two-step method is described in detail elsewhere.[13,15]

For the synthesis of $CH_3NH_3PbI_{3-x}Cl_x$, CH_3NH_3I is reacted with $PbCl_2$ instead of reaction between CH_3NH_3Cl and PbI_2 due to less soluble char-acteristics of CH_3NH_3Cl. Three moles of CH_3NH_3I is recommended to react with one mole of $PbCl_2$ to induce analogue structure with $CH_3NH_3PbI_3$.[12]

7.3   Crystal Structure and Related Properties

In a perovskite structure with the general formula ABX_3 (X = oxygen or halogen), the cation A is stabilized in the cubo-octahedral site and the cation B coordinates with six X anions to form octahedral geometry. The cation A is usually larger than the cation B and is usually too large to occupy the bulky cubo-octahedral site (Figure 7.2). For the case of X = halide such as Cl, Br or I, the cation A is monovalent and the cation B is divalent to satisfy charge neutrality. Crystal stabilization of perovskite ABX_3 can be simply estimated from the Goldschmidt tolerance factor (t),[16] where $t = (r_A + r_X)/[2^{1/2}(r_B + r_X)]$, where r_A, r_B and r_X are the ionic radii of A, B and X, respectively. It has been generally accepted that perovskite can be stabilized when t is in the range between 0.78 and 1.05.[17] However, it has been argued that the perovskite structure is not stable even for the most favorable range of $0.8 < t < 0.9$,[18] which indicates that perovskite stability cannot be predicted solely by tol-erance factor. Additional factor to be considered is octahedral factor (μ), $\mu = r_B/r_X$, for perovskite formability.[19] For instance, formability of alkali metal halide perovskite was determined from the t–μ mapping, where

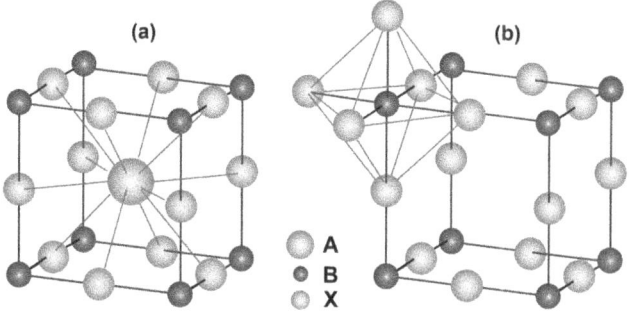

Figure 7.2 (a) The perovskite structure for ABX$_3$. (b) ReO$_3$ structure with the same octahedral (BX$_6$) framework in the perovskite.

Table 7.1 Estimation of A cation radii in AMX$_3$ (M = Pb^{2+} or Sn^{2+}; X = Cl$^-$, Br$^-$ or I$^-$).

B^a	X^a	$r_A{}^b$ (pm) at $t = 0.8$	$r_A{}^b$ (pm) at $t = 1.0$
Pb^{2+} ($r = 119$ pm)	Cl$^-$ ($r = 181$ pm)	158.4	243.3
	Br$^-$ ($r = 196$ pm)	160.4	249.5
	I$^-$ ($r = 220$ pm)	163.5	259.4
Sn^{2+} ($r = 118$ pm)	Cl$^-$ ($r = 181$ pm)	157.3	241.8
	Br$^-$ ($r = 196$ pm)	159.3	248.1
	I$^-$ ($r = 220$ pm)	162.4	258.0

^{a}Effective ionic radii for 6 coordination number.
$^b r_A = t \times \sqrt{2}(r_B + r_X) - r_X$.

perovksite is found in the region between $0.813 < t < 1.107$ and $0.442 < \mu < 0.895$.[19] In Table 7.1, ionic radii of r_A in APbX$_3$ and ASnX$_3$ perovskite are estimated based on effective ionic radii,[20–22] where the cation A is allowed to have one or two C–C (150 pm) or C–N (148 pm) bonds. For this reason, stabilization of methylammonium in the A site can be explained.

Tolerance factors for CH$_3$NH$_3$PbX$_3$ are calculated to be 0.85, 0.84 and 0.83 for X = Cl, Br and I, respectively, based on radii of CH$_3$NH$_3{}^+ = 180$ pm,[23] Pb$^{2+} = 119$ pm, Cl$^- = 181$ pm, Br$^- = 196$ pm and I$^- = 220$ pm. It was found that t of most cubic perovskites are in the range between 0.8 and 0.9.[24] Thus, methylammonium lead halide perovskites are therefore expected to be cubic structure. Crystal phase and temperature-dependent phase transition were studied, where cubic phase was detected at temperature >178.8 K for X = Cl, >236.9 K for X = Br and >327.4 K for X = I.[25] For the iodide perovskite, the tetragonal phase is stabilized at ambient temperature with lattice parameters of a $= 8.855$ Å and c $= 12.659$ Å because the cubic phase is stabilized above 54.4 °C. In Figure 7.3, the unit cell and the 4×4×4 supercell for cubic CH$_3$NH$_3$PbI$_3$ perovskite are displayed.[26]

The basic physical properties of CH$_3$NH$_3$PbX$_3$ are as follows. Dielectric constants are estimated to be 23.9, 25.5 and 28.8, and dipole moments

(a)   **(b)**

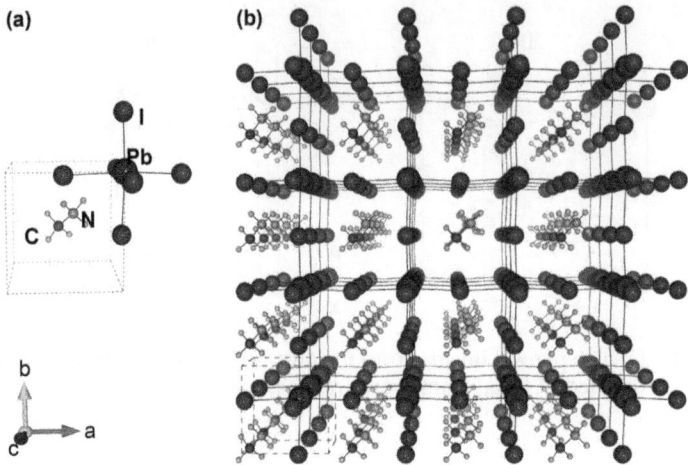

Figure 7.3   (a) The unit cell and (b) the 4×4×4 supercell of cubic CH₃NH₃PbI₃
perovskite.
Reprinted from Giorgi *et al.*[26]

are calculated to be 0.85, 0.766 and 0.854 for X = Cl, Br and I, respectively.[25]
Bandgap energies increase as halogen size decreases from I to Cl. The calculated bandgap is 1.57 eV (X = I), 1.80 eV (X = Br) and 2.34 eV (X = Cl) for
the cubic structure of $CH_3NH_3PbX_3$, which has the same tendency with the
data observed experimentally (*ca.* 1.5 eV for X = I, *ca.* 2.3 eV for X = Br and
ca. 3.1 eV for X = Cl).[27] Physical properties can be tuned by replacing A and/
or B cations while maintaining perovskite structure. For instance, methylammonium in the A site can be replaced with formamidinium
$HC(NH_2)_2^+$,[28] or ethylammonium,[29] which leads to a change in the bandgap
energy from 1.52 eV for $CH_3NH_3^+$ to 1.45 eV for $HC(NH_2)_2^+$ and 2.2 eV for
$CH_3CH_2NH_3^+$ due to mainly structural modification. For instance, replacement of methylammonium with the larger ethylammonium in the A site
leads to a 2H perovskite structure having an orthorhombic crystal phase with
a = 8.7419(2) Å, b = 8.14745(10) Å, and c = 30.3096(6) Å, where infinite chains
of face-sharing (PbI_6) octahedra running along the *b*-axis of the unit cell are
separated from one another by ethylammonium ions.[29] Pb^{2+} in the B site
can be also replaced with divalent element such as Sn^{2+}, in which bandgap
becomes narrower ($\sim$1.5 eV $\rightarrow$ $\sim$1.2 eV) and optical property is modified
(PL emission wavelength shifts from $\sim$750 nm to $\sim$1000 nm) as well.[28]
Bandgap tuning is important in terms of spectral responsivity.

7.4   Opto-electronic Properties

Opto-electronic properties of perovskite organometal halides are also
related to structural dimensionality. The report on semiconductor-to-metal
transition in organotin iodide in 1994 has attracted much attention.[30] The

two-dimensional layered perovskite $(C_4H_9NH_3)_2SnI_4$ is an insulator, which can be transformed to be metallic by increasing the number of corner-sharing two-dimensional SnI_6 octahedral layer. Insertion of $CH_3NH_3^+$ ions between the SnI_6 layers is a method to increase the number of two-dimensional SnI_6 octahedral sheet by forming inner $CH_3NH_3SnI_3$ perovskite layer (Figure 7.4). Resistivity of the $(C_4H_9NH_3)_2(CH_3NH_3)_{n-1}Sn_nI_{3n+1}$ perovskite was found to decrease as n increases[30] and the cubic perovksite $(n \rightarrow \infty)$ $CH_3NH_3SnI_3$ showed a low carrier density p-type metal with a small effective mass of $0.16m_0$ (m_0 is free electron mass) associated with high Hall mobility.[31]

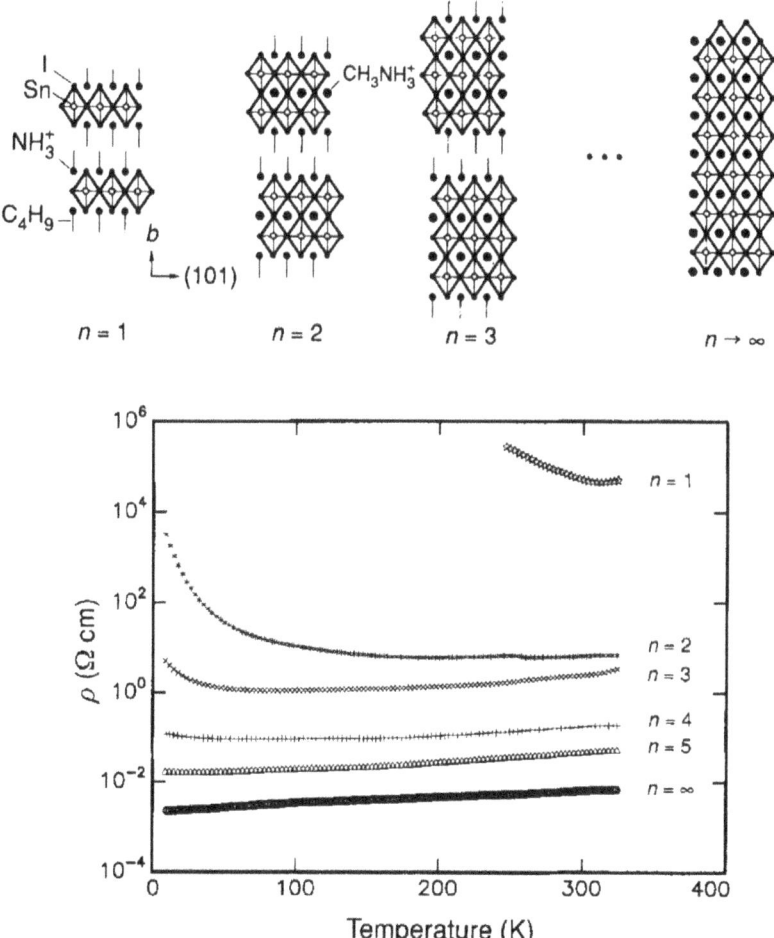

Figure 7.4 (Top) Schematic representation of $(C_4H_9NH_3)_2(CH_3NH_3)_{n-1}Sn_nI_{3n+1}$ and three-dimensional perovskite $(n \rightarrow \infty)$ $CH_3NH_3SnI_3$. (Bottom) Effect of the number of SnI_6 octahedral layers on resistivity. Reprinted from Mitzi *et al.*[30]

Since organometal halide perovskite contains both organic and inorganic part, electronic excitation can lead to Frenkel-type exciton with high exciton binding energy and small Bohr exciton radius, and/or Wannier-type exciton with low exciton binding energy and large Bohr exciton radius. Frenkel exciton arises in general from organic molecules, whereas Wannier exciton is originated from inorganic part, usually metal halide octahedral.[32] Bohr radius, binding energy and reduced mass of the lowest–energy excitons in $CH_3NH_3PbX_3$ are determined to be 20 Å, 76 meV and 0.13 m_0 for X = Br and 22 Å, 50 meV and 0.15 m_0 for X = I.[33] Small photocarrier effective masses of $m_e^* = 0.23m_0$ and $m_h^* = 0.29m_0$ were also estimated for $CH_3NH_3PbI_3$ in the presence of spin–orbit coupling.[26] Since Bohr radii are relatively large and exciton binding energies are small, the excitons from $CH_3NH_3PbX_3$ (X = Br, I) are considered to be three-dimensional Wannier excitons. Contrary to the three-dimensional structure, two-dimensional organometal halide perovskite structures have higher exciton binding energy of more than 200 meV.[34–36]

$CH_3NH_3PbI_3$ and mixed halide $CH_3NH_3PbI_{3-x}Cl_x$ are found to be able to transport both electron and hole. Photoluminescence (PL) quenching measurements revealed that mixed halide perovskite showed both electron and hole diffusion lengths (L_n) as high as 1000 nm (Stranks *et al.*[37]) that was much longer than the absorption depth (100–200 nm), while electron and hole diffusion lengths were around 100 nm for $CH_3NH_3PbI_3$.[38] Table 7.2 summarizes photo-electron and photon-hole behaviors for the $CH_3NH_3PbI_3$ and $CH_3NH_3PbI_{3-x}Cl_x$. Since diffusion coefficients (D) and lengths are known as listed in Table 7.2, time constant for recombination (τ_R) can be roughly estimated based on $\tau_R = L_n^2/D_e$, where D_e is electron diffusion coefficient. If we take $L_n = 130$ nm and $D_e = 0.036$ cm^2 s^{-1} for $CH_3NH_3PbI_3$,[38] τ_R can be estimated to be as short as around 5×10^{-9} s. Although no significant difference in D between tri-iodide and mixed iodide perovskites, τ_R will be reduced for $CH_3NH_3PbI_{3-x}Cl_x$ due to the longer L_n. In addition to charge transport ability, $CH_3NH_3PbI_3$ is also able to accumulate charges because of high density of states,[39] which was first observation in light-absorbing material. This indicates that the perovskite solar cell is a new kind of photovoltaic devices, different from the sensitized solar cell and even the conventional inorganic thin film solar cells.

Table 7.2 Diffusion coefficients (D) and diffusion length (L_n) for $CH_3NH_3PbI_{3-x}I_x$ and $CH_3NH_3PbI_3$.

Perovskite	Carrier	D (cm^2 s^{-1})	L_n (nm)
$CH_3NH_3PbI_{3-x}I_x$	Electron	0.042 ± 0.016	1069 ± 204
	Hole	0.054 ± 0.022	1213 ± 243
$CH_3NH_3PbI_3$	Electron	0.017 ± 0.011 (0.036)	129 ± 41 (130)
	Hole	0.011 ± 0.007 (0.022)	105 ± 32 (110)

The data with errors were based on Samuel *et al.*[37] and data in parenthesis were based on Xing *et al.*[38]

7.5 Perovskite Solar Cell Fabrication

Perovskite solar cell can be prepared by the following structures: (1) sensitized structure, (2) meso-superstructure and (3) planar p–n junction structure. The sensitized structure can be realized by just replacing dye molecule with perovskite in the conventional dye-sensitized structure. The first version of the perovskite solar cell is a sensitized structure, where $CH_3NH_3PbI_3$ perovskite dots forms on the nanocrystalline TiO_2 surface and spiro-MeOTAD HTM is infiltrated in the perovskite coated mesoporous TiO_2 film.[11] In the sensitized structure, photogenerated charges are separated by n-type TiO_2 and p-type HTM, that is, electron injection into TiO_2 should follows photo-induced charge generation. For the meso-superstructure, perovskite layer forms on scaffold layer such as Al_2O_3, where the perovskite thin layer deposited on the surface of the scaffold layer should be continuous in order for electrons to be transported.[12] Since electron injection does not occur in the meso-superstructure, injecting oxides are not necessarily required. Finally, the perovskite solar cell can be fabricated without electron injecting oxide or scaffold layer because of the electron and hole transporting ability in the perovskite light harvester.[14] A pillared structure is also proposed, where the injecting oxide layer, such as mesoporous TiO_2, is first prepared and then the pores are filled with perovskite. In addition to pore filling, a thin perovskite capping layer forms over the mesoporous TiO_2 film.[40] In this case, the TiO_2 matrix also supports a kind of skeleton to build the perovskite. To fabricate high efficiency perovskite solar cells, careful control of thickness, crystallinity and morphology of the perovskite layer is required.

The laboratory perovskite solar cell can be fabricated as follows. First, a transparent conducting oxide (TCO) substrate is etched either chemically or by laser ablation. A compact layer forms on the etched TCO in order to protect direct contact between the HTM layer and TCO. Materials for the compact layer are recommended to be oxides with electron transporting property or hole blocking property. Since the compact layer is aiming to inhibit the TCO/HTM contact and transport electrons, the thinner it is, the better. Less than 100 nm is recommended for this purpose. Once the compact layer has been prepared, mesoporous TiO_2 film is deposited using a screen or spin-coating method. By considering the electron and hole diffusion length of perovskite, mesoporous TiO_2 film should be controlled within the sub-micometer range. If a planar structure is considered, a mesoporous oxide layer is not required but only the compact layer is needed. Perovskite is then infiltrated into the pores in the mesoporous TiO_2 film by either a one-step or two-step method. When coating with perovskite, the coating process is carefully controlled to form a thin perovskite capping layer on the mesoporous TiO_2 film. The capping layer is recommended not to exceed the electron and hole diffusion length of the perovskite materials, such as less than 100 nm for $CH_3NH_3PbI_3$ and less than 1000 nm for $CH_3NH_3PbI_{3-x}Cl_x$. The surface of the perovskite capping layer is important in terms of junction formation with the HTM layer. A flat surface is highly recommended to induce the best contact with the HTM and reduce series resistance as well. HTM layer thickness is also important, where the HTM layer is thin enough to extract and transport holes. Finally, metal electrode forms on the HTM layer, in which several metals are candidates for the positive electrode but work function and chemical inertness are considered when choosing the metal materials. Au is commonly recommended.

7.6 Device Structures and Performances

7.6.1 CH$_3$NH$_3$PbI$_3$-based Perovskite Solar Cells

7.6.1.1 Combination with TiO$_2$

The first report on a high efficiency and long-term durable perovskite solar cell in 2012 was based on a sub-micrometer thick mesoporous TiO$_2$ layer whose surface was decorated with CH$_3$NH$_3$PbI$_3$ nanodots.[11] TEM confirmed that the perovskite nanodots were sparsely deposited and did not have interconnection. Thus this structure allows electron injection from perovskite to TiO$_2$. In this sensitized type, HTM infiltration is important in order to induce contact between HTM and perovskite-sensitized TiO$_2$. In spite of a relatively low concentration of the deposited CH$_3$NH$_3$PbI$_3$, a PCE of 9.7% was demonstrated because of the high absorption coefficient of CH$_3$NH$_3$PbI$_3$. The morphology of the deposited CH$_3$NH$_3$PbI$_3$ can be significantly influenced by a spin-coating condition or deposition technique. Infiltration of perovskite into mesopores of TiO$_2$ film was proposed instead of nanodot or extremely thin absorber layer structure, where over-layer (or capping layer) of CH$_3$NH$_3$PbI$_3$ formed on top of the mesoporous TiO$_2$ film.[40] Several HTM materials were tested and polytriarylamine (PTAA) showed the highest efficiency (12.0%) among the studied HTMs, along with high fill factor of 0.727. It has been an important issue how to infiltrate well the perovskite into mesopores. A two-step sequential coating method was proposed,[13] which was originally developed by Liang *et al.*[41] Two-step deposition led to better infiltration of CH$_3$NH$_3$PbI$_3$ and thereby delivered a PCE as high as 15%.[13] Figure 7.5 shows the transformation of PbI$_2$ to CH$_3$NH$_3$PbI$_3$ by dipping into CH$_3$NH$_3$I solution monitored by X-ray diffraction.

One-dimensional nanorods and nanofibers have been used for the perovskite solar cells. Hydrothermally grown rutile TiO$_2$ nanorods were found to be a good material for perovskite solar cells, where a PCE of 9.4% was demonstrated with a 0.56 μm thick nanorod film.[42] Electrospun nanofibrous TiO$_2$ film was also demonstrated PCE as high as 9.8%.[43] Such one-dimensional structures might be better in an electron transport and porous structure than in a nanoparticulated structure.

Hole transport in CH$_3$NH$_3$PbI$_3$ was utilized using a TiO$_2$/CH$_3$NH$_3$PbI$_3$ heterojunction structure without HTM layer.[44] A PCE of 5.5% was first

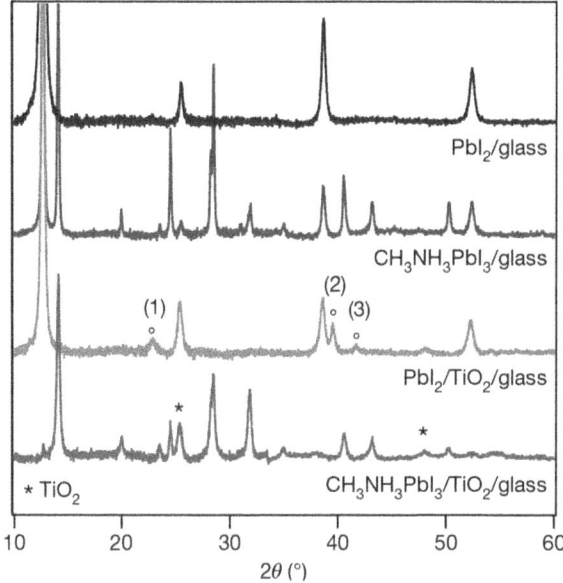

Figure 7.5 X-ray diffraction patterns of PbI$_2$ on glass and TiO$_2$-coated glass before and after transformation to CH$_3$NH$_3$PbI$_3$ perovskite by dipping in CH$_3$NH$_3$I solution.
Reprinted from Burschka *et al.*[13]

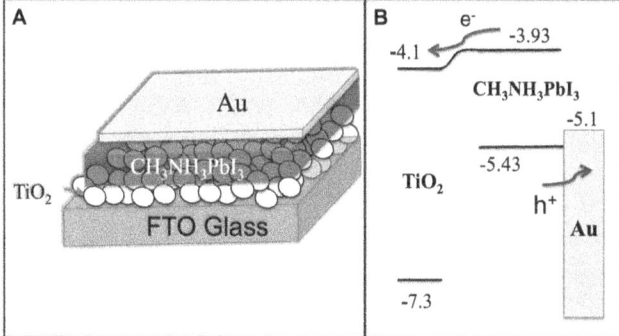

Figure 7.6 (A) Schematic structure of depleted HTM-free perovskite heterojunction solar cell and (B) its band diagram.
Reprinted from Abu Laban and Etgar.[45]

demonstrated with the structure of FTO/100 nm thick TiO$_2$ blocking layer/ 500 nm thick nanosheet TiO$_2$/CH$_3$NH$_3$PbI$_3$/Au, which was improved to more than 8%.[45] Mott–Schottky analysis confirmed that a depletion zone between TiO$_2$ and CH$_3$NH$_3$PbI$_3$ existed, which was therefore called a depleted heterojunction solar cell (Figure 7.6).

7.6.1.2   Combination with ZnO

ZnO is expected to be good candidate because of better electron transport than TiO_2. However, most of organic dye-sensitized ZnO particle or nanorod failed to show high conversion efficiency. This is because ZnO is unstable toward acidic dye solution. $CH_3NH_3PbI_3$ was first attempted to ZnO nanrod, where a PCE of 5% was demonstrated with good stability.[46] Since J_{SC} was less than 13 mA cm^{-2}, a much lower value compared to the theoretical one, which exceeds 20 mA cm^{-2}, further improvement is a challenge.

7.6.1.3   Combination with ZrO₂

Nanostructured ZrO_2 was applied to perovskite solar cell to investigate photovoltaic property. $CH_3NH_3PbI_3$ deposited ZrO_2 mesoporous layer exhibited substantially photovoltaic activity with V_{OC} of ~900 mV although its performance was lower than TiO_2[39] Impedance spectroscopic study using a three-electrode electrochemical cell confirmed that ZrO_2 was not charged under bias voltage up to 0.9 V, whereas TiO_2 was charged.[39] This indicates that a photogenerated charge in $CH_3NH_3PbI_3$ perovskite is not allowed to be injected into ZrO_2. Higher efficiency of 10.8% was achieved from ZrO_2 scaffold layer with $CH_3NH_3PbI_3$, where V_{OC} approached 1.07 V due to prolonged electron life time.[47]

7.6.2   Mixed Halide and Non-iodide Perovskite Solar Cells

7.6.2.1 $CH_3NH_3PbI_{3-x}(Cl$ or $Br)_x$ Meso-superstructure

It was confirmed that the mesoporous Al_2O_3 thin film coated with mixed halide perovskite showed a PCE of 10.9% upon contacting spiro-MeOTAD.[12] Since the used Al_2O_3 merely acted as a scaffold layer, photo-excited electrons should be transported through the perovskite without being injected into Al_2O_3. The first trial to form a mixed halide perovskite $CH_3NH_3PbI_2Cl$ on a Al_2O_3 surface was performed by spin-coating the DMF solution containing 3 M CH_3NH_3I and 1 M $PbCl_2$. Compared to $CH_3NH_3PbI_3$ the iodide–chloride mixed halide perovskite was mentioned to be more stable to processing in air. Although the crystal structure has not been well resolved, the prominent (hk0) peaks observed in an X-ray diffraction pattern indicates that two chloride ions might be located at the axial position in the PbI_4Cl_2 octahedron. The non-injecting Al_2O_3-$CH_3NH_3PbI_2Cl$ system showed higher V_{OC} than the injecting TiO_2-$CH_3NH_3PbI_2Cl$ system because of a higher conduction band of the perovskite.

For the case of a scaffold layer such as Al_2O_3, particle inter-connection may not be required since neither electron injection to the insulating layer nor electron transport in this scaffold layer occur. Thus, high-temperature annealing may not necessarily be required. The low-temperature processed perovskite solar cell exhibited a PCE as high as 12.3%.[48] Al_2O_3 mesoporous layers dried at 150 °C were prepared using a binder-free nanoparticlulate (diameter ~ 20 nm) Al_2O_3 paste, where the thickness of the Al_2O_3 scaffold layer was varied from 0 (no scaffold) to *ca.* 400 nm. A perovskite $CH_3NH_3PbI_{3-x}Cl_x$ capping layer formed on the 80 nm-thick Al_2O_3 film but no capping layer formed on the 400 nm-thick Al_2O_3 film. The highest short-circuit photocurrent density (J_{SC}) was observed from the 80 nm Al_2O_3 with the perovskite capping layer, while the open-circuit voltage (V_{OC}) and fill factor were highest for the 400 nm Al_2O_3 without capping layer, which suggests that highly crystalline perovskite is important for obtaining a high J_{SC} and pin-hole (shunting path) free perovskite is required to have a high V_{OC} and fill factor. For the Al_2O_3–perovskite system, a one-step deposition process was proposed using a composite solution containing Al_2O_3 nano-particle suspension and the perovskite precursor in DMF, where 5 wt% Al_2O_3 showed 7.16%.[49]

Molecular level engineering on the TiO_2 surface with self-assembled monolayer of C_{60} (C_{60}SAM) was found to improve significantly photovoltaic performance of TiO_2/$CH_3NH_3PbI_{3-x}Cl_x$/P3HT structure, in which PCE was improved from 3.8% without C_{60}SAM to 6.7% with C_{60}SAM due to significant increase in both J_{SC} and V_{OC}.[50] The improvement was interpreted by C_{60}SAM function that inhibits electron injection from the perovskite to TiO_2.

7.6.2.2   High V_{OC} Bromide Perovskite $CH_3NH_3PbBr_3$

Based on the meso-superstructured concept,[12] $CH_3NH_3PbBr_3$ was studied as both a light harvester and electron transporter although its bandgap ($E_g = 2.3$ eV) is larger than $CH_3NH_3PbI_3$. The Al_2O_3/$CH_3NH_3PbBr_3$/N,N'-dialkyl perylenediimide (PDI) structure showed V_{OC} as high as 1.3 V.[51] Such a high V_{OC} was likely to be due to non-occurrence of electron injection to the insulating Al_2O_3 and a large difference between the conduction band of $CH_3NH_3PbBr_3$ (4.2 eV with respect to vacuum level) and the HOMO level of PDI (5.8 eV). $CH_3NH_3PbBr_3$ was also applied to mesoporous TiO_2 layer and this bromide perovskite sensitized TiO_2 was in contact with poly[N-9-hepta-decanyl-2,7-carbazole-alt-3,6-bis(thiophen-5-yl)-2,5-dioctyl-2,5-di-hydro-pyrrolo[3,4-]pyrrole-1,4-dione] (PCBTDPP), which resulted in a surprisingly high V_{OC}, being close to 1.2 eV.[52] However, when P3HT (HOMO = 5.2 eV) was used as HTM, a much lower V_{OC} of about 0.5 V was observed. When con-sidering electron injection from the bromide provksite to TiO_2, V_{OC} will be determined by difference between the Fermi level of TiO_2 and the HOMO level of HTM (PCBTDPP: 5.4 eV and P3HT: 5.2 eV). Despite small differences in HOMO levels, the large difference in V_{OC} between PCBTDPP and P3HT was presumably ascribed to the degree of chemical interaction and strength of light filtering effect.[52] A weaker light filtering effect along with stronger interaction in PCBTDDP-based device rendered significant suppression of charge recombination and upward shift of Fermi level as well, resulting in high V_{OC}. Electron lifetime was measured for P3HT, which was one order of magnitude shorter than spiro-MeOTAD in the TiO_2/$CH_3NH_3PbI_3$/HTM system because of much faster recombination rather than electron transport.[53] The above results suggest that other factors except for energy levels are carefully considered in selecting HTM.

7.6.3 Planar Heterojunction Without Mesoporous Oxide Layers

Since organolead halide perovskites have both electron and hole transporting properties as shown in the Table 7.2, mesoporous oxide may not be necessarily required. $CH_3NH_3PbI_{3-x}Cl_3$ was deposited directly on FTO with thin compact TiO_2 layer either by co-evaporation of CH_3NH_3I and $PbCl_2$ using thermal evaporator.[14] Vapor-deposited perovskite showed an average PCE of 12.3% and maximum PCE of 15.4%, which was much higher than a solution-processed one, with a PCE of 8.6%. The higher PCE from the vapor-deposited perovskite is due to a homogeneous layer structure as can be seen in Figure 7.7, which implies a highly crystalline and homogeneous layer thickness. The highest efficiency was delivered at around 300 nm-thick

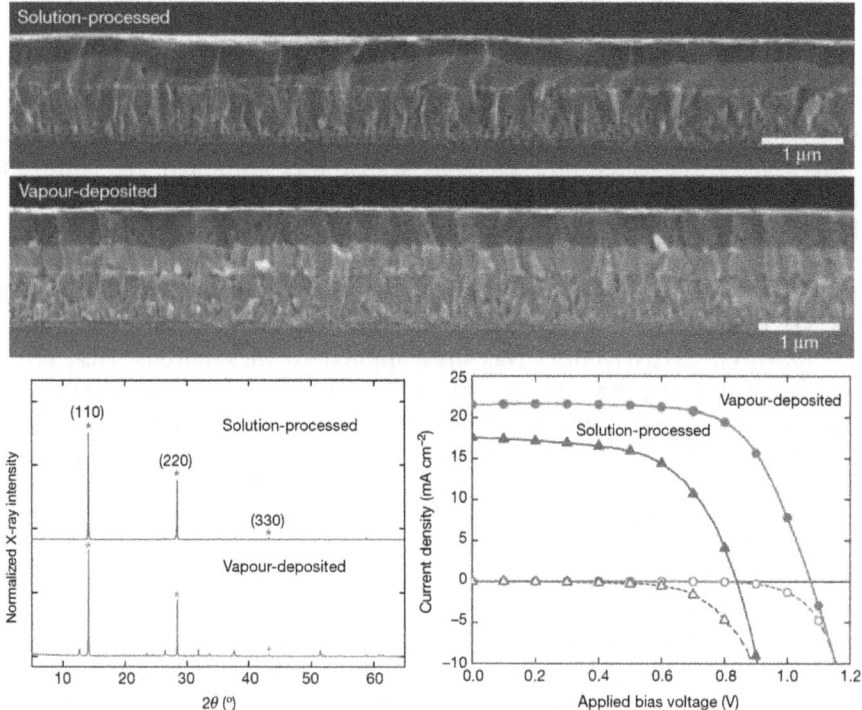

Figure 7.7 Comparison of cross-sectional SEM images, X-ray diffraction patterns and *I–V* curves between solution-processed and vapor-deposited $CH_3NH_3PbI_{3-x}Cl_x$.
Reprinted from Liu *et al.*[14]

CH3NH3PbI$_{3-x}$Cl$_x$. Such high efficiency in the planar heterojunction structure is likely to be related to improved charge transport within the perovskite layer by low-level doping of chloride ions (3–4%).[54]

7.7 Summary

Organolead halide perovskite has been found to have light harvesting property with high absorption coefficient and charge transporting and accumulation properties as well. Thus, perovskite can be used as either sensitizer or p-type or n-type light harvester. An all-solid-state perovskite solar cell based on organolead iodide demonstrated 9.7% in 2012 and 1 year later its efficiency increased to more than 15%, which implies that organolead halide perovskite is a promising solar cell material. The bandgap of 1.5 eV for CH$_3$NH$_3$PbI$_3$ can be tuned by replacing A or B or X ions in the ABX$_3$ perovskite structure within the allowed tolerance factors, which can further improve photovoltaic performance more than 20%. Since the perovskite layer is as thin as sub-micrometer levels, a perovskite solar cell can be classified as a new type of thin-film solar cell technology. Moreover, a perovskite solar cell will deliver low levelized solar power costs because the material and processing costs are expected to be 'dirt cheap'.

Acknowledgements

This work was supported by the National Research Foundation of Korea (NRF) grants funded by the Ministry of Science, ICT & Future Planning (MSIP) of Korea under contracts No. NRF-2010-0014992, NRF-2012M1A2A2671721, NRF-2012M3A6A7054861 (Nano Material Technology Development Program) and NRF-2012M3A6A7054861 (Global Frontier R&D Program on Center for Multiscale Energy System).

References

1. M. Helen, *Nature*, 1945, **155**, 484–485.
2. B. Wul, *Nature*, 1946, **157**, 808.
3. R. Cohen, *Nature*, 1992, **358**, 136–138.
4. C. N. R. Rao, P. Ganguly, A. K. Raychaudhuri, R. A. Mohan Ram and K. Sreedhar, *Nature*, 1987, **326**, 856–857.
5. A. Schilling, M. Cantoni, J. D. Guo and H. R. Ott, *Nature*, 1993, **363**, 56–58.
6. D. B. Mitzi, C. A. Feild, W. T. A. Harrison and A. M. Guloy, *Nature*, 1994, **369**, 467–469.
7. R. E. Wasylishen, O. Knop and J. B. Macdonald, *Solid State Commun.*, 1985, **56**, 581–582.
8. A. Poglitsch and D. Weber, *J. Chem. Phys.*, 1987, **87**, 6373–6378.
9. A. Kojima, K. Teshima, Y. Shirai and T. Miyasaka, *J. Am. Chem. Soc.*, 2009, **131**, 6050–6051.

10. J.-H. Im, C.-R. Lee, J.-W. Lee, S.-W. Park and N.-G. Park, *Nanoscale*, 2011, **3**, 4088–4093.

11. H.-S. Kim, C.-R. Lee, J.-H. Im, K.-B. Lee, T. Moehl, A. Marchioro, S.-J. Moon, R. Humphry-Baker, J.-H. Yum, J. E. Moser, M. Grätzel and N.-G. Park, *Sci. Rep.*, 2012, **2**, 591.

12. M. M. Lee, J. Teuscher, T. Miyasaka, T. N. Murakami and H. J. Snaith, *Science*, 2012, **338**, 643–647.

13. J. Burschka, N. Pellet1, S.-J. Moon, R. Humphry-Baker, P. Gao1, M. K. Nazeeruddin and M. Gratzel, *Nature*, 2013, **499**, 316–319.

14. M. Liu, M. B. Johnston and H. J. Snaith, *Nature*, 2013, **501**, 395–398.

15. K. Liang, D. B. Mitzi and M. T. Prikas, *Chem. Mater.*, 1998, **10**, 403–411.

16. V. M. Goldschmidt, *Ber. Dtsch. Chem. Ges.*, 1927, **60**, 1263–1268.

17. P. M. Woodward, *Acta Crystallogr., Sect. B: Struct. Sci.*, 1997, **53**, 44–66.

18. C. Li, K. C. K. Soh and P. Wu, *J. Alloys Compd.*, 2004, **372**, 40–48.

19. C. Li, X. Lu, W. Ding, L. Feng, Y. Gao and Z. Guo, *Acta Cryst.*, 2008, **B64**, 702–707.

20. G. Alagona, C. Ghio and P. Kollman, *J. Am. Chem. Soc.*, 1986, **108**, 185–191.

21. R. D. Shannon, *Acta Cryst.*, 1976, **A32**, 751–767.

22. E. J. Gabe, *Acta Cryst.*, 1961, **14**, 1296.

23. B. N. Cohen, C. Labarca, N. Davidson and H. A. Lester, *J. Gen. Physiol.*, 1992, **100**, 373–400.

24. L. Q. Jianga, J. K. Guob, H. B. Liua, M. Zhua, X. Zhoua, P. Wuc and C. H. Li, *J. Phys. Chem. Solids*, 2006, **67**, 1531–1536.

25. A. Poglitsch and D. Weber, *J. Chem. Phys.*, 1987, **87**, 6373–6378.

26. G. Giorgi, J.-I. Fujisawa, H. Segawa and K. Yamashita, *J. Phys. Chem. Lett.*, 2013, **4**, 4213–4216.

27. E. Mosconi, A. Amat, Md. K. Nazeeruddin, M. Grätzel and F. De Angelis, *J. Phys. Chem. C*, 2013, **117**, 13902–13913.

28. C. C. Stoumpos, C. D. Malliakas and M. G. Kanatzidis, *Inorg. Chem.*, 2013, **52**, 9019–9038.

29. J.-H. Im, J. Chung, S.-J. Kim and N.-G. Park, *Nanoscale Res. Lett.*, 2012, **7**, 353.

30. D. B. Mitzi, C. A. Feild, W. T. A. Harrison and A. M. Guloy, *Nature*, 1994, **369**, 467–469.

31. D. B. Mitzi, C. A. Feild, Z. Schlesinger and R. B. Laibowitz, *J. Solid State Chem.*, 1995, **114**, 159–163.

32. G. C. Papavassiliou, G. A. Mousdis and I. B. Koutselas, *Adv. Mater. Opt. Electron.*, 1999, **9**, 265–271.

33. K. Tanaka, T. Takahashi, T. Ban, T. Kondo, K. Uchida and N. Miura, *Solid State Commun.*, 2003, **127**, 619–623.

34. G. C. Papavassiliou, *Prog. Solid State. Chem.*, 1997, **25**, 125–270.

35. G. A. Mousdis, V. Gionis, G. C. Papavassiliou, C. P. Raptopoulou and A. Terzis, *J. Mater. Chem.*, 1998, **8**, 2259–2262.

36. S. Elleuch, T. Dammak, Y. Abid, A. Mlayah and H. Boughzala, *J. Lumin.*, 2010, **130**, 531–535.

37. S. D. Stranks, G. E. Eperon, G. Grancini, C. Menelaou, M. J. P. Alcocer, T. Leijtens, L. M. Herz, A. Petrozza and H. J. Snaith, *Absorber, Science*, 2013, **342**, 341–344.

38. G. Xing, N. Mathews, S. Sun, S. S. Lim, Y. M. Lam, M. Grätzel, S. Mhaisalkar and T. C. Sum, *Science*, 2013, **342**, 344–347.

39. H.-S. Kim, I. Mora-Sero, V. Gonzalez-Pedro, F. Fabregat-Santiago, E. J. Juarez-Perez, N.-G. Park and J. Bisquert, *Nat. Commun.*, 2013, **4**, 2242.

40. J. H. Heo, S. H. Im, J. H. Noh, T. N. Mandal, C.-S. Lim, J. A. Chang, Y. H. Lee, H.-j. Kim, A. Sarkar, M. K. Nazeeruddin, M. Grätzel and S. I. Seok, *Nat. Photonics.*, 2013, **7**, 486–491.

41. K. Liang, D. B. Mitzi and M. T. Prikas, *Chem. Mater.*, 1998, **10**, 403–411.

42. H.-S. Kim, J.-W. Lee, N. Yantara, P. P. Boix, S. A. Kulkarni, S. Mhaisalkar, M. Grätzel and N.-G. Park, *Nano Lett.*, 2013, **13**, 2412–2417.

43. S. Dharani, H. K. Mulmudi, N. Yantara, P. T. T. Thrang, N. G. Park, M. Graetzel, S. Mhaisalkar, N. Mathews and P. P. Boix, *Nanoscale*, 2013, **6**, 1675–1679.

44. L. Etgar, P. Gao, Z. Xue, Q. Peng, A. K. Chandiran, B. Liu, M. K. Nazeeruddin and M. Grätzel, *J. Am. Chem. Soc.*, 2012, **134**, 17396–17399.

45. W. Abu Laban and L. Etgar, Depleted hole conductor-free lead halide iodide heterojunction solar cells, *Energy Environ. Sci.*, 2013, **6**, 3249–3253.

46. D. Bi, G. Boschloo, S. Schwarzmuller, L. Yang, E. M. J. Johansson and A. Hagfeldt, *Nanoscale*, 2013, **5**, 11686–11691.

47. D. Bi, S.-J. Moon, L. Haggman, G. Boschloo, L. Yang, E. M. J. Johansson, M. K. Nazeeruddin, M. Grätzel and A. Hagfeldt, *RSC Adv.*, 2013, **3**, 18762–18766.

48. J. M. Ball, M. M. Lee, A. Hey and H. J. Snaith, *Energy Environ. Sci.*, 2013, **6**, 1739–1743.

49. M. J. Carnie, C. Charbonneau, M. L. Davies, J. Troughton, T. M. Watson, K. Wojciechowski, H. Snaith and D. A. Worsley, *Chem. Commun.*, 2013, **49**, 7893–7895.

50. A. Abrusci, S. D. Stranks, P. Docampo, H.-L. Yip, A. K.-Y. Jen and H. J. Snaith, *Nano Lett.*, 2013, **13**, 3124–3128.

51. E. Edri, S. Kirmayer, D. Cahen and G. Hodes, *J. Phys. Chem. Lett.*, 2013, **4**, 897–902.

52. B. Cai, Y. Xing, Z. Yang, W.-H. Zhang and J. Qiu, *Energy Environ. Sci.*, 2013, **6**, 1480–1485.

53. D. Bi, L. Yang, G. Boschloo, A. Hagfeldt and E. M. J. Johansson, *J. Phys. Chem. Lett.*, 2013, **4**, 1532–1536.

54. S. Colella, E. Mosconi, P. Fedeli, A. Listorti, F. Gazza, F. Orlandi, P. Ferro, T. Besagni, A. Rizzo, G. Calestani, G. Gigli, F. De Angelis and R. Mosca, *Chem. Mater.*, 2013, **25**, 4613–4618.

CHAPTER 8

All-oxide Photovoltaics

SVEN RÜHLE* AND ARIE ZABAN*

Center for Nanotechnology & Advanced Materials, Department of
Chemistry, Bar Ilan University, Ramat Gan 52900, Israel
*Email: sven.ruhle@gmail.com; Arie.Zaban@biu.ac.il

8.1 Introduction to All-oxide Photovoltaics

To reach grid parity further price reductions for photovoltaic (PV) systems
are required, calling for continued upscaling of the production processes or
for novel PV cell concepts based on high efficiency and cheap materials in
combination with low cost deposition methods. Metal oxide (MO) semiconductors are very attractive to achieve that goal, many of them show great
chemical stability, are non-toxic, abundant, fulfil the requirements for low
cost manufacturing methods at ambient conditions, and could show sufficiently high conversion efficiency. Consequently, devices made of MO
semiconductors can be very inexpensive, stable and environmentally safe,
which are besides the conversion efficiency the most important requirements for photovoltaics.

This chapter presents photovoltaic cells entirely based on metal oxides
except for the metal back contact. We discuss basic material requirements
for photovoltaic cells and provide guidelines for the search for new PV materials. Metal oxides are reviewed that are used today as electron and hole
transport layers and an overview is given of metal oxide semiconductors with
a bandgap in the visible part of the solar spectrum, suitable as light absorbers for PV cells. Existing all-oxide photovoltaic cells concepts are discussed with the focus on Cu_2O-based devices, which are the most

RSC Energy and Environment Series No. 11
Advanced Concepts in Photovoltaics
Edited by Arthur J Nozik, Gavin Conibeer and Matthew C Beard
© The Royal Society of Chemistry 2014
Published by the Royal Society of Chemistry, www.rsc.org

investigated all-oxide PV system. We review recently proposed oxide photovoltaics such as $BiFeO_3$ or Bi_2O_3 and take a brief look at alternative light absorber such as Fe_2O_3, which today is mainly investigated for solar water splitting. Finally, we propose strategies for searching for novel multi-component MO composites using computational methods, which can direct experimentalist, as well as using combinatorial material science as an efficient tool for material screening.

8.2 Solar Cell Design Rules

A photovoltaic cell in general consists of a light absorber that converts photons into charge carriers, a driving force responsible for charge transport and selective contacts that collect only one type of charge carrier. To achieve high light-to-electric power conversion efficiencies energy losses have to be minimized that occur when charge carriers are transferred from the metastable excited states into the contacts.

8.2.1 Light Absorption

The conversion of light into electronic energy is schematically shown in Figure 8.1. In thermal equilibrium the density of electrons and holes in the conduction band (CB) and valence band (VB) can be estimated by the thermal energy k_BT, the energy difference between the Fermi level E_F and the conduction and valence band edge (E_{CB} and E_{VB}) and the effective density of electronic states (N_{CB} and N_{VB}) at the respective band edges. The Fermi distribution can be approximated by a Boltzmann distribution if the energy difference between the band edges and E_F is larger than a few k_BT:[1]

$$n_{e,0} = N_{CB} \exp[-(E_{CB} - E_F)/k_BT] \tag{8.1}$$

$$n_{h,0} = N_{VB} \exp[(E_{VB} - E_F)/k_BT] \tag{8.2}$$

Upon light absorption electron–hole pairs are generated (Figure 8.1b). The lifetime of charge carriers far from the band edges is extremely short (order of fs to ps) and energy is released *via* phonon emission until the carriers occupy metastable states close to the band edges (Figure 8.1c). The large difference in life times of the excited states deep in the band compared to metastable states close to the band edges leads to a quasi-thermal equilibrium charge carrier distribution, which is described by Fermi statistics and can under the restrictions mentioned above be approximated by a Boltzmann distribution. The occupation probability for electrons and holes close to the band edges is now described by two separate quasi-Fermi levels for electrons and holes, E_{Fn} and E_{Fp}, respectively. In contrast to E_F the quasi-Fermi levels are not constant throughout the

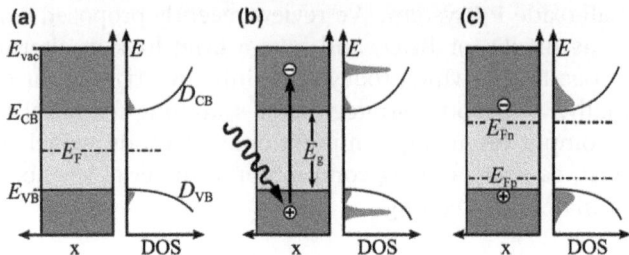

Figure 8.1 Energy band diagram of an intrinsic light absorber showing the local vacuum level E_{vac}, conduction and valence band edges (E_{CB} and E_{VB}) and the respective density of electronic states (DOS) as a function of energy E. (a) In thermal equilibrium the Fermi energy E_F is at mid-gap and the density of occupied electron and hole states is shown in dark grey. (b) Electron–hole generation far away from the band edges with a very short life time. (c) Electrons and holes after phonon emission and electron–phonon equlibration. The occupation probabilities can be described by quasi-Fermi levels E_{Fn} and E_{Fp}.

device and can be expressed as a function of the local charge carrier densities n_e and n_h:

$$E_{Fn} = E_{CB} + k_B T \ln(n_e/N_{CB}) \tag{8.3}$$

$$E_{Fp} = E_{VB} - k_B T \ln(n_h/N_{VB}) \tag{8.4}$$

The maximum photovoltage that can be generated in a PV cells is defined by the energy difference of the quasi-Fermi levels, *e.g.*

$$V_{OC} = q^{-1}(E_{Fn} - E_{Fp}) \tag{8.5}$$

A basic requirement for PV light absorbers in efficient single junction solar cells is a suitable bandgap for the standard solar spectrum on Earth (AM 1.5G). The bandgap should on the one hand be narrow enough to convert many photons into electron–hole pairs while on the other hand it should be large enough to generate high photovoltages. The maximum light-to-electric power conversion efficiency as a function of the bandgap can be calculated assuming detailed balance and this result is known as the Shockley–Queisser limit.[2] The calculations assume that all photons with an energy above the bandgap are converted into charge carriers (100% quantum efficiency), photons with sub-bandgap energy do not contribute to electrical charge generation (0% quantum efficiency), and radiative recombination is assumed to be the only electron–hole recombination process. The calculations reveal that the highest single junction cell conversion efficiencies can be achieved with a light absorber bandgap between 0.9 to 1.7 eV, this is shown in Figure 8.2 and a list of relevant metal oxide semiconductor bandgaps is presented in section 8.3.2.

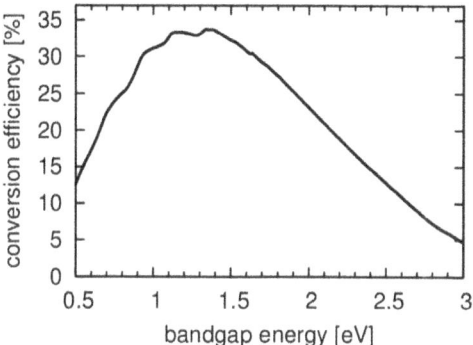

Figure 8.2 Detailed-balance calculations for an ideal single junction PV cell under AM 1.5G illumination showing the maximum conversion efficiency as a function of the absorber bandgap. The maximum conversion efficiency of $\sim 33\%$ is known as the Shockley–Queisser limit.

8.2.2 Charge Transport

In a light absorber the generation of electron–hole pairs is proportional to the light intensity inside the semiconductor, described by Beer–Lambert law:

$$I(x) = I_0 \exp(-\alpha x) \tag{8.6}$$

where I_0 is the incident light intensity, I is the light intensity as a function of penetration depth x and α is the absorption coefficient. To absorb 95% of the incident light an absorber thickness of $3/\alpha$ is required[3] and to extract the charges a driving force is needed, which is defined by the gradient of the quasi-Fermi levels. The electron and hole current densities are defined as:

$$J_e = n_e \eta_e \frac{d}{dx} E_{Fn} \tag{8.7}$$

$$J_h = n_h \eta_h \frac{d}{dx} E_{Fp} \tag{8.8}$$

where η_e and η_h are the electron and hole mobilities, respectively. The driving force can be explicitly expressed by the band structure,[4] replacing the band edge energies E_{CB} and E_{VB} by the local vacuum level E_{vac}, electron affinity χ and bandgap E_g:

$$E_{CB} = E_{vac} - \chi \tag{8.9}$$

$$E_{VB} = E_{vac} - \chi - E_g \tag{8.10}$$

and inserting the expressions for E_{Fn} and E_{Fp} [Equation (8.3) and Equation (8.4)] into Equation (8.7) and Equation (8.8), leading to:

$$J_e = q D_e \frac{d}{dx} n_e + n_e \eta_e \left(\frac{d}{dx} E_{vac} - \frac{d}{dx} \chi - k_B T \frac{d}{dx} \ln N_{CB} \right) \tag{8.11}$$

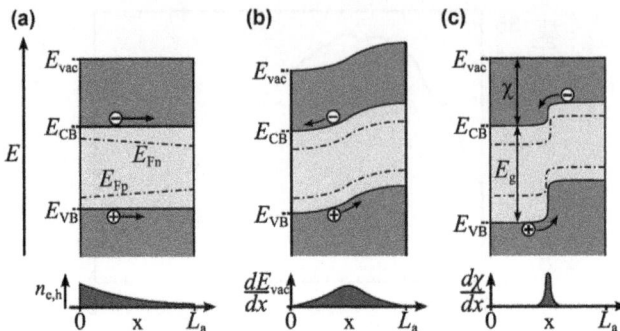

Figure 8.3   Energy band diagrams of a light absorber showing the different driving forces for charge transport. (a) Diffusion currents caused for example by the exponentially decreasing generation of electrons and holes. (b) Drift currents caused by electric fields, shown as gradients in the local vacuum level E_{vac}. (c) Electron and hole currents caused by gradients of the electron affinity χ and bandgap E_g, which can occur at interfaces or in materials with a continuously changing composition.

$$J_h = -qD_h \frac{d}{dx} n_h + n_h \eta_h \left(\frac{d}{dx} E_{vac} - \frac{d}{dx}(\chi + E_g) + k_B T \frac{d}{dx} \ln N_{VB} \right) \qquad (8.12)$$

with the elementary charge q and the diffusion coefficients D_e and D_h for electrons and holes according to Einstein relation ($D_e = \eta_e k_B T/q$ and $D_h = \eta_h k_B T/q$). Besides diffusion (1st term) and drift current (2nd term) Equation (8.11) and Equation (8.12) explicitly show how gradients of the band edges (expressed by $d\chi/dx$ and dE_g/dx) and density of states contribute to charge transport (Figure 8.3) or can block transport for one type of carrier (selective contacts, see below).

Furthermore, the life time τ of photo-excited charge carriers has to be larger than the travel time into the contacts for efficient charge collection. In a homogeneous absorber with constant χ, N_{CB}, N_{VB} and E_g the only driving force at flat band potential is diffusion, with a diffusion length L_D that should be larger than the absorber thickness L_a:

$$L_D = (D\tau)^{\frac{1}{2}} = L_a \qquad (8.13)$$

8.2.3   Selective Contacts

At interfaces where χ, E_g, N_{CB} and N_{VB} are not constant, strong gradients occur over length scales of few atomic layers. Consequently, the band edges in energy band diagrams are often shown with a discrete step instead of a continuous transition. Interface gradients can be used to suppress transport for one type of carrier to create a selective contact,[5] which is often realized using electron and hole transport layers (ETL and HTL, respectively) with a larger bandgap compared to the absorber.[6] For an electron selective contact it is required that electrons from the absorber can move through the conduction band of the ETL towards the contact while hole injection is prohibited by

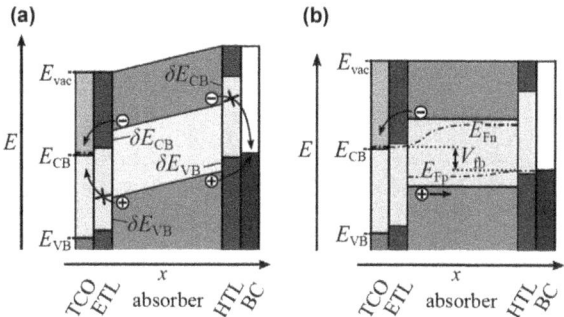

Figure 8.4 Selective contacts shown for a p–i–n junction solar cells with a n-type electron transport layer (ETL) and a p-type hole transport layer (HTL). (a) The valence band offset δE_{VB} at the ETL–absorber interface and the conduction band offset δE_{CB} at the absorber–HTL interface create a barrier for hole and electron transfer, respectively, thus creating selective interfaces. (b) A δE_{CB} at the ETL–absorber interface and a δE_{VB} at the absorber–HTL interface lead to a flat band potential V_{fb} that remains below the splitting of the quasi-Fermi levels.

a substantial valence band offset δE_{VB} that creates a strong gradient $[d(\chi + E_g)/dx]$ at the interface blocking hole transport (Figure 8.4a). For the hole selective contact δE_{CB} creates the barrier for electron transfer while the E_{VB} at the interface has to permit hole transport towards the contact. Contact selectivity is of particular importance at the front side of the absorber where the light enters, due to the high electron–hole generation rate that decays exponentially as a function of penetration depth [Equation (8.6)].

Experimentally the energy band alignment at interfaces can be measured using photoelectron spectroscopy.[7,8] Besides the relative position of the band edges it is important that the interfaces have a low trap density due to undesired trap assisted recombination which can cancel the beneficial effect of the selective transport layer. Today MOs play an important role as ETLs or HTLs in organic–inorganic and purely inorganic photovoltaics. However, substantial research is dedicated to the investigation of MO–MO interfaces, which can show additional outstanding effects that are not present in the bulk of the interface forming MOs. Examples are high mobility electron gas,[9] interface superconductivity,[10] magneto-electric coupling or quantum Hall effect in oxide hetero-structures.[11]

8.2.4   Optimized Energy Levels at Interfaces

For the design of efficient PV cells the open circuit voltage V_{OC} measured between the contacts should be maximized and approach the theoretical limit defined in Equation (8.5). This requires that conduction band offsets δE_{CB} for electron transport and valence band offsets δE_{VB} for hole transport from the absorber to the contacts should be minimized. In an ideal solar cell the E_{CB} and E_{VB} should be continuous at the absorber/ETL, ETL/contact and

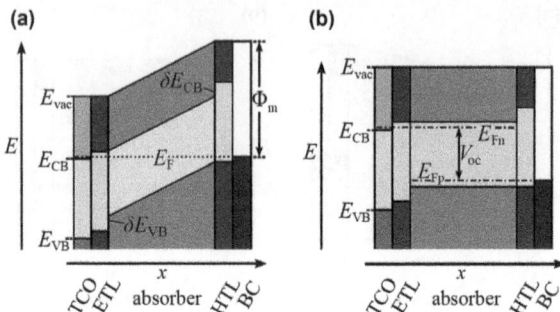

Figure 8.5   Selective contacts in a p–i–n solar cell with optimized energy band
design. (a) The conduction band is nearly continuous from the absorber to the TCO front contact as well as the valence band between the
absorber and back contact (BC). (b) At open circuit no energy loss
occurs at the selective interface such that the V_{OC} corresponds to the
quasi-Fermi level splitting.

absorber/HTL, HTL/BC interfaces, which allows the quasi-Fermi levels to be
horizontal throughout the device such that the V_{OC} can approach its maximum (Figure 8.5). High work function (Φ_m) back contact materials are required to realize such an ideal band structure and research for p-type TCOs
is directed to find such materials.

In summary, for the development of efficient all-oxide PV cells the following requirements should be fulfilled:

- Absorber bandgap between 0.9 eV $\leq E_g \leq$ 1.7 eV
- Absorber thickness $L_a \geq 3/\alpha$
- Diffusion length $L_D \geq L_a$
- Selective contacts with a minimized δE_{CB} and δE_{CB} at the absorber/ETL
  and absorber/HTL interface, respectively.

8.3   Metal Oxides for All-oxide Photovoltaics

8.3.1   Electronic Properties

Metal oxides cover the range from metallic compounds (*e.g.*, RuO_2), semiconductors with a bandgap in the visible part of the light spectrum (*e.g.*,
Fe_2O_3, Cu_2O) to transparent wide bandgap compounds which are highly
insulating (*e.g.*, MgO, $BaTiO_3$). In metal oxides the oxygen has a strong
tendency to fill its octet due to its high electronegativity, leaving the metal
atoms positively charged. Oxidation numbers give a rough estimate
about the charge on the individual atoms though computational methods
such as Bader charge analysis show that the calculated charge can be considerably smaller than the oxidation numbers. Transition metal oxides, especially, can have outstanding properties due to their partially filled *d*-shell,
such as ferromagnetism. Figure 8.6 shows the periodic table of elements

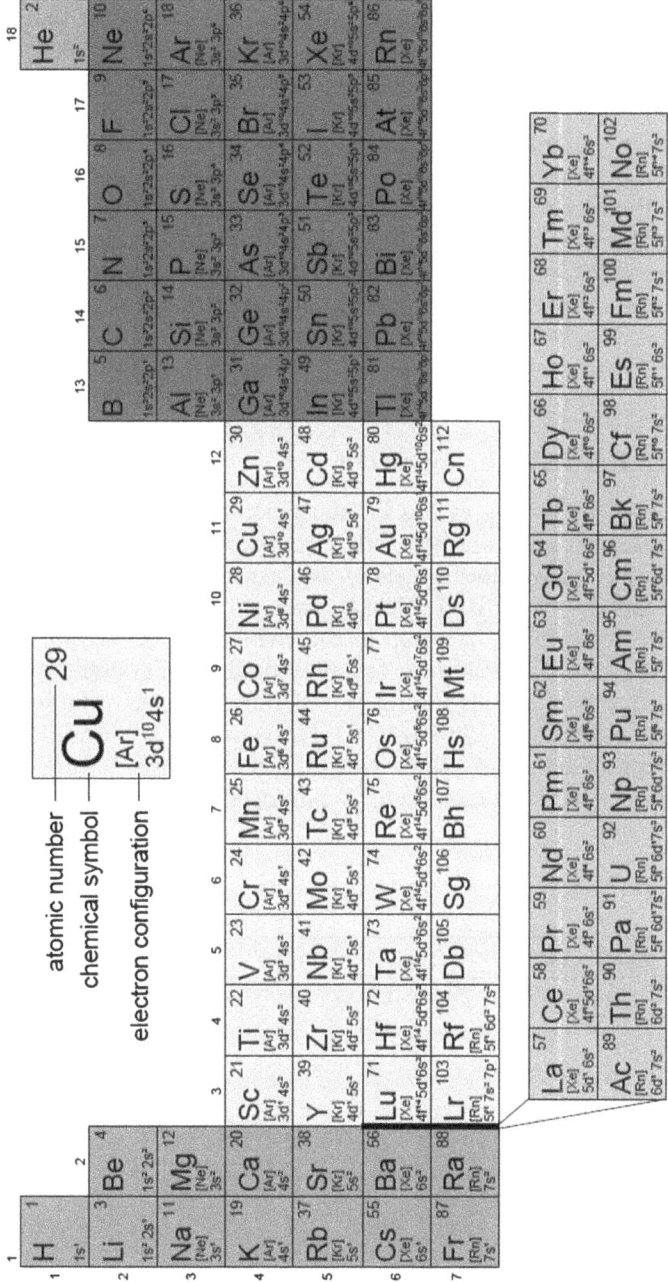

Figure 8.6 Periodic Table with the electronic configuration of the elements. Different grey scales show the blocks in which the *s*-, *p*-, *d*- and *f*-shell are filled. Transition metals are shown in light grey.

including the electronic configurations with the transition metals in light grey.

In MO semiconductors the metal-oxide bonds vary from strongly polar to ionic and consequently the electronic properties of the individual atoms have a strong impact on the electronic band structure. The upper valence band edge has typically a strong character of either oxygen 2*p*-states or occupied *d*-states in transition metals oxides. The lower conduction band edge has predominant character of unoccupied metal *s*- or transition metal *d*-states. The *d*-orbitals are located closer to the core of the metal atom compared to the *s*-orbitals of the next higher shell. Consequently excited electrons or holes that are moving in bands with a pronounced *d*-character are strongly localized, which leads to low charge carrier mobilities and short diffusion lengths. However, with increasing complexity of the metal oxide composition materials have been developed with excellent transport properties such as high temperature superconductors.[12]

8.3.2 Metal Oxide Light Absorber

For single absorber PV cells with a bandgap between 0.9 and 1.7 eV high conversion efficiencies around 30% are theoretically possible (Figure 8.6). Bandgaps outside this window can be of interest for multi-junction cells or for semi-transparent PV cells. Table 8.1 gives an overview of binary and ternary metal oxides with a bandgap in the visible or near IR range of the optical spectrum, showing that there is very limited number of binary oxides with an optimized bandgap. It should be noted that literature values of the bandgap often scatter substantially and are sometimes contradictive regarding the type of bandgap (direct or indirect). This can be due to different characterization methods such as photoconductivity, photoelectron

Table 8.1 Energy bandgap of binary and ternary metal oxide light absorber at 300 K.

Metal oxide	Bandgap (eV)	Reference
Ag_2O	1.2–1.5	13
Bi_2O_3	2.8–3.3	14
CoO	1.6	15
Co_3O_4	0.9–1.5	15,16
Cu_2O	2.0	17
CuO	0.9–1.6	18–20
FeO	2.4	21
Fe_2O_3	2.1	22
VO_2	0.7	23
V_2O_5	2.5	24
Cr_2O_3	2.6	25
$CuNb_3O_8$	1.3	26
$CuFeO_2$	1.6	27
$BiFeO_3$	2.7	28

spectroscopy combined with inverse photoelectron spectroscopy or optical transmission and reflection measurements in conjunction with Tauc plots. Especially the latter is commonly used, where $(\alpha h\nu)^n$ is linearly dependent on the photon energy $h\nu$ with $n = 2$ for direct and $n = \frac{1}{2}$ for indirect bandgap semiconductors. The intercept of a linear fit with the x-axis defines E_g. In reality, Tauc plots often do not show a clear linear regime, forcing the researcher to choose an appropriate fitting range, which consequently leads to uncertainties of the derived bandgap energy. Difficulties identifying a linear regime are often due to an exponential distribution of bandgap states below the conduction band edge or multiple reflections in thin films which often are not appropriately taken into account for the calculation of the absorption coefficient α.

The second requirement for a suitable absorber material besides the bandgap is a minority carrier diffusion length $L_D \geq 3/\alpha$. For a number of MOs this requirement is difficult to fulfil due to a low charge carrier mobility caused by a partially filled d-band. Additionally synthesis specific parameters such as crystallinity, grain size, bulk and interface defect concentration at grain boundaries can have a major impact on charge transport. Especially defect states within the bandgap can be involved in trapping/detrapping events that slow down the transport and act as recombination sites, thus reducing the lifetime and the diffusion length.

8.3.3 Wide Bandgap Metal Oxides

Wide bandgap metal oxides are important components for all-oxide photovoltaics, either as transparent conducting front electrodes, as electron and hole transport layers that provide the required charge selectivity between absorber and the contacts, or as back electrodes.

8.3.3.1 Transparent Conducting Oxides

MOs are today widely used as components in photovoltaic cells and modules, mainly as n-type transparent conducting oxides (TCO).[29,30] Fluorine-doped tin oxide (FTO),[31] indium tin oxide (ITO)[32] or aluminium-doped zinc oxide (AZO)[33] are the most common TCOs and widely applied in commercially available thin film solar cells such as CdTe, Cu(In,Ga)Se$_2$ (CIGS), amorphous or microcrystalline Si. For all-oxide photovoltaic cells TCOs are required as front electrodes. With the progressing development of p-type TCOs high work function contact materials are in reach, which will allow improving the device structure to achieve high photovoltages.[34]

8.3.3.2 Electron Transport Layers

Wide bandgap electron conducting layers such as TiO$_2$, ZnO or SnO$_2$ are typically n-type due to oxygen vacancies[35] with typical doping densities in the order of 10^{16}–10^{19} cm^{-3}. Wide bandgap MOs are today used as electron transport layers in thin film Cu(In,Ga)Se$_2$ (CIGS) solar cells, where an in-trinsic ZnO layer on top of an AZO front electrode is suppressing hole injection from the adjacent CdS layer.[36] The ability of wide bandgap MOs to accept electrons while blocking hole injection has widely been used in nano-composite photovoltaics such as dye-sensitized solar cells (DSC),[37] quantum dot sensitized solar cells (QDSC),[38,39] extreme thin absorber solar cells (ETA cells)[40] or in recently reported perovskite solar cells,[41] which are all based on mesoporous MO films covered with a thin absorber layer while the pores are filled with a redox electrolyte or a solid state hole conductor for the transport of positive charges. TiO$_2$ is very popular due to its chemical stability in acidic and basic environment and can be produced as mesoporous films or rod and tube-shaped nano-structures. A higher versatility in terms of shapes has been achieved with ZnO which has been grown as rods, wires, whiskers, tubes, sheets, and more.[42] Compact thin film ZnO and TiO$_2$ layers have been used in polymer based thin film solar cells to achieve a selective contact.[43–45]

The requirements for an optimized selective contact using an n-type wide bandgap MO as an ETL are a minimized energy step δE_{CB} at the absorber/ETL interface as outlined in section 8.2.4. Following this way of thinking a number of wide bandgap MOs have been investigated in nano-composite solar cells to increase the photovoltage, optimizing the location of the E_{CB} with respect to the excited absorber state. In DSCs mesoporous films of Nb$_2$O$_5$,[46] SnO$_2$,[47] SrTiO$_3$ [48] as well as core/shell systems based on mesoporous TiO$_2$ electrodes coated with Al$_2$O$_3$, MgO, Nb$_2$O$_5$, SiO$_2$, SrTiO$_3$ or ZnO were investigated.[49–51] In QDSCs electron injection from CdS quantum dots into ZrO$_2$ was observed[52] in contrast to DSCs where ZrO$_2$ has been used as a reference system that does not permit electron injection from the dye. For nano-composite cells it was pointed out that an exchange of the TiO$_2$ surface by a different wide bandgap MO does not only change the position of E_{CB}, it changes the entire interface with the absorber which includes the density of QDs or dye molecules and thus the photo current density, the density of surface states that affect the recombination kinetics, or the surface pH, which is responsible for a shift of the flat band potential in an ideal system by 59 mV/pH.[53]

For all-oxide PV a modification of the wide bandgap ETL does not only change the δE_{CB} at the interface with the absorber, it can also change the density of interface states and their relative location in energy with respect to the band edges, this plays a critical role for device performance if such states act as recombination centres.[54] In conjunction with a Cu$_2$O absorber Ga$_2$O$_3$ has been recently successfully tested to replace the commonly used ZnO ETL in heterojunction solar cells, leading to significant higher photovoltages (see section 8.4.4).

8.3.3.3   Hole Transport Layers

As hole transport materials V$_2$O$_5$,[55] WO$_3$,[56] MoO$_3$,[57] and NiO[58] have been used in organic solar cells.[59,60] MoO$_3$, V$_2$O$_5$ and WO$_3$ are n-type semiconductors with a high work function ranging from 6.7 to 7.0 eV that act as HTL due to the proximity of their E_{CB} to the E_{VB} of the organic semiconductor, leading to the injection of holes into the HTLs CB. However, the injected holes could recombine with electrons from the absorber CB since n-type HTLs do not block electron injection and thus do not provide charge injection selectivity like p-type HTLs. Hence these MOs are not ideal HTLs.

NiO, in contrast, is a wide bandgap p-type semiconductor with a valence band edge $E_{VB} \sim 5.4$ eV, a bandgap $E_g \sim 3.6$ eV and a CB edge E_{CB} around 1.8 eV. The work function for undoped NiO depends on the deposition conditions and ranges from 3.8 to 5.4 eV. In PV cells with an organic P3HT light absorber NiO has been used as a hole transport and electron blocking layer. It provides contact selectivity as defined in section 8.2.3 due to the large energy barrier for electron injection from the P3HT into the NiO and a small valence band offset that does not create a large Schottky barrier providing an essentially ohmic contact that allows free hole transfer into the NiO.[61]

In the search for more p-type wide bandgap MOs ternary compounds have been investigated, such as $CuAlO_2$, which is a direct bandgap MO with $E_g \sim 3.5$ eV and conductivity values of around 1 S cm^{-1}.[62] $CuGaO_2$ has a similar bandgap of ~ 3.4 eV[63] and is also a p-type MO, however the electrical conductivity of 5.6 mS cm^{-1} is significantly lower compared to $CuAlO_2$.

Quaternary MOs such as Zn–Ni–Co–O; compounds have been investigated using combinatorial material science with continuous compositional spreads. Over a wide compositional range a work function of 5.8 ± 0.1 eV was observed while p-type conductivity of ~ 100 S cm^{-1} and a wide-bandgap was measured over a broad compositional range near Co_2NiO_4.[34]

8.4 Cu_2O-based Photovoltaics

Cu_2O is the best investigated metal oxide light absorber for photovoltaics with extensive research going back to the 1970s.[64,65] As a single junction device its electronic bandgap of around 2.0 eV is not ideal for the AM1.5G spectrum leading to a maximum theoretical conversion efficiency of $\sim 23\%$ compared to $\sim 33\%$ for an optimized bandgap of 1.34 eV.[2] However, as a top cell in a multi-junction stack consisting of three or more junctions its bandgap is nearly optimized,[66] furthermore Cu_2O can be a very attractive absorber for semi-transparent photovoltaic cells.

8.4.1 Cu_2O Synthesis

Cu_2O is typically p-type and can be produced by a number of methods including electrochemical deposition, anodic oxidation, spray pyrolysis, chemical vapour deposition, pulsed laser deposition, sputtering and thermal oxidation.

Electrochemical deposition of p-type Cu_2O, schematically shown in Figure 8.7a, is typically carried out from basic solution (pH 8–13) where $CuSO_4$ is dissolved in H_2O followed by the addition of lactic acid that forms a

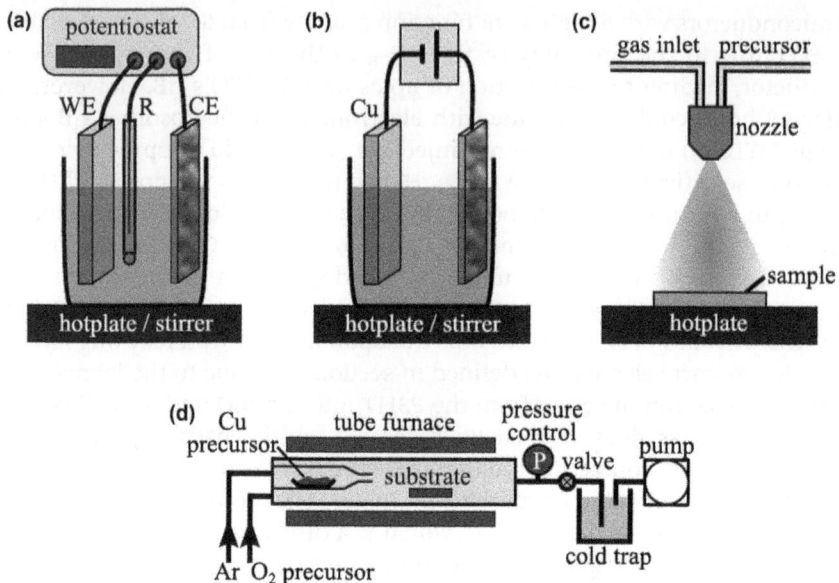

Figure 8.7 Experimental set-up used for (a) electrochemical deposition, (b) anodic oxidation, (c) spray pyrolysis, and (d) chemical vapour deposition.

complex with the Cu^{2+} ion to prevent the precipitation of $Cu(OH)_2$ when NaOH is added to adjust the solution pH.[67] Electro-deposition can be carried out under galvanostatic as well as potentiostatic control at solution temperatures between 10–80 °C, and compact Cu_2O films with a thickness of several µm can be achieved.[17,68] Electrodeposition at pH = 7.5 using $CuSO_4$ or $Cu(acetate)_2$ solutions, has been reported as a synthesis route for undoped n-type Cu_2O,[69] but these results are disputed[70] and the origin of the majority carriers so far cannot be explained.[71]

Under anodic oxidation Cu_2O can be formed on metallic copper surfaces (Figure 8.7b). In cyclic voltammetry the Cu_2O layer is formed by oxidation of Cu during the forward sweep while the layer keeps on growing during the reverse sweep due to the reduction of an $Cu(OH)_2$ or CuO overlayer.[72] Corrosion studies have shown that Cu_2O layers with a thickness in the order of nanometres are created in weakly alkaline and acidic solutions and that the layer thickness depends on the type of electrolyte.[72,73] It was furthermore shown that Cu_2O nano-structures can be synthesized by anodic oxidation of copper, at applied potentials between 2 and 10 V using plain water.[74]

Spray pyrolysis was successfully applied to produce Cu_2O films using an aqueous precursor solution containing $Cu(acetate)_2$, glucose, and 2-propanol which was sprayed by a nozzle onto a substrate located on a hotplate with air as a carrier gas (Figure 8.7c). Depending on the precursor composition and hotplate temperature the formation of metallic copper, cupric oxide or cuprous oxide was observed.[75,76]

Chemical vapour deposition (CVD) is particularly suited for the fabrication of high purity crystals with crystalline and optical properties of high quality. The deposition of Cu_2O can be carried out in a reactor where a copper containing precursor such as CuI,[77] copper(II) acetylacetonate $[Cu(C_5H_7O_2)_2]$[78] or bis(1,1,1,5,5,5-hexafluoropentane-2,4-dionato)copper(II)[79] is thermally decomposed while a process gas containing O_2, N_2O or H_2O provides the required oxygen. Figure 8.7d shows a system where the Cu-precursor is decomposed under Ar flow and released into the reactor to react with the process gas.

Cu_2O films have been produced by reactive RF[80] and DC[81] magnetron sputtering from a high purity Cu target on soda-lime glass or ITO.[82] Epitaxial growth was achieved on MgO substrates at substrate temperatures between 300 °C and 650 °C and an oxygen partial pressure between 0.003 mTorr and 1 mTorr.[80–82]

Pulsed laser deposition has been used to produce Cu_2O onto Si, MgO, Y-ZrO_2 and $SrTiO_3$ substrates using an KrF laser with a wavelength of 248 nm at an oxygen pressure below 1 mTorr and a substrate temperature above 600 °C using Cu_2O target.[83,84]

Thermal oxidation of high purity copper metal (99.999%) at low oxygen partial pressure in a furnace at elevated temperature (750–1100 °C) leads to the formation of micrometer sized Cu_2O crystals of high crystalline quality.[85,86] Thermal oxidation is the method by which the highest light to electric power conversion efficiencies of Cu_2O based photovoltaic cells have thus far been reported.[87]

8.4.2 Electronic and Optical Properties of Cu_2O

Besides CuO and Cu_4O_3, Cu_2O is one of the stable phases of the copper oxides. It has a cubic structure and the unit cell contains four copper and two oxygen atoms. It has an electron affinity of $\sim$3.1 eV and an electronic bandgap of 2.18 eV at 4.2 K. Four direct optical transitions are distinguished, called yellow, green, blue and indigo according to the spectral order of appearance, with interband transition energies of 2.17 eV, 2.30 eV, 2.62 eV and 2.76 eV, respectively.[70] The yellow and green transitions are parity forbidden and theoretically for direct forbidden transitions $(\alpha h\nu)^{2/3}$ is a linear function of the photon energy $h\nu$. Based on this relation a bandgap of 1.96 eV was experimentally derived from optical measurements at room temperature for parity forbidden transitions[17] while direct allowed transitions at room temperature typically showed an onset at around 2.5 eV, derived from $(\alpha h\nu)^2$ *versus* $h\nu$ plots.[75] It was pointed out recently that above bandgap the absorption coefficient is given by a superposition of absorption mechanisms such that $(\alpha h\nu)^n$ *versus* $h\nu$ plots do not provide accurate data, neither for $n = \frac{2}{3}$ nor $n = 2$.[88] Experimental data show that α changes strongly with energy from 10^3 cm^{-1} at $\sim$2.1 eV to 10^4 cm^{-1} at $\sim$2.3 eV, thus requiring an absorber thickness $L_a \sim 3$ μm to convert at least 95% of photons with an energy above 2.3 eV.

Cu₂O is intrinsically p-type due to Cu vacancies that dominate under most growth conditions.[71] For Cu_2O prepared by thermal oxidation a hole mobility of 75 cm² V⁻¹ s⁻¹ was measured, which was further improved to 130 cm² V⁻¹ s⁻¹ by Cl-doping, using an annealing step in the presence of HCl vapour.[89] However, Cl-doping reduced the diffusion length and thus might not be beneficial for solar cells.[86] The doping of Cu_2O with impurities such as F, P, N, In, Mg, Si or Na on the other hand did not show any improvement of the conductivity.[86]

8.4.3 Cu₂O Schottky Junction Cells

The first Cu₂O-based PV cells with a light to electric energy conversion efficiency above 1% were Schottky junctions, where a thin metallic front electrode was deposited onto the Cu₂O absorber. Figure 8.8 shows an energy band diagram of a Schottky junction that is formed between a thin semi-transparent metal front contact (FC) with a work function Φ_m and a p-type Cu₂O semiconductor with the work function Φ_{sc}. Junction formation leads to electron transfer from the low work function front contact to the high work function semiconductor, which creates a space charge layer (SCL) with a built-in potential φ_{bi}. The Schottky barrier height Φ_b for p-type semi-conductor/metal junctions is defined as:

$$\Phi_b = (\chi + E_g) - \Phi_m \qquad (8.14)$$

The Schottky barrier height defines the energy barrier for hole transfer from the front contact into the semiconductor. Under illumination

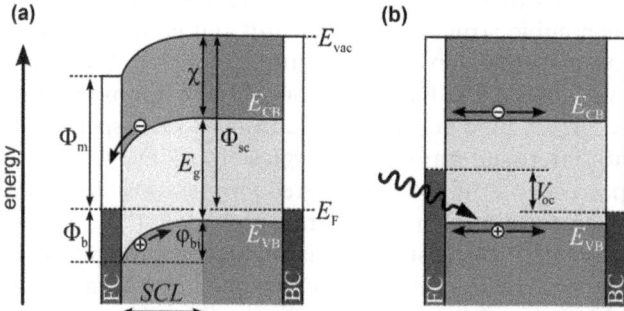

Figure 8.8   (a) Energy band diagram of a Schottky junction showing the difference in work function Φ_m of the metal front contact (FC) and a p-type semiconductor Φ_{sc} that leads to the formation of a space charge layer (SCL). The Schottky barrier height Φ_b is a function of Φ_m, the electron affinity χ and bandgap E_g [Equations (8.1) and (8.2)]. The local vacuum level (E_{vac}), conduction and valence band edge (E_{CB} and E_{VB}) depict the ohmic nature of the back contact (BC). (b) Under open circuit conditions the built-in potential φ_{bi} is compensated by the open circuit voltage V_{OC}.

electron–hole pairs are generated within the semiconductor and separated in the electric field of the SCL, electrons move to the metal front contact while holes migrate towards the back contact. At flat band potential the electrostatic driving force for chare separation is compensated by the photovoltage such that the open circuit voltage V_{OC} is approximately limited by the built-in potential φ_{bi} (see Figure 8.8).

In the 1970s Cu/Cu$_2$O Schottky junctions based on single and poly-crystalline Cu$_2$O were investigated. Single crystals of Cu$_2$O were grown by a crucible-free Czochralski technique[90] with subsequent annealing at 750–960 °C under well-defined oxygen partial pressure before Cu metal contacts were formed by surface reduction under hydrogen.[91] Polycrystalline Cu$_2$O was grown by thermal oxidation of high purity Cu plates.[91] Ohmic back contacts were achieved using Au,[92] Ag[92] or colloidal graphite paint,[91] allowing back-side illumination through the graphite. Photo-response measurements of single and polycrystalline Cu$_2$O/Cu junctions showed an onset substantially below the Cu$_2$O bandgap at around 0.8 eV, corresponding to the Cu/Cu$_2$O barrier height, which was attributed to hole injection from the Cu front contact into the Cu$_2$O valence band.[91] Furthermore, for single crystalline Cu/Cu$_2$O junctions a reduction of the Schottky barrier hight was observed as a function of increasing O$_2$ partial pressure during the Cu$_2$O annealing process, which was discussed in terms of defects imparted to the Cu$_2$O crystals prior to contact formation.[91] Alternative growth methods of Cu$_2$O onto Cu metal such as anodic oxidation showed a lower performance compared to devices produced by thermal oxidation, which was attributed to undesired side reactions during anodic oxidation.[93]

The open circuit voltage of Cu/Cu$_2$O Schottky junction solar cell remained 0.6V,[92,96] which is too low to achieve light to electric power conversion efficiencies above 10%. To increase the built-in potential and consequently the V_{OC} the Cu front contact was replaced by low work-function metals. Measurements of metal/Cu$_2$O junctions however showed that Φ_b did not scale with Φ_m[92] as predicted by eq. 8.9. Figure 8.9 shows the measured barrier height Φ_b as a function of the front contact work function Φ_m while the dashed line describes the correlation between Φ_b and Φ_m according to Equation (8.14) for a metal/Cu$_2$O junction.[92] Depth concentration profiles using Auger electron spectroscopy revealed a Cu rich phase at the junction, indicating that Cu$_2$O Schottky barriers made with low work function metals are essentially Cu/Cu$_2$O cells due to the reduction of the Cu$_2$O surface (Figure 8.10).[97] It was concluded that the fabrication of efficient Cu$_2$O solar cells requires a metal/insulator/semiconductor structure (MIS) to prevent the reduction of Cu$^+$ to Cu at the interface, using a wide bandgap semiconductor layer which is sufficiently thin to permit efficient electron tunnelling through eventual barriers at the interface. Furthermore, it was suggested to replace the metal front contact by TCOs to circumvent the reduction of Cu$^+$ and to reduce light reflection that occurs at thin semitransparent metal front contacts. For ITO/Cu$_2$O junctions barrier heights between 1.0 and 1.25 eV were reported[95] and a V_{OC} up to 400 mV was

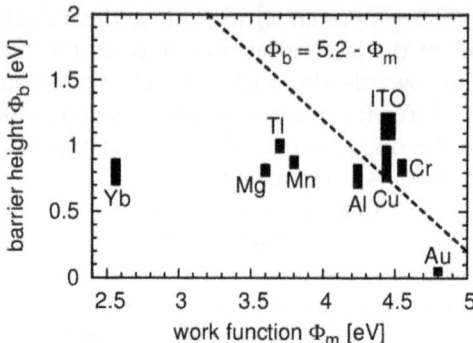

Figure 8.9 Barrier height Φ_b as a function of the front contact work function Φ_m. The dashed line shows the correlation expected for an ideal Cu_2O–metal junction.
Data taken from Olsen *et al.*,[92] Park *et al.*[94] and Sears *et al.*[95]

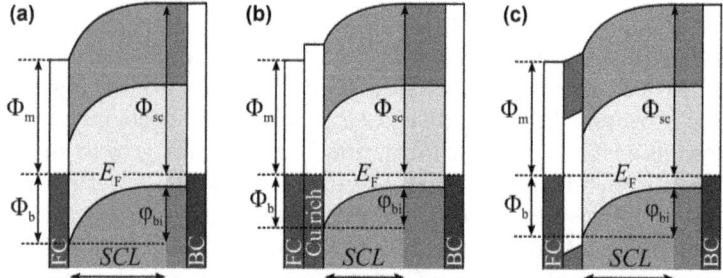

Figure 8.10 (a) Energy band diagram of an ideal Schottky junction showing an increase of the barrier height Φ_b and built-in potential φ_{bi} when a front contact with a low work function Φ_m is used. (b) Real Schottky barrier based on a Cu_2O p-type absorber. The deposition of a low work function front contact leads to the reductions of Cu_2O to metallic Cu such that the barrier height Φ_b is essentially a Cu–Cu_2O junction. (c) Metal–insulator–semiconductor (MIS) device that prevents Cu_2O reduction at the junction, using an insulator thickness that is thin enough to allow electron tunnelling from the Cu_2O into the FC.

achieved.[86] With Al-doped ZnO/Cu_2O junctions, conversion efficiencies above 2% were achieved with a V_{OC} of 500 mV.[99]

8.4.4 Cu_2O-based Heterojunction Cells

Heterojunction solar cells were proposed to improve the light to electric power conversion efficiency beyond the limitations of Cu_2O-based Schottky junctions. Heterojunction cells are mostly based on a TCO front contact, covered with an n-type MO followed by the p-type Cu_2O and a metal back contact (Figure 8.11). It was shown that a ZnO window layer increased the V_{OC} by ~50 % compared to a Schottky junction where the TCO was

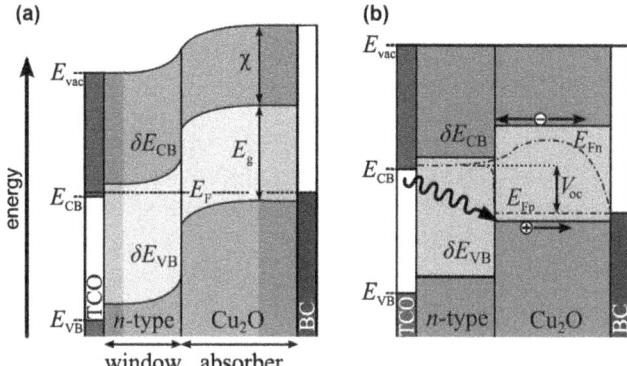

Figure 8.11 Cu_2O heterojunction solar cells. (a) Cell at short circuit showing a large energy difference δE_{CB} between the window layer and the absorber. (b) At open circuit the V_{OC} remains considerable below the maximum quasi-Fermi level splitting due to the large δE_{CB} and the missing back contact selectivity.

deposited directly onto the Cu_2O.[86] Even though early work suggested that Cu enrichment takes also place at the ZnO/Cu_2O interface[100] ZnO remained the most investigated intrinsic n-type window layer for heterojunction Cu_2O cells and the device performance was strongly dependent on the synthesis method. ZnO/Cu_2O heterojunctions have been entirely produced by RF magnetron sputtering[80] or electrochemical deposition,[101,102] with the latter showing a J_{SC} of ~ 2.7 mA cm^{-2} and a V_{OC} above 300 mV.[102] The highest conversion efficiencies have been achieved with heterojunction cells based on thermally oxidized Cu_2O layers onto which ZnO was deposited by sputtering[86] or pulsed laser deposition (PLD)[103] leading to power conversion efficiencies above 4%.[104] Even though there has been an encouraging increase of the power conversion efficiency over the years it was pointed out that the band alignment in a ZnO/Cu_2O junction is not ideal with a CB offset (δE_{CB}) of 0.97 eV, responsible for a significant loss in photovoltage.[105]

To reduce δE_{CB} alternative n-type materials were investigated. PV cells of In_2O_3, SnO_2, mixtures of CdO and SnO_2 in conjunction with Cu_2O revealed that a similar Cu enrichment was found at the interface like in Schottky junctions cells with low work function metals.[106] Thus, the energetics at these heterojunctions are determined by a Cu/Cu_2O layer and not by energy levels of the wide bandgap MO. Only pure CdO/Cu_2O heterojunctions seemed to be free of metallic Cu.[106] Heterojunctions produced by electrophoretic deposition of TiO_2 followed by electrodeposition of Cu_2O showed a J_{SC} above 1 mA cm^{-2} with a V_{OC} below 250 mV, demonstrating the fabrication of ZnO/Cu_2O alternatives by low cost methods, but without a gain in performance.[98] Furthermore, heterojunctions based on cupric and cuprous oxide (CuO/Cu_2O) were produced by a two-step electrochemical deposition on a Ti substrate with an annealing step at 500 °C in the middle to convert Cu_2O into CuO. The conversion efficiency of the final device structure of Ti/CuO/Cu$_2$O/Au, however, was very low with $J_{SC} \sim 0.3$ mA cm^{-2} and $V_{OC} \sim 300$ mV.[107] High conversion efficiencies of $\sim 5.4\%$ were recently achieved with p-type Cu$_2$O produced by thermal oxidation in conjunction with a Ga$_2$O$_3$ ETL produced by PLD. This efficiency increase was mainly due to a reduction of δE_{CB} at the Ga$_2$O$_3$/Cu$_2$O interface, which resulted in an increase of V_{OC} from 700 to 800 mV compared to a ZnO/Cu$_2$O device produced by the same processes, while the J_{SC} of almost 10 mA cm^{-2} did not differ significantly for both devices.[87]

8.4.5 Cu$_2$O Homojunction Cells

The reported synthesis of intrinsic n-type Cu$_2$O triggered the fabrication of Cu$_2$O based p–n homojunctions,[69] though the origin for the n-type character is under debate.[71] Sequential electrochemical deposition at different solution pH was used to produce p–n junctions with Al contacts connected to the n-type and Cu connected to the p-type side, which showed rectification up to −6 V in reverse direction followed by irreversible junction breakdown.[69] Photovoltaic cells were produced onto ITO substrates by electrochemical deposition of p-type Cu$_2$O, followed by the n-type layer and an Au/Pd, Cu, Al or sputtered ITO back contact.[108] Illumination through the ITO substrate showed that the cells with the ITO back contact provided the best performing solar cells with $J_{Sc} \sim 3.6$ mA cm^{-2} and $V_{OC} \sim 600$ mV.[109] Devices with an inverse junction starting from an FTO substrate, an n-type layer and then the p-type layer with a Au back contact also showed J_{SC} up to 3.5 mA cm^{-2}, but a significantly lower $V_{OC} < 120$ mV.[110]

8.4.6 Nano-structured Cu$_2$O-based Photovoltaic Cells

Depending on the preparation method the minority carrier diffusion length L_D in Cu$_2$O was found to be substantially smaller then $3/\alpha$ and for electrochemically deposited Cu$_2$O it was argued that L_D is even smaller than $1/\alpha$, causing substantial recombination within the light absorber. Nano-composite solar cell geometries were proposed to circumvent this limitation using ZnO nanowires,[111] mesoporous TiO$_2$ films[17] or arrays of TiO$_2$ nanotubes[112] as ETLs in conjunction with electrochemically grown Cu$_2$O to fill the voids in between the nano-structures. In comparison to flat bilayer solar cells it was shown that nano-structures can increase the J_{SC}, but the conversion efficiency remained in all cases below 1%,[113–116] and thus significantly below the performance of heterojunction cells where the Cu$_2$O was produced by thermal oxidation of high purity copper. So far, nano-structures have not been able to compensate for the losses in electrochemically grown Cu$_2$O due to its lower crystal quality (Figure 8.12).

Even though Cu$_2$O has a non-ideal bandgap for single junction cells it has been the working horse materials for all-oxide photovoltaics over the last years. Finding efficient MO light absorbers with a narrower bandgap remains one of the challenges. Ag$_2$O is a structural analogue to Cu$_2$O with a

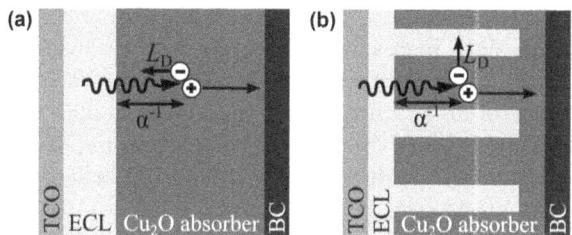

Figure 8.12 (a) Schematic drawing of a thin film Cu_2O heterojunction cell. L_D is shorter than the penetration depth α^{-1} of the light leading to strong recombination. (b) Nano-structures decouple both length scales, providing an interface for charge separation within L_D while α^{-1} is still larger than L_D.

direct bandgap of about 1.4 eV, which is perfectly matching the solar spectrum. Ag_2O can be prepared by sputtering and is stable at room temperature, but begins to dissociate above 60 °C in vacuum due to the weakness of the silver–oxygen bond. Ag doping of Cu_2O or AgCuO alloys might provide a way to shift the Cu_2O into the visible without losing its good charge transport properties.

8.5 Further Metal Oxide-based Photovoltaics

8.5.1 ZnO–Fe_2O_3 Heterojunction Solar Cells

The α-phase of Fe_2O_3 is under heavy investigation for solar water splitting due to its chemical stability which is comparable to TiO_2. The optical bandgap is in the visible at around 2 eV which allows harvesting by far more photons from the solar spectrum compared to TiO_2 with its bandgap of 3.2 eV. Based on the Fe_2O_3 bandgap light to electric power conversion efficiencies comparable to those of Cu_2O could be achieved. However, Fe_2O_3 is intrinsically n-type with a half filled d-band ($3d^5$ configuration), leading to a short minority carrier diffusion length in contrast to p-type Cu_2O, which has a filled d-band and a much larger diffusion length. A photovoltaic ZnO/Fe_2O_3 heterojunction cell has been demonstrated with a low J_{SC} of a few $\mu A\ cm^{-2}$ and a fill factor of approximately 25%. Limiting factors were the low carrier mobility and the non-optimized device structure consisting of an n-type ZnO window layer in combination with an n-type Fe_2O_3 light absorber.[18]

8.5.2 Bi_2O_3 Solar Cells

Bi_2O_3 has been investigated as a light absorber in a simple PV cell configuration consisting of an ITO front contact, the Bi_2O_3 layer as a light absorber and a RuO_2 or Au back contact. Bi_2O_3 was produced by sputtering showing a bandgap of 2.8–3.3 eV depending on the substrate temperature and oxygen partial pressure during deposition. Sub-bandgap light absorption observed in some films was attributed to residual metallic Bi phase in the films. With increasing oxygen content a downward shift of the Fermi level towards the valence band was observed, indicating a more p-type character of the material. At a substrate temperature of 200 °C the crystalline β-Bi_2O_3 phase was identified while room temperature deposition resulted in an amorphous phase with wide optical bandgap. PV cells based on β-Bi_2O_3 with an ITO front and Au back contact showed a V_{OC} up to 680 mV and a J_{SC} of 0.3 mA cm^{-2}, while the fill factor reached not more than $\sim 28\%$.[8] Even though these values are low for PV cells, they are still remarkable, considering the wide bandgap of the absorber, which is far from being optimized and the absence of charge selective layers adjacent to the contacts.

8.5.3   Ferro-electric $BiFeO_3$ Solar Cells

Photovoltaic cells based on ferroelectric oxides such as $BiFeO_3$ have been reported.[117-120] Solar cells with a lateral design were presented, where two Pt contact stripes were deposited on top of a $BiFeO_3$ film parallel to the domain boundaries. Even though the device had a symmetric structure it was explained that the required asymmetry to generate a photocurrent and voltage originated from the electric field within the domains and at the domain boundaries.[117] The domains were electrically connected in series while illuminated in parallel with the same light intensity. The band diagram of the $BiFeO_3$ PV system at short circuit is schematically shown in Figure 8.13a. An electric field in the domain generates a drift current such that

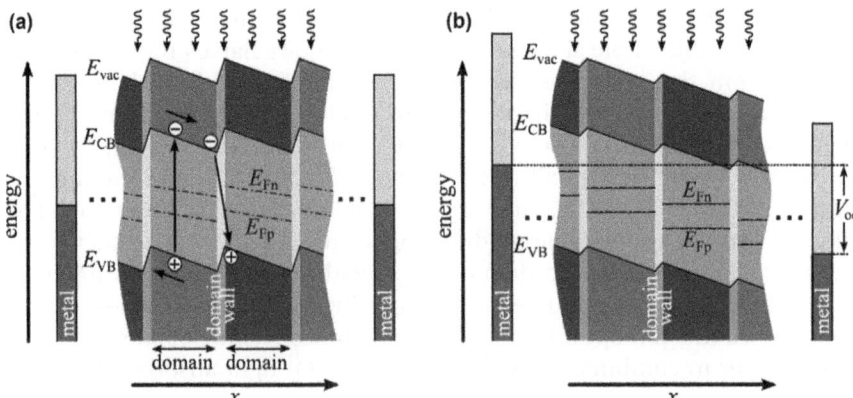

Figure 8.13   Energy band diagrams for $BiFeO_3$-based solar cells. (a) At short circuit electron–hole pairs are generated within a ferroelectric domain. Due to the lateral design each domain is exposed directly to the light. Electron–hole pairs migrate towards domain walls where they recombine with carriers from the neighboring domain. Domain walls are shown with exaggerated width for better visibility. (b) At open circuit high photovoltages are achieved due to the electronic series connection of the domains.

photo-generated electrons and holes move towards opposite domain boundaries where they recombine with charge carriers of opposite sign from the neighbouring ferroelectric domain.[117,119] The maximum photovoltage generated at a single ferroelectric junction was determined to be in order of 10 mV.[119] With tens to hundreds of ferroelectric domains connected in series photovoltages up to 15 V have been reported, which is schematically shown in Figure 8.13b.[119] It was furthermore shown that the direction of the photocurrent was switchable by an electrostatic potential applied before illumination.[118,120] Ferroelectric $BiFeO_3$ based PV cells have generated considerable interest, however high light to electric power conversion efficiencies cannot be expected due to the large $BiFeO_3$ bandgap of 2.7 eV.[118]

8.6 Combinatorial Material Science for Novel Metal Oxides

All-oxide photovoltaics have potentially important and significant advantages over existing PV technologies, it can be based on non-hazardous, chemically stable, abundant and cheap materials[121] which can be processed under ambient conditions. However, the number of suitable MOs, especially as light absorbers, is very limited and far from optimum as shown in section 8.3. Until today Cu_2O is clearly the work-horse material for all-oxide PV. It can be produced in high crystallinity and quality, leading to low defect state density and providing the required diffusion length, but its bandgap is too high for single junction devices. Other MO semiconductors with an absorption edge in the visible suffer from a short L_D, calling for the development of better materials. The development of novel ternary, quaternary and higher order multi-component metal oxides has been proposed to provide the desired optoelectronic properties while keeping the advantages of low cost processability and chemical stability.

High temperature superconductors are a good example where the approach of increased material complexity has led to an increase of the critical temperature T_c above which the material loses its superconducting properties. $HgBa_2Ca_2Cu_3O_8$, $TlBa_2Ca_3Cu_4O_{11}$ or $Bi_2Sr_2Ca_2Cu_3O_6$ are multi-component MOs which have reached $T_c > 100$ K, demonstrating that MO can show exceptional electronic properties.[122] Another example are dielectric materials such as $CaCu_3Ti_4O_{12}$ or $Ba_{0.94}Bi_{0.06}(FeTa)_{0.5}O_3$ with dielectric constants in the order of 10^5, which is orders of magnitude larger compared to binary MOs.[123,124] Today powerful experimental and computational methods are available which will accelerate the development of novel materials for new application such as all-oxide PV cells.[18,20]

8.6.1 Density Functional Theory

Computational methods to calculate the optical and electronic properties of semiconductors have become popular due to increased computational power that is today available within the financial boundaries of research budgets. Density functional theory (DFT) is used to calculate the electronic band structure of semiconductors based on a known crystal structure, however especially for metal oxides and transition metal oxides the calculated bandgap can differ substantially from the experimentally measured one. Semi-empirical extensions have been developed to adjust the calculated bandgap to experimental values.[125] Based on such models the impact of doping can be investigated, using tens to hundreds of unit cells which create a super cell and in which one metal atom can be replaced by a dopant atom. For haematite (α-Fe_2O_3), for example, it was calculated that the replacement of one iron atom within a 120-atom super cell by a transition metal atom from the 4th, 5th or 6th period can cause a drastic reduction of the electronic bandgap.[126] These results call for experimental verification and most likely require further refinement of the computational model to provide solid data for the search of doped MOs with improved optical bandgap.

Furthermore, DFT calculations have been used to investigate the potential of transition metal oxide alloys for light harvesting in solar cells such as $Fe_{1-x}M_xO$ (M = Mg, Mn, Ni or Zn). Especially for $Fe_{1-x}Zn_xO$ the calculations predicted a bandgap between 1.1 and 1.5 eV for $0.5 < x < 0.8$, which is nearly optimized for the AM 1.5G spectrum.[28] Similar computational methods have been used to calculate the electronic properties of heterostructures containing two simple band insulators, for example $LaAlO_3$ and $SrTiO_3$ where unanticipated properties from conductivity to magnetism were predicted and experimentally confirmed.[127,128] Heterostructures of $LaVO_3$ grown on $SrTiO_3$ were proposed as a new class of transition metal oxide absorbers for PV applications, for which DFT calculations predicted a direct bandgap or ~ 1.1 eV.[129]

8.6.2 Combinatorial Material and Device Fabrication

Combinatorial materials science recently is moving into the focus of interest for the development of novel multi-component electronic and functional materials.[130] Combinatorial materials science requires a fabrication method in which the material composition is changed in a systematic fashion across a substrate to create a material library. Such libraries can be produced by simultaneous deposition from two or more sources leading to continuous compositional spreads (CCS). Alternatively sequential deposition of thin layers from different sources can be used with layer thicknesses in the order of Angstroms to nanometres in conjunction with substrate heating to produce a homogeneous material composition. Sputtering and pulsed laser deposition are the most common deposition methods for thin film CCS libraries. In conjunction with high throughput characterization methods structural, compositional and optical information can be extracted for a manifold of compositions and thus help to synthesize and identify novel multi-component metal oxides as light absorbers, low work function ETL or high work function HTL.

The combinatorial approach has been extended to investigate the performance of homogeneous materials as well as CCS using combinatorial solar cell device libraries.[18] In this approach the individual solar cell layers are deposited onto a TCO covered glass substrate where the layer thickness of the ETL, absorber and HTL are varied in systematic fashion. The FTO serves as joint front electrode while a grid of 13×13 round back contacts defines a grid of 169 solar cells.[131] An automated scanning system characterizes all solar cells under simulated AM 1.5G light and in conjunction with transmission/reflection measurements information about the absorption coefficient, bandgap, spectrally integrated quantum efficiency can be extracted.[20]

With the fast development of computational and experimental methods for combinatorial materials science, fast progress is expected for the development of novel MOs for photovoltaic applications with improved optical and electronic properties. In the coming years we expect that the number of new PV materials and concepts which can achieve conversion efficiencies above 10% will increase drastically and lead to a price reduction of PV generated electricity.

References

1. C. Kittel, *Introduction to Solid State Physics*, John Wiley & Sons, 7th edn, 1996, p. 216.
2. W. Shockley and H. J. Queisser, *J. Appl. Phys.*, 1961, **32**, 510.
3. W. Jaegermann, A. Klein and T. Mayer, *Adv. Mater.*, 2009, **21**, 4196.
4. J. S. Fonash and S. Ashok, *Appl. Phys. Lett.*, 1979, **35**, 535.
5. J. Bisquert, D. Cahen, G. Hodes, S. Rühle and A. Zaban, *J. Phys. Chem. B*, 2004, **108**, 8106.
6. P. Würfel, *Physics of Solar Cells – From Principles to New Concepts*, Wiley-VCH edn., 2005.
7. A. Klein, *Thin Solid Films*, 2012, **520**, 3721.
8. V. Pfeifer, P. Erhart, S. Li, K. Rachut, J. Morasch, J. Brötz, P. Reckers, T. Mayer, S. Rühle, A. Zaban, I. Mora Seró, J. Bisquert, W. Jaegermann and A. Klein, *J. Phys. Chem. Lett.*, 2013, **4**, 4182.
9. P. Brinks, W. Siemons, J. E. Kleibeuker, G. Koster, G. Rijnders and M. Huijben, *Appl. Phys. Lett.*, 2011, 98.
10. N. Reyren, S. Thiel, A. Caviglia, L. F. Kourkoutis, G. Hammerl, C. Richter, C. Schneider, T. Kopp, A.-S. Rüetschi and D. Jaccard, *Science*, 2007, **317**, 1196.
11. H. Y. Hwang, Y. Iwasa, M. Kawasaki, B. Keimer, N. Nagaosa and Y. Tokura, *Nat. Mater.*, 2012, **11**, 103.
12. J. G. Bednorz and K. A. Müller, *Z. Phys. B*, 1986, **64**, 189.
13. E. Fortin and F. L. Weichman, *Phys. Status Solidi B*, 1964, **5**, 515.
14. J. Morasch, S. Li, J. Brötz, W. Jaegermann and A. Klein, *Phys. Status Solidi A*, 2013, n/a.

15. N. A. M. Barakat, M. S. Khil, F. A. Sheikh and H. Y. Kim, *J. Phys. Chem. C*, 2008, **112**, 12225.

16. A. Gulino, P. Dapporto, P. Rossi and I. Fragalà, *Chem. Mater.*, 2003, **15**, 3748.

17. P. E. de Jongh, D. Vanmaekelbergh and J. J. Kelly, *Chem. Mater.*, 1999, **11**, 3512.

18. S. Rühle, A. Y. Anderson, H.-N. Barad, B. Kupfer, Y. Bouhadana, E. Rosh-Hodesh and A. Zaban, *J. Phys. Chem. Lett.*, 2012, **3**, 3755.

19. H. Kidowaki, T. Oku, T. Akiyama, A. Suzuki, B. Jeyadevan and J. Cuya, *J. Mater. Sci. Res.*, 2012, **1**, 138.

20. A. Y. Anderson, Y. Bouhadana, H.-N. Barad, B. Kupfer, E. Rosh-Hodesh, H. Aviv, Y. R. Tischler, S. Rühle and A. Zaban, *ACS Combi. Sci.*, 2014.

21. Y. Xu and M. A. A. Schoonen, *Am. Mineral.*, 2000, **85**, 543.

22. H. Dotan, O. Kfir, E. Sharlin, O. Blank, M. Gross, I. Dumchin, G. Ankonina and A. Rothschild, *Nat. Mater.*, 2013, **12**, 158.

23. D. Fu, K. Liu, T. Tao, K. Lo, C. Cheng, B. Liu, R. Zhang, H. A. Bechtel and J. Wu, *J. Appl. Phys.*, 2013, **113**, 043707.

24. D. Zhang, R. Huang, T. Zhang, Y. Li, Y. Chen, Y. Zhong, P. Fan and J. Huang, *Phys. Status Solidi A*, 2012, **209**, 2229.

25. M. F. Al-Kuhaili and S. M. A. Durrani, *Opt. Mater.*, 2007, **29**, 709.

26. U. A. Joshi and P. A. Maggard, *J. Phys. Chem. Lett.*, 2012, **3**, 1577.

27. C. G. Read, Y. Park and K.-S. Choi, *J. Phys. Chem. Lett.*, 2012, **3**, 1872.

28. M. C. Toroker and E. A. Carter, *J. Mater. Chem. A*, 2013, **1**, 2474.

29. E. Fortunato, D. Ginley, H. Hosono and D. C. Paine, *MRS Bull.*, 2007, **32**, 242.

30. D. S. Ginley and C. Bright, *MRS Bull.*, 2000, **25**, 15.

31. A. E. Rakhshani, Y. Makdisi and H. A. Ramazaniyan, *J. Appl. Phys.*, 1998, **83**, 1049.

32. D. J. Milliron, I. G. Hill, C. Shen, A. Kahn and J. Schwartz, *J. Appl. Phys.*, 2000, **87**, 572.

33. E. Fortunato, P. Nunes, D. Costa, D. Brida, I. Ferreira and R. Martins, *Vacuum*, 2002, **64**, 233.

34. A. Zakutayev, J. D. Perkins, P. A. Parilla, N. E. Widjonarko, A. K. Sigdel, J. J. Berry and D. S. Ginley, *MRS Commun.*, 2011, **1**, 23.

35. S. Lany, J. Osorio-Guillén and A. Zunger, *Phys. Rev. B*, 2007, **75**, 241203.

36. E. Wallin, U. Malm, T. Jarmar, O. L. M. Edoff and L. Stolt, *Prog. Photovoltaics*, 2012, **20**, 851.

37. L. M. Peter, *J. Phys. Chem. Lett.*, 2011, **2**, 1861.

38. S. Rühle, M. Shalom and A. Zaban, *ChemPhysChem*, 2010, **11**, 2290.

39. I. Mora-Seró and J. Bisquert, *J. Phys. Chem. Lett.*, 2010, **1**, 3046.

40. T. Dittrich, A. Belaidi and A. Ennaoui, *Sol. Energy Mater. Sol. Cells*, 2011, **95**, 1527.

41. H.-S. Kim, C.-R. Lee, J.-H. Im, K.-B. Lee, T. Moehl, A. Marchioro, S.-J. Moon, R. Humphry-Baker, J.-H. Yum, J. E. Moser, M. Gratzel and N.-G. Park, *Sci. Rep.*, 2012, **2**.

42. C. Klingshirn, *ChemPhysChem*, 2007, **8**, 782.

43. P. Ravirajan, A. M. Peiro, M. K. Nazeeruddin, M. Grätzel, D. D. C. Bradley, J. R. Durrant and J. Nelson, *J. Phys. Chem. B*, 2006, **110**, 7635.
44. J. Boucle, P. Ravirajan and J. Nelson, *J. Mater. Chem.*, 2007, **17**, 3141.
45. P. Atienzar, T. Ishwara, B. N. Illy, M. P. Ryan, B. C. O'Regan, J. R. Durrant and J. Nelson, *J. Phys. Chem. Lett.*, 2010, **1**, 708.
46. F. Lenzmann, J. Krueger, S. Burnside, K. Brooks, M. Grätzel, D. Gal, S. Rühle and D. Cahen, *J. Phys. Chem. B*, 2001, **105**, 6347.
47. S. Ferrere, A. Zaban and B. A. Gregg, *J. Phys. Chem. B*, 1997, **101**, 4490.
48. S. Burnside, J.-E. Moser, K. Brooks, M. Grätzel and D. Cahen, *J. Phys. Chem. B*, 1999, **103**, 9328.
49. S. G. Chen, S. Chappel, Y. Diamant and A. Zaban, *Chem. Mater.*, 2001, **13**, 4629.
50. R. Jose, V. Thavasi and S. Ramakrishna, *J. Am. Ceram. Soc.*, 2009, **92**, 289.
51. L. Grinis, S. Kotlyar, S. Rühle, J. Grinblat and A. Zaban, *Adv. Funct. Mater.*, 2010, **20**, 282.
52. S. Greenwald, S. Rühle, M. Shalom, S. Yahav and A. Zaban, *Phys. Chem. Chem. Phys.*, 2011, **13**, 19302.
53. H. Gerischer, *Electrochim. Acta*, 1989, **34**, 1005.
54. S. Saraf, M. Markovich and A. Rothschild, *Phys. Rev. B*, 2010, **82**, 245208.
55. N. Espinosa, H. F. Dam, D. M. Tanenbaum, J. W. Andreasen, M. Jørgensen and F. C. Krebs, *Materials*, 2011, **4**, 169.
56. F. Li, S. Ruan, Y. Xu, F. Meng, J. Wang, W. Chen and L. Shen, *Sol. Energy Mater. Sol. Cells*, 2011, **95**, 877.
57. D. Y. Kim, J. Subbiah, G. Sarasqueta, F. So, H. Ding and Y. Gao, *Appl. Phys. Lett.*, 2009, **95**, 093304.
58. K. X. Steirer, J. P. Chesin, N. E. Widjonarko, J. J. Berry, A. Miedaner, D. S. Ginley and D. C. Olson, *Org. Electron.*, 2010, **11**, 1414.
59. J. Meyer, S. Hamwi, M. Kröger, W. Kowalsky, T. Riedl and A. Kahn, *Adv. Mater.*, 2012, **24**, 5408.
60. E. L. Ratcliff, B. Zacher and N. R. Armstrong, *J. Phys. Chem. Lett.*, 2011, **2**, 1337.
61. M. D. Irwin, D. B. Buchholz, A. W. Hains, R. P. H. Chang and T. J. Marks, *Proc. Natl. Acad. Sci. U. S. A.*, 2008, **105**, 2783.
62. H. Kawazoe, M. Yasukawa, H. Hyodo, M. Kurita, H. Yanagi and H. Hosono, *Nature*, 1997, **389**, 939.
63. H. Yanagi, H. Kawazoe, A. Kudo, M. Yasukawa and H. Hosono, *J. Electroceram.*, 2000, **4**, 407.
64. B. P. Rai, *Sol. Cells*, 1988, **25**, 265.
65. A. E. Rakhshani, *Solid-State Electron.*, 1986, **29**, 7.
66. A. S. Brown and M. A. Green, *Physica E*, 2002, **14**, 96.
67. T. D. Golden, M. G. Shumsky, Y. Zhou, R. A. VanderWerf, R. A. Van Leeuwen and J. A. Switzer, *Chem. Mater.*, 1996, **8**, 2499.
68. K. Han and M. Tao, *Sol. Energy Mater. Sol. Cells*, 2009, **93**, 153.
69. L. Wang and M. Tao, *Electrochem. Solid-State Lett.*, 2007, **10**, H248.

70. B. K. Meyer, A. Polity, D. Reppin, M. Becker, P. Hering, P. J. Klar, T. Sander, C. Reindl, J. Benz, M. Eickhoff, C. Heiliger, M. Heinemann, J. Bläsing, A. Krost, S. Shokovets, C. Müller and C. Ronning, *Phys. Status Solidi B*, 2012, **249**, 1487.
71. D. O. Scanlon and G. W. Watson, *J. Phys. Chem. Lett.*, 2010, **1**, 2582.
72. L. M. Abrantes, L. M. Castillo, C. Norman and L. M. Peter, *J. Electroanal. Chem. Interfacial Electrochem.*, 1984, **163**, 209.
73. H. H. Strehblow and B. Titze, *Electrochim. Acta*, 1980, **25**, 839.
74. D. P. Singh, N. R. Neti, A. S. K. Sinha and O. N. Srivastava, *J. Phys. Chem. C*, 2007, **111**, 1638.
75. T. Kosugi and S. Kaneko, *J. Am. Ceram. Soc.*, 1998, **81**, 3117.
76. P. Pattanasattayavong, S. Thomas, G. Adamopoulos, M. A. McLachlan and T. D. Anthopoulos, *Appl. Phys. Lett.*, 2013, **102**, 163505.
77. M. Ottosson and J.-O. Carlsson, *Surf. Coat. Technol.*, 1996, **78**, 263.
78. S. Eisermann, A. Kronenberger, A. Laufer, J. Bieber, G. Haas, S. Lautenschläger, G. Homm, P. J. Klar and B. K. Meyer, *Phys. Status Solidi A*, 2012, **209**, 531.
79. J. Pinkas, J. C. Huffman, D. V. Baxter, M. H. Chisholm and K. G. Caulton, *Chem. Mater.*, 1995, **7**, 1589.
80. S. Ishizuka, K. Suzuki, Y. Okamoto, M. Yanagita, T. Sakurai, K. Akimoto, N. Fujiwara, H. Kobayashi, K. Matsubara and S. Niki, *Phys, Status Solidi C*, 2004, **1**, 1067.
81. D. J. Miller, J. D. Hettinger, R. P. Chiarello and H. K. Kim, *J. Mater. Res.*, 1992, **7**, 2828.
82. J. Deuermeier, J. Gassmann, J. Brotz and A. Klein, *J. Appl. Phys.*, 2011, **109**, 113704.
83. I. Pallecchi, E. Bellingeri, C. Bernini, L. Pellegrino, A. S. Siri and D. Marré, *J. Phys. D: Appl. Phys.*, 2008, **41**, 125407.
84. S. B. Ogale, P. G. Bilurkar, N. Mate, S. M. Kanetkar, N. Parikh and B. Patnaik, *J. Appl. Phys.*, 1992, **72**, 3765.
85. G. P. Pollack and D. Trivich, *J. Appl. Phys.*, 1975, **46**, 163.
86. A. Mittiga, E. Salza, F. Sarto, M. Tucci and R. Vasanthi, *Appl. Phys. Lett.*, 2006, **88**, 163502.
87. T. Minami, Y. Nishi and T. Miyata, *Appl. Phys. Express*, 2013, 6.
88. C. Malerba, F. Biccari, C. L. A. Ricardo, M. D'Incau, P. Scardi and A. Mittiga, *Sol. Energy Mater. Sol. Cells*, 2011, **95**, 2848.
89. A. O. Musa, T. Akomolafe and M. J. Carter, *Sol. Energy Mater. Sol. Cells*, 1998, **51**, 305.
90. D. Trivich and G. P. Pollack, *J. Electrochem. Soc.*, 1970, **117**, 344.
91. J. A. Assimos and D. Trivich, *J. Appl. Phys.*, 1973, **44**, 1687.
92. L. Olsen, F. Addis and W. Miller, *Sol. Cells*, 1982, **7**, 247.
93. W. M. Sears and E. Fortin, *Sol. Energy Mater*, 1984, **10**, 93.
94. Y. Park, V. Choong, Y. Gao, B. R. Hsieh and C. W. Tang, *Appl. Phys. Lett.*, 1996, **68**, 2699.
95. W. M. Sears, E. Fortin and J. B. Webb, *Thin Solid Films*, 1983, **103**, 303.
96. R. J. Iwanowski and D. Trivich, *Phys. Status Solidi A*, 1986, **95**, 735.

97. L. C. Olsen, R. C. Bohara and M. W. Urie, *Appl. Phys. Lett.*, 1979, **34**, 47.
98. M. Ichimura and Y. Kato, *Mater. Sci. Semicond. Process.*, 2013, **16**, 1538.
99. Y. Nishi, T. Miyata, J.-i. Nomoto and T. Minami, *Thin Solid Films*, 2012, **520**, 3819.
100. J. Herion, E. A. Niekisch and G. Scharl, *Sol. Energy Mater.*, 1980, **4**, 101.
101. J. Katayama, K. Ito, M. Matsuoka and J. Tamaki, *J. Appl. Electrochem.*, 2004, **34**, 687.
102. S. S. Jeong, A. Mittiga, E. Salza, A. Masci and S. Passerini, *Electrochim. Acta*, 2008, **53**, 2226.
103. T. Minami, Y. Nishi, T. Miyata and J.-i. Nomoto, *Appl. Phys. Express*, 2011, **4**, 062301.
104. Y. Nishi, T. Miyata and T. Minami, *J. Vac. Sci. Technol. A*, 2012, **30**, 04D103.
105. B. Kramm, A. Laufer, D. Reppin, A. Kronenberger, P. Hering, A. Polity and B. K. Meyer, *Appl. Phys. Lett.*, 2012, **100**, 094102.
106. L. Papadimitriou, N. A. Economou and D. Trivich, *Sol. Cells*, 1981, **3**, 73.
107. R. P. Wijesundera, *Semicond. Sci. Technol.*, 2010, **25**, 045015.
108. C. M. McShane, W. P. Siripala and K.-S. Choi, *J. Phys. Chem. Lett.*, 2010, **1**, 2666.
109. C. M. McShane and K.-S. Choi, *Phys. Chem. Chem. Phys.*, 2012, **14**, 6112.
110. H. M. Wei, H. B. Gong, L. Chen, M. Zi and B. Q. Cao, *J. Phys. Chem. C*, 2012, **116**, 10510.
111. K. P. Musselman, A. Marin, L. Schmidt-Mende and J. L. MacManus-Driscoll, *Adv. Funct. Mater.*, 2012, **22**, 2202.
112. D. Li, C.-J. Chien, S. Deora, P.-C. Chang, E. Moulin and J. G. Lu, *Chem. Phys. Lett.*, 2011, **501**, 446.
113. B. D. Yuhas and P. Yang, *J. Am. Chem. Soc.*, 2009, **131**, 3756.
114. K. P. Musselman, A. Wisnet, D. C. Iza, H. C. Hesse, C. Scheu, J. L. MacManus-Driscoll and L. Schmidt-Mende, *Adv. Mater.*, 2010, **22**, E254.
115. K. P. Musselman, A. Marin, A. Wisnet, C. Scheu, J. L. MacManus-Driscoll and L. Schmidt-Mende, *Adv. Funct. Mater.*, 2011, **21**, 573.
116. J. Cui and U. J. Gibson, *J. Phys. Chem. C*, 2010, **114**, 6408.
117. J. Seidel, D. Fu, S.-Y. Yang, E. Alarcón-Lladó, J. Wu, R. Ramesh and J. W. Ager, III, *Phys. Rev. Lett.*, 2011, **107**, 126805.
118. S. Y. Yang, L. W. Martin, S. J. Byrnes, T. E. Conry, S. R. Basu, D. Paran, L. Reichertz, J. Ihlefeld, C. Adamo, A. Melville, Y. H. Chu, C. H. Yang, J. L. Musfeldt, D. G. Schlom, J. W. Ager III and R. Ramesh, *Appl. Phys. Lett.*, 2009, **95**, 062909.
119. S. Y. Yang, J. Seidel, S. J. Byrnes, P. Shafer, C. H. Yang, M. D. Rossell, P. Yu, Y. H. Chu, J. F. Scott, J. W. Ager, L. W. Martin and R. Ramesh, *Nat. Nano.*, 2010, **5**, 143.
120. T. Choi, S. Lee, Y. J. Choi, V. Kiryukhin and S.-W. Cheong, *Science*, 2009, **324**, 63.
121. C. Wadia, A. P. Alivisatos and D. M. Kammen, *Environ. Sci. Technol.*, 2009, **43**, 2072.

122. B. A. Parkinson, *Photoelectrochemical Hydrogen Production*, Springer, 2012, p. 173.
123. C. Homes, T. Vogt, S. Shapiro, S. Wakimoto, M. Subramanian and A. Ramirez, *Phys. Rev. B*, 2003, **67**, 092106.
124. Y.-L. Chai, C.-S. His, Y.-T. Lin and Y.-S. Chang, *J. Alloys Compd.*, 2014, **588**, 248.
125. R. Gillen and J. Robertson, *J. Phys.: Condens. Matter*, 2013, **25**, 165502.
126. Z. D. Pozun and G. Henkelman, *J. Chem. Phys.*, 2011, **134**, 224706.
127. R. Pentcheva and W. E. Pickett, *J. Phys.: Condens. Matter*, 2010, **22**, 043001.
128. B. Kalisky, J. A. Bert, B. B. Klopfer, C. Bell, H. K. Sato, M. Hosoda, Y. Hikita, H. Y. Hwang and K. A. Moler, *Nat. Commun.*, 2012, **3**, 922.
129. E. Assmann, P. Blaha, R. Laskowski, K. Held, S. Okamoto and G. Sangiovanni, *Phys. Rev. Lett.*, 2013, **110**, 078701.
130. R. Potyrailo, K. Rajan, K. Stoewe, I. Takeuchi, B. Chisholm and H. Lam, *ACS Combi. Sci.*, 2011, **13**, 579.
131. S. Rühle, H. N. Barad, Y. Bouhadana, D. A. Keller, A. Ginsburg, K. Shimanovich, K. Majhi, R. Lovrincic, A. Y. Anderson and A. Zaban, *Phys. Chem. Chem. Phys.*, 2014, **16**, 7066.

CHAPTER 9

Active Layer Limitations and Non-geminate Recombination in Polymer–Fullerene Bulk Heterojunction Solar Cells

TRACEY M. CLARKE,* GUANRAN ZHANG AND
ATTILA J. MOZER*

ARC Centre of Excellence for Electromaterials Science, Intelligent Polymer
Research Institute, University of Wollongong, North Wollongong,
NSW 2500, Australia
*Email: tclarke@uow.edu.au; attila@uow.edu.au

9.1 Introduction

Organic photovoltaics (OPV) has become an increasingly popular field of research in recent years. The race to reach power conversion efficiencies of 10% and beyond has led to substantial improvements in materials design and device architectures, coupled with an improved understanding of the operating mechanism of bulk-heterojunction OPV devices.

Many of the early improvements in polymer–fullerene photovoltaic devices were accomplished by optimizing the nanomorphology of the bicontinuous, interpenetrating network of donor and acceptor phases. For example, the landmark 2.5% for MDMO-PPV:PCBM was achieved in 2001[1] by changing the solvent to chlorobenzene. The serendipitous discovery of thermal annealing in P3HT:PCBM[2] in 2003 proved to be another milestone in the OPV

RSC Energy and Environment Series No. 11
Advanced Concepts in Photovoltaics
Edited by Arthur J Nozik, Gavin Conibeer and Matthew C Beard
© The Royal Society of Chemistry 2014
Published by the Royal Society of Chemistry, www.rsc.org

journey, initially boosting the efficiencies of such devices to 2.5%. Further work on the P3HT:PCBM system, which has rapidly become the 'standard' in OPV research[3] (despite its numerous idiosyncrasies, such as very high crystallinity relative to most polymers), has led to efficiencies of 5%.[4]

Around the same time interest started to grow in a new class of polymers: donor–acceptor or push–pull polymers, which possess a partial intra-molecular charge transfer character and as such have a lower band gap compared to P3HT and MDMO-PPV. The reason for this particular interest is that low band gap polymers are able to absorb the red–near-infrared component of the solar spectrum (where the sun's maximum photon flux is), thereby substantially increasing photon harvesting and the consequent short circuit current, J_{SC}. One polymer that attracted particular attention was PCPDTBT, which has an absorption maximum at 775 nm and achieved efficiencies of 3% in blends with $PC_{70}BM$.[5]

In 2007, Peet *et al.* reported a considerable increase in power conversion efficiency of PCPDTBT:$PC_{70}BM$ photovoltaic devices if the additive octane-dithiol was added to the solution during processing.[6] This enhancement was due to increases in J_{SC} and fill factor (FF). Further research by numerous groups indicated that this increase in efficiency was due to an improved phase separation and morphology,[7] higher charge photogeneration yields,[8] lower triplet yields,[9] less geminate recombination,[10-12] and higher charge carrier mobility.[13]

This interest in low band gaps highlights the importance of the energy levels of the polymer, but those of the acceptor – often a fullerene – are equally vital. A schematic illustration of the energy levels of a donor–acceptor heterojunction illustrates this point in Figure 9.1. On the most fundamental level, a 'LUMO–LUMO' offset (ΔE_{LUMO}) is required between the photo-excited donor (polymer) and acceptor (fullerene) in order for electron transfer and thus charge separation to take place. A 'HOMO–HOMO' offset (ΔE_{HOMO}) is required for a hole transfer from the photo-excited acceptor to

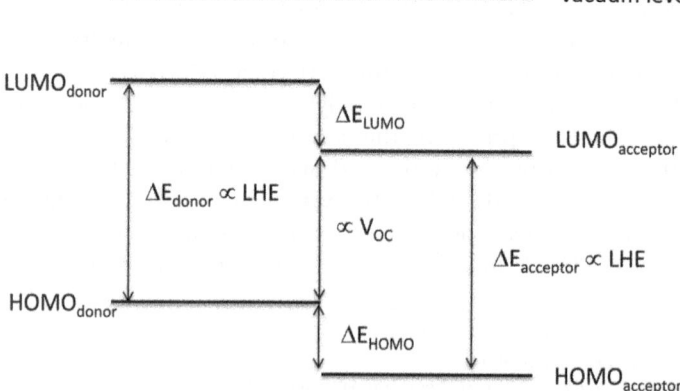

Figure 9.1   Schematic illustration of the energy levels in a donor–acceptor solar cell.

the donor. The required magnitude of this driving force has been much debated in the literature: if the energy offsets are too large then it is considered that much of the photon's energy is wasted. It has been suggested that a ΔE_{LUMO} of 0.3 eV is sufficient for efficient charge separation,[14] but it is noted that experimental data to support the validity of this 0.3 eV LUMO level offset requirement is very limited.[15] Another school of thought on this topic is that the magnitude of the energy offset – and thus the driving force for charge separation – is proportional to the yield of full separated charge carriers.[16] P3HT:PCBM, for example, has a very high LUMO offset and also produces a high free charge carrier yield. It was observed that the push–pull type polymers do not follow the trend of increasing charge photogeneration yield with increasing driving force for charge separation,[8] possibly because their partial intramolecular charge transfer character assists in the charge separation process.

The difference in quasi-Fermi levels at the electron and hole contacts determines the open circuit voltage, V_{OC}, which in turn is related to the energy levels of the donor and acceptor moieties. It has been shown that the V_{OC} is proportional to the energy difference between the donor's ionization potential (HOMO) and the acceptor's electron affinity (LUMO).[14] In a quest to enhance open-circuit voltages, many groups have utilized the strategy of a deeper polymer HOMO. However, this can provide an additional complication since a polymer ionization potential that is too high may raise the energy of the charge-separated state to the point where it is no longer thermodynamically accessible and alternative pathways to lower energy triplet states may be activated instead.[17–19] Furthermore, photo-induced hole transfer may be turned off if ΔE_{HOMO} is too low. The above trade-off between increasing light harvesting efficiency (LHE) without decreasing the V_{OC} is the reason that when designing low bandgap polymers, the LUMO is lowered rather than the HOMO raised.

In 2006, Scharber *et al.*[14] presented an empirical limit to the power conversion efficiency of a hypothetical donor–$PC_{60}BM$ solar cell by assuming (1) an external quantum efficiency (EQE) of 0.65 for photon energies equal to or larger than the donor bandgap ΔE_{donor}; (2) an electrical fill factor of 0.65; and (3) a V_{OC} proportional to the energy difference between the $PC_{60}BM$ LUMO level (assumed to be 4.3 eV) and the donor HOMO, adjusted by a 0.3 eV empirical factor. A maximum power conversion efficiency of 10% was predicted in the case of $\Delta E_{LUMO} = 0.3$ eV and a bandgap smaller than 1.74 eV. This empirical calculation provided important insights into energy level optimization and has led to a plethora of new co-polymers with performance at or above 5%.

Table 9.1 summarizes the energy levels and hole mobilities of approximately 30 polymers, and Table 9.2 summarizes their photovoltaic responses, as reported by October 2013. Table 9.3 shows selected molecular structures of several of these polymers. One of the highest efficiencies for a single junction polymer–fullerene bulk heterojunction solar cell was reported in 2012.[20] The 9.2% efficiency was achieved using an inverted geometry with a

Table 9.1 Energy level, band gap and hole carrier mobility for some semiconducting polymers.

Polymer	HOMO (CV, eV)	LUMO (CV, eV)	Band gap (optical, eV)	Hole mobility (cm^2 V^{-1} s^{-1})
P3HT	-5.2[21]	-3.2[21]	1.9[22]	2×10^{-4} (annealed)[23]
PSiF-DBT[24]	-5.39	-3.57	1.82	$\sim1\times10^{-3\,b}$
PCPDTBT[5]	-5.3	-3.57	1.46	$1\times10^{-3\,b}$
PBDTP-DTBT[25]	-5.35	-3.34	1.7	$8.89\times10^{-2\,c}$
Si-PCPDTBT[26,27]	-5.05	-3.27	1.45	$1\times10^{-3\,b}$
P1[28]	-5.36	-3.55	1.37	
PBDTTT-C[29]	-5.12	-3.55	1.61	2×10^{-4}
PBDTTT-CF[30]	-5.22	-3.45	1.77	$7\times10^{-4\,c}$
PTB7[31]	-5.15	-3.31	1.84	$5.8\times10^{-4\,c}$
PBDTT-DPP[32]	-5.3	-3.63	1.44	$3.1\times10^{-4\,c}$
PCDTBT[33,34]	-5.5	-3.6	1.9	$1\times10^{-3\,a}$
PBDTTBT[35]	-5.31	-3.44	1.75	
PBnDT-FTAZ[36]	-5.36	-3.05	2	$1.03\times10^{-3\,c}$
P2[37]	-5.4		1.73	
PBDTTPD[38]	-5.56	-3.75	1.8	
P3[39]	-5.66	-3.86	1.8	
DT-PDPP2T-TT[40]	-5.1	-3.68	1.35	0.8^a
PDPP3T[41]	-5.17	-3.61	1.3	0.04^c
PDPPTPT[42]	-5.35	-3.53	1.53	0.04 ± 0.01^a
PDTP-DFBT[43]	-5.26	-3.61	1.38	$3.2\times10^{-3\,c}$
PBDTT-SeDPP[44]	-5.25	-3.7	1.38	$6.9\times10^{-4\,c}$
PMDPP3T[45]			1.3	10^{-2} to $10^{-3\,a}$
P4[46]	-5.5	-3.7	1.75	
P(Se)[47]	-5.49	-3.82	1.67	0.017^b
PDTGTPD[48]	-5.6	-3.5	1.69	
PDTSTPD[49]	-5.57	-3.38	1.73	$1\times10^{-4\,a}$
PTDBD2[50]	-5.24		1.68	$1.69\times10^{-4\,c}$
PTAT-3[51]	-5.04	-3.28	1.76	$1.69\times10^{-4\,c}$
PTBF1[52]	-5.15	-3.31	1.68	$4.1\times10^{-4\,c}$
PSeB2[53]	-5.04	-3.26	1.78	$1.35\times10^{-3\,c}$
PBDT-DTNT[54]	-5.19	-3.26	1.58	$1.3\times10^{-5\,c}$
PSiF-DBT[24]	-5.39		1.82	$\sim1\times10^{-3\,a}$
PBDTP-DTBT[25]	-5.35	-3.34	1.7	$8.9\times10^{-2\,c}$
PTPD3T[55]	-5.55	-3.73	1.82	$5.87\times10^{-2\,b}$ $1.2\times10^{-3\,c}$
PBTI3T[55]	-5.58	-3.77	1.81	$2.74\times10^{-3\,b}$ $1.5\times10^{-3\,c}$
PDT-S-T[56]	-5.21	-1.08	1.59	
KP115[57] (PDTSiTTz)			1.83	$1\times10^{-2\,b}$
PDTSTTz[58]	-5.06	-2.81	1.81	$3.56\times10^{-3\,c}$

^{a}Hole mobility obtained *via* FET measurement using pristine polymer.
^{b}Hole mobility obtained *via* FET measurement using polymer:PCBM blend.
^{c}Hole mobility obtained *via* space-charge-limited-current (SCLC) method using a hole only device.

80 nm thick photoactive layer comprising of the low band gap benzo-dithiophene-based polymer PTB7 blended with PC70BM. An additional polyfluorene was incorporated as a surface modification for the ITO, lowering the work function of the ITO such that it formed an ohmic contact with the active layer to facilitate charge carrier collection.

Table 9.2   Structure and photovoltaic performances of some polymer solar cell devices.

Polymer	Device structure	Active layer thickness (nm)	J_{SC} (mA cm^{-2})	V_{OC} (V)	FF (%)	PCE (%)	Comments
PCPDTBT[6]	ITO/PEDOT:PSS/polymer:PC$_{71}$BM/Al	110	16.2	0.62	55	5.5	Alkane dithiol treatment was used. Without dithiol treatment, PCE of 2.8% was reported.
Si-PCPDTBT[26]	ITO/PEDOT:PSS/polymer:PC$_{71}$BM(1:1, w/w)/Ca/Al	80	12.7	0.68	55	4.7	
PCDTBT[34]	ITO/PEDOT:PSS/polymer:PC$_{71}$BM(1:4, w/w)/TiO$_x$/Al	80	10.6	0.88	64	6.0	A $\sim$10 nm thick TiO$_x$ layer was introduced as optical spacer, and an IQE of nearly 100% was reported for the device with a TiO$_x$ layer.
P[28]	ITO/PEDOT:PSS/polymer:PC$_{71}$BM(1:1, w/w)/Al		17.3	0.57	61	5.9	
PBDTTT-C[29]	ITO/PEDOT:PSS/polymer:PC$_{71}$BM(1:1.5, w/w)/Ca/Al	80	14.7	0.7	64	6.6	Average PCE of 6.47% was also reported with inverted structure (ITO/ZnO/polymer:PC$_{71}$BM/MoO$_3$/Ag)
PBDTTT-CF[30]	ITO/PEDOT:PSS/polymer:PC$_{71}$BM(1:1.5, w/w)/Ca/Al	N/A	15.2	0.76	67	7.4	
PTB7[20]	ITO/PFN/polymer:PC$_{71}$BM(1:1.5, w/w)/MoO$_3$/Al	80	17.2	0.754	72	9.2	Inverted structure with a 10 nm PFN polymer layer as ITO surface modifier
PBDTT-DPP[32]	ITO/PEDOT:PSS/polymer:PC$_{71}$BM(1:2, w/w)/Ca/Al	100	13.5	0.74	65	6.5	Both regular and inverted structure devices were fabricated and no obvious difference in PCE was observed
PBDTTBT[35]	ITO/PEDOT:PSS/polymer:PC$_{71}$BM(1:2, w/w)/Ca/Al	80	10.7	0.92	58	5.5	IQE reported was above 80%, with broad response from 350 nm to 700 nm wavelength
PBnDT-FTAZ[36]	ITO/PEDOT:PSS/polymer:PC$_{61}$BM(1:2, w/w)/Ca/Al	160 250 310 400 1000	11.5 11.8 12.2 13.3 14.0	0.74 0.79 0.79 0.74 0.74	70 73 67 58 54	6.0 6.8 6.5 5.8 5.6	PCE listed was the reported average, while the highest PCE given was 7.10%

Table 9.2 (*Continued*)

Polymer	Device structure	Active layer thickness (nm)	J_{SC} (mA cm^{-2})	V_{OC} (V)	FF (%)	PCE (%)	Comments
P2[37]	ITO/PEDOT-PSS/polymer:PC$_{61}$BM(1:1.5, w/w)/Ca/Al	N/A	11.5	0.85	68	6.6	The active layer was casted using a chlorobenzene/1,8-diiodooctane (DIO) mixing solvent
PBDTTTPD[59]	ITO/PEDOT-PSS/polymer:PC$_{61}$BM(1:1.5, w/w)/Ca/Al	100	11.2	0.94	69	7.3	
PDTSTPD[49]	ITO/PEDOT-PSS/polymer:PC$_{71}$BM(1:2, w/w)/BCP/Al	90	12.2	0.88	68	7.3	Optimized performance was achieved by using a CB/DIO solvent for active layer casting. A 5 nm thick BCP layer was used for hole blocking
		220	13.3	0.85	54	6.1	
PDTGTPD[48]	ITO/ZnO/polymer:PC$_{71}$BM/MoO$_3$/Ag	N/A	12.6	0.85	68	7.3	
PDPPTPT[42]	ITO/PEDOT:PSS/polyer:PC$_{71}$BM(1:2, w/w)/LiF/Al	80–90	10.8	0.8	65	5.5	
PDTP-DFBT[43]	ITO/ZnO/polymer:PC$_{71}$BM/MoO$_3$/Ag	~100	17.8	0.68	65	7.9	
PBDTT-SeDPP[44]	ITO/PEDOT:PSS/polymer:PC$_{71}$BM/Ca/Al	100	16.8	0.69	62	7	
PMDPP3T[45]	ITO/PEDOT:PSS/polymer:PC$_{71}$BM(1:3, w/w)/PFN/Al	160	17.8	0.6	66	7	
	PMDPP3T:PC$_{61}$BM(1:3, w/w)	84	9.8	0.62	70	4.3	
		108	14	0.61	66	5.7	
		135	14.8	0.61	65	5.8	
		177	15.7	0.59	59	5.6	
		230	16.9	0.58	56	5.5	
P3[39]	ITO/PEDOT:PSS/polymer:PC$_{71}$BM(1:1, w/w)/TiO$_x$/Al	105	10.51	0.92	63	6.1	3 vol% of 1-chloronaphthalene was added to optimize BHJ nanomorphology, resulting in significant increase of Jsc and FF

Polymer	Device structure	Thickness	J_{sc}	V_{oc}	FF	PCE	Comments
P(Se)[47]	ITO/PEDOT:PSS/polymer: PC$_{71}$BM(1:1, w/w)/Al	75 95 130 200	9.71 10.74 10.24 9.79	0.87 0.88 0.86 0.84	63 62 60 50	5.35 5.79 5.28 4.07	
PTDBD2[50]	ITO/PEDOT:PSS/polymer: PC$_{71}$BM(1:1.2, w/w)/Ca/Al	N/A	13	0.89	65.3	7.6	The active layer was casted using chlorobenzene/DIO mixing solvent. Changing the alkyl side chain on FTT comonomer, they also synthesized PTDBD3 which showed efficiency of 4.9% in polymer:PC$_{71}$BM blend
PTAT-3[51]	ITO/PEDOT:PSS/polymer: PC$_{61}$BM(1:1, w/w)/Ca/Al	100	15	0.66	58	5.62	The active layer was processed from chloroform/DIO mixture
PTBF1[52]	ITO/PEDOT:PSS/polymer: PC$_{71}$BM(1:1.5, w/w)/Ca/Al	70–100	14.1	0.74	68.9	7.2	The active layer was casted in DCB/DIO solvent. Using pure DCB solvent, a PCE of 6.2% was reported
PSeB2[53]	ITO/PEDOT:PSS/polymer: PC$_{71}$BM(1:1.2, w/w)/Ca/Al	100	16.8	0.64	64	6.46	The authors substituted S on PTB9 with Se on different backbone positions, all substituted polymers showed lower band gap than PTB8 and two polymers, PSeB2 and PSeB3, showed enhanced performance compared to PTB9
PBDT-DTNT[54]	ITO/PEDOT:PSS/polymer: PC$_{71}$BM(1:1, w/w)/Ca/Al	80–90	11.71	0.8	61	6	
PSiF-DBT[24]	ITO/PEDOT:PSS/polymer: PC$_{61}$BM(1:2, w/w)/Al	70	9.5	0.9	50.7	5.4	
PBDTP-DTBT[25]	ITO/PEDOT:PSS/polymer: PC$_{71}$BM(1:1.5, w/w)/Ca/Al	102	12.94	0.88	70.9	7.92	Optimized performance was achieved by adding 0.5% DIO, resulting in significant increase in FF (from 48.2% without DIO to 70.9) and PCE

Table 9.2 (Continued)

Polymer	Device structure	Active layer thickness (nm)	J_{SC} (mA cm^{-2})	V_{OC} (V)	FF (%)	PCE (%)	Comments
P4[46]	ITO/PEDOT:PSS/polymer:PC$_{71}$BM(1:2, w/w)/Al	100	13	0.68	55	4.9	The PCE listed was the average performance, while the best performing device had a PCE of 5.4%. The performance was achieved by adding 1-chloronaphthalene (CN, 2% by volume) in the polymer:PCBM chlorobenzene solution. Without CN, the devices have 1.6 ± 0.2% PCE
DT-PDPP2T-TT[40]	ITO/PEDOT:PSS/polymer:PC$_{71}$BM(1:3, w/w)/LiF/Al	84	11.6	0.68	74	5.9	Devices using PC$_{61}$BM as acceptor were also fabricated with different active layer thickness and it was found that, although at low thickness PC$_{71}$BM devices outperform PC$_{61}$BM ones, at 370 nm thickness the PC61BM device showed higher PCE of 5.8%
		93	12.3	0.68	74	6.1	
		137	12.3	0.67	73	6	
		154	13.1	0.66	71	6.2	
		167	12.7	0.66	72	6.1	
		209	14.8	0.66	69	6.7	
		220	14.8	0.66	70	6.9	
		250	15.5	0.67	62	6.4	
		300	15.5	0.67	61	6.3	
		370	15	0.66	53	5.3	
PDPP3T[60]	ITO/PEDOT:PSS/polymer:PC$_{71}$BM(1:2, w/w)/Ca/Al	120	15.41	0.66	66	6.71	For optimized performance, ternary blend of DCB/CF/DIO was used as processing solvent for active layer mixture
PTPD3T[55]	ITO/ZnO/polymer:PC$_{71}$BM(1:2, w/w)/MoO$_x$/Ag	65	10.3	0.795	78.6	6.44	Inverted structure. Chloroform was used as solvent with 2% DIO as processing additive
		90	11.5	0.795	78.5	7.18	
		130	12.3	0.792	78.7	7.72	
		180	12.3	0.786	76.1	7.61	
		250	12.1	0.768	73.9	6.87	
		300	12.1	0.745	71	6.4	

Polymer	Device structure						Comments
PBTI3T[55]	ITO/PEDOT:PSS/polymer: PC71BM(1:2, w/w)/LiF/Al		11.1	0.79	73	6.38	
	ITO/ZnO/polymer: PC71BM(1:2, w/w)/MoOx/Ag	70	10.1	0.86	75.9	6.59	Inverted structure. Chloroform was used as solvent with 2% DIO as processing additive
		100	10.9	0.86	76	7.14	
		120	12.6	0.87	76.6	8.35	
		150	12.1	0.86	74.7	7.73	
		200	11.5	0.84	64.4	6.24	
		270	11.7	0.82	57.6	5.55	
			11.3	0.85	73.2	7.02	
PDT-S-T[56]	ITO/PEDOT:PSS/polymer: PC71BM(1:2, w/w)/LiF/Al	110	16.63	0.73	64	7.79	o-DCB was used as solvent
PDTSTTz[58]	ITO/PEDOT:PSS/polymer: PC71BM(1:1.5, w/w)/Ca/Al ITO/PEDOT:PSS/polymer: PC71BM(1:1, w/w)/Ca/Al	80	11.9	0.77	61	5.59	The active layer was annealed at 100 °C for 15 min to reach reported efficiency. A lower PCE of 4.68% was reported for devices without annealing
KP115[57,61]	ITO/electron-injecting layer/polymer:PC61BM(1:2, w/w)/hole-injecting layer/Ag	50			66	2.4	
		75			69	2.6	
		100			65	4.2	
		140	11.8	0.64	59	4.4	
		150			68	3.8	
		180			68	4.0	
		380			67	4.1	
P3HT	ITO/PEDOT:PSS/polymer: PC61BM(1:0.8, w/w)/Al[62]	170	8.6	0.57	51	2.5	
		880	11	0.57	53	3.3	
	ITO/PEDOT:PSS/polymer: PC61BM(1:1, w/w)/LiF/Al[63]	150	5.36	0.60	65	2.1	
		320	5.47	0.58	51	1.6	
		800	4.31	0.55	38	0.9	
	ITO/ZnO/polymer: PC61BM(1:1, w/w)/Ag[64]	100	8.27	0.49	50	2.0	
		150	9.23	0.55	52	2.7	
		250	10.84	0.56	48	2.9	
		320	11.22	0.56	48	3.0	

Table 9.3 Structures for the selected polymers referred to in Tables 9.1 and 9.2

P3HT

MDMO-PPV

A = C PCPDTBT
A = Si Si-PCPDTBT

PCDTBT

PDTSiTTz (KP115)

PBnDT-FTAZ

PMDPP3T

P4

P(Se)

PTB7

P3

PBDTTBT

Table 9.3 (*Continued*)

DT-PDPP2T-TT

PBTI3T

PTPD3T

PDTP-DFBT

PC60BM

Another example of the highest performing photovoltaic systems was reported by Guo *et al.* in 2013.[55] Here two semicrystalline donor–acceptor polythiophenes were used in inverted bulk heterojunction devices to give devices of up to 8.7% with ~120 nm active layer thicknesses. The primary novelty of these two polymers, PTPD3T and PBTI3T, was that extraordinarily high fill factors exceeding 76% were observed. This was attributed to their nanomorphology, which showed evidence of high order, close packing, and optimal phase separation and vertical gradients.

These highest device efficiencies of approximately 9%[20,55] that have now been achieved for single-junction polymer:fullerene based solar cells are close to the predicted 10% empirical limit. Third generation concepts including tandem (double junction) solar cells offer a viable route towards 15%, using an optimized donor for each sub-cell. Highest efficiencies to date have reached 10.6% for a ITO/ZnO/P3HT:ICBA (220 nm)/PEDOT:PSS/ZnO/PDTP-DFBT:PCBM(100 nm)/MoO$_3$/Ag structure.[43] Advanced charge generation concepts using high dielectric materials[65] or field enhancement may also provide avenues to go beyond the above empirical limit of donor–acceptor solar cells by minimizing ΔE_{LUMO} and ΔE_{HOMO} offsets.

The focus of this chapter is how to increase the power conversion efficiency of single junction organic solar cells by improving charge transport and recombination. For example, increasing the EQE in the above empirical calculation to 0.9 while maintaining a FF of 0.7 should result in ~15% efficiency.[66] The significant challenge is how to increase the EQE for photon energies equal to or larger than the bandgap while maintaining a high FF, and therefore how to achieve high FFs using significantly larger active layer thicknesses (>300 nm).

9.2 Active Layer Limitations

The requirement for greater active layer thicknesses means that a large body of OPV research has been devoted to understanding the dependence of active layer thickness on device parameters.[67] For most systems there exists an optimal device thickness to maximize device efficiency. This optimal thickness is system-dependent but is typically less than 100 nm. This is often too thin in terms of maximal light absorption, as it has been reported that in such cases these solar cells absorb as little as half of the photons available in their absorption bandwidth.[12]

For example, the dependence of the number of photons absorbed by the active layer (and thus J_{SC}) on the active layer thickness reveals three maxima for P3HT:PCBM, due to constructive light interference effects (Figure 9.2).[12,57] The exact positions of these maxima depend on whether the device geometry is standard or inverted. A standard geometry device (where the transparent conductive oxide side is the more positive terminal such that hole extraction occurs), for example, has photon absorption maxima at approximately 80, 220 and 380 nm active layer thicknesses. Conversely, three photon absorption maxima occur at 40, 120 and 300 nm thicknesses for the inverted device with a 10 nm TiO$_x$ n-type window layer. The exact positions of these maxima will

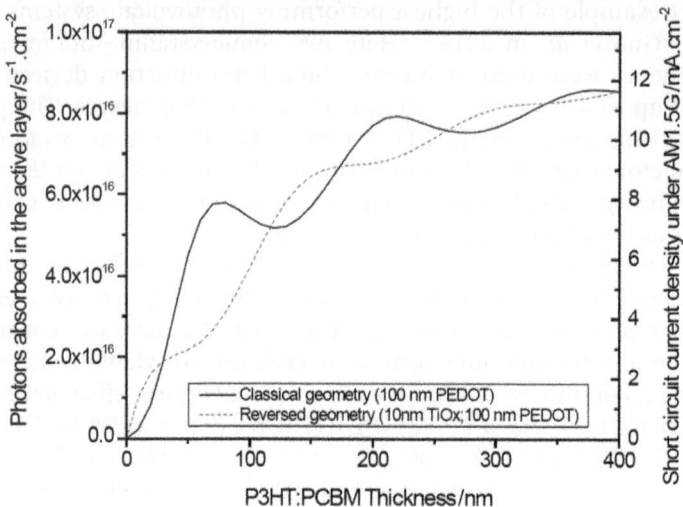

Figure 9.2 The number of photons absorbed in the active layer *versus* film thickness for both the traditional and inverted device geometries. Reprinted with permission from J. Peet, *et al.*, *Appl. Phys. Lett.*, 2011, **98**, 1. Copyright 2011, AIP Publishing LLC.

depend on the particular polymer–fullerene system being examined. For example, lowering the band gap of conjugated polymers not only leads to enhanced absorption but also increases the optimal active layer thickness due to these interference effects occurring at longer wavelengths.[68]

It would be advantageous to target the thicker active layer maxima, in terms of both enhancing light absorption and J_{SC} and also in terms of a manufacturing standpoint. This poses a challenge, especially for inverted devices, as most high performance bulk heterojunction devices cannot maintain a high fill factor with active layer thicknesses beyond 200 nm. This can be observed from an examination of Table 9.2. The third column in Table 9.2 shows the active layer thicknesses of bulk heterojunction solar cells comprised of some of the best performing polymer donors reported to date, while the sixth column details the fill factor (FF). The following general observations can be made:

- The reported active layer thickness of the majority of high performance devices is between 80 to 100 nm.
- With increasing active layer thickness, the FF drops rather significantly, reaching values between 0.5 and 0.58 at thicknesses close to or above 200 nm, *e.g.*, PDTSTPD, PMDPP3T, PBTI3T, and P(Se).
- For some donors, relatively large FF (>0.65) can be achieved using 200 nm, *e.g.*, PBnDT-FTAZ, DT-PDPP2T-TT, PTPD3T and PDTSiTTz.
- For two donors, up to 1 μm active layers are possible with respectable (>0.5) FF, *e.g.*, PBnDT-FTAZ and P3HT.

There are some important practical reasons behind the observed limitation of active layer thickness, including limitations imposed by the deposition technique itself, the difficulty to control morphology with increasing active layer thicknesses, and the solubility of the materials. Another aspect is that space charge – caused by doping and–or asymmetric mobilities – becomes increasingly important with greater active layer thicknesses because it creates field-free regions with low charge carrier collection efficiency.[12] It was noted by Kirchartz *et al.*[69] that this effect was small in thin cells close to the first interference maximum, but that the space charge has a more significant effect on device performance in thicker cells around the second interference maximum. Small *et al.*[70] also attributed their declining fill factor with increasing active layer thickness to space-charge accumulation. The possibility of morphological factors, such as vertical phase segregation[71] and its effect on contact selectivity[67] should also be considered.

The apparent thickness limitation of several high performance OPV blends in Table 9.2 above poses a challenge to the commercialization of low-cost organic photovoltaic cells, which assumes a rather simple industrial-scale fabrication method such as roll-to-roll printing. A trade-off exists between keeping the material cost down by using thinner active layers, while increasing light harvesting and thus increasing EQE at wavelengths where the extinction coefficient of the donor acceptor blend is small. In addition, thin active layers around 100 nm may lead to a much larger variation in power conversion efficiency due to films variations and the formation of pinhole-type defects. Furthermore, the coating speed may also suffer since roll-to-roll printing is more difficult to control with such thin layers. An active layer thickness of $\sim$ 200–300 nm was suggest to be the ideal for high throughput roll-to-roll methods,[72] subject to the optical properties of the particular photoactive blend employed.

A well-known example to illustrate this point is the polymer PCDTBT. One of the first to achieve over 7% efficiency, its internal quantum efficiencies can also approach 100%.[34] However, this performance is only possible with active layer thicknesses of significantly less than 100 nm. In the next section we will examine the reason behind this limitation by studying the charge transport and recombination properties of this blend.

9.3   Charge Transport and Recombination

For many polymer:PCBM systems, as the active layer thickness is increased, the overall device efficiency increases to reach a peak then declines. This efficiency decrease at greater device thicknesses is often a result of a declining fill factor, while the V_{OC} is less affected.[12,70,73–75] This can be observed for the well-known system PCDTBT:PCBM,[76] as shown in Figure 9.3. Here two devices are compared, one with an active layer thickness of 65 nm and the other 165 nm. The thicker device does indeed have a higher J_{SC} as a result of greater light absorption, but also has substantially lower fill factor (and unchanged V_{OC}).

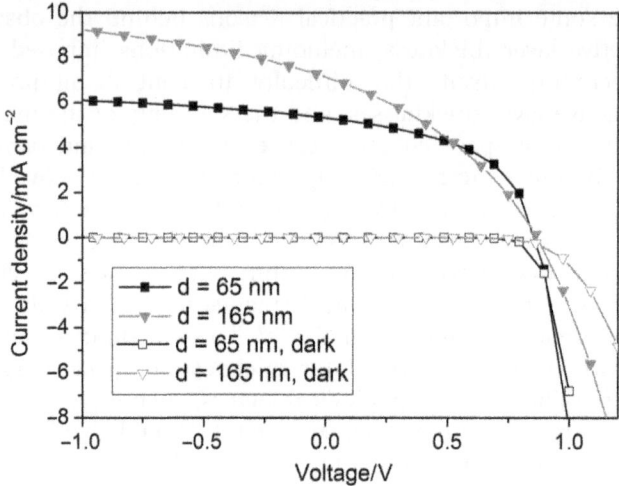

Figure 9.3   Current–voltage curves of thin-film PCDTBT:PCBM 1:2 inverted solar
cells of two different active layer thicknesses, 65 nm and 165 nm,
measured under 100 mW cm^{-2} simulated AM 1.5 illumination (filled
symbols) and in the dark (empty symbols).

Fill factors are inherently linked with bimolecular recombination, such
that a high fill factor is only possible if the photogenerated charges are efficiently extracted from the device with minimal recombination loss. The fill
factor is therefore limited by the charge carrier drift length, L_d,[77] the
distance over which the charge carriers can travel without significant
recombination occurring:

$$L_d = \mu \tau E \tag{9.1}$$

where μ is the charge carrier mobility, τ is the charge carrier lifetime and E is
the electric field. In order to minimize recombination losses, L_d must be
greater than the active layer thickness. A thin active layer, therefore, allows
efficient charge extraction and a higher fill factor. This limitation of the
active layer thickness caused by bimolecular recombination has been noted
by several groups.[73,78] This equation also indicates the key importance of the
mobility–lifetime, $\mu\tau$, product. In order to have a long drift length, the primary requirements are a high charge carrier mobility, a long charge carrier
lifetime, or both. However, the $\mu\tau$ product of organic solar cells is typically
low.[79–81]

From Equation (9.1), τ can be expressed as:

$$\tau = L_d^2 (\mu U_0)^{-1} \tag{9.2}$$

since the electric field is given by:

$$E = U_0 / L_d \tag{9.3}$$

The drift current density J_d can be calculated as:

$$J_d = ne\mu E = e\mu U_0 (\tau \beta L_d)^{-1} = \mu^2 U_0^2 e (L_d^3 \beta)^{-1} \qquad (9.4)$$

by using the expression for charge density n using the bimolecular lifetime τ and the bimolecular recombination coefficient β:

$$n = (\tau \beta)^{-1} \qquad (9.5)$$

The drift length can be calculated from Equation (9.4) as:

$$L_d^3 = \mu^2 U_0^2 e (J_d \beta)^{-1} \qquad (9.6)$$

Figure 9.4 shows the calculated drift lengths by assuming a drift current of $J_d = 18$ mA cm^{-2}, $U_0 = 0.8$ V and by varying the charge mobility μ between 1 and 10^{-6} cm^2 V^{-1} s^{-1} and the bimolecular recombination coefficient β between 10^{-14} and 10^{-9} cm^3 s^{-1}. The empty squares show calculations assuming diffusion-controlled, Langevin recombination, in which case β is given by:

$$\beta_L = e(\mu_n + \mu_p)(\varepsilon_r \varepsilon_0)^{-1} \qquad (9.7)$$

where e is the elementary charge, μ_n and μ_p are the mobilities of the electron and hole respectively, ε_r is the dielectric constant of the surrounding medium, and ε_0 is the permittivity of vacuum. This equation indicates that β_L and thus the resultant recombination rate are directly proportional to the charge carrier mobility. This classical Langevin model[82] assumes that recombination is controlled not by the electron transfer rate between the recombining electron and hole, but by the probability of the two opposite charges encountering one another. This diffusion-limited recombination is

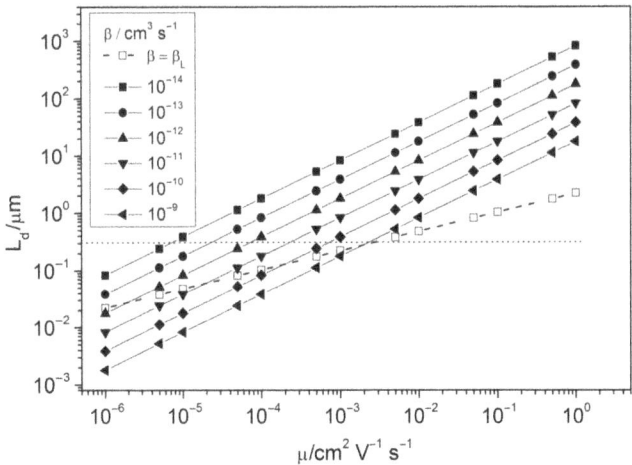

Figure 9.4 Simple calculation of charge carrier drift length as a function of charge carrier mobility for different bimolecular recombination coefficients, using Equation (9.6).

attributed to the rather slow hopping transport typical of disordered organic materials, the timescale of this process being much slower than the interfacial recombination reaction.

Figure 9.4 shows that the calculated drift length increases by either improving the mobility or by reducing the bimolecular recombination coefficient. Furthermore, reaching drift lengths beyond 300 nm in this simple calculation (dotted horizontal line) requires rather large charge mobilities ($>10^{-3}$ cm^2 V^{-1} s^{-1}) when Langevin-type recombination operates.

While simulating the current–voltage curves of bulk heterojunction solar cells involving charge carrier trapping, unbalanced mobilities, space charge effects and diffusion current is much more complex, the above simple model is useful to demonstrate the effect of Langevin-type diffusion-controlled bimolecular recombination, often observed in polymer:fullerene bulk heterojunction solar cells, on drift distances.

In 2005, it was discovered that the well-known polymer:fullerene blend annealed P3HT:PCBM has non-Langevin (reduced) bimolecular recombination,[82] providing a very long-lived charge-separated state on the order of milliseconds. This is unusual since diffusion-controlled recombination predicts a microsecond-scale charge carrier lifetime. Such a long lifetime hence cannot be explained by diffusion-limited (Langevin) recombination.

The primary technique used to probe the extent of this non-Langevin behavior is bulk-generation time-of-flight (TOF), which must be used in combination with other methodologies to establish the charge carrier mobility. Surface-generation TOF, a commonly used technique to determine charge carrier mobilities, must be performed on micrometer-thick active layers and thus the results may not be applicable to very thin layers, which may have a different nanomorphology. Under surface generation conditions, a sheet of photogenerated charge carriers travel through this thick active layer with a characteristic transit time of $t_{tr} = d^2/\mu U_0$ (where d is the film thickness and U_0 is the applied voltage). In a dispersive system, characterized by a continuous decrease in photocurrent versus time and the lack of a clear photocurrent plateau, the transit time can be observed as a 'kink' on a log–log plot. This is illustrated for PCDTBT:PCBM and pristine PCDTBT in Figure 9.5. Advantages of this method include identification of the sign of the charge carrier, elimination of RC (resistance-capacitance time constant) concerns, and straightforward electric field dependence analyses.

Photo-induced charge extraction by linearly increasing voltage (photo-CELIV), in contrast, offers the advantage of examining charge carrier mobility in the exact same device as would then be used for bulk-generation TOF. In photo-CELIV photogenerated charges are extracted after an adjustable delay time by a linearly increasing voltage pulse, $A = \Delta U/\Delta t$. The mobility is estimated from the time at which the maximum current response, t_{max}, occurs:

$$\mu = \frac{2d^2}{3At_{max}^2 \left[1 + 0.36\frac{\Delta j}{j(0)}\right]} \tag{9.8}$$

where d is the active layer thickness, A characterizes the voltage pulse as $\Delta U/\Delta t$, Δj is the photogenerated current response and $j(0) = AC$ (the capacitive current response), where $\Delta j \ll j(0)$ is required for accurate mobility values. A disadvantage of this technique is that a small RC constant is necessary for accurate determination of t_{max} and thus mobility. This can prove a particular limitation if thin film photovoltaic devices with high charge carrier mobility are employed. Representative photo-CELIV results are

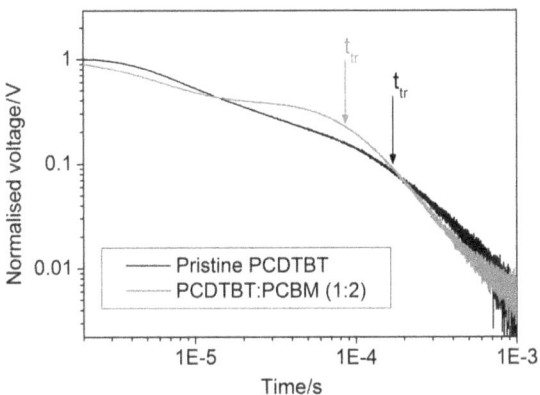

Figure 9.5 The surface generation TOF performed on pristine PCDTBT and PCDTBT:PCBM (1:2) devices with thick active layers of 5.3 μm and 4.0 μm respectively, both with a similar applied electric field of $E = 320 \ (V \ cm^{-1})^{\frac{1}{2}}$ and under low light conditions ($< 1 \ \mu J \ cm^{-2}$). The transit time is shown in each case by an arrow.
Reprinted from T. M. Clarke, *et al.*, *Organic Electronics*, 2012, **13**, 2639, with permission from Elsevier.

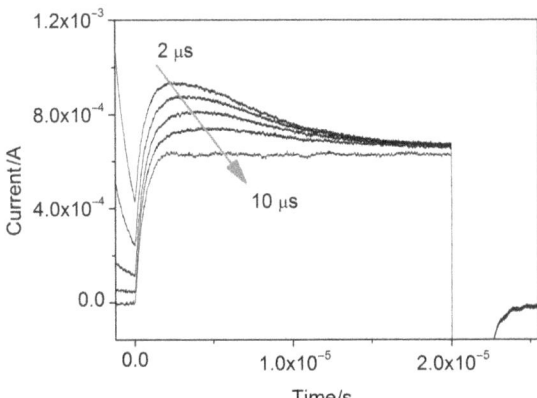

Figure 9.6 Photo-CELIV traces for a PCDTBT:PCBM device with a 165 nm active layer thickness, as a function of the delay time between laser and voltage pulses, using a 3 V, 20 μs voltage pulse offset by 0.70 V, with 532 nm, 1 μJ cm^{-2} excitation. The arrow shows the increase in t_{max} over time, finishing at the dark response.
Adapted from T. M. Clarke, *et al.*, *Organic Electronics*, 2012, **13**, 2639, with permission from Elsevier.

shown in Figure 9.6 for PCDTBT:PCBM. The t_{max} can be seen to move to longer times as the delay time between the laser and voltage pulses increases, denoting a decrease in charge carrier mobility with time.

Bulk photogeneration time-of-flight (TOF) employs the previously determined charge carrier mobility to determine the bimolecular recombination coefficient, β. This technique involves the photovoltaic device being illuminated using a short laser pulse whilst applying a constant DC bias, thereby extracting the photogenerated charges. At low light intensities, when the photogenerated charge carrier density is much smaller than the capacitive charge CU, the photocurrent transient is controlled by the transit time t_{tr} of the charge carriers under the influence of the electric field. As the laser intensity is increased, the photogenerated charge carrier density matches and then exceeds CU, which leads to screening of the external electric field and the appearance of a space charge limited photocurrent transient. When bimolecular – or even higher order[83] – recombination is present, the extracted charge saturates at high laser light intensities.

The charge carrier density N in the device at a delay time t after photoexcitation is given by the difference in charge photogeneration and charge recombination/charge extraction (the continuity equation) as:[82]

$$N(t) = \int_0^d \left(\frac{1}{\alpha L exp(-\alpha x)} + \beta t \right)^{-1} dx - \frac{1}{e} \int_0^t j(t)dt$$

(9.9)

$$= \frac{1}{\alpha \beta t} ln \frac{1 + \beta \alpha L t}{1 + \beta \alpha L t exp(-\alpha d)} - \frac{1}{e} \int_0^t j(t)dt$$

where L and j are incident photon densities and photocurrent densities respectively and α is the absorption coefficient. At high light intensities $(L \uparrow \infty)$ and bulk photogeneration mode $(\alpha d \ll 1)$, the denominator of the first term $ln((1 + \beta \alpha L t)/(1 + \beta \alpha L t exp(-\alpha d)))$ becomes αd, and the following simplified equation for the extracted charge is derived by substituting $j_e = j \times S$, where S is the electrode area:

$$Q_e = \int_0^\infty j_e dt = \frac{edS}{\beta t_e}$$

(9.10)

Since $\beta_L = e\mu/\varepsilon\varepsilon_0$ and $t_{tr} = d^2/\mu U$, then:

$$\frac{CU}{Q_e} = \frac{\varepsilon\varepsilon_0 SU}{d^2 eS} \beta t_e = \frac{\mu U}{d^2 \beta_L} \beta t_e = \frac{\beta t_e}{\beta_L t_{tr}}$$

(9.11)

The ratio of bimolecular recombination coefficient to the Langevin recombination coefficient is therefore obtained as:[82,84]

$$\frac{\beta}{\beta_L} = \frac{CU}{Q_e} \frac{t_{tr}}{t_e}$$

(9.12)

where C is the geometric capacitance, Q_e is the extracted charge and t_e is the extraction time (defined as the difference between the $t_{\frac{1}{2}}$ values at low and high excitation densities). Importantly, this equation is valid irrespective of the *RC* time constant of the TOF measurement conditions. According to the above equation, the magnitude of reduction in bimolecular recombination coefficient is proportional to increased extracted charge density and increased extraction lifetime.

If the parameter β/β_L [Equation (9.12)] approaches 1, then Langevin bimolecular recombination is present. In the case of Langevin recombination, the lifetime of the photogenerated charges is comparable to the extraction time t_e. All the additional photogenerated charges recombine during charge extraction due to their short lifetime. As such, irrespective of the excitation density and thus the initial charge carrier density, the charge extracted from the device (Q_e) does not exceed the capacitive charge (CU) in Langevin recombination: Q_e/CU approaches unity. However, Pivrikas *et al.* noted that it has been shown numerically that for some experimental conditions, such as low circuit resistance, Q_e can surpass CU even in case of Langevin recombination. Secondly, Langevin recombination is present if the $t_{\frac{1}{2}}$ values are independent of (or only weakly dependent on) the excitation density, indicating that the charge carriers are recombining in the device within the transit time, even at the highest excitation densities.

In non-Langevin systems, on the other hand, the lifetime of the photogenerated charges is typically longer than the transit time, causing a delayed charge extraction. As a result, a substantially higher quantity of charge than *CU* can be extracted due to the accumulation of charge within the device. Together, these two attributes lead to a clear gauge of the recombination behavior: the shape of the current decay over time. The charge carrier reservoir in non-Langevin systems causes a distinct photocurrent plateau region prior to the main current decay, which has a characteristic extraction time (and thus a strong dependence of $t_{\frac{1}{2}}$ on light intensity). Conversely, a decay approaching exponential behavior is observed for purely Langevin systems. Since only charge equal to *CU* can be extracted per transit time, the remainder of the charges recombine prior to extraction and no reservoir/plateau is observed.

Bulk generation TOF results are shown in Figure 9.7 for PCDTBT:PCBM. From the saturated photocurrent decay over time, it is apparent from the lack of a photocurrent plateau that this system is dominated by Langevin recombination. Further investigations utilizing the dependence of the extracted charge (Q_e) and $t_{\frac{1}{2}}$ on light intensity provide a β/β_L of 1 when a 165 nm active layer thickness is employed, thereby denoting fully Langevin behavior.

Other methods exist to determine the bimolecular recombination coefficient, β. One is to measure charge carrier decay dynamics and calculate the β from this. For example, the polymer–fullerene system PDTSiTTz:PCBM displays biphasic charge carrier decay kinetics, with both bimolecular and power law components. The kinetics of the fast phase can be fitted to a simple bimolecular equation:

$$n(t) = \frac{n_0}{1 + n_0 \beta t} \tag{9.13}$$

where $n(t)$ is the charge density at time t, n_0 is the initial charge density, and β is the bimolecular recombination coefficient. This is shown in Figure 9.8a,

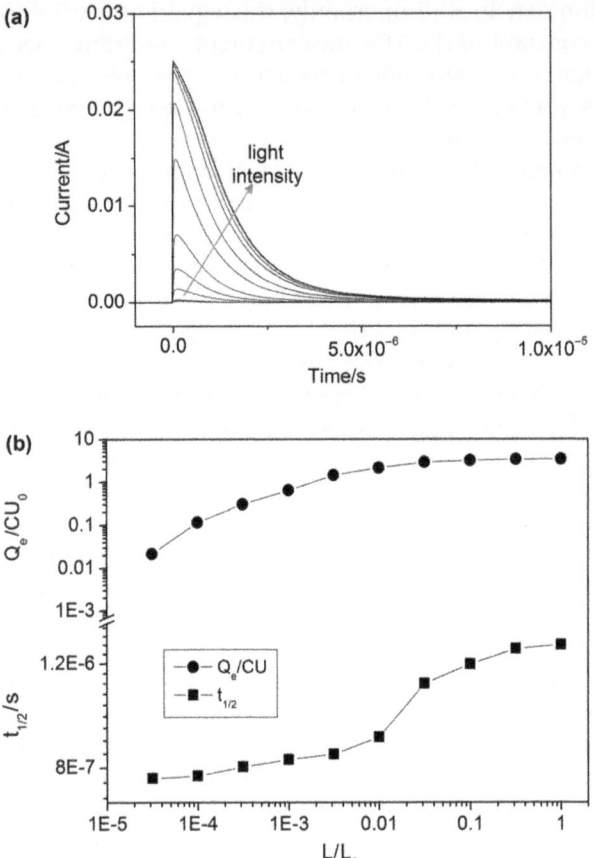

Figure 9.7   (a) TOF trace measured for a PCDTBT:PCBM device with an active layer
thickness of 165 nm as a function of light intensity (from 0.003 to
100 μJ cm^{-2}), using 2 V constant bias and 50 Ω. (b) The resultant
analysis of Q_e/CU and $t_\frac{1}{2}$ (the time at which the current falls to half of its
initial value), from which β/β_L is established.
Adapted from T. M. Clarke, *et al.*, *Organic Electronics*, 2012, **13**, 2639,
with permission from Elsevier.

where the bimolecular recombination coefficient β acquired by this fitting
technique is $\sim 2 \times 10^{-12}$ cm^3 s^{-1} at the highest excitation density, which is
very close to the β value obtained using TOF at 50 Ω. Alternatively, β can be
obtained by calculating the derivative dn/dt at each delay time, assuming
pure bimolecular recombination:

$$\beta(t) = -\frac{dn}{dt}\frac{1}{n^2} \qquad (9.14)$$

At early times and thus high charge densities, β is independent of time
(Figure 9.8b), with a value of $\sim 2 \times 10^{-12}$ cm^{-3} s^{-1}, consistent with the previous methods. At later times and thus lower charge densities, β shows a

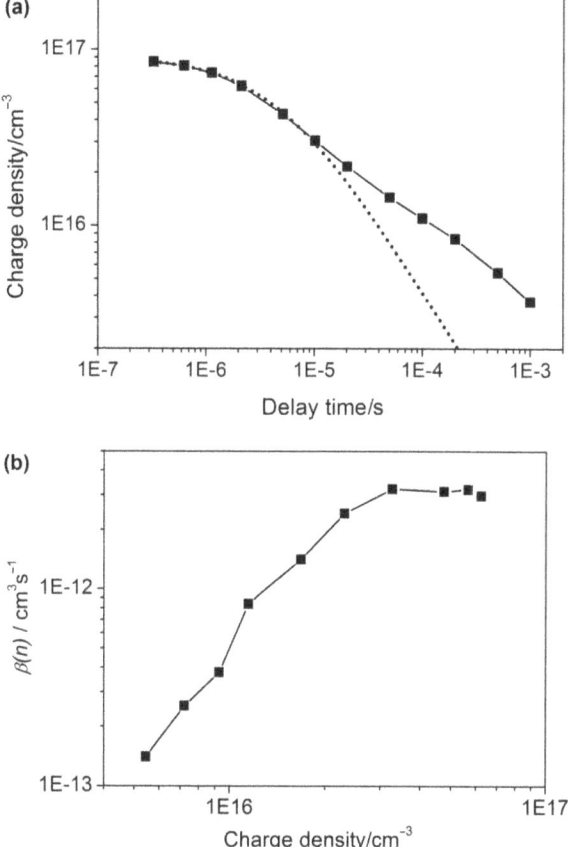

Figure 9.8 (a) The charge density decay dynamics of a PDTSiTTz:PCBM device, where the fast phase has been fitted to a bimolecular equation (dotted line). (b) The bimolecular recombination coefficient as a function of charge carrier density, calculated using Equation (9.14).

significant time dependence and decreases by an order of magnitude. Transient absorption spectroscopy has also been used to investigate β, with plots of β versus charge carrier density published for P3HT:PCBM.[85] These TAS results bear significant resemblance to those presented here for PDTSiTTz:PCBM using time-resolved charge extraction.

9.4 Non-Langevin Bimolecular Recombination

The report of non-Langevin bimolecular recombination in annealed P3HT:PCBM in 2005[82] utilized the TOF technique outlined in the previous section. The authors also numerically calculated photocurrent decays that clearly showed the effect that decreasing β/β_L has on the TOF curves, with an increasingly prominent photocurrent plateau. The measured TOF curves of

P3HT:PCBM showed quite similar behavior to that theoretically expected for non-Langevin recombination, with the observation of a distinct photo-current plateau. In both experimental and theoretical cases, the photo-currents saturated with increasing excitation density, thereby indicating bimolecular recombination. If monomolecular recombination is present instead, t_e does not saturate as a function of light intensity, but has a logarithmic intensity dependence of $t_e \propto \ln L$ instead, where L is the light intensity. The β/β_L parameter was determined to be 10^{-4}, indicating sub-stantially reduced bimolecular recombination. This was attributed to the nanomorphology of P3HT:PCBM, where the bicontinuous interpenetrating network creates separate pathways for electrons and holes, reducing the probability of them meeting in space.

Photo-CELIV has been used to not only provide charge carrier mobilities, but also provide insight into the extent of non-Langevin behavior. Non-Langevin behavior is manifested in photo-CELIV as an extremely large photo-response at early times relative to the capacitive response ($\Delta j \gg j_0$), which shifts to earlier t_{max} values as the delay time is increased. This be-havior indicates that under such high charge carrier density conditions an extraction time is being measured rather than a transit time. For example, Juška *et al.* presented a method for determination of β/β_L using photo-CELIV,[86] such that $\beta/\beta_L = j_{(0)}/\Delta j_{sat}$, where Δj_{sat} is the maximum extraction current density. It was also noted that the time necessary to reach maximum of the CELIV current transient varies with β/β_L.

Further work in 2010 by Hamilton *et al.*[87] used transient photovoltage and differential charging to determine charge carrier densities and recombination rates for P3HT:PCBM before and after thermal annealing, as summarized in Figure 9.9. It was discovered that charge carrier decay dynamics were over an order of magnitude faster in unannealed devices as compared to annealed ones. After annealing the recombination rate of P3HT:PCBM was reduced by approximately two orders of magnitude compared to that expected for Langevin recombination, which, while a smaller reduction, is consistent with what Pivrikas observed.[82] However, prior to thermal annealing the situation is quite different. In this case, the recombination is much closer to Langevin recombination: within one order of magnitude. Hence it is only the thermally annealed P3HT:PCBM which exhibits non-Langevin recombination. Once again this was attributed to the improvement in phase segregation upon annealing, where Monte Carlo simulations were used to show that re-combination is less frequent when the domains are larger.

Also in 2010, Koppe *et al.*[88] investigated the recombination behavior of the ternary blend P3HT:PCPDTBT:PCBM using TOF. Alone, PCPDTBT:PCBM exhibits clear Langevin bimolecular recombination, irrespective of whether it is thermally annealed or not. It was also demonstrated that, as observed previously, thermally annealing of P3HT:PCBM induces non-Langevin behavior. For the ternary blend, when utilising red laser light such that only the PCPDTBT is excited, the TOF behavior is very similar to that of P3HT:PCBM alone: Langevin recombination prior to annealing, which shifts

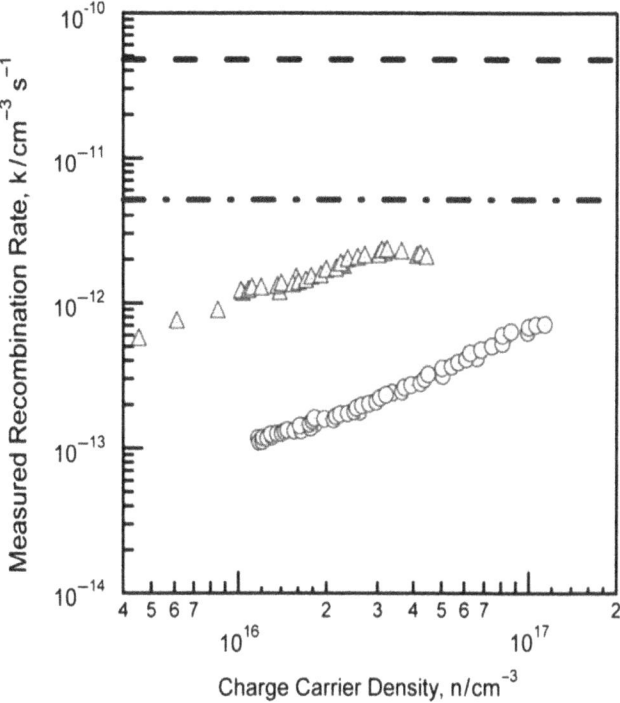

Figure 9.9 The measured recombination rate estimated from charge density and charge lifetime measurements for annealed (circles) and non-annealed (triangles) cells in comparison with the Langevin recombination rate for annealed (dashed line) and non-annealed (dash–dot line) hole mobility values.
Reprinted with permission from R. Hamilton, *et al.*, *J. Phys. Chem. Lett.*, 2010, **1**, 1432. Copyright 2010, American Chemical Society.

to non-Langevin after annealing. The authors were therefore able to conclude that hole transport in the ternary blends is primarily influenced by the P3HT phase, since the charge carriers photogenerated in the PCPDTBT phase are transferred to and subsequently transported through the P3HT.

Another example of the use of P3HT:PCBM's non-Langevin properties in investigating various aspects of photovoltaic behavior was reported by Wetzelaer *et al.* in 2012.[89] Charge transfer (CT) state electroluminescence (EL) was studied in polymer:fullerene bulk heterojunction solar cells. It was noted from the lack of dependence of applied bias on EL quantum efficiency that CT emission in most polymer:fullerene solar cells originates from free-carrier bimolecular recombination. For P3HT:PCBM, however, a voltage-dependent CT EL efficiency was measured, suggesting the presence of additional non-radiative trap-mediated recombination. The authors attributed this to the low, supressed competing bimolecular recombination rate, allowing the non-radiative trap-mediated recombination to dominate the EL characteristics.

The general theme in most scientific reports invoking non-Langevin recombination is that, until recently, they all utilized P3HT:PCBM. This is because P3HT:PCBM was, to the best of our knowledge, the only system clearly and unequivocally demonstrated to possess non-Langevin bimolecular recombination. In 2011, however, non-Langevin recombination behavior was reported in a new polymer, poly[(4,4'-bis(2-ethylhexyl)dithieno[3,2-b:2',3'-d]silole)-2,6-diyl-alt-(2,5-bis 3-tetradecylthiophen-2-yl thiazolo 5,4-d thiazole)-2,5-diyl] (PDTSiTTz), blended with PCBM.[78] It was first noted by Peet *et al.*[57] to maintain a high fill factor of >65% over the entire active layer thickness range to above 300 nm, similar behavior to P3HT:PCBM.

Subsequent charge transport studies on this PDTSiTTz:PCBM blend system were performed and compared to Si-PCPDTBT, which has the same donor unit but a different acceptor unit. Photo-CELIV showed that PDTSiTTz:PCBM has a much higher charge carrier mobility compared to Si-PCPDTBT:PCBM and, furthermore, showed the hallmarks of non-Langevin recombination outlined by Juška *et al.* These included a substantially greater Δj compared to $j_{(0)}$ and a t_{max} that shifts to shorter times under these high charge density conditions. These photo-CELIV results are shown in Figure 9.10. Bulk-generation TOF results clearly showed the distinct photo-current plateau characteristic of non-Langevin recombination for PDTSiTTz:PCBM, but not Si-PCPDTBT:PCBM (Figure 9.11a). Light intensity dependence analysis, as displayed in Figure 9.11b, revealed a much stronger dependence of $t_{\frac{1}{2}}$ on L for PDTSiTTz:PCBM compared to Si-PCPDTBT:PCBM in addition to a greater extracted charge density. This led to a β/β_L of 0.05 for PDTSiTTz:PCBM and a β/β_L much closer to one for Si-PCPDTBT:PCBM ($\beta/\beta_L = 0.5$). Using a simple illustrative equation, the charge carrier drift length for PDTSiTTz:PCBM was established to be substantially longer than in Si-PCPDTBT:PCBM, as a result of both its higher mobility and longer charge carrier lifetime.

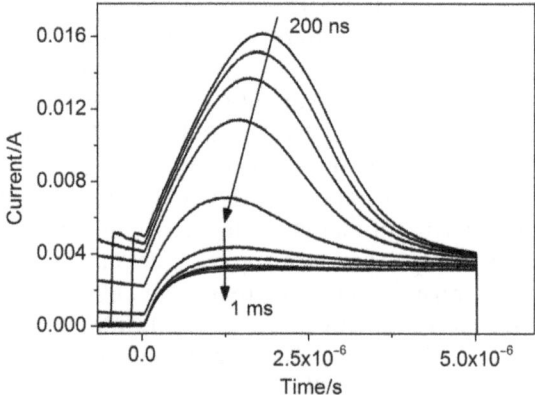

Figure 9.10   Photo-CELIV traces for a PDTSiTTz:PCBM device as a function of the delay time between laser and voltage pulses, using a 6V 5 μs voltage pulse, with 532 nm, 1 μJ cm^{-2} excitation.

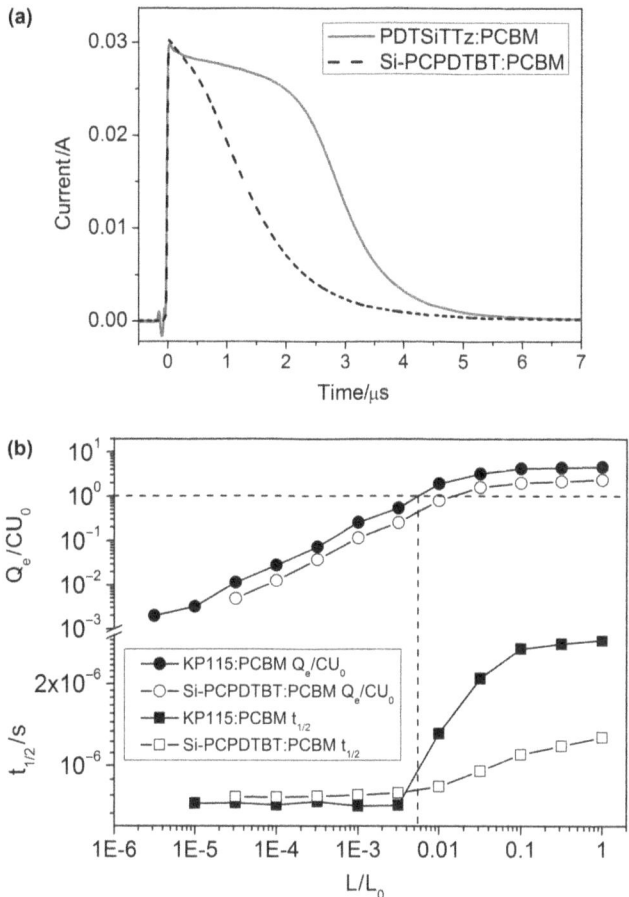

Figure 9.11 (a) Comparison of the saturated TOF traces measured for PDTSiTTz:PCBM and Si-PCPDTBT:PCBM devices as a function of light intensity, using 3 V constant bias and 50 Ω. (b) The resultant analysis of Q_e/CU and $t_{\frac{1}{2}}$ (the time at which the current falls to half of its initial value), from which β/β_L is established.
Reproduced, with permission, from T. M. Clarke, *et al.*, *Adv. Energy Mater.*, 2011, **1**, 1062. Copyright 2011, John Wiley and Sons.

Interestingly, energy-filtered transmission electron microscopy on both PDTSiTTz:PCBM and Si-PCPDTBT:PCBM blend systems revealed that only minor differences in the morphology were observed. The considerable morphological differences between as-cast and thermally annealed P3HT:PCBM films[77,90–92] are absent in these blends. The morphologies of the four blends are compared in Figure 9.12. The lack of clear correlation between the observed nanomorphology and the recombination kinetics of PDTSiTTz:PCBM and Si-PCPDTBT:PCBM demonstrates that non-Langevin recombination could have a more complex origin than previously thought and may not be correlated to morphology in all organic photovoltaic

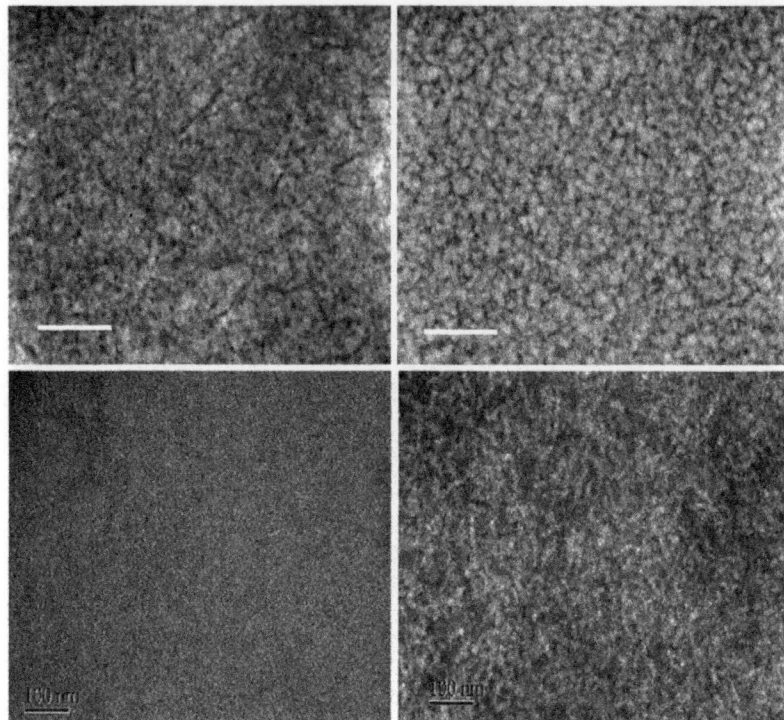

Figure 9.12 The TEM morphologies of Si-PCPDTBT:PCBM (top left) and PDTSiTTz:PCBM (top right), where the scale bar is 0.5 mm, compared to the morphologies of P3HT:PCBM[90] before (bottom left) and after (bottom right) thermal annealing.
P3HT:PCBM images reproduced from P. Vanlaeke *et al.*, *Solar Energy Materials and Solar Cells*, 2006, **90**, 2150, with permission from Elsevier. PDTSiTTz:PCBM and Si-PCPDTBT:PCBM images with permission from T. M. Clarke, *et al.*, *Adv. Energy Mater.*, 2011, **1**, 1062. Copyright 2011, John Wiley and Sons.

blend systems. As such, even in a morphology apparently conducive to recombination, not every encounter between two charge carriers of opposite sign may result in a recombination event (this could be due to either an energetic or mechanical barrier). Furthermore, it is possible that an interfacial structural feature beyond the resolution of current instrumentation exists.

9.5   Mechanism of Reduced Recombination

Several models to explain non-Langevin recombination behavior have been proposed. In 1996, a model by Adriaenssens and Arkhipov[93] introduces the concept of random spatial fluctuations in the potential landscape to explain non-Langevin recombination. In disordered semiconductors, strong local electric fields at the interface coupled to potential energy fluctuations lead to spatially separated percolation networks for electrons and holes. If the amplitude of the potential fluctuations is sufficiently high, this will contribute to the supressed recombination.

Koster *et al.*[94] proposed in 2006 that the supposedly reduced bimolecular recombination in polymer:fullerene blends was due to the fact that the rate of recombination does *not* depend on the sum of the mobilities of both carriers, as per the Langevin expression. Instead, the recombination rate is limited by the slowest carrier mobility, as evidenced by the voltage dependence of the photocurrent and is a consequence of the confinement of each carrier type to its respective phase. In the case of strongly mismatched mobilities, therefore, the Langevin expression gives an erroneously high β value.

In 2008, Groves and Greenham[95] used theoretical methods to tackle the issue of explaining the origin of non-Langevin recombination. They used Monte Carlo simulations, considering the effects of constraining the charge carriers to their respective material phases and also the effect of energetic disorder and deep trapping. With particular regard to the last point, they noted that the Langevin equation assumes a uniform distribution of charge carriers, which is typically not the case in materials with a high degree of energetic disorder, as has often been observed in organic photovoltaic materials. The authors calculated the rate constant for Langevin-type bimolecular recombination and found that donor–acceptor blends (rather than an isotropic, homogeneous medium, as assumed by Langevin) do not necessarily induce significant suppression of the recombination rate. Altering domain size or charge carrier mobility mismatch, for example, reduced the rate constant by less than an order of magnitude. Instead, the authors concluded that the reduced recombination observed in strongly non-Langevin systems such as annealed P3HT:PCBM devices is due to deep carrier trapping.

In 2009 Juška *et al.*[96] proposed a morphology-based model for the annealed P3HT:PCBM blend, with its well-defined phase separation and lamellar structure. They suggested that these lamellae formation contribute to the reduced bimolecular recombination by limiting recombination events to only two dimensions, reducing the probability of the charge carriers to meet and thus recombine.

Deibel *et al.*,[97] on the other hand, suggested in 2009 that steep charge carrier concentration gradients through the active layer, which form due to the interaction of the active layer with the electrodes, are the cause of the non-Langevin recombination. The electron concentration in a photovoltaic device under standard operational conditions is higher at the cathode than at the anode and *vice versa* for holes. A spatially dependent bimolecular recombination rate, which would be proportional to the local electron and hole concentrations, is significantly lower compared to the rate based on the extracted (averaged) charge carrier density. The temperature dependence of the reduction in bimolecular recombination coefficient is well-described by this model, unlike Koster's[94] or Arkhipov's[93] models. In addition to the above contribution to non-Langevin recombination, Deibel *et al.* propose an additional constant, static contribution that could have any number of origins. They suggest energetic disorder, charge carrier mobility mismatch, spatial confinement of charges to their respective domains and/or differing dielectric constants.

Szmytkowski has published a number of papers on this topic. The first, in 2009,[98] invokes unequal charge carrier concentrations in the bulk to explain non-Langevin bimolecular recombination. The foundation of this model is that the Langevin equation depends only on a single, spatially averaged dielectric constant. However, the polymer:fullerene blend typically consists of materials that possesses different dielectric constants. P3HT, for example has $\varepsilon \sim 3$ while the ε of PCBM is closer to 4. Theoretical analyses based on electrostatic polarization suggest that interfacial recombination of charge carriers of one sign located in the lower dielectric constant material competes with bimolecular recombination of charge carriers in the bulk. Charge carrier densities in the bulk are therefore not the equal, leading to the reduction of the Langevin-type bimolecular recombination.

Hilczer and Tachiya presented a unified theory to explain both geminate and bimolecular recombination in 2010.[99] They extended Langevin theory in the way Onsager theory has previously been built upon by taking into account that recombination between an electron and a hole occurs at a non-zero separation with a finite intrinsic recombination rate. This theoretical model includes effects from the intrinsic recombination rate, the diffusion coefficients of each charge carrier, the Onsager radius, and the external electric field. Their analysis suggested that, in the absence of this external electric field, bimolecular recombination is reduced by decreasing an 'intrinsic reactivity parameter' in conjunction with a smaller Onsager radius and an optimized diffusion coefficient. Their theory was successful in modeling Deibel *et al.*[100] and Juška's[83] data on the negative dependence of β/β_L on temperature.

In 2011,[101] Szmytkowski expanded on his 2009 model by exploring the electric field and temperature dependences of the bimolecular recombination coefficients. In particular, the experimental ratio of the interfacial recombination constant to the charge carrier mobility as a function of the electric field was able to be fitted by the proposed theoretical model, the first explanation of the electric field dependence of β.

An additional paper by Szmytkowski in 2012[102] proposed a new theoretical model whereby exciton-polaron annihilation at the donor–acceptor interface competes with bimolecular recombination, thus contributing to the suppression of Langevin recombination. An analytical formula was used to demonstrate an increase of the recombination order when exciton annihilation on coulombically bound electron–hole pairs at the donor–acceptor interface is taken into account.

A study by Hirade *et al.*[103] in 2013 on small molecule bulk heterojunction OPV devices focused on the effect of the dielectric constant. Donor materials comprised of different electron-rich donor and electron-poor acceptor units were blended with PCBM and used to assess the dependence of the strength of the intramolecular donor–acceptor character on the bimolecular recombination constant. It was found that as the dielectric constant increased (a larger intramolecular donor–acceptor interaction), the bimolecular recombination coefficient, as established using TOF, decreased. The end result was an increase in fill factor and efficiency of the devices. It is possible that these results are extendable to polymer materials as well.

Murthy *et al.*[104] used photo-CELIV and time-resolved microwave conductance, which measures the conductance of photogenerated free charge carriers on a nanosecond timescale, to investigate reduced bimolecular recombination in a quinoxaline–thiophene-based co-polymer. The proposed theory to explain non-Langevin behavior in general revolved around the properties of the 'encounter complex' between the two recombining charges: a greater charge delocalization reduces the binding energy between the charges, enhancing the dissociation rate of the encounter complex. It was noted that this encounter complex and the charge transfer state have very similar energetic properties. The authors proposed that increased π–π stacking and order enhance this charge delocalization, and thus more crystalline systems are more likely to be non-Langevin.

Another recent report in 2013, by Ripolles-Sanchis *et al.*[105] applied Marcus charge transfer theory to a variety of polymer systems that vary in recombination rates. They calculated a lower electronic coupling parameter, V_{if}, for P3HT:PCBM than for the other systems, thereby slowing the recombination rate.

The above models can be broadly categorized into three groups:

- *Models based on nanomorphology.* Such morphology-based arguments are the most commonly cited and specifically consider the highly crystalline, fiber-like morphology of annealed P3HT:PCBM.
- *Models focused on spatial inhomogeneity.* Several models, including Arkhipov's,[93] Deibel's[97] and Szmytkowski's[98] 2009 models focus on the spatial inhomogeneity of the parameters governing the bimolecular equation; for example the charge mobility (μ), charge density (n), and dielectric constant (ε).
- *The Langevin model.* Some models fundamentally question the applicability of a simple Langevin model for donor–acceptor heterojunctions and aim to provide more advanced insights into charge recombination, *e.g.*, the minority carrier model of Koster's[94] and Hilczer and Tachiya's.[99]

Models suggesting that the interfacial charge transfer rate, rather than the probability of diffusional encounters controlled by the mobility of recombining charges, dominates the recombination rate have started to emerge. Formation of an interfacial charge transfer 'encounter complex'[104] that can subsequently separate, Marcus charge transfer theory[105] and, to some extent, Hirade's[103] model suggest that the nature of the molecular interface between the donor and the acceptor is important in influencing the interfacial recombination rate.

Jamieson *et al.*[106] has proposed that the increased electron affinity of crystalline PCBM compared to that of PCBM intermixed within the polymer/ PCBM phase provides a driving force for charge separation and prevents the recombination of electrons and holes due to an energy offset. Bartelt *et al.*[73] suggested that holes can also migrate from the well-intermixed polymer/ PCBM regions to polymer aggregates with lower ionization potential, which should further contribute to an energetic barrier to bimolecular recombination at the interface.

To summarize, developing a unified, encompassing model for reduced bimolecular recombination in donor–acceptor bulk heterojunctions is very challenging. This is due to the rather limited number of systems that clearly demonstrate reduced recombination, in addition to the differences in experimental techniques and sample preparation used, thus leading to a rather large variation in the reduction factor β/β_L even for nominally similar samples.

9.6   Summary and Outlook

This chapter has focused on the apparent thickness limitations of most high performance donor–acceptor bulk heterojunction solar cells, which pose significant challenges to achieving 15% conversion efficiency. Efficiencies approaching this value require high external quantum efficiencies and high fill factors using thick photoactive layers. An extensive literature review showed that only a handful of high performance donor–acceptor bulk heterojunction solar cells maintain high fill factor at film thicknesses larger than 200 nm, with PBnDT-FTAZ, DT-PDPP2T-TT, PtPD3T and PDTSiTTz being notable exceptions. Apart from difficulties in controlling active layer morphologies with increasing layer thickness and space charge layer formation in (often inadvertently) doped bulk heterojunctions, Langevin-type bimolecular recombination is shown to limit charge extraction efficiency in thick devices, thereby leading to low fill factors.

Using time-resolved charge extraction techniques such as photo-CELIV, TOF and time-resolved charge extraction, reduced (non-Langevin) bimolecular recombination is demonstrated for PDTSiTTz:PCBM heterojunctions. A brief review of 12 models offering possible explanations for such reduced recombination is presented. This highlights the fact that developing a one-size-fits-all, unified model of reduced recombination is challenging due to the limited number of reduced recombination donor–acceptor systems and the differences in experimental techniques and sample preparation between various reports. Resolving this rather intriguing, fundamentally important issue by establishing design rules for advanced donor–acceptor bulk heterojunction solar cells will be an important milestone affecting not only the performance of laboratory-based devices, but also the commercialization of this low-cost photovoltaic technology.

References

1. S. E. Shaheen, C. J. Brabec, N. S. Sariciftci, F. Padinger, T. Fromherz and J. C. Hummelen, *Appl. Phys. Lett.*, 2001, **78**, 841.
2. F. Padinger, R. S. Rittberger and N. S. Sariciftci, *Adv. Funct. Mater.*, 2003, **13**, 85.
3. M. T. Dang, L. Hirsch and G. Wantz, *Adv. Mater.*, 2011, **23**, 3597.
4. M. D. Irwin, D. B. Buchholz, A. W. Hains, R. P. H. Chang and T. J. Marks, *Proc. Natl. Acad. Sci. U. S. A.*, 2008, **105**, 2783.
5. D. Muhlbacher, M. Scharber, M. Morana, Z. G. Zhu, D. Waller, R. Gaudiana and C. Brabec, *Adv. Mater.*, 2006, **18**, 2884.
6. J. Peet, J. Y. Kim, N. E. Coates, W. L. Ma, D. Moses, A. J. Heeger and G. C. Bazan, *Nat. Mater.*, 2007, **6**, 497.
7. J. T. Rogers, K. Schmidt, M. F. Toney, E. J. Kramer and G. C. Bazan, *Adv. Mater.*, 2011, **23**, 2284.
8. T. Clarke, A. Ballantyne, F. Jamieson, C. Brabec, J. Nelson and J. Durrant, *Chem. Commun.*, 2009, **1**, 89.
9. D. Di Nuzzo, A. Aguirre, M. Shahid, V. S. Gevaerts, S. C. J. Meskers and R. A. J. Janssen, *Adv. Mater.*, 2010, **22**, 4321.
10. G. Grancini, N. Martino, M. R. Antognazza, M. Celebrano, H.-J. Egelhaaf and G. Lanzani, *J. Phys. Chem. C*, 2012, **116**, 9838.
11. S. Yamamoto, H. Ohkita, H. Benten and S. Ito, *J. Phys. Chem. C*, 2012, **116**, 14804.
12. D. J. D. Moet, M. Lenes, M. Morana, H. Azimi, C. J. Brabec and P. W. M. Blom, *Appl. Phys. Lett.*, 2010, **96**, 213506.
13. I.-W. Hwang, S. Cho, J. Y. Kim, K. Lee, N. E. Coates, D. Moses and A. J. Heeger, *J. Appl. Phys.*, 2008, **104**, 033706.
14. M. C. Scharber, D. Mühlbacher, M. Koppe, P. Denk, C. Waldauf, A. J. Heeger and C. J. Brabec, *Adv. Mater.*, 2006, **18**, 789.
15. T. M. Clarke and J. R. Durrant, *Chem. Rev.*, 2010, **110**, 6736.
16. H. Ohkita, S. Cook, Y. Astuti, W. Duffy, S. Tierney, W. Zhang, M. Heeney, I. McCulloch, J. Nelson, D. D. C. Bradley and J. R. Durrant, *J. Am. Chem. Soc.*, 2008, **130**, 3030.
17. J. J. Benson-Smith, L. Goris, K. Vandewal, K. Haenen, J. V. Manca, D. Vanderzande, D. D. C. Bradley and J. Nelson, *Adv. Funct. Mater.*, 2007, **17**, 451.
18. J. J. Benson-Smith, H. Ohkita, S. Cook, J. R. Durrant, D. D. C. Bradley and J. Nelson, *Dalton Trans.*, 2009, 10000.
19. S. Cook, H. Ohkita, J. R. Durrant, Y. Kim, J. J. Benson-Smith, J. Nelson and D. D. C. Bradley, *Appl. Phys. Lett.*, 2006, **89**, 101128.
20. Z. He, C. Zhong, S. Su, M. Xu, H. Wu and Y. Cao, *Nat. Photonics*, 2012, **6**, 591.
21. B. C. Thompson and J. M. J. Frechet, *Angew. Chem. Int. Ed.*, 2008, **47**, 58.
22. V. Shrotriya, J. Ouyang, R. J. Tseng, G. Li and Y. Yang, *Chem. Phys. Lett.*, 2005, **411**, 138.

23. V. D. Mihailetchi, L. J. A. Koster, P. W. M. Blom, C. Melzer, B. de Boer, J. K. J. van Duren and R. A. J. Janssen, *Adv. Funct. Mater.*, 2005, **15**, 795.

24. E. G. Wang, L. Wang, L. F. Lan, C. Luo, W. L. Zhuang, J. B. Peng and Y. Cao, *Appl. Phys. Lett.*, 2008, **92**, 033307.

25. M. Zhang, Y. Gu, X. Guo, F. Liu, S. Zhang, L. Huo, T. P. Russell and J. Hou, *Adv. Mater.*, 2013, **25**, 5.

26. J. H. Hou, H. Y. Chen, S. Q. Zhang, G. Li and Y. Yang, *J. Am. Chem. Soc.*, 2008, **130**, 16144.

27. M. C. Scharber, M. Koppe, J. Gao, F. Cordella, M. A. Loi, P. Denk, M. Morana, H. J. Egelhaaf, K. Forberich, G. Dennler, R. Gaudiana, D. Waller, Z. G. Zhu, X. B. Shi and C. J. Brabec, *Adv. Mater.*, 2010, **22**, 367.

28. R. C. Coffin, J. Peet, J. Rogers and G. C. Bazan, *Nat. Chem.*, 2009, **1**, 657.

29. J. H. Hou, H. Y. Chen, S. Q. Zhang, R. I. Chen, Y. Yang, Y. Wu and G. Li, *J. Am. Chem. Soc.*, 2009, **131**, 15586.

30. H. Y. Chen, J. H. Hou, S. Q. Zhang, Y. Y. Liang, G. W. Yang, Y. Yang, L. P. Yu, Y. Wu and G. Li, *Nat. Photonics*, 2009, **3**, 649.

31. Y. Liang, Z. Xu, J. Xia, S.-T. Tsai, Y. Wu, G. Li, C. Ray and L. Yu, *Adv. Mater.*, 2010, **22**, E135.

32. L. T. Dou, J. B. You, J. Yang, C. C. Chen, Y. J. He, S. Murase, T. Moriarty, K. Emery, G. Li and Y. Yang, *Nat. Photonics*, 2012, **6**, 180.

33. N. Blouin, A. Michaud, D. Gendron, S. Wakim, E. Blair, R. Neagu-Plesu, M. Belletete, G. Durocher, Y. Tao and M. Leclerc, *J. Am. Chem. Soc.*, 2008, **130**, 732.

34. S. H. Park, A. Roy, S. Beaupre, S. Cho, N. Coates, J. S. Moon, D. Moses, M. Leclerc, K. Lee and A. J. Heeger, *Nat. Photonics*, 2009, **3**, 297.

35. L. J. Huo, J. H. Hou, S. Q. Zhang, H. Y. Chen and Y. Yang, *Angew. Chem. Int. Ed.*, 2010, **49**, 1500.

36. S. C. Price, A. C. Stuart, L. Yang, H. Zhou and W. You, *J. Am. Chem. Soc.*, 2011, **133**, 4625.

37. C. Piliego, T. W. Holcombe, J. D. Douglas, C. H. Woo, P. M. Beaujuge and J. M. J. Frechet, *J. Am. Chem. Soc.*, 2010, **132**, 7595.

38. Y. P. Zou, A. Najari, P. Berrouard, S. Beaupre, B. R. Aich, Y. Tao and M. Leclerc, *J. Am. Chem. Soc.*, 2010, **132**, 5330.

39. J. Jo, A. Pron, P. Berrouard, W. L. Leong, J. D. Yuen, J. S. Moon, M. Leclerc and A. J. Heeger, *Adv. Energy Mater.*, 2012, **2**, 1397.

40. W. W. Li, K. H. Hendriks, W. S. C. Roelofs, Y. Kim, M. M. Wienk and R. A. J. Janssen, *Adv. Mater.*, 2013, **25**, 3182.

41. J. C. Bijleveld, A. P. Zoombelt, S. G. J. Mathijssen, M. M. Wienk, M. Turbiez, D. M. de Leeuw and R. A. J. Janssen, *J. Am. Chem. Soc.*, 2009, **131**, 16616.

42. J. C. Bijleveld, V. S. Gevaerts, D. Di Nuzzo, M. Turbiez, S. G. J. Mathijssen, D. M. de Leeuw, M. M. Wienk and R. A. J. Janssen, *Adv. Mater.*, 2010, **22**, E242.

43. J. B. You, L. T. Dou, K. Yoshimura, T. Kato, K. Ohya, T. Moriarty, K. Emery, C. C. Chen, J. Gao, G. Li and Y. Yang, *Nat. Commun.*, 2013, **4**, 1446.

44. L. T. Dou, W. H. Chang, J. Gao, C. C. Chen, J. B. You and Y. Yang, *Adv. Mater.*, 2013, **25**, 825.

45. W. W. Li, A. Furlan, K. H. Hendriks, M. M. Wienk and R. A. J. Janssen, *J. Am. Chem. Soc.*, 2013, **135**, 5529.

46. C. V. Hoven, X. D. Dang, R. C. Coffin, J. Peet, T. Q. Nguyen and G. C. Bazan, *Adv. Mater.*, 2010, **22**, E63.

47. D. H. Wang, A. Pron, M. Leclerc and A. J. Heeger, *Adv. Funct. Mater.*, 2013, **23**, 1297.

48. C. M. Amb, S. Chen, K. R. Graham, J. Subbiah, C. E. Small, F. So and J. R. Reynolds, *J. Am. Chem. Soc.*, 2011, **133**, 10062.

49. T. Y. Chu, J. P. Lu, S. Beaupre, Y. G. Zhang, J. R. Pouliot, S. Wakim, J. Y. Zhou, M. Leclerc, Z. Li, J. F. Ding and Y. Tao, *J. Am. Chem. Soc.*, 2011, **133**, 4250.

50. H. J. Son, L. Y. Lu, W. Chen, T. Xu, T. Y. Zheng, B. Carsten, J. Strzalka, S. B. Darling, L. X. Chen and L. P. Yu, *Adv. Mater.*, 2013, **25**, 838.

51. F. He, W. Wang, W. Chen, T. Xu, S. B. Darling, J. Strzalka, Y. Liu and L. P. Yu, *J. Am. Chem. Soc.*, 2011, **133**, 3284.

52. H. J. Son, W. Wang, T. Xu, Y. Y. Liang, Y. E. Wu, G. Li and L. P. Yu, *J. Am. Chem. Soc.*, 2011, **133**, 1885.

53. H. A. Saadeh, L. Lu, F. He, J. E. Bullock, W. Wang, B. Carsten and L. Yu, *ACS Macro Lett.*, 2012, **1**, 361.

54. M. Wang, X. W. Hu, P. Liu, W. Li, X. Gong, F. Huang and Y. Cao, *J. Am. Chem. Soc.*, 2011, **133**, 9638.

55. X. Guo, N. Zhou, S. J. Lou, J. Smith, D. B. Tice, J. W. Hennek, R. P. Ortiz, J. T. L. Navarrete, S. Li and J. Strzalka, *Nat. Photonics*, 2013, **7**, 825.

56. Y. Wu, Z. J. Li, W. Ma, Y. Huang, L. J. Huo, X. Guo, M. J. Zhang, H. Ade and J. H. Hou, *Adv. Mater.*, 2013, **25**, 3449.

57. J. Peet, L. Wen, P. Byrne, S. Rodman, K. Forberich, Y. Shao, N. Drolet, R. Gaudiana, G. Dennler and D. Waller, *Appl. Phys. Lett.*, 2011, **98**, 1.

58. M. Zhang, X. Guo and Y. Li, *Adv. Energy Mater.*, 2011, **1**, 557.

59. E. T. Hoke, K. Vandewal, J. A. Bartelt, W. R. Mateker, J. D. Douglas, R. Noriega, K. R. Graham, J. M. J. Frechet, A. Salleo and M. D. McGehee, *Adv. Energy Mater.*, 2013, **3**, 220.

60. L. Ye, S. Q. Zhang, W. Ma, B. H. Fan, X. Guo, Y. Huang, H. Ade and J. H. Hou, *Adv. Mater.*, 2012, **24**, 6335.

61. T. M. Clarke, C. Lungenschmied, J. Peet, N. Drolet, K. Sunahara, A. Furube and A. J. Mozer, *Adv. Energy Mater.*, 2013, **3**, 1473.

62. G. Dennler, K. Forberich, M. C. Scharber, C. J. Brabec, I. Tomis, K. Hingerl and T. Fromherz, *J. Appl. Phys.*, 2007, **102**.

63. M. S. Kim, B. G. Kim and J. Kim, *ACS Appl. Mater. Interfaces*, 2009, **1**, 1264.

64. M. S. White, D. C. Olson, S. E. Shaheen, N. Kopidakis and D. S. Ginley, *Appl. Phys. Lett.*, 2006, **89**, 143517.

65. L. J. A. Koster, S. E. Shaheen and J. C. Hummelen, *Adv. Energy Mater.*, 2012, **2**, 1246.

66. M. C. Scharber and N. S. Sariciftci, *Prog. Polym. Sci.*, 2013, **38**, 1929.

67. A. Guerrero, N. F. Montcada, J. Ajuria, I. Etxebarria, R. Pacios, G. Garcia-Belmonte and E. Palomares, *J. Mater. Chem. A*, 2013, **1**, 12345.
68. J. D. Kotlarski, D. J. D. Moet and P. W. M. Blom, *J. Polym. Sci. Pol. Phys.*, 2011, **49**, 708.
69. T. Kirchartz, T. Agostinelli, M. Campoy-Quiles, W. Gong and J. Nelson, *J. Phys. Chem. Lett.*, 2012, **3**, 3470.
70. C. E. Small, S.-W. Tsang, S. Chen, S. Baek, C. M. Amb, J. Subbiah, J. R. Reynolds and F. So, *Adv. Energy Mater.*, 2013, **3**, 909.
71. M. Campoy-Quiles, T. Ferenczi, T. Agostinelli, P. G. Etchegoin, Y. Kim, T. D. Anthopoulos, P. N. Stavrinou, D. D. C. Bradley and J. Nelson, *Nat. Mater.*, 2008, **7**, 158.
72. R. Gaudiana, *J. Polym. Sci. Pol. Phys.*, 2012, **50**, 1014.
73. J. A. Bartelt, Z. M. Beiley, E. T. Hoke, W. R. Mateker, J. D. Douglas, B. A. Collins, J. R. Tumbleston, K. R. Graham, A. Amassian, H. Ade, J. M. J. Fréchet, M. F. Toney and M. D. McGehee, *Adv. Energy Mater.*, 2013, **3**, 364.
74. T.-Y. Chu, S. Alem, P. G. Verly, S. Wakim, J. Lu, Y. Tao, S. Beaupre, M. Leclerc, F. Belanger, D. Desilets, S. Rodman, D. Waller and R. Gaudiana, *Appl. Phys. Lett.*, 2009, **95**, 063304.
75. S. Wakim, S. Beaupre, N. Blouin, B.-R. Aich, S. Rodman, R. Gaudiana, Y. Tao and M. Leclerc, *J. Mater. Chem.*, 2009, **19**, 5351.
76. T. M. Clarke, J. Peet, A. Nattestad, N. Drolet, G. Dennler, C. Lungenschmied, M. Leclerc and A. J. Mozer, *Org. Electron.*, 2012, **13**, 2639.
77. W. W. Ma, C. Yang, X. Gong, K. Lee and A. J. Heeger, *Adv. Funct. Mater.*, 2005, **15**, 1617.
78. T. M. Clarke, D. B. Rodovsky, A. A. Herzing, J. Peet, G. Dennler, D. DeLongchamp, C. Lungenschmied and A. J. Mozer, *Adv. Energy Mater.*, 2011, **1**, 1062.
79. R. A. Street, A. Krakaris and S. R. Cowan, *Adv. Funct. Mater.*, 2012, **22**, 4608.
80. R. A. Street, M. Schoendorf, A. Roy and J. H. Lee, *Phys. Rev. B*, 2010, **81**, 205307.
81. J. R. Tumbleston, Y. Liu, E. T. Samulski and R. Lopez, *Adv. Energy Mater.*, 2012, **2**, 477.
82. A. Pivrikas, G. Juška, A. J. Mozer, M. Scharber, K. Arlauskas, N. S. Sariciftci, H. Stubb and R. Österbacka, *Phys. Rev. Lett.*, 2005, **94**, 176806.
83. G. Juška, K. Genevičius, N. Nekrašas, G. Sliaužys and G. Dennler, *Appl. Phys. Lett.*, 2008, **93**, 143303.
84. A. Pivrikas, N. S. Sariciftci, G. Juška and R. Österbacka, *Prog. Photovoltaics*, 2007, **15**, 677.
85. T. M. Clarke, F. C. Jamieson and J. R. Durrant, *J. Phys. Chem. C*, 2009, **113**, 20936.
86. N. Nekrašas, K. Genevičius, M. Viliūnas and G. Juška, *Chem. Phys.*, 2012, **404**, 56.

87. R. Hamilton, C. G. Shuttle, B. O'Regan, T. C. Hammant, J. Nelson and J. R. Durrant, *J. Phys. Chem. Lett.*, 2010, **1**, 1432.
88. M. Koppe, H.-J. Egelhaaf, G. Dennler, M. C. Scharber, C. J. Brabec, P. Schilinsky and C. N. Hoth, *Adv. Funct. Mater.*, 2010, **20**, 338.
89. G.-J. A. H. Wetzelaer, M. Kuik and P. W. M. Blom, *Adv. Energy Mater.*, 2012, **2**, 1232.
90. T. Erb, U. Zhokhavets, G. Gobsch, S. Raleva, B. Stühn, P. Schilinsky, C. Waldauf and C. J. Brabec, *Adv. Funct. Mater.*, 2005, **15**, 1193.
91. X. Yang, J. Loos, S. C. Veenstra, W. J. H. Verhees, M. M. Wienk, J. M. Kroon, M. A. J. Michels and R. A. J. Janssen, *Nano Lett.*, 2005, **5**, 579.
92. P. Vanlaeke, A. Swinnen, I. Haeldermans, G. Vanhoyland, T. Aernouts, D. Cheyns, C. Deibel, J. D'Haen, P. Heremans, J. Poortmans and J. V. Manca, *Sol. Energy Mater. Sol. Cells*, 2006, **90**, 2150.
93. G. J. Adriaenssens and V. I. Arkhipov, *Solid State Commun.*, 1997, **103**, 541.
94. L. J. A. Koster, V. D. Mihailetchi and P. W. M. Blom, *Appl. Phys. Lett.*, 2006, **88**, 052104.
95. C. Groves and N. C. Greenham, *Phys. Rev. B*, 2008, **78**, 155205.
96. G. Juška, K. Genevičius, N. Nekrašas, G. Sliaužys and R. Österbacka, *Appl. Phys. Lett.*, 2009, **95**, 013303.
97. C. Deibel, A. Wagenpfahl and V. Dyakonov, *Phys. Rev. B*, 2009, **80**, 075203.
98. J. Szmytkowski, *Chem. Phys. Lett.*, 2009, **470**, 123.
99. M. Hilczer and M. Tachiya, *J. Phys. Chem. C*, 2010, **114**, 6808.
100. C. Deibel, A. Baumann and V. Dyakonov, *Appl. Phys. Lett.*, 2008, **93**, 163303.
101. J. Szmytkowski, *Semicond. Sci. Technol.*, 2011, **26**, 105012.
102. J. Szmytkowski, *Phys. Status Solidi RRL*, 2012, **6**, 300.
103. M. Hirade, T. Yasuda and C. Adachi, *J. Phys. Chem. C*, 2013, **117**, 4986.
104. D. H. K. Murthy, A. Melianas, Z. Tang, G. Juška, K. Arlauskas, F. Zhang, L. D. A. Siebbeles, O. Inganäs and T. J. Savenije, *Adv. Funct. Mater.*, 2013, **23**, 4262.
105. T. Ripolles-Sanchis, S. R. Raga, A. Guerrero, M. Welker, M. Turbiez, J. Bisquert and G. Garcia-Belmonte, *J. Phys. Chem. C*, 2013, **117**, 8719.
106. F. C. Jamieson, E. B. Domingo, T. McCarthy-Ward, M. Heeney, N. Stingelin and J. R. Durrant, *Chem. Sci.*, 2012, **3**, 485.

CHAPTER 10

Singlet Fission and 1,3-Diphenylisobenzofuran as a Model Chromophore

JUSTIN C. JOHNSON[a] AND JOSEF MICHL*[b,c]

[a] National Renewable Energy Laboratory, Golden, CO 80401, USA;
[b] Department of Chemistry and Biochemistry, University of Colorado at Boulder, Boulder, CO 80309, USA; [c] Institute of Organic Chemistry and Biochemistry, Academy of Sciences of the Czech Republic, 16610 Prague 6, Czech Republic
*Email: michl@eefus.colorado.edu

10.1 Introduction

10.1.1 Singlet Fission

In the simplest description of singlet fission (SF),[1] a singlet excited S_1 chromophore shares some of its energy with an S_0 ground-state neighbor and both end up in their T_1 triplet states, coupled into an overall singlet. The process was discovered half a century ago,[2] the principles were elucidated,[3] and little attention was paid to it subsequently.

In addition to providing a mechanism for converting excited singlets into triplets in a spin-allowed fashion, *i.e.*, without having to rely on spin–orbit coupling induced spin flips, SF provides an opportunity to produce two excited chromophores using a single sufficiently energetic photon, thus potentially doubling the number of electron–hole pairs resulting from photon absorption in a photovoltaic cell.[4] Obviously, this is only achieved at

RSC Energy and Environment Series No. 11
Advanced Concepts in Photovoltaics
Edited by Arthur J Nozik, Gavin Conibeer and Matthew C Beard

the expense of halving the open-circuit voltage in principle available from such energetic photons. It can still be advantageous if the lower energy photons for which the SF chromophore is transparent are used at the same junction by a different lower bandgap chromophore to produce a single electron–hole pair each (Figure 10.1). A detailed analysis[5] showed that the theoretical efficiency limit for such a single-junction SF photovoltaic cell is nearly 1/2, well above the Shockley–Queisser[6] limit of 1/3. This realization has awakened renewed interest in SF in recent years.[7]

Although efficient SF has been observed in a diffusively formed collision dimer in pentacene solution,[8] and very inefficient SF is known even in certain covalent dimers,[9–12] the process has been mostly studied in organic molecular crystals. The following more detailed description therefore applies to a crystal. There, the initial singlet excitation is often delocalized over a group of chromophores, the two triplet excitons are not necessarily located on neighboring molecules when initially formed, and they are able to move apart by hopping to other molecules. They are at first coherently coupled into an overall doubly excited pure singlet state $^1(T_1T_1)$:

$$S_0 + S_1 \rightleftharpoons {}^1(T_1T_1) \rightleftharpoons T_1 + T_1 \tag{10.1}$$

The description of the SF process provided by Equation (10.1) involves four ordinary kinetic rate constants but this is actually a severe oversimplification. There is evidence[13,14] that at least in some cases, the initial excitation can generate a coherent superposition of the S_0S_1 and $^1(T_1T_1)$ states, which then loses coherence on a slower time scale. Moreover, although the doubly triplet excited singlet state $^1(T_1T_1)$ is an eigenstate of the usual electrostatic Hamiltonian, it is not an eigenstate of the total Hamiltonian, which also contains small spin-dependent terms. Therefore, $^1(T_1T_1)$ develops in time into an initially still coherent mixture of the nine sublevels

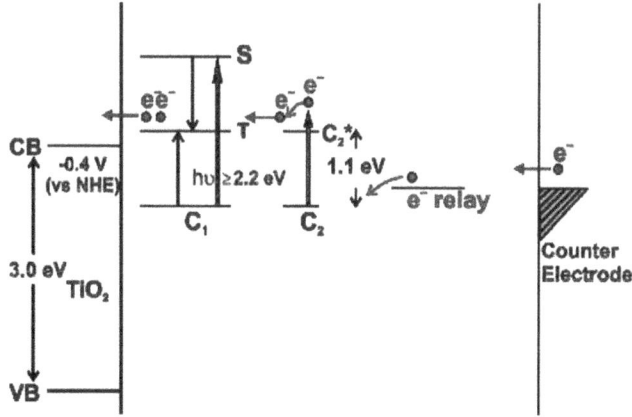

Figure 10.1   Processes occurring in a singlet fission solar cell (schematic). Reprinted with permission from A. F. Schwerin, *et al., J. Phys. Chem. A,* 2010, **114**, 1457. Copyright 2010 American Chemical Society.

spanned by $^1(T_1T_1)$, $^3(T_1T_1)$, and $^5(T_1T_1)$. Most important of the small mixing terms are the spin dipole–spin dipole interaction familiar from triplet EPR spectroscopy (D and E tensor) and, if an external magnetic field is present, also the Zeeman interaction. The initial coherence is gradually lost due to perturbations by the environment. Simultaneously, if the $^1(T_1T_1)$, $^3(T_1T_1)$, and $^5(T_1T_1)$ levels are close in energy, *i.e.*, if the two triplets interact only weakly, as is likely in a molecular crystal, opportunities for triplet energy transfer to other chromophores present will be used. The two triplets will then diffuse apart to yield two free triplet excitons, $2 \times T_1$. The excitons may undergo a geminal or an uncorrelated re-encounter, and such encounters may permit a re-conversion to an excited state located on a single molecule. This does not need to be only the initial S_1 state, but may in principle be any energetically close singlet (S), a triplet (T), or a quintet (Q) molecular state, with probabilities dependent on the weights of the $^1(T_1T_1)$, $^3(T_1T_1)$, and $^5(T_1T_1)$ levels in the triplet pair at the time of encounter. Conversion to the energetically farther separated S_0 and T_1 states is normally not competitive (energy gap law). Conversion to the Q_1 state is generally too endoergic, but in addition to return to S_1, there may be good opportunities for conversion to the T_2 state or even to higher molecular triplet states (*e.g.*, in polyenes). Recombination into S_1 merely restores the initial situation and does not represent an efficiency loss, but conversion to T_2 is likely to be followed by internal conversion into T_1 and thus to result in a loss of one of the two excitations.

The outcome of the complicated dance indirectly implied by Equation (10.1) is described approximately by a density matrix based theory due to Merrifield and Johnson[15] and more quantitatively by the far more complex theory developed by Suna,[16] who pointed out that two of the four rate constants implied by Equation (10.1) do not have independent significance and only their ratio is physically meaningful. These are the constants k_{-1} for the dissociation of the initially formed triplet pair $^1(TT)$ and k_2 for the return of $^1(TT)$ to an $S_1 + S_0$ pair. The branching ratio ε is defined variously as $\varepsilon = k_2/k_{-1}$ or $\varepsilon = 2k_2/(2k_2 + k_{-1})$.

The outcome depends on energetics of the situation:

1. If the free energy of the initial S_1 state is sufficiently higher than that of the final $2 \times T_1$ state, *e.g.*, in pentacene, the formation of two triplets by SF will be irreversible. If the free energy of the molecular T_2 and Q_1 states is also higher than that of the $2 \times T_1$ state, it can be hoped that there will be no T_1–T_1 annihilation and that the triplets will all ultimately produce the desired electron–hole pairs.

2. If the free energy of the initial S_1 state is approximately equal to that of the final $2 \times T_1$ state, as is the case in tetracene, triplet formation by SF and triplet recombination that re-forms the S_1 state will both take place and SF will be reversible. In this instance, delayed fluorescence is likely to appear.

3. If the free energy of the initial S_1 state is sufficiently lower than that of the final $2 \times T_1$ state, triplet formation by SF will be thermally activated

and possibly quite slow. When slow enough, *e.g.*, in anthracene, SF will not be competitive with other processes that deactivate the S_1 state, such as ordinary (prompt) fluorescence. Then, delayed fluorescence resulting from SF will also become inobservable.

In case (3), it may be possible to supply sufficient energy for overcoming the endoergicity by using initial excitation into a vibrationally hot S_1 state or into a higher singlet state. Another means of producing hot S_1 states is polaron recombination. However, reliance on hot states usually does not lead to high triplet yields, because of efficient competition by fast internal conversion.

An additional possible source of efficiency loss in an SF-based photovoltaic cell is inherent to the operation of the cell. The splitting of excitons into charge carriers (polarons) involves production of spin one-half species, and these have been long recognized as quenchers of triplet excitation.[1] It is therefore desirable to minimize triplet–polaron encounters.

In summary, in a search for efficient SF chromophores, the requirement of approximate isoergicity is of paramount importance and the fulfilment of the energy conditions $\Delta E(S_1)$, $\Delta E(T_2) \geq 2 \Delta E(T_1)$ is a well recognized first hurdle for a chromophore choice to overcome.[17] Excessive exoergicity is not desirable, even though it might produce fast SF, because it would entail conversion of electronic into vibrational energy and heat and thus a loss of overall photovoltaic efficiency.

10.1.2 Singlet Fission Chromophores

The energy conditions for an SF chromophore are difficult to meet and only a handful of organic structures are known to qualify. One unfavorable and one favorable circumstance need to be mentioned. First, the excitation energies in question are not those of an isolated chromophore, but of a chromophore in the crystal environment. Since intermolecular interactions usually lower the energy of S_1 more than the energy of T_1, even chromophores that look very promising as isolated species in solution actually need not fulfill the energy criteria in the crystal (*e.g.*, cibalackrot[18]). Second, the chromophore excitation energies obtained from spectra are potential energies. They are normally considered approximately equal to free energies, but this does not hold when two excitations are generated from one. Free energy difference for the SF process from S_1 to $^1(T_1T_1)$ and $2 \times T_1$ are therefore more favorable than would appear to be the case from spectral energies alone, as has been discovered in the case of tetracene.[13]

Although perhaps most critical, the energy criteria are by far not the only demands that a chromophore needs to satisfy in order to qualify as a candidate for a SF system of interest in photovoltaics. Some of the other requirements are a correctly situated absorption edge (ideally, about 2.2 eV), high absorption coefficients throughout the visible, redox potential matched to the other component of the photovoltaic junction, good charge-separation properties from the triplet state, absence of photophysical and photochemical processes capable of competing with SF, extraordinary chemical and photochemical stability, low cost, and low toxicity.

Simple theoretical considerations have been used to identify two categories of π-electron systems that represent promising candidates for meeting the SF energy requirements.[1,17]

10.1.2.1 Hydrocarbons and Derivatives

The first category comprises large alternant hydrocarbons and their simple derivatives, and it includes all of the good SF chromophores that were discovered more or less accidentally before 2010. Most of these, particularly tetracene and pentacene, have been discussed in considerable detail in one or more articles on SF that appeared in a recent dedicated issue of *Acc. Chem. Res.*[14,19–23] and we will not review them again here. Instead, we will exemplify the photophysics of an SF chromophore using 1,3-diphenylisobenzofuran (**1**), which was reported in 2010 as the first chromophore specifically designed for SF and as the first member of the second category, to be discussed next.

10.1.2.2 Biradicaloids

The second category comprises biradicaloids, *i.e.*, compounds that can be formally derived from perfect biradicals (whose triplet state carries two singly occupied exactly degenerate orbitals) by a structural perturbation that removes orbital degeneracy.[17] The two categories are not mutually exclusive.[24] Another example of a biradicaloid candidate is indigo, which satisfies the energy criteria, but has another problem, namely fast photochemical deactivation. In an indigo derivative known as cibalackrot this difficulty is eliminated, and it looks promising when considered in isolation, but strong Davydov interactions in its crystalline forms make SF endoergic.[18] Other candidates have been identified computationally,[25] but have yet been examined in the laboratory.

10.1.3 Chromophore Coupling

When a chromophore is found to satisfy the energy criteria for efficient SF, the next essential question arises: Is its crystal (or amorphous) structure suitable for promoting fast SF, or if it is not, can it be modified by appropriate structure engineering to become suitable? Alternatively, can a dimer or an aggregate be designed in which SF will proceed fast? An approximate theory has been developed starting with diabatic states that resembles standard theories of energy and charge transfer and allows predictions of favorable and unfavorable inter-chromophore geometries.[1,7,19] It assumes that the S_1 and T_1 states are related to the S_0 state by promotion of an electron from the highest occupied molecular orbital (HOMO) to the lowest unoccupied molecular orbital (LUMO). Using mutually Löwdin-orthogonalized Hartree–Fock molecular orbitals HOMO (h_A, h_B) and LUMO (l_A, l_B) on two identical weakly interacting chromophore molecules A and B, the following approximations are adopted for the description of singlet electronic states:

- $S_0{}^A S_0{}^B = |h_A \alpha h_A \beta h_B \alpha h_B \beta|$
- $S_1{}^A S_0{}^B = 2^{-1/2}[|h_A \alpha l_A \beta h_B \alpha h_B \beta| - |h_A \beta l_A \alpha h_B \alpha h_B \beta|]$
- $S_0{}^A S_1{}^B = 2^{-1/2}[|h_A \alpha h_A \beta h_B \alpha l_B \beta| - |h_A \alpha h_A \beta h_B \beta l_B \alpha|]$
- ${}^1(T_1{}^A T_1{}^B) = 3^{-1/2}[|h_A \alpha l_A \alpha h_B \beta l_B \beta| + |h_A \beta l_A \beta h_B \alpha l_B \alpha| - \{|h_A \alpha l_A \beta h_B \alpha l_B \beta| + |h_A \alpha l_A \beta h_B \beta l_B \alpha | + |h_A \beta l_A \alpha h_B \alpha l_B \beta| + |h_A \beta l_A \alpha h_B \beta l_B \alpha|\}/2]$
- ${}^1(C^A A^B) = 2^{-1/2}[|h_A \alpha l_B \beta h_B \alpha h_B \beta| - |h_A \beta l_B \alpha h_B \alpha h_B \beta|]$
- ${}^1(A^A C^B) = 2^{-1/2}[|h_A \alpha h_A \beta l_A \alpha h_B \beta | - |h_A \alpha h_A \beta l_A \beta h_B \alpha|]$.

The Fermi golden rule expression for the rate of singlet fission has the form:

$$w(SF) = (2\pi/\hbar)|\langle{}^1(T_1 T_1)|H_{el}|S_1 S_0\rangle - \langle{}^1(T_1 T_1)|H_{el}|CA\rangle$$
$$\times \langle{}^1 CA + {}^1 AC|H_{el}|S_1 S_0\rangle/\Delta E_{CT}|^2 \rho[E] \qquad (10.2)$$

where ΔE_{CT} is the difference between the energy of the charge-transfer states and the energy of the nearly degenerate $S_0 S_1$ and ${}^1(TT)$ states, and $\rho[E]$ is the Franck–Condon weighted density of states. The first term on the right-hand side of Equation (10.2) is known as the direct contribution and it is usually negligible. The second term is known as the mediated contribution since it involves the mixing of virtual charge-transfer states into the initial $S_0 S_1$ and final ${}^1(T_1 T_1)$ states of the conversion process. In a reasonable approximation, equation (2) can be simplified to Equation (10.3):[7]

$$w(SF) = (3\pi/\hbar)|\langle l_A|F|h_B\rangle[\langle l_A|F|l_B\rangle - \langle h_A|F|h_B\rangle]/\Delta E_{CT}|^2 \rho[E] \qquad (10.3)$$

where F is the Fock operator for the ground state.

When the Löwdin orthogonalization is ignored in view of the small intermolecular overlaps involved, the orbitals l and h are expanded in atomic orbitals μ on chromophore A and ν on chromophore B, and the tight-binding approximation is introduced, a simple approximate Equation (10.4) for the SF rate results and represents a simple design rule for the coupling of a pair of SF chromophores:[7]

$$w(SF) = (3\pi/\hbar)\left|\left[\sum_{(\mu\nu)}(c_{\mu l}c_{\nu h} + c_{\nu l}c_{\mu h})\beta_{\mu\nu}\right]\left[2\sum_{(\mu\nu)}(c_{\mu l}c_{\nu l} - c_{\mu h}c_{\nu h})\beta_{\mu\nu}\right]\right/\Delta E_{CT}\right|^2 \rho[E]$$

$$(10.4)$$

The sums run over pairs $(\mu\nu)$ of interacting atomic orbitals, μ on A and ν on B, and $\beta_{\mu\nu}$ is their resonance (hopping) integral. The coefficients $c_{\mu h}$, $c_{\nu h}$, $c_{\mu l}$, and $c_{\nu l}$ are the amplitudes of h and l at these atomic orbitals. Since the resonance integrals are approximately proportional to overlap integrals, a simple determination of overlaps of atomic orbitals on the two partners, often merely by inspection, permits conclusions to be drawn about the relative merits of various relative geometries of the two partners without any computation.[7] This is particularly useful for a comparison of different geometrical arrangements of the same two chromophore molecules, since the $\rho[E]$ factor can then be assumed to remain constant.

The best type of inter-chromophore arrangement that has been discovered so far is a slip-stacked geometry, with the slip in the direction of the HOMO to LUMO transition moment, while perfectly stacked geometries and those slipped perpendicular to the transition moment are predicted to be poor, as are linearly linked geometries. Other types of geometrical arrangements still remain to be investigated.

In sufficiently polar media that can stabilize a charge-transfer intermediate state ^1CA or ^1AC enough to make it a real state on the first excited singlet surface, SF may also occur as a two-step process. In this case, the initial state S_0S_1 is first transformed into a charge-transfer state by electron transfer and then to $^1(TT)$ by a back-electron transfer.[7]

10.2   1,3-Diphenylisobenzofuran (1)

10.2.1   The Chromophore 1

This structure is a biradicaloid since it can be formally derived from a nearly perfect o-quinodimethane biradical both of whose exocyclic double bonds have been decoupled out of conjugation by a 90° twist. This hypothetical perfect biradical is perturbed by untwisting and introduction of an ether link between the two exocyclic positions, followed by double phenyl substitution for further stabilization. Compounds of this type have been known for some time and 1 is commercially available for use as a trap for singlet oxygen.[26] This implies instability in the simultaneous presence of air and light and precludes any practical use in an inexpensive solar cell. Besides, the absorption edge of 1 lies too far to the blue for good efficiency. Nevertheless, 1 is useful as a model system in the laboratory, because dimers with a variety of structures can be synthesized quite easily.[27,28] Its electrochemical reduction and oxidation have been studied and occur reversibly at − 1.84 V in DMF against the saturated calomel electrode and at + 0.35 V in liquid SO_2 against ferrocene/ferricenium, respectively.[28]

1

The lowest excited electronic states of 1 are located at 24 300 cm^{-1} (S_1) and 11 900 cm^{-1} (T_1) in non-polar solutions and at slightly lower energies in a thin polycrystalline layer.[29] The S_0 to S_1 absorption band is purely polarized along the y axis (shown in formula 1) and it has a positive B term in magnetic circular dichroism. Both at the CC2 and TDDFT levels of calculation, the transitions from S_0 to S_1 and to T_1 correspond to an almost pure HOMO to LUMO excitation. The next triplet state T_2 has not been detected experimentally but is calculated to lie well above the S_1 state in energy, suggesting that **1** fully satisfies the energy requirements for SF. The fluorescence spectrum is a mirror image of the first absorption band and its 5–6 ns lifetime depends only weakly on choice of solvent. The fluorescence quantum yield is 0.96 ± 0.03 in a series of solvents, and no phosphorescence or T_1–T_n absorption are detectable after a pulsed excitation.[29]

From single crystal X-ray analysis, the molecule of **1** is nearly planar in its S_0 state, except that the two phenyl substituents are twisted out of plane by about $24°$, either in a conrotatory (the C_s conformer) or a disrotatory (the C_2 conformer) way. The energies and nearly all other properties of the two conformers are calculated to be virtually identical and the only difference between them seems to be a small change in the shape of the Franck–Condon envelopes of their first absorption band. Both known crystal modifications of **1** contain the disrotatory conformer and their structures are very similar.[29,30]

The S_1–S_n absorption spectrum of **1** was obtained from transient absorption measurements and has a weak peak near 14 500 and a strong one near 22 000 cm^{-1}. Its T_1–T_n absorption spectrum was obtained from sensitization experiments. It contains a weak and broad peak at about 13 500 cm^{-1} and a strong peak at 21 800 cm^{-1}. The spectra of the radical cation and radical anion of **1** were obtained from pulse radiolysis experiments. The former has a strong band at 18 500 cm^{-1} and the latter at 15 200 cm^{-1}.[29]

10.2.2   Polycrystalline Layers of 1

Vapor-deposited or solution-cast thin films of **1** can be readily fabricated and provide a system with non-covalent intermolecular coupling that can be used to probe singlet fission by spectroscopic methods. Comparison of thin films with bulk crystals of **1** can also be made, although in the former case a full structure cannot be determined, and in the latter quantification of Φ_T by spectroscopic means is problematic. Nonetheless, two bulk polymorphs of **1** with similar but distinct crystal structures were discovered, and they exhibit a close correlation of crystal habit, fluorescence, and powder X-ray diffraction (p-XRD) features with the two polycrystalline thin film types (Figure 10.2),[30] identified as α-**1** and β-**1**. The former can be converted into the latter by mild thermal annealing, and evidence for this conversion is seen in the change in extinction spectra, with β-**1** exhibiting an apparent shift and broadening (Figure 10.2C). In addition, the composition can be monitored quantitatively *via* the ratios of the unique (002) p-XRD peaks (Figure 10.2B). The β form of **1** is much more strongly fluorescent ($\Phi_F \sim 60\%$) than the α form ($\Phi_F \sim 10\%$). The fluorescence decay of β-**1** lacks the fastest (< 50 ps) and the slowest (< 20 ns) components observed in α-**1** (Figure 10.3A), which can be assigned to singlet fission and delayed fluorescence after triplet–triplet fusion, respectively. These features are mimicked in the corresponding bulk crystals and further indicate the lack of fast singlet fission in β-**1**.

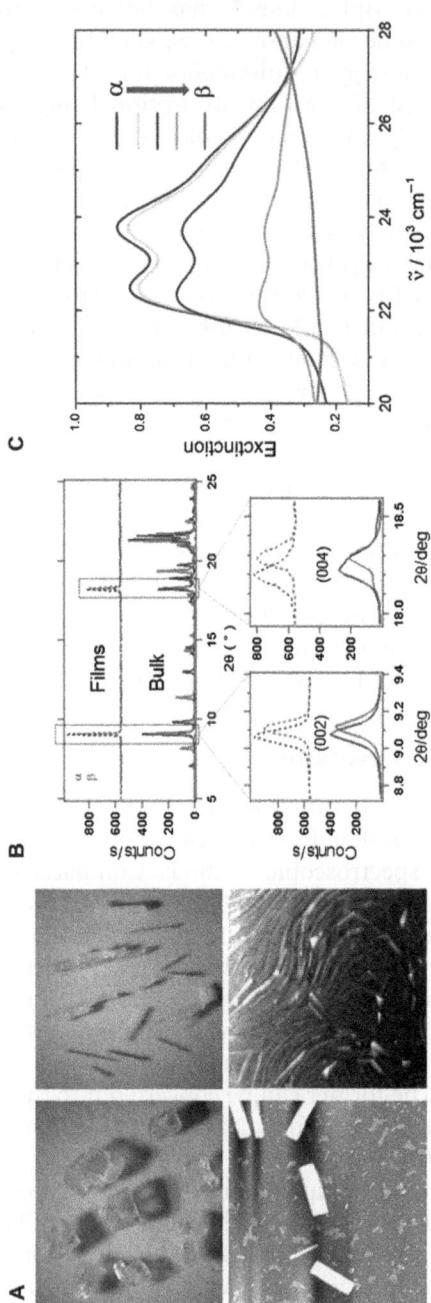

Figure 10.2 (A) Clockwise from upper left: Photographs of bulk crystals, 'block' and 'prism', and atomic force microscopy (AFM) images of α-1 and β-1 films. (B) Powder X-ray diffraction of vapor-deposited thin films and predicted p-XRD from known crystal structures. (C) Extinction spectra of a series of DPIBF films with varying compositions following thermal annealing (accompanied with some sublimation).

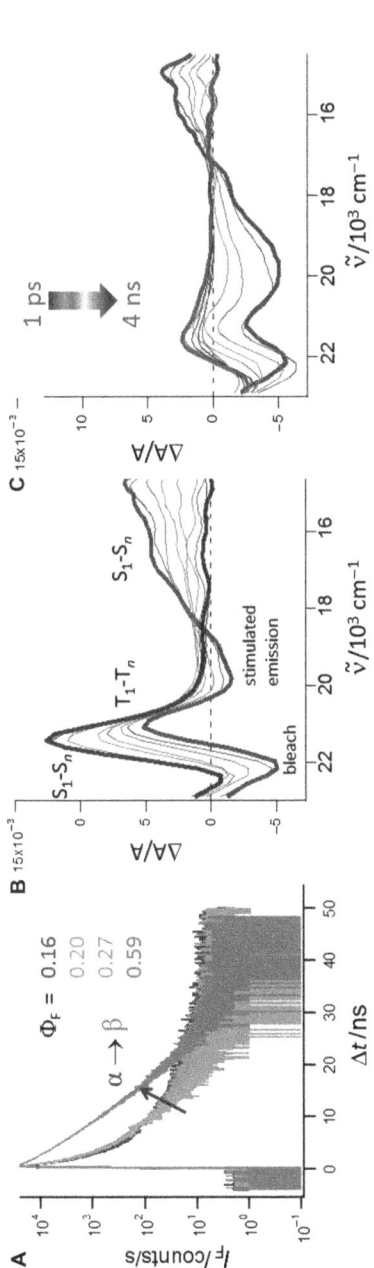

Figure 10.3   (A) Time-resolved fluorescence of α-1 before and after annealing to β-1 and Φ_F for each film along the annealing series. (B) Transient absorption spectra of α-1 at time delays from 1 ps (dark gray) to 4 ns (pale gray). (C) Same as B but for a thermally annealed film, mostly β-1.

Direct monitoring of the triplet population is accomplished by pump–probe transient absorption spectroscopy of the thin films (Figure 10.3B and C).[31,32] The spectra are similar to those of isolated molecules described in section 10.2.1. The most prominent T_1–T_n feature is found at 21 500 cm^{-1}, with a broad S_1–S_n feature near 15 000 cm^{-1} and stimulated emission near 20 000 cm^{-1}. As the relative amount of β-**1** is increased *via* thermal annealing, the T_1–T_n amplitude decreases while the stimulated emission

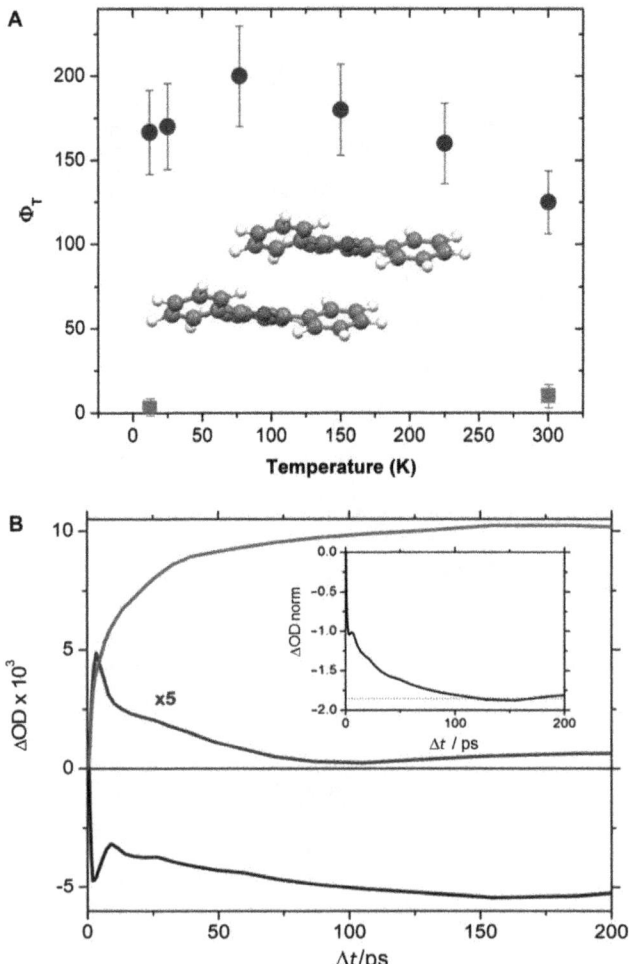

Figure 10.4 (A) Triplet quantum yield *versus* temperature for α-**1** (dark gray circles) and β-**1** (pale gray squares). Inset, slip-stacked structure of closest molecules in the 'block' bulk crystal structure. The structure in 'prisms' is virtually identical. (B) Kinetics of T_1 rise (pale gray), S_1 decay (dark gray), and S_0 bleach (black) in α-**1**. Inset, S_0 bleach rise after correction for spectral shifting at early times and normalization to −1.0 at $\Delta t = 0.2$ ps.
Adapted by permission from J. C. Johnson, *et al.*, *J. Am. Chem. Soc.* **2010**, *132*, 16302. Copyright 2010 American Chemical Society.

increases, indicative of a reduced dominance of the singlet fission pathway. By comparison with the ground state bleach, Φ_T of an α-**1** film was found to be ~200% at 80 K and ~130% at 295 K (Figure 10.4A). As the α:β ratio is reduced and the films become nearly free of α-**1** crystallites, the triplet yield falls to values below 20%. Yet, the slip-stacked columns in each polymorph possess nearly identical slippage angles and distances (Figure 10.4A, inset), and calculations of both Davydov coupling and SF matrix elements yield nearly the same results.

Interconversion of the polymorphs does not change the observed fast (2 ps) and slow (25 ps) rate constants for triplet formation, but instead alters the ratio of their amplitudes. The fast component comprises the largest portion of the overall triplet rise in β-**1** films, while the slower component is dominant in α-**1** films that yield a high Φ_T. The fast component appears to be due to 'hot' singlet fission from a vibrationally excited S_1 state, while the slower component is singlet fission from relaxed S_1. Similar effects have been observed in tetracene thin films.[33] The β form of **1** lacks the slower rise in triplet formation, and the significantly reduced triplet yield for β-**1** appears to be due to the barrier above relaxed S_1 that must be surmounted before SF becomes competitive with other relaxation pathways such as excimer formation. Using photon energies with more than 1000 cm^{-1} excess energy above the origin of S_1, or employing high intensity excitation to induce two-photon absorption also increase the amount of 'hot' singlet fission, consistent with reports of triplet formation in rubrene[34] and perylene.[35,36]

10.2.3   Covalent Dimers of 1

The covalently bound dimers of **1** described in Figure 10.5 and Figure 10.6 (**2–6**) were prepared and investigated spectroscopically in solution.[12,37] They are categorized as weakly or strongly coupled according to the degree of perturbation of spectral properties compared with the monomer **1**.[29]

Weakly coupled dimers **2** and **3** (Figure 10.5A) absorb light with roughly twice the peak absorption coefficient compared with the monomer **1** (~5.7×10^3 and 4.8×10^3 M^{-1} cm^{-1} *versus* 2.4×10^3 M^{-1} cm^{-1}) and a ~100 cm^{-1} red shift in the absorption onset (Figure 10.5B).[12] There is a small increase in the amplitude of the 0–0 vibronic absorption band compared with that of the 0–1 band, but the resemblance to monomer absorption is evident. In non-polar solution the fluorescence also exhibits an increased 0–0 to 1–0 band amplitude ratio for **2** and **3** compared with **1**, but all compounds fluoresce with near unity quantum yield and 5–7 ns mono-exponential decay times (Figure 10.5C). In polar solutions the situation changes dramatically, as **1** retains its high fluorescence quantum yield Φ_F but **2** and **3** exhibit multi-exponential decay kinetics and Φ_F values are reduced by a factor of 5–10 (Figure 10.5D). The fluorescence spectral shape remains unchanged except for solvatochromic shifts, thus the fluorescence is still considered to originate from the vertical S_1 state. Calculations suggest that the excitation is localized on one of the two halves of the dimer. It undoubtedly hops rapidly between the halves.

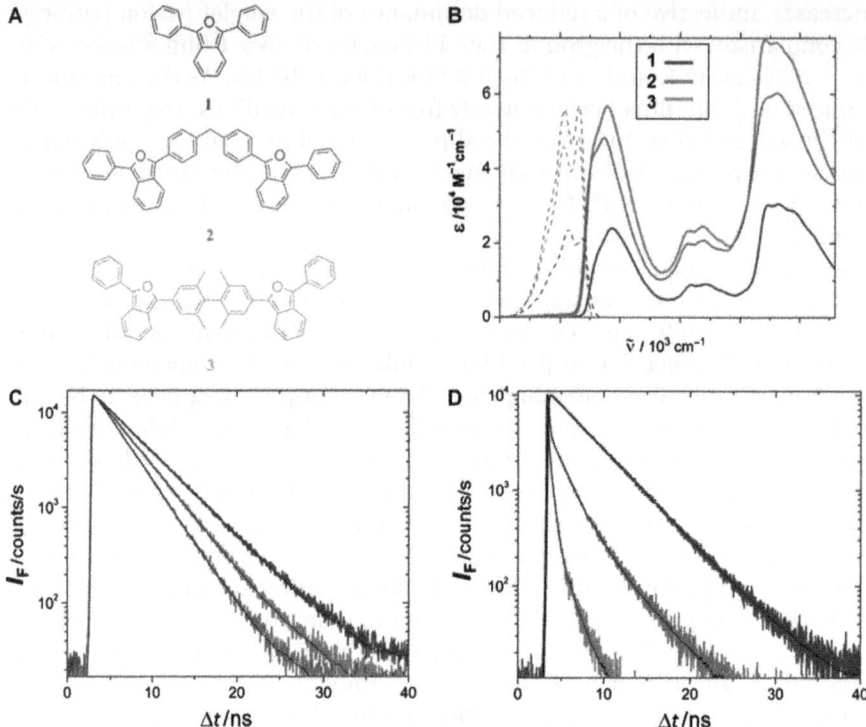

Figure 10.5 (A) Structures of **1** and the weakly coupled dimers **2** and **3**. (B) Absorption spectra in acetonitrile (from above: **2**, **3**, **1**). (C and D) Time-resolved fluorescence decay of **1–3** in toluene (C, from above: **1**, **2**, **3**) and DMSO (D, from above: **1**, **3**, **2**).

Adapted by permission from J. C. Johnson, *et al.*, *J. Phys. Chem. B*, 2013, **117**, 4680. Copyright 2013 American Chemical Society.

Concomitant with the lowering of Φ_F in dimethyl sulfoxide (DMSO) is an increase in the triplet yield Φ_T, from less than 0.5% in non-polar solution to $\sim$3% and $\sim$6% for **2** and **3**, respectively. The T_1–T_n spectra remain mostly unchanged from that of the monomer, again suggesting the formation of one or two localized T_1 states. A global fit of fs–ns transient absorption data at many wavelengths produces T_1–T_n and S_1–S_n spectra that compare favorably with expectations from other experiments and define the rate constants for triplet and singlet formation and decay, which are used to construct the concentration *versus* time profiles shown in Figure 10.7A. Transient absorption spectra also reveal the formation and decay time of a polar intermediate X that lies in the excited state pathway between S_1 and T_1 (Figure 10.7B). The spectrum of X possesses two strong absorption bands in the visible region that are correlated with the absorption spectra of radical cation and anion species determined from pulsed radiolysis experiments on **1** in solution.[29] The sum of the cation and anion spectra (dashed line in Figure 10.7B) matches well with the spectrum of X, which we assign to the

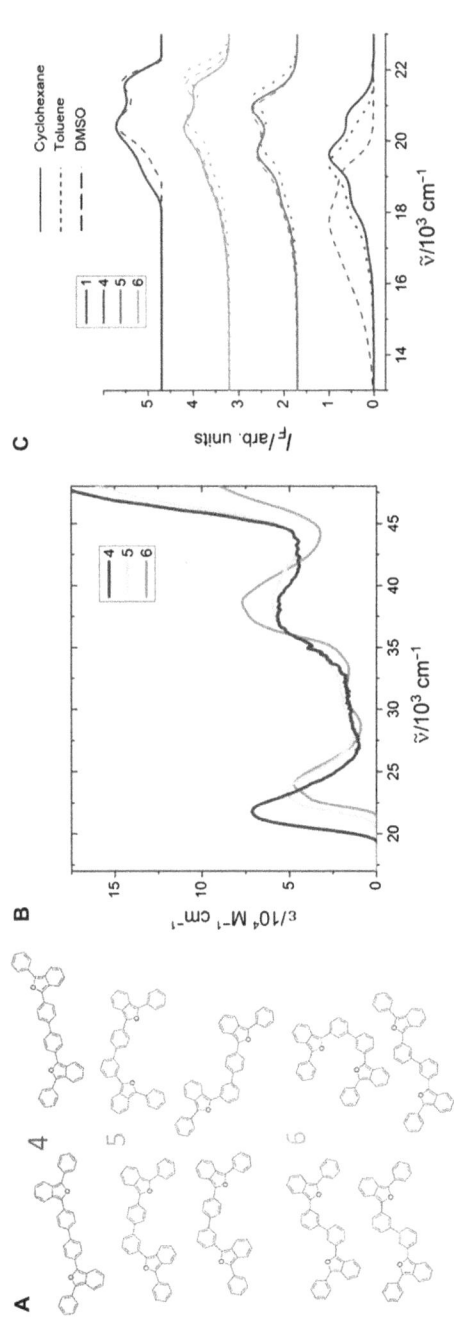

Figure 10.6 (A) Structures of selected conformers of 4–6. (B) Absorption spectra in acetonitrile. (C) Fluorescence spectra (from above, 1, 6, 5, 4).

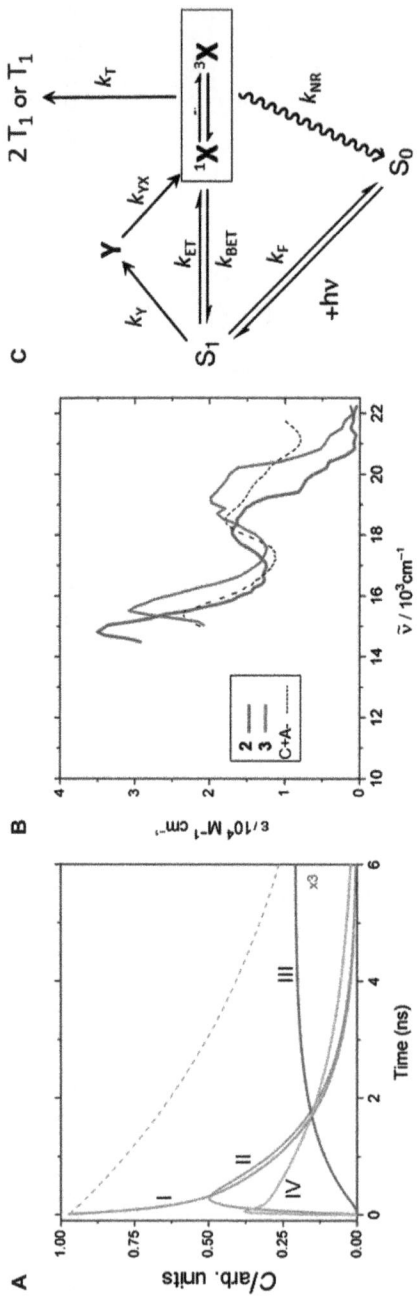

Figure 10.7   (A) Concentration *versus* time profiles for **3** in DMSO (solid lines) and toluene (dashed line). Curve I = S_1, curve II = X, curve III = T_1, and curve IV = Y. (B) Spectra of X derived from global fit of transient absorption data for **2** and **3**, and composite radical cation plus radical anion spectrum from pulsed radiolysis (dashed). (C) Kinetic scheme for photophysics of **2** or **3** (Y is not observed in **2**). Adapted by permission from J. C. Johnson, *et al.*, *J. Phys. Chem. B*, 2013, **117**, 4680. Copyright 2013 American Chemical Society.

C^+A^- state formed *via* intramolecular charge transfer. The formation of C^+A^- is clearly facilitated by the presence of the polar solvent, which is calculated to stabilize C^+A^- by ~ 25–30 kcal mol^{-1} compared with non-polar solutions, which show no C^+A^- formation.

The possible triplet formation routes are presented in Figure 10.7C. For **2** and **3** at room temperature, an excited state equilibrium is formed between population in S_1 and C^+A^-. The variation of the equilibrium constant with temperature favors C^+A^- as the temperature is lowered, increasing also the triplet yield to as high as 10% just above the solvent freezing point. From each $^1(C^+A^-)$ species one T_1 state may form from fast intersystem crossing to $^3(C^+A^-)$ followed by back charge transfer. Alternatively, two T_1 states may form from C^+A^- in a second step of two-step (indirect) singlet fission. A relatively fast internal conversion of C^+A^- to the ground state causes the low observed triplet yields, which prevented a definitive assignment of the true formation path. An additional charge-separated species, Y, possibly a twisted intramolecular CT state,[38] is formed at higher temperatures in **3**. This species forms quickly but does not lead directly to triplets and instead decays to X in less than 200 ps. The formation of Y is not observed in the case of **2**, and this is the only significant difference in the behavior of **2** and **3**.

The directly linked dimers of **1** (**4–6**, Figure 10.6A)[28,37] ostensibly fall under the category of 'strongly coupled'. Steady-state absorption reveals an increasing red shift of the absorption onset as the conjugating link is changed from meta–meta to meta–para to para–para (Figure 10.6B). Strong emission and small solvatochromic shifts characterize fluorescence from S_1 for **5** and **6** in non-polar solvents (Figure 10.6C). A similar change in excited state behavior as was observed in **2** and **3** in polar solvents is found for **5** and **6**, with reduced Φ_F and multi-exponential fluorescence decays. Concomitant with this change is an increase in Φ_T. However, in contrast to **2** and **3**, Φ_T is measurable for **5** and **6** even in solvents of very low polarity, such as toluene (2% and 9%, respectively). A further increase to 9% and 13%, respectively, is measured in DMSO, suggesting that two-step SF occurs with a charge transfer (CT) state as an intermediate. In toluene, the intermediate is not observable by transient absorption for **6** but is for **5**, resembling but not exactly matching the C^+A^- state found for **2** and **3** in highly polar solvents. A state of mixed CT and locally excited character may be accessible from S_1 even in weakly polar solvents and may act as an intermediate in this case. Numerous low-energy conformations are available and complicate the situation. The pathway to triplet formation may then look fairly similar to that shown in Figure 10.7C, with X varying from the truly charge-separated C^+A^- in polar solvents to something closer to charge-neutral in non-polar solvents. The T_1–T_n spectra are similar in shape to that of **1** but shift to lower energy as the linking topology increases the conjugation between the two halves of the dimer (Figure 10.8A). They agree with expectations based on TDDFT calculations, according to which the triplet excitation is localized on one half of the dimer and the two halves are twisted relative to each other.

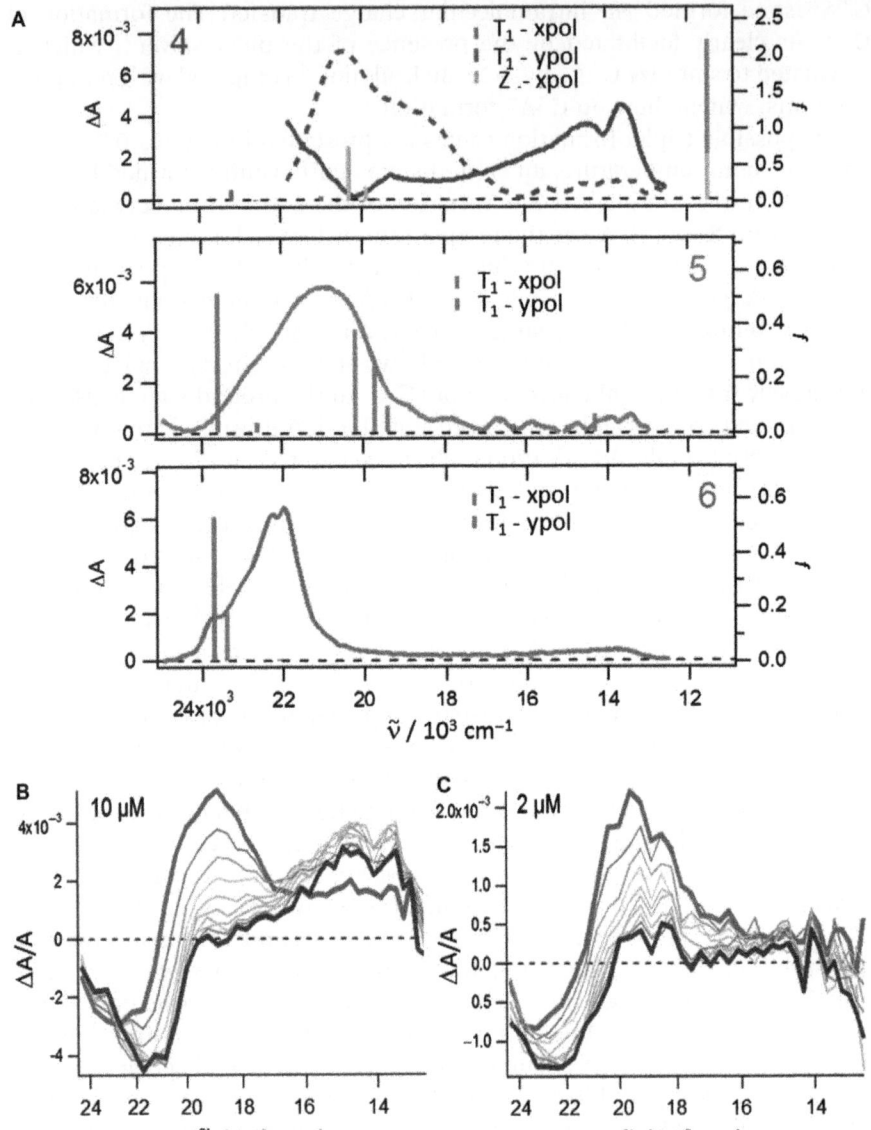

Figure 10.8 (A) T_1–T_n absorption spectra from flash photolysis experiments. Dashed curve is the absorption spectrum of Z. TDDFT calculated transitions for two optimized triplet conformations are shown as sticks: centrally twisted (pale gray) and nearly planar (mid gray). (B) Transient absorption of 10 µM **4** in DMSO at times ranging from 1 µs (pale gray) to 140 µs (dark gray). (C) Same as B but for 2 µM **4**.

Dimer **4** also appears to have states with at least partial CT character, as judged by the strong solvatochromism in fluorescence (Figure 10.6C). However, now Φ_T does not follow the polarity of the solvent as it does in all the other dimers. Instead, Φ_T remains below 4% in all solvents, slightly increasing as solvent polarity decreases. The absorption spectrum of the T_1 transient obtained by the sensitization of **4** with triplet anthracene is quite different from those obtained with **2**, **3**, **5**, and **6**. This is attributed to full delocalization of the triplet excitation through both halves of the molecule, in agreement with DFT optimized geometry, which is nearly planar. The absorption spectrum calculated for this delocalized triplet state agrees with the $T_1 \rightarrow T_n$ spectrum of **4** observed in sensitization experiments, which contains strong bands near 16 000 cm^{-1} (solid line in Figure 10.8A).

Little if any of this delocalized triplet state of **4** is however formed directly from S_1. Instead, in all solvents it is formed from a transient intermediate Z whose absorption spectrum (dashed line in Figure 10.8A) is entirely different from that of the C^+A^- intermediate found for the other dimers and resembles that of the T_1 state of the monomer **1**. This suggests that Z carries localized triplet excitation. Indeed, DFT geometry optimization identified both T_1 and Q_1 conformers of this type, twisted about the central single bond in **4**, and the spectra calculated for these species agree with that of Z (Figure 10.8A, light gray sticks show results for the localized triplet state).

Z is formed in less than 1 ns but decays slowly, in about 50 μs. As shown in Figure 10.8C, the rate of the transformation of Z into T_1 increases with the total concentration of **4**, demonstrating that the conversion is not due to a simple unimolecular isomerization on the triplet surface, but is dominated by a diffusion-controlled bimolecular process, $Z + S_0 \rightarrow S_0 + T_1$.

The action spectrum for the formation of Z is shifted from the steady state absorption spectrum by more than 500 cm^{-1} to higher energies, indicating a threshold for triplet formation that may coincide with an opening of access to a higher energy excited state with a non-vertical geometry. The most likely excited state pathway involving triplet formation is singlet fission or possibly unusually fast intersystem crossing from this unique excited state, perhaps with some CT character, to Z followed by triplet–triplet energy transfer to a ground-state molecule that leads to T_1. The exact nature of Z has not yet been elucidated.

10.3 Current and Future Activities

Further investigation of **1** as a singlet fission chromophore is concerned with optimizing the interchromophore geometry in dimers or thin layer structures[19] and extracting multiple electrons per absorbed photon. Additional dimers are being synthesized for this purpose, including some with the potential to mimic the slip-stacked geometry of α-**1**. We are also attempting to elucidate the details of the formation mechanism and the structure of the intermediate Z in the dimer **4**.

Extraction of a quantum yield of charges that exceeds unity is a certain sign of singlet fission, but requires a delicate balance between the singlet fission rate and charge transfer/recombination rates. Using mid-infrared time-resolved absorption spectroscopy and time-resolved microwave conductivity (TRMC), we have begun to monitor the arrival of electrons into acceptors like TiO_2 and C_{60} after singlet fission in **1**. The various rate constants can be measured and compared with ideal values that first allow singlet fission to complete prior to charge injection from S_1 but that ensure efficient charge injection from T_1 and transport away from the interface before recombination. Full dye-sensitized solar cells (DSSCs) have also been fabricated with **1**/nanoparticle TiO_2/electrolyte architectures that show more than 1% power conversion efficiency and photocurrents exceeding 3 mA cm^{-2} despite limited solar light absorption, implying a high yield of collected charges. The distance between **1** and TiO_2 can be varied with dielectric spacers, and a non-monotonic decrease in photocurrent as a function of **1**/TiO_2 separation has been seen as evidence that charges born from singlet fission are being extracted once injection from S_1 is slowed beyond 10 ps.[39] Further work is aimed at optimizing the local morphology of **1** on the TiO_2 surface such that the SF rate constant is maximized.

Acknowledgements

Our work on singlet fission has been supported by the US Department of Energy, Office of Basic Energy Sciences, Division of Chemical Sciences, Biosciences, and Geosciences (DOE DE-SC0007004 and DE-AC36-08GO28308).

References

1. M. B. Smith and J. Michl, *Chem. Rev.*, 2010, **110**, 6891.
2. S. Singh, W. J. Jones, W. Siebrand, B. P. Stoicheff and W. G. Schneider, *J. Chem. Phys.*, 1965, **42**, 330.
3. C. E. Swenberg and N. E. Geacintov, *Org. Mol. Photophys*, 1973, **18**, 489.
4. A. J. Nozik, R. J. Ellingson, O. I. Mičić, J. L. Blackburn, P. Yu, J. E. Murphy, M. C. Beard and G. Rumbles, Proc. 27th DOE Solar Photochem. Res. Conf., Washington, DC: US Dep. Energy, 2004, 63. Available: http://science.energy.gov/~/media/bes/csgb/pdf/docs/solar_photochemistry_2004.pdf.
5. M. C. Hanna and A. J. Nozik, *J. Appl. Phys.*, 2006, **100**, 074510.
6. W. Shockley and H. J. Queisser, *J. Appl. Phys.*, 1961, **32**, 510.
7. M. B. Smith and J. Michl, *Annu. Rev. Phys. Chem.*, 2013, **64**, 361.
8. B. J. Walker, A. J. Musser, D. Beljonne and R. H. Friend, *Nat. Chem.*, 2013, **5**, 1019.
9. A. M. Müller, Y. A. Avlasevich, K. Müllen and C. J. Bardeen, *Chem. Phys. Lett.*, 2006, **421**, 518.

10. J. Burdett, A. M. Müller, D. Gosztola and C. J. Bardeen, *J. Chem. Phys.*, 2010, **133**, 144506.

11. J. Michl, A. J. Nozik, X. Chen, J. C. Johnson, G. Rana, A. Akdag and A. F. Schwerin, in *Organic Photovoltaics VIII*, eds. Z. H. Kafafi and P. A. Lane. Proc. of SPIE, 2007, Vol. 6656, p. 66560E1.

12. J. C. Johnson, A. Akdag, M. Zamadar, X. Chen, A. F. Schwerin, I. Paci, M. B. Smith, Z. Havlas, J. R. Miller, M. A. Ratner, A. J. Nozik and J. Michl, *J. Phys. Chem. B*, 2013, **117**, 4680.

13. W.-L. Chan, M. Ligges and X.-Y. Zhu, *Nat. Chem.*, 2012, **4**, 840.

14. W.-L. Chan, T. C. Berkelbach, M. R. Provorse, N. R. Monahan, J. R. Tritsch, M. S. Hybertsen, D. R. Reichman, J. Gao and X.-Y. Zhu, *Acc. Chem. Res.*, 2013, **46**, 1321.

15. R. C. Johnson and R. E. Merrifield, *Phys. Rev. B*, 1970, **1**, 896.

16. A. Suna, *Phys. Rev. B*, 1970, **1**, 1716.

17. I. Paci, J. C. Johnson, X. Chen, G. Rana, D. Popović, D. E. David, A. J. Nozik, M. A. Ratner and J. Michl, *J. Am. Chem. Soc.*, 2006, **128**, 16546.

18. J. C. Johnson, B. R. Stepp, J. L. Ryerson, M. B. Smith, A. Akdag, A. J. Nozik, and J. Michl, unpublished results.

19. J. C. Johnson, A. J. Nozik and J. Michl, *Acc. Chem. Res.*, 2013, **46**, 1290.

20. J. Lee, P. Jadhav, P. D. Reusswig, S. R. Yost, N. J. Thompson, D. N. Congreve, E. Hontz, T. Van Voorhis and M. A. Baldo, *Acc. Chem. Res.*, 2013, **46**, 1300.

21. J. J. Burdett and C. J. Bardeen, *Acc. Chem. Res.*, 2013, **46**, 1312.

22. M. W. B. Wilson, A. Rao, B. Ehrler and R. H. Friend, *Acc. Chem. Res.*, 2013, **46**, 1330.

23. P. M. Zimmerman, C. B. Musgrave and M. Head-Gordon, *Acc. Chem. Res.*, 2013, **46**, 1339.

24. T. Minami and M. Nakano, *J. Phys. Chem. Lett.*, 2012, **3**, 145.

25. A. Akdag, Z. Havlas and J. Michl, *J. Am. Chem. Soc.*, 2012, **134**, 14624.

26. J. A. Howard and G. D. Mendenhall, *Can. J. Chem.*, 1975, **53**, 2199.

27. J. C. Johnson, A. Akdag, M. Zamadar, X. Chen, A. F. Schwerin, I. Paci, M. B. Smith, Z. Havlas, J. R. Miller, M. A. Ratner, A. J. Nozik and J. Michl, *J. Phys. Chem. B*, 2013, **117**, 4680.

28. A. Akdag, A. Wahab, P. Beran, L. Rulíšek, P. Dron, J. Ludvík and J. Michl, unpublished results.

29. A. F. Schwerin, J. C. Johnson, M. B. Smith, P. Sreearunothai, D. Popović, J. Černý, Z. Havlas, I. Paci, A. Akdag, M. K. MacLeod, X. Chen, D. E. David, M. A. Ratner, J. R. Miller, A. J. Nozik and J. Michl, *J. Phys. Chem. A*, 2010, **114**, 1457.

30. J. L. Ryerson, J. N. Schrauben, A. J. Ferguson, S. C. Sahoo, P. Naumov, Z. Havlas, J. Michl, A. J. Nozik and J. C. Johnson, *J. Phys. Chem. C*, Article ASAP, DOI: 10.1021/jp502122d.

31. J. C. Johnson, A. J. Nozik and J. Michl, *J. Am. Chem. Soc.*, 2010, **132**, 16302.

32. J. N. Schrauben, J. Ryerson, J. Michl and J. C. Johnson, *J. Am. Chem. Soc.*, Article ASAP, DOI: 10.1021/ja501337b.

33. V. K. Thorsmølle, R. D. Averitt, J. Demsar, D. L. Smith, S. Tretiak, R. L. Martin, X. Chi, B. K. Crone, A. P. Ramirez and A. J. Taylor, *Phys. Rev. Lett.*, 2009, **102**, 017401.
34. L. Ma, K. Zhang, C. Kloc, H. Sun, M. E. Michel-Beyerle and G. G. Gurzadyan, *PCCP*, 2012, **14**, 8307.
35. W. G. Albrecht, M. E. Michel-Beyerle and V. Yakhot, *Chem. Phys.*, 1978, **35**, 193.
36. W. G. Albrecht, H. Coufal, R. Haberkorn and M. E. Michel-Beyerle, *Phys. Status Solidi*, 1978, **89**, 261.
37. J. Schrauben, A. Akdag, J. Wen, Z. Havlas, A. J. Nozik, J. C. Johnson, and J. Michl, unpublished results.
38. J. Jacq, S. Tsekhanovich, M. Orio, C. Einhorn, J. Einhorn, B. Bessieres, J. Chauvin, D. Jouvenot and F. Loiseau, *Photochem. Photobio.*, 2012, **88**, 633.
39. Y. Zhang, J. N. Schrauben, J. C. Johnson and K. Zhu, unpublished results.

Quantum Confined Semiconductors for Enhancing Solar Photoconversion through Multiple Exciton Generation

MATTHEW C. BEARD,*[a] ALEXANDER H. IP,[b]
JOSEPH M. LUTHER,[a] EDWARD H. SARGENT[b] AND
ARTHUR J. NOZIK[a,c]

[a] National Renewable Energy Laboratory, Golden, CO, USA; [b] University of
Toronto, Toronto, ON, Canada; [c] University of Colorado, Boulder, CO, USA
*Email: matt.beard@nrel.gov

11.1 Introduction to Colloidal Quantum Dots

11.1.1 Tuning of Electronic Properties

In their seminal paper on the electronic properties of quantum dots (QDs), Efros and Efros define a 'smallness parameter' ($\lambda = r/a_B$) to describe the degree of quantum confinement,[1] where a_B is the bulk exciton Bohr radius and r is the QD radius. They identified a regime in which the motions of electrons and holes are strongly affected by the physical dimension of the system when $r < a_B$ (*i.e.*, $\lambda < 1$). The Bohr radius can be calculated as $a_B = \varepsilon \hbar^2 / \mu e^2$ (where ε is the dielectric constant of the semiconductor and μ is the reduced effective mass of the electron and hole) and typically ranges between 2 and 100 nm. For PbS a_B is 18–20 nm, for PbSe it is 46 nm, while a_B is 4.7 nm for Si, and

RSC Energy and Environment Series No. 11
Advanced Concepts in Photovoltaics
Edited by Arthur J Nozik, Gavin Conibeer and Matthew C Beard
© The Royal Society of Chemistry 2014
Published by the Royal Society of Chemistry, www.rsc.org

5.4 nm for CdSe. When $\lambda < 1$, electronic states become highly dependent on the QD size and shape, and the electronic states of the QDs no longer form continuous energy bands but rather form discrete quantized energy levels.

Excitons are quasi-particles made of correlated (bound) electron–hole (e–h) pairs that are mutually attracted through the Coulomb potential and in bulk semiconductors give rise to a Rydberg energy series below the free-electron continuum of the conduction band. For most bulk semiconductors, the exciton binding energy is small compared to kT and therefore, electrons and holes do not interact to form excitons at room temperature (or very quickly dissociate). For QDs, however, charge carriers are forced to interact by the physical dimensions of the crystal even at room temperature. The energies of quantized levels are dictated primarily by confinement while the e–h Coulomb attraction provides a relatively small correction. The HOMO-LUMO separation, 1st exciton transition energy, or bandgap, E_g is given, to first order, by $E_g^{\text{NC}} = E_g^{\text{bulk}} + E_{\text{conf}}(r) - E_{\text{Coul}}(r)$, where E_g^{bulk} is the bulk bandgap of the semiconductor, $E_{\text{conf}} = \pi^2\hbar^2/2m^*r^2$ is the confinement energy and $E_{\text{Coul}} = e^2/\varepsilon r$ is the Coulomb energy. Figure 11.1a displays typical absorption spectra of

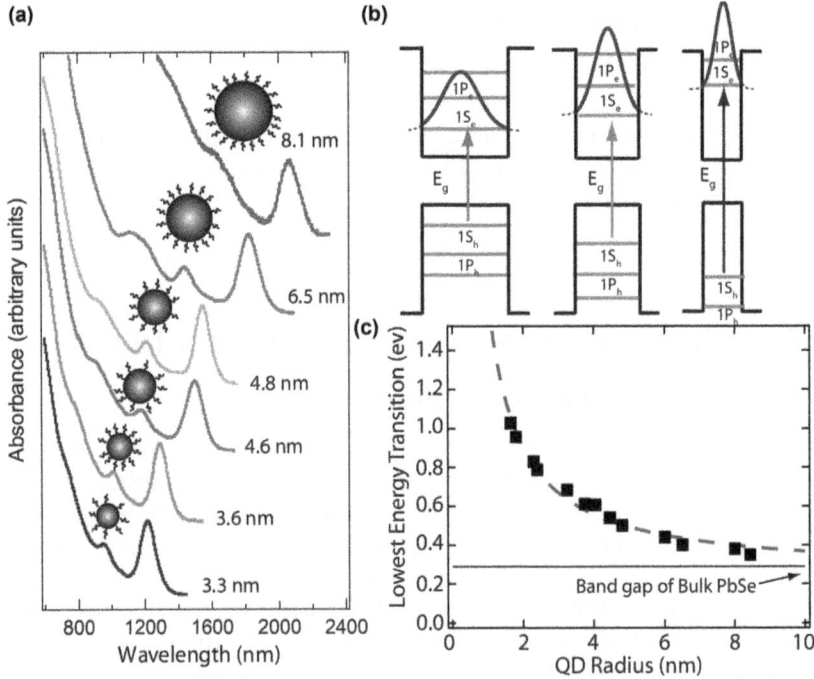

Figure 11.1   (a) Absorption spectra of PbSe QDs with average diameters ranging from 3.3 nm to 8.1 nm. There are several discrete transitions that represent discrete excitonic transitions. (b) The increased quantum confinement of the electronic wavefunction as the particle size decreases. (c) Sizing curve for PbSe QDs.
Figure reproduced with permission from *Materials Today*, 2012, **15**, 508.

PbSe QDs with average radii that vary from about 3 to 8 nm while Figure 11.1b depicts how the lowest energy conduction band electronic wavefunction experiences increased confinement and the separation between the quantized states increase as the QD size decreases (similar to a particle-in-a-box), giving rise to the peaks in the absorption spectra. In Figure 11.1 we label the states according to their orbital angular momentum with quantum number, L, and principal quantum number, n, neglecting spin–orbit interactions.

Sizing curves (Figure 11.1c) relate the QD radius determined *via* TEM analysis to the 1st exciton energy level or E_g. As demonstrated in Figure 11.1 the electronic energy levels are sensitive to the QD size. A change in size of less than one monolayer can have a notable effect on E_g (compare 4.6 and 4.8 nm QDs). Assuming that the Gaussian absorption lineshape of the 1st exciton transition results from a distribution of QD sizes, the size-distribution can be determined by considering the full-width of the Gaussian lineshape and the sizing curve. For PbSe QDs shown in Figure 11.1 the size distribution is typically about 4–6%. For solar cell applications, QDs offer the ability to tune the E_g of the absorber layer by simply changing the size of the QDs and there are at least two benefits from the ability to tune E_g:

- Material systems that have too low an E_g in their bulk form, but would otherwise be attractive candidates can now be utilized in their QD forms. Thus a material system such as PbS with bulk $E_g = 0.4$ eV and strong absorption strength can be considered as a viable PV system under conditions of quantum confinement.
- Multi-junction solar cells can be constructed from the same material system and with fine control over the respective bandgap of each layer.

11.1.2 Competition Between MEG and Hot-carrier Cooling *via* Phonon Emission

Absorption of a photon with energy in excess of the semiconductor's band gap energy results in a 'hot-exciton' with excess energy, $\Delta E_{ex} = h\nu - E_g$. In bulk semiconductors, hot charge carriers interact with the crystal lattice primarily through LO-phonon emission on the ~ 100 fs timescale (electron–phonon scattering rate of $\sim 10^{13}$ s^{-1}), losing energy to the crystal lattice in the form of heat.[2,3] However, if $\Delta E_{ex} > E_g$ ($h\nu > 2E_g$) then a new relaxation channel is available where the excess energy can be used to produce one or more extra e–h pairs (see Figure 11.2a). In bulk semiconductors this relaxation channel is termed impact ionization (I.I.).[4,5] Impact ionization has been investigated as a way to increase the power conversion efficiency of solar cells but the fundamental I.I. process was deemed too inefficient.[6] In bulk semiconductors, there are two factors that limit I.I.: (1) conservation of crystal momentum (Figure 11.2a) reduces the density of final states; and (2) rapid carrier cooling *via* phonon emission to the band edge (~ 1 ps or less). As a result, multiple e–h pairs are not produced *via* I.I. until excess energies of $\Delta E_{ex} > 3$ to $4E_g$ and thus photon energies of >4 to $5E_g$. Due the unique physics offered by

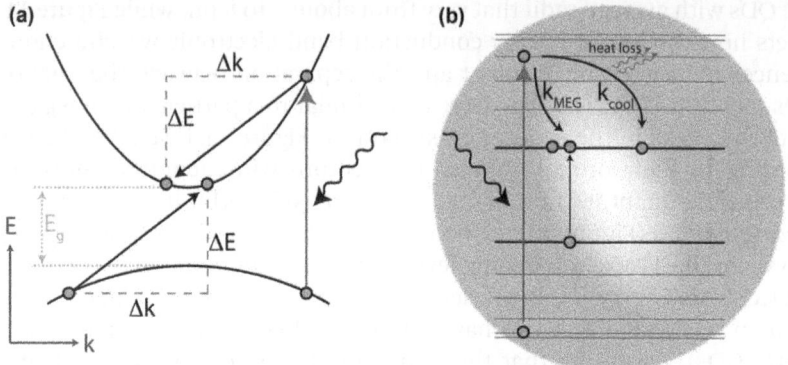

Figure 11.2   (a) Schematic depiction of impact ionization in a bulk semiconductor; both energy (ΔE) and crystal momentum (Δk) must be conserved, as shown. (b) MEG in a quantum-confined structure such as a QD has small dimensions and thus Bloch states with defined momentum are not eigenstates, such that only energy needs to be conserved. Hot carriers undergo one of two competing processes: (1) generation of an extra e–h pair (k_{MEG}) or (2) cooling of the single exciton state (k_{cool}). The same two processes apply for hot holes but are not shown in the diagram.

Reprinted with permission from *Accounts of Chemical Research*, 2013, **46**, 1252. Copyright 2013 American Chemical Society.

quantum confinement, QDs overcome some of these limitations. To distinguish I.I. from the similar yet distinct process in QDs we use the term multiple exciton generation (MEG). In QDs only energy conservation sets fundamental limits on MEG because crystal momentum is a result of long range repeating atomic potentials, which are not present in QDs. Hot-carrier pathways are modified in QD systems[7–9] and carrier–carrier interactions are larger due to the physical confinement within the QD. Thus MEG can better compete with carrier cooling *via* heat generation leading to an enhanced MEG process in QDs compared to I.I. in bulk semiconductors. Finally, the added degrees of freedom afforded by quantum-confined semiconductors (shape, size, composition, surface engineering, *etc.*) allow for a larger parameter space that can be exploited to further improve MEG and thereby, enhance the power conversion efficiencies in QD-based solar cells.

The competition between MEG or I.I. and carrier cooling *via* phonon emission can be expressed by the following equation:[22]

$$QY = 1 + \frac{k_{MEG}}{\left(k_{MEG}^{(1)} + k_{cool}\right)}$$

$$+ \frac{k_{MEG}^{(1)} k_{MEG}^{(2)}}{\left(k_{MEG}^{(1)} + k_{cool}\right)\left(k_{MEG}^{(2)} + k_{cool}\right)} \qquad (11.1)$$

$$+ \cdots$$

where QY is the number of e–h pairs generated per absorbed photon, $k_{MEG}^{(n)}$ is the rate of producing $(n+1)$ excitons from (n) hot excitons and k_{cool} is the cooling rate that is in competition with MEG. Each successive term in Equation (11.1) is only valid when energy conservation is met ($k_{MEG}^{(n)} \geq 0$ when $h\nu < n \times E_g$).[22] To approximate how the QY changes as a function of excess photon energy, we refer to work on impact ionization in bulk semiconductors where the rate of carrier cooling, k_{cool}, is found to be fairly constant with excess energy,[23] while k_{MEG} increases with increasing energy above a threshold photon energy, $h\nu_{th}$.[24,25] A phenomenological expression relates k_{MEG} and k_{cool}:[26]

$$k_{MEG} = Pk_{cool}\left(\frac{h\nu_{ex}}{h\nu_{th}}\right)^2 \qquad (11.2)$$

through a parameter, P, that denotes the competition between k_{MEG} and k_{cool}, where $h\nu_{ex} = h\nu - h\nu_{th}$. Higher values of P indicate higher yields of e–h pairs [the QY in Equation (11.1) is larger]. In Figure 11.3 we plot the QY for various values of P ranging from 0.1 to 10 000. When $P = 10\,000$, the QY approaches the energy conservation limit where two e–h pairs are produced at $2E_g$, three e–h pairs at $3E_g$ and so forth. We also plot in Figure 11.3 the QYs for bulk PbSe and PbSe QDs. We find that P increases from 0.42 to 1.5 when comparing bulk PbSe to PbSe QDs. Thus the competition between MEG and cooling favors MEG in the QD system.

There are two characteristics of the MEG process that are generally reported: (1) the MEG threshold, $h\nu_{th}$, which is the photon energy at which

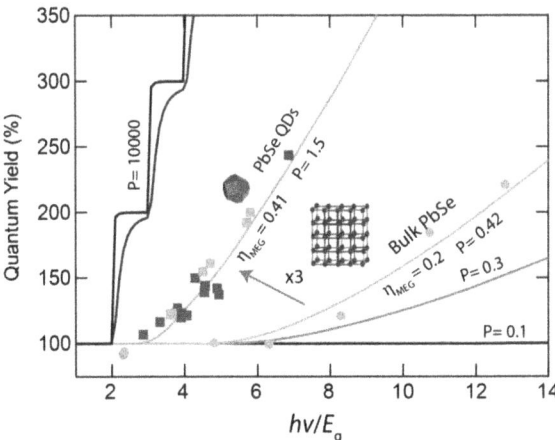

Figure 11.3 Lines are the calculated QY for various values of P (see text). The data points compare the measured QYs for bulk PbSe and PbSe QDs. Reprinted with permission from *Accounts of Chemical Research*, 2013, **46**, 1252. Copyright 2013 American Chemical Society.

the QYs begin to exceed 1; and (2) the electron–hole pair creation energy, ε_{eh},[4,5] which is the required excess energy to produce one additional e–h pair, $\varepsilon_{eh} = \Delta h\nu / \Delta QY$, for $h\nu > h\nu_{th}$. These two characteristics can be directly related by $h\nu_{th} = E_g + \varepsilon_{eh}$ in the limit of $h\nu_{th} \geq 2.2E_g$ and we define a dimensionless MEG efficiency, $\eta_{MEG} = E_g / \varepsilon_{eh}$ (Beard *et al.*[6]) with $h\nu_{th} = E_g(1 + 1/\eta_{MEG})$. η_{MEG} serves as a photon-energy-independent measure of MEG and captures the underlying photophysics regarding the competition between a hot carrier cooling to the band edge and the processes leading to production of additional electron–hole pairs. η_{MEG} increases from 0.2 to 0.4 when comparing bulk PbSe to PbSe QDs

11.1.3   Benefits to Solar Photoconversion

We now discuss how MEG improves the theoretical power conversion efficiency of solar cell with QD absorbers. The Shockley and Queisser[1] (S–Q) analysis of the maximum thermodynamic efficiency for converting sunlight into electrical free energy in a single-junction PV cell assumes:

- Photons with energy less than E_g are not absorbed
- Conduction band electrons and valence band holes created by absorption of high-energy photons immediately relax to their band minima (*i.e.*, the fraction of energy of photons with energy greater than E_g is lost as heat)
- The solar cell radiates light as a black-body above E_g

There are many approaches to increase the theoretical efficiency of the solar cell by using the high photon energies more efficiently (such as the MEG approach) or by recovering the low energy photons. For the former, the multi-junction or tandem junction solar cell is the most successful approach and employs multiple p–n junctions operating in tandem where the layers are designed to each absorb a portion of the total incident sunlight. For an infinite number of junctions in the stack, the theoretical conversion efficiency reaches 68% at 1-sun intensity.[36] Ross and Nozik[7] demonstrate that the same high conversion efficiency possible from a multi-junction solar cell could be obtained by utilizing the total excess kinetic energy of hot photo-generated carriers in a single bandgap device in other ways such as where hot carriers are transported and collected at energy-selective contacts.[37,38]

Figure 11.4 shows detailed balance analysis following the S-Q approach for various MEG characteristics denoted by the value of P (see Figure 11.3). The black line represents the conventional S-Q calculation with just one e-h pair created per photon ($P = 0$). In the figure, curve I for $P = 10\,000$ achieves the maximum multiplication energetically allowed. MEG yields a lower maximum thermodynamic conversion efficiency ($\sim 45\%$, with $E_g = 0.7$ eV) compared to the hot-carrier solar cell, because in the MEG approach there is loss of photon energy between integer multiples of the bandgap energy starting

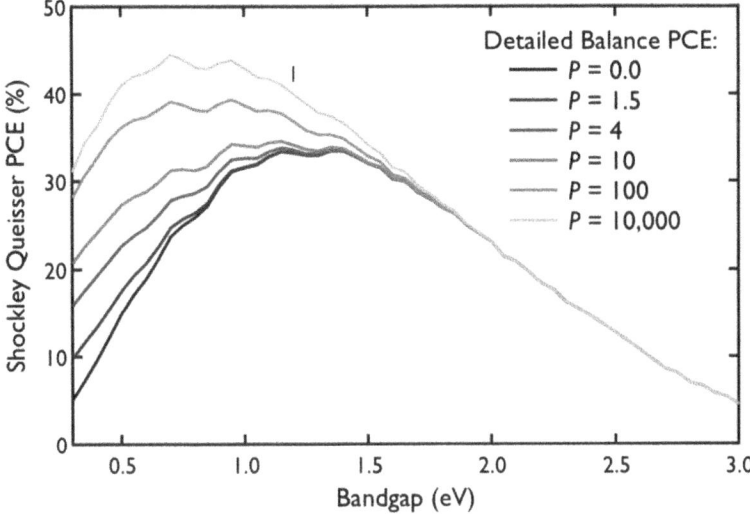

Figure 11.4 The detailed balance-derived maximum allowable PCE.
Reprinted with permission from *Accounts of Chemical Research*, 2013, **46**, 1252. Copyright 2013 American Chemical Society.

at $1E_g$. Nevertheless, the MEG approach is attractive for recouping lost energy because only one absorber layer is needed, conventional carrier contacts can be employed, and the cell can be fabricated mainly using solution-processing techniques.

When concentrated sunlight is used, thermodynamic calculations show that solar cells can convert solar photons into electricity or fuel with higher theoretical power conversion efficiencies.[8] We show in Figure 11.5 how the theoretical efficiency increases under concentration for various MEG characteristics (denoted here by the threshold photon energy where the carrier yield exceeds 1). The black line is for the ideal MEG case ($P = 10\,000$) where the maximum thermodynamic efficiency increases to 75% at 500X (dashed vertical line). For single-junction solar cells that do not exhibit MEG, the theoretical increase in efficiency is relatively small (absolute values of 40% at 500X *versus* 33% at 1X). The hot-carrier solar cell and multi-junction cell approaches have theoretical efficiencies that can exceed 80% (not shown) at maximum solar concentration (46 500X). Similarly, when solar concentration is combined with MEG, the increase in theoretical power conversion efficiency (PCE) exceeds 80%. However, the optimum bandgap shifts to smaller values. If E_g is fixed 0.70 eV, the maximum theoretical efficiency still increases markedly as a function of solar concentration, becoming 62% at 500X for the staircase MEG characteristic, and for a linear MEG characteristic that has a threshold photon energy of $2E_g$, if E_g is fixed at 0.93 eV the PCE is 47% at 500X.

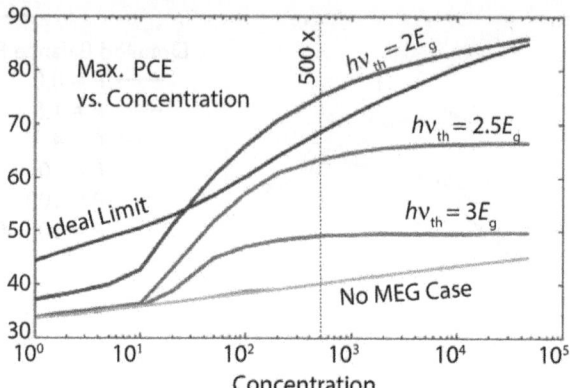

Figure 11.5   Effect of solar concentration on power conversion efficiency (PCE) for various MEG characteristics denoted by the threshold photon energy. For each concentration value the maximum possible PCE is selected from the plot of PCE *versus* bandgap; bandgap is a free variable. Reprinted with permission from *JPCL*, 2012, **19**, 2857. Copyright 2012 American Chemical Society.

11.2   Nanocrystal Synthesis and Physical Properties

11.2.1   Solution Phase Synthesis

Synthesis of colloidal nanomaterials are often performed under air-free conditions using organometallic precursors in organic solvents.[7] Many different quantum-confined semiconductor nanostructures can be made with dimensions on the 1–20 nm length scale. However, QDs of metal chalcogenides, in particular CdSe, PbS and PbSe, have been at the forefront of progress in the development of QD-based solar cells. Accordingly, the synthesis of such materials has been greatly refined.[9–12] There are three main approaches to synthesize colloidal semiconductor nanostructures: hot injection,[13] the heating-up method,[14,15] and the reduction method.[14,16,17] The hot-injection method brings together metal and chalcogenide precursors in a hot (typically 150–300 °C) mixture of organic solvents and surfactants to create conditions analogous to those described decades ago by LaMer and Dinegar for the formation of monodisperse colloids.[18] According to this model, a supersaturation of precursors causes a singular 'nucleation' event of small metal chalcogenide clusters. These nascent QDs then 'grow' by the reactive addition of additional precursors until the precursor concentrations are diminished and reach an equilibrium point, at which the precursor addition occurs at a similar rate to surface redissolution. By manipulating the temperature, solvent, precursors and stabilizing ligand, fine control over the composition and topology is achieved.

A simple non-hot-injection synthetic route (heat-up) that achieves *in situ* halide passivated PbS and PbSe QDs was recently developed. The synthesis mechanism follows a temperature-dependent diffusion growth

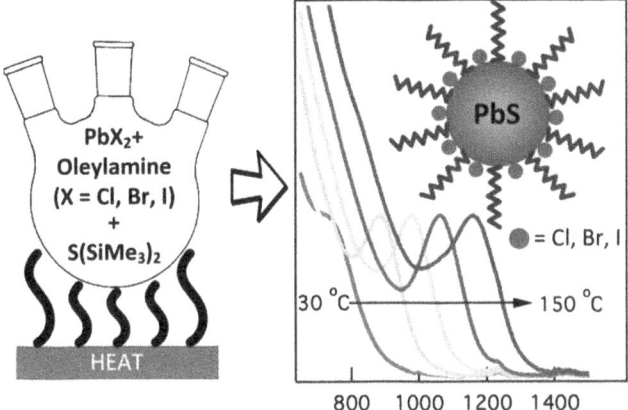

Figure 11.6 Heat-up method of producing PbX QDs. The precursors are added to the reaction flask and then heated to a particular temperature than determines their size.
Reprinted with permission from *ACS Nano*, 2014, ASAP. Copyright 2014 American Chemical Society.

model leading to strategies that can achieve narrow size-distributions for a range of sizes. The precursors are brought together at room temperature and subsequently the temperature is raised to a desired value that then determines the QD size (see Figure 11.6). Narrow size-distributions are obtained by the differences in growth rates of small and large QD that vary with temperature and solvent viscosity. One advantage of the diffusion-controlled synthesis is that scaling up the reaction should in principle be more straightforward than the hot-injection synthesis. To demonstrate the feasibility a 47 g batch of PbS QDs with narrow size-distribution was demonstrated. Another approach to achieve large scale QD synthesis relies on continuous flow synthesis. Bakr and colleagues have reported a continuous flow synthesis method which produces quantum dots of similar quality to those produced using typical batch synthesis methods.[19]

The nanocrystal surfaces are normally terminated by organic molecules bound to surface atoms that control their growth kinetics, allow for dispersion in solvents and generally electronically isolate them from their environment. The as-produced QDs are nearly all single crystals suspended in solution and tend to retain the underlying bulk crystal structure and thus are faceted rather than being spherical. For example, in Figure 11.7 we show TEM pictographs of PbTe and PbSe QDs.[8] Figure 11.7a and b shows smaller QDs that exhibit several facets indicative of the underlying rock salt crystal lattice. In this case, the {100} and {111} facets are the most prominent. As the QDs grow larger, the growth rate along different directions may induce shape changes. Therefore in larger QDs (Figure 11.7c–f)) crystal growth is more prominent in the ⟨111⟩ direction resulting in cubic nanocrystals (NCs) with larger proportion of the {100} facets exposed. The underlying

Figure 11.7   TEM of different sizes of PbSe and PbTe QDs showing both the high
crystallinity (insets are small and wide-angle electron diffraction
patterns). (a and b) 9.0 nm PbTe QDs, (c) 14.3 nm cubic-like PbTe
QDs, (d) ordered array of 15 nm PbSe QDs, (e) PbSe QDs, and
(f) HRTEM of PbSe cubes with 27 nm in size.
Reprinted with permission from *JACS*, 2006, **128**, 3241. Copyright
2006 American Chemical Society.

morphology can impact the packing of QDs in films. Figure 11.7a and b
shows hexagonal close packing while Figure 11.7c–f displays cubic packing
patterns.

11.2.2   Shape and Composition Control

The chemistry and physics of the ligand interaction with different surface
facets can be used to control crystal shape. Figure 11.8 shows TEM pictographs of PbSe NCs grown under varying reaction conditions. The {111}
facets are blocked in Figure 11.8a and b resulting in crystal growth along the
⟨100⟩ direction, thus producing hexapods.[9] Figure 11.8c displays anisotropic
NCs (in this case, QRs) of PbSe which are elongated along the ⟨100⟩ direction
most likely through ligand–ligand and dipolar interactions.[10] Manipulating
the growth mechanism to produce varying shapes and composition is an
ongoing research endeavor. Beyond QRs, platelets or quantum films can be
formed by two-dimensional oriented attachment of QDs into arrays that
recrystallize as single crystals.[11]

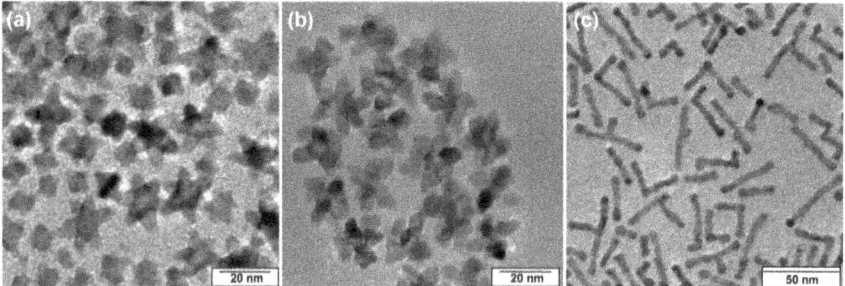

Figure 11.8   TEM pictures of PbSe quantum confined nanostructures showing a few of the many shapes accessible to advanced synthetic techniques. Reprinted with permission from *ACS Nano*, 2012, **6**, 4573. Copyright 2012 American Chemical Society.

11.2.2.1   Stoichiometry and Surface Chemistry

Ligands serve an important role in colloidal QD synthesis, as mentioned previously, and can also influence the QD surface composition. A better understanding of QD–ligand interactions in relation to surface chemistry will promote greater control of photophysical properties. The strength of the interaction between the metal ions and chosen Lewis basic surfactant molecules, particularly for cadmium and lead chalcogenides, typically leads to formation of overall cation-rich stoichiometries in the product particles (Figure 11.9).[20–23] Studies of CdSe QDs have shown that the stoichiometry is heavily dependent on synthetic conditions. It is found to be more sensitive to the purity and amount of surfactant ligands present in the synthesis bath than to the ratio of Cd:Se precursors.[22] Moreover, Weiss has noted a size-dependence in the stoichiometry variation of CdSe QDs[22] that correlates well with the surface-to-volume ratio, supporting Hens' model that colloidal QDs combine a stoichiometric 'bulk-like' core with an outer shell of ions in which faceting leads to a more varied composition.[21] In this model, surface (and near-surface) atoms are more mobile and will reconfigure to produce at least a local minimum surface energy.[24] The Owen group at Columbia University has investigated the details of ligand exchange, including chloride termination of quantum dots, using NMR spectroscopy and reported that the surfaces and stoichiometry of nanocrystals are dynamic and dependent on the composition of the solution.[25,26] Regardless of an imbalance of charges within the strictly inorganic part of QDs, they can achieve overall charge neutrality through binding to ionic ligands, although this can have important ramifications for resulting optical properties.

Interestingly, metal:anion ratios in as-prepared PbSe QDs generally follow the same size-dependent trends observed in CdSe QDs,[20,21,23] but Yu specifically notes that the stoichiometry does *not* depend on reaction conditions, solvents or precursors. This has made direct observation of the effect of surface composition on properties such as PL QY more

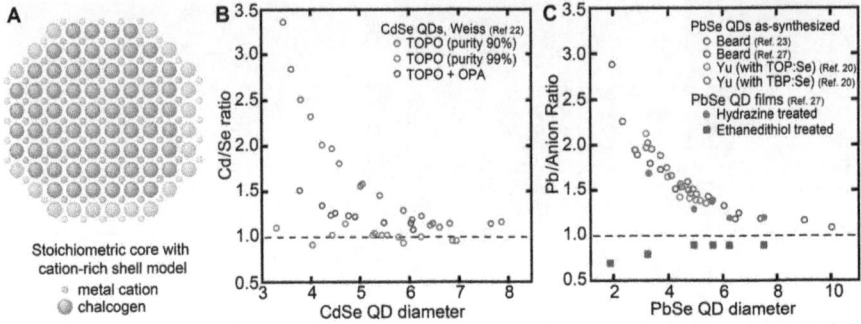

Figure 11.9 (A) Two-dimensional rendering of an atomic QD model showing a cation-rich shell (partially transparent) surrounding a stoichiometric core (solid). This model of a rock-salt structure specifically describes the size-dependent Pb:Se ratio, which trends with the surface:volume ratio. (B) Cd:Se ratios as a function of size from Weiss,[22] showing how the trends depend on the surfactant purity. *n*-Octylphosphonic acid (OPA) is added to high-purity trioctylphosphine oxide (TOPO) to achieve intermediate size-dependent values. (C) Data compiled from literature reports of Pb:Se ratios in PbSe QDs synthesized and measured by varied methods,[20,23,27] including the use of different Se precursors [trioctylphosphine selenide (TOP:Se) and tributlyphosphine selenide (TBP:Se) from Yu[20] are shown]. Squares show data on conductive films of PbSe QDs treated either with hydrazine (circles) or 1,2-ethanedithiol (squares) to enhance electronic coupling. The hydrazine films show Pb:Se ratios similar to as-synthesized QDs, whereas EDT-treated films show anion-rich (Se + S) stoichiometry.[27]
Reprinted with permission from *ACS Nano*, 2013, 7, 1845. Copyright 2013 American Chemical Society.

challenging. Beard has recently reported the treatment of PbSe QDs with a long-chain alkylselenide ligand after synthesis.[27] The alkylselenide binds to Pb ions, particularly at the {111} surface facets, to reduce the overall Pb:Se ratio while still maintaining complete surface passivation. As was found by Jasieniak and Mulvaney in CdSe,[28] surface anions in PbSe directly affect valence band states: the use of the alkylselenide ligand induces hole traps, but also diminishes the propensity of the QDs to oxidize under ambient conditions.[27] In contrast, Klimov and Pietryga found that post-synthesis exposure of PbSe QDs to dilute solutions of chlorine (Cl_2) preferentially removes surface Se ions, leading to an even greater overall enrichment of Pb.[29] Cl_2 oxidizes the Se ions, which are replaced by Cl^- ions to effectively create a thin inorganic passivating layer of $PbCl_2$ that greatly enhances the resistance of these QDs toward oxidation without creating new hole traps, resulting in increased PL QY.[29] As will be discussed below, the stoichiometry also has important implications on electrical properties.

11.2.2.2 Colloidal Nano-heterostructures

Nanoscale and quantum confined materials offer new physics for optical and opto-electronic applications, and has inspired advanced synthetic methods to create multicomponent nanostructures with selectively arranged individual domains for complex functionalities. Material junctions form the basis of modern solid-state technology by controlling electronic current.[30–34] Junctions are key components in digital electronics, electronic switches, signal amplification and processing,[35–38] sensing, light emitting diodes,[39–41] lasers[42,43] and photovoltaics,[44–48] for example. A heterojunction arises when dissimilar semiconductors come into contact and form abrupt interfaces. The junction physics are controlled by the bandgap, work function, electron affinity, and chemical potential of each material as well as how the energy levels or bands align across the material interface (which can be influenced by dipoles or interfacial defects/alloying).[49,50] Some examples of the effects of heterostructuring in nanomaterials include: photoluminescence manipulation in core shell QDs,[51–54] slowed cooling,[55] reduced blinking,[51,54,56] materials with plasmon-assisted absorption enhancement[57] and doping,[58] charge separating interfaces,[59–61] energy funneling,[62,63] and strain[64] effects. Such nano-heterostructures offer promising new materials for modernizing industries such as biological sensing,[65,66] photovoltaics,[67–71] and photocatalysis.[59,72–74]

The electrical properties of heterostructured nanomaterials are governed by the relative alignment of the electron and hole energy levels of the two components at the material interface. When two semiconductor materials form a heterojunction, three possible band alignments can occur: Type I, Type II and Type III. Type III represents the case when the conduction band of one material is deeper than the valence band of the other such that the bands are completely offset. These naming conventions also apply to nanomaterials where the electron and hole states (as opposed to energy bands) can also align in these three configurations. The alignment affects the spatial distribution of the electron and hole wave function within the composite nanostructure[75] and offers distinct control over carrier dynamics.

Incomplete passivation of QD surfaces by the organic ligand shell leads to surface trap states which can affect the carrier dynamics (*e.g.*, reduce the photoluminescence quantum yield).[76–78] Core-shell QDs often have enhanced PLQYs over conventional NCs due to a reduced interaction between the light-emitting core and ligands by way of a shell semiconductor that better passivates and confines carriers to the core. The optical characteristics of core-shell NCs are tunable by tweaking the core diameter, shell thickness and the material compositions and have led to a variety of composite QDs that have use in biological systems[79–81] and devices.[82,83] Recently, advances in the synthesis of multi-component nanomaterials now allow for a variety of shapes and structures. Figure 11.10 shows examples of various nano-heterostructures in a variety of configurations: quantum dot contained

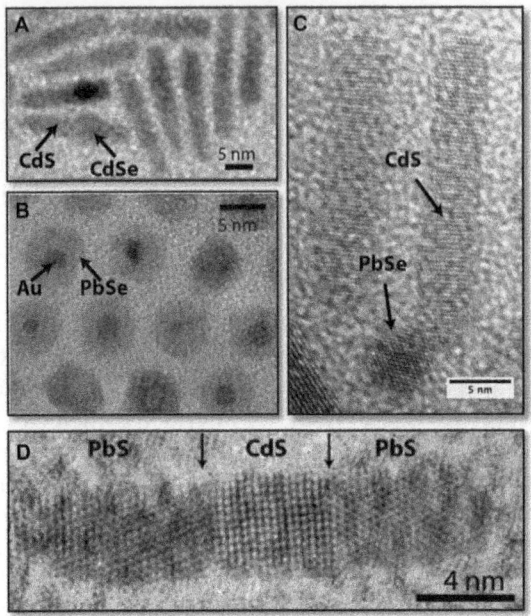

Figure 11.10 TEM pictures of nanocrystals heterostructures made by solution-phase methods. (A) Seeded nanorod growth, resulting in a CdSe QD contained within a CdS nanorods. (B) Core–shell Au-PbSe QDs. (C) PbSe QD grown epitaxially on the end of a CdS nanorods. (D) Segmented nanorods with two axial PbS/CdS interfaces denoted by the arrows.

within a nanorod (a), core–shell spherical quantum dots (b), quantum dot grown on an end facet of a nanorods (c), and end-on (axial) interfaces along the length of a nanorod (d).

11.2.3 Measuring Multiple Exciton Generation

Ultrafast transient absorption (TA) spectroscopy or time-resolved photoluminescence (TRPL) is typically employed to study MEG in colloidal solutions of isolated QDs.[10–13] In 2004, Schaller and Klimov[20] found that when photoexciting a colloidal suspension of PbSe QDs at high photon energies ($h\nu > 3E_g$) a fast component persisted in the transient absorption (TA) dynamics even when the laser fluence was adjusted so that the average number of photoexcited excitons, $\langle N_0 \rangle$, was < 0.1. For pulsed laser excitation of a solution of QDs with pulses of sufficiently short duration the average occupation level follows Poisson statistics,[21] such that the fraction P_m of QDs within the excitation volume with m photogenerated excitons is $P_m = \exp[1 - \langle N_0 \rangle]\langle N_0 \rangle^m/m!$ where $\langle N_0 \rangle$ is the average number of photons absorbed per QD per pulse, given by $\langle N_0 \rangle = \sigma_a \cdot j_p$, with a per-QD absorption cross-section of σ_a (cm^2) and a photon pump fluence, j_p (photons$\times$cm$^{-2}\times$pulse^{-1}). The per-dot absorption

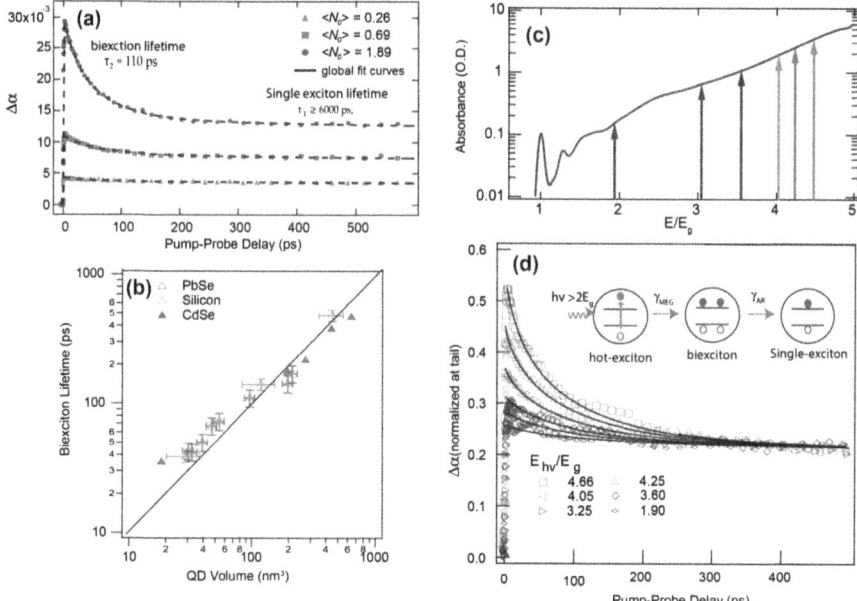

Figure 11.11   MEG transient absorption experiment. (a) The linear absorption. (b) The transients for a constant excitation level but increasing photon energy shown in (a). The amplitude of the fast component increases with increasing photon energy indicative of increased bi-exciton yields.

Reprinted with permission from *Laser & Photonics Review*, 2008, **2**, 377.

cross-section is measured in a separate experiment using ICP-MS[22] or other analytical techniques,[23] and has been measured and tabulated for a variety of QD materials.[22–26] When QDs have more than one exciton, Auger recombination,[27] which describes multi-particle interactions, governs the relaxation dynamics.

Schaller and Klimov analyzed the fast component that they observed for $\langle N_0 \rangle < 0.1$ and $h\nu > 3E_g$ and found that it had the same characteristics of the bi-exciton lifetime, τ_2, measured in a separate experiment for an average occupation level $\langle N_0 \rangle > 1$ and relative photon energy of $h\nu < 2E_g$. The bi-exciton lifetime, τ_2 for typically sized QDs is in the range of ~ 10 to ~ 100 ps and is mainly governed by the volume of the QDs,[28,29] while single exciton lifetimes are at least an order of magnitude longer. The amplitude corresponding to the fast component increased with increasing photon energy after exceeding a photon energy threshold ($h\nu_{th}$) (Figure 11.1) indicating a larger bi-exciton population. Upon investigating QDs with different bandgaps, the fast component tracked the expected linear dependence of τ_2 with QD volume. The appearance of the fast component was ascribed to efficient multi-exciton generation with threshold photon energy of $3E_g$ which is lower than what is required by momentum conservation considerations in bulk

PbSe $\left(\sim 4E_g\right)^{28}$ and $\sim 2X$ lower than what was reported for impact ionization in bulk PbSe, $\sim 6E_g$.[30] The evidence that multiple excitons were produced per absorbed photon, therefore, is the appearance of the fast multi-exciton decay component with the identical time constant as τ_2 when photoexciting above the energy conservation threshold ($>2E_g$) and at low intensity so that each photoexcited QD absorbs at most one photon. The measurement of MEG needs two experiments depicted in Figure 11.11. First (see Figure 11.11a) the intensity dependent dynamics are measured under conditions where MEG is not allowed due to energy constraints ($h\nu < 2E_g$). The bi-exciton lifetime is extracted and it helps to verify that TA dynamics follow Poisson statistics. The bi-exciton lifetime should scale with the QD volume (Figure 11.9b). Next the photon energy is increased while maintaining a constant low fluence (Figure 11.11c and d). The amplitude of the fast component is related to the QY. Reports confirming efficient MEG in PbSe QDs quickly followed[31] and was also reported in PbS[31] and PbTe QDs.[32] MEG has been reported for PbSe,[33,34] CdSe,[35,36] InAs,[37,38] Si,[39] InP,[40] CdTe,[41] SWCNTs,[42] and CdSe/CdTe core-shell QDs.[43] There has been debate[14–19] over TA measurements because of the large spread in reported QYs. Researchers concluded that some discrepancies and variations reported could be attributed to photocharging during the pulsed laser experiments.[19–21] Using lower fluences and rapidly flowing or stirring the solutions during the measurements can minimize or eliminate photocharging effects. We refer interested readers to several reviews of the MEG measurements that detail these complications.[13,18,20]

To obtain accurate QY measurements for each $h\nu_p$ the dynamics are measured at lower and lower values of j_p until they are independent of fluence. The ratio of the early and late time bleaching dynamics, $R_{pop} = \Delta\alpha(\tau_{early})/\Delta\alpha(\tau_{late})$, are related to the as-produced population of excitons, where $\Delta\alpha(\tau)$ is the bleach at pump-probe delay time of τ. The transient bleach at early times captures the total number of excitons created by the pump pulse while the value at late times is related to the total number of QDs excited by the pump pulse. Figure 11.12 plots R_{pop} *versus* j_p for two different pump energies. In the low fluence regime each QD only absorbs

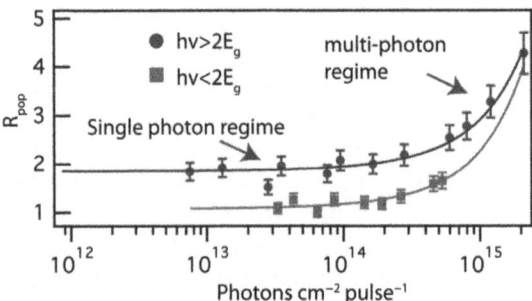

Figure 11.12   Value of R_{pop} *versus* laser fluence for determining σ_a and the QYs.

one photon and R_{pop} reaches a constant value equal to the photon-to-exciton QY while at higher fluences multi-photons are absorbed per QD. The photon fluence where R_{pop} begins to increase is related to the per-dot absorption cross-section and a simple expression can be used to model R_{pop} *versus* j_p. $R_{pop} = QY \cdot j_p \cdot \sigma_a / [1 - \exp(j_p \cdot \sigma_a)]$ as $j_p \rightarrow 0$, $R_{pop} \rightarrow QY$, the model is shown as lines with the circles and squares and can be used to determine σ_a.

11.3 Quantum Dot Solar Cells

Colloidally synthesized quantum dots have been used for photovoltaics for about two decades. In 2002, Nozik proposed three general strategies of incorporating solution grown QDs into solar cells: (1) sensitize wide-bandgap semiconductors with QDs; (2) use electron and/or hole conducting polymers in intimate contact with QDs; and (3) form QD arrays where the QDs are electronically coupled to allow for efficient electron/hole conductivity.[18] The earliest work involved directly forming quantum-confined particles as sensitizers on porous wide-bandgap semiconductors in a QD analog to the dye-sensitized solar cell. At that time, colloidal QD synthesis was still in its infancy and as synthetic techniques progressed to produced monodispersed QD solutions,[19] the concepts of solution casting QDs layers in conjunction with organic polymers led to favorable exciton separation and transport in bulk heterojunction composite films.[20–22] Due to low carrier mobility in hybrid QD films, intimate connection with a charge-separating interface was required for operation of early QD solar cells. Appreciable efficiencies were achieved in 2002 by synthetically elongating the colloidal particle (producing quantum rods and heteropods) to enhance electron transport through the blended polymer/QD film.[23] This concept has been further improved to achieve an efficiency of 3.13% with CdSe nanostructures.[24] Cells based on the dye-sensitized solar cell architecture, using a thin layer of quantum dots to sensitize a mesoporous TiO_2 scaffold, have recently been revisted.[84–87] Santra and Kamat demonstrated the use of a mixture of CdSeS QDs of various bandgaps in order to enhance efficiencies over a single bandgap quantum dot.[88] Alloyed and core-shell CdTe-CdSe quantum dot sensitizers have recently been reported to obtain PCEs greater than 6% in liquid junction cells.[89,90]

11.3.1 Quantum Dot Films

Each of the first two concepts relies on the QDs being in intimate contact with other materials to achieve charge-separation and charge transport, where QDs serve mainly as the light-absorbing component. The third approach requires a highly coupled QD-array that would perform the tasks of light absorption, electron–hole separation, and subsequent charge-carrier transport. However the insulating ligands on QD surfaces prevented efficient charge transport, although photocurrent spectra on an all-QD array with TOPO ligands was recorded in 2000 with 190 nm thick films of CdSe QDs sandwiched between ITO and various metals.[25] In 2003, Guyot-Sionnest and co-workers developed a simple ligand exchange strategy whereby a solid QD film is first formed from the as-made QDs and then soaked in a solution of a different shorter ligand.[26] The conductivity of the resulting QD film increased substantially as the new shorter ligand replaces the original insulating ligands. Bawendi and co-workers demonstrated that ligands as short as butylamine could be used and lead to increased photoconductivity of CdSe QD films. Additionally they showed that an NaOH treatment completely removes the ligands.[27] Talapin and Murray produced field effect transistors (FETs) from PbSe QD solids employing hydrazine (the shortest possible diamine) to remove or replace the oleate ligands, thereby greatly improving electrical coupling.[28] Since exchanging bulky ligands with hydrazine resulted in substantial loss in the film's volume, cracks and pinholes were filled by sequential deposition and treatment which led to the layer-by-layer approach of deposition and ligand removal to create smooth, conformal, pinhole-free films of electronically coupled QDs.[29,30] All-QD PbS and PbSe Schottky junction solar cells[31–33] resulted from these concepts and these QD-layers were later paired with transparent oxide semiconductors to improve the photovoltage and charge-carrier collection.[34,35]

11.3.1.1   Quantum Dot Film Deposition

One challenge to producing efficient solar cells from QD films is the formation of crack-free films of coupled QDs. During film preparation, a QD solution is typically spin- or drop-cast onto a substrate, and the resulting film is then soaked in a suitable chemical reagent such as hydrazine to remove the bulky capping ligands and potentially replace them with a smaller ligand.[91] As a result, the volume of the film has been found to decrease by >40% causing stress within the film that result in cracking and film defects. These defects degrade device performance.[92] However, building up a film with a layer-by-layer process (alternating between dip-coating or spin-coating a thin layer of QDs followed by a chemical treatment), produces uniform, crack-free, and defect-free films.[93] Recently, Fischer et al. reported a solution phase exchange to short thioglycerol ligands, allowing for direct deposition of a well-coupled quantum dot film without requiring post-deposition ligand exchange.[94]

11.3.1.2   Electronic Quantum Dot–Quantum Dot Coupling

The opto-electronic properties of QD solids depend on a delicate interplay between film morphology, interparticle interactions, and nanocrystal surface chemistry. To first approximation the function of the ligand in a QD solid is simply to define the QD–QD spacing. In Figure 11.13 we show that by changing the ligand from oleic acid, to aniline or ethylenediamine, the QD–QD separation can be varied from 1.8 nm to < 0.4 nm.[36] The film

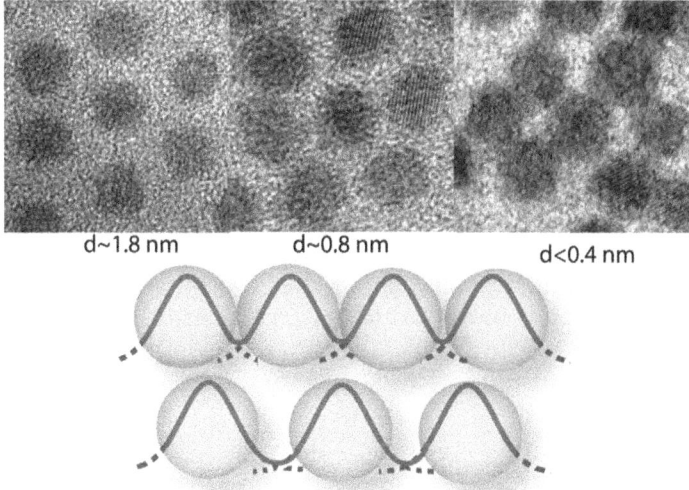

d~1.8 nm   d~0.8 nm   d<0.4 nm

Figure 11.13   The QD–QD distance can be controlled by the choice of QD surface ligand. As produced films with oleic acid capping group have an average QD–QD distance of 1.8 nm while after an ethylenediamine treatment the distance can be reduced to < 0.4 nm. The immediate benefit of a smaller QD–QD distance is greater QD–QD electronic interaction. This is graphically represented in the lower panel where the electronic wavefunctions for excited electrons overlap between neighboring QDs.
Adapted with permission from *JPCB*, 2006, **110**, 25455. Copyright 2006 American Chemical Society.

undergoes a transition from insulating to conducting similar to that of a Mott transition as the QDs become electronically coupled to each other.[37]

However, the ligand exchange can do more than just move the QDs closer to one another. In Talapin and Murray's work on the effect of hydrazine treatment on PbSe QDs films, they found the array displayed n-type behavior but could be switched to p-type when the films were exposed to a vacuum or mild heating.[28] More recently, Zhitomirsky *et al.* have used a halide group ion for surface passivation, which they find to produce n-type PbS QD films.[38] In contrast, conventional thiol treatments produce p-type behavior.[30,39] Recently, Sargent and co-workers built a solar cell where these various passivation strategies have been combined to form homojunction PbS QD solar cells with appreciable efficiencies comparable to the current record 8.6% PbS QD PV devices.[40,41] For bulk semiconductors, intentional doping is at the heart of modern day electronics and provides functionality to the semiconductor's properties. A similar degree of control over doping is still in its infancy for quantum dot films. Ligand exchanges have been explored with hydrazine,[28] thiols,[30] amines,[39] carboxylic acids[42] and recently inorganic ligands. For solar cell applications, thiols, halogens,[43] or combinations of passivants[40,44,45] have proven to be particularly effective.

11.3.1.3 Influence of Stoichiometry on Conductivity

The lead chalcogenides represent an inherently interesting system in which to study the relationship between stoichiometry and electronic properties because in bulk films, the majority carrier type can be manipulated *via* stoichiometry: for instance, Pb (or Se) deficient layers produce p-type (or n-type) films.[95] The concept of applying this type of control to QD arrays in general has recently been modeled,[96] supporting that the effect of stoichiometry on carrier properties might be especially pronounced in PbE QD films. The Talapin and Murray study found that hydrazine-treated PbSe films display strong n-type behavior in field-effect transistors, until a mild heat treatment (that removes hydrazine) switches the majority carrier to p-type.[91] Beard found that hydrazine-treated PbSe arrays retain a Pb-rich stoichiometry, whereas, another popular surface treatment, 1,2-ethanedithiol (EDT), shifts the stoichiometry toward anion-rich, as thiols effectively append S ions to the surface.[27] Oh *et al.* probed the effect of Pb and Se deposition on the energetics of PbSe QD Schottky-junction solar cells.[97] A mild (0.2 Å) deposition of Se results in larger photovoltages and higher current densities, while a similar amount of Pb produces the opposite effect. These effects are attributed to changes in the degree of band-bending at the Schottky junction that result from the shift of the Fermi level in each case. Extrinsic, post-deposition techniques for doping of QD thin films, whether the dopants are delivered as atomic vapor or, perhaps, in solution, may represent facile pathways to p–n and p–i–n structures for efficient charge separation, or even to graded doped layers, *e.g.*, n–n$^+$ layers, for improved carrier extraction at electrodes.[98]

In a separate approach, work by Talapin and co-workers demonstrates that one can take advantage of excellent binding properties of chalcogenides and still exert control over the dominant carrier type in films through the intercession of additional metal cations. For instance, films of CdSe QDs treated with molecular metal chalcogenide complexes[99] exhibit n-type behavior. The use of metal-free chalcogenide complexes as ligands in conjunction with additional metal cations offers even more flexible control over film stoichiometry majority carrier type, particularly in QD films based on materials such as PbE or CdTe that do not exhibit a strong tendency toward one carrier type. In one specific example, K_2Te treatment produces *p*-type CdTe QD films, In_2Te_3 yields ambipolar behavior, whereas $In_2Se_4^{2-}/N_2H_5^+$ enables n-type behavior.[100]

11.3.2 Quantum Dot Material Selection

The bandgap of the QD array is a key factor in selecting material systems for development of QD photovoltaic devices. The systems that have received attention are (bandgap in units of eV): CdTe 1.5, CdSe 1.7, CdS 2.5, PbTe 0.31, PbSe 0.28, PbS 0.41, Cu_2S 1.21, Si 1.12, $CuInSe_2$ 1.0, and $Cu(In_{1-x}Ga_x)(Se_{1-y}S_y)_2$ 1.0–2.4 (depending on Ga and S content). The bandgap of suspended colloidal QDs of these materials range from close to their bulk value to highly blue-shifted bandgaps due to quantum confinement that can be double or triple the bulk bandgap value, depending on the QD diameter. For MEG solar cells CdTe, CdSe, and CdS QDs have little value since their bandgap will be blue-shifted far outside the region of high theoretical efficiency (see Figure 11.4). However, quantum confined PbTe, PbSe, and PbS fall in the region of greatest potential enhancement (0.5 to 1.1 eV, with 0.7 eV being ideal) depending on their size. Other potentially ideal QD systems are III–V materials such as InAs (with bulk bandgap of 0.35 eV), InSb (0.23 eV), and InN (0.8 eV), Ge (0.66 eV), and other IV–VI materials like SnTe (0.18 eV). Group IV materials such as Si (1.12 eV) are also of significant interest despite the fact that the bandgap is a little high for MEG applications. In a separate approach, semiconductor nanocrystals are used as precursor inks for forming thin film bulk semiconductors. Nanocrystals of CdTe, Cu_2S, $CuInSe_2$, $Cu(In_{1-x}Ga_x)(Se_{1-y}S_y)_2$ and Si have been explored.

In addition to matching the absorber bandgap to the AM1.5G spectrum and selecting syntheses with high photoluminescence quantum yield, the stability to oxidation, the stability of the junction, the material and synthesis cost, the global material availability, and potential environmental and health impacts are also important factors to consider. Some materials such as PbTe QDs are very air-sensitive, and devices based on these degrade substantially within seconds outside of an oxygen-free glovebox. On the other hand devices made from $CuInSe_2$ nanocrystals are quite stable and can be prepared without a glovebox (non-encapsulated devices exposed to humid air are stable for years). Regarding material availability and cost, a landmark study was recently performed[101] in which nine semiconductor material systems were identified that could in principle supply worldwide electricity demand *via* photovoltaic devices at material costs substantially lower than silicon. FeS_2 was the highest ranked in this regard, but Cu_2S and PbS also ranked fairly well. PbS naturally occurs in abundance as the mineral galena.

11.3.3   p–n Heterojunction Quantum Dot Solar Cells

The performance of thin film optoelectronic devices comprised of lead chalcogenide (PbX) quantum dots (QDs) has seen rapid development since 2005. Photovoltaic cells incorporating QDs of PbS, PbSe, and their alloy, PbSSe, were first constructed with a simple back-contact Schottky junction between a p-type QD film and a low work function metal electrode.[102–105] Depositing the QD film onto ITO forms an ohmic contact and evaporating a metal with a low work function, such as Ca, Mg, or Al, forms the Schottky contact.

Several reports followed that employ n-type materials which form a heterojunction with the p-type QD film. ZnO, amorphous-Si, and C_{60} have been paired with PbS and PbSe to form heterojunctions.[106–110] Choi *et al.* showed that ZnO nanocrystals (NCs) can be spin coated on top of a PbSe film prior to deposition of the evaporated metal contact.[106] The NC ZnO layer helps to increase the voltage by creating a pseudo p-n junction, since PbSe QD films treated with thiols show p-type behavior under illumination,[93] and ZnO is a well-known n-type material. Leschkies *et al.* has shown ZnO/PbSe heterojunction devices where the ZnO is deposited below the QD film.[107,108] A depleted-heterojunction solar cell was demonstrated by the Sargent group at the University of Toronto, consisting of a layer PbS quantum dots deposited on a TiO$_2$ electrode; this architecture benefits from drift transport within the depletion region to collect photogenerated charge carriers.[111] In 2011, Kramer *et al.* expanded on this architecture with a quantum funnel consisting of layers of various bandgap quantum dots appropriately graded to sweep electrons towards the TiO$_2$ electrode.[112]

The relatively short diffusion lengths of current quantum dot films constrain the thickness of the absorber layers to ensure mostly complete depletion and therefore efficient charge collection. This thickness is often not enough to enable complete absorption, particularly of long wavelength photons. Therefore, strategies have been explored to either improve the collection from a thick absorber, or enhance the absorption within a thinner layer.

For the former approach, cues have been taken from the bulk heterojunction approach of organic solar cells, in which donor and acceptor layers are intermixed in order to efficiently separate electrons and holes. Barkhouse *et al.* used a porous electrode of TiO$_2$ crystals hundreds of nanometers in diameter that was fully infiltrated with PbS quantum dots.[113] The quantum dots within this porous structure were always within a depletion width of an electron extracting interface. This allowed construction of a thicker layer with greater absorption, culminating in an enhancement of short circuit current approaching 30% compared with a planar approach. Rath *et al.* reported in 2012 a bulk nano-heterojunction structure using n-type Bi$_2$S$_3$ nanocrystals mixed with PbS quantum dots in solution and simultaneously cast into a film.[114] Nanowire TiO$_2$ or ZnO structures have also been reported to allow thicker quantum dot layers while maintaining complete depletion, with best reported efficiencies exceeding 7%.[115–118] By varying the ligands used for each layer and creating a p$^+$-p-n structure, Yuan *et al.* demonstrated a heterojunction cell with improved charge extraction at maximum power point conditions.[119]

Enhancing absorption of depleted quantum dot films has followed a few approaches. Cells incorporating a folded-light-path have been shown to allow multiple passes of incident light within a thin, depleted quantum dot film, leading to large enhancements in infrared quantum efficiencies.[120] A periodically etched substrate was used with semi-conformal coating of the active material to produce a nanostructured back electrode which scattered light into the PbS film. This led to a broadband enhancement of the absorption.[121] Absorption has also been increased using plasmonic enhancers. Researchers from the University of Toronto used gold nanoshells specifically tuned to have a strong scattering cross section in the wavelength range most weakly absorbed by the PbS quantum dots. These plasmonic nanoshells were embedded within the quantum dot film to maximize the enhancement of absorption and photocurrent.[122]

Other strategies for improving the efficiencies of quantum dot solar cells have focused on the charge collecting electrodes and their interfaces.[123] Modification of the n-type electrode has been explored to fine tune band alignment, reduce interfacial recombination, and increase the depletion width within the quantum dot absorber.[124–126] Improved hole-extraction has been achieved using metal oxides as selective transport layers between the quantum dot film and the back metal contact.[127–129]

11.3.3.1   Record Cells

A table containing confirmed PV cell efficiencies is maintained by NREL and published online (http://www.nrel.gov/ncpv/images/efficiency_chart.jpg). These devices must be submitted to and measured by independent labs from which they were fabricated following standardized measurement practices, solar simulators, and conditions. The first 'all-QD' solar cell to be included as a new technology platform on this chart appeared in 2010 with an efficiency of 2.94%. This cell was comprised of a heterojunction between a ZnO NC layer and an array of PbS QDs with bandgap of 1.3 eV. The contacts to this cell were ITO on the ZnO side and Au to the PbS QD film. In 2011, a modified treatment [EDT and 3-mercaptoproprionic acid (MPA)] to the PbS QD array and a MoO_x/Al contact to the PbS array resulted in an increase to 4.4% (see Figure 11.14).

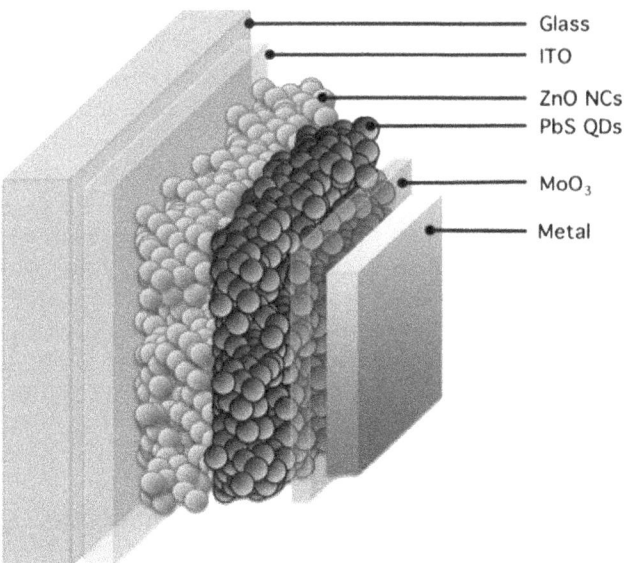

Figure 11.14   An example of PbS based QD solar cell. The PbS QD layer is p-type and forms a p–n heterojunction with the n-type ZnO NCs.

The Sargent group at the University of Toronto then demonstrated a 5.1% efficient device with a PbS QD array passivated with bromide ligands in a junction with TiO$_2$ on FTO rather than ZnO on ITO.[130] In 2012, the (independently verified) record efficiency increased to 7.0% using a hybrid passivation scheme on the PbS QD array.[131] For this solar cell, during PbS QD synthesis, an additional step was performed where CdCl$_2$ and tetradecylphosphonic acid were dissolved in oleylamine and this solution was injected into the QD reaction bath. This Cl$^-$ termination, when used in conjunction with cross-linking MPA ligands during film formation, reduces surface associated trap states within the bandgap of the PbS array. The cell consisted of this PbS array on a ZnO/TiO$_2$ composite layer and the hole contact to the PbS film was MoO$_x$/Au/Ag. Most recently, in 2013, researchers at MIT have demonstrated 8.6% with details of this latest development forthcoming.

11.3.4 Quantum Junction Solar Cells

While heterojunction solar cells have achieved increased efficiencies in recent years, there remain some limitations since proper band alignment to the n-type electrode is required. Thus, even though the quantum dots may be tuned to a smaller bandgap, the change in band alignment may result in a large barrier to charge extraction without re-optimization of the n-type electrode. In 2012, the Sargent group reported a new architecture consisting of a junction formed between n-type and p-type quantum dots, dubbed a quantum junction solar cell.[132] Similar to a homojunction architecture, the same material (PbS QDs) is used on each side of the junction, but uniquely, the bandgap of each side can be tuned independently. This structure was shown to obtain nearly triple the short circuit current density with 0.6 eV bandgap QDs compared with the same bandgap QDs in a heterojunction architecture. Halide ligand exchange in an inert environment was used to enable n-type quantum dot films, with the doping density determined by the halide ion used.[133,134] Optimization of the contact to the n-type layer resulted in efficiencies over 6% and stable performance over tens of hours.[135] Further tuning of the electric field within the active film was achieved through bandgap engineering[136] or graded doping (using bromide and iodide ligands to form a p–n–n$^+$ structure).[137] The latter approach was used to achieve 7.4% PCE, the current record for this architecture. Recently, researchers at ICFO in Spain reported homojunction solar cells using bismuth-doped PbS quantum dots as the n-type layer, with film fabrication occurring under ambient atmosphere.[138]

11.3.5 Multiple Exciton Generation in a Quantum Dot Solar Cell

Two early reports hinted at MEG being active in QD opto-electronic devices. Sambur *et al.* found an internal quantum efficiency (IQE) greater than 100%

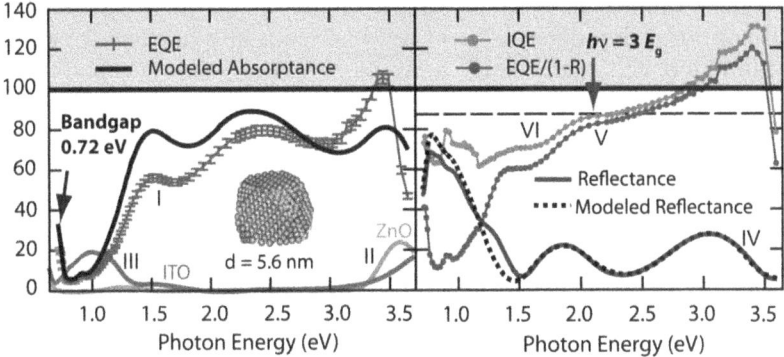

Figure 11.15 Spectral characterization of a PbSe QD solar cell exhibiting MEG. In the left panel, we plot external quantum efficiency (EQE, curve I), and the modeled absorptance of the active layer of the cell (black), ZnO (curve II), and ITO (curve III). On the right panel, we plot the measured (curve IV) and modeled (dashed black) reflectance, the EQE normalized by reflectance (curve V), and the extracted internal quantum efficiency (IQE, curve IV).
Reprinted with permission from AAAS.

in a photo-electrochemical cell that consisted of a monolayer of PbS QDs strongly coupled to an atomically flat anatase surface.[139] Similarly, MEG was invoked as an explanation for increased UV responsivity in PbS QD photo-conductors[140] measured under a large external bias. In 2011, Semonin *et al.* demonstrated the generation of multiple e–h pairs in a PbSe QD based p–n heterojunction solar cell.[141] They found that the external quantum efficiency (EQE) or the number of electron–hole pairs generated per photon *incident* on the solar cell (Curve I, Figure 11.15) exceed 100% for certain photon energies. They reported a PbSe QD (with $E_g = 0.72$ eV) based solar cell with an EQE peaking at 106% (which increased to 114% by application of an anti-reflection coating), confirming that single photons generated multiple excited electrons that were collected at the device electrodes. Through the use of an optical model, the IQE was found to be 130%. Importantly, the IQE data was consistent with previous spectroscopic measurements of MEG. While there are no other reports of an EQE greater than 100% in a solar cell, there are several reports of IQE greater than 100% for solar cells based on bulk semiconductors. For bulk silicon-based solar cells, the photon energy threshold for impact ionization occurs around 3.9 eV, or $3.5E_g$,[142] and at 2.8 eV, or $4.1E_g$, in germanium.[143] The onset for PbSe QD solar cell is $\sim 2.8E_g$ demonstrating that QDs can be used to enhance the MEG process in working solar cells.

11.3.6 Multi-junction Solar Cells

A multi-junction approach is another method that can be used to exceed the single junction Shockley–Queisser limit. A cascading interconnected stack of solar cells with progressively smaller bandgaps is used to reduce losses due to thermalization of carriers. The size tunability of quantum dots allows for precise optimization of constituent cell bandgaps and provides a path for a multi-junction cell made from a single absorber material. Notably, PbS quantum dots can be tuned to the ideal bandgaps for a triple junction cell: 0.71 eV, 1.16 eV and 1.83 eV.[144]

Two reports of tandem (or double junction) quantum dot solar cells were published in 2011. Choi *et al.* reported a tandem cell using an interconnecting layer of ZnO/Au/PEDOT:PSS between the 1 eV and 1.6 eV PbS QD cells.[145] The thin gold layer was found to be crucial for minimization of series resistance from the interlayer. Wang, Koleilat and colleagues reported a depleted heterojunction tandem using a graded recombination layer consisting of a stack of metal oxides with a progression of work functions.[146] This layer allowed efficient recombination of electrons from the back cell with holes from the front cell and led to cells with PCE of 4.2% and open circuit voltages of 1.06 V (Figure 11.16). Besides the design of the interconnecting recombination layer, optimization of the individual subcells was also important for realization of the tandem cell.[147,148]

Extending methods used in recent efficiency advances in single junction cells will allow for further progress in tandem or multi-junction quantum dot solar cells. The precise bandgap tunability and infrared absorption of quantum dots also provide a unique opportunity towards design of a back cell in a multi-junction approach with other materials systems that are unable to harvest long wavelength photons.

11.4   Conclusions and Future Directions

Multiple exciton generation in QDs is a very important process that if harnessed could lead to a new solar conversion efficiency limit. The fact that MEG is enhanced in PbSe QDs is encouraging and a better understanding of the factors that influence MEG will lead to further improvements. To make the largest impact on solar energy technologies the MEG efficiency needs to be further improved so that the onset for MEG occurs as close to $2E_g$ as possible.[19,58] Recent measurements of MEG in PbSe QRs,[29] InP QDs, SWCNT,[49] and Si QDs[50] have shown MEG characteristics that if implemented in PV devices could benefit PCEs. The degrees of freedom afforded by the chemistry and physics of quantum-confined semiconductors, as reviewed here, could produce a material that exhibits MEG at or very near the energy conservation limit.

Beyond MEG, solar energy conversion strategies that incorporate QDs is an emerging field of research with many different approaches. We have reviewed here the use of colloidal QDs in simple photovoltaic devices that consists of p–n heterojunctions as well as emerging multi-junction cells produced from QD absorber layers. The device power conversion efficiencies

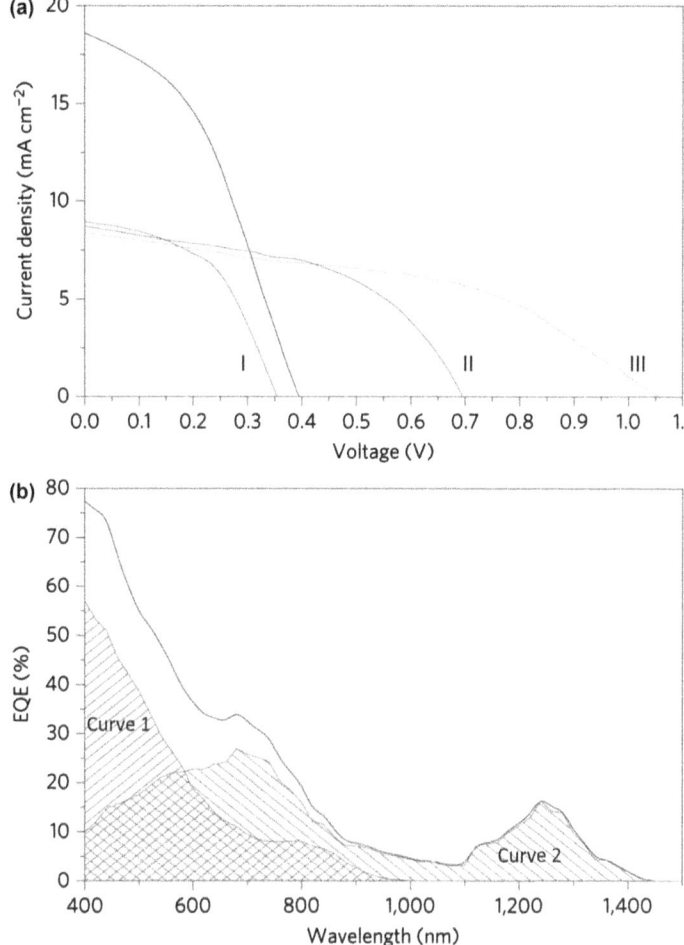

Figure 11.16 (a) Measured current-voltage characteristics under AM1.5 simulated sunlight for large-bandgap (curve III) and small-bandgap (black) cells. The characteristics of the small-bandgap cell using the large-bandgap cell as a filter is shown in curve I. The performance of the tandem cell is shown in curve II, demonstrating addition of the open-circuit voltages of the component cells. (b) The external quantum efficiency of the wide-bandgap cell (curve I) and the small-bandgap cell with (curve II) and without (black) the wide-bandgap film as a filter. The estimated short circuit currents of each component cell are matched. Reprinted by permission from Macmillan Publishers Ltd. (*Nature Photonics* 2011, **5**, 480–484, doi:10.1038/nphoton.2011.123).

have increased from 2% to over 8% in a few years. Further improvements will result from better understand of QD-film junction physics, reduced recombination and improvements in the charge carrier mobility within the QD-film.

Acknowledgements

M.C.B, J.M.L, and A.J.N. gratefully acknowledge funding from the Basic Energy Sciences (BES) division within DOE, the QD work was funded from the Solar Photochemistry program while the QD photovoltaic work is funded by the Center for Advanced Solar Photophysics (CASP) an Energy Frontier Research Center. DOE funding was provided to NREL through contract DE-AC36—08G028308. A.H.I. and E.H.S. acknowledge support from Award KUS—11—009—21 made by King Abdullah University of Science and Technology (KAUST), and funding by the Ontario Research Fund Research Excellence Program, and by the Natural Sciences and Engineering Research Council (NSERC) of Canada.

References

1. A. L. Efros and A. L. Efros, *Sov. Phys. Semicond.*, 1982, **16**, 772–775.
2. W. S. Pelouch, R. J. Ellingson, P. E. Powers, C. L. Tang, D. M. Szmyd and A. J. Nozik, *Phys. Rev. B.*, 1992, **45**, 1450–1453.
3. W. S. Pelouch, R. J. Ellingson, P. E. Powers, C. L. Tang, D. M. Szmyd and A. J. Nozik, *Semicond. Sci. Technol.*, 1992, **7**, B337–B339.
4. J. A. McGuire, J. Joo, J. M. Pietryga, R. D. Schaller and V. I. Klimov, *Acc. Chem. Res.*, 2008, **41**, 1810–1819.
5. R. C. Alig and S. Bloom, *Phys. Rev. Lett.*, 1975, **35**, 1522–1525.
6. M. C. Beard, A. G. Midgett, M. C. Hanna, J. M. Luther, B. K. Hughes and A. J. Nozik, *Nano Lett.*, 2010, **10**, 3019–3027.
7. R. T. Ross and A. J. Nozik, *J. Appl. Phys.*, 1982, **53**, 3813–3818.
8. M. C. Hanna, M. C. Beard and A. J. Nozik, *J. Phys. Chem. Lett.*, 2012, **3**, 2857–2862.
9. C. B. Murray, D. J. Norris and M. G. Bawendi, *J. Am. Chem. Soc.*, 1993, **115**, 8706–8715.
10. C. B. Murray, S. H. Sun, W. Gaschler, H. Doyle, T. A. Betley and C. R. Kagan, *IBM J. Res. Dev*, 2001, **45**, 47–56.
11. M. A. Hines and G. D. Scholes, *Adv. Mater.*, 2003, **15**, 1844–1849.
12. Z. A. Peng and X. Peng, *J. Am. Chem. Soc.*, 2000, **123**, 183–184.
13. Y. Yin and P. Alivisatos, *Nature*, 2005, **437**, 664–670.
14. S. G. Kwon and T. Hyeon, *Acc. Chem. Res.*, 2008, **41**, 1696.
15. J. Joo, H. B. Na, T. Yu, J. H. Yu, Y. W. Kim, F. Wu, J. Z. Zhang and T. Hyeon, *J. Am. Chem. Soc.*, 2003, **125**, 11100–11105.
16. G. Krylova, L. J. Giovanetti, F. G. Requejo, N. M. Dimitrijevic, A. Prakapenka and E. V. Shevchenko, *J. Am. Chem. Soc.*, 2012, **134**, 4384–4392.
17. D. A. Ruddy, J. C. Johnson, E. R. Smith and N. R. Neale, *ACS Nano*, 2010, **4**, 7459–7466.
18. V. K. LaMer and R. H. Dinegar, *J. Am. Chem. Soc.*, 1950, **72**, 4847–4854.
19. J. Pan, A. O. El-Ballouli, L. Rollny, O. Voznyy, V. M. Burlakov, A. Goriely, E. H. Sargent and O. M. Bakr, *ACS Nano*, 2013, **7**, 10158–10166.

20. Q. Q. Dai, Y. N. Wang, X. B. Li, Y. Zhang, M. X. Zhao, B. Zou, J. Seo, Y. D. Wang and W. W. Yu, *ACS Nano*, 2009, **3**, 1518–1524.
21. I. Moreels, K. Lambert, D. De Muynck, F. Vanhaecke, D. Poelman, J. C. Martins, G. Allan and Z. Hens, *Chem. Mater.*, 2007, **19**, 6101–6106.
22. A. J. Morris-Cohen, M. T. Frederick, G. D. Lilly, E. A. McArthur and E. A. Weiss, *J. Phys. Chem. Lett.*, 2010, **1**, 1078–1081.
23. D. K. Smith, J. M. Luther, O. E. Semonin, A. J. Nozik and M. C. Beard, *ACS Nano*, 2011, **5**, 183–190.
24. C. Fang, M. A. van Huis, D. l. Vanmaekelbergh and H. W. Zandbergen, *ACS Nano*, 2009, **4**, 211–218.
25. N. C. Anderson and J. S. Owen, *Chem. Mater.*, 2013, **25**, 69–76.
26. N. C. Anderson, M. P. Hendricks, J. J. Choi and J. S. Owen, *J. Am. Chem. Soc.*, 2013, **135**, 18536–18548.
27. B. K. Hughes, D. A. Ruddy, J. L. Blackburn, D. K. Smith, M. R. Bergren, A. J. Nozik, J. C. Johnson and M. C. Beard, *ACS Nano*, 2012, **6**, 5498–5506.
28. J. Jasieniak and P. Mulvaney, *J. Am. Chem. Soc.*, 2007, **129**, 2841–2848.
29. W. K. Bae, J. Joo, L. A. Padilha, J. Won, D. C. Lee, Q. L. Lin, W. K. Koh, H. M. Luo, V. I. Klimov and J. M. Pietryga, *J. Am. Chem. Soc.*, 2012, **134**, 20160–20168.
30. Z. I. Alferov, *Rev. Mod. Phys.*, 2001, **73**, 767–782.
31. A. I. Gubanov, *Zh. Tekh. Fiz*, 1950, **20**, 1287.
32. A. I. Gubanov, *Zh. Tekh. Fiz*, 1951, **21**, 304.
33. H. Kroemer, *Proc. Inst. Rad. Eng.*, 1957, **45**, 1535.
34. H. Kroemer, *Rad. Corp. Am. Rev.*, 1957, 28.
35. H. Beneking, *Elec. Lett*, 1980, **16**, 602.
36. M. Feng, N. Holonyak and R. Chan, *Appl. Phys. Lett.*, 2004, **84**, 1952–1954.
37. M. Feng, N. Holonyak and W. Hafez, *Appl. Phys. Lett.*, 2004, **84**, 151–153.
38. H. Xu, E. W. Iverson, K. Y. Cheng and M. Feng, *Appl. Phys. Lett.*, 2012, **100**, 113508.
39. A. P. Kulkarni, C. J. Tonzola, A. Babel and S. A. Jenekhe, *Chem. Mater.*, 2004, **16**, 4556–4573.
40. F. So, J. Kido and P. Burrows, *MRS Bull.*, 2008, **33**, 663–669.
41. Q.-H. Wu, *Crit. Rev. Solid State Mater. Sci*, 2013, **38**, 318.
42. H. Kroemer, *Proc. IEEE.*, 1963, **51**, 1782.
43. G. Walter, N. Holonyak, M. Feng and R. Chan, *Appl. Phys. Lett.*, 2004, **85**, 4768–4794.
44. T. L. Chu and S. S. Chu, *Solid State Electron.*, 1995, **38**, 533–549.
45. V. Fthenakis, *Renew. Sustain. Energy Rev.*, 2009, **13**, 2746–2750.
46. A. Hagfeldt and M. Gratzel, *Acc. Chem Res.*, 2000, **33**, 269–277.
47. A. Romeo, A. Terheggen, D. Abou-Ras, D. L. Batzner, F. J. Haug, M. Kalin, D. Rudmann and A. N. Tiwari, *Prog. Photovoltaics*, 2004, **12**, 93–111.
48. A. Shah, P. Torres, R. Tscharner, N. Wyrsch and H. Keppner, *Science*, 1999, **285**, 692–698.

49. G. Margaritondo, *J. Vac. Sci. Technol. B*, 1993, **11**, 1362–1369.
50. C. Tejedor and F. Flores, *J. Phys. C: Solid State Phys.*, 1978, **11**, L19–L23.
51. B. Mahler, P. Spinicelli, S. Buil, X. Quelin, J. P. Hermier and B. Dubertret, *Nat. Mater.*, 2008, **7**, 659–664.
52. P. Reiss, M. Protiere and L. Li, *Small*, 2009, **5**, 154–168.
53. P. Spinicelli, S. Buil, X. Quelin, B. Mahler, B. Dubertret and J. P. Hermier, *Phys. Rev. Lett.*, 2009, **102**, 136801.
54. X. Y. Wang, X. F. Ren, K. Kahen, M. A. Hahn, M. Rajeswaran, S. Maccagnano-Zacher, J. Silcox, G. E. Cragg, A. L. Efros and T. D. Krauss, *Nature*, 2009, **459**, 686–689.
55. A. Pandey and P. Guyot-Sionnest, *Science*, 2008, **322**, 929–932.
56. F. Garcia-Santamaria, Y. F. Chen, J. Vela, R. D. Schaller, J. A. Hollingsworth and V. I. Klimov, *Nano Lett.*, 2009, **9**, 3482–3488.
57. J. M. Luther and J. L. Blackburn, *Nat. Photonics*, 2013, **7**, 675–677.
58. J. S. Lee, E. V. Shevchenko and D. V. Talapin, *J. Am. Chem. Soc.*, 2008, **130**, 9673–9675.
59. L. Amirav and A. P. Alivisatos, *J. Phys. Chem. Lett.*, 2010, **1**, 1051–1054.
60. S. U. Nanayakkara, G. Cohen, C.-S. Jiang, M. J. Romero, K. Maturova, M. M. Al-Jassim, J. van de Lagemaat, Y. Rosenwaks and J. M. Luther, *Nano Lett.*, 2013, **13**, 1278–1284.
61. J. B. Rivest, S. L. Swisher, L. K. Fong, H. M. Zheng and A. P. Alivisatos, *ACS Nano*, 2011, **5**, 3811–3816.
62. T. A. Klar, T. Franzl, A. L. Rogach and J. Feldmann, *Adv. Mater.*, 2005, **17**, 769–773.
63. A. L. Rogach, T. A. Klar, J. M. Lupton, A. Meijerink and J. Feldmann, *J. Mater. Chem.*, 2009, **19**, 1208–1221.
64. A. M. Smith, A. M. Mohs and S. Nie, *Nat. Nanotechnol*, 2009, **4**, 56–63.
65. J. S. Choi, Y. W. Jun, S. I. Yeon, H. C. Kim, J. S. Shin and J. Cheon, *J. Am. Chem. Soc.*, 2006, **128**, 15982–15983.
66. C. Fang and M. Q. Zhang, *J. Mater. Chem.*, 2009, **19**, 6258–6266.
67. M. C. Beard, J. M. Luther, O. E. Semonin and A. J. Nozik, *Acc. Chem. Res.*, 2013, **46**, 1252–1260.
68. S. Kumar, M. Jones, S. S. Lo and G. D. Scholes, *Small*, 2007, **3**, 1633–1639.
69. H. Lee, S. W. Yoon, J. P. Ahn, Y. D. Suh, J. S. Lee, H. Lim and D. Kim, *Sol. Energy Mater. Sol. Cells*, 2009, **93**, 779–782.
70. J. Schrier, D. O. Demchenko and L. W. Wang, *Nano Lett.*, 2007, **7**, 2377–2382.
71. O. E. Semonin, J. M. Luther and M. C. Beard, *Mater. Today*, 2012, **15**, 508–515.
72. K. P. Acharya, R. S. Khnayzer, T. O'Connor, G. Diederich, M. Kirsanova, A. Klinkova, D. Roth, E. Kinder, M. Imboden and M. Zamkov, *Nano Lett.*, 2011, **11**, 2919–2926.
73. C. J. Wang, K. W. Kwon, M. L. Odlyzko, B. H. Lee and M. Shim, *J. Phys. Chem. C*, 2007, **111**, 11734–11741.

74. Y. H. Zheng, L. R. Zheng, Y. Y. Zhan, X. Y. Lin, Q. Zheng and K. M. Wei, *Inorg. Chem.*, 2007, **46**, 6980.
75. D. Steiner, D. Dorfs, U. Banin, F. Della Sala, L. Manna and O. Millo, *Nano Lett.*, 2008, **8**, 2954–2958.
76. B. O. Dabbousi, J. RodriguezViejo, F. V. Mikulec, J. R. Heine, H. Mattoussi, R. Ober, K. F. Jensen and M. G. Bawendi, *J. Phys. Chem. B*, 1997, **101**, 9463–9475.
77. S. Kim and M. G. Bawendi, *J. Am. Chem. Soc.*, 2003, **125**, 14652–14653.
78. X. G. Peng, M. C. Schlamp, A. V. Kadavanich and A. P. Alivisatos, *J. Am. Chem. Soc.*, 1997, **119**, 7019–7029.
79. S. W. Kim, J. P. Zimmer, S. Ohnishi, J. B. Tracy, J. V. Frangioni and M. G. Bawendi, *J. Am. Chem. Soc.*, 2005, **127**, 10526–10532.
80. D. S. Wang, J. B. He, N. Rosenzweig and Z. Rosenzweig, *Nano Lett.*, 2004, **4**, 409–413.
81. J. P. Zimmer, S. W. Kim, S. Ohnishi, E. Tanaka, J. V. Frangioni and M. G. Bawendi, *J. Am. Chem. Soc.*, 2006, **128**, 2526–2527.
82. M. C. Schlamp, X. G. Peng and A. P. Alivisatos, *J. Appl. Phys.*, 1997, **82**, 5837–5842.
83. H. S. Yang and P. H. Holloway, *J. Phys. Chem. B*, 2003, **107**, 9705–9710.
84. H. J. Lee, J.-H. Yum, H. C. Leventis, S. M. Zakeeruddin, S. A. Haque, P. Chen, S. I. Seok, M. Grätzel and M. K. Nazeeruddin, *J. Phys. Chem. C.*, 2008, **112**, 11600–11608.
85. H. Lee, H. C. Leventis, S. J. Moon, P. Chen, S. Ito, S. A. Haque, T. Torres, F. Nüesch, T. Geiger, S. M. Zakeeruddin, M. Grätzel and M. K. Nazeeruddin, *Adv. Funct. Mater.*, 2009, **19**, 2735–2742.
86. S. Giménez, I. Mora-Seró, L. Macor, N. Guijarro, T. Lana-Villarreal, R. Gómez, L. J. Diguna, Q. Shen, T. Toyoda and J. Bisquert, *Nanotechnology*, 2009, **20**, 295204.
87. I. Hod and A. Zaban, *Langmuir*, 2013, Article ASAP.
88. P. K. Santra and P. V. Kamat, *Journal of the American Chemical Society*, 2013, **135**, 877–885.
89. Z. Pan, K. Zhao, J. Wang, H. Zhang, Y. Feng and X. Zhong, *ACS Nano*, 2013, **7**, 5215–5222.
90. J. Wang, I. Mora-Seró, Z. Pan, K. Zhao, H. Zhang, Y. Feng, G. Yang, X. Zhong and J. Bisquert, *Journal of the American Chemical Society*, 2013, **135**, 15913–15922.
91. D. V. Talapin and C. B. Murray, *Science*, 2005, **310**, 86–89.
92. M. Law, J. M. Luther, O. Song, B. K. Hughes, C. L. Perkins and A. J. Nozik, *J. Am. Chem. Soc.*, 2008, **130**, 5974–5985.
93. J. M. Luther, M. Law, Q. Song, C. L. Perkins, M. C. Beard and A. J. Nozik, *ACS Nano*, 2008, **2**, 271–280.
94. A. Fischer, L. Rollny, J. Pan, G. H. Carey, S. M. Thon, S. Hoogland, O. Voznyy, D. Zhitomirsky, J. Y. Kim, O. M. Bakr and E. H. Sargent, *Advanced Materials*, 2013, **25**, 5742–5749.
95. R. S. Allgaier and W. W. Scanlon, *Phys. Rev*, 1958, **111**, 1029.

96. O. Voznyy, D. Zhitomirsky, P. Stadler, Z. Ning, S. Hoogland and E. H. Sargent, *ACS Nano*, 2012, **6**, 8448–8455.

97. S. J. Oh, N. E. Berry, J.-H. Choi, E. A. Gaulding, T. Paik, S.-H. Hong, C. B. Murray and C. R. Kagan, *ACS Nano*, 2013, 7, 2413–2421.

98. Z. Ning, D. Zhitomirsky, V. Adinolfi, B. Sutherland, J. Xu, O. Voznyy, P. Maraghechi, X. Lan, S. Hoogland, Y. Ren and E. H. Sargent, *Advanced Materials*, 2013, **25**, 1719–1723.

99. M. V. Kovalenko, M. Scheele and D. V. Talapin, *Science*, 2009, **324**, 1417–1420.

100. A. Nag, D. S. Chung, D. S. Dolzhnikov, N. M. Dimitrijevic, S. Chattopadhyay, T. Shibata and D. V. Talapin, *J. Am. Chem. Soc.*, 2012, **134**, 13604–13615.

101. C. Wadia, A. P. Alivisatos and D. M. Kammen, *Environ. Sci. Technol.*, 2009, **43**, 2072–2077.

102. E. J. D. Klem, D. D. MacNeil, P. W. Cyr, L. Levina and E. H. Sargent, *Appl. Phys. Lett.*, 2007, **90**, 183113.

103. G. I. Koleilat, L. Levina, H. Shukla, S. H. Myrskog, S. Hinds, A. G. Pattantyus-Abraham and E. H. Sargent, *ACS Nano*, 2008, **2**, 833–840.

104. J. M. Luther, M. Law, M. C. Beard, Q. Song, M. O. Reese, R. J. Ellingson and A. J. Nozik, *Nano Lett.*, 2008, **8**, 3488–3492.

105. W. Ma, J. M. Luther, H. M. Zheng, Y. Wu and A. P. Alivisatos, *Nano Lett.*, 2009, **9**, 1699–1703.

106. J. J. Choi, Y. F. Lim, M. B. Santiago-Berrios, M. Oh, B. R. Hyun, L. Sun, A. C. Bartnik, A. Goedhart, G. G. Malliaras, H. D. Abruña, F. W. Wise and T. Hanrath, *Nano Lett.*, 2009, **9**, 3749–3755.

107. K. S. Leschkies, T. J. Beatty, M. S. Kang, D. J. Norris and E. S. Aydil, *ACS Nano*, 2009, **3**, 3638–3648.

108. K. S. Leschkies, A. G. Jacobs, D. J. Norris and E. S. Aydil, *Appl. Phys. Lett.*, 2009, **95**.

109. B. Sun, A. T. Findikoglu, M. Sykora, D. J. Werder and V. I. Klimov, *Nano Lett.*, 2009, **9**, 1235–1241.

110. S. W. Tsang, H. Fu, R. Wang, J. Lu, K. Yu and Y. Tao, *Appl. Phys. Lett.*, 2009, **95**, 183505.

111. A. G. Pattantyus-Abraham, I. J. Kramer, A. R. Barkhouse, X. Wang, G. Konstantatos, R. Debnath, L. Levina, I. Raabe, M. K. Nazeeruddin, M. Grätzel and E. H. Sargent, *ACS Nano*, 2010, **4**, 3374–3380.

112. I. J. Kramer, L. Levina, R. Debnath, D. Zhitomirsky and E. H. Sargent, *Nano Lett.*, 2011, **11**, 3701–3706.

113. D. A. R. Barkhouse, R. Debnath, I. J. Kramer, D. Zhitomirsky, A. G. Pattantyus-Abraham, L. Levina, L. Etgar, M. Grätzel and E. H. Sargent, *Advanced Materials*, 2011, **23**, 3134–3138.

114. A. K. Rath, M. Bernechea, L. Martinez, F. P. G. de Arquer, J. Osmond and G. Konstantatos, *Nat. Phot.*, 2012, **6**, 529–534.

115. I. J. Kramer, D. Zhitomirsky, J. D. Bass, P. M. Rice, T. Topuria, L. Krupp, S. M. Thon, A. H. Ip, R. Debnath, H. C. Kim and E. H. Sargent, *Advanced Materials*, 2012, **24**, 2315–2319.

116. X. Lan, J. Bai, S. Masala, S. M. Thon, Y. Ren, I. J. Kramer, S. Hoogland, A. Simchi, G. I. Koleilat, D. Paz-Soldan, Z. Ning, A. J. Labelle, J. Y. Kim, G. Jabbour and E. H. Sargent, *Advanced Materials*, 2013, **25**, 1769–1773.

117. J. Jean, S. Chang, P. R. Brown, J. J. Cheng, P. H. Rekemeyer, M. G. Bawendi, S. Gradečak and V. Bulovič, *Adv. Mat.*, 2013, **25**, 2790–2796.

118. H. Wang, T. Kubo, J. Nakazaki, T. Kinoshita and H. Segawa, *J. Phys. Chem. Lett.*, 2013, **4**, 2455–2460.

119. M. Yuan, D. Zhitomirsky, V. Adinolfi, O. Voznyy, K. W. Kemp, Z. Ning, X. Lan, J. Xu, J. Y. Kim, H. Dong and E. H. Sargent, *Advanced Materials*, 2013, **25**, 5586–5592.

120. G. I. Koleilat, I. J. Kramer, C. T. O. Wong, S. M. Thon, A. J. Labelle, S. Hoogland and E. H. Sargent, Scientific Reports, 2013, 3.

121. M. M. Adachi, A. J. Labelle, S. M. Thon, X. Lan, S. Hoogland, and E. H. Sargent, Scientific Reports, 2013, 3.

122. D. Paz-Soldan, A. Lee, S. M. Thon, M. M. Adachi, H. Dong, P. Maraghechi, M. Yuan, A. J. Labelle, S. Hoogland, K. Liu, E. Kumacheva and E. H. Sargent, *Nano Letters*, 2013, **13**, 1502–1508.

123. I. Mora-Sero, L. Bertoluzzi, V. Gonzalez-Pedro, S. Gimenez, F. Fabregat-Santiago, K. W. Kemp, E. H. Sargent and J. Bisquert, *Nat. Comms.*, 2013, 4.

124. H. Liu, J. Tang, I. J. Kramer, R. Debnath, G. I. Koleilat, X. Wang, A. Fisher, R. Li, L. Brzozowkski, L. Levina and E. H. Sargent, *Advanced Materials*, 2011, **23**, 3832–3837.

125. K. W. Kemp, A. J. Labelle, S. M. Thon, A. H. Ip, I. J. Kramer, S. Hoogland and E. H. Sargent, *Advanced Energy Materials*, 2013, **3**, 917–922.

126. P. Maraghechi, A. J. Labelle, A. R. Kirmani, X. Lan, M. M. Adachi, S. M. Thon, S. Hoogland, A. Lee, Z. Ning, A. Fischer, A. Amassian and E. H. Sargent, *ACS nano*, 2013, **7**, 6111–6116.

127. J. B. Gao, C. L. Perkins, J. M. Luther, M. C. Hanna, H. Y. Chen, O. E. Semonin, A. J. Nozik, R. J. Ellingson and M. C. Beard, *Nano Lett.*, 2011, **11**, 3263–3266.

128. P. R. Brown, R. R. Lunt, N. Zhao, T. P. Osedach, D. D. Wanger, L.-Y. Chang, M. G. Bawendi and V. Bulović, *Nano Lett.*, 2011, **11**, 2955–2961.

129. B.-R. Hyun, J. J. Choi, K. L. Seyler, T. Hanrath and F. W. Wise, *ACS Nano*, 2013, **12**, 10938–10947.

130. J. Tang, K. W. Kemp, S. Hoogland, K. S. Jeong, H. Liu, L. Levina, M. Furukawa, X. Wang, R. Debnath, D. Cha, K. W. Chou, A. Fischer, A. Amassian, J. B. Asbury and E. H. Sargent, *Nat. Mater.*, 2011, **10**, 765–771.

131. A. H. Ip, S. M. Thon, S. Hoogland, O. Voznyy, D. Zhitomirsky, R. Debnath, L. Levina, L. R. Rollny, G. H. Carey, A. Fischer, K. W. Kemp, I. J. Kramer, Z. Ning, A. J. Labelle, K. W. Chou, A. Amassian and E. H. Sargent, *Nat. Nanotechnol.*, 2012, **7**, 577–582.

132. J. Tang, H. Liu, D. Zhitomirsky, S. Hoogland, X. Wang, M. Furukawa, L. Levina and E. H. Sargent, *Nano Lett.*, 2012, **12**, 4889–4894.

133. D. Zhitomirsky, M. Furukawa, J. Tang, P. Stadler, S. Hoogland, O. Voznyy, H. Liu and E. H. Sargent, *Adv. Mater.*, 2012, **24**, 6181.

134. Z. Ning, Y. Ren, S. Hoogland, O. Voznyy, L. Levina, P. Stadler, X. Lan, D. Zhitomirsky and E. H. Sargent, *Adv. Mat.*, 2012, **24**, 6295–6299.

135. H. Liu, D. Zhitomirsky, S. Hoogland, J. Tang, I. J. Kramer, Z. Ning and E. H. Sargent, *Appl. Phys. Lett.*, 2012, **101**.

136. V. Adinolfi, Z. Ning, J. Xu, S. Masala, D. Zhitomirsky, S. M. Thon and E. H. Sargent, *Appl. Phys. Lett.*, 2013, **103**.

137. See reference 98.

138. A. Stavrinadis, A. K. Rath, F. P. G. de Arquer, S. L. Diedenhofen, C. Magén, L. Martinez, D. So and G. Konstantatos, *Nat. Comms.*, 2013, **4**.

139. J. B. Sambur, T. Novet and B. A. Parkinson, *Science*, 2010, **330**, 63–66.

140. V. Sukhovatkin, S. Hinds, L. Brzozowski and E. H. Sargent, *Science*, 2009, **324**, 1542–1544.

141. O. E. Semonin, J. M. Luther, S. Choi, H. Y. Chen, J. B. Gao, A. J. Nozik and M. C. Beard, *Science*, 2011, **334**, 1530–1553.

142. L. R. Canfield, R. E. Vest, R. Korde, H. Schmidtke and R. Desor, *Metrologia*, 1998, **35**, 329–334.

143. M. Wolf, R. Brendel, J. H. Werner and H. J. Queisser, *J. Appl. Phys.*, 1998, **83**, 4213–4221.

144. E. H. Sargent, *Nat. Phot.*, 2009, **3**, 325–331.

145. J. J. Choi, W. N. Wenger, R. S. Hoffman, Y.-F. Lim, J. Luria, J. Jasieniak, J. A. Marohn and T. Hanrath, *Adv. Mat.*, 2011, **23**, 3144–3148.

146. X. Wang, G. I. Koleilat, J. Tang, H. Liu, I. J. Kramer, R. Debnath, L. Brzozowski, D. A. R. Barkhouse, L. Levina, S. Hoogland and E. H. Sargent, *Nature Photonics*, 2011, **5**, 480–484.

147. X. Wang, G. I. Koleilat, A. Fischer, J. Tang, R. Debnath, L. Levina and E. H. Sargent, *ACS Appl. Mater. Interfaces*, 2011, **3**, 3792–3795.

148. G. I. Koleilat, X. Wang, A. J. Labelle, A. H. Ip, G. H. Carey, A. Fischer, L. Levina, L. Brzozowski and E. H. Sargent, *Nano Lett.*, 2011, **11**, 5173–5178.

CHAPTER 12

Hot Carrier Solar Cells

GAVIN CONIBEER,*[a] JEAN-FRANÇOIS GUILLEMOLES,[b]
FENG YU[a] AND HUGO LEVARD[b]

[a] School of Photovoltaic and Renewable Energy Engineering and Centre
for Advanced Photovoltaics, The University of New South Wales, Sydney,
2052 Australia; [b] Institut de Recherche et Developpement de l'Energie
Photovoltaique, Chatou 78401, France
*Email: g.conibeer@unsw.edu.au

12.1 Introduction to Hot Carrier cells

The hot carrier solar cell is a solar energy converter that utilizes the excess thermal energy of photo-excited carriers to generate DC electric power. Unlike conventional solar cells, the hot carrier cell maintains the carrier hot population by inhibiting ultra-fast cooling processes. The excess carrier energies above the respective band edges thus contribute to a higher conversion efficiency than that of the conventional cells, which is restricted by the detailed-balance (Shockley–Queisser) limit for a single junction cell to 33%. The hot carrier cell has a limiting efficiency of 66% for unconcentrated sunlight rising to 85% at maximum concentration (46 200 suns), assuming ideal operation under detailed balance.[1-3]

An ideal hot carrier cell would absorb a wide range of photon energies and extract a large fraction of the energy to give very high efficiencies by extracting 'hot' carriers before they thermalize to the band edges. Hence an important property of a hot carrier cell is to slow the rate of carrier cooling to allow hot carriers to be collected whilst they are still at elevated energies ('hot'), and thus allowing higher voltages to be achieved from the cell and hence higher efficiency. A hot carrier cell must also only allow extraction of carriers from

RSC Energy and Environment Series No. 11
Advanced Concepts in Photovoltaics
Edited by Arthur J Nozik, Gavin Conibeer and Matthew C Beard
© The Royal Society of Chemistry 2014
Published by the Royal Society of Chemistry, www.rsc.org

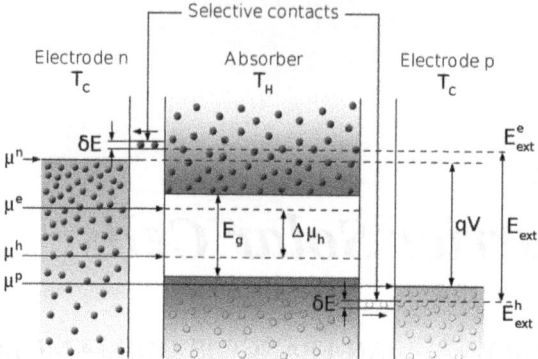

Figure 12.1   Band diagram of the hot carrier cell. The device has four stringent requirements: (a) to absorb a wide range of photon energies; (b) to slow the rate of photogenerated carrier cooling in the absorber; (c) to extract these 'hot carriers' over a narrow range of energies, such that excess carrier energy is not lost to the cold contacts; and (d) to allow efficient renormalisation of carrier energy *via* carrier–carrier scattering. From LeBris.[5]

the device through contacts that accept only a very narrow range of energies (energy selective contacts or ESCs). This is necessary in order to prevent cold carriers in the contact from cooling the hot carriers, *i.e.*, the increase in entropy on carrier extraction is minimized.[4] Figure 12.1 is a schematic band diagram of a hot carrier cell illustrating these requirements.[5]

The microscopic processes which occur in semiconductors determine the carrier dynamics under light excitation. In the case of conventional cells, the photo-generated carriers will first go through an ultra-fast elastic scattering, which renormalizes the carrier energy distribution to a thermal distribution at high temperature. These carriers at the same time gradually lose their energy by colliding with atoms in the material to produce phonons (lattice vibrations at specific resonant frequency). This energy dissipation occurs on the timescale of picoseconds, after which the carriers have energies very close to the band edges. These thermalized carriers then transport towards the respective contact and contribute to the current. Some of the electrons can recombine with holes during the process, leading to nonradiative recombination, re-emission, or photon-recycling.

12.2   Modelling of Hot Carrier Solar Cells

12.2.1   Thermodynamic Analysis for the Hot Carrier Cell

Classically, the limiting efficiency of an energy conversion process is obtained when the internal entropy generation is negligible and given by:

$$\eta_{\text{Carnot}} = \frac{W}{E_s} = 1 - \frac{T}{T_s} \tag{12.1}$$

This is the Carnot efficiency of the process, with a value of 95%, where T_s is the source temperature (sun temperature, about 6000 K) and T is the sink temperature (*i.e.*, cell environment, about 300 K). The issue here is that to have negligible internal entropy generation, the converter and the sun's surface should have almost the same temperature, otherwise the radiative transfer between the two would result in finite entropy generation. But then, the amount of work produced is infinitesimally small: most of the sun's power is internally recycled, which defeats the purpose (see Green[1] for a discussion).

There is yet another way for solar energy conversion using a Carnot engine using a reciprocal system, closer to what is expected for solar cell operation. The sun could heat up a black-body absorber to a temperature T_a, and a Carnot engine could then extract work reversibly from heat at T_a and the heat sink at temperature T. This is a thermodynamic analysis of an endor-eversible system,[6] which leads to the formula:

$$\eta_{\text{endo}} = \left(1 - \frac{C}{C_m}\frac{T_a^4}{T_s^4}\right)\left(1 - \frac{T}{T_a}\right) \tag{12.2}$$

where C is the solar concentration and C_m is the maximal possible solar concentration ($C_m = 46\,200$).

Maximal power is extracted at maximal concentration for an optimal temperature of $T_a = 2480$ K. The efficiency is then 85.4%. The endo-reversible efficiency is quite dependent on the concentration ratio as illustrated on Figure 12.2. As we will see later, an ideal hot carrier solar cell in the energy conservation limit with zero bandgap would operate in this limit.

12.2.2   Models for Ideal Hot Carrier Cells

The concept of a hot carrier solar cell was first introduced for photoelectrolysis cells by Williams and Nozik in 1978.[8] and by Boudreaux *et al.* 1980[9] and then more generally for solar photon conversion by Ross and Nozik in 1982.[3] In standard devices, the photon energy in excess of the threshold absorption energy is given to the photogenerated carrier population, and then to the lattice as heat, with only the bandgap energy being converted as electrical energy. If the carrier thermalization (thermal equilibrium of carriers with the lattice) is reduced sufficiently, then the conversion of the total available energy into potential energy is possible and leads to higher conversion efficiency.

Immediately after absorption, the energy distribution of the photo-generated carriers is a non-equilibrium distribution that depends on the energy distribution of the incident photons, and on the electron and hole effective masses and respective density of states.

The carrier–carrier scattering time constant is usually much less than one picosecond, and decreases with an increasing carrier density, while the time constant for the carrier–phonon interaction is of the order of several picoseconds. Considering these time constants and without carrier extraction, the dynamics of the carrier after a pulsed monochromatic excitation is

(a)

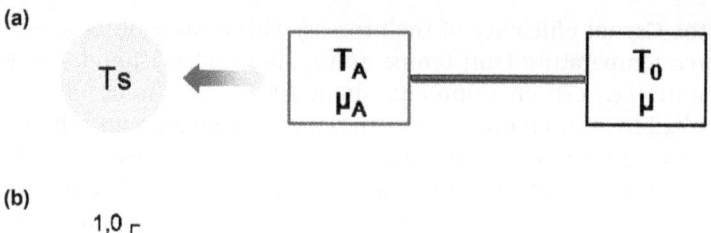

(b)

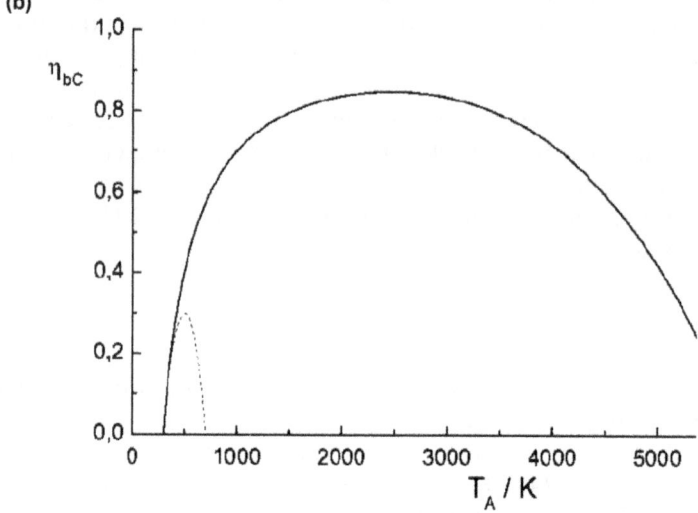

Figure 12.2 Efficiency of an endo-reversible solar cell. (a) Schematics of the device, showing that solar heat is radiatively exchanged with an absorber at temperature T_A while a Carnot engine operating between T_A and T_0 produces work. T_0 is connected to a cold reservoir. (b) The efficiency is computed for various absorber temperatures, displaying a maximum. The full curve corresponds to full concentration and the dotted one to no concentration of solar radiation.

described in Figure 12.3 from Green.[1] Immediately after absorption at $t = 0$, carriers are generated in a narrow energy range (2). Carrier–carrier scattering then occurs and carriers are redistributed in a hot thermal distribution (3)–(4), within a picosecond. This distribution is then cooled towards the lattice temperature because of interaction with phonons (5). Finally, carriers recombine with a nanosecond to sub-microsecond time constant (7) or are extracted through contacts.

If the excitation rate is higher than the thermalization rate, and lower than the carrier–carrier scattering rate, a steady state hot distribution can be established, and the carrier kinetic energy, which is usually lost as heat, can be used and contributes to the conversion efficiency.

Under steady state conditions there are various possible carrier distributions:

- *Hot non-equilibration*: the carriers are not even equilibrated among themselves, a common carrier temperature and chemical potential cannot be defined.

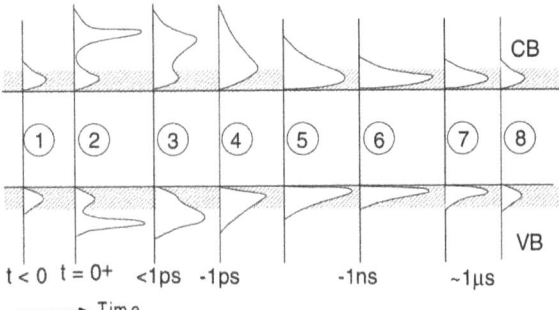

Figure 12.3 Time evolution of electron and hole population after a laser excitation.[1]

- *Hot equilibration*: the electron population equilibrates thermally and the hole population equilibrates thermally but separately, but neither equilibrate with the lattice; thus resulting in a hot carrier population at $T_H > T_C$, where T_C is the temperature of the lattice. A chemical potential (quasi-Fermi level) can be defined for carriers in each band.
- *Full thermalization*: the common situation where carriers are at thermal equilibrium with the environment at ambient temperature. A chemical potential can be defined for carriers in each band.

Assuming such steady state hot carrier population is achieved, specific care is required regarding carrier extraction. If the photogenerated electrons and holes are withdrawn at electrodes, dissipation of the excess kinetic energy of an excited carrier to the lattice temperature will occur rather than its conversion to useful work. In other words, the carrier's kinetic energy has to be converted into potential energy, or voltage. This can be achieved with energy selective contacts[2] that are described below in more detail.

Figure 12.1 is a schematic of the principle of a hot carrier solar cell. Photons are absorbed in the absorber having a bandgap E_g, where carriers are in thermal and chemical disequilibrium with the lattice at temperature T_C, characterized by a temperature $T_H > T_C$ and a quasi-Fermi level splitting $\Delta\mu_H = \mu^e - \mu^h \neq 0$. Carriers are extracted through selective contacts with a transmission energy range δE, electrons at energy E^e_{ext}, and holes at energy E^h_{ext}, towards electrodes with an applied voltage, $V, = (\mu^n - \mu^p)/q$. The extraction energy is defined as $E_{ext} = E^e_{ext} - E^h_{ext}$. Normally $E_{ext} > E_G =$ the bandgap.

12.2.3 Detailed Balance Models and Limit of Efficiency

In order to evaluate the potential of the hot carrier solar cell concept, models have been developed to evaluate its maximal achievable efficiency.

These models are based on a detailed balance approach that defines the electrical power output of a device as the difference between the power absorbed and the different types of losses.

In 1961, a model was proposed by Shockley and Queisser,[10] where the current is not only determined by photon absorption, but is equal to the difference between absorption and radiative recombination:

$$J = q\left[\dot{N}_{abs} - \dot{N}_{em}(V)\right] \tag{12.3}$$

where J is the electric current density in A m^{-2}, N_{abs} is the absorbed photon current density in m^{-2} s^{-1}, N_{em} is the emitted photon current density in m^{-2} s^{-1} depending on the terminal voltage, V, of the device, and q is the charge of an electron.

Assuming the sun emits a black body spectrum at their respective temperature T_S, the photon flux incident on the cell per unit area, per unit solid angle and per photon energy is:

$$dN_{ph} = \frac{1}{4\pi^3\hbar^3 c^2}\frac{E^2}{\exp(E/k_B T_S) - 1}\cos\theta\,\sin\theta\,d\theta\,\varphi\,dE \tag{12.4}$$

where E is the photon energy, θ is the inclination angle, and φ is the azimuthal angle.

Integrating over directions and energies, the absorbed photon current density is then given by Planck's law:

$$\dot{N}_{abs} = \frac{1}{4\pi^2\hbar^3 c^2}\int_{E_G}^{\infty}\frac{E^2 dE}{\exp(E/k_B T_S) - 1} \tag{12.5}$$

where f is a geometrical factor that accounts for the solid angle for photo-absorption. If incident photons comes from a cone with half angle θ_{max}, f is given by $f = 2\pi\int\cos\theta\sin\theta\,d\theta = \pi\sin^2\theta_{max}$. It is 1 for maximally concentrated radiation ($\theta_{max} = \pi/2$) and 6.7×10^{-4} for non-concentrated light (the sun is viewed from a solid angle $\Omega = 6.8 \times 10^{-5}$). The absorptivity is assumed to be 1 above the bandgap energy, and zero below.

In 1982, Ross and Nozik[3] suggested a device where the carriers reach a thermal equilibrium amongst themselves at a temperature T_H, but are completely thermally insulated from the lattice temperature T_C ($<T_H$). Equation (12.3) becomes:

$$J = \frac{q}{4\pi^2\hbar^3 c^2}\int_{E_G}^{\infty}\left[\frac{fE^2}{\exp(E/k_B T_S) - 1} - \frac{E^2}{\exp(E - \Delta\mu_H/k_B T_C) - 1}\right]dE \tag{12.6}$$

where for simplicity conduction band electrons and valence band holes have reached thermal equilibrium at T_H and have a chemical potential μ^e and μ^h respectively. $\Delta\mu_H = \mu^e - \mu^h$ is the chemical potential difference between electrons and holes. If a pathway is introduced for carrier extraction, where electrons and holes are extracted at specific energies E^e and E^h, this provides a limited interaction between the outside world and the hot system, which continues to be thermally insulated. Then it holds that according to the Ross and Nozik model:[3]

$$qV = E_{\text{ext}}\left(1 - \frac{T_C}{T_H}\right) + \Delta\mu_H \frac{T_C}{T_H} \qquad (12.7)$$

Equation (12.6) is not sufficient to determine the current voltage characteristic because of the two unknowns, $\Delta\mu_H$ and T_H. An additional equation describing the energy balance in the system is required. The absorbed and emitted energy current are simply obtained from the photon current expressions by multiplying the integrand in Equation (12.6) by the energy E of the photons and must equal the energy current extracted through the contacts, that is:

$$J \cdot E_{\text{ext}} = \frac{q}{4\pi^2\hbar^3c^2}\int_{E_G}^{\infty} E\left[\frac{fE^2}{\exp(E/k_BT_S) - 1} - \frac{E^2}{\exp(E-\mu_H/k_BT_C) - 1}\right]dE \qquad (12.8)$$

Combining Equations (12.6) and (12.8), the current is determined for a given value of E_G, E_{ext} and f. The voltage is then obtained from Equation (12.7). The current–voltage characteristic and the cell efficiency are obtained, with a predicted optimal limit of 66% under AM1.5 illumination and a bandgap approaching zero.

In this model, particle conservation was assumed (*i.e.*, one net photon in is one net electron out), which means that only radiative recombinations are considered, but also that Auger recombination and impact ionization are neglected. In 1997, Würfel[2] proposed another approach where impact ionization is assumed to be dominant over all other recombination processes and faster than carrier extraction. In these conditions, the electrons and holes are at chemical equilibrium ($\Delta\mu_H$) and are described by a Fermi–Dirac distribution at a temperature $T_H > T_C$. The limit of efficiency in that case is 53% under non-concentrated 6000 K black-body illumination, and 85% under full concentration.

These particle conservation (Ross and Nozik)[3] and energy conservation (Wurfel)[2] models represent two extreme conditions for ideal materials; namely zero electron and hole interaction (zero Auger coefficients) and infinitely fast Auger processes (instantaneous impact ionisation) respectively. They give very similar limiting efficiencies at maximum concentration but differ markedly under 1 sun. For any real material the Auger coefficients lie somewhere between these extremes, with partial impact ionization primarily for high energy carriers well above the bandgap.[11] Much more realistic models, which are necessarily more specific and complex, are considered in section 12.2.5.

12.2.4 The Mechanisms of Carrier Thermalization

In the above determination of achievable efficiencies, it was stated that the electron–hole plasma was thermally insulated from the environment, which is not the case in practical conditions. The carriers thermalize with the environment mainly because of interactions with the lattice vibration modes. This process usually occurs in few picoseconds, which is much faster than the carrier extraction time in conventional cells. In order to allow a hot carrier population regime, the carriers have to be extracted before they thermalize, which means that the thermalization rate has to be controlled and reduced.[12]

12.2.4.1 Carrier–Carrier Scattering

Immediately after absorption, the photogenerated electron–hole plasma distribution is not an equilibrium Fermi-Dirac distribution. In the case where a laser excites the carriers, for instance, electrons and holes are generated in a very narrow region in the energy–momentum space. This population, in chemical and thermal disequilibrium with the environment, is subject to different scattering processes that bring it to equilibrium.

12.2.4.1.1 Intraband Scattering. The intraband scattering comes from Coulomb elastic interaction between free carriers.[13,14] Electron-electron, electron–hole and hole–hole scattering occur with exchange of energy and momentum to more uniformly distribute the excess kinetic energy amongst carriers. Here the carriers stay in their respective energy band (no recombination), and this interaction is called intraband scattering.

The intraband scattering is a very fast two-particle process that depends quadratically on the carrier density. Its typical time scale is in the femtosecond range.

Considering this process, the free carrier population naturally evolves towards a Fermi–Dirac distribution defined by a temperature that is common to electrons and holes (thermal equilibrium), which can be much higher than the environment temperature, since no interaction with the lattice has occurred. In the case of different time constant for electron–hole interaction and electron–electron or hole–hole interaction, different temperatures for electrons and holes can be obtained.[15]

12.2.4.1.2 Interband Scattering. The collisions between carriers can also result in electron–hole pair generation or recombination in non-radiative processes. Two cases are possible:

- *Auger recombination*: an electron and a hole recombine, and give their energy to a third free carrier (electron or hole).
- *Impact ionization*: a high energy free electron (or hole) gives a part of its energy to give rise to an exciton.

Table 12.1 Comparison of limits of efficiency ($\eta_{\max}$), optimal band gap ($E_{G,\,opt}$), and carrier temperature in optimal operating conditions (T_H) for conventional single junction and different hot carrier models under non-concentrated ($\Omega_S = 6.8 \times 10^{-5}$) and fully concentrated ($\Omega_S = 2\pi$) 6000 K black-body spectrum.

Model	Ω_S	$\eta_{\max}$	$E_{G,\,opt}$	T_H
Conventional single junction	6.8×10^{-5}	31%	1.4 eV	300 K
	2π	41%	1.1 eV	300 K
Particle conservation	6.8×10^{-5}	66%	0 eV	3600 K
	2π	86%	0 eV	4200 K
Impact ionization	6.8×10^{-5}	53%	0.9 eV	348 K
	2π	85%	0 eV	2470 K

Here again the free carrier plasma total energy is conserved, but the number of particles is not. It is a three particle process, and therefore less probable than intraband scattering. A typical time constant is 100 ps, given in Table 12.1. It also depends on the bandgap, and can become a very efficient process in small bandgap materials. The exchange between electrons and holes results in an equilibration of electron and hole chemical potentials. In the limit of very fast Auger recombination and impact ionization, electrons and holes can be considered in chemical equilibrium with quasi-Fermi level splitting approaching zero.[2]

12.2.5 Modelling of Hot Carrier Solar Cell Efficiency

The preceding discussion deals with the ideal situation without the constraint of real material properties. The calculation can be carried for a HCSC closer to reality based on an Indium Nitride (InN) absorber layer. InN has been chosen as a potential material because of its narrow electronic bandgap for absorption of a wide range of photon energies. This band has only fairly recently been found to be at 0.7 eV,[16,17] previously the widely held value at about 1.7 eV was due to difficulty in removing oxygen impurities. Importantly InN also has a very wide phonon bandgap, significantly larger than its acoustic phonon energy. As discussed below, this is good for suppression of phonon decay, and hence for slowing carrier cooling.

Figure 12.4a is a schematic representation defining the energy and particle fluxes into and out of a hot carrier cell. Calculation of limiting efficiency was performed taking into account real optical and electronic properties of InN, removing most of the ideality assumptions used in other models.[18–21] The detailed band structure of wurtzite bulk InN has been considered in performing computation of carrier densities, pseudo-Fermi potentials and II–AR time constants.[22] Results have been calculated considering ideal energy selective contacts for the HCSC, which means that contacts have a very high conductivity and a discrete energy transmission level. The results of this calculation are shown in Figure 12.4b. The maximum limiting efficiency is found to be 52.0% at maximum concentration, 43.6% at 1000 suns and only

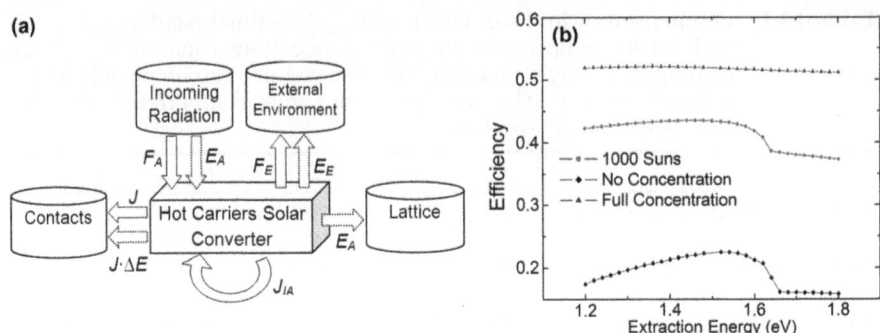

Figure 12.4   (a)Schematic representation of energy and particle fluxes interactions used in the model (particle fluxes – full line arrow, energy fluxes – dotted line arrow). (b) HCSC efficiency as a function of carrier extraction energy level. Parameters used are: thermalization time = 100 ps, concentration = 1000, lattice temperature = 300 K and absorber layer thickness = 50 nm. (*J*, *F* and *E* are current density and particle and energy fluxes as denoted by subscripts for *A* absorption, *E* emission and *IA* Auger processes.)

22.5% at 1 sun. Higher efficiency occurs at higher concentration primarily because of the increasingly dominant effect of II-AR processes over thermalisation as the density of carriers increases. The degree to which the optimum extraction energy is important increases with decrease in concentration (*i.e.* the curve gets less flat). This again is due to the predominance of II-AR at high concentrations. At lower concentration the greater potential for thermalisation means there will be a better defined optimum extraction. There is also a weakly increasing dependence of optimum extraction energy from 1.38 to 1.44 to 1.52 eV as concentration decreases, because lower concentrations drive greater advantage in extracting carriers at a higher energy, but that there is a sharp drop off in this once carrier energy is so high that there are few hot carriers to extract and hence a lower current.

12.2.6   Modelling of Non-ideal ESCs

The limiting efficiency for the hot carrier InN solar cell has been calculated considering non-ideal energy selective contacts (ESCs). In this case the carriers are not extracted on a single energy level, but over a finite energy window. Calculations have been performed taking into account contact resistance and entropy generation effects.

The flux of current travelling through the ESCs towards the cold metal electrodes can be described using the following relation:

$$J_{e,h}(\varepsilon) = T_{e,h}(\varepsilon) \times \left[f_{T_C,\mu_{e,h}}(\varepsilon) - f_{T_{rt},V_{e,h}}(\varepsilon) \right]$$

$$\times \frac{e}{\pi^3 \hbar} \sum_{\min \varepsilon} \left(\int_0^\varepsilon \int_0^{\varepsilon - \varepsilon_z} \frac{dk_y\, dk_z}{d\varepsilon_y\, d\varepsilon_z} d\varepsilon_y\, d\varepsilon_z \right) \tag{12.9}$$

The current density in this case is proportional to the occupation probability at the two sides of the ESC. Equation (12.9) has been derived assuming no correlation of energy of electrons in three different directions as shown in Equation (12.10). This assumption is acceptable if there is a parabolic dispersion relation at minimum energy point along the three different directions:

$$\bar{\varepsilon} = \varepsilon_x \hat{x} + \varepsilon_y \hat{y} + \varepsilon_z \hat{z} ; \quad \bar{k} = k_x \hat{x} + k_y \hat{y} + k_z \hat{z} \qquad (12.10)$$

Based on the energy and carrier conservation, $\Delta\mu$ and T_C at steady state are calculated:

$$F_A - F_E + F_{IA} - \frac{J}{e} = 0 ; \quad E_A - E_E - E_{TH} - \frac{E_J}{e} = 0 \qquad (12.11)$$

The quantities are as defined in Figure 12.4a.

The maximum efficiency has been found for a ΔE between 1.15 eV and 1.2 eV with a transmission energy window δE of 0.02 eV. The value of limiting efficiency is 39.6% compared to 43.6% calculated in the previous section using ideal ESCs. The drop in efficiency is mostly due to the decrease of open circuit voltage related to the decreased extraction level, Equation (12.11). This is partially compensated by an increase in extracted current due to increased II rate.

Figure 12.5a shows calculated efficiency as a function of δE for several values of extraction energy. In all the curves two different trends can be identified. If the value of δE is too close to zero, the efficiency is very low due to low carrier extraction, thus a very small value of short circuit current. The conductivity of the contact in this case is indefinitely large. Enlarging δE, the number of carriers available for extraction increases, improving J_{SC}, and so the maximum efficiency. In general the efficiency peak has been found for values of δE from 0.02 eV to 0.1 eV depending on the extraction energy ΔE. For the configurations which show higher efficiencies, $\Delta E < 1.35$ eV, the optimum value of δE goes from 0.02 eV to 0.05 eV. This optimum value for δE of between is very close to kT_{RT}. This represents the variation in energy in the contacts such that approximately the kT_{RT} will inevitably be lost anyway by carriers thermalizing within the contacts. Thus it sets a lower limit on a reasonable δE. Therefore this result indicates that the transmission energy range has to be very small and confirms once again the high selectivity requirements of ESCs for HCSC.[23]

In Figure 12.5b the value of maximum efficiency as a function of ΔE is reported for different values of δE. It can be observed that for small transmission energy window the extraction energy which allows maximum efficiency is lower compared to the one calculated using ideal ESCs. This effect is related to the higher occupancy at lower energies, which increases the value of J_{SC} for contacts with a small transmission window.

Modelling and computation of HCSC efficiencies with non-ideal selective energy contacts has been carried out (Figure 12.5).[22,24] These are contacts

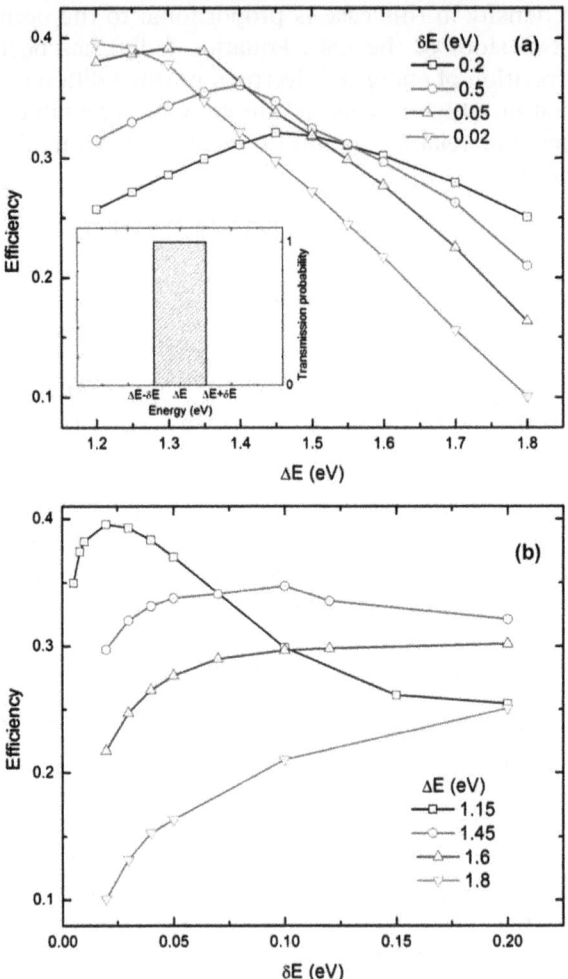

Figure 12.5 (a) Calculated HCSC efficiency *versus* extraction energies, ΔE, under 1000-sun illumination, for different extraction energy window width, δE. (b) Calculated HCSC efficiency *versus* extraction energy window, δE, under 1000-sun illumination, for different extraction energies, ΔE.

extracting electrons at one side and extracting holes at the other side with a narrow transmission window. Normally selective energy contacts are proposed to be realized with resonant tunnelling diodes, since this structure gives confined energy levels, at which the transmission probability peaks. The transmission probabilities, as a function of energy levels, are modelled with one-dimensional quantum transmission treatment. The resultant transmission probabilities are involved in the device model, by relating the carrier statistics inside the absorber to the currents, *i.e.*, carrier current and heat current thought the contacts. By equating the incoming and outgoing fluxes current corresponding to a given voltage is obtained, resulting in an MPP efficiency.

From the computation the optimal efficiency is close to 40% for an InN absorber assuming complete optical absorption, even with a specific contact configuration. The contact configuration proposed is InGaN/InN/InGaN resonant tunnelling diode, with the advantage of uniform nitride deposition with the absorber. Also variation of indium fraction gives an extra dimension to optimize the transmission peaks, leading to good cell performance. The results are shown in Figure 12.6.

12.2.7 Monte Carlo Modelling of Real Material Systems

The Ensemble Monte Carlo (EMC) method has been successfully used to solve the Boltzmann transport equation and calculate charge carrier transport properties for many years in a variety of materials and devices.[25] Here, this method is extended to account for optical absorption, radiative recombination, Auger recombination processes and impact ionization in a non-parameterized way. This has been achieved by the use of the sampled carrier distribution (characteristic of the EMC) to calculate all carrier–carrier interactions consistently (the corresponding scattering rate being updated with the evolving carrier distribution), and by the use of matrix elements and overlap factors calculated in a Non-local Empirical Pseudopotential Method for the semiconductor's electronic structure. Thus, the illumination of the HCSC absorber and the energy selective contact energy position and selectivity are the only parameters left to study the hot carrier solar device operation. This new model has been applied to a 100 nm thick $In_{0.53}Ga_{0.47}As$ absorber under a fully concentrated solar illumination.[26] At the bottom of the absorber, an ideal reflector is modelled to enhance the absorption in a numerically easy way (all solar rays are assumed to be normal to the surface). Energy selective contacts (ESC) are modelled by transmission windows, in which the carrier extraction is treated as a scattering process. Details about how the hot phonon effect is included in the calculation can be found in Tea *et al.*[26]

According to the calculations, 34% of the incoming photon flux is absorbed. Radiative recombination and impact ionization are found to be negligible in all calulations. Figure 12.7 shows the calculated photocurrent versus the electron and hole ESC position within the corresponding band. The maximum calculated photocurrent is 820 A cm^{-2}, and corresponds to the configuration where the ESCs are located at the band edges ($E_{ext} = 0$). When the ESC enters the conduction and valence bands the photocurrent drops due to Auger recombination (also shown in Figure 12.5). Indeed, ESC far from the band edges enable the electron and hole population to accumulate. This favours Auger recombinations which have been determined to be the main current loss process. Examination of Figure 12.5 also reveals that the extracted current is more sensitive to the hole contact energy position than the electron one.

The same calculations have been performed including the phonon bottleneck effect (hot LO phonons) using the relaxation time approximation (see Tea *et al.*[26] for more details about the implementation), in the configuration

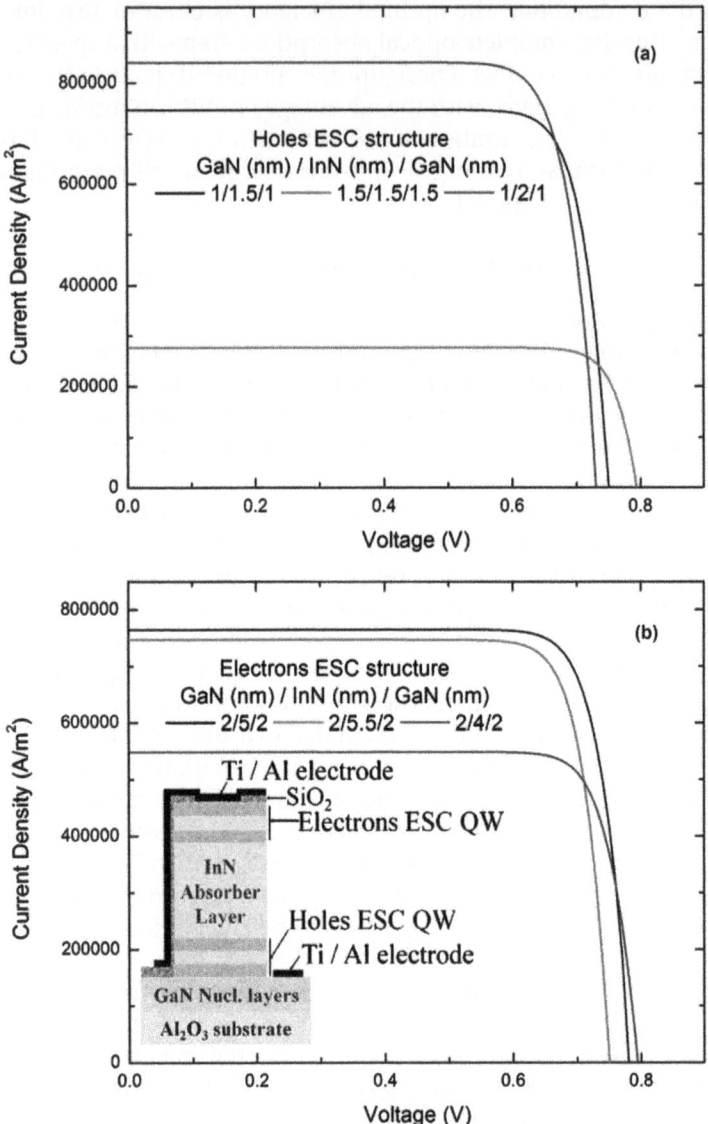

Figure 12.6 Current–voltage characteristics of the HCSC with In$_x$Ga$_{1-x}$N/InN/ In$_x$Ga$_{1-x}$N quantum wells as ESCs. Different configurations for electron contacts and hole contacts have been taken into account.

where the electron and hole contacts are located at 0.2 and 0.1 eV from the band edges respectively. The LO phonon lifetime in In$_{0.53}$Ga$_{0.47}$As is still unknown, so the LO phonon lifetime in GaAs ($\tau_{LO} = 2.5$ ps at 300 K) have been used instead. The electron and hole distributions did not show any significant heating features. This suggests that a raw semiconductor slab may not be a

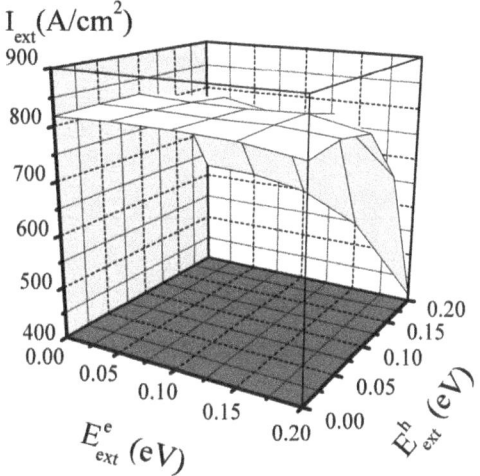

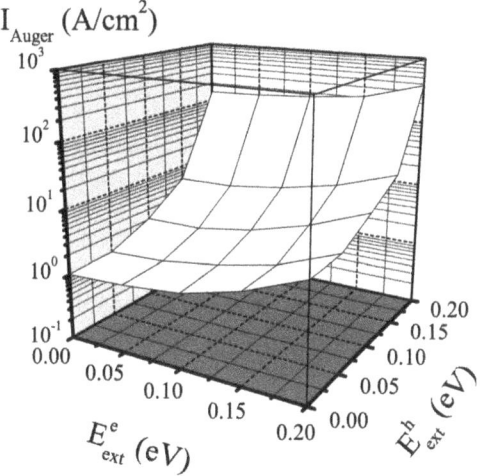

Figure 12.7   Extracted current (top) and lost current by Auger processes (bottom) *versus* energy selective contact. Energies are given within the conduction band (electron contact) and the valence bands (hole contact) relative to the respective band edges.[26]

suitable candidate for a HCSC absorber.[26] However, using achievable longer LO lifetimes, as *e.g.*, in quantum wells, the steady-state electron distribution can be maintained heated as shown in Table 12.2. This is a promising modelling result in accordance with recent experimental results where the carrier temperature spatial distribution has been evidenced by hyperspectral imaging in multiple III–V quantum wells (measured temperature 600 K).[27] The calculated extracted currents are not significantly impacted by the hot phonon effect.

Table 12.2 Average distribution of electron energy and temperature *versus* LO phonon lifetime.

τ_{LO} (ps)	$\langle E \rangle$ (eV)	T (K)
2.5	0.060	320
5	0.099	486
10	0.137	497
20	0.173	544

12.2.8 Summary of Modelling Section

Introduction of an increasing number of parameters to realize more realistic situations leads to both increased complexity of the models and to their applicability only to more specific applications. The thermodynamic approach embodied in first the particle conservation and then the energy conservation models gives the limiting efficiencies for hot carrier solar cells under ideal conditions. Increasing reality first with a hybrid model in which the Auger coefficients are neither zero nor infinite but have more realistic intermediate values chosen from literature, leads to a more specific calculation and to efficiencies very strongly dependent on the thermalization time used. Introduction of an increasing number of material parameters for specific materials leads to results which are closer to what can be fabricated in reality but also which are necessarily much more specific to individual situations. In these models it is necessary to assume a certain level of thermalization time constant. In general this is chosen as 100 ps as this is perhaps accessible with some materials such as InN, but all the modelling agrees that the longer this time constant the higher the efficiencies that can be obtained. Hence we will now look at the mechanisms of carrier cooling and methods and mechanisms aimed to reduce the thermalization rate.

12.3 Hot Carrier Absorbers: Slowing of Carrier Cooling

The energy dissipation of photo-generated carriers is a multi-step process. The inelastic scattering of carriers takes place in the first few tens of femtoseconds after their generation. This process normalizes momentum and leads to a distribution of electron energies which can be described by Boltzman distribution and a single high temperature, *i.e.*, a thermal population, and a separate (generally lower temperature) thermal population of holes.[28] Then on a longer timescale (typically several picoseconds) carriers scatter inelastically with phonons, predominantly emitting optical phonons in a series of discrete hops in each of which energy and momentum are conserved in the combination of electron and emitted phonon. For polar semiconductors the major scattering process is with longitudinal optical phonon modes. These optical phonons emitted by the excited carriers then interact with other phonons due to the anharmonic nature of the crystal potential. Through various routes these over-populated optical phonons decay into low-energy acoustic phonons. The final step is the transport of these acoustic phonons, macroscopically illustrated as the heat dissipation to the environment. The carrier cooling process can be slowed down by blocking any of these three processes. Other processes, such as direct emission of acoustic phonons and diffusion of optical phonons, are not significant for polar semiconductors.

The cooling of carriers by emission of optical phonons leads to the build-up of a non-equilibrium 'hot' population of optical phonons which, if it remains hot, will drive a reverse reaction to re-heat the carrier population, thus slowing further carrier cooling. Therefore the critical factor is the mechanism by which these optical phonons decay into acoustic phonons, or heat in the lattice. The principal mechanism by which this can occur is the Klemens mechanism, in which the optical phonon decays into two acoustic phonons of half its energy and of equal and opposite momenta.[29] The build up of emitted optical phonons is strongly peaked at zone centre both for compound semiconductor due to the Fröhlich interaction (strong quadratic dependence on momentum) and for elemental semiconductors due to the deformation potential interaction (less strong linear dependence on momentum). The strong coupling of the Fröhlich interaction also means that high energy optical phonons are also constrained to near zone centre even if parabolicity of the bands is no longer valid as is the case for high energy carriers well above the band minima.[30] This zone centre optical phonon population determines that the dominant optical phonon decay mechanism is this pure Klemens decay.

12.3.1 Electron–Phonon Interactions

The major interaction in polar materials is the Fröhlich interaction between hot electrons and longitude optical phonon modes. In non-polar materials the deformation potential interaction dominates, in which transitory charges on net covalent bonds lead to weak electron–phonon interactions an order of magnitude lower than for polar bonds. The importance of investigating the fundamental particle interactions occurring in the device is demonstrated by the significant dependence between the carrier energy relaxation time and the output efficiency. The thermalization of carriers occurs in three stages each of which could be interrupted in order to slow cooling. The first of these is the Fröhlich interaction between electrons and optical phonons, which can potentially be blocked by modifying bond polarization. The second is decay of optical phonons decay into acoustic modes, which can be modified by a phonon bottleneck effect. And the third is the build-up of non-equilibrium acoustic phonons, which could also contribute to reducing carrier energy relaxations. In order to understand the mechanism of slow relaxation rates in different types of materials the three interaction processes need to be considered.

The Fröhlich interaction occurs in polar semiconductors and mediates energy exchange between electrons and longitudinal phonons. It affects the total energy and state transition rates by an interaction Hamiltonian. The Fröhlich interaction Hamiltonian for super-lattice structures involves an overlap integral over one dimension (the direction perpendicular to planes). The derivation of the Hamiltonian involves the interaction between electrons and polarization fields induced by longitudinal mode vibrations. The non-uniform spatial distribution of free carriers also contributes to an additional screening effect on the polarization fields.[31] By obtaining the divergence of the polarization field an effective phonon envelope function is developed and incorporated in the overlap integral.

The dispersion relations of phonon modes are regarded as important for the purpose of realizing the hot carrier cell. The so-called phonon-bottleneck effect has an important role for slowing down the carrier cooling processes, especially for polar semiconductors. This is because prevention of further phonon decay leads to a non-equilibrium phonon population and an increased probability of the back reaction occurring to transfer the phonon energy back to electrons or holes. Hence by restricting the energy dissipation of optical phonons, they feed energy back to electrons and essentially impede the process of carrier energy loss.

12.3.2 Phonon Decay Mechanisms

An optical phonon will decay into multiple lower energy phonons. The processes require the conservation of energy and momentum. However, additionally, the principal decay path must be into two LA phonons only, which are of equal energy and opposite momenta, *via* an anharmonicity in the lattice, this was suggested by Klemens[7] and has been shown to apply to a wide range of materials.

12.3.2.1 *Wide Phononic Gaps in III–Vs and Analogues*

A large energy gap between high-lying optical phonon energies and low-lying acoustic phonon energies can impede the decay of optical phonons. This 'phononic bandgap' could possibly prevent the anharmonic decay of high-lying phonons, especially the simplest case of Klemens decay.

For some compounds in which there is a large difference in masses of the constituent elements, there exists a large gap in the phonon dispersion between acoustic and optical phonon energies. If large enough this phonon bandgap can prevent Klemens decay of optical phonons, because no allowed states at half the LO phonon energy exist. InN is an example of such a material with a very large phonon gap.

The second desirable property is a flat dispersion relation. As the group velocity for phonon transport is the tangential slop of the dispersion curve, a flat dispersion relation can contribute to small heat conductivity. This inhibits the energy dissipation into the environment.

The prevention of the Klemens mechanism forces optical phonon decay *via* the next most likely, Ridley mechanism, of emission of one TO and one low energy LA phonon. Such a mechanism only has appreciable energy loss (although still much less than Klemens decay) if there is a wide range of optical phonon energies at zone centre. This is only the case for lower symmetry structures such as hexagonal. For a high symmetry cubic structure, LO and TO modes are close to degenerate at zone centre and the Ridley mechanism is severely restricted or forbidden. Unfortunately cubic InN is very difficult to fabricate precisely because of the large difference in masses that give it its interesting phononic dispersion.

A flat dispersion relation is advantageous for other reasons. As the group velocity for phonon transport is the tangential slop of the dispersion curve, a flat dispersion relation can contribute to small heat conductivity. This inhibits the energy dissipation into the environment. In addition to the small dispersion in cubic structures dispersionless structures can be achieved by engineering nanostructures.

Slowed cooling has been observed in some III–V compounds in which there is a large difference in atomic mass. This has been shown in slowed carrier cooling for InN[32] and for slowed carrier cooling in InP compared to the small mass ratio GaAs.[33]

Analogues of InN with abundant elements, but also with narrow E_g are discussed below in section 12.4.

12.3.3 Nanostructures for the Absorber

Nanostructures are good options to fulfill the requirements of the hot carrier absorber. Two types of nanostructures have been suggested: nanowell structure and nanodot structures. The nanowell structure has very thin layers of narrow bandgap materials stacked in layers separated by potential barriers, creating oscillating potentials for the electrons. The electronic dispersion relations become flat for energies below the barrier height. This can potentially reduce the rate of electron–phonon polar interactions.

12.3.3.1 *Slowed Carrier Cooling in Multiple Quantum Wells*

Low-dimensional multiple quantum well (MQW) systems have also been shown to have lower carrier cooling rates. Comparison of bulk and MQW materials has shown significantly slower carrier cooling in the latter. Figure 12.8 shows data for bulk GaAs as compared to MQW GaAs/AlGaAs materials as measured using time resolved transient absorption by Rosenwaks *et al.*[34] recalculated to show effective carrier temperature as a function of carrier lifetime by Guillemoles *et al.*[35] It clearly shows that the carriers stay hotter for significantly longer times in the MQW samples, particularly at the higher injection levels by $1\frac{1}{2}$ orders of magnitude. This is due to an enhanced 'phonon bottleneck' in the MQWs allowing the threshold intensity at which a certain ratio of LO phonon re-absorption to

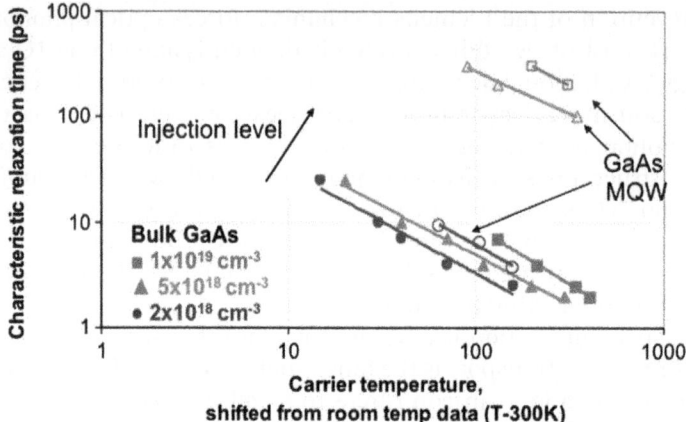

Figure 12.8 Effective carrier temperature as a function of carrier lifetime for bulk GaAs as compared to GaAs/AlGaAs MQWs: time resolved transient absorption data for different injection levels, from Rosenwaks,[34] recalculated by Guillemoles.[35]

emission is reached which allows maintenance of a hot carrier population, to be reached at a much lower illumination level. More recent work on strain balanced InGaAs/GaAsP MQWs by Hirst *et al.*[36] has also shown carrier temperatures significantly above ambient, as measured by photo-luminescence. An increase in In content to make the wells deeper and to reduce the degree of confinement is seen to increase the effective carrier temperatures.[37]

The mechanisms for the reduced carrier cooling rate in these MQW systems are not yet clear. However there are at least three possible effects that are likely to contribute, more than one of which could well occur in parallel. The first of these effects is that in bulk material photogenerated hot carriers are free to diffuse deeper into the material and hence to reduce the hot carrier concentration at a give depth. This will also decrease the density of LO phonons emitted by hot carriers as they cool and make a phonon bottleneck more difficult to achieve at a given illumination intensity. Whereas in a MQW there are physical barriers to the diffusion of hot carriers generated in a well and hence a much greater local concentration of carriers and therefore also of emitted optical phonons. Thus the phonon bottleneck condition is achieved at lower intensity.

The second effect is that for the materials systems which show this slowed cooling, there is very little or no overlap between the optical phonon energies of the well and barrier materials. For instance the optical phonon energy ranges for the GaAs wells and AlGaAs barriers used in Feng *et al.*[31] at 210–285 cm^{-1} and 280–350 cm^{-1}, respectively, exhibit very little overlap in energy, with zero overlap for the zone centre LO phonon energies of 285 and 350 cm^{-1}.[38] Consequently the predominantly zone centre LO phonons emitted by carriers cooling in the wells will be reflected from the interfaces and will remain confined in the wells, thus enhancing the phonon bottleneck at a given illumination intensity.

Thirdly, if there is a coherent spacing between the nanowells (as there is for these MQW or super-lattice systems) a coherent Bragg reflection of phonon modes can be established which blocks certain phonon energies perpendicular to the wells, opening up one dimensional phononic bandgaps (analogous to photonic bandgaps in modulated refractive index structures. For specific ranges of nanowell and barrier thickness these forbidden energies can be at just those energies required for phonon decay. This coherent Bragg reflection should have an even stronger effect than the incoherent scattering of the second mechanism above at preventing emission of phonons and phonon decay in the direction perpendicular to the nanowells.

It is likely that all three of these effects will reduce carrier cooling rates. None depend on electronic quantum confinement and hence should be exhibited in wells that are not thin enough to be quantized but are still quite thin (perhaps termed 'nanowells'). In fact it may well be that the effects are enhanced in such nanowells as compared to full QWs due to the former's greater density of states and in particular their greater ratio of density of electronic to phonon states which will enhance the phonon bottleneck for emitted phonons. The fact that the deeper and hence less confined wells in Clady *et al.*[33] show higher carrier temperatures is tentative evidence to support the hypothesis that nanowells without quantum confinement are all that are required. Whilst several other effects might well be present in these MQW systems, further work on variation of nanowell and barrier width and comparison between material systems, will distinguish which of these reduced carrier diffusion, phonon confinement or phonon folding mechanisms might be dominant.

12.3.4 Hot Carrier Cell Absorber Requisite Properties

The above discussion allows us to estimate the major properties required for a good hot carrier absorber material. These are listed below in approximate order of priority, although their relative importance may well change in light of future research:

1. Large phononic bandgap $(E_{O,min} - E_{LA})$, to suppress Klemens decay, requires large mass difference between elements
2. Narrow optical phonon energy dispersion $(E_{LO} - E_{O,min})$, to minimize the loss by Ridley decay, requires a high symmetry
3. Small $E_g < 1$ eV, to allow a broad range of strong photon absorption
4. A small LO optical phonon energy (E_{LO}), to reduce the amount of energy lost per LO phonon emission but not so small as to increase the occupancy of LO modes unduly: $E_{LO} \times \exp(-E_{LO}/kT)$

5. A small maximum acoustic phonon energy (E_{LA}). This maximizes ($E_{O,min} - E_{LA}$) and is important if E_{LO} is also small. It is achieved for a flat dispersion

6. Good renormalization rates in the material, *i.e.*, good e–e and h–h scattering. This condition is met in most semiconductors quite easily, with e–e scattering rates of less than 100 fs, but it may be comprised in nanostructures

7. Good carrier transport in order to allow transport of hot carriers to the contacts

8. Ability to make good quality, ordered, low defect material

9. Earth-abundant and readily processable materials

10. No, or low, toxicity of elements, compounds and processes

12.4 Hot Carrier Absorber: Choice of Materials

InN has most of the properties mentioned above, except 4, 8 and 9, and is therefore a good model material for a hot carrier cell absorber. InN is investigated below in combination with its alloys with GaN. However the abundance of In is very low, so it is difficult to see InN as a long-term material suitable for large-scale implementation. Hence analogues of InN, which retain its interesting property of a large phonon bandgap, are also being investigated.

12.4.1 Analogues of InN

As InN is a model material, but has the problems of abundance and bad material quality, another approach is to use analogues of InN to attempt to emulate its near ideal properties. These analogues can be II–IV–nitride compounds, large mass anion III–Vs, group IV compounds/alloys or nanostructures.

12.4.1.1   *II–IV–Vs: ZnSnN, ZnPbN, HgSnN, HgPbN*

With reference to Figure 12.9, it can be seen that replacement of In on the III sub-lattice with II–IV compounds is analogous and is now quite widely being investigated in the Cu_2ZnSnS_4 analogue to $CuInS_2$.[39]

ZnGeN can be fabricated[40,41] and is most directly analogous with Si and GaAs. However, its bandgap is large at 1.9 eV. It also has a small calculated phononic bandgap.[42] ZnSnN has a smaller electronic gap (1 eV) and larger calculated phononic gap[38,39] It is however difficult to fabricate, and also its phononic gap is not as large as the acoustic phonon energy making it difficult to block Klemens decay completely. HgSnN or HgPbN should both have smaller E_g and larger phononic gaps. These materials have not yet been fabricated and have the problem that Hg has about half the abundance of In and is toxic.[43]

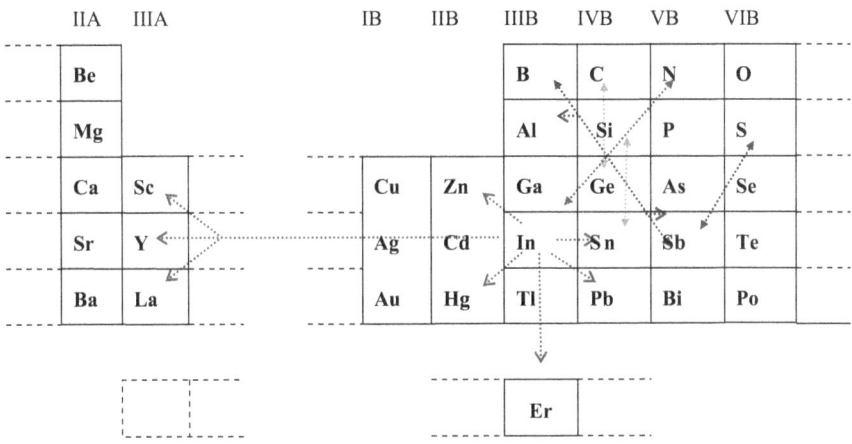

Figure 12.9   Use of the periodic table to analyse possible analogue compounds of InN based on atomic mass combination and electro-negativity.

12.4.1.2   Large Mass Anion

Bi and Sb are heavy elements and hence their compounds with B and Al have large predicted phononic gaps.[44] BiB has the largest phononic gap but AlBi, Bi_2S_5, Bi_2O_3 (bismuthine) are also attractive. AlSb has a calculated and measured phononic gap the same size as its acoustic phonon energy,[45] with an E_g of 1.5 eV. It also has a very high calculated phonon lifetime probably because of its large phononic gap.[46] This, and the fact that it is of reasonably widespread use and can be made of reasonable quality, makes AlSb marginal in suitability as an absorber material and similar to InP. BSb has a large elemental mass difference giving it a very large phonon bandgap more than twice the acoustic phonon energy, as calculated for a 1D force constant model, sufficient to block Klemens decay completely if the material quality is high enough. It also can be fabricated by co-evaporation and has a measured E_g of 0.59 eV.[47] This bandgap is indirect and hence not ideal, but the absorption coefficient remains high even out to 1700 nm. Bi is a rare element with an earth abundance about twice that of In, with Sb about five times the abundance of In.[20] Nonetheless their large phonon gaps make particularly BSb and AlSb interesting to pursue further for proof of concept.

12.4.1.3   Transition Metal Nitrides

The Group IIIA nitrides LaN, YN and ScN would have been thought to have large phononic gaps, but as they form in a NaCl structure rather than the zincblende structure of the Group IIIB nitrides, they actually have zero phonon bandgap.[48,49] However, hafnium and zirconium nitrides both have large predicted and measured phononic gaps,[26] both bigger than the acoustic phonon energy, whilst that for TaN is rather smaller. Hf and Ta are about the same abundance as Ge and As, with Zr 100 times more abundant and all are widely used metals.[20] Again because of their large mass cations these oxides and nitrides all have large phonon gaps. That for Hf_3N_4 shows a large gap (although still too small to block Klemens decay completely)[50] but also a narrow optical phonon dispersion, which is important for blocking secondary Ridley decay. They can also readily be fabricated and have existing applications. Interestingly both HfN and ZrN have zero electronic bandgap, leading to the surprising conclusion that these semi-metals may be good absorbers for a hot carrier cell.

The lanthanides can also form III–nitrides. ErN and other RE nitrides can be grown by MBE. The phononic bandgaps of the Er compounds are predicted to be large, because of the heavy Er cation, but its discrete energy levels make it not useful as an absorber, although the combination of properties in a nanostructure could be advantageous.

12.4.1.4   Group IV Alloys/Compounds

All of the combinations Si/Sn, Ge/C or Sn/C look attractive with large gaps predicted in 1D models. However being all group IVs they only form weak largely covalent bonds. Unfortunately SiC, whether 3C, 4H or 6H, has too narrow a phononic gap. Nonetheless GeC and SnSi are of significant interest and the former does form a compound, although stoichiometric quality as yet is very low.[51]

The IV–VI PbS also has a large mass ratio and is interesting, but its measured phonon dispersion shows an additional TO branch which fills the phonon gap,[42] thus likely making it unsuitable for an absorber in bulk form. But initial modelling indicates that phonon folding in a nanostructure could reflect these modes and make PbS nano-crystals highly suitable.

There are several other inherent advantages of group IV compounds/alloys all of which are associated with the four valence electrons of the group IVs that result in predominantly covalent bonding:

12.4.1.4.1   Covalently bonded crystals.   The elements form completely co-valently bonded crystals pr imarily in a diamond structure (tetragonal is also possible as in βSn). However for group IV compounds the decreasing electronegativity down the group results in partially ionic bonding. This is not strong in SiC and whilst it tends to give co-ordination numbers of 4, can nonetheless result in several allotropes of decreasing symmetry: 4c, 4h, 6h. However, as the difference in period increases for the as yet theoretical group IV compounds, so too does the difference in electronegativity and hence also the bond ionicity and hence the degree of ordering. For a hot carrier absorber this trend is in the right direction because it is just such a large difference in the period which is needed to give the large mass difference and hence large phononic gaps. All of GeC, SnSi, SnC (and the Pb compounds) have computed phononic gaps large enough to block Klemens decay, and should also tend to form ordered diamond structure compounds.

12.4.1.4.2 Small Electronic Bandgap. Because of their covalent bonding, the group IV elements have relatively small electronic bandgaps as compared to their more ionic III–V and much more ionic II–VI analogues in the same period: *e.g.*, Sn 0.15 eV, InSb 0.4 eV, CdTe 1.5 eV. In fact to achieve approximately the same electronic bandgap one must go down one period from group IV to III–V and down another period from III–V to II–VI: *e.g.*, Si 1.1 eV, GaAs 1.45 eV, CdTe 1.5 eV. This means that for group IV compounds there is greater scope for large mass difference compounds whilst still maintaining small electronic bandgaps. A small bandgap of course being important for broadband absorption in an absorber, property 3 in the desirable properties for hot carrier absorbers listed above.

12.4.1.4.3 Smaller E_g. The smaller E_g would tend to be for the larger mass compounds of Pb or Sn. Which, to give large mass difference, would be compounded with Si or Ge. This trend towards the lower periods of group IV also means that the maximum optical phonon and maximum acoustic phonon energies will be smaller for a given mass ratio, the desirable properties 4 and 5 listed above.

12.4.1.4.4 Abundance. Furthermore, unlike most groups, the group IV elements remain abundant for the higher mass number elements, desirable property 9; in fact, it is Ge in the middle of the period which has the lowest moderate abundance.[20] Property 10 is also satisfied because the group IVs have low toxicity.

12.4.1.5 Nanostructures

As discussed in section 12.4.3 the phonon dispersion of quantum dot (QD) nanostructures can be calculated in a similar way to compounds. Their phononic properties can be estimated from consideration of their combination force constants. Hence it is possible to 'engineer' phononic properties in a wider range of nanostructure combinations. Of the materials discussed above the Group IVs lend themselves most readily to formation of nanostructures instead of compounds due to their predominantly covalent bonding, which allows variation in the coordination number. Therefore the nanostructure approaches of section 12.4.3 are consistent with a similar description as analogues of InN, whether it be III–V QDs, colloidally dispersed QDs or for core shell QDs.

12.4.2 Modelling Phonon Properties in Group IV and III–V compounds

The group IV compounds or alloys look very promising as large phonic bandgap materials and have a number of other advantages s discussed above. The drastic improvement of computational power over the last 10 years allows us to revisit the two-phonon final states of the LO-phonon decay with a high degree of accuracy in various semiconductors. In the present work we computed the three-phonons decay process of the optical phonon produced by electron cooling, for different group III–V and IV semiconductors in the zincblend phase, within the density functional perturbation theory (DFPT).[52] These semiconductors were chosen so to cover a wide diversity of impacting parameter on phonon lifetime, such as the mass difference between atoms or the LO-TO splitting. We derive an accurate complete representation of the two-phonon final states in the reciprocal space.

12.4.2.1 Computational Methodology

12.4.2.1.1 LO–Phonon Decay and Two-phonon Final States. The optical phonon decay processes has already been extensively described in the literature.[18,19,23] In a pure semiconductor, the main mechanism at stake for optical phonon (oph) relaxation is a three-phonon process, in which it decays into two lower energy phonons following the energy and momentum conservation rules:

$$\hbar\omega_\alpha + \hbar\omega_\beta = \hbar\omega_{oph} \left(1 \pm \delta_\omega\right) \tag{12.12}$$

$$\mathbf{k}_\alpha + \mathbf{k}_\beta = \mathbf{k}_{oph} \left(1 \pm \delta_{\mathbf{k}}\right) \tag{12.13}$$

where α and β label the two final phonons polarization branch. Considering that the emitted phonon has a momentum very close to the zone centre,[15] it is often very convenient to set the second term in Equation (12.13) to **0**, so that the two final phonons have exactly opposite wave vectors. This is called the 'zone-centre' approximation (Figure 12.10).

Computing the dispersion relation of bulk semiconductors was done within the Density Functional Perturbation Theory, using the linear response method implemented in the Quantum Espresso package.[49] Pseudopotentials generated within the Local Density Approximation using the Troullier–Martins formalism were used. For each semiconductor, a very good accordance was expected and found with experimental phonon dispersion relation data whenever they exist,[53–57] as can be seen in the insets of Figure 12.11 and Figure 12.12. The dispersion relation was computed on a very tight mesh into the whole irreducible Brillouin zone (1/48th of the whole BZ for a bulk unstrained FCC crystal), so that the volume of each k-point is 0.125×10^{-6} in units of $(2\pi/a)^3$, where a is the lattice parameter. This is equivalent to mesh the ΓX axis in the reciprocal space into 200 k-points. The energy and momentum conservation rules for a three-phonon process [Equations (12.12) and (12.13)] are then applied. The delta function in Equation (12.12) has to be replaced by a non-zero width. Haro-Poniatowski *et al.*[33] chose it as 1% of the optical phonon frequency. Considering the density of the mesh and the agreement with experiment, we choose 0.1%, so that the small density of

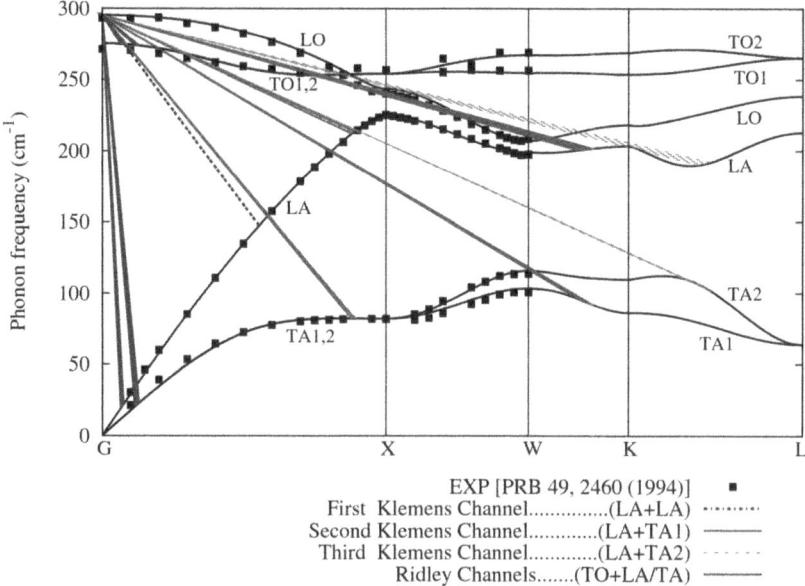

Figure 12.10 LO–phonon decay two-phonon final states in GaAs along high symmetry lines.

final states in k-space is light enough to allow the following analysis. An example is shown on Figure (12.10), where we applied this methodology on GaAs on a particular path in the BZ (the negative k region is folded onto the positive one for clarity). Klemens[17] expected the two-phonon final states of GaAs to be the ensemble of the LA modes of opposite momentum whose energies are exactly half of the LO phonon energy (the so-called 'Klemens channel'). In 1996, Ridley[58] suggested that in GaN this decay must be governed by a three-phonon process involving the LO phonon, a LA and a TO phonon ('Ridley' channel). Vallée and Boganni[59] reinvestigated the LO relaxation in GaAs and assessed that the LO decay is lead by the creation of one LA and one zone-edge LO near the L point ('Vallée–Bogani' channel). These bulk processes will be here extended and accurately distinguished as follow. The Klemens final states are divided into three relevant subspaces for the (upcoming) analysis: the two-phonon final states that involve two LA phonons will be referred to as the 'first Klemens surface', and similarly, when they involve one LA and one TA phonon, it will be referred to as the 'second' and 'third Klemens surface', one for each TA branch. The case where the two resulting phonons are one TO and one acoustic (either LA or TA), is labelled the 'Ridley surfaces'; distinction between the different Ridley surfaces (TOs + LA/TAs) does not bring here any relevant information. Finally, the 'Vallée–Bogani' channel is defined as above.

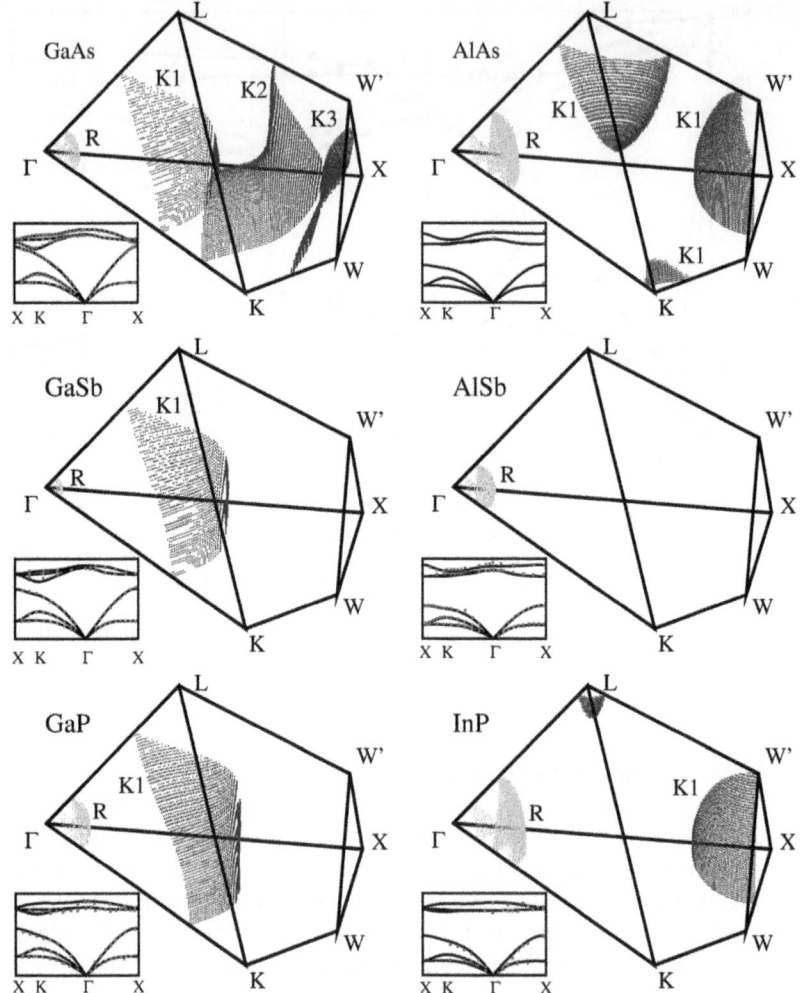

Figure 12.11 LO decay two-phonon final states in III–V semiconductors. K1, K2 and K3 stands for 'first', 'second' and 'third Klemens surface', respectively, while R stands for 'Ridley surfaces'.

12.4.2.1.2 III–V Binaries. GaAs is a case state since it has been the most studied semiconductor of the III–V group, both experimentally and theoretically. In addition to the Ridley channels, the three different Klemens channels are allowed. The contribution of each channel in Table 12.3 is in contradiction with some previous studies.[35] Indeed, Vallée and Bogani[52] claimed that the LO-decay in GaAs is led by another two-phonon process, involving one LO-phonon near L point of the Brillouin zone and one LA-phonon. This channel does not appear in our result for the chosen energy conservation function width of 0.1%. Trying to relax this latter parameter, we find that at 7%, a similar channel appears, but involving

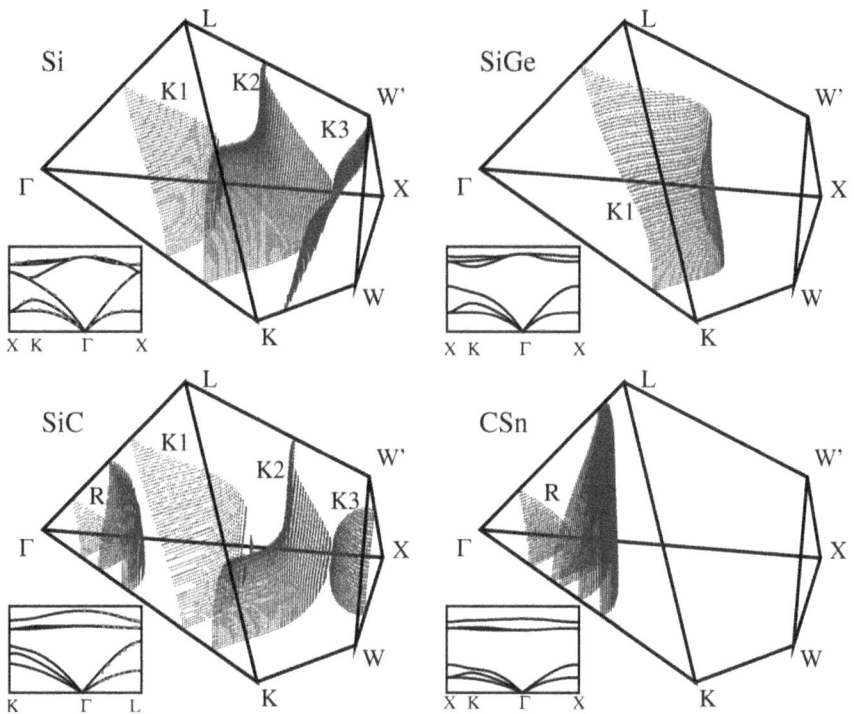

Figure 12.12 LO decay two-phonon final states in IV–IV semiconductors. K1, K2 and K3 stands for 'first', 'second' and 'third Klemens surface', respectively, while R stands for 'Ridley surfaces'.

Table 12.3 Contributions of the different channels in the two-phonon final states.

	Klemens channels (%)	K1 : LA + LA	K2 : LA + TA1	K3 : LA + TA2	Ridley channels (%)
GaAs	93	10	31	59	7
AlAs	67	100	0	0	33
AlSb	0	0	0	0	100
GaSb	94	100	0	0	6
GaP	74	100	0	0	26
InP	16	100	0	0	84
Si	100	12	34	54	0
SiGe	100	100	0	0	0
SiC	64	18	21	61	36
CSn	0	0	0	0	100
SiSn	87	100	0	0	13

a zone edge LO-phonon near the K-point of the Brillouin zone. The Vallée–Bogani channel, as it is described above, only appears at 7.2%. In both cases, the tolerance parameter is too large to seem reasonable for the authors: in a phonon lifetime calculation, the Gaussian linewidth in the energy conservation relation is taken at most for simple semiconductor to be 5 cm^{-1} (Barman *et al.*[60]) which is four times lower than the required value for Vallée–Bogani channel to be observed. This channel is considered not to take place within the current approximations. This conclusion is consistent with recent Boltzmann transport equation solving *via* Monte Carlo calculation.[20] When increasing the mass difference, the second and third Klemens surfaces disappear, as in GaSb and GaP. Note that the former case in of great interest in the research for HCSC absorber candidates. The remaining allowed channels leads to an almost spherical first Klemens surface. This is a of great interest when considering creating a gap in the dispersion relation in order to block the channel, because the gap has to remain at a constant energy in all reciprocal directions, which can be directly associated with a quantum dot type of nanostructure. The narrow splitting between LO and TO branches is also responsible in GaSb for the Ridley states to be confined very close to the Γ point. As the mass difference between the two atoms increases, the first Klemens surface is split into three parts in the vicinity of particular high symmetry point, K, L and X, as in AlAs. This behaviour is emphasized in InP, where even the first Klemens states near K are not allowed. AlSb is the ultimate case since the Klemens channels are forbidden everywhere because all acoustic branches remain below half of the LO-phonon energy. Unfortunately, AlSb has an indirect bandgap, in contradiction with condition (4).

12.4.2.1.3 Group IV Binaries. The same calculation is performed for Si, SiGe, SiC, CSn and SiSn, and the two-phonon final states are depicted on Figure 12.12. Si is the case state for non-polar group-IV semiconductors. The three Klemens channels are energetically allowed, and the non-polar feature makes the Ridley channel not relevant. As in GaAs, we can see from Table 12.3 that in this case the first Klemens process is not dominant in the purely acoustic final states. It is clear also when considering the relative size of the three Klemens surfaces on Figure 12.12. SiGe is another example for which the mass difference is high enough to prevent the second and third Klemens channels, leaving only possibilities for the first Klemens process. Interestingly, the first Klemens surface differs from the pure Si case. This is an edge case: the first Klemens surface is bowing in the centre, which for higher mass difference like in AlAs, leads to the case where the same surface is split in three. The LO and TO branches are energetically so closed around Gamma that no Ridley channel is observed within an energy conservation function width of 0.1%. In SiC, the high electronegative feature of the Carbon atom is responsible for the large and complete LO–TO splitting, so that the gap between the LO and the TO branch is effective in the whole Brillouin zone. As a consequence, we find a large number of Ridley final states. In addition, considering that the mass ratio between the Si atom and the C atom is rather weak, all the Klemens channel are energetically allowed. In the CSn case, the mass ratio is close to 10, which is much bigger than any of the other semiconductors treated here. In addition, the LO–TO splitting is even wider than in SiC. While no Klemens channel decay occurs, the Ridley surfaces are stretched even farther from the Γ point (Figure 12.12) than in SiC.

12.4.3 Phonon Modulation in Quantum Dot Nanostructure Arrays for Absorbers

Nanostructures offer the possibility of modification of the phonon dispersion of a composite material. III–V compounds or indeed most of the cubic and hexagonal compounds can be considered as very fine nanostructures consisting of 'quantum dots' of only one atom (say In) in a matrix (say N) with only one atom separating each 'QD' and arranged in two inter-penetrating fcc lattices. Modelling of the one-dimensional phonon dispersion in this way gives a close agreement with the phonon dispersion for zincblende InN extracted from real measured data for wurtzite material.[61]

Similar 'phonon bandgaps' should appear in good quality nanostructure super-lattices, through coherent Bragg reflection of modes such that gaps in the super-lattice dispersion open up.[62] There is a close analogy with *photonic* structures in which modulation of the *refractive index* in a periodic system opens up gaps of disallowed *photon* energies. Here modulation of the ease with which phonons are transmitted (the *acoustic impedance*) opens up gaps of disallowed *phonon* energies.

12.4.3.1 Force Constant Modelling of III–V QD Materials

Three-dimensional force constant modelling, using the reasonable assumption of simple harmonic motion of atoms in a matrix around their rest or lowest energy position, reveals such phononic gaps.[63] The model calculates longitudinal and transverse modes and can be used to calculate dispersions in a variety of symmetry directions and for different combinations of QD sub-lattice structure and super-lattice structure.

III–V Stranski–Krastinov grown QD arrays of InAs in InGaAs and InGaAlAs matrices are fabricated at the University of Tokyo using MBE. These are investigated for evidence of phonon dispersion modulation and potential slowed carrier cooling. In order to understand the expected phonon dispersions these are being modelled using the three-dimensional force constant technique.

Lattice matched and strain compensated material pairs that may produce large phonon bandgaps are of interest. Previous iterations in the design of these structures indicated the importance of separating 'light' and 'heavy' atoms to different parts of the nanostructure. Initially, the lightest atom in the system, As, was present in both the QD and the matrix. This meant that the reduced mass of both regions (proportional to the sum of the inverses of each atomic mass) was very similar as the light element dominates in this case.

Structures with an $In_{0.5}Ga_{0.5-x}Al_xAs$ matrix (with $x = 0.4$) and InAs QDs were grown. Significant Al content was introduced into these structures with

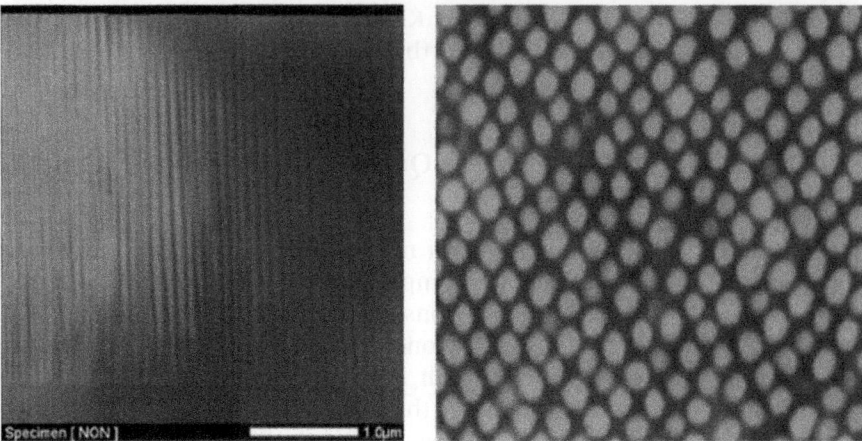

Specimen [NON]   1.0µm

Figure 12.13   Images of the structure as grown and modelled. (Above) the near-
perfect stacking of the structure is shown. (Above right) in-plane
InAs QD arrangement. (Right) simple hexagonal superlattice structure showing stacking. The QDs are similar to flattened disks with in
plane dimensions of 40–50 nm and heights of 7 nm.

the expectation that this light element, segregated to the matrix material,
might produce appreciable phonon bandgaps. Some images derived from
characterisation of the structure are presented in Figure 12.13. The super-
lattice of QDs has a simple hexagonal structure. Extraordinary periodic
out-of-plane stacking is achievable and largely defect free structures can be
grown on the order of microns.

Force constant modelling of this structure predicts an appreciable phonon
bandgap, as shown in phonon energies in Figure 12.14 and in DOS in
Figure 12.15. This bandgap is due almost entirely to the significant fraction
of Al in the system which reduces the mass of the matrix as compared to the
QDs considerably. A small bandgap is present due to the mass difference
between In, Ga and As, but it is less than a quarter of the size shown in
Figure 12.14 and Figure 12.15. Due to computational constraints the size of
the modelled QDs is small, only about 1nm in diameter. While actual sizes
for these structures are too computationally intensive to model without extreme effort, recent modelling with gradually increasing size suggests that
the dispersion relations scale linearly with size. That is, once the discrete
distances (bond lengths) are small relative to the super-lattice unit cell dimension, the dispersion should look exactly the same when scaled such that
the relative dimensions are preserved. This requires confirmation in greater
detail in future work. Figure 12.14 also shows that the phonon bandgap
extends in the three major reciprocal space directions and hence appears to
be complete in reciprocal space.

This same InAs QD array structure has been analysed using low wavenumber high resolution Raman at the University of Chemnitz in order to
investigate the phonon energies and any evidence of Brillouin zone folding.[64]

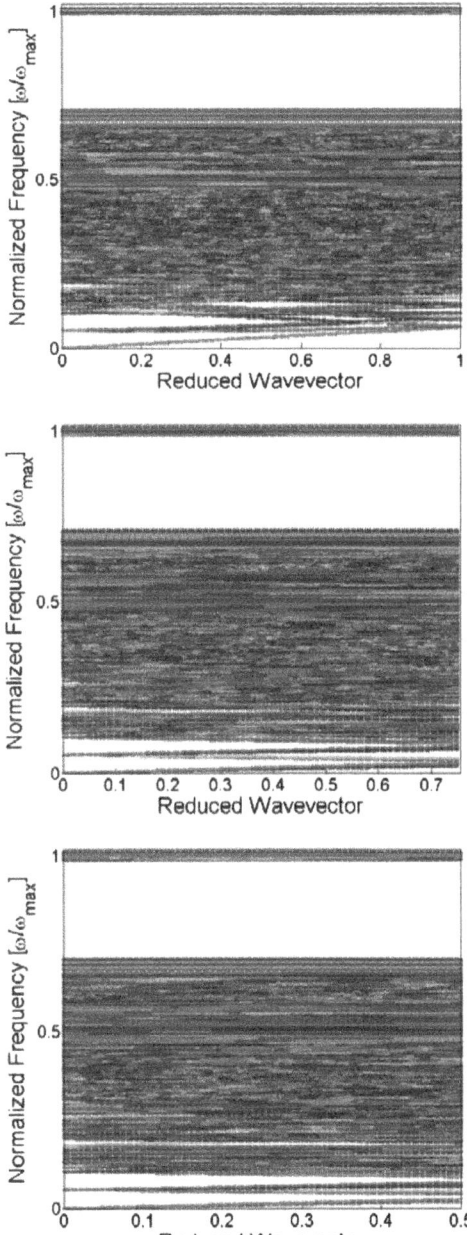

Figure 12.14 Dispersion curves for InAs QDs in a simple hexagonal SL. The matrix is $In_{0.5}Ga_{0.5-x}Al_xAs$ with $(x = 0.4)$. There is a large phonon bandgap present in the system due to the presence of Al. The gap seems very tolerant to Al location. (Top): <100>, (Middle): <110>, (Bottom): <111> directions.

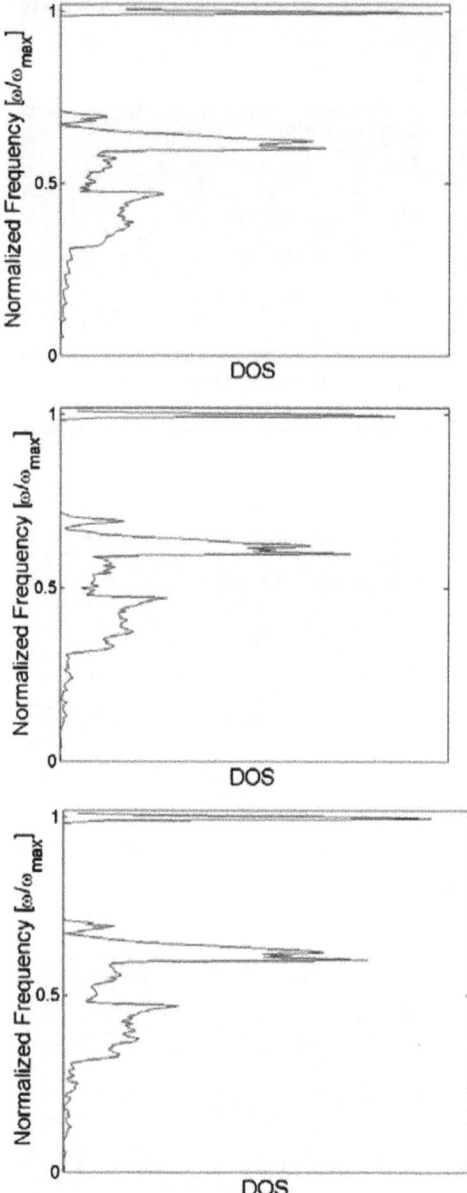

Figure 12.15 Densities of states (DOS) for the dispersion curve in Figure 12.14. Note small changes in the DOS could be due to pseudo-random locations of the Ga and Al particles or to differences in crystalline direction. (Top): <100>, (Middle): <110>, (Bottom): <111> directions respectively.

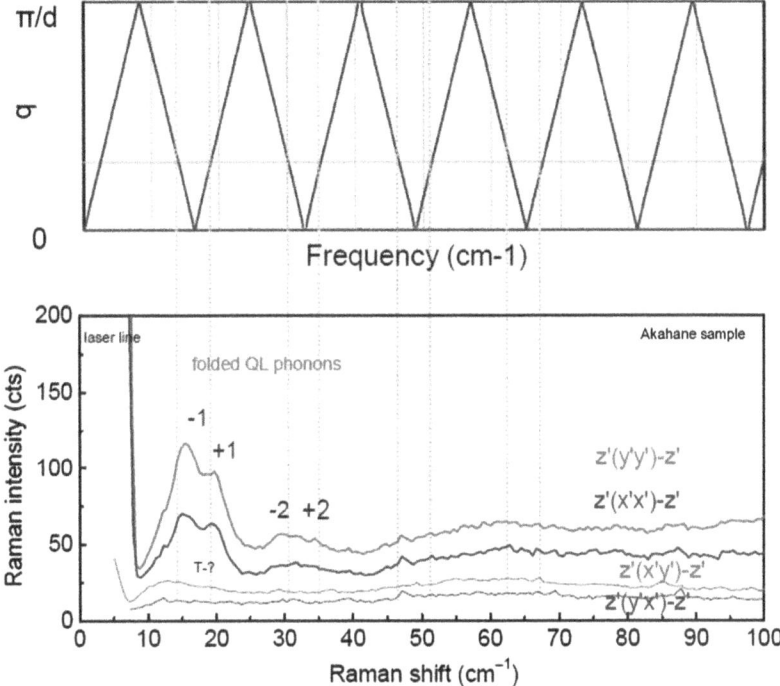

Figure 12.16 Optically active phonon energies for InAs QD in InGaAlAs matrix.[33] (Top) folded phonon mode calculated using a 1D Rytov force constant model in which the acoustic impedance in the superstructure is modulated. These modes fold back to the mini-Brillouin zone centre. The appropriate momentum, q, for the photon energy of the incident laser is indicated at about $\pi/3d$. (Bottom) low wavenumber Raman showing optically active phonons at very low wavenumber. The peaks and peak doublets align well with the calculated values for the appropriate photon momentum for the first to values of m, the number of phonon folds. The decrease in peak doublet intensity with increasing m number is due to the reduced scattering of photons with phonons as the number of folds increases and hence the optical character of the phonon mode also decreases.

As shown in Figure 12.16 the phonon modes are folded in reciprocal space such that even very low wavenumber acoustic phonons at 15 and 20 cm^{-1} become Raman active, illustrating that they have taken on an optical character. Folding from the QD interfaces allows this by crating standing waves at long wavelengths consistent with the super-lattice dimensions. This requires high periodicity in the super-lattice and is hence an indication of its quality, as also shown in the AFM image in Figure 12.13.

This modelling of phonon bandgaps and experimental evidence for phonon modulation in real QD super-lattices is an indication of the potential of these structures to modulate phonon energies in a hot carrier absorber. With appropriate QD material, size, spacing and super-lattice spacegroup a phonon bandgap can be opened at half the maximum optical phonon energy and thus block Klemens decay to acoustic phonons and hence enhance phonon bottleneck, which should slow carrier cooling.

12.5 Contacting Hot Carrier Cells

The requirement for extraction of energies of carriers from the hot carrier absorber only over narrow energy ranges poses significant practical difficulties. Energy selective contacts can be realized by the purely quantum phenomenon of resonant tunnelling. This effect exists if electron-reflecting interfaces are closely spaced, *i.e.*, with separations comparable to the electron wavelengths. Interference between different electron waves leads to discrete transmission peaks.

Hence one practical implementation of the requirement for a narrow range of contact energies is an energy selective contact (ESC) based on double barrier resonant tunnelling. Tunnelling to the confined energy levels in a quantum dot layer embedded between two dielectric barrier layers, can give a conductance sharply peaked at the line up of the Fermi level on the 'hot' absorber side of the contact with the QD confined energy level. Conductance both below this energy and above it should be very significantly lower. This is the basis of the current work on double barrier resonant tunnelling ESCs.

12.5.1 Modelling Optimized Materials for Energy Selective Contacts

Energy selective contacts (ESCs) for a hot carrier solar cell can be implemented using different materials and structures. The main property that needs to be achieved is precise energy selectivity in carrier extraction. Different models have been developed to identify possible materials for ESCs and to optimize their properties. Amongst the different possible material configurations, group IV and group III–V double barrier resonant tunnelling structures look to be the most promising.

Modelling of group IV materials has shown that the structure with the highest potential to be used as ESC consists of double barrier structure (DBS) with quantum dots (QDs) embedded in a dielectric matrix.

Modelling of electrical properties for the structure shown in Figure 12.17 has been performed for Si QDs in Si based dielectric materials. The reasons for focusing on Si in the first instance for group IV materials are abundance and well established growth techniques. Structures consisting of Si QDs and the following dielectric barriers have been modelled: SiO_2, Si_3N_4 or SiC. Results show that optimal energy selectivity properties are obtained with higher lateral confinement within the middle layer of the structure and higher vertical conductivity. Amongst the material taken into account in the simulations, structures consisting of Si QDs in a SiO_2 with SiC barriers have shown the best overall energy extraction properties.

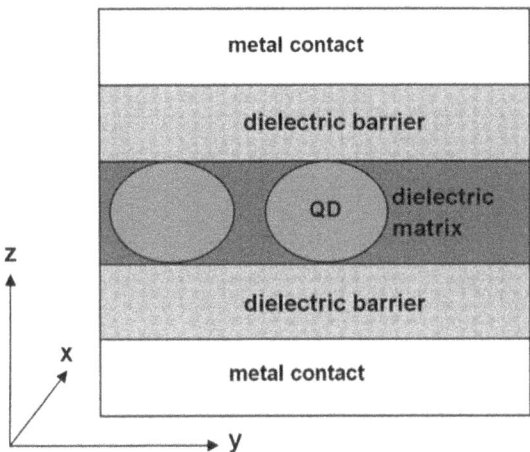

Figure 12.17 Schematic diagram of a double barrier structure. Silicon QDs are formed in a dielectric matrix that may not be the same as the barrier material. The growth direction of the structure is along the z axis.

ESCs for a III–V based HC solar cell can be realized using a QW structure in a DBS, probably requiring either MBE or MOVPE growth. Modelling results have shown that, for a HC cell based on an InN absorber, the optimum configuration for ESCs is constituted by an InN/In$_x$Ga$_{1-x}$N DBS.[65] This structure allows modulation of the extraction energy and shape of the transmission probability function independently for the electrons and holes contacts. The value of extraction energy can be modified by engineering the stoichiometry and the thickness of the QW structure. In general the optimal extraction energy for electrons and holes is different, due to the different values of effective mass. Thus, the hole ESC QW should be physically thinner and have a higher barrier than the electron ESC QW to achieve similar currents at reasonable extraction energies, as illustrated in Figure 12.18.

The transmission coefficient of a double-barrier semiconductor structure has been numerically calculated using a tight-binding method. The phonon scattering during the transmission has been incorporated with a Greens function method. This constructs a framework for simulating a real resonant tunnelling system with ultra-thin layers. It will help in optimizing the material system for the ESCs. Some sampling results for the DBS of AlAs/GaAs/AlAs are shown in Figure 12.19.

Here we mainly aim to utilize the 1st energy peak. From the transmission calculation of an AlAs/GaAs/AlAs double-barrier structure, this peak is visible and reasonably significant if only normally incident electrons are considered. (These constitute the majority of energy transmission in any case.) With the consideration of phonon scattering, side-peaks appear which are not favourable for energy-selective extraction. This indicates a

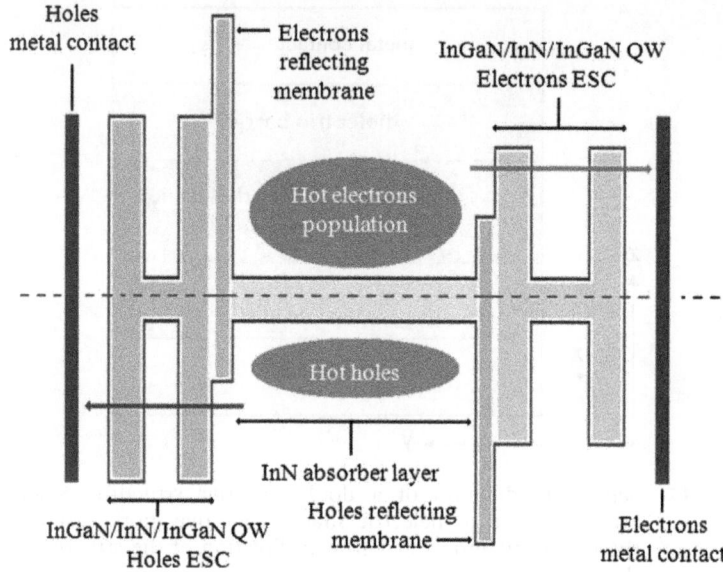

Figure 12.18   Schematic of a HC solar cell based on an InN absorber and InN/InGaN ESCs.

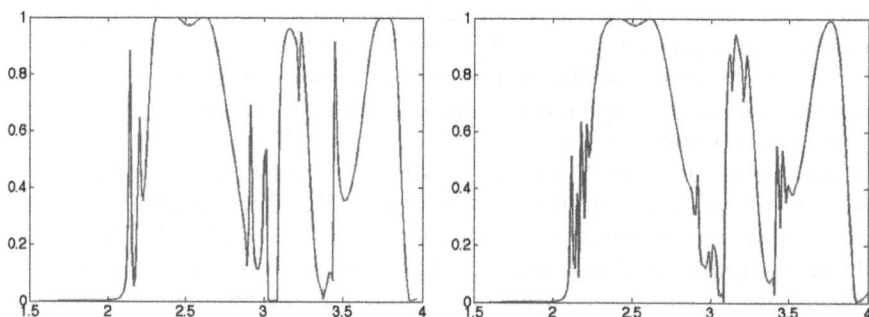

Figure 12.19   The total transmission (*y* axis: 0–1) profile as a function of the normally penetrating electron energy (*x* axis: eV) through the zinc-blende DBS AlAs (3 unit cells)/GaAs (3 unit cells)/AlAs (3 unit cells), with different phonon scattering properties. Left: No phonon scattering considered; Right: with phonon scattering and the initial phonon occupation number $= 0$.

preference of choosing materials that are less polar. Besides, the addition of arbitrary directions off normal may broaden the transmission peak with a little loss on the selectivity. Such a Greens function model can treat any type of atomic system in terms of the scattering problem. It is ready for use as a tool to choose the optimal material system for the energy-selective transmission.

12.5.2 Triple Barrier Resonant Tunnelling Structures for Carrier Selection and Rectification

A normal Energy Selective Contact (ESC) using double barrier resonant tunnelling is symmetric and does not rectify electrons from holes. An ESC which uses triple barrier resonant tunnelling through two separately tuned quantum well or quantum dot layers can give a much greater energy selectivity and also is a filter to separate electrons and holes in opposite directions (*i.e.*, rectification). The mini-band line-ups of the confined levels need to be tuned by the well thicknesses, effective mass of electrons and holes and barrier heights such as illustrated schematically in Figure 12.20 With appropriate choice of these for the left contact there can be an alignment of the confined levels such as to give a resonant channel in the conduction band for electrons but misalignment of levels in the valence band such that there is no channel for holes. Similarly for the right contact the valence band confined levels are aligned to allow holes to pass but misaligned in the conduction band so as to block electrons. For non-equal electron and hole effective masses (as is the case for most materials) this condition will in general be met.[66,67]

The collection channels for electrons and holes would ideally be aligned with the peak occupancies of energy of the hot electron and hot hole populations respectively, which in turn will depend on the respective

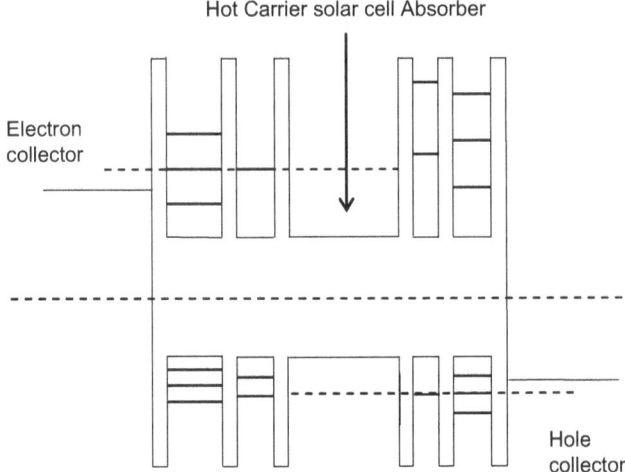

Figure 12.20 Triple barrier double well resonant tunnelling energy selective contacts. Wells are of different width designed to give alignment of 1st and 2nd confined levels in the conduction band on the left and for 1st and 2nd confined levels in the valence band on the right. In general for non-equal effective masses allignment of first and second confined energy levels for electrons will not give the same condition for holes thus giving electron/hole rectification and a preferred collection direction in the external circuit.

temperatures of these populations. This condition will give the highest efficiency, but as discussed in section 12.5.1, for high electron–electron and hole–hole carrier scattering rates, it is not very critical.

Ideally there would be two of these triple barrier double QW/QD devices on either side of the absorber, but as there is much less energy in the high effective mass hole population a simpler double barrier or simple band contact on the hole side is also possible, without much loss in collected energy, as investigated in section 12.2.6. Various materials combinations could be used for such a design, but specifically III–V triple barrier QWs based on the InGaN/GaN system or Al_2O_3/Si QW/Al_2O_3 look particularly promising.

The exact thickness of QWs or size of QDs and barrier thicknesses and contact work functions will have to be tuned for the particular confined energy levels involved in each material. Use of more than two QW or QD layers on either side is also possible. This will increase energy selection by requiring even more careful alignment.

This is the first application of this concept to energy selective contacts for hot carrier, Hot Lattice or Thermoelectric cells, but the concept itself was first suggested by Brennan and Summers[68] and has been used for energy filtering in quantum cascade lasers for several years. It is also not the first application of the concept to solar cells as, Barnham *et al.*[69] at Imperial College, London, have used the concept for a multi-QW solar cell design. But it is the first application to hot carrier or hot lattice devices and its ability to rectify is likely to be a very significant advantage.

12.5.3 Optical Coupling for Hot Carrier Cells

The difficulty of engineering a material both good for slowed carrier cooling and with good transport to contacts, followed by an appropriate ESC structure with little loss, can perhaps be overcome by operation of the hot carrier absorber in open circuit such as to allow recombination and re-emission of photons which are then incident on a high efficiency conventional solar cell. This concept was first suggested by Farrel *et al.*,[70] (although in its ideal form it is very close to a thermo-photovoltaic converter with an optical filter[1]). Farrell suggested either a down converter configuration with the absorber in front of the solar cell or an up-conversion one with the absorber behind a bifacial cell. It is now being developed with extension to a combination of both up- and down-conversion before illuminating the cell and with the addition of an optically selective distributed Bragg reflector tuned to only pass photons with energy just above the bandgap of the solar cell. Longer and shorter wavelength photons are reflected back to the absorber where they are re-absorbed and help to maintain the high carrier temperature. This concept is illustrated in Figure 12.21.

Such a selective emitter can only be placed on the rear side of the absorber as if on the front it would block most incoming sunlight, but this is not a large limitation as any solar cell must similarly allow both absorption and emission over a wide range from its front surface. The amount of loss from

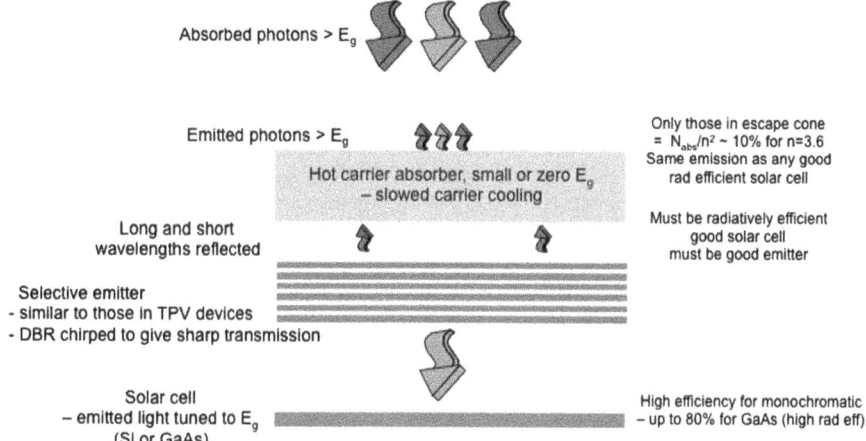

Figure 12.21 Optically coupled hot carrier cell. The absorber is in open circuit. It has appropriate mechanisms for slowing carrier cooling and emits photons over a wide range of energies. On the back surface a DBR selective emitter allows only photons just above its band gap to be incident on an efficient conventional Si or GaAs cell.

re-emission is limited to about 8% for an absorber of refractive index n about that of silicon at 3.6, $1/(1 + n^2) = 8\%$.

It is still required that very significant slowing of carrier cooling is achieved in the absorber. In fact it might need to be even longer than in an electrically coupled device, because the length of time hot carriers remain in the absorber is increased. But importantly the need to allow electrical transport of hot carriers to contacts is removed and so highly localized designs in which carriers do not move very far are possible. These can include isolated QD systems or arrays with emphasis on phonon modulation to slow cooling rather than having to also be concerned for close spacing for carrier transport.

DBRs are well known in layered dielectric materials and can be applied readily as selective emitters. Just one example of such a tunable photonic structure in porous silicon is given by Johnson *et al.*[71]

The advantage of tuning the selective emitter as a band pass for a specific wavelength means that a highly efficient solar cell can be used. This could be a PERL Silicon cell at 25.0% under 1 sun[72] or an Alta Devices GaAs at 28.8% efficiency under 1 sun[73] Illuminated by a nearly monochromatic emission from the selective emitter such a GaAs cell's efficiency for that stage of the conversion can be up to 80%.

12.6 Summary and Conclusion

The hot carrier solar cell has the potential to achieve very high efficiencies in a device that is essentially a single junction. Detailed balance calculations indicate limiting efficiencies as high as 65% under 1 sun and 85% under maximum concentration. However a series of modelling developments has shown that as real material parameters are introduced the predicted efficiencies decrease. What emerges is that maximization of the thermalization time constant for hot carriers is critical to improved efficiency.

The carrier cooling mechanisms are investigated and depend primarily on emission of optical phonons by cooling carriers, predominantly electrons. Under some circumstances these optical phonons can be produced at such a high density that they cannot decay away fast enough and a 'phonon bottleneck' is formed that allows the phonon energy to scatter back with the electron ensemble thus re-heating it. Creating the conditions for this phonon bottleneck seems the most fruitful route for significantly increasing the thermalization time constant.

Quantum well nanostructures exhibit such phonon confinement with significantly hot carrier temperatures. The reasons for this are not completely clear but are affected by the restriction of hot carriers diffusing in the direction perpendicular to the wells and by confinement of phonons in the wells. Prevention of decay of optical phonons into acoustic phonons is another method for maximizing phonon bottleneck. Materials with a large difference in acoustic and optical phonon energies can block this Klemens route for phonon decay. A range of materials are identified as having these properties with the principle requirement that there is a large mass difference between their constituent atoms. Some of the most promising are III-nitrides, especially InN, and their analogues, which include transition metal nitrides (of which HfN and ZrN are most interesting) and group IV compounds (of which SnSi has the most impressive modelled properties). Experimental demonstrations of these effects are very limited at present although there are encouraging signs that these properties will soon be demonstrated in several material groups.

Contacting to hot carrier cells requires specific contacts which only allow transmission of a narrow range of energies. This is so that cold carriers in the contacts do not cool carriers in the absorber. The most promising route to such contacts at present is the double barrier resonant tunnelling structure which can be tuned to specific energies. Such structures in high quality have been made in III–Vs and demonstration of resonant tunnelling achieved. Thin film structures involving silicon and oxides have also shown promising proof of concept. An alternative to electrically contacting is to allow the hot carrier absorber to stay at open circuit and re-radiate photons from hot carriers recombining. Such an approach requires an optically selective filter to illuminate a high efficiency conventional solar cell and has the advantage that optical and electrical properties can be optimized in separate structures.

Combination of absorbers and contacts in full devices has yet to be realized. But there are now a number of designs for such combinations and the next few years should see their fabrication and demonstration of full proof of concept of these challenging but highly promising hot carrier devices.

References

1. M. A. Green, *Third Generation Photovoltaics: Ultra-High Efficiency at Low Cost*, Springer-Verlag, 2003.
2. P. Würfel, *Sol. Energy Mats Sol. Cells*, 1997, **46**, 43.
3. R. Ross and A. J. Nozik, *J. Appl. Phys.*, 1982, **53**, 3318.
4. S. K. Shrestha, P. Aliberti and G. J. Conibeer, *Sol. Energy Mater. Sol. Cells*, 2010, 1546–1550.
5. A. Le Bris and J.-F. Guillemoles, *Appl. Phys. Lett.*, 2010, **97**, 113506; A. LeBris, PhD Thesis, (UPMC, Paris 2010).
6. A. De Vos, *Endoreversible Thermodynamics of Solar Energy Conversion*, Oxford University Press, Oxford, 1992.
7. P. Wurfel1, A. S. Brown, T. E. Humphrey and M. A. Green, *Prog. Photovoltaics: Res. Appl.*, 2005, **13**, 277–285.
8. F. E. Williams and A. J. Nozk, *Nature*, 1978, **271**, 137.
9. D. S. Boudreaux, F. E. Williams and A. J. Nozik, *J. Appl. Phys.*, 1980, **51**, 2158.
10. W. Shockley and H. J. Queisser, *J. Appl. Phys.*, 1961, **32**, 10–519.
11. H. K. Jung, K. Taniguchi and C. Hamaguchi, *J. Appl. Phys.*, 1996, **79**, 2473.
12. M. Neges, K. Schwarzburg and F. Willig, *Sol. Energy Mater. Sol. Cells*, 2006, **90**, 2107–2128.
13. P. Y. Yu and M. Cardona, *Fundamentals of Semiconductors*, Springer, 1996.
14. B. K. Ridley, *Quantum Processes in Semiconductors*, Oxford University Press, Oxford, 1999.
15. J. Shah, IEEE *J. Quantum Electron.*, 1986, **QE-22**, 1728; P. Lugli and S. M. Goodnick, *Phys. Rev. Lett.*, 1987, **59**, 716.
16. V. Y. Davydov, V. V. Emtsev, I. N. Goncharuk, A. N. Smirnov, V. D. Petrikov, V. V. Mamutin, V. A. Vekshin, S. V. Ivanov, M. B. Smirnov and T. Inushima, *Appl. Phys. Lett.*, 1999, **75**, 3297.
17. D. Fritsch, H. Schmidt and M. Grundmann, *Phys. Rev. B*, 2004, **69**, 165204.
18. P. Würfel, *Sol. Energy Mater. Sol. Cells*, 1997, **46**, 43.
19. R. Ross and A. J. Nozik, *J. Appl. Phys.*, 1982, **53**, 3318.
20. Y. Takeda, T. Ito, T. Motohiro, D. König, S. K. Shrestha and G. Conibeer, *J. Appl. Phys.*, 2009, **105**, 074905–074910.
21. P. Wurfel1, A. S. Brown, T. E. Humphrey and M. A. Green, *Prog. Photovoltaics: Res. Appl.*, 2005, **13**, 277–285.
22. P. Aliberti, Y. Feng, Y. Takeda, S. K. Shrestha, M. A. Green and G. J. Conibeer, *J. Appl. Phys.*, 2010, **108**, 094507–094510.
23. G. Conibeer, C. W. Jiang, D. König, S. K. Shrestha, T. Walsh and M. A. Green, *Thin Solid Films*, 2008, **216**, 6968–6973.
24. Y. Y. Feng, P. Aliberti, B. P. Veettil, R. Patterson, S. Shrestha, M. A. Green and G. Conibeer, *Appl. Phys. Lett.*, 2012, **100**, 053502.

25. C. Jacoboni and L. Reggiani, *Rev. Mod. Phys.*, 1983, **55**, 645; M. Saraniti and S. M. Goodnick, *IEEE Trans. Electron Devices*, 2000, **47**, 1909.

26. (a) E. Tea, H. Levard, H. Hamzeh, A. Le Bris, S. Laribi, F. Aniel and J. F. Guillemoles, Proceedings of the 27th European Photovoltaic Solar Energy Conference and Exhibition, Hambourg 2011, Frankfurt, sept 2012, p. 516; (b) J.-F. Guillemoles, A. Le Bris, L. Lombez, S. Laribi, E. Tea and F. Aniel, in: Numerical Simulation of Optoelectronic Devices (NUSOD), 2011 11th International Conference on 5-8 Sept. 2011, pp. 17–18; (c) E. Tea, PhD thesis, University Paris Sud, 2011, http://tel.archives-ouvertes.fr/tel-00670433.

27. (a) J. Rodiére, H. Levard, A. Le Bris, S. Laribi, L. Lombez, C. Colin, S. Collin, J. L. Pelouard, P. Christol, J. F. Guillemoles, Proceedings 27th European Photovoltaic Solar Energy Conference and Exhibition Proceedings, Hambourg 2011, Frankfurt, Sept 2012, p. 89; (b) J. Rodiére, L. Lombez, J. F. Guillemoles, H. Folliot, O. Durand, 28th European Photovoltaic Solar Energy Conference and Exhibition, pp. 127–128, DOI: 10.4229/28thEUPVSEC2013-1DO.4.3.

28. Y. Feng, R. Patterson, S. Lin, S. Shrestha, S. Huang, M. Green and G. Conibeer, *Appl. Phys. Lett.*, accepted May 2013. 10.1063/1.4811263 Date: 7 June 2013.

29. P. G. Klemens, *Phys. Rev.*, 1966, **148**, 845.

30. G. Conibeer, R. Patterson, P. Aliberti, L. Huang, J-F. Guillemoles, D. König, S. Shrestha, R. Clady, M. Tayebjee, T. Schmidt and M. A. Green, 24th European Photovoltaic Solar Energy Conference, (Hamburg, 2009) Hot Carrier solar cell Absorbers.

31. Y. Feng, S. Lin, M. A. Green and G. Conibeer, *J. Appl. Phys.*, 2013, **113**, 024317.

32. F. Chen and A. N. Cartwright, *Appl. Phys. Lett.*, 2003, **83**, 4984.

33. R. Clady, M. J. Y. Tayebjee, P. Aliberti, D. Konig, N. J. Ekins-Daukes, G. J. Conibeer, T. W. Schmidt and M. A. Green, *Prog. Photovoltaics: Res. Appl.*, 2012, **20**, 82–92.

34. Y. Rosenwaks, M. Hanna, D. Levi, D. Szmyd, R. Ahrenkiel and A. Nozik, *Phys. Rev. B*, 1993, **48**, 14675–14678.

35. J-F. Guillemoles, G. Conibeer and M. A. Green, Proc. *21st European Photovoltaic Solar Energy Conference* (Dresden Germany, 2006) pp. 234–237.

36. L. Hirst, M. Führer, A. LeBris, J-F. Guillemoles, M. Tayebjee, R. Clady, T. Schmidt, Y. Wang, M. Sugiyama and N. Ekins-Daukes, Proc. 37th IEEE Photovoltaics Specialists Conference, (Seattle, 2011) talk 927.

37. L. C. Hirst, M. Fuhrer, D. J. Farrell, A. Le Bris, J-F. Guillemoles, M. J. Y. Tayebjee, R. Clady, T. W. Schmidt, M. Sugiyama, Y. Wang, H. Fujii, N. J. Ekins-Daukes, Proc. SPIE Photonics West conf. (San Francisco 2012) InGaAs/GaAsP quantum wells for hot carrier solar cells.

38. C. Colvard, T. A. Gant and M. V. Klein, *Phys. Rev. B*, 1985, **31**, 2080–2091.

39. T. K. Todorov, K. B. Reuter and D. B. Mitzi, *Adv. Mater.*, 2010, **22**, 1–4.

40. W. R. L. Lambrecht, *Phys. Rev. B*, 2005, **72**, 155202.

41. S. C. Erwin and I. Zutic, *Nat. Mater.*, 2004, **3**, 410–414.
42. T. R. Paudel and W. R. L. Lambrecht, *Phys. Rev. B*, 2009, **79**, 245205.
43. (a) P. A. Cox, in *The Elements: Their Origin, Abundance, and Distribution*, Oxford University Press, 1989; (b) P. A. Cox *McGraw-Hill Encyclopedia of Science & Technology*, 7th edn., McGraw-Hill, New York, 1992; (c) A. M. James and M. P. Lord, in *Macmillan's Chemical and Physical Data*, Macmillan, London, 1992.
44. D. König, K. Casalenuovo, Y. Takeda, G. Conibeer, J-F. Guillemoles, R. Patterson, L. Huang and M. A. Green, *Physica E*, 2010, **42**, 2862–2866.
45. H. Bilz and W. Kress, *Phonon Dispersion Relations in Insulators*, Springer, 1979.
46. S. Barman and G. P. Shrivestava, *Phys. Rev.*, 2004, **69**, 235208.
47. (a) S. Dalui, S. N. Das, S. Hussain, D. Paramanik, S. Verma and A. K. Pal, *J. Cryst. Growth*, 2007, **305**, 149–155; (b) S. Hussain, S. Dalui, R. K. Roy and A. K. Pal, *J. Phys. D: Appl. Phys.*, 2006, **39**, 2053; (c) Y. Yao, D. Konig and M. A. Green, *Solar Energy Materials and Solar Cells*, 2013, **111**, 123–126.
48. B. Saha, T. D. Sands and U. V. Waghmare, *J. Appl. Phys.*, 2011, **109**, 073720.
49. B. Saha, J. Acharya, T. D. Sands and U. V. Waghmare, *J. Appl. Phys.*, 2010, **107**, 033715.
50. A. Norlund-Christensen, W. Kress and M. Miura, *Phys. Rev. B*, 1983, **28**, 977.
51. Z. T. Liu, J. Z. Zhu, N. K. Xu and X. L. Zheng, *Jpn. J. Appl. Phys.*, 1997, **36**, 3625.
52. P. Giannozzi, S. de Gironcoli, P. Pavone and S. Baroni, *Phys. Rev. B*, 1991, **43**, 7231.
53. R. Patterson, M. Kirkengen, B. P. Veettil, D. König, M. A. Green and G. Conibeer, *Sol. Energy Mater. Sol. Cells*, 2010, **94**, 1931.
54. A. Debernardi, C. Ulrich, K. Syassen and M. Cardona, *Phys. Rev. B*, 1999, **59**, 6774.
55. F. Vallée, *Phys. Rev. B*, 1994, **49**, 2460.
56. P. Maly, A. C. Maciel, J.-F. Ryan, N. J. Mason and P. J. Walker, *Semicond. Sci. Technol.*, 1994, **9**, 719.
57. E. Haro-Poniatowski, J. L. Escamilla-Reyes and K. H. Wanser, *Phys. Rev. B*, 1996, **53**, 23232.
58. B. K. Ridley, *J. Phys.: Condens. Matter*, 1996, **8**, L511.
59. F. Valle and F. Bogani, *Phys. Rev. B*, 1991, **43**, 12049.
60. (a) S. Barman and G. P. Strivastava, *Appl. Phys. Lett.*, 2002, **81**, 3395; (b) A. Debernardi, *Solid State Commun.*, 2000, **113**, 1; (c) M. Canonico, C. Poweleit, J. Menéndez, A. Debernardi, S. R. Johnson and Y.-H. Zhang, *Phys. Rev. Lett.*, 2002, **88**, 215502; (d) L. Paulatto, F. Mauri and M. Lazzeri, *Phys. Rev. B*, 2013, **87**, 214303.
61. G. Conibeer, R. Patterson, L. Huang, J.-F. Guillemoles, D. Konig, S. Shrestha and M. A. Green, *Sol. Energy Mater. Sol. Cells*, 2010, **94**, 1516–1521.

62. G. Conibeer, N. J. Ekins-Daukes, J-F. Guillemoles, D. König, E-C. Cho, C-W. Jiang, S. Shrestha and M. A. Green, *Sol. Energy Mater. Sol. Cells*, 2009, **93**, 713.

63. R. Patterson, M. Kirkengen, B. Puthen Veettil, D. Konig, M. A. Green and G. Conibeer, *Sol. Energy Mater. Sol. Cells*, 2010, **94**, 1931–1935.

64. Y. Kamikawa-Shimizu, R. Patterson, A. Milekhin, K. Akahane, Y. Shoji, D. R. T. Zahn, Y. Okada and G. Conibeer, Low dimensional structures and nanostructures, Physica E, accepted Oct 13, in press, available online 31 October 2013.

65. Y. Feng, P. Aliberti, B. P. Veettil, R. Patterson, S. Shrestha, M. A. Green and G. Conibeer, *Appl. Phys. Lett.*, 2012, **100**, 053502.

66. G. Conibeer, C. W. Jiang, D. König, S. K. Shrestha, T. Walsh and M. A. Green, *Thin Solid Films*, 2008, **216**, 6968–6973.

67. A. Hsieh, Investigation of Si/SiO2 and Si/SiOx Quantum Well Structures for Applications as Energy Selective Contacts and All-Silicon Tandem Cells, PhD Thesis, UNSW, 2014.

68. K. F. Brennan and C. J. Summers, *J. Appl. Phys. Lett.*, 1987, **61**, 614–623.

69. K. Barnham and G. Duggan, *J. Appl. Phys.*, 1990, **67**, 3490.

70. M. Farrell, Y. Takeda, K. Nishikawa, T. Nagashima, T. Motohiro and N. J. Ekins-Daukes, *Appl. Phys. Lett.*, 2011, **99**, 111102.

71. C. Johnson, P. J. Reece and G. J. Conibeer, *Sol. Energy Mater. Sol. Cells*, 2013, **112**, 168.

72. J. Zhao, A. Wang and M. A. Green, *Appl. Phys. Lett.*, 1998, **73**, 1991–1993.

73. B. M. Kayes, H. Nie, R. Twist, S. G. Spruytte, F. Reinhardt, I. C. Kizilyalli and G. S. Higashi, Proc. 37th IEEE Photovoltaics Specialists Conference, Seattle, 2011.

CHAPTER 13

Intermediate Band Solar Cells

YOSHITAKA OKADA,* TOMAH SOGABE AND YASUSHI SHOJI

Research Center for Advanced Science and Technology (RCAST), The University of Tokyo, 4-6-1 Komaba, Meguro-ku, Tokyo, 153-8904, Japan
*Email: okada@mbe.rcast.u-tokyo.ac.jp

13.1 Introduction

For an ideal case, in which all the non-radiative recombination processes within solar cell were negligibly small and hence ignored, the theoretical maximum energy conversion efficiency in thermodynamic upper limit becomes as high as ~85% by assuming a fully concentrated black-body radiation at 5800 K.[1,2] The maximum efficiency of a single-junction solar cell, however, is limited to the Shockley–Queisser limit of 31–32% for air mass (AM) 1.5 spectrum.[3] The main physical processes that limit the efficiency of solar cell are the losses by thermal dissipation or thermalization, and non-absorption of low-energy below-bandgap photons. These losses could add up to as much as 40–50% of incident solar energy. Thus improving the cell efficiency directly means developing methods to reduce these losses.

One of the concepts that are well established today is to split the solar spectrum among multiple bandgap absorbers or sub-cells, *e.g.*, tandem or multi-junction cells. The other approaches employ more advanced techniques such as hot carriers, multi-exciton generation (MEG), and intermediate-band (IB) absorption in low-dimensional nanostructures such as semiconductor quantum dots.[4–6] Among various approaches studied, this chapter is devoted to describing the basic principles and development of the state-of-the-art technologies for quantum dot (QD)-based IB solar cells.

RSC Energy and Environment Series No. 11
Advanced Concepts in Photovoltaics
Edited by Arthur J Nozik, Gavin Conibeer and Matthew C Beard
© The Royal Society of Chemistry 2014
Published by the Royal Society of Chemistry, www.rsc.org

A high-density array of QDs incorporated in the active region of a p-i-n single-junction solar cell has attracted significant interest as a potential IB solar cell which utilizes the sub-bandgap infrared photons to generate additional photocurrent other than that corresponding to the valence-to-conduction band (VB-CB) optical transition, through extra absorption *via* superlattice miniband states.[4-6] The QDs also offer possibility for reducing thermal dissipation loss. If such a nanostructured solar cell were truly realized, the conversion efficiency of such IB solar cell can exceed the Shockley–Queisser limit of conventional single-junction solar cell. The calculated maximum efficiency for solar cell incorporated with one IB is $> \sim 47\%$ under 1 sun, and $\sim 63\%$ under full solar concentration.[6]

In a QD-IB solar cell, QDs are required to be homogeneous and small in size, and are regularly and tightly placed. This structural configuration then leads to formation of an intermediate-band (IB) or a superlattice miniband that is well separated in energy from the higher-order states.[7] Secondly, IB states should be partially filled, or ideally be half-filled, with electrons in order to ensure an efficient pumping of electrons by providing both the empty states to receive electrons being photo-excited from VB, and filled states to promote electrons to CB *via* absorption of second sub-bandgap photons.[8] This implies that an independent quasi-Fermi level can be defined within IB, as schematically shown in Figure 13.1.

Proposed implementation of QD-IB solar cell must accompany two-step carrier generation *via* IB states as shown in Figure 13.1, but it has been difficult to clearly demonstrate this operation concept *at room temperature*, which is due to, as will be discussed in more detail in the following, relatively small optical generation rates from QD-IB states to CB (photons indicated in Figure 13.1).

The rate of thermal escape of electrons from IB to CB continuum as well as recombination from CB to IB transitions (reverses process) also increase significantly at room temperature.

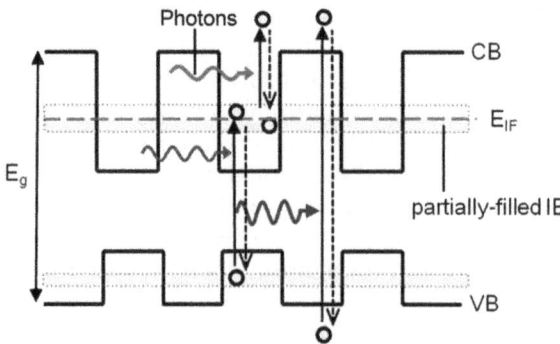

Figure 13.1   Schematic energy band diagram of QD-IB solar cell shown with possible photo-absorption and recombination processes involved. The energy bandgap of host material and quasi-Fermi level of IB (QDs) are given by E_g and E_{IF}, respectively.

The demonstration of QD-IB solar cell is presently undergoing two research stages. The first is to develop technology to fabricate high-density QDs arrays or superlattices with low interface defect densities and sufficiently long carrier lifetimes in the optimized QD/host material band-gap system. The fabrication of QDs arrays is most commonly achieved by taking advantage of spontaneous self-assembly of coherent three-dimensional (3D) islands in lattice-mismatched epitaxy, well known as Stranski–Krastanov (S-K) growth in molecular beam epitaxy (MBE) and metalorganic vapor phase epitaxy (MOVPE). However, the number of QDs is severely limited by lattice strain accumulation in single crystalline material as the number of stacked QD layers is increased. In S-K growth of InAs/(Al)GaAs system, which is one of the most popularly investigated and well established QD material system, misfit dislocations are generated after typically 10–15 layers of stacking.[9] Recently, strain-compensation or strain-balanced growth technique have shown a significantly improved QD quality as well as characteristics of QD solar cells even after stacking of 50–100 QD layers by S-K growth. To date, InAs QDs in AlGaInAs matrix on InP substrate,[10,11] InAs QDs in GaAsP,[12] and in GaP[13] matrices on GaAs substrate, and InAs QDs in GaNAs on GaAs(001),[14–16] and on (311)B substrate[17,18] of high material quality have been reported.

The second stage needed to implement a high efficiency QD-IB solar cell is to realize partially filled or ideally half-filled IB states in order to maximize the photocurrent generation by two-step photon absorption, which is the central operating principle of IB solar cell. Martí *et al.*[19] have used a *modulation doping* technique to fabricate a QD-IB solar cell, in which GaAs barrier was delta (δ)-doped with Si with a sheet density equaling the InAs QD areal density. Due to spin degeneration, this condition equals half the density of electronic states in IB. On the other hand, Strandberg and Reenaas[20] have calculated that IB must be partially filled by means of doping within a reasonable optical length in order to achieve high efficiency, if QD-IB solar cell were to be operated under 1 sun conditions. However, it becomes possible to sustain a reasonable population of photogenerated electrons even in a non-doped QD-IB solar cell if it was operated under concentrated sunlight, typically 100–1000 suns. Yoshida *et al.*[21] have also recently reported similar calculated results based on a self-consistent device simulation method.

The sub-bandgap photons can optically pump the electrons both from VB to IB states and from IB to CB states as depicted in Figure 13.1. However, both the rate of thermal escape, as opposed to optical excitation, of electrons from IB to CB continuum and the recombination rate from CB to IB transitions increase significantly at room temperature. If all the carriers in IB states are in electrochemical equilibrium, which is a condition defined by an independent quasi-Fermi level E_{IF}, then a clear detection of photo-excitation of electrons from IB to CB giving rise to an increase in the photocurrent at room temperature should become possible. A photocurrent production as a direct result of two-step photon absorption has been successfully detected at a cryogenic temperature by Martí *et al.*[19] and more recently at room temperature by Okada *et al.*[22] and Sablon *et al.*,[23] though the rate of this two-step current generation is still small in both reports and it requires further improvements.

In this chapter, we will first review the analysis for various cell designs where QD technology could be implemented to configure high-efficiency IB solar cell. Then the self-organized growth of QDs, control of size and density of QDs, and direct doping of QDs will be reviewed. Last we summarize with specific experimental results for proof of operation with two-step photon absorption from the current state-of-the-art IB solar cells implementing QD superlattice.

13.2 Numerical Analysis of QD-IB Solar Cell Characteristics

Multi-stacked or QD superlattice solar cells can operate as intermediate band (IB) solar cells. To fabricate a high-efficiency QD-IB solar cell, not only a high crystalline quality but also a structural optimization is crucial. In this respect, device-domain simulation of IB solar cell would play an important role. Device-domain simulations have been reported in recent references.[20,21,24-26] In this section, we summarize our results of one-dimensional (1D) device-domain simulation for IB solar cell[21,26] which is based on the drift-diffusion method.[27] In particular, the effects of doping in the IB region and incident sunlight concentration on the cell characteristics will be discussed.

For device simulation, ideal ohmic contacts are assumed for the majority carriers and zero surface recombination velocity for the minority carriers. The zero surface recombination velocity refers to an ideal passivation of surface or equivalently a lossless window layer. The dark current can then be determined by the total amount of recombination processes inside the cell itself. Two cases of IBSC, as schematically shown in Figure 13.2, are modeled; IB region of 1.0 μm thickness is (1) non-doped (w/o doping), and (2) n-doped (w doping) such that a half of IB states are filled with electrons. The material parameters used are from the database available for GaAs at 300 K except for

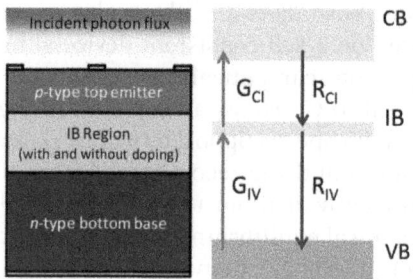

Figure 13.2 Schematic drawing of IBSC structure employed for device simulation. The IB region is assumed to be 1.0 μm thick and either non-doped (w/o doping) or doped (w doping) such that IB is half-filled.
(Reprinted from K. Yoshida, Y. Okada and N. Sano, *J. Appl. Phys.*, 2012, 112, 084510.)

the values of absorption coefficients and refractive index. To understand the intrinsic effects of impurity doping in IB and light concentration on operation characteristics of IBSC, a simple model for the absorption coefficients is employed, in which all the absorption coefficients for VB to CB, VB to IB and IB to CB optical transitions are constant at 10^4 cm^{-1}, and the overlap of absorption coefficients among various transitions is ignored.

Figure 13.3 shows the current–voltage (*I–V*) characteristics calculated for IB solar cells with non-doped (w/o doping) and doped (w doping) IB region under 1 and 1000 sun illuminations, respectively. It is noted that the current density is normalized by the light concentration factor *X*. For reference, *I–V* characteristics for GaAs control cell under the same illumination conditions are also plotted. The GaAs control cell shows a smaller short-circuit current (I_{SC}), but a larger open-circuit voltage (V_{OC}) than those of IB cells under both 1 sun and 1000 suns. The fill factors for both IB cells also degrade compared to GaAs control cell. This is qualitatively explained as follows: Since IB can be regarded as a carrier generation center, I_{SC} in IB cells increases. But at the same time, IB also works as a recombination center and thus, it degrades V_{OC} and fill factor. Introducing doping in the IB region remarkably improves I_{SC} under 1 sun also shown in Figure 13.3. This is because a large occupation probability of IB states enhances the net optical generation rate *via* IB. Recent experiments observe that donors introduced in multi-stacked quantum dot layers indeed increase I_{SC}.[19,22,23] The experimental results reported therein are consistent with our device simulation and it demonstrates the important role of doping in the IB region.

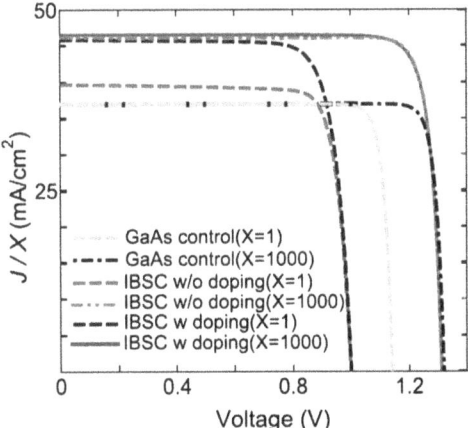

Figure 13.3 Calculated current–voltage (*I–V*) characteristics of IB solar cells with non-doped (w/o doping), doped (w doping) IB region, and without IB (GaAs control) cases, under 1 sun and 1000 suns illuminations, respectively. The I_{SC} is normalized by the light concentration factor *X*. (Reprinted from K. Yoshida, Y. Okada, and N. Sano, *J. Appl. Phys.*, 2012, **112**, 084510.)

On the other hand, as the sunlight concentration is increased, while I_{SC} in GaAs control cell increases linearly with respect to concentration factor X, I_{SC} in IB cells shows a non-linear dependence with respect to X. This dependence arises from the complexity of carrier generation rate associated with electron occupancy in IB. As sunlight concentration increases, the photofilling effect comes into play and the difference in I_{SC} between non-doped and doped IB cells becomes negligible under high concentrations. In addition, V_{OC} and fill factor for IB cells both improve as sunlight concentration is increased because a larger voltage is now required to reduce the larger net carrier generation *via* IB.

The I_{SC}, V_{OC} and cell efficiency η is plotted as a function of light concentration factor X for IB cells and GaAs control cell as shown in Figure 13.4. As shown in Figure 13.4a, I_{SC} in IB cell with non-doped IB (w/o doping) shows a strong non-linear dependence on X, compared to IB cell with doped IB (w doping) and GaAs control cell. Further, I_{SC} in non-doped IB cell approaches that of doped IB cell as X becomes larger than 1000 suns as a result of photofilling effect. Figure 13.4b shows that V_{OC} in both IB cells, non-doped and doped IB is strongly dependent on X, while that for GaAs control cell is logarithmically dependent, which is typical in single junction solar cells.[2] This is because the recombination processes between CB and VB dominate in low bias range under large X, compared with recombination rates *via* IB. Thus, the operation of IB cells becomes similar to that of GaAs control cell under high sunlight concentration and results in a value of V_{OC} which is similar to V_{OC} of GaAs control cell. Last, Figure 13.4c shows that the cell efficiencies as a function of X. Due to smaller V_{OC} and fill factor in IB cells observed under weak illumination, the efficiency of IB cells, especially for non-doped IB cell becomes smaller than that of control cell under weak illumination. However, the conversion efficiency of IB cells recovers fast due to the photofilling of IB states, which in turn increases all of I_{SC}, V_{OC} and fill factor. In order to evaluate the expected performance of

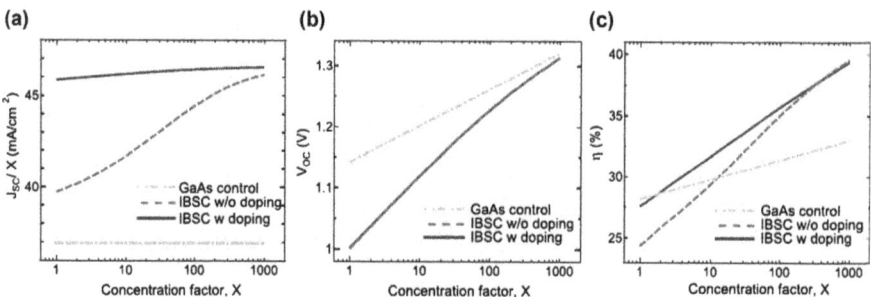

Figure 13.4   (a) I_{SC}, (b) V_{OC}, and (c) cell efficiency η calculated for IB solar cells with non-doped (w/o doping) and doped (w doping) IB region, and for GaAs control cell (reference) as a function of the sunlight concentration factor X, respectively. The I_{SC} is normalized by the concentration factor X. (Reprinted from K. Yoshida, Y. Okada and N. Sano, *J. Appl. Phys.*, 2012, **112**, 084510.)

QD-IB solar cells in more detail, Tomić has carried out a rigorous theoretical study to calculate the absorption coefficients for each optical transition process in InAs-based QDs.[28]

13.3 Fabrication of QD-IB Solar Cells

In a QD-IB solar cell, QDs are required to be homogeneous and small in size, and are regularly and tightly placed. This structural configuration then leads to formation of an intermediate-band (IB) or a superlattice miniband that is well separated in energy from the higher-order states.[7,28] Therefore, advanced crystal growth technology must be developed.

13.3.1 Growth and Properties of High-density InAs QD Arrays on High-index Substrate

The most popular fabrication technique of QDs is to take advantage of spontaneous self-assembly or self-organization of coherent 3D islands in lattice-mismatched heteroepitaxy known as S-K growth mode in molecular beam epitaxy (MBE) or metal-organic vapor phase epitaxy (MOVPE). The heteroepitaxy of InAs grown on GaAs (001) substrate is one of the most investigated QDs formed by S-K mode today. The lattice mismatch between InAs and GaAs bulk is about 7.2%, and lattice strain introduced by the large mismatch will drive into S-K growth. Consequently, coherent InAs 3D islands are formed after the growth of a thin wetting layer. The QD size and density are dependent on the growth conditions such as the growth temperature, deposition thickness, and V/III flux ratio. Kakuda *et al.* have grown uniform and extremely high-density *in-plane* InAs QDs on GaAs (100) substrate by using Sb-mediated molecular beam epitaxy (MBE).[29]

On the other hand, InAs-based QD structures grown on a high-index GaAs (311)B substrate can be made higher in density under equivalent growth conditions, and more significantly, exhibit a superior in-plane spatial ordering characteristics.[30] This high degree of in-plane ordering of QDs formed on (311)B surface could facilitate the formation of a superlattice miniband,[17,31] which is favorable for QD-IB solar cells. Figure 13.5 shows the atomic force microscope (AFM) image of the topmost InGaAs QDs layer grown on GaAs (311)B substrate by MBE at 480 °C.[31] The In composition is 40%. The inset shows the two-dimensional fast Fourier transformation (2D-FFT) image which is a good indicator of the degree of in-plane spatial ordering. The mean diameter, height, and in-plane density of QDs are determined to be 60.8 nm, 10.1 nm, and 3.1×10^{10} cm^{-2}, respectively. The clear fine peaks shown in 2D-FFT image indicate that a high degree of in-plane ordering with a six-fold symmetry has been obtained.

The InGaAs QDs grown on GaAs (311)B show some unique structural features. Figure 13.6a shows an enlarged AFM image of a single InGaAs QD grown on GaAs(311)B substrate, and Figure 13.6b and c show the bright-field (BF) cross-sectional scanning transmission electron microscope (STEM)

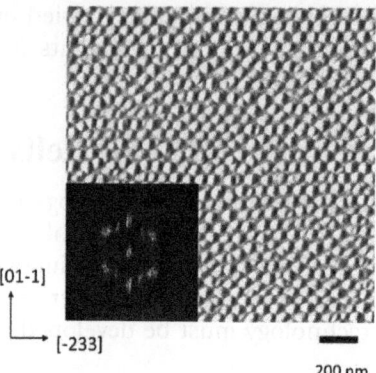

Figure 13.5 AFM image of InGaAs QDs grown on GaAs (311)B substrate. The inset
shows 2D-FFT image.
(Reprinted from Y. Shoji, K. Akimoto and Y. Okada, *J. Phys. D: Appl.
Phys.*, 2013, **46**, 024002.)

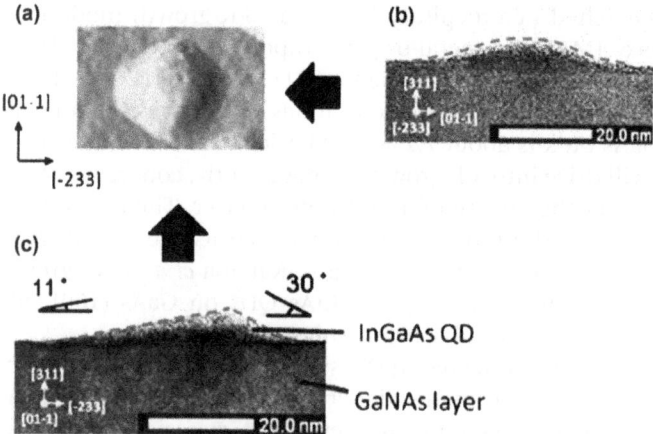

Figure 13.6 (a) Enlarged AFM image of single QD. (b) and (c) show the bright-field
cross-sectional STEM images for the uncapped QD viewed along
(b) [01-1] and (c) [-233], respectively.
(Reprinted from Y. Shoji, K. Akimoto, and Y. Okada, *J. Phys. D: Appl.
Phys.*, 2013, **46**, 024002.)

images for the uncapped QD viewed along (b) [01-1] and (c) [-233], respectively. Symmetrical lens-shaped QDs are observed across [01-1], while their
shape is asymmetric along [-233]. The facet angle for each side wall of QD
with respect to the underlying buffer layer horizontal surface is about 11°
and 30°, respectively as also shown in the Figure 13.6c. This suggests that
two different facets are dominantly formed along [-233], and this is thought
to be caused by an elastic anisotropy of GaAs (311)B surface itself.[30–32] The
anisotropic shape is then expected to affect the final alignment of QDs in the
growth direction.

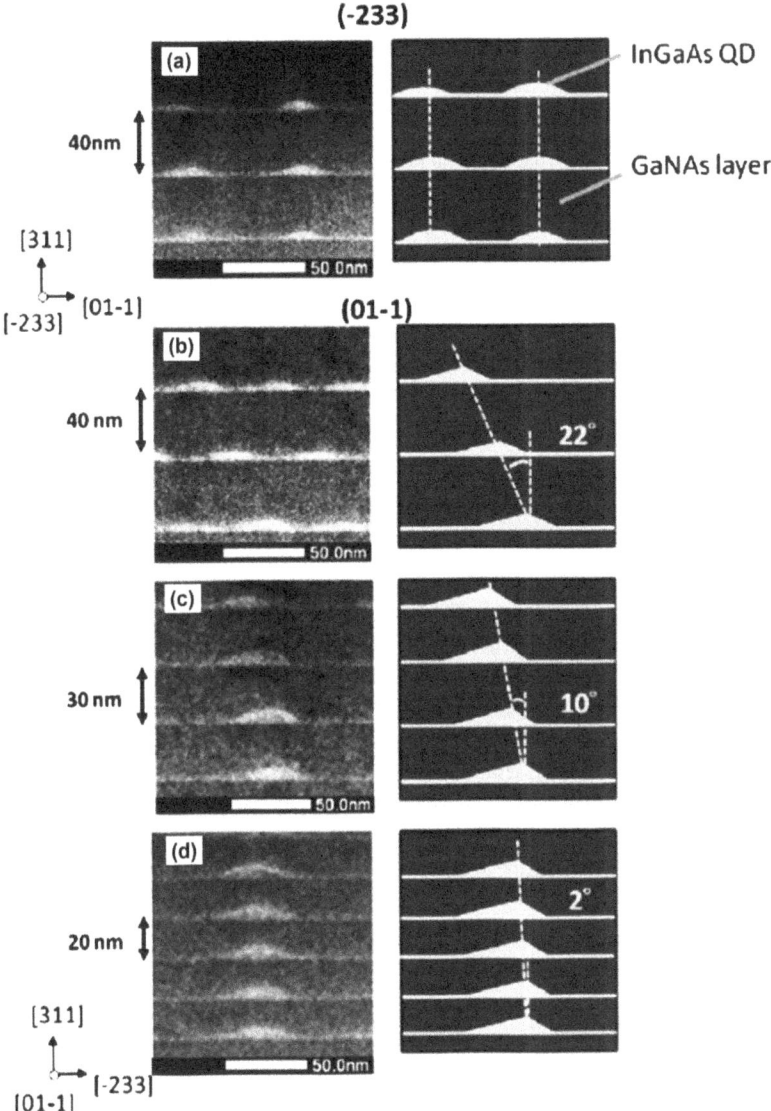

Figure 13.7 (a) Cross-sectional STEM image for the embedded InGaAs/GaNAs strain-compensated QD structure with $d_{SL} = 40$ nm viewed along [01-1]. (b), (c) and (d) show images for the embedded QD structure with (b) $d_{SL} = 40$ nm, (c) 30 nm, and (d) 20 nm viewed along [-233], respectively. (Reprinted from Y. Shoji, K. Akimoto and Y. Okada, *J. Phys. D: Appl. Phys.*, 2013, **46**, 024002.)

Figure 13.7a shows the high-angle annular dark-field (HAADF) STEM and illustrated images for the portion of embedded InGaAs/GaNAs *strain-compensated* QDs structure with 40 nm-thick spacer layers (SLs) viewed along [01-1]. Figure 13.7b–d show the STEM and illustrated images for the embedded QD structure with (b) 40 nm-thick, (c) 30 nm-thick and (d) 20 nm-thick SLs viewed along [-233], respectively. These images provide sound information on the QD configuration because the contrast in this image is roughly proportional to $Z^{1.7}$, where Z is the atomic number of the scattering atoms.[33] The facet angle of embedded QD determined from Figure 13.7b is in good agreement with that of uncapped QD as of Figure 13.6c though the interface between the embedded QDs and SLs become less clear owing to out-diffusion of In. The embedded QDs have almost the same shape as the uncapped topmost QDs, and are vertically aligned in the growth direction when viewed along [01-1] as in (a). On the other hand, the QD alignment for the same 40 nm-thick SL sample is inclined at an angle of 22° with respect to the growth direction when viewed along [-233] as shown in (b).

In order to clarify the origin of inclined alignment of QDs, HAADF and BF STEM images of the embedded QD structure were compared. Figure 13.8 show the cross-sectional (a) HAADF and (b) BF STEM images for the embedded multi-stacked InGaAs/GaNAs QDs structure viewed along [-233], respectively. The BF image provides the local strained state of individual QD. The boundary areas associated with individual QD along [-233] direction are not identical between Figure 13.8a and b as also illustrated in the schematic image. This indicates that the local strain around QD extends further outward from the lower-angle facet, and hence the vertical alignment of QD is tilted along the direction of a stronger strain field. This suggests that an asymmetric extension of local strain around QDs in the lower layer strongly affects the formation of QDs in the upper layers.

In addition, the inclination angle of vertical alignment QDs becomes smaller from 22° to 2° with decreasing SL thickness from $d_{SL} = 40$ nm to 20 nm. These results suggest that the strain field extends asymmetrically resulting in vertically tilted alignment of QDs for sample with thick SLs,

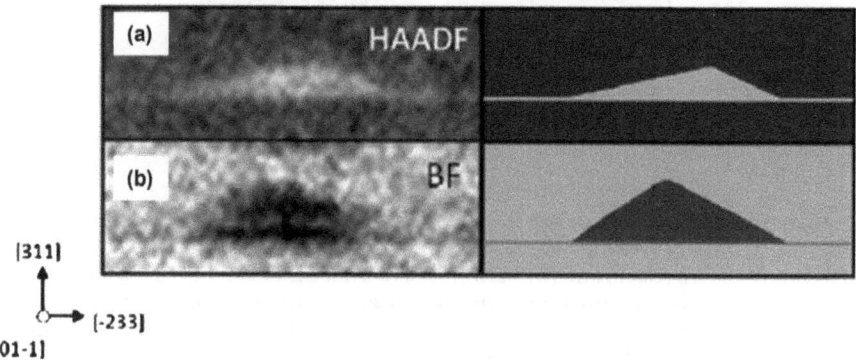

[311]

[-233]

[01-1]

Figure 13.8 (a) Cross-sectional (a) HAADF and (b) BF STEM images for the embedded InGaAs/GaNAs strain-compensated QD structure viewed along [-233], respectively.
(Reprinted from Y. Shoji, K. Akimoto and Y. Okada, *J. Phys. D: Appl. Phys.*, 2013, **46**, 024002.)

while the propagated local strain field is strong enough to generate the nucleation site of QD formation just above the lower QD in the sample with thinner SLs. From the viewpoint of optical absorption, a QD superlattice with an inclined lattice structure may exhibit different absorption cross-section, coefficient and polarization dependence, which will be of great interest to solar cell application and yet wait to be clarified.

The photoluminescence (PL) spectra are shown in Figure 13.9 measured for 10-layer stacked $In_{0.4}Ga_{0.6}As/GaAs$ QDs with (a) $d_{SL} = 10$ nm, (b) 20 nm, and (c) 30 nm, respectively. The QDs are lattice strained as GaAs is used as SLs instead of GaNAs. Further, Figure 13.9d shows the spectrum for a single layer of $In_{0.4}Ga_{0.6}As/GaAs$ QDs. The measurements are done at 16 K within a closed cycle helium cryostat. The PL spectrum for single layer InGaAs QDs as of (d) shows a single peak at around wavelength of 955 nm. On the other, PL spectra for 10-layer stacked QDs consist of two components, and the lower-energy peak becomes dominant with a decrease in the SL thickness. This observation has been well documented[34] and the PL peak observed at around wavelength of 990 nm results from emission of uncoupled QDs and the lower-energy signal arises due to an electronic coupling between the stacked QDs.

In order to evaluate the effect of possible electronic coupling or equivalently superlattice miniband (IB) formation between the stacked QDs, the dependence of PL decay time on SL thickness has been evaluated. Shown in Figure 13.10 are the results of time-resolved PL measurements for 10-layer stacked strained $In_{0.4}Ga_{0.6}As/GaAs$ QDs with $d_{SL} = 40$, 30, 20, and 10 nm, respectively. The detection energy has been set at the peak position of PL spectrum for each sample. The PL decay time τ estimated by fitting with a single exponential function is 0.80, 0.94, 1.11 and 1.86 ns for $d_{SL} = 40$, 30, 20,

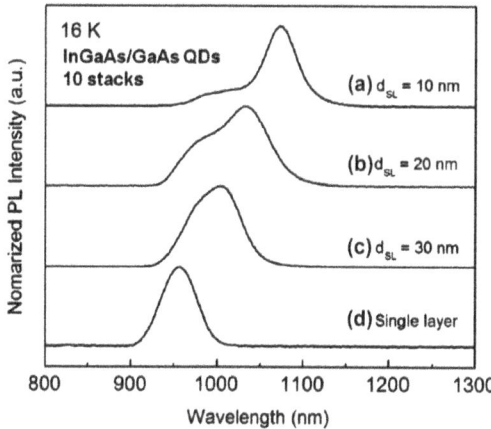

Figure 13.9 PL spectra taken at 16 K for the 10-layer stacked InGaAs/GaAs strained QDs with (a) $d_{SL} = 10$ nm, (b) 20 nm and (c) 30 nm, respectively. (d) shows the PL spectrum for a single layer of InGaAs/GaAs QDs. (Reprinted from Y. Shoji, K. Akimoto and Y. Okada, *J. Phys. D: Appl. Phys.*, 2013, **46**, 024002.)

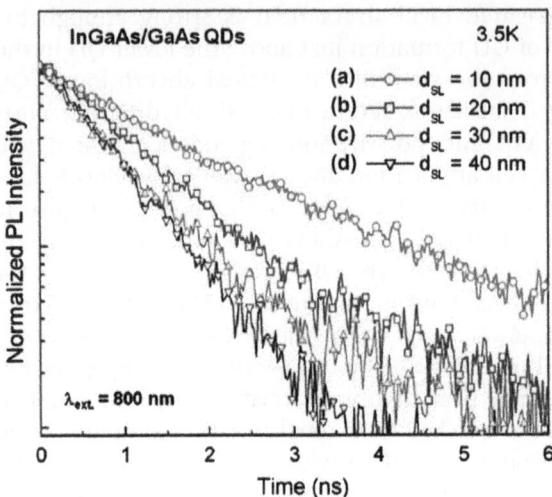

Figure 13.10   PL decay profiles for 10 layer-stacked InGaAs/GaAs QDs with $d_{SL} = 10$ nm to 40 nm.
(Reprinted from Y. Shoji, K. Akimoto and Y. Okada, *J. Phys. D: Appl. Phys.*, 2013, **46**, 024002.)

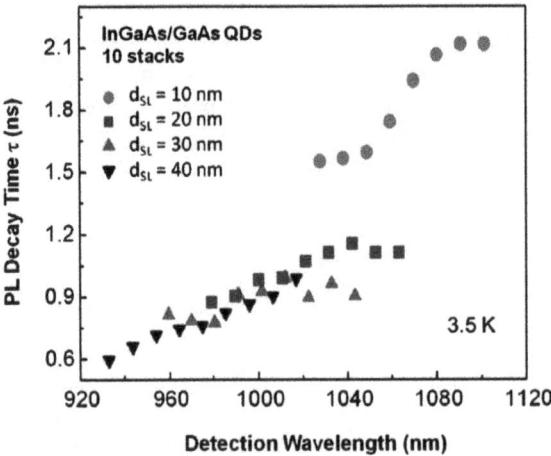

Figure 13.11   PL decay times as a function of detection wavelength measured for 10 layer-stacked InGaAs/GaAs QDs with $d_{SL} = 10$ nm to 40 nm.
(Reprinted from Y. Shoji, K. Akimoto and Y. Okada, *J. Phys. D: Appl. Phys.*, 2013, **46**, 024002.)

and 10 nm, respectively. It is clear that the carrier lifetime τ increases with decreasing SL thickness, and this result is considered to be influenced by the QD size or electronic coupling.[35] Figure 13.11 shows the detection wavelength dependence of PL decay time τ for the 10-layer stacked InGaAs/GaAs QDs with $d_{SL} = 40$, 30, 20, and 10 nm, respectively. From $d_{SL} = 40$ nm down to 20 nm, τ monotonously increases with increasing detection wavelength. This is due to a reduction of the oscillator strength with increase in QD size arising from size inhomogeneity.[36] However, the increasing rate of τ changes drastically for $d_{SL} = 10$ nm. This could suggest that the carrier lifetime increases because electronic coupling between the stacked QDs facilitates the separation of generated excitons in samples with thin SLs.

13.3.2 InAs/GaAs QD-IB Solar Cells Fabricated on High-index Substrate

Shoji *et al.* have fabricated p-i-n QD solar cell structures on GaAs (311)B substrate as illustrated in Figure 13.12.[31] The 10 stacked pairs of $In_{0.4}Ga_{0.6}As$ QDs with GaAs SLs are inserted into the i-layer. For the top p-type emitter side, AuZn Ohmic contact is deposited, and In is used for the bottom n-type ohmic contact. Figure 13.13 shows the external quantum efficiency (EQE) spectrum measured for 10-layer stacked InGaAs/GaAs QD solar cell with $d_{SL} = 15$ nm grown on GaAs (311)B substrate. The EQE is degraded in the 400–600 nm wavelength region but this is mostly due to the absence of an anti-reflection coating (ARC) layer as will be discussed later. It is evident that by embedding InGaAs/GaAs QD structure in the i-layer, EQE extends into the longer wavelength range beyond the bandgap of GaAs (880 nm). Thus EQE response in the wavelength of > 880 nm is attributed to the contribution from InGaAs/GaAs QDs.

In order to detect the production of a photocurrent as a direct result of optical transition of electrons from IB to CB, a measurement system such as

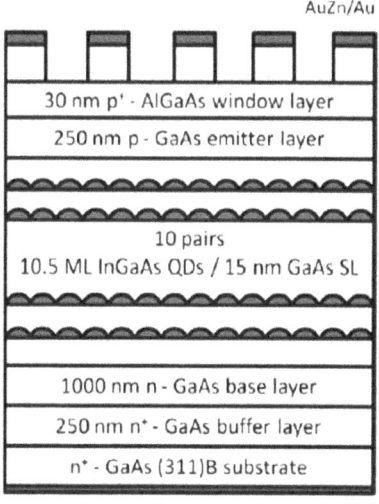

Figure 13.12   Schematic representation of InGaAs/GaAs QD-IB solar cell structure fabricated on GaAs (311)B high-index substrate.
(Reprinted from Y. Shoji, K. Akimoto and Y. Okada, *J. Phys. D: Appl. Phys.*, 2013, **46**, 024002.)

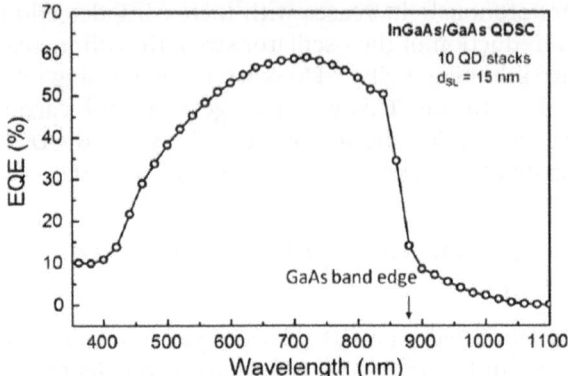

Figure 13.13   External quantum efficiency (EQE) spectrum taken at short-circuit condition for InGaAs/GaAs QD-IB solar cell with GaAs SLs of 15 nm thickness grown on GaAs (311)B.
(Reprinted from Y. Shoji, K. Akimoto and Y. Okada, *J. Phys. D: Appl. Phys.*, 2013, **46**, 024002.)

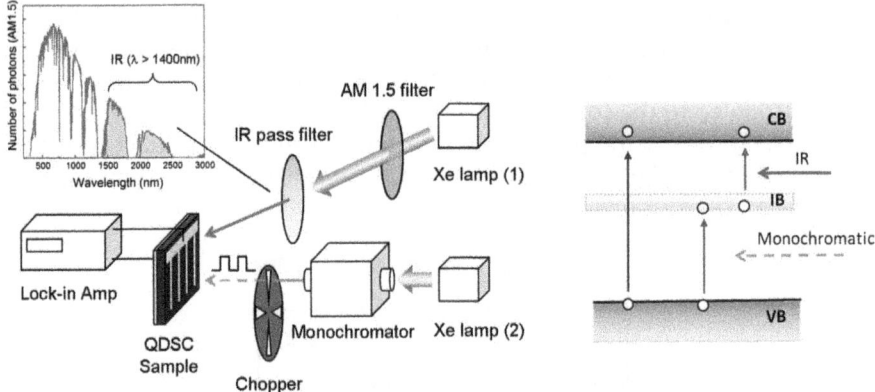

Figure 13.14   Measurement setup used to characterize photocurrent production as a direct result of optical transitions of electrons from IB to CB. Light from a Xe lamp with AM 1.5 filter placed at the exit passes through a set of filters that altogether allow only IR photons of λ > 1400 nm to be transmitted [Light source (1)]. The illumination of low energy photons from this IR source can then only pump the electrons from IB to CB.
(Reprinted from Y. Okada, T. Morioka, K. Yoshida, R. Oshima, Y. Shoji, T. Inoue and T. Kita, *J. Appl. Phys.*, 2011, **109**, 024301.)

the one shown in Figure 13.14 is frequently used.[22] Light from a Xe lamp with an AM 1.5 filter placed in front passes through an appropriate set of filters that altogether allow only the infrared (IR) region of l >1400 nm, as indicated as light source (2) in Figure 13.14. The illumination of low energy photons from this IR source can therefore only pump electrons from IB to CB, but do not have sufficient energies to pump electrons from VB to IB or directly from VB to CB. The QE curves are then measured, with light

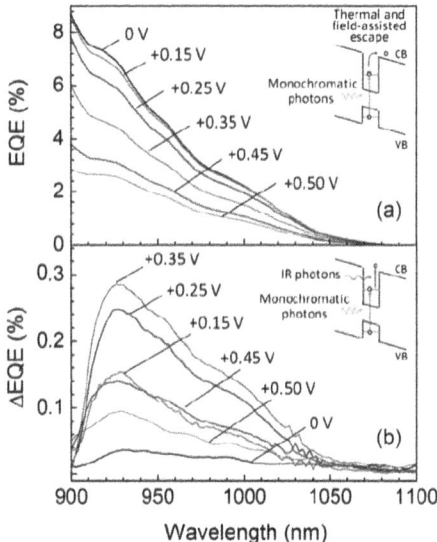

Figure 13.15   Bias dependence of (a) EQE, and (b) difference in EQE values, ΔEQE, between with (EQE$_{IR\_on}$) and without (EQE$_{IR\_off}$) IR light illumination, respectively.
(Reprinted from Y. Shoji, K. Akimoto and Y. Okada, *J. Phys. D: Appl. Phys.*, 2013, **46**, 024002.)

source (1) in the figure, first *without* and then *with* a continuous illumination of IR light.

Figure 13.15 shows the bias dependence of (a) EQE, and (b) difference in EQE value, DEQE, taken *with* (EQE$_{IR\_on}$) and *without* (EQE$_{IR\_off}$) IR light illumination, respectively. The forward bias is varied from $V = 0$ to $+0.50$ V. The scan range is from 900 to 1100 nm in wavelength, in order to estimate the contribution of absorption *via* IB. A production of photocurrent due to absorption of IR photons producing IB to CB optical transition is clearly seen at room temperature as shown in Figure 13.15b. However, the optical transition rate from QD-IB layers to CB at room temperature is still small at around 0.1% at ∼920 nm. This result indicates that the electrons pumped from VB to IB in the InGaAs QDs layers can escape to CB relatively easily by thermal excitation and/or built-in electric filed at the short-circuit condition as illustrated in the inset of Figure 13.15(a). The efficient thermal processes would increase thermal injection of carriers into QDs as well, which subsequently increases the recombination loss leading to a reduction in V_{OC}.

Further, it is most probable that the built-in electric filed across the intrinsic i-layer would de-couple the miniband structure of stacked QDs. Nonetheless, with increasing forward bias, despite the EQE is reduced, the photocurrent production due to two-step absorption actually increases in the range of 0 to $+0.35$V as shown in Figure 13.15b. DEQE increases to ∼0.28% at ∼920 nm. Figure 13.16 show the calculated energy band diagram for the

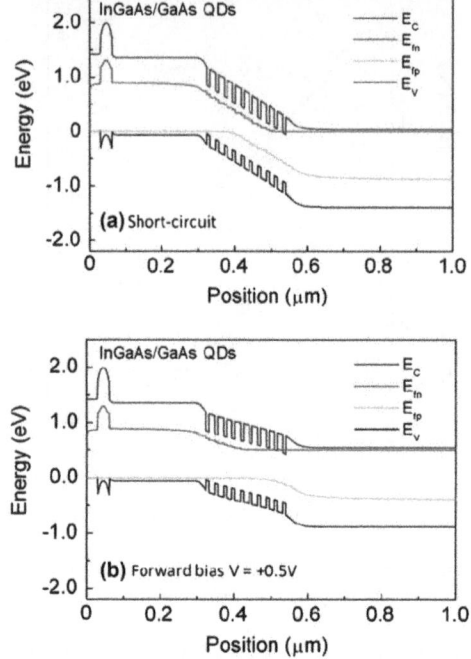

Figure 13.16 Calculate energy band structure for 10-layer stacked InGaAs/GaAs QD solar cell with $d_{SL} = 15$ nm under (a) short-circuit condition, and (b) forward bias of $V = +0.5$ V, respectively.
(Reprinted from Y. Shoji, K. Akimoto and Y. Okada, *J. Phys. D: Appl. Phys.*, 2013, **46**, 024002.)

10-layer stacked InGaAs/GaAs QDSC with $d_{SL} = 15$ nm as of Figure 13.12, under (a) short-circuit condition, and (b) forward bias of $V = +0.5$ V, respectively. A forward bias injects carriers into IB state and thus causes a partial filling of QDs, which in turn changes the energy band structure into a near flat-band condition. Consequently, the optical transition rate from IB to CB is increased, which in turn increases the photocurrent production by two-step optical transitions. This is a strongly required mode of operation for successful IB solar cells. As the forward bias is increased to $V > +0.35$ V, the photocurrent production due to two-step transition is decreased, because the recombination rate now becomes dominantly large in comparison.

Finally, Figure 13.17 plots the current–voltage curve measured for an 8-layer stacked InGaAs/GaAsSb QD solar cell with a much thicker SL of $d_{SL} = 60$ nm grown on GaAs (311)B substrate. The total QD density is 3.2×10^{11} cm^{-2}. The short-circuit current density (I_{SC}), open-circuit voltage (V_{OC}) and fill factor (FF) are $I_{SC} = 28.1$ mA cm^{-2}, $V_{OC} = 0.824$ V, and FF $= 0.747$, and the efficiency is $\eta = 17.3\%$. While an increase in I_{SC} is clearly visible, V_{OC} on the other hand is commonly reduced by ~ 0.2 V with respect to reference GaAs cell without QDs. This drop in voltage is due to the fact that the *optical transition rate from QD-IB layers to CB at room temperature is still small as*

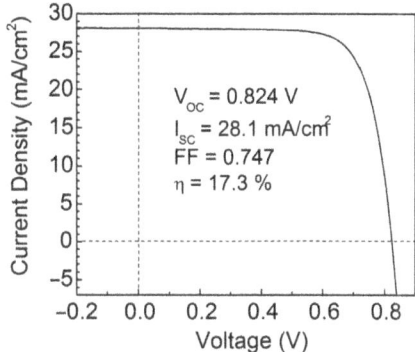

Figure 13.17 Current–voltage curve measured under AM 1.5 irradiation for an 8-layer stacked InGaAs/GaAsSb QD solar cell on GaAs (311)B with a much thicker GaAs spacer $d_{SL} = 60$ nm. ARC coating is deposited.

discussed above and that most electrons in QD-IB are efficiently escaping in and out of QDs by *thermal* and *field-assisted* processes.[37] Increasing the barrier height or band offset would reduce the thermal escape and capturing rates and to and from QD-IB and CB thereby reducing the loss of voltage. The recovery of V_{OC} has been proven with an InAs/GaAs QD solar cell by lowering the cell temperature which in a similar way reduces any thermal processes involved within QDs.[38]

13.3.3 Growth and Properties of InAs/GaAsSb QDs with Type-II Band Alignment

In order to further increase the conversion efficiency beside optimizing the band structure and increasing the QDs density as discussed above, (1) use of type-II energy band alignment leading to longer carrier lifetimes,[39,40] (2) doping of QDs in order to partially fill the IB with electrons, and (3) operation under concentrated sunlight are considered also essential. In this section, QD-IB solar cells with InAs/GaAsSb type-II heterostructure are reviewed.

The InAs QDs embedded in GaAs fabricated by S-K growth mode are the most extensively investigated QD system. This material combination has a type-I energy band alignment with a radiative recombination lifetime of τ_{rad} of a few nanoseconds, typically as shown in Figures 13.10 and 13.11 in the previous section. However, adding Sb into GaAs shifts the valence-band maximum (VBM) which then leads to a looser confinement of holes, and increasing Sb composition in GaAsSb alloy eventually changes the band alignment to type-II. As a result, τ_{rad} becomes significantly longer than that for type-I structure. Indeed, τ_{rad} as long as 65 ns has been demonstrated in InAs/GaAs$_{0.78}$Sb$_{0.22}$ QDs.[41]

Nishikawa *et al.* have recently analyzed in detail on the dependence of Sb composition on τ_{rad} in InAs/GaAsSb QD system.[39,40] The InAs/GaAs$_{1-x}$Sb$_x$ QD samples as schematically drawn in Figure 13.18 of various Sb compositions: 10% $(x = 0.1)$, 14% $(x = 0.14)$, and 18% $(x = 0.18)$ are prepared by

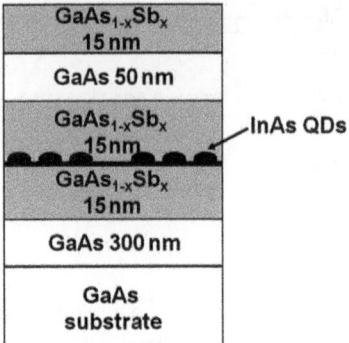

Figure 13.18   Schematic structure of the QD samples with GaAs$_{1-x}$Sb$_x$ layers inserted between InAs QDs and GaAs layers.
(Reprinted from K. Nishikawa, Y. Takeda, K. Yamanaka, T. Motohiro, D. Sato, J. Ota, N. Miyashita and Y. Okada, *J. Appl. Phys.*, 2012, **111**, 044325.)

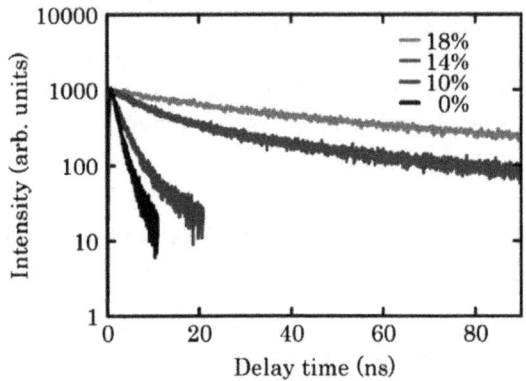

Figure 13.19   Dependence of PL decay curves on Sb compositions in InAs/GaAs$_{1-x}$Sb$_x$ QD samples measured at 77 K. The excitation intensity is 38 mW cm^{-2}. The detected wavelengths are 1140 nm, 1150 nm, and 1180 nm for Sb composition of 10%, 14%, and 18% samples, respectively.
(Reprinted from K. Nishikawa, Y. Takeda, K. Yamanaka, T. Motohiro, D. Sato, J. Ota, N. Miyashita and Y. Okada, *J. Appl. Phys.*, 2012, **111**, 044325.)

MBE. After growing a 15 nm-thick GaAs$_{1-x}$Sb$_x$ layer on GaAs buffer layer on GaAs (100) substrate, two monolayers of InAs are deposited to form QDs by S-K growth mode. The QDs are subsequently capped by another 15 nm-thick GaAs$_{1-x}$Sb$_x$ layer and 50 nm-thick GaAs layer. The QDs are hemispherical shaped with a diameter of around 20 nm and a height of 5 nm, and the areal density was $\sim 7 \times 10^{10}$ cm^{-2}.

Figure 13.19 shows the photoluminescence (PL) decay curves observed at each peak wavelength in the PL spectra for the three samples. It can be

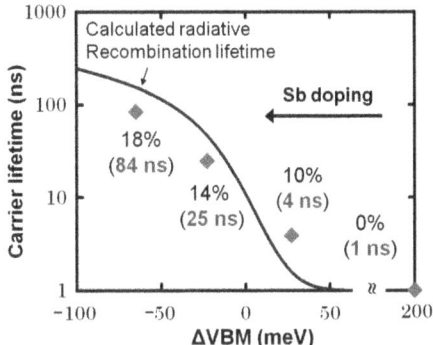

Figure 13.20 Comparison for calculated τ_{rad} (solid line) vs. experimentally determined PL decay time τ_{PL} $(t=0)$ (squares) at 77 K under weak excitation of 38 mW cm^{-2}. ΔVBM $=-65$, -23, and 27 meV correspond to the Sb compositions of 18%, 14%, and 10%, respectively.
(Reprinted from K. Nishikawa, Y. Takeda, K. Yamanaka, T. Motohiro, D. Sato, J. Ota, N. Miyashita and Y. Okada, *J. Appl. Phys.*, 2012, **111**, 044325.)

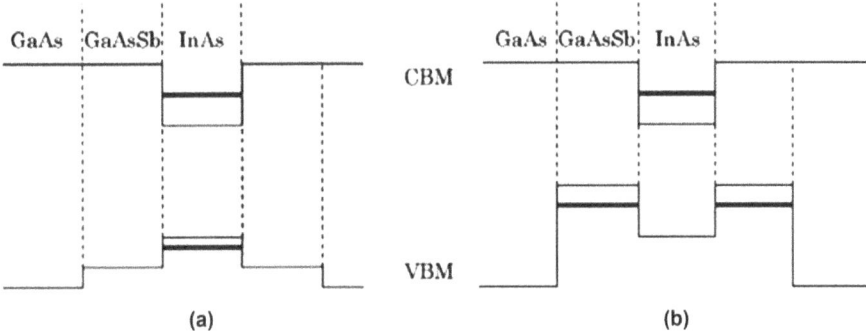

Figure 13.21 Energy band diagrams of InAs/GaAs$_{1-x}$Sb$_x$/GaAs samples, for type-I (left) and type-II (right) alignment. Thick lines schematically represent the positions of confined energy levels for electrons and holes.
(Reprinted from K. Nishikawa, Y. Takeda, K. Yamanaka, T. Motohiro, D. Sato, J. Ota, N. Miyashita and Y. Okada, *J. Appl. Phys.*, 2012, **111**, 044325.)

clearly seen that with increasing Sb composition, PL decay time becomes longer. Though not shown, with decreasing excitation intensity, the decay time also becomes longer. The 18% Sb QD sample shows PL decay time of $\tau_{\text{PL}} > 100$ ns under weak excitation as shown in Figure 13.20. The dependence of the PL decay profile on Sb composition shown in Figure 13.19 is well interpreted by the band offset relationship between the GaAs$_{1-x}$Sb$_x$ layers and InAs QDs. Figure 13.21 illustrates the proposed energy band diagrams of the samples. When InAs QDs are embedded in GaAs (no Sb), both electrons and holes are confined in the QDs (type-I alignment). Increasing the Sb composition shifts the VBM of the GaAs$_{1-x}$Sb$_x$ to higher energy, whereas hardly changes the location of the conduction band minimum (CBM). This results in a loose confinement and widely spread wavefunctions of holes, and hence the decay time becomes longer, because it is basically determined from the overlap between the electron and hole wavefunctions. With higher Sb composition, eventually, a type-II alignment is formed, in which the holes are located in the GaAs$_{1-x}$Sb$_x$ layers whereas the electrons are still confined in InAs QDs, leading to a very small overlap of the wavefunctions. Therefore, the decay time becomes significantly longer than those for type-I samples. Thus the observed very long decay times of the 18% and 14% Sb composition samples are due to the formation of type-II alignment. From the view point of QD-IB solar cells, these long carrier lifetimes in QD states are favorable in order to increase the optical generation rate and hence photocurrent production from IB to CB.

13.3.4   InAs/GaAsSb QD-IB Solar Cells with Type-II Band Alignment

Shoji *et al.* have fabricated three different types of QD-IB solar cells on GaAs (311)B substrate by MBE.[42] Figure 13.22a shows the schematic structure of p-i-n QD solar cell. The eight stacked pairs of In$_{0.4}$Ga$_{0.6}$As QDs/GaAs or GaAsSb spacer layer as shown in Figure 13.22b–c are inserted into the i-layer. The i-layer is grown at 480 °C, and the growth rates of QDs, GaAs, and GaAsSb layers were 1.0 mm h^{-1}, 0.6 mm h^{-1}, and 0.6 mm h^{-1}, respectively. ARC films have not been deposited.

Figure 13.23 shows the AFM image of In$_{0.4}$Ga$_{0.6}$As QDs grown on GaAs(311)B substrate, and the QD mean diameter, height, and sheet density, can be determined as 54.1 nm, 5.6 nm, and 3.2 × 10^{10} cm^{-2}, respectively.

The excitation-power-dependent PL has been investigated in order to evaluate the energy band structure of QDs. Figure 13.24 shows the excitation power dependence of PL peak energy. The PL peak of InGaAs QDs capped with GaAs layer shows a negligible change in the peak energy with increasing excitation density. For a type-I QD structure, it is well known that the PL intensity increases with excitation density without changes in the peak position, until when the emission peak from the excited states becomes predominant as the excitation power density is further increased.[43] For the InGaAs QDs capped with GaAsSb layer, the peak shows a blueshift with the cube root of excitation power. The value of the blueshift of PL spectrum for the QDs capped with GaAsSb is 49.1 meV. This result is consistent with luminescence of type-II band alignment.[44] Chiu *et al.* explained that the blueshift is due to the band bending effect induced by the spatially excited carriers in a type-II band structure.[45] The strong localized carriers around the interface of type-II band structure form a charged plane and hence, produce an approximated triangular potential well as illustrated in Figure 13.25.

The electric field strength of the interface is then written as:

$$\varepsilon = \frac{2\pi e n_w}{\varepsilon_0} \propto I^{1/2}$$

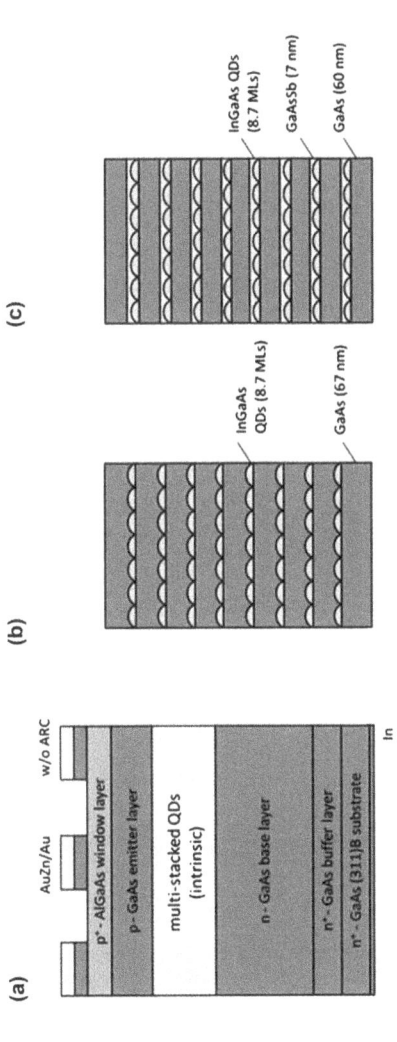

Figure 13.22 A(a) Schematic structure of p-i-n QD-IB solar cell grown on GaAs (311)B substrate. The eight stacked pairs of $In_{0.4}Ga_{0.6}As$ QDs/GaAs or GaAsSb spacer layer as shown in (b) to (c) are inserted into the i-layer.

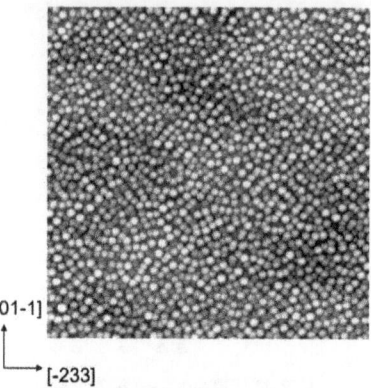

Figure 13.23 AFM image of InGaAs QDs grown on GaAs (311)B substrate. (a) Schematic structure of p-i-n QD-IB solar cell grown on GaAs (311)B substrate. The scan size is 2 μm × 2 μm. The QD mean diameter, height, and sheet density are 54.1 nm, 5.6 nm, and 3.2×10^{10} cm^{-2}, respectively.

(Reprinted from Y. Shoji, K. Akimoto and Y. Okada, 'InGaAs/GaAsSb Type-II Quantum Dots for Intermediate-Band Solar Cell', Proceedings of the 38th IEEE Photovoltaic Specialists Conference, Austin, June 2012.)

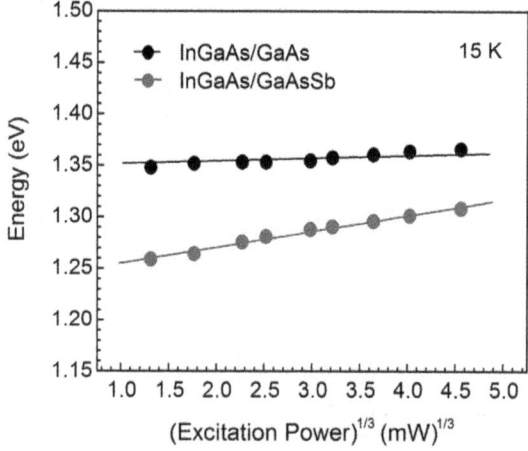

Figure 13.24 The peak energy of PL spectrum as a function of excitation power for InGaAs QDs embedded in GaAs and GaAsSb layers.

(Reprinted from Y. Shoji, K. Akimoto, and Y. Okada, 'InGaAs/GaAsSb Type-II Quantum Dots for Intermediate-Band Solar Cell', Proceedings of the 38th IEEE Photovoltaic Specialists Conference, Austin, June 2012.)

where n_w is the electron density and I is the excitation photon flux. The electron ground state E_{gs} in a triangular potential well is given by:[46]

$$E_{gs} = \text{const.} \times \varepsilon^{2/3} \propto I^{1/3}$$

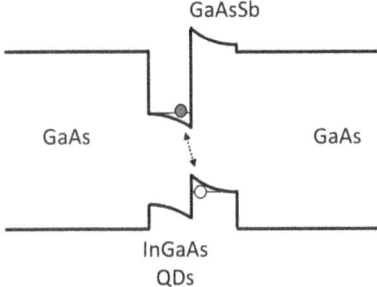

Figure 13.25 Schematic energy band structure for the InGaAs QDs capped with GaAsSb layer.
(Reprinted from Y. Shoji, K. Akimoto, and Y. Okada, 'InGaAs/GaAsSb Type-II Quantum Dots for Intermediate-Band Solar Cell', Proceedings of the 38th IEEE Photovoltaic Specialists Conference, Austin, June 2012.)

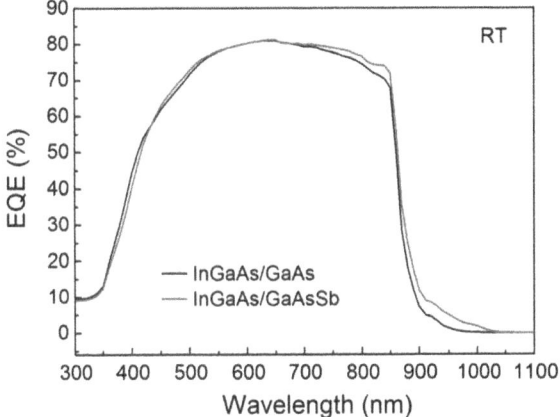

Figure 13.26 EQE spectrum of InGaAs/GaAs and InGaAs/GaAsSb QD solar cells grown on GaAs(311)B substrate. ARC films are deposited.
(Reprinted from Y. Shoji, K. Akimoto, and Y. Okada, 'InGaAs/GaAsSb Type-II Quantum Dots for Intermediate-Band Solar Cell', Proceedings of the 38th IEEE Photovoltaic Specialists Conference, Austin, June 2012.)

The quantum state is thus expected to increase proportionally with the cube root of the excitation power. Accordingly, the result obtained in Figure 13.24 indicates that type-II structure is indeed formed between InGaAs QDs and GaAsSb layers.

Figure 13.26 shows the external quantum efficiency (EQE) measured for the eight-layer stacked InGaAs/GaAs QD cell (type-I band alignment) and InGaAs/GaAsSb QD cell with type-II band alignment. It is clearly seen that by embedding InGaAs QDs, EQE extends into the longer wavelength range beyond the bandgap of GaAs (880nm) for both samples. Thus EQE response in the wavelength of >880 nm is attributed to the contribution from InGaAs

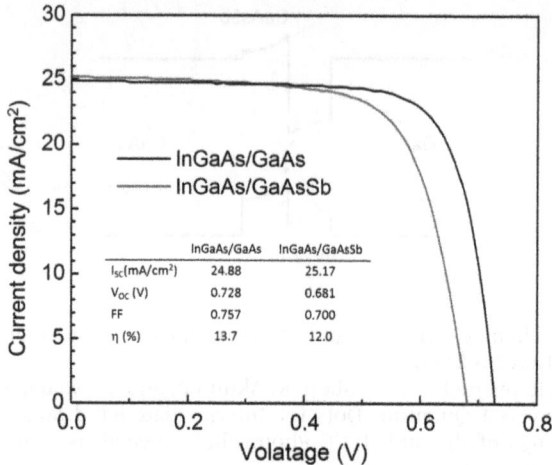

Figure 13.27   Current–voltage curves measured for eight-layer stacked InGaAs/
GaAsSb and InGaAs/GaAs QD solar cells grown on GaAs (311)B
substrate. ARC films are deposited.
(Reprinted from Y. Shoji, K. Akimoto, and Y. Okada, 'InGaAs/GaAsSb
Type-II Quantum Dots for Intermediate-Band Solar Cell', Proceed-
ings of the 38th IEEE Photovoltaic Specialists Conference, Austin,
June 2012.)

QDs. Additionally, the EQE response of InGaAs QDs is improved by embedding them with GaAsSb layers compared to InGaAs/GaAs type-I QD cell.
This observation could be due to effect of reduction of recombination loss or
longer carrier lifetime, which is a direct consequence of formation of type-II
band alignment between InGaAs QDs and GaAsSb layer as discussed in the
previous section.[39–41]

Finally, Figure 13.27 shows the AM1.5 current–voltage curves measured
for the eight-layer stacked InGaAs/GaAs QD cell and InGaAs/GaAsSb QD cell
with type-II band alignment. For InGaAs/GaAsSb QD cell, I_{SC} is increased
compared to InGaAs/GaAs QD cell while V_{OC} of InGaAs/GaAsSb QD cell is
degraded. This is attributed to an increased dark current induced by the
defects formed due to accumulation of lattice strain by GaAsSb layer and
thus material quality needs improvement.

13.3.5   Characteristics of QD-IB Solar Cells under Concentrated Sunlight

As we have pointed out repeatedly, QD-IB solar cells which commonly suffer
from relatively small absorption by QDs leading to lowering of V_{OC} and efficiency are expected to recover fast and perform better under concentrated
illumination operation. In a well-developed cell, a 50-layer stacked
InAs/GaNAs QD cell with ~15.7% efficiency at 1 sun has been shown to
improve to efficiency of ~20.3% at 100 suns and ~21.2% at 1000 suns

QD-SC   GaAs pin
(4 cells) (4 cells)

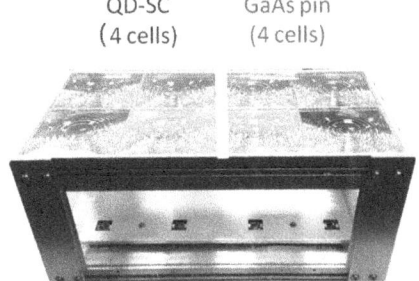

Figure 13.28 Photograph of InAs/GaAs QD-IB CPV module. Each four sub-modules are connected in series and mounted on Al heat sink plate. (Reprinted from T. Sogabe, Y. Shoji, M. Ohba, H-F. Hong, C-H. Wu, C-T. Kuo, and Y. Okada, 'First Demonstration of InAs/GaAs Quantum Dot CPV Module', Proceedings of the 9th International Conference on Concentrator PV Systems, Miyazaki, April 2013.)

illumination.[47] Thus the evaluation of carrier dynamics *via* IB for IB solar cells under a concentrating photovoltaic (CPV) module configuration is another important step toward future commercialization. Recently, Sogabe *et al.* have reported the first demonstration and testing of InAs/GaAs QD-IB CPV module.[48] Figure 13.28 shows a picture of the fabricated CPV module, which contains two separate parts with four series-connected GaAs control sub-modules on one side and four InAs/GaAs QD-IB CPV sub-modules on the other side. The InAs/GaAs QD-IB CPV and GaAs control cells are fabricated by adopting a p-i-n cell structure on n-type GaAs (001) substrate using MBE. The QD-IB CPV module characterization has been performed at room temperature using HELIOS 3198, a standard CPV measurement system with the light source from a flash lamp. The peak conversion efficiency of 19.2% has been obtained for GaAs reference module and 15.3% for a 10-layer stacked InAs/GaAs QD-IB CPV module at 116 suns under IEC62108 standard as shown in Figure 13.29.

13.4 Conclusion and Future Research

In this chapter, we have attempted to review theoretical analysis and QD technologies that could be implemented to realize high-efficiency intermediate-band (IB) solar cell.

From self-consistent drift-diffusion method adopted for IB solar cell, it is shown that doping of the IB region can increase the optical carrier generation *via* IB states, especially for low concentrated illuminations. However, optimal doping density must be carefully estimated not to degrade the material quality and the cell performance. Under higher concentrated illumination operation, photofilling effect can control the occupation rate of IB states and hence this leads favorably to higher I_{sc}, V_{OC} and cell efficiency η.

We have also discussed self-organized growth of QDs and control the *in-plane* density of QDs. Though not mentioned in detail here, compensating

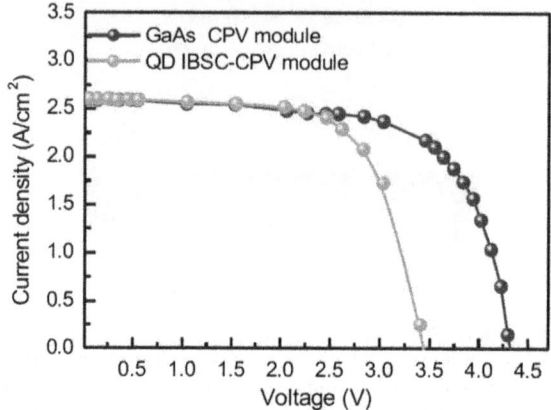

Figure 13.29   Current–voltage characteristics of QD-IB CPV module and GaAs reference module measured at 116 suns. Each module consists of four sub-modules.
(Reprinted from T. Sogabe, Y. Shoji, M. Ohba, H-F. Hong, C-H. Wu, C-T. Kuo, and Y. Okada, 'First Demonstration of InAs/GaAs Quantum Dot CPV Module', Proceedings of the 9th International Conference on Concentrator PV Systems, Miyazaki, April 2013.)

for or balancing out the lattice strain induced by self-assembled QDs with spacer layer that exerts an opposite biaxial strain works remarkably well in order to achieve improved size uniformity and to avoid generation of defects and dislocations in *multiple-stacked* QD superlattice structure.[10–17] Up to 100 layers of multi-stacking of InAs/GaNAs QDs are reproducibly achieved with the present technology.[49] The junction quality is good and the fundamentals of solar cell performance can be studied and analyzed clearly without the need to be concerned about the crystalline quality.

However, if the IB region is not partially filled with carriers, one can expect to observe a large mismatch between the rates of optical excitation of carriers from VB to IB in QD superlattice, and from QD-IB into CB, particularly in III-V based QD material system. Of the two absorption processes, the optical absorption rate from QD-IB to CB is presently the limiting factor which must be improved in order to maximize the photocurrent generation by two-step photon absorption. For this, fast recombination rate in QDs is the dominant factor in limiting V_{OC} in most of the QD solar cells reported to date.

The *modulation doping* technique has been applied to fabricate QD-IB solar cell,[19,23] in which the barrier layers are δ-doped with sheet densities equaling to or up to few times the QD areal density. *Photofilling* is another approach to effectively achieve partial filling of QD-IB states.[20] It is expected that a reasonable population of photogenerated electrons is sustained even in non-doped QD-IB solar cells if operated under concentrated sunlight, typically 100–1000 suns illumination. *Direct doping* of QDs is a new evolving technology to control the quasi-Fermi level of QD-IB states in solar cell.[22] In direct Si-doping of InAs QDs in GaNAs strain-compensating matrix, Si atoms are evenly incorporated into QDs during the assembling stage of growth and a uniform array of partially filled QDs has been obtained.[50,51] A clear photocurrent production due to two-step photon absorption has been observed at room temperature in such QD-IB solar cells with direct Si-doped InAs QDs.[20] This is evidence for the proof of concept of QD-IB solar cells and it opens up a pathway for fast development of high-efficiency IB solar cells with efficiencies exceeding the Shockley–Queisser limit. In order to evaluate and predict the expected and actual performance of working QD-IB solar cells more accurately, further studies both on the band theory[7,28,52] and spectroscopic characterization[53] of QD-IB structures are expected.

For high-efficiency IB solar cells, operation under sunlight concentration is not only an effective way to reduce the cell cost but also it is a favorable way to ensure the principle of operation in which the carrier photogeneration rate outperforms the recombination *via* IB. The QD-IB solar cells which commonly suffer from small absorption by QDs leading to lowering of V_{OC} and efficiency are expected to recover fast and perform better under concentrated illumination operation. In the first CPV module using InAs/GaAs QD-IB solar cells reported by Sogabe *et al.*,[48] 15.3% efficiency has been obtained for InAs/GaAs QD-IB CPV module at 116 suns illumination under IEC62108 standard, which appears to be a very promising step towards commercialization.

Acknowledgements

We would like to gratefully acknowledge Professors A. Luque and A. Martí of Universidad Politécnica de Madrid, Professor K. Barnham and Dr N. Ekins-Daukes of Imperial College London, Professor S. Tomić of University of Salford, Dr R. Oshima of National Institute of Advanced Industrial Science and Technology (AIST), Professor N. Sano and Dr K. Yoshida of University of Tsukuba, Professor K. Miyano, Dr R. Tamaki and Dr N. Miyashita of RCAST, The University of Tokyo, and Professor T. Kita of Kobe University for their collaboration and direct contribution to this work. We are also grateful to Professor K. Yamaguchi of University of Electro-Communications, Dr K. Akahane of National Institute of Information and Communications Technology (NICT), and Dr T. Motohiro, Y. Takeda, D. Sato, K. Nishikawa, and J. Ota of Toyota Motor Co. for their valuable comments and discussion. We also thank Dr C.-T. Kuo, C.-H. Wu and H-F. Hong of Institute of Nuclear Energy Research (INER) for technical support in the assembly and testing of CPV modules.

This work is supported by New Energy and Industrial Technology Development Organization (NEDO), and Ministry of Economy, Trade and Industry (METI), Japan. The Strategic International Cooperative Program by Japan Science and technology Agency (JST) is also gratefully acknowledged.

References

1. M. A. Green, *Third Generation Photovoltaics: Advanced Solar Energy Conversion*. Springer Series in Photonics, 2003.
2. P. Würfel, *Physics of Solar Cells: From Basic Principles to Advanced Concepts*. Wiley-VCH, 2009.
3. W. Shockley and H. J. Queisser, *J. Appl. Phys.*, 1961, **32**, 510.
4. A. J. Nozik, *Physica E*, 2002, **14**, 115.
5. A. J. Nozik, *Chem. Phys. Lett.*, 2008, **457**, 3.
6. A. Luque and A. Martí, *Phys. Rev. Lett.*, 1997, **78**, 5014.
7. S. Tomić, T. S. Jones and N. M. Harrison, *Appl. Phys. Lett.*, 2008, **93**, 263105.
8. A. Martí, L. Cuadra and A. Luque, *IEEE Trans. Electron Devices*, 2001, **48**, 2394.
9. G. S. Solomon, J. A. Trezza, A. F. Marshall and J. S. Harris Jr., *Phys. Rev. Lett.*, 1996, **76**, 952.
10. Y. Okada, N. Shiotsuka, H. Komiyama, K. Akahane and N. Ohtani, Multi-Stacking of Highly Uniform Self-Organized Quantum Dots for Solar Cell Applications, *Proceedings of the 20th European Photovoltaic Solar Energy Conference*, (WIP, 2005) 51.
11. K. Akahane, N. Yamamoto and M. Tsuchiya, *Appl. Phys. Lett.*, 2008, **93**, 041121.
12. V. Popescu, G. Bester, M. C. Hanna, A. G. Norman and A. Zunger, *Phys. Rev. B*, 2008, **78**, 205321.
13. S. M. Hubbard, C. D. Cress, C. G. Bailey, R. P. Raffaelle, S. G. Bailey and D. M. Wilt, *Appl. Phys. Lett.*, 2008, **92**, 123512.
14. R. Oshima, A. Takata and Y. Okada, *Appl. Phys. Lett.*, 2008, **93**, 083111.
15. Y. Okada, R. Oshima and A. Takata, *J. Appl. Phys.*, 2009, **106**, 024306.
16. R. Oshima, Y. Okada, A. Takata, Y. Yagi, K. Akahane, R. Tamaki and K. Miyano, *Physica Status Solidi C*, 2011, **8**, 619.
17. Y. Shoji, R. Oshima, A. Takata and Y. Okada, *J. Crystal Growth*, 2010, **312**, 226.
18. Y. Shoji, K. Narahara, H. Tanaka, T. Kita, K. Akimoto and Y. Okada, *J. Appl. Phys.*, 2012, **111**, 074305.
19. A. Martí, E. Antolín, C. R. Stanley, C. D. Farmer, N. López, P. Días, E. Cánovas, P. García-Linares and A. Luque, *Phys. Rev. Lett.*, 2006, **97**, 247701.
20. R. Strandberg and T. W. Reenaas, *Prog. Photovoltaics Res. Appl.*, 2011, **19**, 21.
21. K. Yoshida, Y. Okada and N. Sano, *Appl. Phys. Lett.*, 2010, **97**, 133503.
22. Y. Okada, T. Morioka, K. Yoshida, R. Oshima, Y. Shoji, T. Inoue and T. Kita, *J. Appl. Phys.*, 2011, **109**, 024301.
23. K. A. Sablon, J. W. Little, V. Mitin, A. Sergeev, N. Vagidov and K. Reinhardt, *Nano Lett.*, 2011, **11**, 2311.
24. A. S. Lin and J. D. Phillips, *IEEE Trans. Electron Devices*, 2009, **56**, 3168.

25. I. Tobías, A. Luque and A. Martí, *Semicond. Sci. Technol.*, 2011, **26**, 014031.
26. K. Yoshida, Y. Okada and N. Sano, *J. Appl. Phys.*, 2012, **112**, 084510.
27. S. Selberherr, *Analysis and Simulation of Semiconductor Device Simulation.* Springer-Verlag, 1984.
28. S. Tomić, *Phys. Rev. B*, 2010, **82**, 195321.
29. N. Kakuda, T. Yoshida and K. Yamaguchi, *Appl. Surf. Sci.*, 2008, **254**, 8050.
30. K. Akahane, H. Xu, Y. Okada and M. Kawabe, *Physica E*, 2001, **11**, 94.
31. Y. Shoji, K. Akimoto and Y. Okada, *J. Phys. D: Appl. Phys.*, 2013, **46**, 024002.
32. G. Springholz, V. Holy, M. Pnczolits and G. Bauer, *Science*, 1998, **282**, 734.
33. Z. Yu, P. E. Batson and J. Silcox, *Ultramicroscopy*, 2003, **96**, 275.
34. M. K. Zundel, P. Specht, K. Eberi, N. Y. Jin-Phillipp and F. Phillipp, *Appl. Phys. Lett.*, 1997, **71**, 2972.
35. O. Kojima, H. Nakatani, T. Kita, O. Wada, K. Akahane and M. Tsuchiya, *J. Appl. Phys.*, 2008, **103**, 113504.
36. T. Takagahara, *Phys. Rev. B*, 1987, **36**, 9293.
37. G. Jolley, H-F. Lu, L. Fu, H-H. Tan and C. Jagadish, *Appl. Phys. Lett.*, 2010, **97**, 123505.
38. A. Luque, A. Martí and C. Stanley, *Nat. Photonics*, 2012, **6**, 146.
39. K. Nishikawa, Y. Takeda, T. Motohiro, D. Sato, J. Ota, N. Miyashita and Y. Okada, *Appl. Phys. Lett.*, 2012, **100**, 113105.
40. K. Nishikawa, Y. Takeda, K. Yamanaka, T. Motohiro, D. Sato, J. Ota, N. Miyashita and Y. Okada, *J. Appl. Phys.*, 2012, **111**, 044325.
41. Y. D. Jang, T. J. Badcock, D. J. Mowbray, M. S. Skolnick, J. Park, D. Lee, H. Y. Liu, M. J. Steer and M. Hopkinson, *Appl. Phys. Lett.*, 2008, **92**, 251905.
42. Y. Shoji, K. Akimoto and Y. Okada, InGaAs/GaAsSb Type-II Quantum Dots for Intermediate-Band Solar Cell, *Proceedings of the 38th IEEE Photovoltaic Specialists Conference*, Austin (June 2012).
43. T. T. Chen, C. L. Cheng, Y. F. Chen, F. Y. Chang, H. H. Lin, C-T. Wu and C-H. Chen, *Phys. Rev. B*, 2007, **75**, 024301.
44. C. Y. Jin, H. Y. Liu, S. Y. Zhang, Q. Jiang, S. L. Liew, M. Hopkinson, T. J. Badcock, E. Nabavi and D. J. Mowbray, *Appl. Phys. Lett.*, 2007, **91**, 021102.
45. Y. S. Chiu, M. H. Ya, W. S. Su and Y. F. Chen, *J. Appl. Phys.*, 2002, **92**, 5810.
46. C. Weisbuch and B. Vinter, *Quantum Semiconductor Structures.* Academic Press, 1991.
47. Y. Okada, K. Yoshida, Y. Shoji, A. Ogura, P. Garcia-Linares, A. Marti and A. Luque, AIP Conference Proceedings 1477, pp. 10–13 2012; doi: 10.1063/1.4753822.
48. T. Sogabe, Y. Shoji, M. Ohba, H-F. Hong, C-H. Wu, C-T. Kuo and Y. Okada, First Demonstration of InAs/GaAs Quantum Dot CPV Module,

Presented at the 9th International Conference on Concentrator PV Systems, Miyazaki, 2013 p. 79.

49. A. Takata, R. Oshima, Y. Shoji, K. Akahane and Y. Okada, Fabrication of 100 layer-stacked InAs/GaNAs strain-compensated quantum dots on GaAs (001) for application to intermediate band solar cell, *Proceedings of the 35th IEEE Photovoltaic Specialists Conference*, Hawaii (2010).

50. T. Inoue, S. Kido, K. Sasayama, T. Kita and O. Wada, *J. Appl. Phys.*, 2010, **108**, 063524.

51. J. Phillips, K. Kamath, X. Zhou, N. Chervela and P. Bhattacharya, *Appl. Phys. Lett.*, 1997, **71**, 2079.

52. T. Sogabe, T. Kaizu, Y. Okada and S. Tomić, *J. Renew. Sustain. Energy*, 2014, **6**, 011206.

53. A. Takahashi, T. Ueda, Y. Bessho, Y. Harada, T. Kita, E. Taguchi and H. Yasuda, *Phys. Rev. B*, 2013, **87**, 235323.

CHAPTER 14

Spectral Conversion for Thin Film Solar Cells and Luminescent Solar Concentrators

WILFRIED VAN SARK,[*][a] JESSICA DE WILD,[b,†]
ZACHAR KRUMER,[c] CELSO DE MELLO DONEGÁ[c]
AND RUUD SCHROPP[d,e]

[a] Energy and Resources, Copernicus Institute of Sustainable Development, Utrecht University, Heidelberglaan 2, 3584 CS Utrecht, the Netherlands; [b] Physics of Devices, Debye Institute for Nanomaterials Science, P.O. Box 80000, 3508 TA Utrecht, the Netherlands; [c] Condensed Matter and Interfaces, Debye Institute for Nanomaterials Science, P.O. Box 80000, 3508 TA Utrecht, the Netherlands; [d] Solar Energy, Energy research Centre of the Netherlands (ECN), High Tech Campus Building 5, p-057 (WAY), 5656 AE Eindhoven, The Netherlands; [e] Plasma & Materials Processing, Department of Applied Physics, Eindhoven University of Technology (TU/e), P.O. Box 513, 5600 MB Eindhoven, the Netherlands
*Email: w.g.j.h.m.vansark@uu.nl

[†] J. de Wild is now at Luxembourg University, Campus Belval (Gabriel Lippmann Institute), 41 Rue du Brill, L-4422 Belvaux, Luxembourg

RSC Energy and Environment Series No. 11
Advanced Concepts in Photovoltaics
Edited by Arthur J Nozik, Gavin Conibeer and Matthew C Beard
© The Royal Society of Chemistry 2014
Published by the Royal Society of Chemistry, www.rsc.org

14.1   Introduction

Photovoltaic (PV) solar energy systems have seen a steep decrease in cost recently, predominantly due to much lower PV module prices. Still, drivers in research and development remain to further lower the cost to bulk electricity levels while at the same time increasing the conversion efficiency: next or third generation PV, or, perhaps, 'Solar 3.0'. Commercial modules can be purchased with module efficiency higher than 20% nowadays, and laboratory solar cells are getting closer and closer[1] to the Shockley–Queisser limit of 31% for single junction cells.[2] A fundamental limitation is that conventional single-junction semiconductor solar cells only optimally convert photons of energy close to the bandgap (E_g). Other photons are not converted efficiently or not at all. This is generally known as the mismatch between the incident solar spectrum and the spectral absorption properties of the material.[3] Photons with energy (E_{ph}) smaller than the bandgap are not absorbed and their energy is not used for carrier generation. Photons with energy E_{ph} larger than the bandgap can be absorbed, but the excess energy, $E_{ph} - E_g$, is lost due to thermalization of the excited electrons. These fundamental spectral losses amount to about 50%.[4] Several approaches have been suggested to overcome these losses and some of these are treated in this book, *e.g.*, multiple stacked cells,[5] intermediate bandgaps,[6] multiple exciton generation,[7] quantum dot concentrators[8,9] and spectral converters, the latter being down- and up-converters,[10,11] and down-shifters.[12,13]

14.1.1   Spectral Conversion

Spectral conversion aims at modifying the incident solar spectrum such that a better match is obtained with the wavelength dependent conversion efficiency of the solar cell. It can be applied to existing solar cells and optimization of the solar cell and spectral converter can be done separately. Different types of spectral conversion can be distinguished (Figure 14.1):

- *Down-conversion or quantum cutting*, in which one high energy photon is transformed into two lower energy photons
- *Down-shifting or luminescence*, in which one high energy photon is transformed into one lower energy photon
- *Up-conversion*, in which two low energy (sub-bandgap) photons are combined to give one high energy photon

Down-shifting can give an efficiency increase by shifting photons to a spectral region where the solar cell has a higher external quantum efficiency, *i.e.*, basically improving the blue response of the solar cell, and improvements of up to 10% relative efficiency increase have been predicted,[13] as well as efficiencies beyond the SQ limit.[10,11] For example, Richards[12] has shown for c-Si that the potential relative gain in efficiency could be 32% and 35% for down-conversion and up-conversion, respectively, both calculated for the

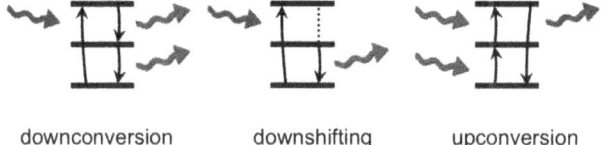

downconversion   downshifting   upconversion

Figure 14.1   Schematic energy diagrams showing photon absorption and subsequent downconversion, downshifting, and up-conversion.
Reprinted, with permission, from W. G. J. H. M. van Sark, A. Meijerink, R. E. I. Schropp, Nanoparticles for solar spectrum conversion, Chapter 10 in *Nanotechnology for Photovoltaics* (L. Tsakalakos, ed.), Taylor and Francis, Boca Raton, FL, USA, 2010, pp. 351–390.

standard 1000 W m^{-2} air mass (AM) 1.5 solar spectrum. Note, this efficiency is highly dependent on the efficiency of the up-converter and its absorption bands.[14]

Research on spectral conversion is focused on organic dyes, quantum dots, lanthanide ions, and transition metal ion systems for up- and down-conversion.[15,16] An up-conversion layer is to be placed at the back of the solar cells where the transmitted photons are converted to wavelengths that can be absorbed and are reflected back into the solar cell. Down-converters and down-shifters have to be placed at the front of the solar cell and any absorption loss (*i.e.*, absorption not leading to down-shifting) will reduce the overall efficiency of the system. An alternative use of down-shifters is found in the luminescent solar concentrator.[17,18] These consist of a highly transparent plastic plate, in which luminescent species are dispersed, which absorb incident light and emit light at a red-shifted wavelength, with high quantum efficiency. Internal reflection ensures collection of part of the emitted light in solar cell(s) that are located at the edge(s) of the plastic plate.

Down-conversion with close to 200% internal quantum efficiency has been demonstrated, but the actual quantum efficiency is lower due to concentration quenching and parasitic absorption processes.[19,20] Even for a perfect 200% quantum yield system a higher solar cell response requires a reflective coating to reflect the isotropically emitted photons from the down-conversion layer back towards the solar cell. However, no proof-of-principle experiments have been reported to demonstrate an overall efficiency gain using down-conversion materials. An up-converter also emits isotropically but since it is placed at the back of the solar cells, the up-converted photons can easily be directed into the solar cell by placing a reflector behind the up-converter layer.

The usefulness of down- and up-conversion and down-shifting depends on the incident spectrum and intensity. While solar cells are commonly designed and tested according to the ASTM standard,[21] these conditions are rarely met outdoors. Spectral conditions for solar cells vary from AM0 (extraterrestrial) via AM1 (equator, summer and winter solstice) to AM10 (sunrise, sunset). The weighted average photon energy (APE)[22] can be used to parameterize this; the APE (using the range 300–1400 nm) of AM1.5G is 1.674 eV, while the APE of AM0 and AM10 is 1.697 eV and 1.307 eV, respectively. Further, overcast skies cause higher scattering leading to diffuse spectra, which are blue-rich: the APE of the AM1.5 diffuse spectrum is 2.005 eV, *i.e.*, much larger than the APE of the AM1.5 direct spectrum of 1.610 eV. As down-conversion and down-shifting effectively red-shift the incident spectrum, the more relative energy an incident spectrum contains in the blue part of the spectrum (high APE) the more gain can be expected.[13,23] Application of down-conversion/shifting layers will therefore be more beneficial for regions with high diffuse irradiation fraction, such as northwestern Europe, where this fraction can be 50% or higher on an annual basis. In contrast, the relative gain of solar cells due to with up-converter layers will be higher in countries with high direct irradiation fractions or in early morning and evening due to the high air mass resulting in low APE, albeit that the non-linear response to intensity may be limiting. Up- and down-conversion layers could be combined on the same solar cell to overcome regionally dependent efficiencies. Optimization of either up- or down-conversion layers could be very effective if the solar cell bandgap is a free design parameter. For various semiconductor types (III–V, II–VI, IV–IV) the bandgap can be tuned by changing the composition. Further bandgap freedom can potentially be achieved using quantum confinement.

14.1.2 This Chapter

In this chapter, two examples of spectral conversion for solar cells will be treated: up-conversion for thin film silicon solar cells with an up-converter based on lanthanides and down-shifting in luminescent solar concentrators.

Full spectrum absorption combined with effective generation and collection of charge carriers is a prerequisite for attaining high efficiency solar cells. In this chapter, we will describe up-conversion for thin film silicon solar cells. The high bandgap of amorphous silicon of ~ 1.8 eV implies that the material is transparent for sub-bandgap, near infrared (NIR) light, constituting a high photon loss. Up-conversion at the back of the cell may enhance the response of solar cells in the infrared,[11] as is demonstrated in several papers.[24–28] Recent studies show increased up-conversion by broadening the NIR absorption[29] or choosing a suitable matrix for the up-converter.[30] We will show improvements in efficiency of amorphous silicon cells using monochromatic laser light,[26] as well as by using broadband excitation.[27]

The development of luminescent solar concentrators (LSCs) began in the 1970s as an alternative approach to lower the costs of PV.[17,18] Unlike standard solar concentrators both direct and diffuse light is concentrated by a factor of typically 5–10, without the need for expensive tracking systems. Also smaller silicon (or other more expensive) solar cells can be used. As the cost of the transparent plastic is expected to be much lower than the area cost of the solar cell, the cost per watt-peak (W_p) of an LSC is lower than that of a planar silicon solar cell.[31] With present day silicon cell cost of below

1 \$/W$_p$, it is harder than envisaged some years ago for LSCs to be competitive with standard silicon modules. However, note that the market for LSCs is different from that of standard PV as building integrated PV applications (facades, sound barriers) require extreme flexibility of design.[31,32] In this chapter, we will address various issues that should be addressed to increase the efficiency beyond the present efficiency record of 7.1%,[33] in particular the approach to reduce self-absorption of emitted photons by the luminescent species.

14.2 Up-conversion for Thin Film Silicon

14.2.1 Introduction

Up-conversion in lanthanide ions is extensively investigated since the 1960s.[34] Lanthanides are most commonly found in the trivalent ionized state and have a $4f^n5s^25p^6$ electron configuration, where n varies from 0 to 14. Interactions between the 4f electrons in the partly filled inner $4f^n$ shell give rise to a rich energy level structure with absorption and emission ranging from UV to NIR. Because the 4f electrons are shielded by the outer $5p^6$ and $5s^2$ shells, the energy level structure and optical properties are barely influenced by the surrounding host lattice.

The up-converter material used is crucial for obtaining any significant increase in solar energy conversion efficiency. In this chapter, results on the up-converter materials β-NaYF$_4$:Yb^{3+}, Er^{3+} and Gd$_2$O$_2$S:Yb^{3+}, Er^{3+} are described; in both materials Yb^{3+} absorbs light around 980 nm and Er^{3+} emits in the visible spectrum (400–700 nm), see Figure 14.2. These absorption and emission wavelengths are very suitable for use with a-Si:H single junction solar cells, as the absorption edge of a-Si:H lies between the wavelengths for absorption and emission and the spectral response is very high in that emission range.

Different mechanisms are responsible for the up-converter luminescence. The dominant up-conversion mechanism in Gd$_2$O$_2$S and NaYF$_4$: Yb^{3+}, Er^{3+} is energy transfer up-conversion (ETU). The Yb^{3+} ion has only one excited state and is an ideal sensitizer for Er^{3+} because of the relatively high oscillator strength of the $^2F_{7/2} \rightarrow {}^2F_{5/2}$ transition and the fact that Er^{3+} has a state with similar energy ($^4I_{11/2}$) which is populated by energy transfer from Yb^{3+}, see Figure 14.2. Population of the first excited state of Er^{3+} ($^4I_{11/2}$) is therefore directly proportional to the incoming light intensity. When up-conversion is the main route, energy transfer from the first excited state ($^4I_{11/2}$) to the second excited ($^4F_{7/2}$) state follows. After some small energy-relaxation steps, emission is observed from $^4S_{3/2}$, $^2H_{11/2}$ (green) and $^4F_{9/2}$ (red) states. The $^4F_{9/2}$ state can also be reached after energy transfer from the $^4I_{13/2}$ state, see Figure 14.2. Because two or more photons are required for up-converted emission, a higher order dependence of the incoming light intensity is expected:

$$N_n \propto N_{n-1}N_s \propto (N_s)^n \propto P_{in}^{\ n} \tag{14.1}$$

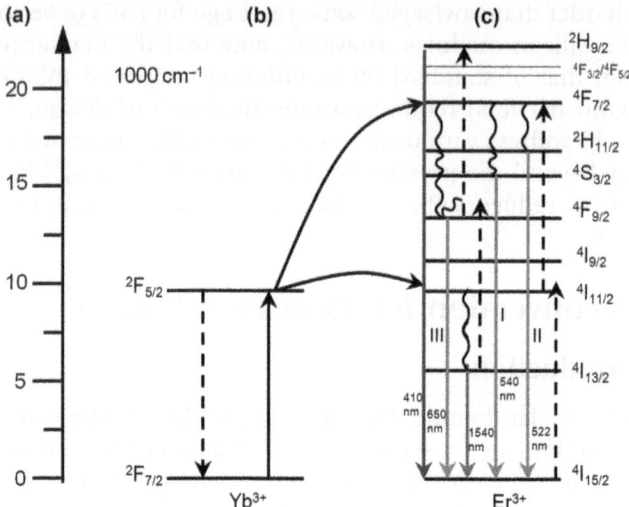

Figure 14.2   Up-conversion in the (Yb^{3+}, Er^{3+}) couple. The dashed lines represent energy transfer, the full lines in part (c) represent the radiative decay, and the curly lines in part (c) indicate multiphonon relaxation processes. The main route is a two-step energy transfer that leads to excitation to the $^4F_{7/2}$ state of the Er^{3+} ion. After relaxation from this state emission is observed from the $^2H_{11/2}$, the $^4S_{3/2}$ level (I and II), and the $^4F_{9/2}$ level (III).
Reprinted, with permission, from J. de Wild, T. F. Duindam, J. K. Rath, A. Meijerink, W. G. J. H. M. van Sark, R. E. I. Schropp, *IEEE J Photovoltaics*, 2013, 3, 17–21. © 2013 IEEE.

where N_n is the n-th excited state in the Er^{3+} ion, N_s the excited state of the sensitizer ion Yb^{3+}, and n the number of photons needed to excite the up-converted state. When a higher energy level saturates, other processes like non-radiative relaxation to lower energy states occur and as a consequence, deviations from the expected power law dependence are observed.[35,36]

The up-converted emission intensity is thus proportional to the population of the higher excited state N_n. When an up-converter is applied to the back of a solar cell, the increased photogenerated current is due to this emission, therefore:

$$I_{SC}^{UC} \propto P_{in}{}^n \qquad (14.2)$$

where I_{SC}^{UC} is the photogenerated short circuit current in the solar cell due to up-conversion, and P_{in} the incoming light intensity. Hence, for the current increase due to up-conversion we expect a quadratic power dependence on the concentration factor, because the up-converted emission occurs after absorption of two photons.

14.2.2 Up-conversion Results

14.2.2.1 Monochromatic Light

A near-infrared to visible up-conversion phosphor [β-NaYF$_4$:Yb^{3+}(18%), Er^{3+}(2%)] has been applied at the back of a thin film hydrogenated amorphous silicon (a-Si:H) solar cell in combination with a white back reflector to investigate its response to sub-bandgap infrared irradiation. The up-converter NaYF$_4$:Yb,Er is a very efficient up-converting material. The maximum up-conversion efficiency, defined as power out, P_{out}/P_{in} for 520–580 nm light has been reported to be 5.5%, which was reached at an intensity of 20 W cm^{-2}.[37,38] Based on power dependent emission spectra it was also concluded that for high excitation powers 50% of all NIR photons absorbed by the material are converted to visible photons, *i.e.*, blue (405 nm), green (540 nm) and red (650 nm) emission.[39]

β-NaYF$_4$:Er^{3+} (2%),Yb^{3+} (18%) phosphors were synthesized in house.[26] The up-converter powder mixture was applied to the rear of the solar cells by first dissolving it in a solution of polymethylmethacrylate (PMMA) in chloroform after which it was drop cast, resulting in an up-converter layer thickness of 200–300 µm. As a back reflector, white foil[40] was used.

Standard p-i-n amorphous silicon solar cells were made by 13.56 MHz plasma enhanced chemical vapour deposition (PECVD) with an area of 0.16 cm^2 and an intrinsic layer thickness of 500 nm. As a back contact 1 µm sputtered aluminum doped zinc oxide from a ZnO:Al$_2$O$_3$ 0.5% target was used. Front side illumination with AM1.5 light yields an efficiency of 8%; for backside illumination (through the n-layer) an efficiency of 5% was determined. The external quantum efficiency peaked at 0.77 for a wavelength of 530 nm.

The solar cells were illuminated with a NIR diode laser (981/986 nm), with a maximum power of 28 mW. The laser beam was not focused; the beam area was 1 mm^2, *i.e.*, power density of 3 W cm^{-2}. Current–voltage characteristics are shown in Figure 14.3, in which a cell with up-converter is compared to a cell without up-converter. There is a clear three-fold improvement due to up-conversion: the cell with up-converter shows short circuit current I_{SC} of 6.2 µA (0.039 mA cm^{-2}), compared to I_{SC} of 2.1 µA (0.013 mA cm^{-2}) for the cell without up-converter. The response to NIR laser light without up-converter, with energy lower that the bandgap of a-Si:H, is due to sub-bandgap absorption that arises from a continuous density of localized states. Therefore, the cell absorbs part of the NIR radiation before it reaches the backside of the cell where it can be up-converted. With EQE calculated as:

$$EQE = \frac{I_{SC}}{P_{in}q/h\nu} \qquad (14.3)$$

where I_{SC} is the short circuit current, P_{in} the input power, q the electron charge, $h\nu$ the energy of the photon, a maximum value of 0.03% is determined at NIR wavelength.

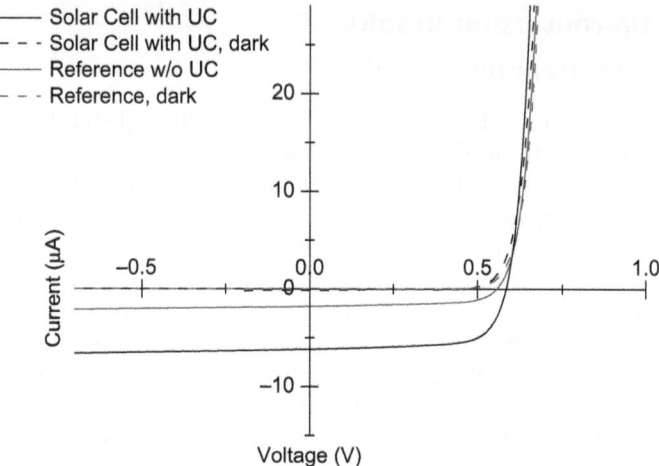

Figure 14.3 *I–V* curves of a-Si:H solar cells with and without up-converter at the back side. The reference cell has the white back reflector only. In the dark the solar cells have the same *I–V* curve; however, when illuminated with the NIR diode laser ($\sim$ 980 nm), current is generated in both solar cells. The photocurrent in the cell with the up-conversion phosphor is a factor of three larger.
Reprinted from J. de Wild, J. K. Rath, A. Meijerink, W. G. J. H. M. van Sark, R. E. I. Schropp, *Solar Energy Materials and Solar Cells*, 2010, **94**, 2395–2398, Copyright (2010), with permission from Elsevier.

The intensity of solar illumination in the NIR in practice is several orders of magnitude lower than that of a laser. Therefore, a study was performed on the relation between emission and excitation density. Figure 14.4 shows this dependence for the red (653 nm) and green (522 and 540 nm) emissions as a function of excitation intensity. The slopes are determined to be 1.9 ± 0.1 (red emissions, solid line) and 1.65 ± 0.06 (green emissions, dashed line), respectively, which is consistent with a two-photon absorption process. Note that the deviation from the expected slope of 2 is commonly observed and marks the transition to a slope of 1 in the high power regime.[36] The red emission is at least 10 times less intense than the green one. For the higher excitation densities the relative intensity of the red emission increases since feeding through a three-photon process starts to contribute at higher powers. This also explains the difference in slope for the two emissions.

Also the external quantum efficiency of solar cells due to non-linear processes such as up-conversion is strongly dependent on the illumination intensity. As EQE is proportional to I_{SC}/P_{in} [Equation (14.3)], and I_{SC} itself is proportional to $P_{out} \propto P_{in}{}^n$, the power dependence of the external quantum efficiency is:

$$\text{EQE} \propto \frac{P_{out}}{P_{in}} = P_{in}{}^{n-1} \tag{14.4}$$

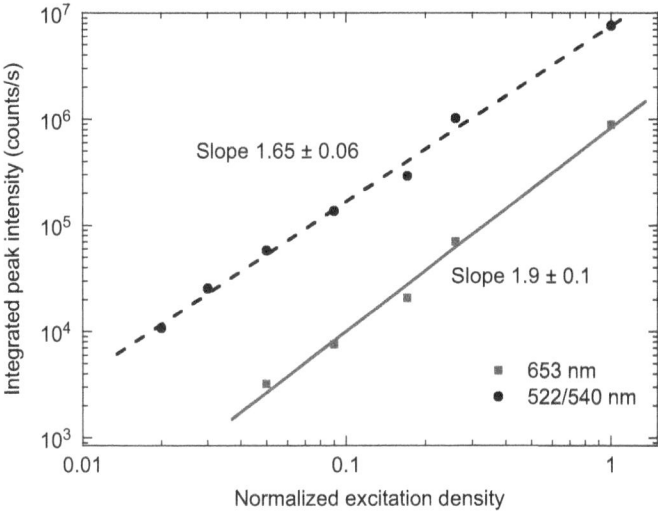

Figure 14.4   Power dependence of the UC emission intensity in β-NaYF$_4$. The slope in a double logarithmic plot gives the dependence of the emitted light on the excitation power density. The slope of the 525/550 nm emission is 1.65 ± 0.06 (dashed line) and of the 650 nm emission is 1.9 ± 0.1 (solid line) revealing a quadratic dependence of the emitted power on the excitation density.
Reprinted from J. de Wild, J. K. Rath, A. Meijerink, W. G. J. H. M. van Sark, R. E. I. Schropp, *Solar Energy Materials and Solar Cells*, 2010, **94**, 2395–2398, Copyright (2010), with permission from Elsevier.

From Figure 14.5, in which EQE is plotted as a function of input power a slope of 0.72 ± 0.02 is determined, and hence $n = 1.72$. This is in very good agreement with the slope of 1.65 found earlier for the green emission, which is the most significant part of the up-converted light. For the cell without up-converter a very weak dependence on the illumination intensity was found, inferring that the sub-bandgap response is almost a linear process and the relative contribution of up-conversion becomes more significant at higher excitation powers.

The efficiency of the up-converter can be calculated from the measured EQE while comparing with expected efficiencies for lower excitation densities. Of importance here is that the incoming light has a wavelength of 980 nm, whereas the up-converted light has a wavelength of predominantly 540 nm, and up-converted light enters the solar cell through the back of the cell. As the back reflector has high reflectivity,[40] all light is assumed to reflect back into the solar cell. Furthermore, only green emission is considered to simplify the calculation, but this is reasonable as at lower light intensities the green emission is at least ten times stronger than emission at other wavelengths. Finally, as EQE due to illumination from the back by the up-converted photons does not depend on the up-conversion efficiency, EQE

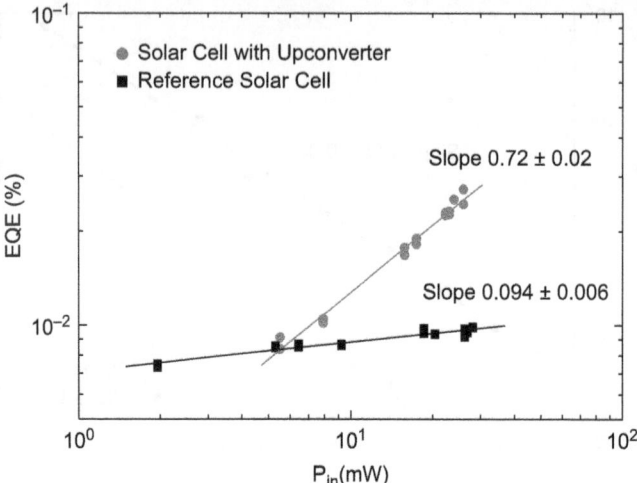

Figure 14.5   Double logarithmic plot of EQE as a function of excitation power. The slope in the case of the solar cell with up converter (0.72, line with circles) matches the expected power dependence of P_{in}^{n-1}, with n equalling 1.65 as found above (Figure 14.4). The EQE of the reference cell is approximately constant which is expected for a one-photon process. Reprinted from J. de Wild, J. K. Rath, A. Meijerink, W. G. J. H. M. van Sark, R. E. I. Schropp, *Solar Energy Materials and Solar Cells*, 2010, **94**, 2395–2398, Copyright (2010), with permission from Elsevier.

can simply be measured by spectral response measurements. The EQE from backside illumination at 540 nm was measured to be 0.62.

The efficiency of the up-converter is defined as P_{out}/P_{in}. Recalling Equation (14.1), and rewriting I_{sc} as $(P_{out}q/h\nu_{bs})EQE_{bs}$, where EQE_{bs} is the measured quantum efficiency at the specific wavelength (here 540 nm) for backside illumination, the efficiency of the up-converter P_{out}/P_{in} can then be written as:

$$\eta_{UC} = \frac{EQE_{UC}q/h\nu_{bs}}{EQE_{bs}q/h\nu_{UC}} \quad (14.5)$$

where $h\nu_{bs}$, $h\nu_{UC}$ is the energy of the emitted and incoming photons from the up-converter, respectively. An up-conversion efficiency of 0.05% is found, whereby EQE_{UC} is corrected for the sub-bandgap response without up-converter.

To compare with the expected efficiency of the up-converter it is important to know the power *density*. The incident power density on the cells was measured to be 3 W cm^{-2}. However, not all light reaches the up-converter layer due to absorption in the localized states of the semiconductor layers and in the front and back TCO layers. Therefore, only 40% of the incoming light is transmitted and reaches the up-converter layer, leading to a power density of 1.2 W cm^{-2}. A decrease of the up-converter efficiency of

$\eta = \eta_{max}(P_{in}/P_{max})^{1.65}$, with η_{max} and P_{max} taken from,[37,38] leads to an up-converter efficiency of $\sim 0.05\%$, consistent with the measured EQE.

This very close match of experimentally determined efficiency and expected efficiency of the up-conversion process means that the approximation that all light is reflected back into the solar cell is a good approximation. The fact that this match is so close may be due to the fact that the EQE_{bs} is higher than measured. This can be explained by the diffuse back reflector that diffusively scatters the light back into the solar cell, which leads to a longer path length of the incoming light and consequently a higher chance that the light will be absorbed.

14.2.2.2 Broadband Light

In this section the up-converter material used was Gd_2O_2S: Er^{3+} (5%),Yb^{3+} (10%). As above, the material was applied to the rear of the solar cells by first dissolving it in a solution of PMMA in chloroform after which it was drop cast. A layer of 200–300 μm was obtained. As a back reflector white foil[40] was used. Standard p-i-n amorphous silicon solar cells were made by PECVD with an area of 0.16 cm^2 and an intrinsic layer thickness of 350 nm. As back contact a 1.5 μm ZnO:Al layer from a $ZnO:Al_2O_3$ 0.5% target was applied.

To show the enhancement of solar cell performance due to up-conversion, current–voltage measurements were performed. A set-up was made, which concentrated the light coming from a Wacom solar simulator. This type of simulator operates using a halogen and xenon lamp. The typical xenon lamp peaks are suppressed with a dichroic mirror and most near infrared light is coming from the halogen lamp. With a convex glass lens, the broadband light could be concentrated up to 25 times. A long pass 900 nm filter was added so that only sub-bandgap light reached the solar cell. The spectral match of the produced spectrum to the AM1.5 standard spectrum is +14% for the 900–1100 nm region.[41]

Two types of p-i-n a-Si:H solar cells were made, one on Asahi glass and one on flat ZnO:Al 0.5% superstrate. The efficiency obtained for the cells is 8% for textured and 5% for flat solar cells, both without any back reflector. Backside illumination yields an efficiency of 5% for textured solar cells and 4% for flat solar cells. These low values for backside illumination are mainly caused by current losses in the blue part, due to bad carrier collection. Blue light is absorbed close to the top of the cell in the n-layer and holes are hardly collected at the p side. The spectral response measured through the n-layer, shows a quantum efficiency of 0.7 for both textured and flat solar cells at 550 nm; the spectral response at 660 nm is lower, *i.e.*, 0.4 for textured cells and 0.15 for flat cells. The transmission for 900–1040 nm was 40–45% for the textured solar cells and between 60 and 80% for the flat solar cell. The thickness of the i-layer was chosen such that an interference maximum occurred at 950 nm, increasing the transmission at this wavelength. Therefore the up-converter layer in the flat solar cell configuration will absorb more

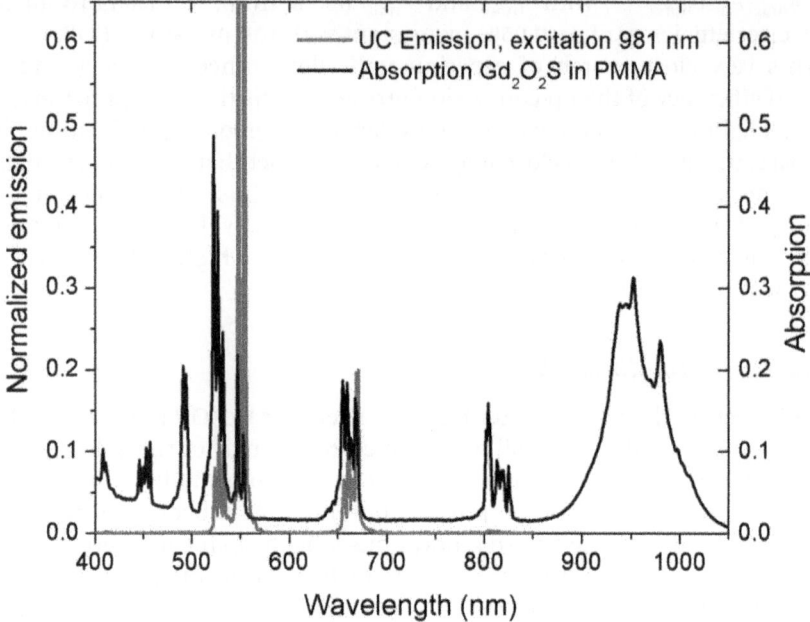

Figure 14.6   Up-converted emission and absorption spectra of the up-converter in PMMA layer. The emission spectrum is obtained when the up-converter shows no saturation and only emission peaks from the $^4S_{3/2}$, $^2H_{11/2}$ (510–560 nm) and $^4F_{9/2}$ (650–680 nm) states are observed. Reprinted, with permission, from J. de Wild, T. F. Duindam, J. K. Rath, A. Meijerink, W. G. J. H. M. van Sark, R. E. I. Schropp, *IEEE J Photovoltaics*, 2013, **3**, 17–21. © 2013 IEEE.

light. The back reflector adds $\sim 11\%$ to the current and thus to the efficiency. The absorption in the PMMA layer was 1.5%.

The absorption and emission spectra for the up-converter Gd_2O_2S: Er^{3+} (5%),Yb^{3+} (10%) are shown in Figure 14.6. The absorption is highest around 950 nm. The up-converter was excited with a xenon lamp at 950 ± 10 nm and 980 ± 10 nm. The $^4F_{7/2}$ state at 2.52 eV is reached after two times energy transfer from Yb to Er. The up-converter was already shown to be very efficient at low light intensities. Although the absorption at 950 nm (1.31 eV) is higher, excitation at 980 nm (1.26 eV) nm leads to a two times higher up-converted emission intensity. This may be attributed to the perfectly resonant energy transfer step of 980 nm (1.26 eV), since the $^4F_{7/2}$ state is at 2.52 eV.

As up-conversion is a two-photon process, the efficiency should be quadratically dependent on the excitation power density. This was investigated with a 980 nm NIR laser, and intensity was varied with neutral density filters. Up-conversion spectra were recorded in the range of 400 to 850 nm under identical conditions with varying excitation power. Varying the intensity shows that for low light intensities the red part is less than 6% of the total

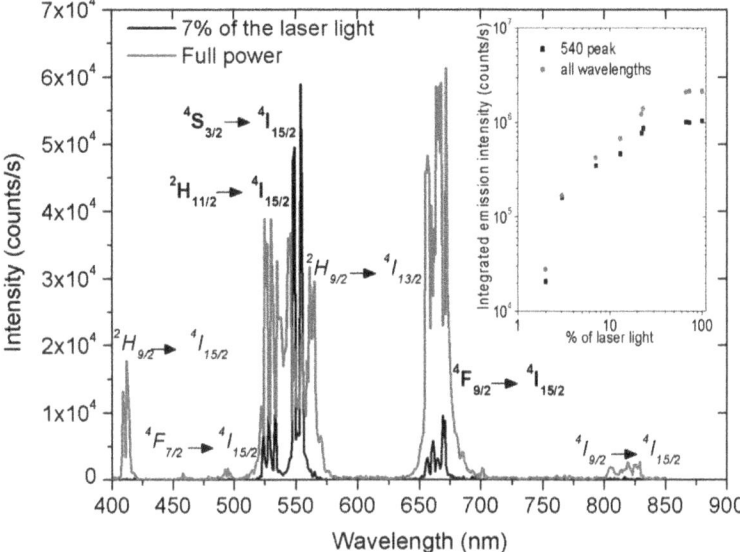

Figure 14.7 Up-converted emission spectra under low and high excitation density. For the low excitation power the green state was not yet saturated. The intensities may be compared. New peaks (italic) are assigned: $^2H_{9/2} \rightarrow {^4I_{15/2}}$ transition at 410 nm, $^4I_{9/2} \rightarrow {^4I_{15/2}}$ transition at 815 nm and the intermediate transition $^2H_{9/2} \rightarrow {^4I_{13/2}}$ at 560 nm.
Reprinted, with permission, from J. de Wild, T. F. Duindam, J. K. Rath, A. Meijerink, W. G. J. H. M. van Sark, R. E. I. Schropp, *IEEE J Photovoltaics*, 2013, **3**, 17–21. © 2013 IEEE.

emission, see Figures 14.6 and 14.7. Only when the emission from the green emitting states becomes saturated the red emission becomes more significant and even blue emission from $^2H_{9/2}$ state is measured, see Figure 14.7. Emission intensity is clearly not increasing quadratically with the excitation power density. Instead emission from higher and lower energy states is visible. The inset depicts the integrated emission peaks for the green and total emission, and it is shown that at very high laser intensities the total emission is saturated.

As shown above, sub-bandgap response in the near infrared due to the band tails of a-Si:H solar cells cannot be neglected. Therefore, in order to distinguish between up-converter response and sub-bandgap response, intensity dependent measurements are performed on solar cells with and without up-converter. *I–V* curves are measured under concentrated solar light, with wavelengths longer than 900 nm and I_{SC} is extracted from these measurements. Intrinsic response of the band tails is linearly dependent on the light intensity, while response due to up-converted light is expected to be quadratically increasing with the concentration. Figure 14.8 shows the current measured of the different solar cells for the different concentration factor of the sub-bandgap light. The slope of the line fitted to the data yields

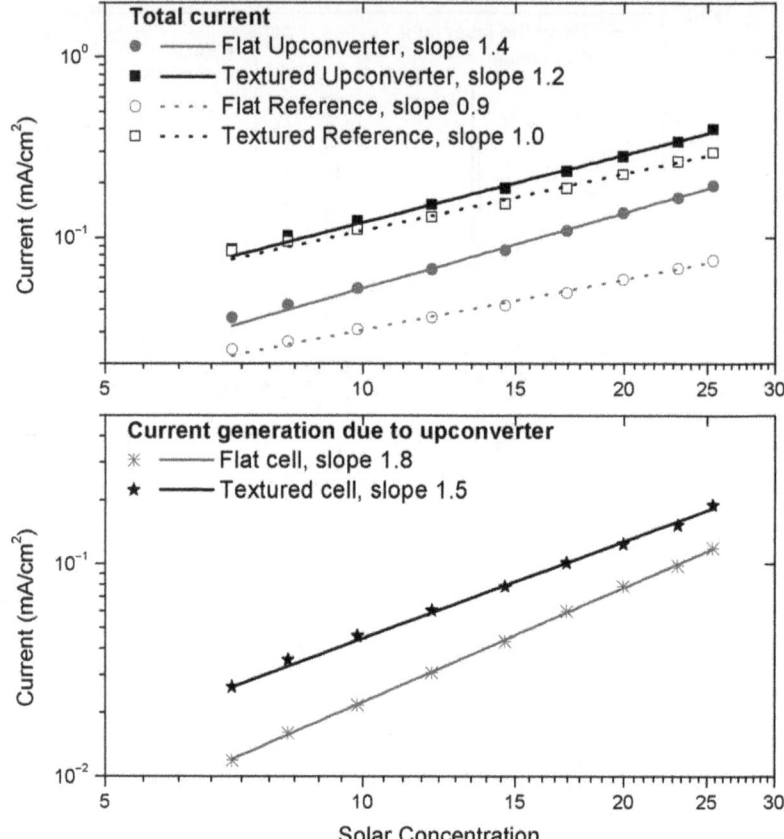

Figure 14.8 Current measured in the solar cells under illumination of sub-bandgap light. In the upper figure the total current of the reference and up-converter cells are plotted as a function of the concentration factor, in the lower graph the current generated by the up-converter is shown. The slope for sub-bandgap response is 1 for flat and textured solar cells. The contribution of the up-converter increases the slope slightly, when corrected for the sub-bandgap response the slope is 1.5 for the textured and 1.8 for the flat solar cells.

Reprinted, with permission, from J. de Wild, T. F. Duindam, J. K. Rath, A. Meijerink, W. G. J. H. M. van Sark, R. E. I. Schropp, *IEEE J Photovoltaics*, 2013, **3**, 17–21. © 2013 IEEE.

the value n, as given by Equation (14.1). As expected the sub-bandgap response is linearly increasing with the light intensity and values of n higher than 1 are measured for the solar cells with up-converter (for short: 'up-converter solar cells'). The value is close to 1 because a large part of the total current is due to the sub-bandgap response (see Figure 14.8, upper graph). When the total current measured for the up-converter solar cells is corrected for the sub-bandgap response, the current due to up-conversion only shows a higher value for n (see Figure 14.8 lower graph). Values of $n = 1.5$ for textured solar cells and $n = 1.8$ for flat solar cells are determined.

It is noted that the current is not increasing quadratically with increasing concentration. It is unlikely that the up-converter is saturated, because the power density is far below the saturation level of 0.6 W cm^{-2}.[42] It is therefore more likely that the deviations are due to decreasing carrier collection efficiency with increasing concentration. This effect would play a larger role in textured solar cells, because they have a higher defect density than flat solar cells. This may explain why for flat solar cells the value n is closer to 2, than the value n for the textured solar cells.

The textured solar cells were also measured using laser light excitation. Monochromatic laser light with wavelength at 981 nm and power density of 0.2 W cm^{-2} gave a current density of 0.14 mA cm^{-2} for the up-converter solar cells and 0.04 mA cm^{-2} for the reference solar cells. With monochromatic laser light the contribution of sub-bandgap absorption is much smaller, because for broadband excitation all wavelengths longer than 900 nm add to the up-converted light, and thus, to the photocurrent.

The current due to up-conversion is comparable to the current measured under 20 sun: ~ 0.1 mA cm^{-2} (Figure 14.8). First, this is in contrast with previously reported experiments with broad band excitation of c-Si solar cells,[43] where the current under broad band excitation was much smaller than under laser light excitation. However, another up-converter host was applied (NaYF$_4$) and different processes are taking place in the up-converter, namely excited state absorption. Using the up-converter host in this work (Gd$_2$O$_2$S), ETU is the main up-converter path and the broadband absorption of Yb^{3+} may increase the transfer between Yb^{3+} and Er^{3+}. Second, the power that is absorbed by Yb^{3+} is 3.44 mW cm^{-2},[44] which yields a broadband power density of 70 mW cm^{-2} under a concentration of 20 suns. This is three times less than the power density of the laser. A large difference here is that light of the solar simulator extends to further than 1600 nm, thus the Er^{3+}:$^2I_{13/2}$ state is also excited directly. Addition of other paths that lead to up-converted light may contribute to the current. These paths may be non-resonant excited state absorption between the energy levels of Er^{3+} or three-photon absorption around 1540 nm at Er^{3+}:$^2I_{13/2}$ (Figure 14.2). Direct excitation of the Er^{3+} $^2I_{13/2}$ state followed by excited state absorption (ESA) from $^2I_{13/2}$ to $^2F_{9/2}$ results in a visible photon around 650 nm, while three-photon absorption around 1540 nm results in emission from the $^2F_{9/2}$ state too. Wavelengths required for these transitions are around 1540 nm and 1200 nm, which are present within the broad spectrum that the solar cell receives.

14.3 Luminescent Solar Concentrators

A luminescent solar concentrator (LSC) consists of a highly transparent plastic plate, in which luminescent species are dispersed. These species absorb incident light and emit it at a red-shifted wavelength, with high quantum efficiency. Internal reflection ensures that emitted light is directed

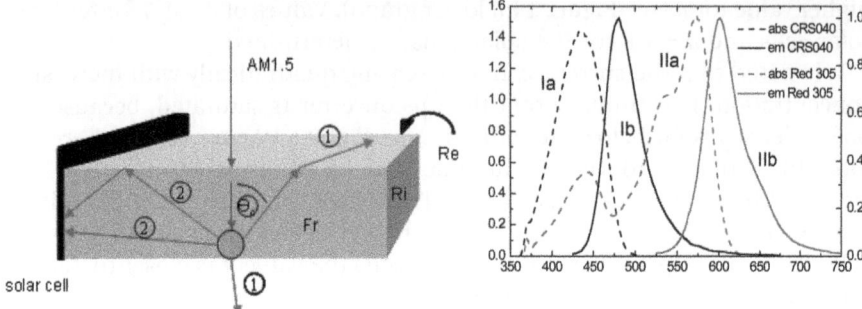

Figure 14.9 (left) Schematic 3D view of a luminescent concentrator. AM1.5 light is incident at the top. The light is absorbed by luminescent species, and its luminescence is randomly emitted. Part of the emission falls within the escape cone (critical angle θ_c and is lost from the luminescent concentrator at the surfaces (1). The other part (2) is guided to the solar cell by total internal reflection.[9] (right) Absorption cross-section and normalized emission spectra of two fluorescent dyes (CRS040, green fluorescence, curves Ia and Ib; and Red305, red fluorescence, curves IIa and IIb), illustrating the Stokes' shift.
Reprinted from W. G. J. H. M. van Sark, R. E. I. Schropp, *Renewable Energy*, 2013, **49**, 207–210, Copyright (2013), with permission from Elsevier.

towards solar cell(s) that are located at the side(s) of the plastic plate, see Figure 14.9. The LSC is an example of the use of down-shifting.

Down-shifting was already suggested in the 1970s to be used for LSCs.[17,18] In those early times, organic dye molecules were employed as luminescent species to absorb incident light and re-emit it at a red-shifted wavelength. The expected efficiency of $\sim 30\%$[45,46] was not reached in practice due to the inability of the used organic dye molecules to meet stringent requirements, such as high quantum efficiency and stability, and the limited transparency of collector materials in which the dye molecules were dispersed.[17,47] Today, new organic and inorganic luminophores have been developed that have very high luminescence quantum efficiency (LQE) (near unity) and are available in a wide range of colours with better re-absorption properties while providing improved UV stability.

14.3.1 Operating Principles

The operating principle of the LSC is illustrated in Figure 14.9. Light is incident on the top of a plastic plate with refractive index n. In this plate luminescent species are present, which absorb light and isotropically emit it at a wavelength that is, in general, red shifted as a result of the Stokes' shift of the particular luminescent species (see also Figure 14.9). The Stokes' shift is defined as the difference between maxima in absorption and emission spectra, related to the same transition. Part of the emission falls within the escape cone [determined by the critical angle (θ_c)] and is lost from the luminescent concentrator at the surfaces (1 in Figure 14.9. The critical angle is defined by $\theta_c = \sin^{-1}(1/n)$. The other part (2) of the luminescence is guided to the solar cell by total internal reflection, which is lossless (except for absorption in the volume of the plate) within the angular range. The other sides of the plastic plate should be covered with mirrors to reflect photons outside the angular range.

The efficiency of the complete LSC device including solar cell η_{LSC} can be expressed as the product of the optical efficiency η_{opt} of the LSC plate and the efficiency of the solar cell η_{PV}. The optical efficiency is expressed as:[48]

$$\eta_{opt} = (1 - R)\eta_{abs}\eta_{LQE}\eta_S\eta_{trap}\eta_{mat}(1 - \eta_{self})\eta_{TIR} \tag{14.6}$$

with R the surface reflection coefficient, η_{abs} the absorption efficiency (the fraction of sunlight absorbed by the luminescent species), η_{LQE} the quantum efficiency of the luminescent species, η_S the Stokes' efficiency, which is the energy loss between absorption and emission, η_{trap} the efficiency of trapping of light (fraction of light outside the escape cone), η_{mat} the efficiency of transmission of light in the plastic matrix, η_{self} the efficiency of self absorption of the luminescent species, η_{TIR} the efficiency of total internal reflection.

The Fresnel reflection loss equals $[(n-1)/(n+1)]^2$. For a plastic with $n = 1.49$ (*e.g.*, polymethylmethacrylate, PMMA) this loss amounts to 0.0387, which may be lowered by applying an anti-reflection coating. The fraction of sunlight that can be absorbed by the luminescent species depends on the specific material; organic dyes have narrow absorption bands. As an example, the absorption and emission spectra for Fluorescence Yellow CRS040 (coumarine dye, from Radiant Color) and Lumogen F Red305 (perylene, from BASF) are depicted in Figure 14.9. The loss due to non-unity luminescent quantum efficiency is typically low for organic dyes. Stokes' shift between absorption and emission is required to be small to minimize energy loss; on the other hand it should be large to minimize the overlap between the emission and absorption spectrum causing re-absorption (or self-absorption). The light trapping efficiency can be written as $(1/n)(n^2 - 1)^{\frac{1}{2}}$, which equals 0.741 for PMMA; the critical angle being $\theta_c = 42.2°$. This efficiency can be enhanced by the use of selective mirrors or mirrors based on photonic materials.[31] The transmission efficiency in the matrix depends on scattering or absorption in the matrix; an absorption coefficient of 1 m^{-1} is typical. Self-absorption in general leads to a red shift of the emission spectrum, as well as intensity loss. Total internal reflection loss depends on surface quality. Taking into account all these losses and including the PV efficiency, typical LSC device efficiencies range from 3% to 8%.

14.3.2 Efficiency

The efficacy of an LSC is evaluated with the achieved concentration factor and its power conversion efficiency. The concentration factor or ratio is defined as the ratio of radiative power delivered to the solar cell attached to

the LSC plate and the radiative power delivered to the bare solar cell illuminated under the same (STC) conditions. The power conversion efficiency
is defined as the ratio of energy delivered at the terminal of the solar cell
while attached to the LSC and the energy of incident light that the entire
device was illuminated with. Note that the wavelength of emitted light is
close to the maximum of the spectral response of the cell, hence in this case
the cell efficiency is much higher than the broadband (STC) efficiency. As an
example, the efficiency of an LSC with an aperture area 10 times as large as
the solar cell delivering five times as much current as the bare solar cell
would be half the efficiency of the attached solar cell. From the principle of
detailed balance as discussed by Shockley and Queisser (SQ), it has been
derived that maximally 33% of the energy in the solar spectrum can be
converted with LSCs assuming the presence of a perfect band pass mirror at
the top surface and perfect absorption of the luminophore.[46] The maximum
thermodynamically possible concentration (C) strongly depends on Stokes'
shift of the luminescent species.[45,46] For the Lumogen F Red 300 dye it can
be calculated that $C = 119$, as absorption and emission maxima are 578 and
613 nm, respectively (Smestad *et al.*[45] arrive at $C = 102$). This maximum is
one order of magnitude larger than typically obtained in practical LSCs.

   Experimental work to surpass the long-standing 4% efficiency record of
LSCs,[48–50] albeit for smaller area size, has been successful. For example,
Goldschmidt *et al.*[51,52] showed a conversion efficiency of 6.7% for a stack of
two plates with different dyes, to which four GaInP solar cells were placed at
the sides. The plate was small (4 cm^2), and the concentration ratio was only
0.8. It was argued that the conversion efficiency was limited by the spectral
range of the organic dyes used, and that if the same quantum efficiency that
was reached for dyes emitting in the 450–600 nm spectral range could be
realized for those emitting in the 650–1050 nm range an efficiency of 13.5%
could be feasible. They also discuss the benefits of a photonic structure on
top of the plate, to reduce the escape cone loss.[51] The proposed structure is a
so-called rugate filter; this is characterized by a varying refractive index in
contrast to standard Bragg reflectors. Rugate filters suppress the side loops
that could lead to unwanted reflections. The use of these filters would increase the efficiency by $\sim 20\%$, as was determined for an LSC consisting of
one plate and dye. Slooff *et al.*[33] presented results on $50 \times 50 \times 5$ mm^3 PMMA
plates in which both CRS040 and Red305 dyes were dispersed at 0.003 and
0.01 wt%, respectively. The plates were attached to either multicrystalline Si
(mc-Si), GaAs or InGaP cells, and a diffuse reflector (97% reflectivity) was
used at the rear side of the plate. The highest efficiency measured was 7.1%
for four GaAs cells connected in parallel (7% if connected in series), and to
date this is the record efficiency. Desmet *et al.*[53] reached 4.2% efficiency
using crystalline silicon cells, using a red and blue dye, which is higher than
the 2.7% reported by Slooff *et al.*[33] Using thermodynamic as well as ray-trace
models, LSC device efficiencies have been calculated between 2.4% for an
LSC with mc-Si cell at certain mirror configurations and 9.1% for an LSC
with InGaP cell for improved specifications.[9]

In an attempt to classify operating regimes of luminescent solar concentrators Farrell and Yoshida introduced the concept of first and second generation LSCs.[54] They argue that the first generation consists of the 'traditional' flat plate LSCs containing isotropic emitters, whereas in the second generation device directional emitters or photonic filters are employed that enhance the waveguiding mechanism. They predict the LSC efficiency to depend on geometric concentration, and in fact demonstrate that experimentally attained efficiency values can be well described by their model: low geometric concentration yields high efficiency, and *vice versa*. Second generation LSCs may be able to attain efficiencies approaching the SQ limit.

The ideal LSC should be based on material systems that:[31,55,56]

- Absorb all photons with wavelength >950 nm, and emit them red-shifted at ~1000 nm, for use with c-Si solar cells
- Have as low as possible spectral overlap between absorption and emission spectra to minimize re-absorption losses
- Have near unity luminescence quantum yield
- Have low escape cone losses
- Be stable outdoors for longer than 10 years
- Be easy to manufacture at low cost

Presently, there are no luminescent species that fulfil all requirements, and research predominantly is focused on finding new suitable species. In the following some aspects of this research will be highlighted, in particular re-absorption loss.

14.3.3 Alternative Luminescent Species

14.3.3.1 Quantum Dots

Quantum dots have been proposed as an alternative for organic dye molecules in luminescent solar concentrators.[57–60] Quantum dots (QDs) are nanometre-sized semiconductor crystals of which the emission wavelength can be tuned by their size, as a result of quantum confinement.[61,62] QDs have advantages over dyes in that:

- Their absorption spectra are far broader, extending into the UV
- Their absorption properties may be tuned simply by the choice of nanocrystal size
- They are inherently more stable than organic dyes[63]

Moreover, there is a further advantage in that the red-shift between absorption and luminescence is quantitatively related to the spread of QD sizes, which may be determined during the growth process, providing an additional strategy for minimizing losses due to re-absorption.[57] A recent review shows a wealth of colloidal QDs of which material candidates for use in LSCs can be selected.[64] Spherical QDs have been reported with LQE larger than 80%;[65] however, their Stokes' shift is small and of the same order as that of organic dyes. Type-I CdSe/ZnS core/shell quantum dots, which absorb photons up to a wavelength of ~600 nm and re-emit them energy-shifted at ~625 nm, have been used in LSCs.[66] Type-I denotes a band alignment between the core and shell materials that results in the formation of direct excitons. Bomm *et al.*[66] have addressed several problems regarding incorporation these QDs in an organic polymer matrix, *viz.* phase separation, QD agglomeration leading to turbidity, and luminescence quenching due to exciton energy transfer.[67] Quantum dot luminescent solar concentrators (QDLSCs) have been fabricated using CdSe core/multishell QDs with LQE of 60%[68] that were dispersed in laurylmethacrylate (LMA).[69] UV-polymerization was employed to yield transparent PLMA plates with QDs without any sign of agglomeration. A mc-Si cell was glued to one side of this plate; all other sides were covered with aluminium mirrors. Compared to the bare cell (5×0.5 cm^2) that generated a current density of 40.28 mA cm^{-2} at 1000 W AM1.5G spectrum, the best QDLSC made generated a current of 77.14 mA cm^{-2}, nearly twice as much. However, the QDLSC efficiency was 3.5%. This can be explained by the fact that the power conversion efficiency of an LSC is calculated with respect to the area of the top surface of the LSC, while the efficiency of a solar cell is calculated with respect to solar cell area, which is much smaller. Thus, twice the current from a five times larger device results in two fifths of the bare cell efficiency. In addition, exposure to a 1000 W sulfur lamp for 280 h continuously showed very good stability: the current density decreased by 4% only, on average. However, re-absorption is still an issue, as was demonstrated by the observation of a small red shift in the emission spectrum for long photon pathways. In another effort, Bomm *et al.*[70] have dispersed nanorods (NRs) in PLMA, showing excellent transmittance of 93%; for long rods (aspect ratio of 6) a QE of 70% is observed, which is only slightly smaller than the QE in solution, implying that these rods are stable throughout the polymerization process. In addition, these NRs have also been dispersed in cellulose triacetate (CTA), and a ~10 μm thin film on a glass substrate was made showing bright orange luminescence.[70]

Schüler *et al.*[71] reported on the successful fabrication of thin silicon oxide films containing CdS QDs by using a sol–gel dip-coating process, whereby the 1–2 nm sized CdS QDs are formed during thermal treatment after dip-coating. Depending on the anneal temperature, the colors of the LSC ranged from green for 250 °C to yellow for 350 °C and orange for 450 °C. Reda[72] also found that absorption and emission spectra of CdS QDLSCs, made using sol–gel spin coating followed by annealing, are affected by annealing temperature. Also, luminescence intensity and Stokes' shift both decreased after 4 weeks outdoors exposure to sunlight, which was probably caused by aggregation and oxidation. Oxidation of QDs leads to blue shift in emission,[73] and blue-shifts have also been observed by Gallagher *et al.*[74] who dispersed CdSe QDs in several types of resins (urethane, PMMA, epoxy), for fabrication of LSC plates. Commercially available CdSe/ZnS core/shell QDs with LQE of 57% were used in LSCs, both liquid (QDs dissolved in toluene, between two $6.2 \times 6.2 \times 0.3$ cm^3 glass plates) and solid (QDs dispersed in epoxy):[75] an efficiency of 3.98% and 1.97% was obtained, respectively, using 18.3% c-Si solar cells. For comparison, using Lumogen F Red300 in toluene, an efficiency of 2.6% was determined, *i.e.*, smaller than using QDs.

14.3.3.2   Luminescent Ions

Other alternatives for dye molecules in LSCs are luminescent ions. Traditionally, efficient luminescent materials rely on the luminescence of transition metal ions and lanthanide ions. In case of transition metal ions intraconfigurational $3d^n$ transitions are responsible for the luminescence, while in case of lanthanide ions both intraconfigurational $4f^n$–$4f^n$ transitions and interconfigurational $4f^n$ to $4f^{n-1}5d$ transitions are capable of efficient emission. In most applications efficient emission in the visible is required and emission from lanthanide ions and transition metal ions is responsible for almost all the light from artificial light sources (*e.g.*, fluorescent tubes, flat displays and cathode ray tubes) and white light emitting diodes (LEDs).[76] For LSCs to be used in combination with c-Si solar cells efficient emission in the NIR is needed. The optimum wavelength is between 700 and 1000 nm, which is close to the bandgap of c-Si and in the spectral region where c-Si solar cells have their optimum conversion efficiency. Two types of schemes can be utilized to achieve efficient conversion of visible light into narrow band NIR emission. A single ion can be used if the ion shows a strong broad band absorption in the visible spectral range followed by relaxation to the lowest excited state from which efficient narrow band or line emission in the NIR occurs. Alternatively, a combination of two ions can be used where one ion (the sensitizer) absorbs the light and subsequently transfers the energy to a second ion (the activator), which emits efficiently in the NIR. Both concepts have been investigated for LSCs by incorporating luminescent lanthanides and transition metal ions in glass matrices. The stability of these systems is good, in contrast to LSCs based on dye molecules, however, the quantum efficiency of luminescent ions in glasses is much lower than in crystalline compounds, especially in the infrared, thus hampering their use for LSCs,[77–79] see also Debije and Verbunt.[31] As an example, the europium ion has been used for LSCs; however, as the absorption spectrum is very narrow, the attained LSC device efficiencies are low.[80–84] In contrast, De Boer *et al.*[85] have demonstrated that SrB$_4$O$_7$: 5% Sm^{2+}, 5% Eu^{2+} may be a good luminescent candidate material that has high absorption, little re-absorption, and high quantum efficiency. This Sm^{2+}-based phosphor has a broad absorption between 350 and 600 nm, and has an emission peak around 700 nm. Using this layer as LSC with 50×5 mm^2 solar cells, it was shown that the LSC efficiency was only slightly lower compared to an LSC with organic dye.[53]

14.3.4   Re-absorption

Luminescent species in the LSC may re-absorb previously emitted photons before they can reach the solar cell. Upon every re-absorption event the photon energy can be dissipated again through any of the loss mechanisms in the LSC. Even in an idealized LSC with only escape cone losses, the trapping probability for every emitted photon is $P_{trap} = 1 - P_{cone} = 0.75$. Consequently for a given input energy flux Φ_{in} the output flux Φ_{out} decreases after an average of $N - 1$ re-absorptions per photon to:[86]

$$\Phi_{out} = \Phi_{in}(\eta_{LQE})^N(1 - P_{cone})^N \qquad (14.7)$$

Clearly, re-absorption maximizes the occurring losses as the light output into the solar cell decreases exponentially with the number of re-absorption events. This limits the size of LSC plates and thus the maximum concentration ratio.

Using luminescent species that have a negligible absorption cross-section at their emission energy re-absorption can be minimized. This implies that their emission spectrum should be well separated from their absorption spectrum: Stokes' shift should be reasonably large. Besides luminescent ions, colloidal Type-II semiconductor hetero-nanocrystals fulfil this requirement. Semiconductor hetero-nanocrystals (HNCs) consist of two (or more) different materials joined in the same particle by hetero-interfaces. Depending on the energy offsets between the valence and conduction band levels of the materials that are combined at the hetero-interface, different carrier localization regimes will be observed after photo-excitation.[64] In Type-I HNCs both carriers are primarily localized in the same material, whereas in Type-II HNCs electrons and holes are spatially separated, creating a spatially indirect exciton. The photon emitted upon radiative recombination of the spatially indirect exciton will have a lower energy than the bandgap of both components of the HNC, and thereby its re-absorption probability is greatly reduced.

In this chapter, we will compare the self-absorption behaviour of several organic dyes and CdSe QDs with that of CdTe/CdSe Type-II HNCs. Two different samples of CdTe/CdSe HNCs were used: nearly spherical concentric core/shell QDs (hereafter referred to as Type-II iso) and dot core/rod shell nanorods (hereafter referred to as Type-II an). In order to do so, we developed a liquid model LSC device, consisting of a quartz cell in which the studied luminescent solution can be quickly and easily exchanged. The loss mechanisms in this model LSC are the same as in a real LSC, and therefore our experiments offer a convenient and reliable way to evaluate the potential of luminescent species for application in LSCs.

Self-absorption occurs when there is a finite probability for a luminescent species to absorb the light emitted by the same species. This is visualized in an overlap between the spectra of absorption $A(\lambda)$ and emission $F(\lambda)$. This overlap σ_{SA} quantifies the self-absorption cross-section per 1 cm optical path as:[86]

$$\sigma_{SA} = \frac{\int_0^\infty F(\lambda)A(\lambda)d\lambda}{\int_0^\infty F(\lambda)d\lambda} \tag{14.8}$$

It gives the fraction of the emission that is absorbed by the luminescent species itself while travelling through 1 cm of the medium. This optical pathlength was chosen because it corresponds to that used for the absorption measurements in solution. The self-absorption cross-section can be used to characterize luminescent species from their emission spectra and absorption spectra.

The measurements of self-absorption effects were conducted with a position-adjustable excitation source that illuminates different spots in the material so that the optical pathlength travelled by the emitted light inside the luminescent solution can be varied (Figure 14.10). A Spectral Products ASBN-D2-W100F-L Halogen lamp (power 100 W, colour temperature 3000 K) filtered through a band filter at (520 ± 25) nm was used as excitation source. The optical densities of the various luminescent solutions were adjusted to 0.17 ± 0.04 at 520 nm, in order to ensure that the amount of light absorbed by different samples was comparable.

An optical fibre with a slit exit aperture and a lens were used to focus the excitation light on a quartz cell (the liquid model LSC device). The quartz cell was screw capped to prevent oxidation of the luminescent solution inside. The internal volume of the cell was $10 \times 10 \times 35$ mm^3. The bottom of the quartz cell was mounted in firm contact with a second optical fibre, which collected the emitted light and guided it to a monochromator and subsequently to a CCD (Spectra Pro 300i from Action Research) to record the emission spectra. The cell with the collection fibre could be moved along

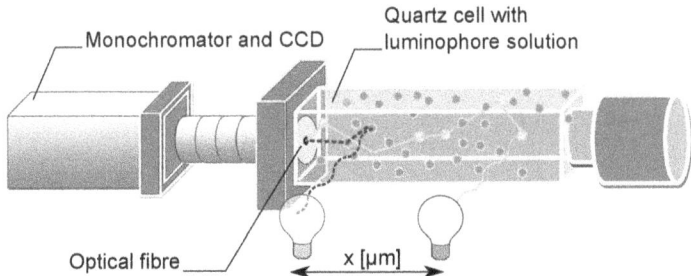

Figure 14.10   Schematic representation of the setup for intensity profile measurements.
Reprinted from Z. Krumer, S. J. Pera, R. J. A. van Dijk-Moes, Y. Zhao, A. F. P. de Brouwer, E. Groeneveld, W. G. J. H. M. van Sark, R. E. I. Schropp, C. de Mello-Donegá, *Solar Energy Materials and Solar Cells*, 2013, **111**, 57–65, Copyright (2013), with permission from Elsevier.

the *x*-axis with a step-size of several micrometres. When operating with air-sensitive samples, the cells were filled under nitrogen atmosphere. The *x*-axis was chosen to point upwards to collect eventual gas bubbles in the bottleneck of the quartz cell, such that they could not influence the measurement. Toluene was employed as solvent for all the samples but Rhodamine 6G, which was dissolved in ethanol.

The absorption and emission spectra are provided in Figure 14.11. The self-absorption cross section of Rhodamine 6G was calculated to be 8.5% and its quantum yield was taken to be 95%.[87] The quantum yield of Lumogen Orange was found to be 95%. The self-absorption cross section was calculated to be 6.7%. The absorption spectra of the semiconductor QDs (Figure 14.11) are much wider than those of the dyes, since they consist of a large number of different exciton transitions that partially overlap producing a spectrally broad absorption spanning from the band edge to the UV.

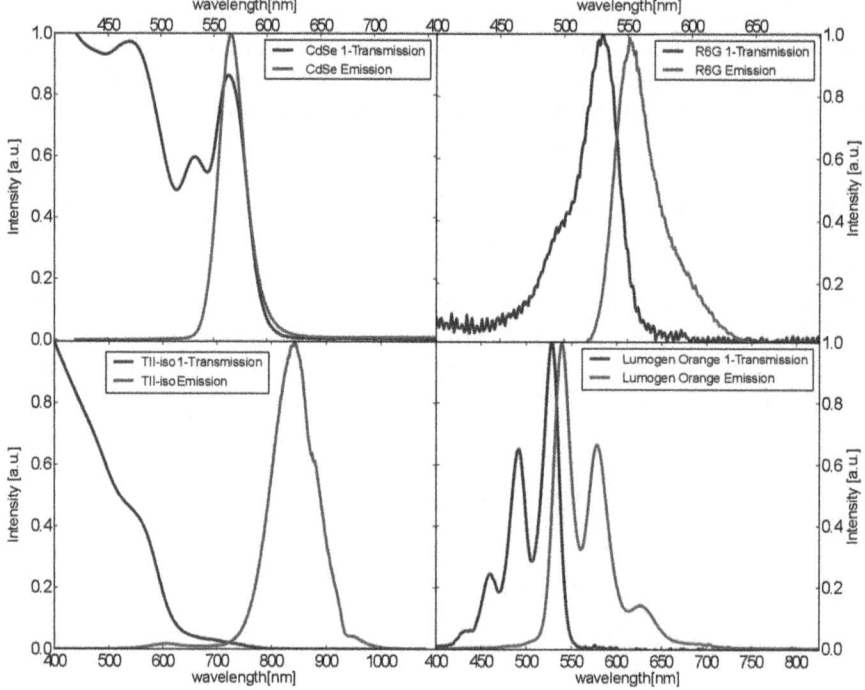

Figure 14.11   Normalized absorption and emission spectra of diluted solutions of selected dyes (Rhodamine 6G, top right, and Lumogen Orange, bottom right) and colloidal semiconductor nanocrystals (CdSe, top left, and CdTe/CdSe/ZnS core/multishell Type-II HNCs, bottom left). Reprinted from Z. Krumer, S. J. Pera, R. J. A. van Dijk-Moes, Y. Zhao, A. F. P. de Brouwer, E. Groeneveld, W. G. J. H. M. van Sark, R. E. I. Schropp, C. de Mello-Donegá, *Solar Energy Materials and Solar Cells*, 2013, **111**, 57–65, Copyright (2013), with permission from Elsevier.

The self-absorption cross-sections of the semiconductor QDs were calculated to be 52.2%, 2.5% and 0.2%, for CdSe, Type-II-iso and Type-II-an QDs, respectively. The quantum yields of the semiconductor QDs were found to be 27% and 48%, for CdSe, and Type-II-iso QDs, respectively. The absorption and emission spectra of Type-II-an QDs are similar to those of Type-II-iso QDs (not shown).

The emitted light can subsequently be absorbed by the emitting medium itself. Following Lambert–Beer's law, the probability for absorption increases exponentially with increasing path of the light through the medium, the longer the path between the spot of excitation and the detector, the higher the probability for this process occurring and repeating itself. In a weakly luminescent sample, such as CdSe QDs, the measured emission peak-intensity drops rapidly with increasing optical path (Figure 14.12). Although a spectral shift of the peak is observed, the position of the red band edge of the peak remains unchanged.

For single composition QDs (*e.g.*, CdSe QDs) and Type-I HNCs the energy of every emitted photon is up to a few (<20) meV lower than the bandgap of the emitting quantum dot:[64] intrinsic (or resonant) Stokes' shift. The so-called global (or non-resonant) Stokes' shift is larger which is due to the fact that real samples always consist of an ensemble of QDs of slightly different sizes. The size distribution of the ensemble of QDs is reflected in the line width of the optical transitions, which becomes broader due to inhomogeneous broadening, and in the non-resonant Stokes shift, which increases for larger size dispersions and higher concentrations. The increase in the non-resonant Stokes shift is due to the fact that the absorption cross-sections at energies far

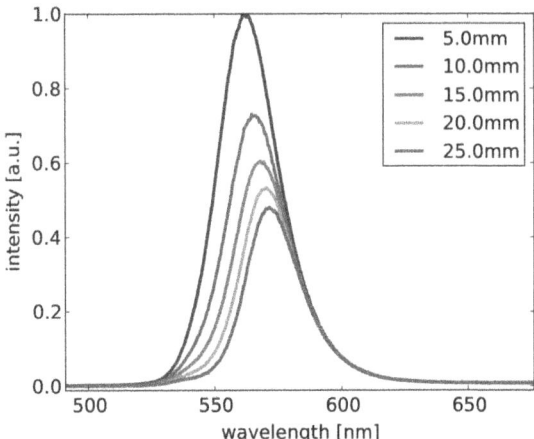

Figure 14.12 Emission spectra of CdSe quantum dots recorded at different distances from the excitation spot.
Reprinted from Z. Krumer, S. J. Pera, R. J. A. van Dijk-Moes, Y. Zhao, A. F. P. de Brouwer, E. Groeneveld, W. G. J. H. M. van Sark, R. E. I. Schropp, C. de Mello-Donegá, *Solar Energy Materials and Solar Cells*, 2013, **111**, 57–65, Copyright (2013), with permission from Elsevier.

above the band-edge scale with the volume. Therefore, larger QDs will absorb relatively more light, resulting in a red-shift of the ensemble PL spectrum from the statistically weighted maximum at excitation energies far above the band-edge. Moreover, self-absorption also leads to a concentration- and optical path-length-dependent increase in the observed non-resonant Stokes shift. Figure 14.12 shows that this increase is primarily due to the shift of the emission peak to lower energies with increasing number of self-absorption events, which can occur as a result of increasing optical path lengths and/or increasing concentrations. A requirement for the self-absorption process is that the photon energy must be equal to or larger than the bandgap of the absorbing QD. Therefore, photons emitted by the largest QDs in the ensemble do not have sufficient energy to be re-absorbed by most of the QDs in the ensemble, and therefore have a higher probability of reaching the detector. In contrast, photons emitted by smaller QDs carry sufficient energy to be absorbed by most members of the QD-ensemble and therefore are more likely to be absorbed before reaching the detector. This is consistent with the observed loss in intensity in the range 530–570 nm and the almost unchanged red shoulder starting from 580 nm, as the optical path increased. The overall effect is the loss in total intensity and a total red shift as shown in Figure 14.13 for CdSe QDs.

The integration of the recorded emission spectra provides a measure for the amount of light that would reach the solar cell at the side-face of a luminescent solar concentrator. We have performed the same analysis on a selection of dyes and colloidal semiconductor QDs and HNCs. In order to allow the self-absorption characteristics of the different samples to be properly compared, intensities have been normalized with respect to that acquired at the closest distance ($d = 5$ mm) from the detector (Figure 14.14).

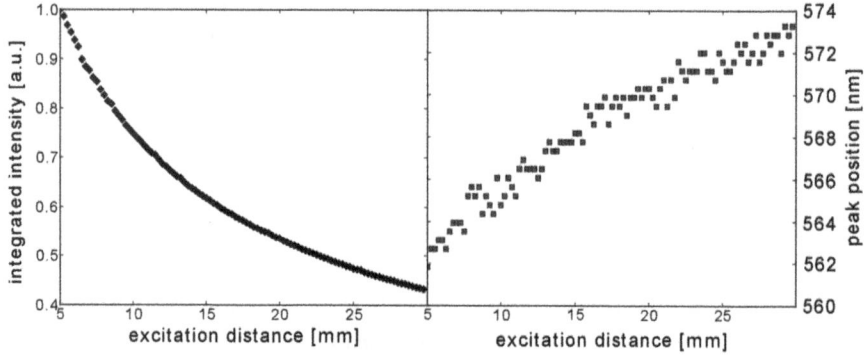

Figure 14.13   Effects of self-absorption on intensity (left) and on peak position (right) of colloidal CdSe QDs.
Reprinted from Z. Krumer, S. J. Pera, R. J. A. van Dijk-Moes, Y. Zhao, A. F. P. de Brouwer, E. Groeneveld, W. G. J. H. M. van Sark, R. E. I. Schropp, C. de Mello-Donegá, *Solar Energy Materials and Solar Cells*, 2013, **111**, 57–65, Copyright (2013), with permission from Elsevier.

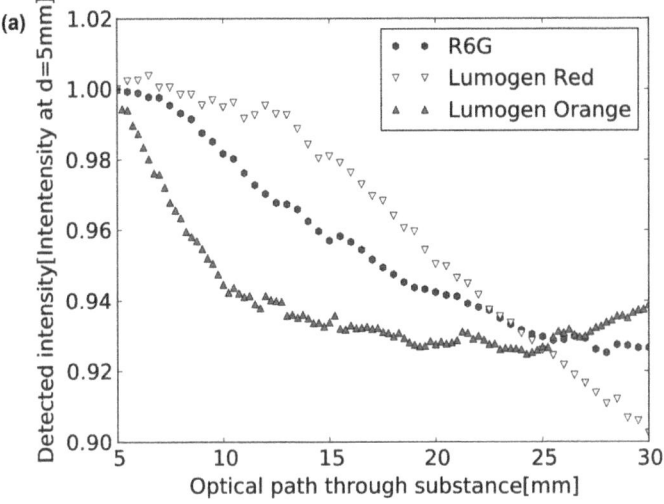

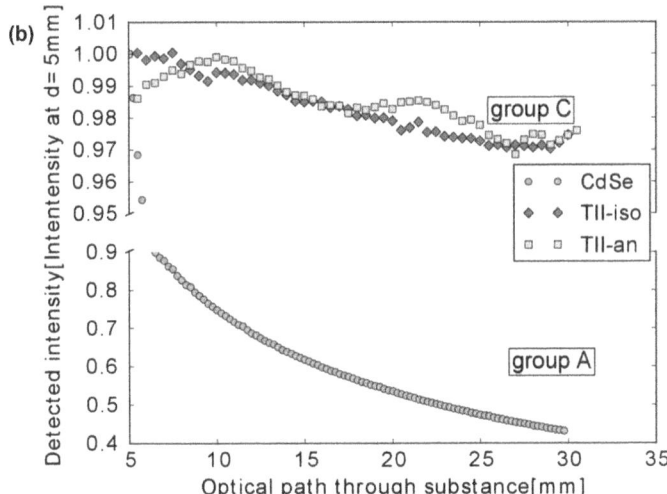

Figure 14.14   Normalized intensity profiles of a selection of dyes (top, classified as group B in the text) and semiconductor nanocrystals (bottom, classified as groups A and C in the text). Normalized intensity profiles for dyes (Lumogen Orange, Lumogen Red, Rhodamine 6G) and QDs (CdSe QDs, Type-II-an QDs).
Reprinted from Z. Krumer, S. J. Pera, R. J. A. van Dijk-Moes, Y. Zhao, A. F. P. de Brouwer, E. Groeneveld, W. G. J. H. M. van Sark, R. E. I. Schropp, C. de Mello-Donegá, *Solar Energy Materials and Solar Cells*, 2013, **111**, 57–65, Copyright (2013), with permission from Elsevier.

The same experiments were also carried out on non-luminescent silica nanoparticles of comparable size (25 nm) to the QDs (5–15 nm) in order to provide a reference for potential losses due to light scattering. The results show that the impact of light scattering is negligible.

Table 14.1   Summary of the key properties for the luminescent species CdSe QDs, Rhodamine 6G (R6G), Lumogen Orange (Lumo O), CdTe/CdSe/ZnS-core/multishell Type-II QDs (TII-iso), and CdTe/CdSe dot-core/rod-shell Type-II nanorods (TII-an).

Sample	η_{LQE} (%)	σ_{SA} (%)	$\Delta\lambda$ (nm)	Loss (%)	ζ
CdSe QDs	27	52.2	11.0	57	4.04
R6G	95	8.5	4.0	7.0	1.31
Lumo O	95	6.7	2.2	7.5	1.0
TII-iso	—	2.5	−0.6	3	4.13
TII-an	48	0.2	−0.6	3	4.02

η_{LQE}, luminescence quantum yield; σ_{SA} self-absorption cross section; $\Delta\lambda$, spectral shift after 30 mm optical path length; Loss, relative intensity loss after 30 mm optical path length; ζ, integrated absorption spectra from 300 to 800 nm normalized with respect to Lumogen Orange. Reprinted from Z. Krumer, S. J. Pera, R. J. A. van Dijk-Moes, Y. Zhao, A. F. P. de Brouwer, E. Groeneveld, W. G. J. H. M. van Sark, R. E. I. Schropp, C. de Mello-Donegá, *Solar Energy Materials and Solar Cells* 2013, **111**, 57–65, Copyright (2013), with permission from Elsevier.

All samples show intensity losses over the optical path. The observed losses are summarised in Table 14.1. The quantitative differences allow the classification of the investigated luminescent species in three distinct groups: those with a strong intensity loss (group A: CdSe QDs), those for which the output is moderately reduced (group B: Lumogen and Rhodamine dyes), and those for which the output is only slightly reduced (group C: both Type-II HNCs). The increase in intensities at longer optical paths observed in some cases can be explained by the increasingly stronger reflection of excitation light as the bottleneck of the quartz cell is approached.

The observed intensity loss in groups A and B can be confidently attributed to self-absorption as it is accompanied by a red-shift of the peak, which is also correlated to the intensity drop: the luminescent species with the strongest intensity loss (CdSe QDs) also reveals the largest red-shift (and lowest quantum yield) (see Table 14.1).

For both Type-II HNCs a mean red-shift of −0.6 nm was found (Table 14.1). As this is of the same order of magnitude as the resolution of the monochromator used (0.5 nm), we conclude that no detectable red shift is apparent. Together with the minimal intensity loss this indicates the absence of self-absorption in Type-II samples. From all the investigated samples the Type-II HNCs show the least self-absorption effects: the photons emitted by the Type-II HNC samples seem almost unaffected by the amount of material they travelled through.

In this study monochromatic excitation was used to properly compare self-absorption properties of different luminescent species. However, as colloidal semiconductor NCs are characterized by a broad absorption spectrum compared to dyes, it is interesting to assess the effect of using broadband excitation (closer to sunlight). To this end, we have integrated absorption spectra from 300 to 800 nm, and normalised the result with respect to the total absorption cross section of Lumogen Orange (quantity ζ in Table 14.1). The amount of emitted light F is then given by:

$$F = \zeta\eta_{LQE} \tag{14.9}$$

If this is taken into account one can conclude that under uniform broad band excitation the light output of Type-II-iso HNCs at 1 cm optical path length would be about two times larger than that of the Lumogen Orange, despite its lower LQE (48% *versus* 95%, see Table 14.1). However, LQEs as high as 80% have been reported in the literature for CdTe/CdSe Type-II HNCs similar to those investigated in the present work,[88] demonstrating that the high outputs reported here for Type-II HNCs can be even better with proper optimization of the sample preparation. In conclusion, Type-II HNCs would be suitable candidates to surpass the 10% LSC efficiency barrier.

14.4 Conclusion and Outlook

In this chapter two examples of spectral conversion for use in solar cells have been discussed: up-conversion for thin film silicon solar cells and down-shifting for luminescent solar concentrators. As thin film silicon has a high bandgap, application of up-converter materials at the back of these cells may enhance the solar cell response by up-converting unabsorbed NIR light. We have reviewed existing work, as up-conversion for solar cells is an emerging field and proof of concept experiments have been reported. In our work, we have demonstrated efficiency improvement for both mono-chromatic as well as broadband light for the lanthanide ions Er and Yb in two different host matrices: $NaYF_4$ and Gd_2O_2S. This field of research is now challenged to improve up-conversion efficiency, which is difficult considering the low excitation densities that are typical for solar illumination. Also, the design of the solar cell/up-converter system must be optimized for maximum efficiency. The use of plasmonic resonance effects may be especially promising to increase the up-conversion efficiency, or the use of organic molecules showing triplet–triplet annihilation.[89] We estimate that a combination of up-converters with sensitizers for increased absorption in combination with plasmonic resonators to enhance the excitation strength, may lead to up-conversion efficiencies well above 1% for excitation densities of several mW cm^{-2}.[44]

Down-shifting in luminescent concentrators has been developed some 40 years ago, and has found a revival of interest due to the development of better luminescent species, which has lead to a record efficiency of 7.1% in 2008. Since then, several routes have been pursued as separate endeavours while a concerted effort has been missing.[31] Nevertheless, new luminescent species are under development that should absorb in a wide (up to 850–900 nm) range, have a wide enough Stokes' shift to prevent re-absorption, have near-unity luminescent quantum efficiency, and can be easily processed at low cost. In this chapter we have reviewed operating principles as well as critical issues that hamper further efficiency improvement. We have focused on re-absorption, and have shown that a new class of quantum dots, *i.e.*, Type-II hetero-nanocrystals may be good candidates as luminescent species as they are not suffering from re-absorption: a luminescent solar concentrator at >10% efficiency with c-Si solar cells should be within reach.

Alternatively, the originally proposed three-plate stack[17] could be further developed using perhaps a combination of organic dyes, nanocrystals, or rare earth ions,[55] with optimized dedicated thin film or organic solar cells for each spectral region.

Acknowledgements

The authors gratefully acknowledge numerous colleagues at Utrecht University and elsewhere who contributed to the presented work. This work was partly financially supported by the European Commission as part of the Framework 6 integrated project FULLSPECTRUM (contract SES6-CT-2003-502620), AgentschapNL as part of their Netherlands Nieuw Energie Onderzoek (New Energy Research) programme, Netherlands Foundation for Fundamental Research on Matter (FOM) and HyET Solar as part of their Joint Solar Program 2, Netherlands Organization for Scientific Research (NWO), and Utrecht University as part of its Focus and Mass programme.

References

1. M. A. Green, K. Emery, Y. Hishikawa, W. Warta and E. D. Dunlop, *Prog. Photovoltaics: Res. Appl.*, 2014, **22**, 1–9.
2. W. Shockley and H. J. Queisser, *J. Appl. Phys.*, 1961, **32**, 510–519.
3. M. A. Green, *Solar Cells: Operating Principles, Technology and Systems Application*. Englewood Cliffs: Prentice-Hall; 1982.
4. M. Wolf, *Energy Convers.*, 1971, **11**, 63–73.
5. D. C. Law, R. R. King, H. Yoon, M. J. Archer, A. Boca, C. M. Fetzer, S. Mesropian, T. Isshiki, M. Haddad, K. M. Edmondson, D. Bhusari, J. Yen, R. A. Sherif, H. A. Atwater and N. H. Karam, *Sol. Energy Mater. Sol. Cells*, 2010, **94**, 1314–1318.
6. A. Luque and A. Marti, *Phys. Rev. Lett.*, 1997, **78**, 5014–5017.
7. V. I. Klimov, *J. Phys. Chem. B*, 2006, **110**, 16827–16845.
8. A. J. Chatten, K. W. J. Barnham, B. F. Buxton, N. J. Ekins-Daukes and M. A. Malik, *Sol. Energy Mater. Sol. Cells*, 2003, **75**, 363–371.
9. W. G. J. H. M. Van Sark, K. W. J. Barnham, L. H. Slooff, A. J. Chatten, A. Büchtemann, A. Meyer, S. J. McCormack, R. Koole, D. J. Farrell, R. Bose, E. E. Bende, A. R. Burgers, T. Budel, J. Quilitz, M. Kennedy, T. Meyer, D. C. De Mello, A. Meijerink and D. Vanmaekelbergh, *Opt. Express*, 2008, **16**, 21773–21792.
10. T. Trupke, M. A. Green and P. Würfel, *J. Appl. Phys.*, 2002, **92**, 1668–1674.
11. T. Trupke, M. A. Green and P. Würfel, *J. Appl. Phys.*, 2002, **92**, 4117–4122.
12. B. S. Richards, *Sol. Energy Mater. Sol. Cells*, 2006, **90**, 2329–2337.
13. W. G. J. H. M. Van Sark, A. Meijerink, R. E. I. Schropp, J. A. M. Van Roosmalen and E. H. Lysen, *Sol. Energy Mater. Sol. Cells*, 2005, **87**, 395–409.
14. J. A. Briggs, A. C. Atre and J. A. Dionne, *J. Appl. Phys.*, 2013, **113**, 124509.
15. B. M. Van der Ende, L. Aarts and A. Meijerink, *Phys. Chem. Chem. Phys.*, 2009, **11**, 11081–11095.

16. W. G. J. H. M. Van Sark, A. Meijerink and R. E. I. Schropp, in *Nanotechnology for Photovoltaics*, ed. L. Tsakalakos, Taylor & Francis, Boca Raton, 2010, pp. 351–390.
17. A. Goetzberger and W. Greubel, *Appl. Phys.*, 1977, **14**, 123–139.
18. W. H. Weber and J. Lambe, *Appl. Opt.*, 1976, **15**, 2299–2300.
19. R. T. Wegh, H. Donker, K. D. Oskam and A. Meijerink, *Science*, 1999, **283**, 663–666.
20. A. Meijerink, R. Wegh, P. Vergeer and T. Vlugt, *Opt. Mater.*, 2006, **28**, 575–581.
21. American Society for Testing and Materials (ASTM), Standard Tables for Reference Solar Spectral Irradiances: Direct Normal and Hemispherical on 37_ Tilted Surface, Standard G173-03(2008). ASTM, West Conshohocken, 2008.
22. T. Minemoto, M. Toda, S. Nagae, M. Gotoh, A. Nakajima, K. Yamamoto, H. Takakura and Y. Hamakawa, *Sol. Energy Mater. Sol. Cells*, 2007, **91**, 120–122.
23. W. G. J. H. M. Van Sark, *Thin Solid Films*, 2008, **516**, 6808–6812.
24. P. Gibart, F. Auzel, J. Guillaume and K. Zahraman, *Jpn. J. Appl. Phys.*, 1996, **35**, 4401–4402.
25. A. Shalav, B. S. Richards, T. Trupke, K. W. Krämer and H. U. Güdel, *Appl. Phys. Lett.*, 2005, **86**, 013505.
26. J. De Wild, J. K. Rath, A. Meijerink, W. G. J. H. M. Van Sark and R. E. I. Schropp, *Sol. Energy Mater. Sol. Cell*, 2010, **94**, 2395–2398.
27. J. De Wild, T. F. Duindam, J. K. Rath, A. Meijerink, W. G. J. H. M. Van Sark and R. E. I. Schropp, *IEEE J. Photovoltaics*, 2013, **3**, 17–21.
28. G. Shan and G. P. Demopoulos, *Adv. Mater.*, 2010, **22**, 4373–4377.
29. W. Zou, C. Visser, J. A. Maduro, M. S. Pshenichnikov and J. C. Hummelen, *Nat. Photon*, 2012, **6**, 560.
30. A. Ivaturi, S. K. W. MacDougall, R. Martin-Rodrìguez, M. Quintanilla, J. Marques-Hueso, K. W. Krämer, A. Meijerink and B. S. Richards, *J. Appl. Phys.*, 2013, **114**, 013505.
31. M. G. Debije and P. P. C. Verbunt, *Adv. Energy Mater.*, 2012, **2**(12–35).
32. W. G. J. H. M. Van Sark, *Renew. Energy*, 2013, **49**, 207–210.
33. L. H. Slooff, E. E. Bende, A. R. Burgers, T. Budel, M. Pravettoni, R. P. Kenny, E. D. Dunlop and A. Büchtemann, *Physica Status Solidi (RRL)*, 2008, **2**, 257–259.
34. F. Auzel, *Chem. Rev.*, 2004, **104**, 139–173.
35. M. Pollnau, D. R. Gamelin, S. R. Lüthi, H. U. Güdel and M. P. Hehlen, *Phys. Rev. B*, 2000, **61**, 3337–3346.
36. J. F. Suyver, A. Aebischer, S. García-Revilla, P. Gerner and H. U. Güdel, *Phys. Rev. B*, 2005, **71**, 125123.
37. M. P. Hehlen, M. L. F. Philips, N. J. Cockroft and H. U. Güdel, *Encyclopedia of Materials: Science of Technology*, Pergamon, New York, 2001, vol. 10, p. 9458.
38. R. H. Page, K. I. Schaffers, P. A. Waide, J. B. Tassano, S. A. Payne, W. F. Krupke and W. K. Bischel, *J. Opt. Soc. Am. B*, 1998, **15**, 996.

39. J. F. Suyver, J. Grimm, K. W. Krämer and H. U. Güdel, *J. Lumin.*, 2005, **114**, 53.

40. H. Knauss, E. L. Salabas, M. Fecioru, H. D. Goldbach, J. Hötzel, S. Krull, O. Kluth, R. Kravets, P. A. Losio, J. Reinhardt, J. Sutterlüti and T. Eisenhammer, Proceedings of the 25th European Photovoltaic Solar Energy Conference, 2010, pp. 3064–3067.

41. Voss Electronic GmbH., Manual solar simulator Model: WXS-140-SUPER. Munchen, 2011.

42. J. De Wild, A. Meijerink, J. K. Rath, W. G. J. H. M. Van Sark and R. E. I. Schropp, Proceedings of the 25th European Photovoltaic Solar Energy Conference, 2010, pp. 255–259.

43. J. C. Goldschmidt, S. Fischer, P. Löper, K. W. Krämer, D. Biner and M. Hermle, *Sol. Energy Mater. Sol. Cells*, 2011, **95**, 1960–1963.

44. J. De Wild, A. Meijerink, J. K. Rath, W. G. J. H. M. Van Sark and R. E. I. Schropp, *Energy Environ. Sci.*, 2011, **4**, 4835–4848.

45. G. Smestad, H. Ries, R. Winston and E. Yablonovitch, *Sol. Energy Mater.*, 1990, **21**, 99–111.

46. E. Yablonovitch, *J. Opt. Soc. Am.*, 1980, **70**, 1362–1363.

47. R. L. Garwin, *Rev. Sci. Instrum.*, 1960, **31**, 1010–1011.

48. A. Goetzberger, *Fluorescent* Solar Energy Concentrators: Principle and Present State of Development, in *High-Efficient Low-Cost Photovoltaics – Recent Developments*, ed. V. Petrova-Koch, R. Hezel and A. Goetzberger, Springer, Heidelberg, 2008, pp. 159–176.

49. V. Wittwer, W. Stahl and A. Goetzberger, *Sol. Energy Mater.*, 1984, **11**, 187–197.

50. A. Zastrow, The physics and applications of fluorescent concentrators: A review. Proceedings of SPIE 2255 (1994) pp. 534–547.

51. J. C. Goldschmidt, M. Peters, A. Bösch, H. Helmers, F. Dimroth, S. W. Glunz and G. Willeke, *Sol. Energy Mater. Sol. Cells*, 2009, **93**, 176–182.

52. J. C. Goldschmidt, Chapter 1.27 in *Photovoltaic Solar Energy Technology*, ed. W. G. J. H. M. Van Sark; vol. 1 *Comprehensive Renewable Energy*, ed. A. Sayigh, Elsevier, Oxford, 2012, pp. 587–601.

53. L. Desmet, A. J. M. Ras, D. K. G. De Boer and M. G. Debije, *Opt. Lett.*, 2012, **37**, 3087–3089.

54. D. J. Farrell and M. Yoshida, *Prog. Photovoltaics*, 2012, **20**, 93–99.

55. B. C. Rowan, L. R. Wilson and B. S. Richards, *IEEE J. Quantum Electron.*, 2008, **14**, 1312–1322.

56. R. Reisfeld, *Opt. Mater.*, 2010, **32**, 850–856.

57. K. W. J. Barnham, J. L. Marques, J. Hassard and P. O'Brien, *Appl. Phys. Lett.*, 2000, **76**, 1197–1199.

58. A. J. Chatten, K. W. J. Barnham, B. F. Buxton, N. J. Ekins-Daukes and M. A. Malik, *Sol. Energy Mater. Sol. Cells*, 2003, **75**, 363–371.

59. A. J. Chatten, K. W. J. Barnham, B. F. Buxton, N. J. Ekins-Daukes and M. A. Malik, The Quantum Dot Concentrator: Theory and Results. In Proceedings of Third World Congress on Photovoltaic Energy

Conversion (WPEC-3) (ed. K. Kurokawa, L. Kazmerski, B. McNelis, M. Yamaguchi, C. Wronski and W. C. Sinke), 2003, pp. 2657–2660.

60. S. J. Gallagher, B. C. Rowan, J. Doran and B. Norton, *Sol. Energy*, 2007, **81**, 540.
61. A. P. Alivisatos, *J. Phys. Chem.*, 1996, **100**, 13226–13239.
62. S. V. Gaponenko, *Optical Properties of Semiconductor Nanocrystals*. Cambridge University Press, Cambridge, 1998.
63. M. Bruchez Jr, M. Moronne, P. Gin, S. Weiss and A. P. Alivisatos, *Science*, 1998, **281**, 2013–2016.
64. C. De Mello Donegá, *Chem. Soc. Rev.*, 2011, **40**, 1512–1546.
65. X. Peng, M. C. Schlamp, A. V. Kadavanich and A. P. Alivisatos, *J. Am. Chem. Soc.*, 1997, **119**, 7019–7029.
66. J. Bomm, A. Büchtemann, A. J. Chatten, R. Bose, D. J. Farrell, N. L. A. Chan, Y. Xiao, L. H. Slooff, T. Meyer, A Meyer, W. G. J. H. M. Van Sark and R. Koole, *Sol. Energy Mater. Sol. Cells*, 2011, **95**, 2087–2094.
67. R. Koole, P. Liljeroth, C. De Mello Donegá, D. Vanmaekelbergh and A. Meijerink, *J. Am. Chem. Soc.*, 2006, **128**, 10436–10441.
68. R. Koole, M. Van Schooneveld, J. Hilhorst, C. De Mello Donegá, D. C. 't Hart, A. Van Blaaderen, D. Vanmaekelbergh and A. Meijerink, *Chem. Mater.*, 2008, **20**, 2503–2512.
69. J. Lee, V. C. Sundar, J. R. Heine, M. G. Bawendi and K. F. Jensen, *Adv. Mater.*, 2000, **12**, 1102–1105.
70. J. Bomm, A. Büchtemann, A. Fiore, L. Manna, J. H. Nelson and W. G. J. H. M. Van Sark, *Beilstein J. Nanotechnol.*, 2010, **1**, 94–100.
71. A. Schüler, M. Python, M. Valle del Olmo and E. De Chambrier, *Sol. Energy*, 2007, **81**, 1159–1165.
72. S. M. Reda, *Acta Material.*, 2008, **56**, 259–264.
73. W. G. J. H. M. Van Sark, P. L. T. M. Frederix, A. A. Bol, H. C. Gerritsen and A. Meijerink, *ChemPhysChem*, 2002, **3**, 871–879.
74. S. J. Gallagher, B. C. Rowan, J. Doran and B. Norton, *Sol. Energy*, 2007, **81**, 540–547.
75. M. G. Hyldahl, S. T. Bailey and B. P. Wittmershaus, *Sol. Energy*, 2009, **83**, 566–573.
76. G. Blasse and B. C. Grabmaier, *Luminescent Materials*. Springer, Berlin, 1995.
77. J. A. Levitt and W. H. Weber, *Appl. Opt.*, 1977, **16**, 2684–2689.
78. R. Reisfeld and Y. Kalisky, *Chem. Phys. Lett.*, 1981, **80**, 178–183.
79. V. V. Popov and V. N. Yakimenko, *J. Appl. Spectrosc.*, 1995, **61**, 573–579.
80. O. Moudam, B. C. Rowan, M. Alamiry, P. Richardson, B. S. Richards, A. C. Jones and N. Robertson, *Chem. Commun.*, 2009, **43**, 6649–6651.
81. G. Katsagounos, E. Stathatos, N. B. Arabatzis, A. D. Keramidas and P. Lianos, *J. Lumin.*, 2011, **131**, 1776–1781.
82. X. Wang, T. Wang, X. Tian, L. Wang, W. Wu, Y. Luo and Q. Zhang, *Sol. Energy*, 2011, **85**, 2179–2184.
83. T. Wang, J. Zhang, W. Ma, Y. Luo, L. Wang, Z. Hu, W. Wu, X. Wang, G. Zou and Q. Zhang, *Sol. Energy*, 2011, **85**, 2571–2579.

84. J. W. E. Wiegman and E. Van der Kolk, *Sol. Energy Mater. Sol. Cells*, 2011, **103**, 41–47.

85. D. K. G. De Boer, D. J. Broer, M. G. Debije, W. Keur, A. Meijerink, C. R. Ronda and P. P. C. Verbunt, *Opt. Express*, 2012, **20**, A395–A405.

86. Z. Krumer, S. J. Pera, R. J. A. Van Dijk-Moes, Y. Zhao, A. F. P. De Brouwer, E. Groeneveld, W. G. J. H. M. Van Sark, R. E. I. Schropp and C. De Mello-Donegá, *Sol. Energy Mater. Sol. Cells*, 2013, **111**, 57–65.

87. R. Kubin and A. Fletcher, *J. Lumin.*, 1983, **27**, 455–462.

88. C. De Mello Donegá, *Phys. Rev. B*, 2010, **81**, 165303.

89. T. F. Schulze, J. Czolk, Y-Y. Cheng, B. Fückel, R. W. MacQueen, T. Khoury, M. J. Crossley, B. Stannowski, K. Lips, U. Lemmer, A. Colsmann and T. W. Schmidt, *J. Phys. Chem. C*, 2012, **116**, 22794–22801.

CHAPTER 15

Triplet–triplet Annihilation Up-conversion

TIMOTHY W. SCHMIDT*[a] AND MURAD J. Y. TAYEBJEE[b]

[a] School of Chemistry, UNSW, Sydney 2052, Australia; [b] School of Photovoltaic and Renewable Energy Engineering, UNSW, Sydney 2052, Australia
*Email: timothy.schmidt@unsw.edu.au

15.1   Introduction

Most commercially available photovoltaic (PV) devices are single threshold solar cells. Crystalline silicon,[1] CdTe,[2] organic photovoltaics,[3] dye-sensitized solar cells[4] and the emerging perovskite solar cells[5] all exploit a single absorption threshold to bring about their photovoltage.

Consequently, all of these devices suffer from several inherent losses that arise from either inefficient use of the solar spectrum or the unavoidable generation of entropy, and are attributed to several macroscopic mechanisms.[6] The most important for the purposes of this chapter are:

- Photons of energy less than the threshold are transmitted
- Energy in excess of the threshold is lost as heat
- Radiative losses

One strategy to address the first of the items listed above is up-conversion.[7] Up-conversion is the general process whereby low energy light is converted to higher energy light by a process of conjoining the energy of several photons, or otherwise exploiting entropy generation.[8] By harvesting

RSC Energy and Environment Series No. 11
Advanced Concepts in Photovoltaics
Edited by Arthur J Nozik, Gavin Conibeer and Matthew C Beard
© The Royal Society of Chemistry 2014
Published by the Royal Society of Chemistry, www.rsc.org

sub-threshold light, and re-radiating this at a higher energy, above the threshold (bandgap), the energy conversion efficiency of a solar cell is increased. Up-conversion using rare earth ions has been researched for many years, and there have been several reports of its application to photovoltaics.[9–13] In this chapter, we review developments in the field of triplet–triplet annihilation up-conversion (TTA-UC).[14–18]

To gauge the potential effectiveness of this strategy, we need to first consider the single-threshold cell and revisit the Shockley–Queisser limit.[19] We then consider the limiting efficiency of an up-converting solar cell with a single lower threshold, a situation to which TTA-UC corresponds. Following, we describe the physical detail of TTA-UC, and experiments which gauge its effectiveness. The chapter will be concluded with a perspective on the future directions of research in this field.

15.2   The Limiting Efficiency of a Single Threshold Solar Cell

In 1961, Shockley and Queisser (SQ) published a detailed balance study on the efficiency limit of single threshold devices that has retained its utility to this day.[19] Their underlying assumptions were that:

- All recombination is radiative
- Every photon with energy greater than the bandgap is absorbed
- There is infinite carrier mobility
- All absorption occurs at the surface of the cell

This detailed balance approach has been used in several theoretical limiting efficiency studies for photovoltaic devices by altering either the cell architecture, irradiation spectrum, or the level of solar concentration. For instance, single threshold devices under global solar radiation,[20] intermediate band solar cells,[21] and annihilation based up-converting solar cells,[22] have all been modelled.

15.2.1   Photon Ratchet Model

The Shockley–Queisser model can be couched in terms of a photon-driven ratchet[23] (Figure 15.1). The ratchet is attached to a weight such that a single turn of the ratchet does work E_1, against both the weight and the pawl. So that the work is not undone, an energy sacrifice of Q is made. The stored energy per turn is then $W = E_1 - Q$. The ratchet is taken to be photon-driven, such that it makes a turn upon absorption of photons of energy $h\nu > E_1$. The rate of absorption, k, is given by the integral:

$$k = \sigma \int_{E_1}^{\infty} \mathscr{E}_E^S(E)\mathrm{d}E, \qquad (15.1)$$

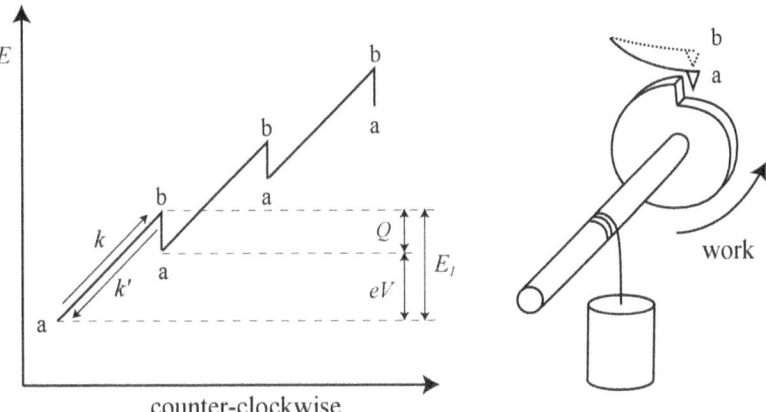

counter-clockwise

Figure 15.1 Left: The energy landscape for the photon-driven ratchet pictured right. One turn of the ratchet does work eV on the weight, and Q on the pawl. The energy Q is dissipated each turn to prevent the ratchet slipping backwards, equivalent to radiative losses in the solar cell considered here.
Adapted with permission from *J. Phys. Chem. Lett.*, 2012, **3**, 2749–2754. Copyright 2012 American Chemical Society.

where $\mathscr{E}_E^S(A)$ is the solar photon spectral density in $s^{-1}\,m^{-2}\,J^{-1}$ as a function of photon energy. The absorption cross-section, σ, is equal to the cross-section of the absorber over the spectral region of interest, $E > E_1$ and is otherwise zero.

To obey the laws of thermodynamics, if the device absorbs light, it must also emit light. The rate of emission, k', can be derived through considerations of thermodynamics:

$$k' = \frac{\sigma}{4\pi^2 c^2 \hbar^3} \int\limits_{E_1}^{\infty} \frac{E^2}{\exp(\beta E) - 1}\, dE, \tag{15.2}$$

where the factor of 1/4 accounts for Lambertian emission.

Now, the voltage of the device is the amount of work done per coulomb of electronic charge. This quantity is then given by the stored work per turn of the ratchet, divided by the charge of the electron, $V = W/e$. The current density, J, is the balance of the rates k and k', multiplied by the charge on the electron, and divided by the area of the absorber. The rate of emission is attenuated by a Boltzmann factor, where $\beta = k_B T$:

$$J = \frac{e}{\sigma}(k - k' \exp(-\beta Q)). \tag{15.3}$$

The maximum rate of work (power) is calculated by maximizing the quantity $P = VJ$ as a function of the bandgap, E_1, and operating voltage, V. The limiting efficiency results that Shockley and Queisser,[19] Hanna and Nozik,[24] and Brown and Green[20] calculated under various spectra are all reproduced using the above equations. Single threshold solar cell energy

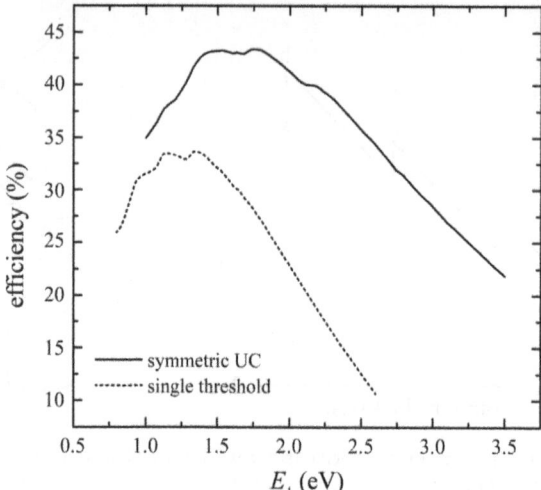

Figure 15.2   The energy conversion efficiency limits for single threshold solar cells and those augmented with symmetric up-conversion as a function of the absorption threshold. The single threshold cell peaks at an efficiency of 33.7% with a bandgap of 1.34 eV. The up-converting cell has an efficiency limit of 43.4% with an upper threshold of 1.76 eV.

conversion efficiency limits are 31.1% under the 6000 K black-body spectrum (1 kW m^{-2}), and 33.7% under AM1.5G illumination.[25] The variation of energy conversion efficiency with bandgap is given in Figure 15.2.

15.3   Up-conversion

The rate k can be increased by augmenting the number of photons impinging on the solar cell with energy above the bandgap. One way to achieve this is to absorb photons down to about half the bandgap energy, and use this power to create photons at the bandgap. The chemical procedure by which this is achieved will be detailed below. For the time being, we will continue to discuss limiting efficiencies, and as such, introduce the equivalent circuit for an up-conversion-enhanced single threshold solar cell. This cell transmits light of energy $h\nu < E_1$ which is then harvested by a pair of lower threshold cells, held in series. The voltages of these identical cells add to drive a light-emitting diode with a bandgap matching that of the top, power-producing cell. The architecture is illustrated in Figure 15.3.

Both of the lower threshold cells may be analyzed as ratchets, with an input solar spectrum attenuated by absorption down to photon energy E_1 by the top, power producing cell. These lower threshold cells, in series, must each have half the surface area of the top cell, and thus their absorption rate is given by:

$$k_2 = \frac{\sigma}{2} \int_{E_2}^{E_1} \mathscr{E}_E^S(E)\, \mathrm{d}E. \qquad (15.4)$$

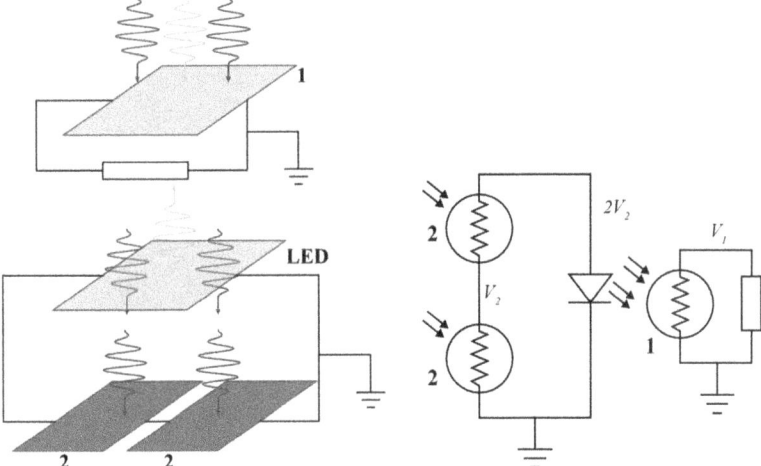

Figure 15.3 The architecture for an ideal up-converting solar cell. The top, power-producing cell has an energy threshold of E_1, and is bifacial. It transmits light of $h\nu < E_1$ through an LED of the same threshold and onto two lower threshold cells of half the area. Photons of $h\nu > E_2$ are absorbed by these cells, whose voltages add to power the LED. The LED radiates at energy E_1 into the top cell. The circuit diagram is given at right.

The rate constant of emission by these cells is:

$$k_2' = \frac{\sigma}{8\pi^2 c^2 \hbar^3} \int_{E_2}^{\infty} \frac{E^2}{\exp(\beta E) - 1} \, \mathrm{d}E \tag{15.5}$$

The light-emitting diode receives a small amount of flux from the top, power producing cell, which is a bifacial version of that used to calculate the Shockley–Queisser limit above, since it must accept radiation from both sides. This cell is governed by the equation:

$$J_1 = \frac{e}{\sigma}(k_{UC} + k_1 - k_1' \exp(-\beta Q_1)), \tag{15.6}$$

where:

$$k_1 = \sigma \int_{E_1}^{\infty} \mathscr{E}_E^S(E) \, \mathrm{d}E, \tag{15.7}$$

and:

$$k_1' = \frac{\sigma}{2\pi^2 c^2 \hbar^3} \int_{E_1}^{\infty} \frac{E^2}{\exp(\beta E) - 1} \, \mathrm{d}E \tag{15.8}$$

The rate of radiation here is twice that of the cell used above, due to its bifacial nature.

Now, k_{UC} is the balance of particle fluxes in the up-convertor:

$$k_{UC} = k_2 + k_1' \exp(-\beta Q_1)/2 - k_2' \exp(-\beta Q_2)$$
$$= k_1' \exp(-\beta(E_1 - 2eV_2))/2 \tag{15.9}$$

The quantities Q_1 and Q_2 are the differences between the voltages and the absorption thresholds of the upper and lower photon absorbing cells:

$$Q_1 = E_1 - eV_1 \tag{15.10}$$

$$Q_2 = E_2 - eV_2 \tag{15.11}$$

To optimize the power output for given upper and lower absorption thresholds, E_1 and E_2, the voltage of the top cell is taken as the independent variable. This sets the rate of emission into the bottom cell. The voltage of the up-convertor, $2V_2$, is solved from Equation (15.9), which provides k_{UC}. The current of the top cell, J_1, is then determined, from which the power output, and the energy conversion efficiency may be determined.

The limiting efficiency of a solar energy conversion device with the architecture as put forward above is given in Figure 15.2. As compared to a single threshold solar cell, with a limit of $\eta = 33.7\%$ the symmetric up-convertor peaks at 43.4%, with upper and lower thresholds optimized to $E_1 = 1.76$ and $E_2 = 1.11$ eV . From Figure 15.4 it is clear that E_2 must exceed

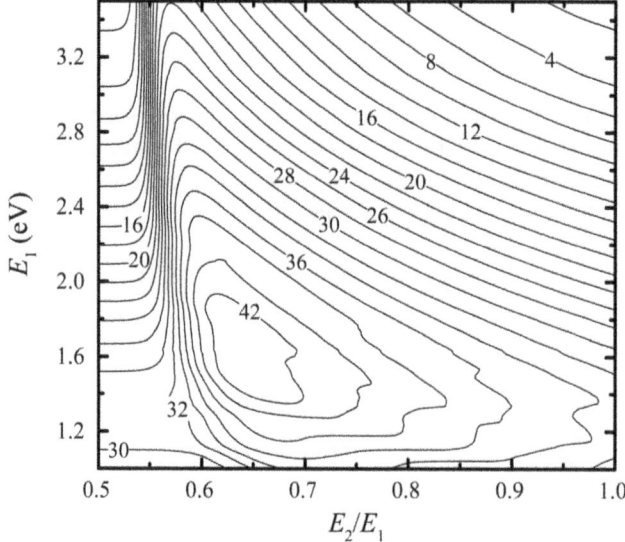

Figure 15.4 A contour plot of the maximum (%) efficiency of an upconverting solar cell with lower threshold E_2 and upper threshold E_1. The maximum is 43.4%, located at $E_1 = 1.76$ and $E_2 = 1.11$ eV. The left hand edge represents ineffectual up-conversion, and the right hand edge is equivalent to the Shockley–Queisser limit.

$0.5E_1$. In order to drive an LED with threshold E_1, the voltage must exceed $(E_1 - Q)/e$. However, each cell with threshold E_2 makes the sacrifice of Q, and thus if $2(E_2 - Q) \sim E_1 - Q$, then $E_2 \sim (E_1 + Q)/2$.

Where E_2 is too low, the peak efficiency is in the vicinity $E_1 \approx 1.3$ eV with $\eta \sim 31\%$, which is similar to the Shockley–Queisser limit. The full 33.7% is not realized since the cell is bifacial, and thus radiates faster than the ideal single threshold device. The maximum efficiency is found at about $E_2 = 0.65E_1$, and then the efficiency degrades back to that of an optimal single threshold device where $E_2 > E_1$.

15.3.1 Summary

Theoretical considerations show that solar cells with bandgaps of 1.5–2.0 eV are ideal for coupling to an up-convertor, while lower bandgap cells, such as c-Si, have relatively little to gain. In the following section, we will review up-conversion by triplet–triplet annihilation in organic molecules.

15.4 Triplet–triplet Annihilation

Up-conversion by triplet–triplet annihilation (TTA-UC) in organic molecules has a long history. In 1962, Parker and Hatchard published a report entitled 'Sensitized Anti-Stokes Delayed Fluorescence'.[26] In the following year, Parker proposed the mechanism accepted today. In this process, illustrated in Figure 15.5, sensitizer molecules absorb photons to undergo transitions from S_0 to S_1. Subsequently, intersystem crossing (ISC) occurs, and the

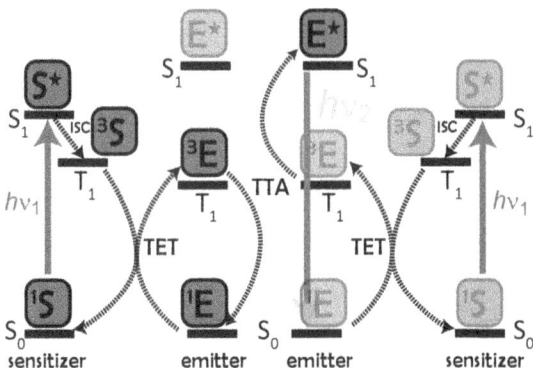

Figure 15.5 A schematic of triplet–triplet annihilation up-conversion. Ground state sensitizer molecules (^1S) absorb low energy photons ($h\nu_1$) to become excited to the S_1 state, (S^*). These undergo intersystem crossing (ISC) to T_1 (^3S). Triplet energy transfer to emitter molecules (TET) takes place, returning the sensitizer to S_0 (^1S), while exciting the emitter from the S_0 state (^1E) to the T_1 triplet state (^3E). Triplet emitter molecules annihilate (TTA), to bring about one in the S_1 state (E^*) which fluoresces at a higher energy than absorbed ($h\nu_2 > h\nu_1$).

molecules transition to the T_1 state.[27] The energy is then transferred to another species, termed emitters, by triplet energy transfer (TET). Upon two triplet emitters meeting, they may undergo triplet–triplet annihilation to a state where one emitter is placed in the fluorescent S_1 state, while the other is quenched to S_0. The ensuing fluorescence is at a shorter wavelength than the originally absorbed photons, and thus up-conversion has occurred.

15.4.1 Typical TTA Up-conversion Combinations

The sensitizer should be selected so that it has efficient intersystem crossing to the triplet state. This is usually achieved with the heavy-atom effect. The triplet should not be too much lower in energy than the singlet. Optimally, the gap would be several times k_BT. A further consideration is that the sensitizer should not absorb the finally up-converted light. Porphyrins containing a heavy metal atom, or other metals complexed with heterocyclic ligands such as Ru(bipy)$_3$, are typically used as sensitizers. Platinum acetylides of polycylic aromatic hydrocarbon chromophores have also been studied.

The emitter should be chosen so that it possesses a triplet state a little over half the energy of the first excited singlet. The triplet should be long lived, and thus the intersystem crossing rate should be low. The excited singlet should emit light with near unity quantum yield, which requires that the second triplet state, T_2, be higher in energy than S_1, and preferably $>2E_{T1}$. The emitter triplet should ideally be below or at least similar in energy to that of the sensitizer, though endothermic TET is possible.[28]

The most commonly studied emitter is diphenylanthracene. However, since it emits in the blue, this is not useful for photovoltaic applications. All of the papers which have applied TTA-UC to photovoltaics have used rubrene as the emitter (see, for instance Figure 15.6).[14,16,18]

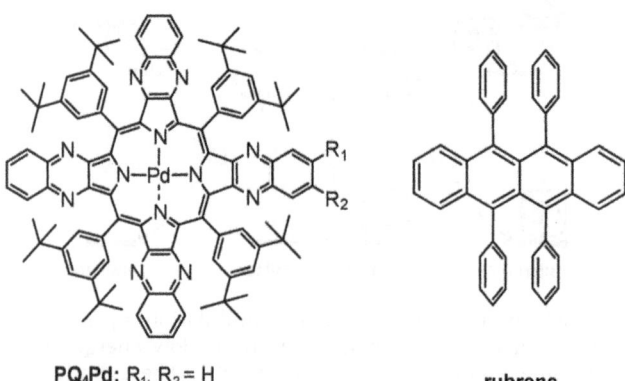

PQ$_4$Pd: R$_1$, R$_2$ = H   **rubrene**
PQ$_4$PdNA: R$_1$ = NO$_2$, R$_2$ = NH$_2$

Figure 15.6   Molecules used in TTA-UC-PV.[14,16,18] The palladated porphyrins serve as sensitizers, with rubrene as the emitter.

15.4.2 Efficiency Considerations

Interest in TTA-UC has increased in recent years due to potential application in photovoltaics.[29–31] TTA-UC has been demonstrated to occur under the incoherent excitation of sunlight,[29] but there remained for some time a question over its eventual efficiency. From a naïve spin-statistical argument, one can make the case that only one-in-nine collisions between two triplets would result in a singlet. This is because there are nine spin-states that result when two triplets meet. A single triplet state has one unit of angular momentum, $1\hbar$, which may be oriented in the laboratory frame three ways, $m_s \pm 1, 0$, hence 'triplet'. The maximum angular momentum achievable with two triplets is thus $2\hbar$, which may be oriented five ways, $m_s = \pm 2, \pm 1, 0$, and is termed a quintet state. However, the angular momenta of two triplets may not be parallel, and thus they can also add to $1\hbar$ (a bimolecular triplet), or cancel altogether to bring about a singlet state, $m_s = 0$. Thus, the nine possible outcomes from the collision of two triplets comprise the five components of the quintet, three of the triplet, and one singlet. The singlet state alone is capable of populating the fluorescent excited state of the emitter. But, if it were the case that only 1/9 (11%) of collisions resulted in a singlet, this alone would not reduce the potential quantum efficiency. Only if some of the other eight collisions (quintet and triplet spin-states) resulted in non-fluorescent products, then the quantum efficiency would be diminished,[32] as energy is wasted.

Less naïvely, one might consider that usually molecular quintet states lie at quite high energies, and would not be accessible, even with the energy of two molecular triplets. As such, the bimolecular quintet is unable to populate a molecular quintet state. Further, annihilation to a high-lying triplet state, T_n, would return one triplet to the reaction mixture, due to rapid internal conversion to T_1. So, if three triplet collisions burn one molecular triplet each, with no resultant photon, and the singlet channel burns two triplets for the production of one photon, then one photon is produced per five triplets lost, which is 40% of the maximum quantum efficiency, rather than 11%.[33] Note that the actual quantum efficiency is half this number.

The argument was settled by a series of kinetic experiments performed by Cheng *et al.*[32,33] These authors showed that instantaneously, after laser pulse excitation, >40% of triplets could be observed contributing to up-conversion. Further, by considering the fluorescence yields due to the those triplets undergoing second order decay, it was found that >60% of triplets could contribute to TTA-UC for the chosen system, likely due to a thermally activated triplet channel.

With no *a priori* reason to reject TTA-UC for application in photovoltaics, one should consider how to render it efficient under 1-sun illumination. In solvents with reasonable viscosities, such as toluene, second order chemical reactions can proceed with rates approaching 10^{10} M^{-1} s^{-1}. However, using the more conservative and quite reasonable rate of 10^9 M^{-1} s^{-1} as a guide, one can calculate the quenching rate of sensitizer triplets as $k_{TET}[E]$. For an emitter concentration of $[E] = 10^{-2}$ M, this amounts to a quenching rate of 10^7 s^{-1}, and thus a quenching lifetime of 100 ns. Thus, a sensitizer triplet lifetime of $\gg 100$ ns will result, in principle, in near unity transfer of triplets from sensitizer to emitter.

Considering rapid and efficient triplet energy transfer, the emitter triplet state kinetics can then be described by the equation:

$$\frac{d[^3E]}{dt} = k_\phi[^1S] - k_1[^3E] - k_2[^3E]^2 \tag{15.12}$$

where $[^1S]$ and $[^3E]$ are the concentrations of ground state sensitizer and excited state emitter (Figure 15.5). The rate constants k_ϕ, k_1 and k_2 in this case are the excitation rate of sensitizer, due to photon flux, and the first and second order decay rates of the emitter triplets.

The excitation rate of the sensitizer is calculated by an integral of the product of photon flux, $\mathscr{E}_E(E)$, and absorber cross-section, $\sigma_S(E)$, over the photon energies:

$$k_\phi = \int_0^\infty \sigma_S(E)\mathscr{E}_E(E)\,dE \tag{15.13}$$

For typical organic dyes, this quantity is of the order 1 s^{-1} under solar illumination.

The values of k_1 and k_2 are typically about 10^4 s^{-1} and 10^9 M^{-1} s^{-1} respectively. Thus, under steady-state conditions, we find that:

$$\frac{d[^3E]}{dt} = k_\phi[^1S] - k_1[^3E] - k_2[^3E]^2 = 0, \tag{15.14}$$

and thus:

$$[^3E] = \frac{-k_1 + \sqrt{k_1^2 + 4k_2k_\phi[^1S]}}{2k_2} \tag{15.15}$$

Where the emitter triplet concentration is low, the first order decay is dominant, and the concentration of emitter triplets is given by:

$$[^3E] \approx \frac{k_\phi[^1S]}{k_1} \tag{15.16}$$

As such, it is clear that increasing the excitation rate, and the concentration of sensitizer will bring about higher triplet concentrations, and thus a greater proportion of triplets undergoing second order decay (f_2), which can bring about up-converted photons:

$$f_2 = \frac{k_2[^3E]}{k_2[^3E] + k_1} \tag{15.17}$$

Where TTA-UC is inefficient $(k_1 \gg k_2[^3E])$, we arrive at the following expression using Equation (15.16):

$$f_2 \approx \frac{k_2 k_\phi [^1S]}{k_1^2}.$$ (15.18)

For typical values, in solution, under 1-sun illumination:

$$f_2 \approx \frac{10^9 \cdot 1 \cdot 10^{-4}}{10^8} = 0.001.$$ (15.19)

The grand challenge of TTA-UC is to lift this quantity by several orders of magnitude. This should be achieved by increasing k_ϕ, concentrating the sensitizer, and finding intrinsically long-lived (low k_1) yet fast reacting emitter triplets (high k_2).

15.5 Application to Photovoltaics

Despite the typical f_2 values being orders of magnitude away from ideal, the field of TTA-UC has been progressed by application to various solar cells which suffer from transmission of long-wavelength light. In 2012, our group published the first report of TTA-UC-PV, where an hydrogenated amorphous silicon solar cell was optically contacted to a cuvette holding a solution of up-converter (Figure 15.7). Though far from optimal, a measurement was made which laid down the rules for characterizing TTA-UC-PV. Of interest, eventually, is the increase in the solar cell short circuit current under one-sun

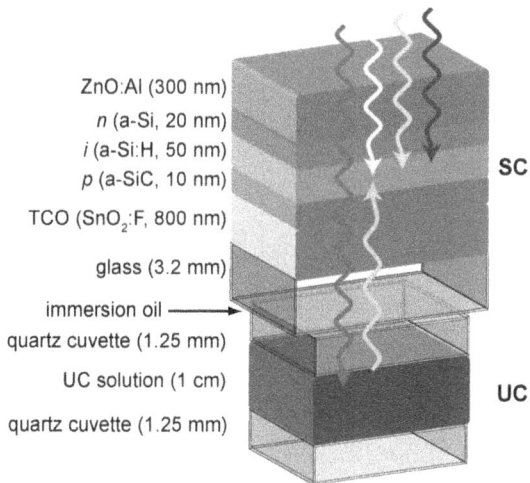

ZnO:Al (300 nm)
n (a-Si, 20 nm)
i (a-Si:H, 50 nm)
p (a-SiC, 10 nm)
TCO (SnO$_2$:F, 800 nm)
glass (3.2 mm)
immersion oil
quartz cuvette (1.25 mm)
UC solution (1 cm)
quartz cuvette (1.25 mm)

SC

UC

Figure 15.7 The architecture of the first application of TTA-UC-PV in 2012.[14] The solar cell (SC) is a p–i–n structure sandwiched between transparent conductive oxides, mounted on glass. The cell is optically contacted, using immersion oil, to a quartz cuvette containing the up-conversion (UC) mixture.

illumination, ΔJ_{SC}. This is calculated by integrating the change in the EQE curve, due to up-conversion, over the incoming photon flux:

$$\Delta J_{SC} = \frac{e}{C^2} \int_0^\infty \Delta EQE(E, C) \mathcal{E}_E^S \, dE \tag{15.20}$$

Here the integral is divided by the solar concentration factor, under which the measurement was performed, squared. This is because inefficient TTA-UC is a quadratic process. As such, this Figure-of-Merit is sometimes quoted in units of mA cm^{-2} sun^{-2}. The solar concentration factor is calculated based on the excitation rate compared to that expected under 1 sun. That is:

$$C = \frac{k_\phi}{k_\phi^\odot} \tag{15.21}$$

where:

$$k_\phi^\odot = \int_0^\infty \sigma_S(E) T(E) \mathcal{E}_E^S \, dE \tag{15.22}$$

where $T(E)$ is the transmission of the solar cell, assuming the up-convertor is at the rear.

15.6 Measurement

The EQE curve of a solar cell is the result of a linear measurement. The increase in current upon increasing the flux at a given wavelength is measured by application of a chopped probe beam, and using lock-in detection of photocurrent (*cf.* Figure 15.8). Since TTA-UC is a non-linear process at low

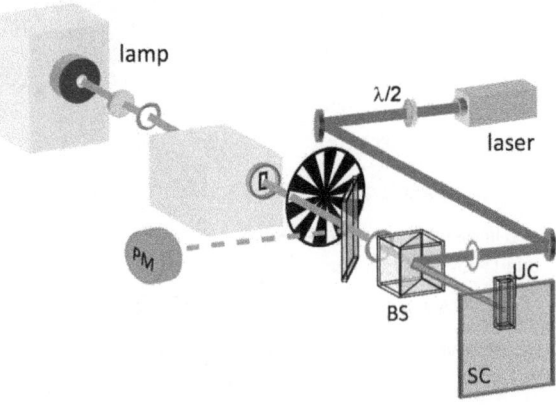

Figure 15.8 The experimental set-up used to characterize TTA-UC-PV. The linear current response of a solar cell (SC) with an optically contacted up-convertor (UC) to a chopped, monochromated incoherent light beam from a lamp, is determined using lock-in current amplification. The up-convertor and solar cell are placed under the conditions of constant illumination by a diode laser, which serves to fill traps in the cell while charging the up-conversion medium with triplets.

flux, this measurement should be made while otherwise holding the cell under a constant state of illumination (bias). This serves the purpose of trap-filling in the case of the solar cell, and charging the up-conversion medium with a background concentration of triplets so that the EQE measurement can be made. The measurement is made with the up-convertor active and inactive, and the ratio between these curves is then plotted and compared to a model for the expected behaviour.

The results from Reference 18 are displayed in Figure 15.9. For OPV cells (top two panels), the EQE ratio reaches as high as 1.12, indicating that the cell converts 12% more photons into carriers at that wavelength with the up-convertor active. The modeled data is from an equation derived in Reference 16:

$$\frac{\text{EQE}_{\text{UC}}(\lambda)}{\text{EQE}_0(\lambda)} = 1 + \text{const.} \times \frac{T_\text{p}(\lambda)}{\text{EQE}_0(\lambda)} \frac{\sigma_\text{p}(\lambda)\sigma_\text{b}}{\sigma_\text{p}(\lambda) + \sigma_\text{b}} \qquad (15.23)$$

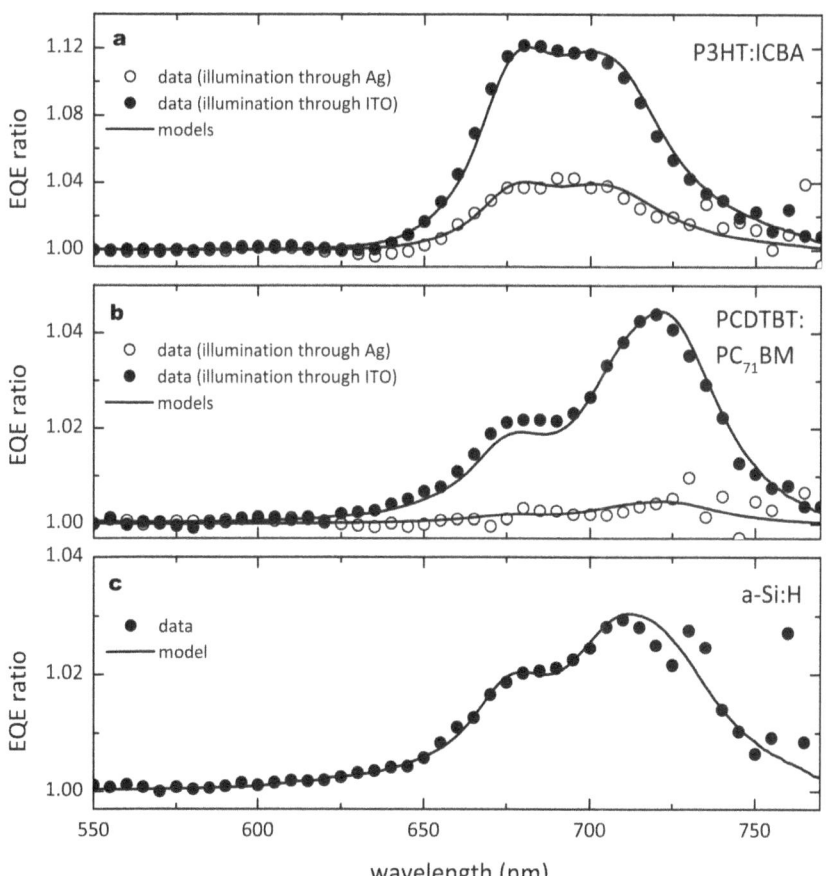

Figure 15.9   Measured EQE ratios from Reference 18. The top two panels are OPV devices, and the bottom is hydrogenated amorphous silicon.
Adapted with permission from *J. Phys. Chem. C*, 2012, **116**, 22794–22801. Copyright 2012 American Chemical Society.

where σ_p and σ_b are the absorption cross-sections of the sensitizer at the pump and bias wavelengths, T_p is the transmission of the solar cell at the probe wavelength, and EQE_{UC} and EQE_0 are the EQEs with and without the up-convertor activated (controlled by alignment of pump and probe beams). The only free parameter is the constant of multiplication, the others being taken from measurement. Inspection of Figure 15.9 reveals a more than satisfactory agreement between the measurement and the model.

15.7   The Figure of Merit

The evolution of the Figure of Merit is shown in Figure 15.10. From the published results on rare-earth up-convertors,[10,11] it was seen that the first attempt by our group to characterize TTA-UC-PV brought about a Figure-of-Merit (FoM) some 200 times better.[14] For both of the samples measured, the FoM was measured in the 10^{-4} mA cm^{-2} sun^{-2} range. This is increased to about 10^{-3} mA cm^{-2} sun^{-2} when an optimally matched solar cell is chosen,[18] and an optical back reflector is used.[16] The effect of the back reflector is to double-pass the incoming light, leading to a near-doubling of the effective value of k_ϕ[16,17] Indeed a Lambertian back-reflector is calculated to enhance the FoM by a factor of 3, since the weakly absorbed wavelengths are reflected into long pathlengths in the transverse direction.

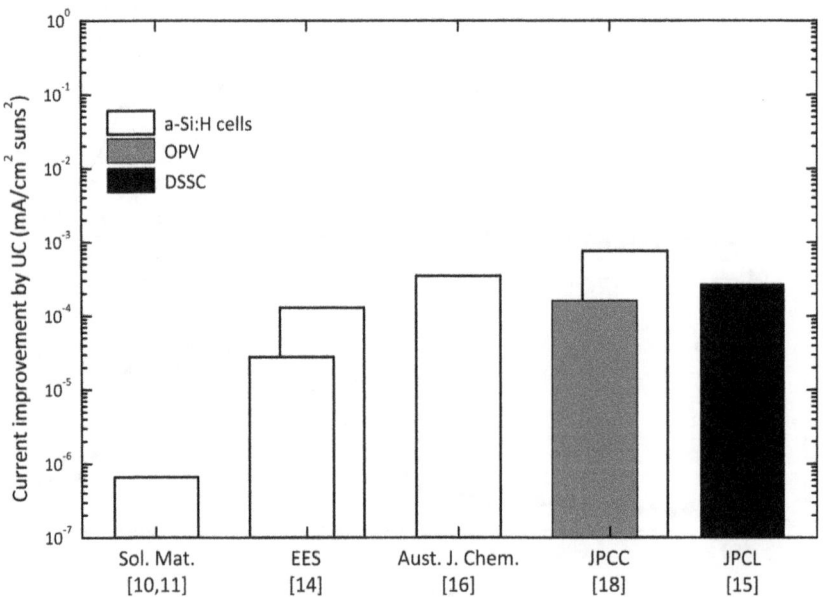

Figure 15.10   The evolution of the Figure of Merit for TTA-UC-PV.

15.8 Prospects

To be device relevant, the up-convertor should provide some increase of >0.1 mA cm^{-2} under 1 sun. It has been shown that an FoM of 10^{-3} mA cm^{-2} sun^{-2} can be achieved with molecules in solution. However, if the quantity $k_\phi^\ominus[^1S]$ can be increased by a factor of 100, then the FoM should follow with a similar scaling. $k_\phi^\ominus$ itself can be increased by locally concentrating the sunlight. This has been demonstrated with a field of spherical mirrors at the rear of the cell, but the improvements are marginal.[17] An exciting prospect is plasmonics, where light can be concentrated into some regions of the up-convertor at the expense of others. This should result in a massive increase in $k_\phi^\ominus$ in the 'hot spots' of the plasmonic concentrator.

The concentration of sensitizer molecules, $[^1S]$, can be increased by incorporation into a support structure. In so doing, the sensitizer molecules will be prevented from coming into contact, which might cause problems in terms of quenching. Dyes can be ligated to nanoparticles at a surface concentration of about 1 nm^{-2}. For close-packed spheres, the pore volume fraction is 0.26. The effective concentration of dye for $R = 10$ nm particles is 0.856 nm^{-3}, since the surface area to volume ratio of a sphere is $3/R$.[34] This corresponds to a concentration exceeding 1 M, which is 1000-fold higher than that employed in the highest FoM measurements.

Eventually, there is enough of an expected enhancement to pursue a combination of sensitizer immobilization on nanoparticle surfaces, and plasmonic structures. There are other problems to solve before TTA-UC becomes a viable PV technology. It is unknown how the performance of TTA-UC degrades with time. While solutions in the lab are stable over years, this is without daily illumination. In the favour of the up-convertors, they will be protected from the harshest of UV and blue photons by virtue of being placed behind the solar cell.

Liquids are a problem, and there have been attempts to replace liquid emitters with conjugated polymers, wherein the exciton can migrate throughout the structure, rather than the chromophore diffusing.[35] However, attempts have thus far failed, in part due to low energy trapping sites hindering exciton diffusion.

One concept to bring about a solid-state up-convertor is illustrated in Figure 15.11. Sensitizer molecules are attached to the surface of insulating silica nanoparticles. The nanoparticles contain embedded, plasmonic gold spheres to enhance the electric fields in their vicinity. The triplets are transferred to a conjugated polymer in which the up-conversion takes place.[35]

Bringing about such a material is a challenging prospect. There are unknowns, such as whether the triplet states will be mobile enough in the highly heterogeneous medium, and what the dominant energy quenching mechanisms are. It remains to be seen whether TTA-UC can be made efficient under one sun illumination. But, if such materials do emerge, they will

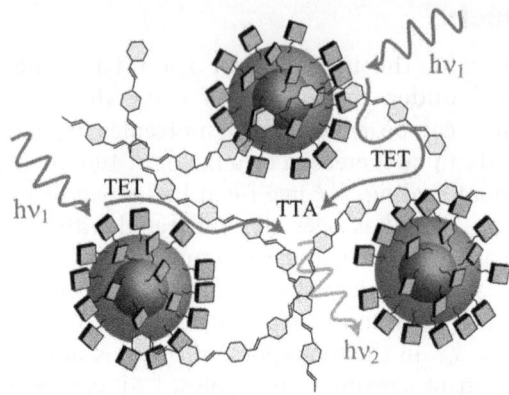

Figure 15.11   Concept of an efficient up-convertor (see text).

be heterogeneous, meso/nanostructured, and include optical concentrators such as plasmonic nanoparticles.

References

1. J. H. Zhao, A. H. Wang, M. A. Green and F. Ferrazza, *Appl. Phys. Lett.*, 1998, **73**, 1991.
2. J. Britt and C. Ferekides, *Appl. Phys. Lett.*, 1993, **62**, 2851.
3. C. J. Brabec, N. S. Sariciftci and J. C. Hummelen, *Adv. Funct. Mater.*, 2001, **11**, 15.
4. B. O'Regan and M. Grätzel, *Nature*, 1991, **353**, 737.
5. M. M. Lee, J. Teuscher, T. Miyasaka, T. N. Murakami and H. J. Snaith, *Science*, 2012, **338**, 643.
6. L. C. Hirst and N. J. Ekins-Daukes, *Prog. Photovoltaics: Research and Applications*, 2010.
7. T. Trupke, M. A. Green and P. Würfel, *J. Appl. Phys.*, 2002, **92**, 4117.
8. N. J. Ekins-Daukes, I. Ballard, C. D. J. Calder, K. W. J. Barnham, G. Hill and J. S. Roberts, *Appl. Phys. Lett.*, 2003, **82**, 1974.
9. B. Ahrens, P. Loper, J. C. Goldschmidt, S. Glunz, B. Henke, P. T. Miclea and S. Schweizer, *Physica Status Solidi (A)*, 2008, **205**, 2822.
10. J. de Wild, A. Meijerink, J. K. Rath, W. G. J. H. M. van Sark and R. E. I. Schropp, *Sol. Energy Mater. Sol. C.*, 2010, **94**, 1919.
11. J. de Wild, J. K. Rath, A. Meijerink, W. G. J. H. M. van Sark and R. E. I. Schropp, *Sol. Energy Mater. Sol. C.*, 2010, **94**, 2395.
12. A. C. Pan, C. del Cañizo and A. Luque, *Mater. Sci. Eng. B*, 2009, **159–160**, 212.
13. A. Shalav, B. S. Richards and M. A. Green, *Sol. Energy Mater. Sol. C.*, 2007, **91**, 829.

14. Y. Y. Cheng, B. Fückel, R. W. MacQueen, T. Khoury, R. G. C. R. Clady, T. F. Schulze, N. J. Ekins-Daukes, M. J. Crossley, B. Stannowski, K. Lips and T. W. Schmidt, *Energy Environ. Sci.*, 2012, **5**, 6953.

15. A. Nattestad, Y. Y. Cheng, R. W. MacQueen, T. F. Schulze, F. W. Thompson, A. J. Mozer, B. Fückel, T. Khoury, M. J. Crossley, K. Lips, G. G. Wallace and T. W. Schmidt, *J. Phys. Chem. Lett.*, 2013, **4**, 2073.

16. T. F. Schulze, Y. Y. Cheng, B. Fückel, R. W. MacQueen, A. Danos, N. J. L. K. Davis, M. J. Y. Tayebjee, T. Khoury, R. G. C. R. Clady, N. J. Ekins-Daukes, M. J. Crossley, B. Stannowski, K. Lips and T. W. Schmidt, *Aust. J. Chem.*, 2012, **65**, 480.

17. T. F. Schulze, Y. Y. Cheng, T. Khoury, M. J. Crossley, B. Stannowski, K. Lips and T. W. Schmidt, *J. Photonics Energy*, 2013, **3**, 034598.

18. T. F. Schulze, J. Czolk, Y. Y. Cheng, B. Fückel, R. W. MacQueen, T. Khoury, M. J. Crossley, B. Stannowski, K. Lips, U. Lemmer, A. Colsmann and T. W. Schmidt, *J. Phys. Chem. C*, 2012, **116**, 22794.

19. W. Shockley and H. J. Queisser, *J. Appl. Phys.*, 1961, **32**, 510.

20. A. S. Brown and M. A. Green, *Physica E*, 2002, **14**, 96.

21. A. Luque and A. Marti, *Phys. Rev. Lett.*, 1997, **78**, 5014.

22. N. J. Ekins-Daukes and T. W. Schmidt, *Appl. Phys. Lett.*, 2008, **93**, 063507.

23. M. J. Y. Tayebjee, A. A. Gray-Weale and T. W. Schmidt, *J. Phys. Chem. Lett.*, 2012, **3**, 2749.

24. M. C. Hanna and A. J. Nozik, *J. Appl. Phys.*, 2006, **100**, 4510.

25. ASTM standard G173 standard tables for reference solar spectral irradiances: Direct normal and hemispherical on 37° tilted surface, ASTM International, West Conshohocken, PA, 2008.

26. C. A. Parker and C. G. Hatchard, *Proc. Chem. Soc. London*, 1962, 386.

27. C. A. Parker, *Proc. R. Soc. London Ser. A*, 1963, **276**, 125.

28. Y. Y. Cheng, B. Fückel, T. Khoury, R. G. C. R. Clady, N. J. Ekins-Daukes, M. J. Crossley and T. W. Schmidt, *J. Phys. Chem. A*, 2011, **115**, 1047.

29. S. Baluschev, V. Yakutkin, T. Miteva, Y. Avlasevich, S. Chernov, S. Aleshchenkov, G. Nelles, A. Cheprakov, A. Yasuda, K. Müllen and G. Wegner, *Angew. Chem. Int. Ed.*, 2007, **46**, 7693.

30. D. V. Kozlov and F. N. Castellano, *Chem. Commun.*, 2004, **24**, 2860.

31. T. N. Singh-Rachford, A. Haefele, R. Ziessel and F. N. Castellano, *J. Am. Chem. Soc.*, 2008, **130**, 16164.

32. Y. Y. Cheng, B. Fückel, T. Khoury, R. G. C. R. Clady, M. J. Y. Tayebjee, N. J. Ekins-Daukes, M. J. Crossley and T. W. Schmidt, *J. Phys. Chem. Lett.*, 2010, **1**, 1795.

33. Y. Y. Cheng, T. Khoury, R. G. C. R. Clady, M. J. V. Tayebjee, N. J. Ekins-Daukes, M. J. Crossley and T. W. Schmidt, *Phys. Chem. Chem. Phys.*, 2010, **12**, 66.

34. R. W. MacQueen, T. F. Schulze, T. Khoury, Y. Y. Cheng, B. Stannowski, K. Lips, M. J. Crossley and T. Schmidt, *Proc. SPIE*, 2013, **8824**, 882408–882409.

35. V. Jankus, E. W. Snedden, D. W. Bright, V. L. Whittle, J. A. G. Williams and A. Monkman, *Adv. Funct. Mater.*, 2012, **23**, 384.

CHAPTER 16

Quantum Rectennas for Photovoltaics

FENG YU,[a] GARRET MODDEL[b] AND RICHARD CORKISH*[a]

[a] School of Photovoltaic and Renewable Energy Engineering and Australia-US Institute for Advanced Photovoltaics, The University of New South Wales, Sydney 2052, Australia; [b] Department of Electrical, Computer, and Energy Engineering, University of Colorado, Boulder CO 80309-0425, USA
*Email: r.corkish@unsw.edu.au

16.1 Introduction

Quantum antennas for photovoltaics are special cases in the rapidly growing field of optical antennas. An optical antenna is 'a device that converts freely propagating optical radiation into localized energy, and *vice versa*'.[1-4] Quantum antennas for photovoltaics are specifically required to couple optical solar radiation to a load, commonly via a rectifier.[5] The combined antenna and rectifier is termed a 'rectenna'.

A rectenna, or rectifying antenna (Figure 16.1) is a device for the conversion of electromagnetic energy propagating through space to direct current electricity in a circuit, available to be delivered to a load or to storage. It has one or more elements, each consisting of an antenna, filter circuits and a rectifying diode or bridge rectifier either for each antenna element or for the power from several elements combined. They are under consideration as alternatives to conventional solar cells.

Conventional solar cells, with the exception of their reflection coatings, are quantum devices, only able to be understood through quantum physics. On the

RSC Energy and Environment Series No. 11
Advanced Concepts in Photovoltaics
Edited by Arthur J Nozik, Gavin Conibeer and Matthew C Beard
© The Royal Society of Chemistry 2014
Published by the Royal Society of Chemistry, www.rsc.org

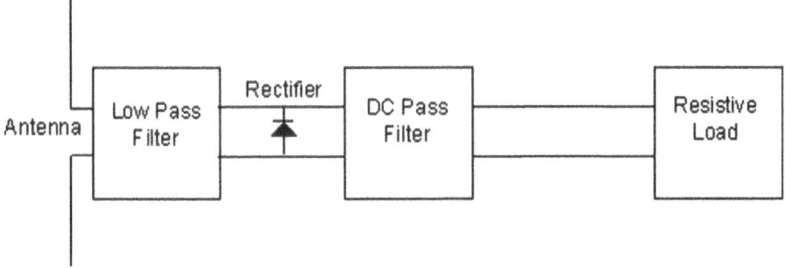

Figure 16.1 Conceptual structure of a rectenna.
Reproduced from Corkish *et al.*[6]

other hand, the wave nature of light is routinely exploited at longer wavelengths in radio and microwave frequency bands for communications, heating and sensing. Photon energies are small at radio frequencies and a large number is required to produce significant power density. Engineers routinely use wave models for radiation in that regime. At shorter wavelengths, fewer photons are required for the same power density and particle models are commonly applied. In principle, there should be no reason why the electromagnetic wave technologies which are so successfully used for radio communications cannot be scaled to optical frequencies, although quantum models may be necessary for at least some aspects. However, there are significant practical issues, especially concerning the sub-millimeter size scales involved.

In parallel, antenna structures used to generate surface plasmons are exciting great interest in the physics and engineering communities, including in the photovoltaics arena.[7,8] In these applications, including the enhancement of light absorption by dye molecules,[9] nanoscale metallic particles absorb and re-emit light and are used to intensify light absorption in conventional solar cells, even beyond the ergodic limit. These applications are beyond the scope of this article and are considered elsewhere.[10,11]

There are significant overlaps and common interests with radioastronomy. A solar rectenna is similar to a simple radiotelescope or radiometer[12] but differs in that the radio telescope needs to measure the intensity of the radiative power received by the antenna, often with a square-law detector which produces an output voltage proportional to the input power, while the rectenna needs to convert that power to useful DC electricity. In terms of a solar cell analogy, the radiotelescope observes the open circuit voltage while the rectenna extracts power at the maximum power point.

16.2 History of Quantum Antennas for Photovoltaics Research

16.2.1 Optical and Infrared Rectennas

This field has been briefly reviewed in the past by Corkish *et al.*,[6,13] Goswami[14] and Rzykov *et al.*[15] and more recently and extensively by

Eliasson,[16] and Moddel and Grover.[17] The concept of using antennas to convert solar energy to rectified electricity first appeared in the literature in the early 1970s when Bailey,[18] Bailey et al.,[19] and Fletcher and Bailey[20] proposed the idea of collecting solar energy with devices based on the wave nature of light. He suggested artificial pyramid or cone structures like those in eyes. He describes pairs of the pyramids as modified dipole antennas, each pair electrically connected to a diode (half-wave rectifier), low-pass filter and load. The antenna elements needed to be several wavelengths long to permit easier fabrication (Figure 16.2).

Marks[21] patented the use of arrays of submicron crossed $\lambda/2$ dipoles on an insulating sheet with fast full-wave rectification (Figure 16.3). Marks' structure is essentially a conventional broadside array antenna with the output signal from several dipoles feeding into a transmission line to convey their combined power to a rectifier. This design requires the oscillations from each dipole to add in phase. Marks also patented devices to collect and convert solar energy using solidified sheets containing oriented metal dipole particles or molecules[22] and his later patent[23] describes a 'submicron metal cylinder with an asymmetric metal—insulator–metal tunnel junction at one end' to absorb and rectify light energy. Marks also proposed a system[24,25] in which a plastic film containing parallel chains of iodine molecules form linear conducting elements for the collection of optical energy. The theoretical conversion efficiency was claimed to be 72%[24] and a recently active but now expired web site[26] stated that the material was being actively developed.

Farber,[27] in addition to proving the concept of single-frequency microwave power reception by pyramidal dielectric antenna elements and rectification, attempted reception of light energy by SiC particles on modified abrasive paper. The results were, however, inconclusive, despite some electrical output being observed. This work was extended by Goswami et al.[14]

Kraus, in the second edition of his text,[28] proposed two orthogonally polarized arrays of λ dipoles, one array above the other on either side of a transparent substrate, with a reflector behind, to receive and rectify sunlight.

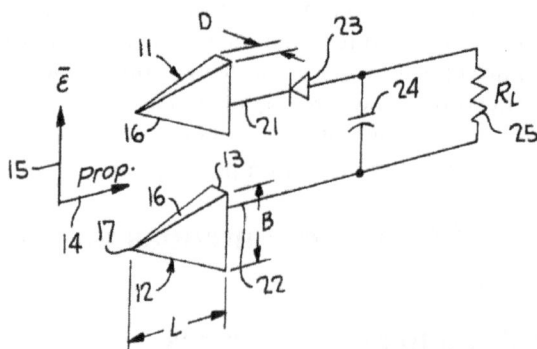

Figure 16.2   Electromagnetic wave energy converter proposed by Fletcher and Bailey. Reproduced from Bailey.[18]

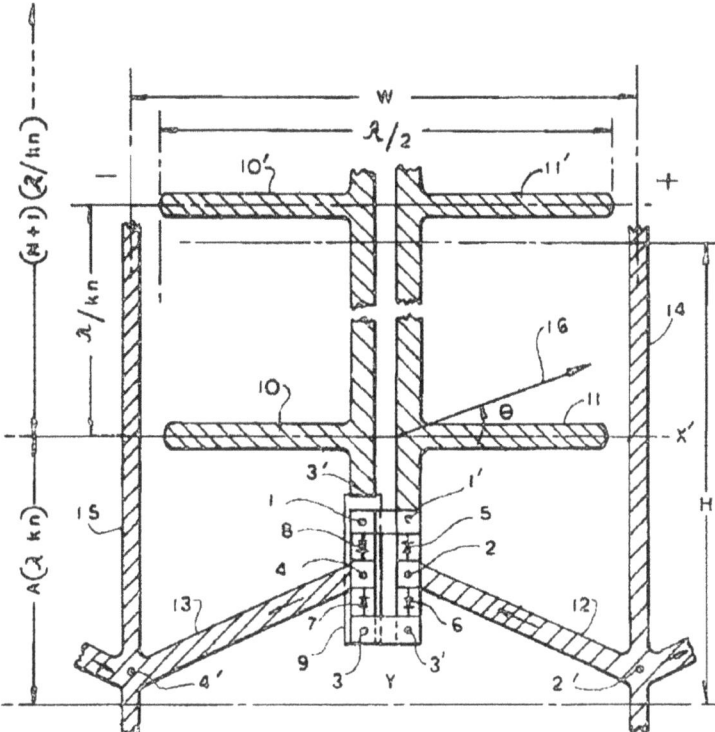

Figure 16.3   Marks' rectenna using array of dipoles and discrete rectifiers.
Reproduced from Bailey *et al.*[19]

There is no mention there of any attempt to realize the device. Kraus stated that 100% aperture efficiency is, in principle, possible. A later textbook states that optical rectification is impractical because the diode electron relaxation time is too slow.[29]

Lin *et al.*[30] reported the first experimental evidence for light absorption in a fabricated resonant nanostructure and rectification at light frequency. The device used grooves and deposited metallic elements to form a parallel dipole antenna array on a silicon substrate and a p–n junction for rectification. They observed an output resonant with the dipole length and dependent on light polarization and angle of the incoming light, indicating that the device possessed antenna-like characteristics.

Berland *et al.*[31,32] undertook extensive development of theoretical and experimental models for optical antennas coupled to fast tunnel diodes, stating a theoretical efficiency for sunlight of up to 85%. They built model dipole rectenna arrays operating at 10 GHz, achieving over 50% conversion efficiency and integrated metal–insulator–metal (MIM) rectifier diodes into a μm-scale antenna.

Eliasson and Moddel proposed the use of metal/double-insulator/metal diodes for rectenna solar cells,[16,33] and demonstrated high responsivity at

60 GHz.[34] To circumvent *RC* time constant limitations of metal/insulator/metal diodes, the group proposed the use of traveling wave diodes[35,36] and graphene geometric diodes, which were demonstrated for infrared radiation.[37]

Wang *et al.*[38] made random arrays of aligned carbon nanotubes and demonstrated the polarization and length-dependence effects in the visible range.

Sarehraz *et al.*[39–41] focussed on the issues of skin effect in metallic antenna elements and the tiny voltage produced by each. Assuming a dipole structure, they calculate the available power and the efficiency of an MIM rectifier and DC power. Diode efficiency increases with input power and the authors conclude that, with their assumptions, about 5000 dipoles would need to feed each rectifier to exceed the efficiency of a silicon photovoltaic cell.

Corkish *et al.*[6] attempted to address the question of the theoretical limit on the efficiency of rectenna collection solar energy. They proposed combining concepts from classical radio-astronomical radiometry[42] with theories for rectification of electrical noise.[43] This work, like most preceding studies of solar rectennas, discussed a proposed extension of classical physical concepts, reliant on the Rayleigh-Jeans approximation and applicable for $hf(kT) \ll 1$. This approach is extended in the present chapter.

Kotter *et al.*[44] modeled, using a general Method-of-Moments software package, periodic arrays of square loop antennas for mid-infrared wavelengths and constructed devices. Modelling predicted a theoretical efficiency of 92% for antenna absorption of solar energy, with peak performance at 10 μm wavelength. Test devices were built on silicon wafers using electron beam lithography and prototype roll-to-roll printed arrays were produced. Peak operation for experimental devices was at 6.5 μm.

Osgood *et al.*[45] observed 1 mV signals from nanorectenna arrays of silver patterned lines coupled to NiO-based rectifier barriers illuminated by 532 nm and 1064 nm laser pulses and Nunzi[46] proposed reception by arrays of metallic, resonant nanoparticles and rectification by molecular diodes,[47] covalently linked to the antennas.

Eliasson[16] suggested that coherent illumination was required for rectennas to avoid cancellation of different components of the current, and that sunlight provides sufficient coherent only over a limited illumination area. Mashaal and Gordon[48] analyzed the coherence radius for broadband solar illumination.

Miskovsky *et al.*[49] proposed a sharp-tip geometry to rectify optical-frequency radiation to circumvent the *RC* time constant limitations of metal–insulator–metal diodes. Choi *et al.*[50] have presented a different approach to forming an asymmetric tunneling diode for high frequency rectification.

Gallo[51] proposed an innovative thermo-photovoltaic system in which the earth absorbs solar energy and the re-emitted thermal radiation is intercepted and converted by an infrared antenna array using printed gold square spirals. This could be of potential interest for stationary aerial platforms for earth observation, surveillance, etc. The antenna array was simulated at 28.3 THz (10.6 μm) for reception and rectification of circularly polarized plane wave. As noted by Grover,[52] in accordance with the second law of thermodynamics the efficiency of such an approach depends on the antenna being at a significantly lower temperature than the earth and is likely to be very low in any conceivable implementation.

Grover *et al.*[53] rejected the use of classical approaches in favor of the semiclassical treatment of photon assisted transport, generalized for tunnel devices. They used this method to derive a piecewise-linear approximation to the current–voltage curve for an optical rectenna under monochromatic illumination. Using this treatment Joshi *et al.*[54] calculated the upper bound for rectenna solar cell conversion efficiency under broadband illumination.

Optical antenna structures, or antenna-coupled detectors, are under development for many actual and potential uses other than for solar energy collection.[4] In many of these other applications, issues discussed in this review as problems for implementation of antennas for energy collection can be seen as advantages. Small physical size, narrow bandwidth and polarization dependence can improve performance in many applications.[4] The research in this field has moved from infrared[55] devices to optical wavelengths as lithography technology has made smaller devices feasible. Here too, more tractable problems than solar energy rectification are commonly addressed. In particular, the difficult question of how to efficiently convert received energy to DC electricity does not always arise. Skigin and Lester[5] reviewed optical antennas under development for a range of purposes, referring to dipole, bow-tie and Yagi-Uda styles. Antenna–load interactions at optical frequencies were investigated Olmon and Rashke.[3] They described the reception process by partitioning into three main steps: excitation of an antenna resonance by a freely propagating mode; its transformation into a nanoscale spatial localization; and near-field coupling to a quantum load. They suggested the necessary extension of antenna theory for the design of impedance-matched optical antenna systems coupled to loads. Vandenbosch and Ma[56] analyzed the upper bounds of the antenna efficiency for different metals in solar rectennas.

16.2.2 Wireless Power Transmission

Another potential application for rectennas is the wireless transmission of electrical power, either terrestrially or between satellites or between satellites and Earth.[57,58] Wireless power transmission is an old dream that is now commonly realized over short distances for battery charging in consumer and industrial devices. Interest in terrestrial wireless power transmission can be traced from Heinrich Hertz in the 19th century.[59] Nicola Tesla was one of the pioneers in its implementation with his experiments at Colorado Springs[60] at the beginning of the 20th century. W.C. Brown, at Raytheon Corp., led a long program of research into microwave power transmission with many technical successes for terrestrial, aerial and space applications, achieving efficiency of 84%.[59] The potential use of ground-based rectennas to collect microwave transmission from orbiting solar power stations has also attracted significant interest in recent decades.[57]

The problem being addressed in those studies is much simpler than that of solar energy reception and rectification since coherent radiation of a single polarization and frequency, commonly in the atmospheric window in the microwave band, may be chosen, excluding most of the more difficulties discussed below.

16.2.3 Radio-powered Devices

A range of low-power electronic devices has been developed to derive operating power from electromagnetic fields. The earliest radio receivers, crystal sets, were self powered by rectification of the incoming radio frequency signal. Several inventors have developed devices to monitor microwave oven leakage or other radiation by powering an indicator from the leaking radiation via a rectenna. A similar method has been used to power active biotelemetry devices, implanted therapeutic devices, radio frequency identification tags and transponders and even the proposed recharging of batteries in microwave ovens. Motjolopane and van Zyl[61] reviewed options for harvesting ambient microwave energy to supply indoor distributed wireless sensor.

16.2.4 Radio Astronomy

Radio astronomy technology provides, perhaps, the closest existing similarity to technology that might eventually permit antenna collection of solar energy and it might provide useful insights for future developments. The field of astronomy has a technological divide between optical and radio astronomy. Practitioners of the former rely on optical instruments such as lenses, reflectors and cameras. Radio astronomers share the extensive use of reflectors but antennas and radio receivers are the core components of radiotelescopes.[62] However, the two are likely to merge and overlap as the instrument technologies for each extend into the intervening gap. Radio astronomers routinely deal with broad bandwidth signals, a range of coherence and polarizations and, at the shorter wavelength end of their range of interest, components of small physical scale. They detect and measure black-body and/or other forms of radiation from celestial bodies, including the sun, but do not seek to maximize the extraction of energy.

16.3 Research Problems Concerning Rectennas for Photovoltaics

16.3.1 Fundamental Problems

16.3.1.1 Partial Coherence

In principle, it is possible to generate electrical power from purely incoherent photon sources, as this does not violate the second law of thermodynamics. In fact non-coherent black-body radiation at an elevated temperature is equivalent to thermally agitated electromagnetic noise. Together with a heat sink, such noise at a higher colour temperature can contribute to DC electricity with the use of a rectifying heat engine. This conversion is limited by thermodynamic efficiencies, *i.e.* the Carnot efficiency or, more strictly, the Landsberg limit.[63] However, if sunlight is inherently partly coherent, higher efficiency figures may be achieved. In the field of traditional radio technologies, the rectification efficiency can even achieve 100% for purely coherent electromagnetic waves.

This consideration introduces a fundamental problem concerning the coherence properties of sunlight. Sunlight is normally regarded as incoherent radiation due to the nature of spontaneous emission. However coherence theory implies that even radiation emitted from incoherent sources still has equal-time partial coherence if the spatial separation is sufficiently small. Verdet[64] studied the coherence problem of sunlight for the first time. Since then researchers developed the van Cittert–Zernike theorem based on far-field assumptions.[65–67] Recently Agarwal *et al.*[68] used a different approach to study the partial coherence of sunlight, leading to expressions in good agreement with the far-field result.

The mutual coherence between two field points is defined by the equal-time mutual coherence function (EMCF),[48] which is the statistical time average of the product of the field at the two positions (Figure 16.4), *i.e.* $\langle E^*(r_1)E(r_2)\rangle$. From the treatment by Agarwal *et al.*, for simplicity, a scalar field U (as one component of E) is used instead of the vector electric field E. The time-dependence of this field is converted into the frequency domain by Fourier transformation, enabling the possibility of dealing with the wide spectrum of sunlight. The mutual coherence is characterized by a cross-spectral density function instead:

$$W(r_1,r_2,\omega) \equiv \langle U^*(r_1,\omega)U(r_2,\omega)\rangle \tag{16.1}$$

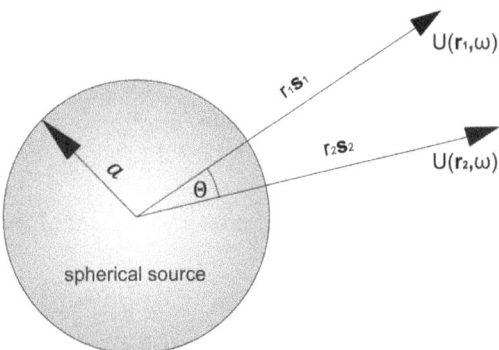

Figure 16.4 Schematic diagram for two light rays emitted from a spherical source. Reproduced from Agarwal and Wolf.[68]

This scalar field, U, as a solution of the scalar Helmholtz equation, can be expanded into a series of spherical harmonics:

$$U(r, \omega) = \sum_{lm} c_{lm} h_l^{(1)}(kr) Y_{lm}(\theta, \phi) \tag{16.2}$$

where $h_l^{(1)}$ is the spherical Hankel function of the first kind and Y_{lm} denotes the spherical harmonics. $k = 2\pi/\lambda = \omega/c$ is the wave-vector of the monochromatic light. θ and ϕ denote the angular coordinates of the field point. The coefficients c_{lm} are random, depending on the statistical properties of the field on the surface of the light source. The cross-spectral density is then expressed by Equation (16.3) after substituting Equation (16.2) into Equation (16.1):

$$W(r_1, r_2, \omega) = \sum_{lm} \sum_{l'm'} c_{lm}^* c_{l'm'} h_l^{(1)*}(kr_1) h_{l'}^{(1)}(kr_2) Y_{lm}^*(\theta_1, \phi_1) Y_{l'm'}(\theta_2, \phi_2) \tag{16.3}$$

The next step is to include appropriate boundary conditions for Equation (16.3). Considering the fields at the surface of the spherical source, its cross-spectral density is delta-function correlated:

$$W(as_1, as_2, \omega) = I_0(\omega) \delta^{(2)}(S_2 - S_1) \tag{16.4}$$

where s_1 and s_2 are the respective unit vectors in the directions of the vectors r_1 and r_2 and $I_0(\omega)$ is the effective intensity of the field on the spherical surface of the source. $\delta^{(2)}$ is the two-dimensional Dirac delta function with respect to the spherical coordinates (θ, ϕ). This delta function can be expanded according to the spherical harmonic closure relation:

$$\delta^{(2)}(s_2 - s_1) = \sum_{lm} Y_{lm}^*(\theta_1, \phi_1) Y_{lm}(\theta_2, \phi_2) \tag{16.5}$$

The boundary condition, Equation (16.4), attributes the physical origin of the partial coherence of sunlight to the geometric correlation at the spherical surface of the light source. By matching Equation (16.3) to the boundary condition, the correlation function of c_{lm} is found to be:

$$\langle c_{lm}^* c_{l'm'} \rangle = \delta_{ll'} \delta_{mm'} \frac{I_0(\omega)}{|h_l^{(1)}(ka)|^2} \tag{16.6}$$

where $\delta_{ll'}$ and $\delta_{mm'}$ are the Kronecker delta functions. By using the spherical harmonic addition theorem the cross-spectral function takes its final form:

$$W(r_1, r_2, \omega) = \sum_l \frac{2l+1}{4\pi} \frac{I_0(\omega)}{|h_l^{(1)}(ka)|^2} h_l^{(1)*}(kr_1) h_l^{(1)}(kr_2) P_l(\cos \Theta) \tag{16.7}$$

where Θ is the angle between r_1 and r_2. P_l is the Legendre polynomial of order l.

The angular dependence of the degree of coherence is obtained by setting $r \equiv r_1 = r_2$, and then normalizing $W(r_1, r_2, \omega)$ according to its peak value (at $\Theta = 0$). The numerical results (Figure 16.5) show that the angular spread

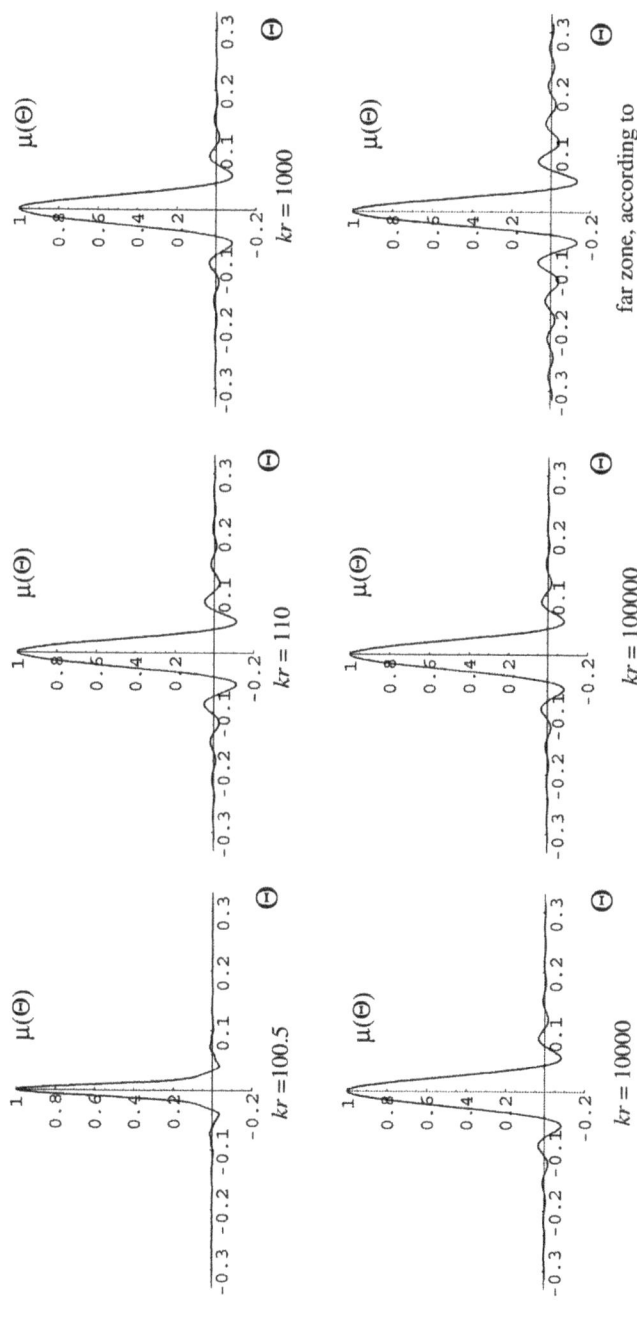

Figure 16.5 Degree of coherence with different angular distances between two field points. Each figure corresponds to a specified radial distance to a spherical source whose radius is kr = 100. Reproduced from Agarwal and Wolf.[68]

of the coherence remains a constant when the field points are more than a few wavelengths away from the source. This confirms that the far-field relation also applies to field points essentially within the near-field regime.

Figure 16.5 also provides a comparison between the expression derived by Agarwal and Wolf[68] and the traditional far-field van Cittert–Zernike theorem:

$$W(rs_1, rs_2, \omega) = \frac{J_1[2ka \sin(\Theta / 2)]}{ka \sin(\Theta / 2)} \tag{16.8}$$

where Θ is still the angular distance between the unit vectors s_1 and s_2 and J_1 is the Bessel function of the first kind and order 1. Numerically, this expression in a very good agreement with the general relation derived here, enabling us to analyze the coherence problem simply with the van Cittert–Zernike relation.

The previous analysis gives expressions for the degree of coherence at a quasi-monochromatic spectrum. To convert the whole spectrum of sunlight by the rectenna system, the partial coherence of the broadband radiation requires investigation. Mashaal and Gordon[48] have provided a quantitative analysis. In their treatment, the far-field van Cittert–Zernike theorem is represented in terms of the solar angular radius Φ and the intensity, I (per solid angle of black-body radiation from the sun):

$$\langle E^*(r_1, \omega) \cdot E(r_2, \omega) \rangle = \pi\Phi^2 I \frac{J_1[2ka \sin(\Theta / 2)]}{ka \sin(\Theta / 2)} \approx \pi\Phi^2 I \frac{J_1(k\Phi|r_1 - r_2|)}{k\Phi|r_1 - r_2| / 2} \tag{16.9}$$

where r_1 and r_2 denote the field point positions in the transverse direction. For a circular antenna placed transverse to the incident sunlight, the maximum separation between r_1 and r_2 is the antenna diameter, $2b$. The whole-band equal-time mutual coherence function (EMCF) is simply an integration of the quasi-monochromatic EMCF, as the coherence between waves of different frequencies always drops to zero:

$$\langle E^*(r_1) \cdot E(r_2) \rangle = 2\pi\Phi^2 \int_{\lambda_{min}}^{\lambda_{max}} I_{BB}(\lambda) \frac{J_1(k\Phi|r_1 - r_2|)}{k\Phi|r_1 - r_2| / 2} d\lambda \tag{16.10}$$

where the spectrum of black-body radiation is:

$$I_{BB}(\lambda) = 2hc^2/\lambda^5 \cdot [\exp(hc/kT\lambda) - 1]^{-1} \tag{16.11}$$

The average power intercepted by the antenna is:[48]

$$\langle P \rangle = \frac{1}{A_{ap}} \int_{A_{ap}} \int_{A_{ap}} \langle E^*(r_1) \cdot E(r_2) \rangle dA dA' = \int_{\lambda_{min}}^{\lambda_{max}} I_{BB}(\lambda)\lambda^2 \left[1 - J_0^2(k\Phi b) - J_1^2(k\Phi b)\right] d\lambda \tag{16.12}$$

where the integrations are over the antenna's full aperture area A_{ap}. From Equation (16.12) the intercepted power is proportional to the aperture area (πb^2), i.e. $\langle P \rangle = \pi^2\Phi^2 b^2 I$, if the antennas size is sufficiently small, $(k\Phi b \ll 1)$.

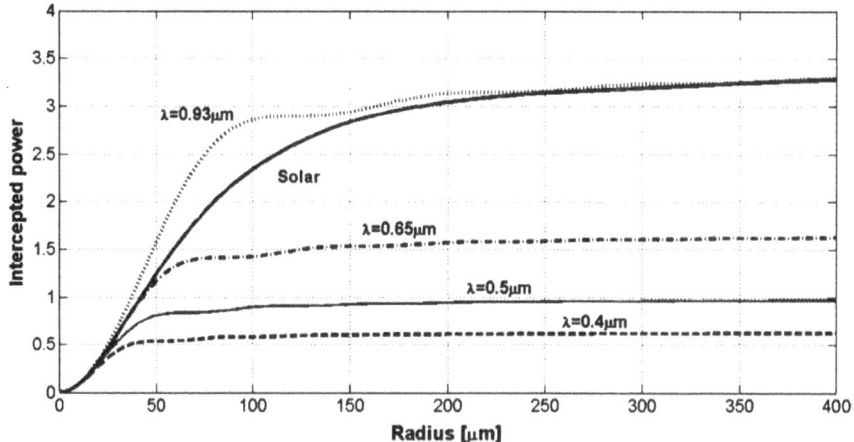

Figure 16.6 Intercepted power (normalized to the asymptotic value for $\lambda = 0.5$ μm) as a function of detector radius. For each curve the asymptotic power is $\overline{\lambda^2}I$, while at small radius, all curves converge to the result for coherence of $\pi^2 \Phi^2 b^2 I$.
Reproduced from Mashaal and Gordon.[48]

This corresponds to the case of pure-coherent interception. On the other hand, the pure-incoherent limit exists for large antennas, $(k\Phi b \gg 1)$. At this limit the intercepted power does not increase with the antenna size $(\langle P \rangle = \overline{\lambda^2}I)$, as it is restricted by the partial coherence of sunlight. Figure 16.6 illustrates the increase of intercepted power with the antenna radius. When the antenna radius exceeds approximately a hundred times of the wavelength, the intercepted power gradually levels off.

Coherence efficiency can be defined as the ratio of the intercepted power to its value at the pure-coherence limit $(\pi^2 \Phi^2 b^2 I)$. This efficiency provides a measure of the loss of collectible power due to the incoherence of sunlight. Figure 16.7 provides the coherence efficiency for whole-spectrum sunlight, as a function of the antenna radius. A 90% coherence efficiency can be achieved with a radius of 19 μm.

16.3.1.2 Polarization

Most traditional antennas only accept a single linear or a single circular polarization, which is insufficient to match the unpolarized nature of directly incident sunlight. Such antennas can at maximum absorb 50% of the total radiation. To overcome this problem, cross-polarized structures have been designed as a combination of two linear or circularly polarized antennas placed orthogonally. With these structures it is possible to provide a 100% aperture efficiency in principle.[28] Prior splitting of sunlight into two orthogonal linearly or circularly polarized components, by use of birefringent crystals for example, would be necessary for some possible converter designs, such as Song's ratchets.[69]

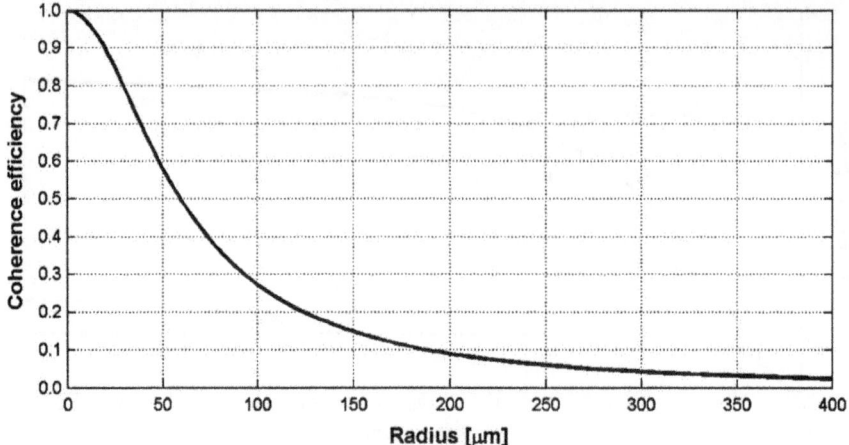

Figure 16.7 Coherence efficiency (intercepted power relative to its value in the pure-coherence limit) as a function of antenna radius for solar radiation. Reproduced from Mashaal and Gordon.[48]

Another practical design adopts a unidirectional conical four-arm form that is effectively two intertwined antennas for both polarizations (Figure 16.18). This sinuous antenna can produce power outputs for either the two linear or the two circular polarizations. It also provides a large bandwidth. None of the frequency-independent antenna designs are very directive (*i.e.*, they are restricted to low solar concentration ratios) but it would be feasible to use them as feed antennas for concentrating reflectors or lenses.[70]

16.3.1.3   *Bandwidth*

The broadband nature of sunlight limits the possibility of accepting its energy by a single antenna. In fact, 60% of the solar spectrum is contained in a fractional bandwidth of 60%.[39] On the other hand, a fractional bandwidth of 15–20% is usually regarded as wide bandwidth for conventional microwave antennas, presenting a problem for the application of antenna techniques to solar power harvesting.

Radio antenna designs, termed 'frequency independent', have been devised to achieve greater bandwidths,[71] but their complexity clearly presents a challenge for fabrication at the scale required for optical reception. In order to make an antenna independent of frequency it is necessary to ensure that the antenna's radiating structures are specified by angles only and to be truly frequency independent, an antenna would need to infinite in size and its feed point would need to be infinitely fine. In practice, antenna engineers can obtain up to 100:1 bandwidth. Multi-arm planar spiral antennas are frequently used to obtain the required frequency range but, undesirably in our case, have radiation lobes on either side of the substrate plane.

Attachment to a transparent dielectric substrate or lens would allow radiation to be better received through the substrate side. Conical spiral antennas (Figure. 16.8), 3D structures with spiral arms wrapped over the surface of a cone, concentrate the radiation pattern in the direction of the cone apex, demonstrating a 10:1 bandwidth.[71] Planar spiral antennas have been made, with lenses, in the THz range[26] but the current authors have not identified examples of infrared or optical conical spiral antennas.

Another design of wide bandwidth antenna[28] is a planar version of the exponential horn antenna (Figure 16.9), which has the advantage of being compatible with printed-circuit fabrications. The horn takes the form of an exponential notch in the conducting surface of a circuit board, and couples to a 50 Ω strip line on the other surface of the board. It can achieve a bandwidth of 5:1, which is still far from the requirement for whole-band sunlight absorption.

Apart from the difficulty of fabricating a frequency-independent antenna for optical wavelengths, the introduction of harmonics is another important issue. These harmonics, allowed by the wide bandwidth, make it more difficult to achieve high rectification efficiency.[6] Furthermore, an optical rectenna operating with terrestrial sunlight intensities cannot efficiently convert the entire solar spectrum,[54] as discussed later in this chapter.

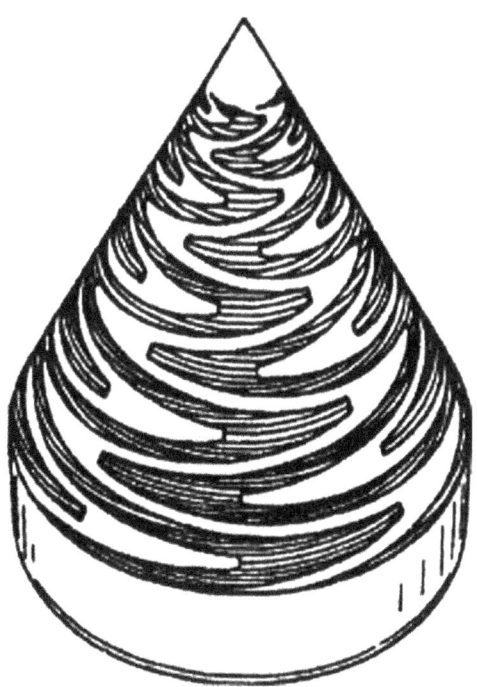

Figure 16.8   Conical sinuous, dual-polarized antenna.
Reproduced from DuHamel and Scherer.[70]

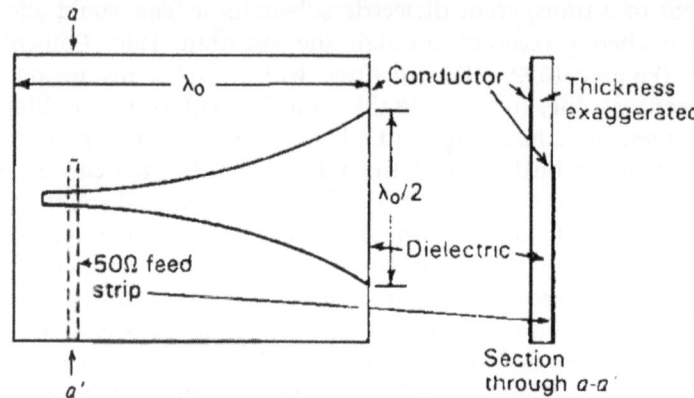

Figure 16.9 Exponential notch broadband antenna with 50 Ω microstrip feed. Reproduced from Kraus.[28]

However, in practice, it is both infeasible and unnecessary to accept the whole solar spectrum using a single antenna. A practical solution may be to absorb sunlight using an approach that resembles the operation of a tandem solar cell. With this approach different rectennas in optical series are used to split the whole spectrum of sunlight and contribute to the power output collectively. Unlike spectral splitting in conventional solar cells, where different materials are required for each spectral range, spectral splitting can be accomplished in rectenna solar cells simply be selecting different operating voltages for each spectral range,[54] along with an appropriately sized antenna. This allows the adoption of narrow-bandwidth antennas. In fact even if each individual antenna has a fractional bandwidth of 20% (typical figure for traditional microwave antennas), only 11 antennas are required to cover the whole wavelength range (0.2–2 μm) of sunlight.

16.3.2 Practical Problems

16.3.2.1 Element Size

Unlike the commonly used microwave antennas that are of one-half the wavelength of the incident light, optical antennas at resonance may have different lengths from that predicted by classical antenna theory. Mühlschlegel *et al.*[7] found that the length of optical antennas at resonance is considerably shorter than one-half the wavelength, if accounting for the finite metallic conductivity at optical frequencies. The excitation of plasmon modes at optical frequencies also plays a fundamental role in this. As demonstrated by Podolskiy, optical excitation of surface plasmons results in strong local field enhancement if the antenna length is around one-half of the plasmon wavelength.[72] These surface plasmons have much shorter wavelengths than free-space radiation at optical frequencies, contributing to a shorter antenna length at resonance. Mühlschlegel *et al.*[7] demonstrated that nanometer-scale gold dipole antennas at resonance, with an optimal antenna length of 255 nm, led to white-light supercontinuum generation. This scale is far below the coherence limit of sunlight so that the incoherence during rectification is no longer an issue.

The dimension of the rectification component is limited by its ultrafast response speed at optical frequencies. Conduction electrons interacting with the electromagnetic field of a coherent light simultaneously undergo two motions.[73] We are interested in estimating the magnitude of the transverse oscillation at optical frequencies. Semchuk *et al.*[73] obtained an expression, valid for non-relativistic motion, for the amplitude of the transverse oscillation:

$$x_{max} = \frac{qE_0}{4\pi^2 f^2 m} \tag{16.13}$$

where f and E_0 are the frequency and the peak amplitude of the external field, respectively and m is the electron mass. This leads to a restriction on the maximum excursion of the electron from its equilibrium position:

$$x_{max} \ll \frac{c}{2\pi f} \ll \frac{\lambda}{2\pi}$$

For light of wavelength 500 nm this results in an amplitude of several nanometers and places a restriction on the dimensions of the structures that could be used as rectifying elements. The most popular option of the high-frequency rectifier is the metal–insulator–metal (MIM) diode, which is under intensive investigation. Such structures need to be extremely thin, *i.e.* several nanometers, partly because the current needs to be sufficiently high to provide a low impedance to match that of the antenna for efficient power coupling.[74]

16.3.2.2 Rectifier Speed

It is a challenging task to rectify electric signals with optical frequencies. The frequency limit of Schottky diodes is in the far-infrared region,[74] beyond which the responsivity drops quickly. The conventional p–n junction diode has an even worse response to high frequencies. In fact the oscillation period for optical frequencies (600 THz is the centre of the visible window) is around 1/600 THz = 1.7 fs. This time scale is even shorter than the electron relaxation time,[77] which implies that any rectification relying on diffusive transport of carriers would be not fast enough to respond to optical frequencies.

However, there are no fundamental restrictions on the feasibility of using rectifiers that rely on quantum transport. A promising option is the metal–insulator–metal (MIM) diode. The nanometer-scale insulator layer sandwiched between two metal layers works as a potential barrier, allowing electrons to tunnel through under positive bias. When the diode is

negatively biased, the conduction band of the insulator becomes flat, contributing to a large effective barrier thickness. This blocks the transport of
electrons as an exponential function of the voltage.

Grover has pointed out that a severe problem preventing existing MIM
diodes to be applied to solar energy conversion is the limitation of its RC
time constant.[74] Even for extremely low resistance MIM diodes, the RC time
constant, which we desire to be below 1 fs, is still too long for visible frequencies. This analysis puts a limit for parallel plate diodes, though it might
be overcome by other potential technologies. Within the regime of quantum
transport, each planar mode can carry at maximum a conductance quantum,
i.e. $(12.9 \text{ k}\Omega)^{-1}$ with spin degeneracy included.[74] If the diode consists of a
solid with a planar periodicity of ~ 5 Å, the density of planar modes would be
around $1/(5\text{Å})^2 = 4 \times 10^{18}$ m^{-2}. If electrons of these modes always have complete transmission, the resistance is $12.9 \text{ k}\Omega/4 \times 10^{18}$ m$^{-2} = 3.2 \times 10^{-15}$ Ω m^2.
To estimate the capacitance of the diode, we consider a scenario: a short
separation of 1nm for ballistic transport, and a material permittivity of 10.
This gives a capacitance of ~ 0.1 F m^{-2} and hence a RC constant of $\sim 10^{-16}$ s.
This sets a fundamental limit for the minimal RC value one could ever get. It
does not change with the diode area and is shorter than the time scale of
visible frequencies. This topic is further discussed, with a different approach,
in section 16.7.1.1.

16.3.2.3 Filtering

The rectenna technology has been intensively investigated for converting
microwave into DC energy.[75-77] The frequency range of these studies is from
1 GHz to 10 GHz, much lower than visible frequencies. However, their
general design structures might still potentially apply to rectennas operating
at higher frequencies. A common idea is that a rectenna should include an
input filter before the rectifier and an output filter after the rectifier
(Figure 16.1).

The input filter can be a band-pass filter[76] or a lowpass filter.[77] Both the
input filter and the output filter are used to store energy during the off
period of the diode. It was found that power flow continuity was able to be
achieved, even with half-wave rectification, by suitable filter design.[78]

More importantly, the input filter restricts the flowing-back and re-radiation
of high-frequency harmonics generated by the rectifier. By correctly setting the
RC time constant of the input filter, it allows the fundamental to pass without
much attenuation, while rejecting all the harmonics. Computer simulation
has revealed that the power loss due to harmonics generated by the diode is
significant, especially when a resonant loop forms between the diode and
transmission lines at a harmonic frequency.[77]

For collection of broadband sunlight, there is an additional problem as
the harmonics generated by rectification of long-wavelength light can have
frequencies within the desired bandwidth for energy acceptance. In fact
the solar spectrum has a frequency range from 150 THz to 1500 THz, being

wide enough to allow harmonics to pass. It will become necessary to split the spectrum and direct different fractions to different rectennas. Another possible method is to use a stack of rectennas, each comprising two orthogonally oriented arrays. Unlike a tandem photovoltaics stack, which effectively works as a series of high-pass filters, the rectennas are arranged in an increasing order of operating frequency, if their input filters are low-pass.

The output filter is basically a DC filter, aiming to block AC power from reaching the load, where it would be lost as heat. It was found that the distance between the rectifier and the output filter might be used to cancel the capacitive reactance of the diode, which is needed to maximize the diode efficiency.[75]

Although the rectenna structure and the filters can be fabricated using conventional transmission-line technology, there is trend in the microwave region of using microstrip-printing techniques.[75,76] In the design by McSpadden *et al.*[75] the output filter is a RF short chip capacitor, while the low pass filters are strips printed on the opposite side of the substrate, as they require a much lower capacitance (Figure 16.10).

16.3.2.4 Element Matching

There are two types of matching problems for rectenna-based solar energy conversion. The first is to match the characteristics of components within one rectenna circuit in order to obtain the maximum conversion efficiency, while the second is to combine the power from different rectenna circuits for a useful DC voltage.

Impedance matching between the antenna and the diode forms the most important of the matching problems. Mismatch between components can lead to reflection or re-radiation, rather than absorption of power. Especially

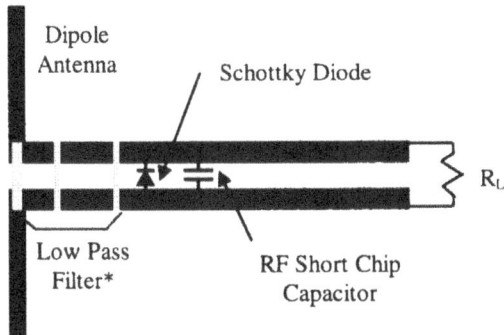

* Capacitive Strips Printed on Opposite Side of Substrate

Figure 16.10 Microwave printed rectenna element. The dipole antenna and transmission line are printed on one side of the substrate and the low-pass filter is formed by strips printed on the opposite side. Reproduced from Semchuk *et al.*[75]

when there is a lack of input filters, the non-linear nature of the rectifier generates significant re-radiation of harmonic frequency energy.[79] The simplest case of rectifying monochromatic radiation is discussed in the following content.

The small-signal circuit model is illustrated in Fig. 16.11. The antenna, as a receiver, is modeled as an ac voltage source $V\cos\omega t$ with an internal resistance, R_A.[74] A resistor and a capacitor connected in parallel represent the differential resistance R_D and the junction capacitance C_D of the diode.

The voltage across the rectifier includes a DC component v_r, as the result of the rectification process, and an ac component v_{ac}, which involves the fundamental transmitted from the antenna and the harmonics generated from its non-linearity. To model the rectifier, consider the Taylor expansion of its I–V characteristic in the neighborhood of v_r:

$$I(V) = I(v_r) + I'(v_r)v_{ac} + \frac{I''(v_r)}{2}v_{ac}^2 + O(v_{ac}^3) \tag{16.14}$$

The ac component is assumed to be so small that the higher-order term $O(v_{ac}^3)$ is regarded as negligible. Using time-averaging technique, the rectified current i_r and the power dissipated in the diode are retrieved:

$$i_r = \langle I \rangle = I(v_r) + \frac{I''(v_r)}{2}\langle v_{ac}^2 \rangle \tag{16.15}$$

$$P_r = \langle Iv \rangle \approx i_r v_r + I'(v_r)\langle v_{ac}^2 \rangle \tag{16.16}$$

It is noted that the differential resistance $R_D = 1/I'(v_r)$. From Equation (16.15) and Equation (16.16) we can define the responsivity of the rectifier as:

$$\beta(v_r) = \frac{i_r - I(v_r)}{P_r - i_r v_r} = \frac{1}{2}\frac{I''(v_r)}{I'(v_r)} \tag{16.17}$$

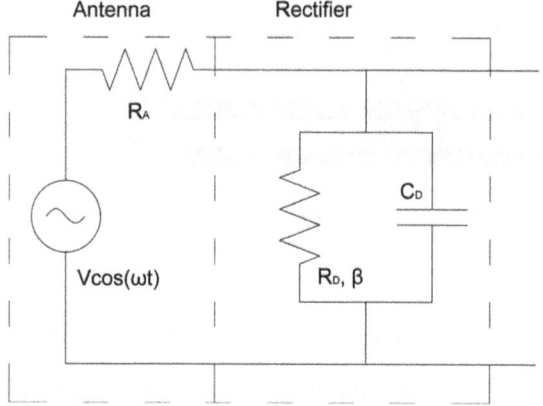

Figure 16.11   Small-signal circuit model of the rectenna. Adapted from Podolskiy *et al.*[74]

The responsivity of a diode reflects its ability to rectify current. It is a constant depending on the material and structure of the rectifier. Detailed circuit analysis on the rectenna system gives expression for the overall conversion efficiency:

$$\eta = \beta^2 PR_A \left\{ \frac{4R_AR_D/(R_A+R_D)^2}{1+[\omega R_AR_D/(R_A+R_D)C_D]^2} \right\}^2 \frac{R_AR_D}{(R_A+R_D)^2} \tag{16.18}$$

The dimensionless term inside the braces is defined as the coupling efficiency, η_C. Its numerator has the same form as the term behind the braces, reaching its maximum when $R_D=R_A$. This corresponds to the condition of impedance matching between the rectifier and the antenna. The denominator has its minimum value at $\omega(R_A\|R_D)C_D=0$. For RC value increases above the time period of the incident light, the overall efficiency drops quickly.

A typical antenna for microwave purposes has an effective impedance of around 100 Ω. This scale of resistance is hard to achieve for the rectifier, unless the junction capacitance is sacrificed.[52] This problem might be avoided by operating the rectenna with a load of low resistance and low capacitance.

With respect to matching aspects of the methods of combining power from different antennas, two basic families of designs have been proposed. Bailey, for example, had individual pairs of antenna elements each supplying its own rectifier and the DC rectifier outputs were combined.[18] Kraus, for example, on the other hand, combined the electrical oscillations from many antenna elements in a particular phase relationship and delivered their combined output to a rectifier.[28] With the former method there is the immediate concern that the tiny power expected from one or two antenna elements may not be enough to produce sufficient voltage to allow proper operation of any conceivable rectifier diode. The latter approach overcomes that problem but at the expense of the need for spatial coherence across all the antenna elements feeding a rectifier. In fact the micro-scale coherence limit for sunlight allows the combination of coherent power from a few nano-scale antennas. In addition, Kraus' suggested antenna structure is not sufficiently broadband for the light spectrum.

16.3.2.5 Materials/Skin Effect

The ac current density in a conductor decreases exponentially from its value at the surface due to the so-called skin effect:

$$J=J_se^{-d/\delta} \tag{16.19}$$

The skin depth δ characterizes the effective thickness of the surface region that conducts ac currents. It is frequency-dependent according to Equation (16.26):[80]

$$\delta = \sqrt{\frac{2\rho}{\omega\mu}}\sqrt{\sqrt{1+(\rho\omega\epsilon)^2}+\rho\omega\epsilon} \tag{16.20}$$

This equation is valid at least up to microwave frequencies. In this frequency range, metallic materials have the relation $\delta \propto \sqrt{1/\omega}$ as $\rho\omega\epsilon \ll 1$. Assuming the skin depth is much smaller than the diameter of the transmission line, the skin depth and, hence, the conductance drops by a factor of 1/10 000 if increasing the frequency from 60 Hz to 6 GHz, which is still far below optical frequencies. This would lead to a significant resistive loss for optical rectennas.

As the time periods of optical frequencies are comparable to or even shorter than the electron relaxation time, Equation (16.20) requires modification due to the ballistic nature of electron transport. Nevertheless, the resistance continues to increase with frequency rise due to the skin effect. Sarehraz *et al.* have indicated that silver at optical frequencies has a skin depth of only 2–3 nm and a resistance of 5–7 Ω per square.[39]

The skin effect[56] can be controlled by replacing metallic materials with dielectrics. If assuming the skin depth δ is much smaller than the wire diameter D, the resistance R is:

$$R = \frac{L\rho}{\pi D \delta} = \frac{L}{\pi D}\left\{\sqrt{\frac{\rho\omega\mu}{2}}\left[\sqrt{\sqrt{1+(\rho\omega\epsilon)^2}+\rho\omega\epsilon}\right]^{-1}\right\} \qquad (16.21)$$

where L denotes the length of the wire. The minimal value of the material-dependent term in the braces can be achieved if the resistivity ρ and the permeability, μ, of the material are both small while the permittivity, ϵ, is as large as possible. For optical frequencies ($\sim 10^{15}$ Hz), the term $\rho\omega\epsilon$ is comparable to 1. This indicates that an increasing permittivity ϵ begins to significantly decrease the resistance ($\sim \epsilon^{-0.5}$) while the material resistivity ρ becomes less relevant. Low resistivity dielectrics could thus potentially benefit the solar application of rectennas.

Another approach is to optimize the geometric design of the antenna. A large surface area is required, which gives advantages to planar rather than wire antennas. Apart from the antenna, other components of the rectenna, including the rectifier and the filters, also requires optimization in order to avoid high resistances.[39]

16.4   Thermodynamics of Rectennas

Strictly speaking the rectenna system is not a thermodynamic system as the excited plasmon polaritons do not equilibrate. In fact at least for microwave frequencies the excitation retains the same spectral density as the incident light. In addition, the spatial coherence of the sunlight makes it possible to have coherent signals, which are different from thermal emissions.

However, in some special cases the excited plasmons can have the spectral density of an equilibrated distribution, *i.e.*, Bose–Einstein statistics. In this situation, with the assumption that partial coherence of incident light is weak, the incident light excites ac signals in the antenna that simulate thermal noise. A thermodynamic analysis can thus improve our understanding on the limitation of such solar energy converters.

16.4.1 Broadband Antenna Modeled as a Resistor

The thermal noise, or the Johnson–Nyquist noise, of a resistor, originates from the thermal agitation arising from electromagnetic energy. This electrical energy is like white noise, with evenly distributed components at all frequencies. Although the power accepted by the antenna and the thermal noise of a hot resistor seem unrelated, they in fact share common properties. A simple way to interpret this is to consider the thermal emission from a hot resistor. The emission, originated from the Johnson–Nyquist noise, has similar properties to black-body radiation.

Following this philosophy, Dicke described a thought experiment in 1946, leading us to the formulae for the voltage excited in the antenna.[42] In Figure 16.12 the antenna is bathed in background radiation emitted from a black-body at temperature T. Regardless of other energy losses, the radiation emitted from the black-body can be either reflected or accepted by the antenna. The antenna is assumed to have effective impedance matched with the transmission line, and the resistive load at the other end of the circuit matches the transmission line too. The whole system is at the same temperature T and is isolated from the environment. This allows detailed balance of the energy transfer between the black-body and the circuit. According to the transmission line theory, the power flowing to the load is not reflected back for the impedance matching configuration. On the other hand, the thermally agitated voltage generated across the resistive load feeds energy back to the antenna for re-emission. These two powers are equal to each other, enabling the substitution of the antenna with a resistor of the antenna's effective impedance (Figure 16.12).

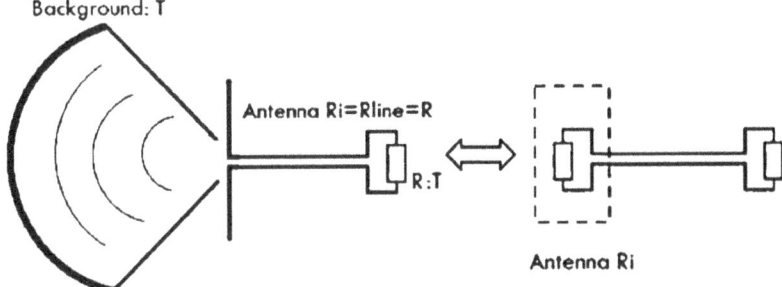

Figure 16.12   An impedance-matched antenna-resistor system bathed in black-body radiation. The thermal equilibrium between the background and the resistor ensures the balance of two powers flowing in opposite directions. The background radiation plus the antenna is thus equivalent to the hot resistor.

A more strict proof of this equivalency is provided as below in terms of their spectral densities. Due to the non-reflecting property of this circuit configuration, the power flowing into the load P_R equals to the power intercepted by the antenna:

$$P_R = \int \frac{1}{2} \frac{\Omega \cdot A_{eff}}{4\pi^3 c^2} \frac{\hbar\omega^3}{e^{\hbar\omega/kT} - 1} d\omega \tag{16.22}$$

The integration is over the whole spectral range of the radiation. Ω and A_{eff} denote the acceptance solid angle and the effective aperture area of the antenna, respectively. The 1/2 term accounts for the fact that the antenna can only intercept one polarization and, thus, absorb only half of the incident power. The rest of the integrand is simply the Planck's equation for the black-body radiation. According to the antenna theory, there is a general relation between the acceptance solid angle and the effective area:

$$\Omega \cdot A_{eff} = \lambda^2 \tag{16.23}$$

where λ is the wavelength of the radiation. By applying this relation to Equation (16.22), the frequency dependence of the integrand changes from ω^3 to ω. This is essentially because the density of electromagnetic modes changes its form from three-dimensional to one-dimensional. If regarding the antenna as a voltage source v_a, the power dissipated by the load is equal to the power absorbed by the antenna:

$$P_R = \int \frac{1}{2\pi} \frac{\hbar\omega}{e^{\hbar\omega/kT} - 1} d\omega = \int \frac{\langle v_a^2 \rangle_\omega}{4R} d\omega \tag{16.24}$$

This indicates that the voltage component (in a frequency range from ω to $\omega + d\omega$) excited by the antenna absorbing the black-body radiation at temperature T is:

$$\langle v_a^2 \rangle_\omega = \frac{2}{\pi} \frac{R\hbar\omega}{e^{\hbar\omega/kT} - 1} \tag{16.25}$$

At the low frequency limit, *i.e.*, $\hbar\omega \ll kT$, there is an approximate expression $\langle v_a^2 \rangle_\omega \approx 2/\pi \cdot kTR$ or $\langle v_a^2 \rangle_f \approx 4kTR$. This is exactly the expression for the spectral density of the Johnson–Nyquist noise.

As a conclusion, an antenna with ideal broadband absorption can generate voltages with the spectral density of the thermal noise across a hot resistor. The thermal noise is at the temperature of the incident black-body radiation, and is excited across a resistor of the antenna's effective impedance. It is noted that for an antenna with incomplete absorption, the excited frequency-dependent voltage is scaled down by the antenna's absorptivity. However, unlike the Johnson–Nyquist noise, which is purely incoherent, the information of partial coherence can be retained in the excited ac voltage.

16.4.2 Energetics of Thermal Rectification

From the previous section, the solar radiation plus the ideal broadband antenna can be substituted by a hot resistor at a fixed temperature, T_C. Thus the rectification process can now be considered as for the thermal noise generated by a hot resistor,[6] with the assumption of complete incoherence. In this section thermal rectification by a cold diode is considered, providing a quantitative picture for the rectenna system.

We assume that the antenna is ideal, *i.e.*, it converts all power into radiation. The equivalent circuitry is shown in Figure 16.13(a), representing a thermal converter between the resistor (heat source) at temperature T_C and the diode (heat sink) at temperature T_A. This converter rectifies Johnson–Nyquist noise generated across the hot resistor and extracts a DC current, i, through the load. It is noted that the hot resistor, R, shown in Figure 16.13(a), which represents the antenna accepting sunlight, is replaced by a noiseless resistor R and an AC voltage source v(t) in Figure 16.13(b). The capacitor C is the output filter for generating a DC voltage $\langle u(t) \rangle$ across the load.

For the purpose of circuit analysis, the hot resistor is replaced by a noiseless resistor and a voltage source v(t) connected in series (Figure 16.13(b)). The correlation between v(t) and the voltage across the load, u(t) will be revealed in the following analysis.

Corresponding to the absorption of the antenna, the power supplied from the voltage source is:

$$S_R = \langle v(t) i_R(t) \rangle = \langle v \cdot (v - u) / R \rangle = \frac{\langle v^2 \rangle - \langle uv \rangle}{R} \tag{16.26}$$

where $i_R(t)$ denotes the current following through the noiseless resistor R. Some of the absorbed power is re-emitted by the antenna. This corresponds to the power dissipated by the resistor R:

$$C_R = \frac{\langle (u - v)^2 \rangle}{R} = \frac{\langle u^2 \rangle + \langle v^2 \rangle - 2 \langle uv \rangle}{R} \tag{16.27}$$

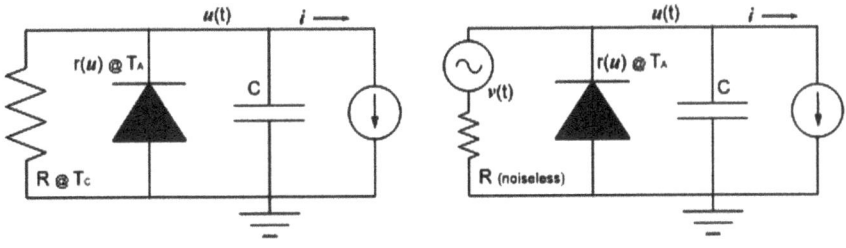

Figure 16.13 (a) Equivalent circuitry of the rectenna system: a hot resistor R at temperature T_C rectified by a diode at a lower temperature T_A; (b) the hot resistor is equivalent to a noiseless resistor and a voltage source v(t) connected in series

The rest is the available power extracted from the antenna:

$$Q_R = S_R - C_R = \frac{\langle uv \rangle}{R} - \frac{\langle u^2 \rangle}{R} \tag{16.28}$$

To evaluate the first term in the expression for Q_R, the relation between u(t) and v(t) is expressed according to the Kirchoff's law:

$$v(t) = \left[1 + \frac{R}{r(u)}\right] u(t) + RC\dot{u}(t) \tag{16.29}$$

where r(u)denotes the non-linear resistance of the diode. By solving this differential equation the time average value of v(t)u(t) is found:

$$\langle uv \rangle = \lim_{T \to \infty} \frac{1}{2TRC} \int_{-T}^{T} v(t_1) e^{-\frac{t_1}{R'C}} dt_1 \int_{-\infty}^{t_1} v(t_2) e^{\frac{t_2}{R'C}} dt_2 \tag{16.30}$$

where $R' = R/[1 + \langle R/r(u) \rangle]$. The integrations can be evaluated in the frequency domain, *i.e.* expanding v(t) with v(ω) using the Fourier transformation. By considering the orthogonality of exp(iωt) components, Equation (16.30) is simplified to a resistance-independent form:

$$\langle uv \rangle = \frac{1}{RC} \int_{0}^{\infty} \frac{v^2(\omega) R' C}{1 + \omega^2 R'^2 C^2} d\omega = \frac{kT}{C} \tag{16.31}$$

where $v^2(\omega) = 2RkT/\pi$ is the Johnson–Nyquist noise at the low-frequency limit. This gives the first term in Equation (16.28). The second term $\langle u^2 \rangle/R$ can be calculated from the statistical distribution of the voltage, u. Its probability density has been calculated from a Markovian diffusion model proposed by Sokolov:[43]

$$p(u; i) = \begin{cases} A \exp\left(-\dfrac{Cu^2}{2kT_C}\right), u > 0 \\ A \exp\left(-\dfrac{Cu^2 + 2iuRC}{2kT_A}\right), u < 0 \end{cases} \tag{16.32}$$

where i is the extracted current through the load and A is a normalisation factor. This distribution is derived for an ideal rectifier, *i.e.* zero resistance for u > 0 while infinite resistance for u < 0. The rectified voltage $\langle u \rangle$ is calculated by the same distribution function, as well as the DC output power:

$$P(i) = i \langle u \rangle = i \int_{-\infty}^{\infty} p(u; i) u \, du \tag{16.33}$$

The efficiency of thermal rectification writes $\eta_{rc}(i, T_C) = P(i)/Q_R(i)$. It varies with the dimensionless current $x = iR\sqrt{C/kT_C}$, as illustrated in Figure 16.14(a). The maximum efficiency increases with the temperature of the hot resistor (Figure 16.14(b)). It is considerably lower than the Carnot efficiency, but it increases faster at higher temperatures.[43] This indicates a maximum thermal rectification efficiency of 49%, corresponding to the sun's temperature at 6000 K.

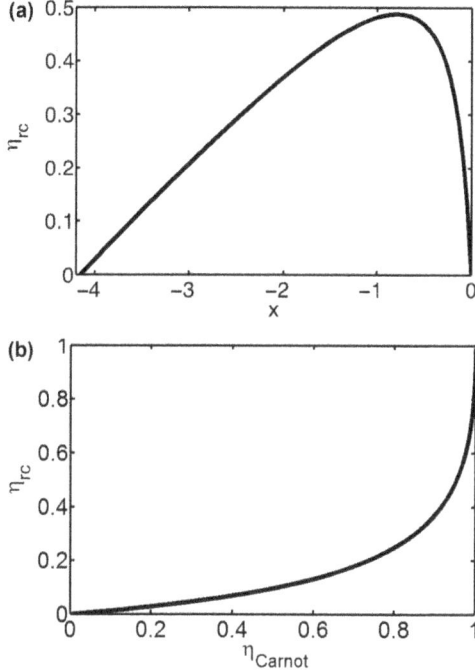

Figure 16.14 (a) The rectification efficiency η_{rc} as a function of the dimensionless current $x = iR\sqrt{C/kT_C}$. The temperature $T_C = 6000$ K $(\eta_{Carnot} = 0.95)$. The maximal efficiency is attained at a finite current; (b) The maximal rectification efficiency η_{rc} as a function of η_{Carnot}. At moderate temperature differences (moderate η_{Carnot}) the efficiency of the engine is considerably lower than η_{Carnot}, but it increases rapidly when η_{Carnot} approaches unity.
Figure 16.14(b) is reproduced from Reference 43.

16.5 Quantum Rectification

In considering the rectification process one usually thinks of applying a time-varying ac voltage to a diode to produce a smoothly time-varying current. Due to the asymmetry in the diode's $I(V)$ characteristics the current flows dominantly in one direction, producing a DC current output. That is not the way that an optical rectenna works – or for that matter, any rectifier working at optical frequencies. We can gain insight into optical frequency rectification by looking at the conduction band profile of a MIM diode at different modulation frequencies. In Figure 16.15 the effect of a sub-optical-frequency AC voltage is shown. The energy difference between the left and right hand Fermi levels is modulated by the applied AC voltage.

At optical frequencies the photon energy divided by the electronic charge, $\hbar\omega/e$, is on the order of the voltage at which there is significant nonlinearity in the $I(V)$ characteristic. A semi-classical (quantum) approach is required to evaluate the tunneling current. Photon-assisted tunneling (PAT) theory was

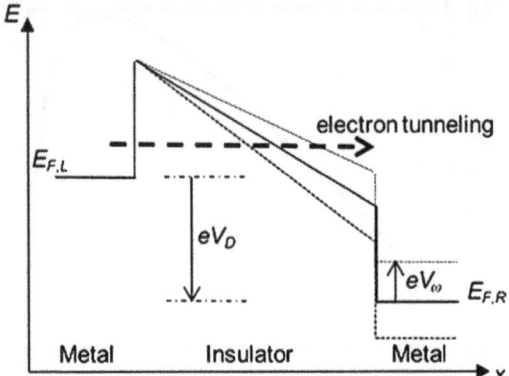

Figure 16.15 Classical model of the conduction band profile of an MIM diode modulated by a sub-optical-frequency ac voltage. The ac signal causes the Fermi level difference between the left and right sides of the tunnel junction to oscillate, which causes a change in the tunneling distance and hence in the tunnel current.
Reproduced from Grover, Joshi and Moddel.[53]

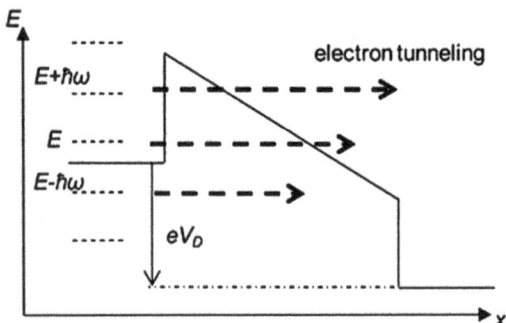

Figure 16.16 Effect of photon-assisted tunnelling (PAT) on electron tunneling through an MIM diode. The semiclassical theory gives the probabilities for electrons at energy E to absorb or emit photons and thus occupy multiple energy levels separated by $\hbar\omega$.
Reproduced from Grover, Joshi and Moddel.[53]

developed by Tien and Gordon[81] to analyze the interaction of photons in a superconducting junction and adapted to tunnel devices by Tucker and Millea.[82] The result can be seen in the effect on the conduction band profile of an MIM diode under optical frequency modulation, shown in Figure 16.16. The sea of electrons below the Fermi level in the metal conduction band now occupy multiple energy levels that are separated by $\pm\hbar\omega$ from the original energy levels, and multiples of $\hbar\omega$ for higher-order, less probable interactions.

The effect of PAT can be seen in the $I(V)$ characteristics that result. To make the illuminated $I(V)$ curve formation clear, we start with a simple piecewise linear $I(V)$ curve, shown in Figure 16.17(a). Scaled PAT components

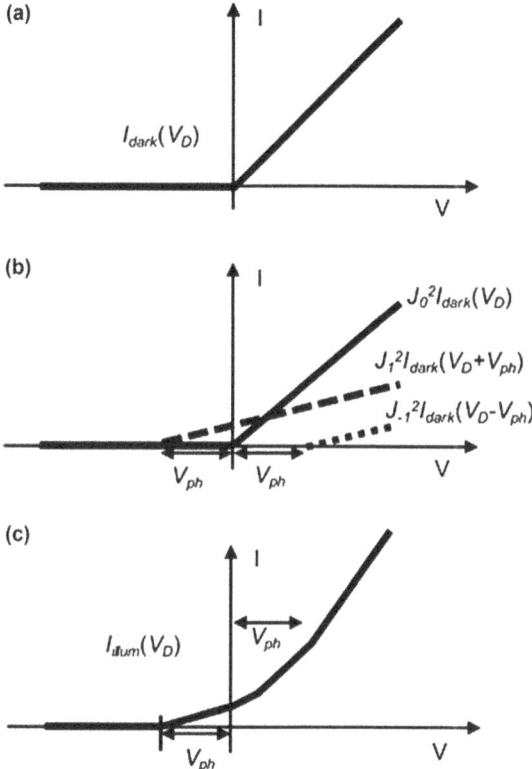

Figure 16.17   (a) Piecewise linear dark $I(V)$ curve. (b) Under high frequency illumination scaled components of the dark $I(V)$ curve shifted by $\pm\hbar\omega$ are
added. (c) The illuminated $I(V)$ curve obtained by adding the components in (b).
Reproduced from Grover, Joshi and Moddel.[53]

of the dark $I(V)$ curve shifted by $\pm\hbar\omega$ are added, as shown in Figure 16.17(b),
and added together to produce the resulting curve shown in Figure 16.17(c).
Unlike conventional solar cells, which produce power in the 4th quadrant of
the $I(V)$ curve, rectenna solar cells produce power in the 2nd quadrant.

The $I(V)$ curve in the 2nd quadrant appears to be triangular in
Figure 16.17. This would give a poor maximum fill factor of 25%. In fact,
when the load is matched to the diode for each voltage, corresponding to
constant incident power, a more rectangular $I(V)$ curve results.[53] A more
realistic set of $I(V)$ curves is shown in Figure 16.18, showing the illuminated
$I(V)$ curve as more rectangular.

The quantized nature of the tunneling process affects not only the rectification process but also the diode ac resistance as seen by the antenna. At
optical frequencies the rectification proceeds by discrete electron energy
shifts, as opposed to the continuous variations shown in Figure 16.15. The
diode ac resistance also becomes a function of the $I(V)$ curve at $\pm\hbar\omega$ about

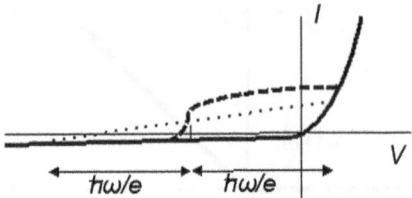

Figure 16.18   Sketch of an $I(V)$ curve for a rectenna diode. The solid curve shows
the $I(V)$ for the rectenna in the dark, and the dashed curve shows the
$I(V)$ under illumination. The operating voltage for the maximum
power point is indicated by a small vertical line on the V axis. The
secant resistance is the reciprocal of the slope of the line connecting
the dark $I(V)$ curve at $\pm\hbar\omega$ about the operating voltage and is shown
as a dotted line.
Reproduced from Moddel.[83]

the operating voltage. This 'secant resistance'[53] is the reciprocal of the slope
of the dotted line shown in Figure 16.18. The secant resistance of the dark
$I(V)$ curve determines the coupling efficiency between the antenna and diode
at optical frequencies, and the conventional resistance of the illuminated
$I(V)$ curve at the operating point determines the DC coupling between the
diode and the load.

The quantum nature of the rectification process at optical frequencies, as
described by PAT theory, has several consequences.[53] It limits the quantum
efficiency of rectennas to unity, *i.e.*, one electron of current for each incident
photon. It reduces the AC resistance of the diode, as compared to the small-
signal differential resistance. This is good because the diode impedance
must match to the low impedance of the antenna, $\sim 100~\Omega$, for optimal
power coupling. This quantum rectification also has severe implications for
the power conversion efficiency, as described in the next section.

16.6   Broadband Rectification Efficiency Limit

Microwave rectennas have demonstrated broadband power conversion efficiencies well in excess of 80%.[84] When Bailey proposed the use of optical
rectennas for solar energy conversion in 1972,[18] the technology was seen as a
way to break through the Shockley–Queisser limit of 34%,[85] which is function of the Trivich–Flinn limit of 44% imposed by a quantum process[86] reduced by thermodynamic considerations. Behind Bailey's proposal was the
implicit assumption that rectenna rectification was not subject to the 44%
limit imposed by a quantum process. Earlier in this chapter we considered
the efficiency limitations based upon thermodynamic considerations. Here
we explore whether Bailey's implicit assumption was correct, and what the
broadband efficiency limit is based upon the actual quantum rectification
process as described by photon-assisted tunneling theory.

For monochromatic illumination, the power conversion efficiency for
optical rectennas can approach 100%, just as with conventional solar cells.

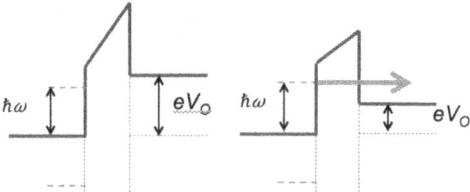

Figure 16.19 Band diagram for an MIM diode under two operating voltages, V0, under monochromatic illumination with photon energy $\hbar\omega$ (a). When $\hbar\omega < eV_0$ the electrons have insufficient energy to tunnel from the left hand metal to an empty state in the right hand metal. (b) When $\hbar\omega > eV_0$ the electrons do have sufficient energy to tunnel to an empty state.
(Unpublished, courtesy of Saumil Joshi.)

The key to achieving high efficiency is the operating voltage, V_0, assuming that other parameters such as antenna efficiency, diode $I(V)$ asymmetry, antenna-diode impedance matching are all perfect. The operating voltage is a self bias that is determined by the load resistance times the DC photo-current. The effect of the operating voltage can be seen in the band diagrams of Figure 16.19, which are special cases of Figure 16.16. In Figure 16.19(a) the operating voltage times the electron charge (eV_0) is higher than the photon energy ($\hbar\omega$). The consequence is that the electrons excited in the left hand metal do not have sufficient energy to tunnel to an unfilled state in the right hand metal. In Figure 16.19(a) $eV_0 < \hbar\omega$ and the excited electrons can tunnel to an empty state. If $eV_0 \ll \hbar\omega$ the efficiency will be poor because only a fraction of the energy of each incident photon is used, and so the optimal efficiency results when $eV_0 \cong \hbar\omega$.

For broadband illumination from the sun optimal efficiency would result only if a different operating voltage could be selected for each photon energy region of the spectrum. Since the photocurrent from the entire spectrum is channelled to a single diode there can be only one operating voltage, and hence the power conversion efficiency is compromised. This efficiency has been calculated based on photon-assisted tunneling theory assuming perfect antenna efficiency, diode $I(V)$ asymmetry, and antenna-diode impedance matching.[54] The result is 44%, identical to the Trivich–Flinn efficiency[86] and the Shockley–Queisser 'ultimate efficiency' limit.[85] The diode operating voltage in rectenna solar cells plays the role that bandgap plays in conventional solar cells.

Bailey's original intention of exceeding the efficiency limits of a quantum process are not realized with optical rectennas. The reason that the beyond 80% power conversion efficiency of microwave rectennas does not apply here is that the microwave rectennas operate in the classical domain whereas the solar rectennas operate in the quantum regime. The classical regime applies when $\hbar\omega/e$ is much smaller than the voltage at which the $I(V)$ curve exhibits significant nonlinearity, and the photon flux is low enough that the AC voltage developed at the diode is much less than $\hbar\omega/e$.

A way to circumvent this efficiency limit would be to use some sort of a 'photon homogenizer'[83] that could process the broad band of photon energies and convert it to a single photon energy. Then a single optimal operating voltage could be used in the rectification process. In theory at least, this could be achieved by the mixing of frequencies in the diode to produce sum, difference, and harmonic frequencies. This would allow a high operating voltage corresponding to the highest photon energies of interest, and photons of lower energies would be mixed together to produce sufficient energies for the electrons to tunnel to empty states. Such mixing would require sufficiently high intensity to engage higher-order rectification processes, and it would, in fact, result in higher efficiency.[54] These higher intensities are not achievable with solar illumination for optical rectennas, even if large antennas or optical concentrators are used. The reason is that the coherence of terrestrial sunlight extends over a diameter of only 19 μm.[48] Gathering the sunlight from a larger area decreases the coherence and results in diminishing returns for the rectified current due to cancellation of out-of-phase components of the current in the diode. Without some innovation to greatly improve the nonlinearity of the diode far beyond what has been achieved for any type of room-temperature diode, or to somehow create coherence in the illumination, the ultimate conversion efficiency limit of 44% remains. As with conventional solar cells, the conversion efficiency of rectenna solar cells could be increased by splitting the spectrum and directing each spectral region to a rectenna solar cell at a different optimal operating voltage.[74] Rectennas have an inherent advantage over conventional solar cell in spectral splitting because they not require materials matched to each spectral range.

16.7   High-frequency Rectifiers

16.7.1   MIM/MIIM Rectifiers

Two types of transducers have commonly been used for IR and optical antenna devices,[4] microbolometers and metal—insulator–metal (MIM) diodes.[74] Both, but especially the microbolometer, are sensitive to the temperature of the surrounding materials. Microbolometers respond more slowly than MIM diodes, with the latter speed being limited, in theory, to about 10^{-15} s by the speed of electron tunneling through the junction. Experimental devices respond more slowly.

16.7.1.1   RC Time Constant Limitation of MIM Diodes

Although the transit time of electrons through the insulator is sufficiently short to allow optical frequency rectification, other constraints severely limit the response time of MIM diodes. In particular, the *RC* time constant for the antenna/diode system is the culprit.[74] In the usual rectenna circuit, the diode and the antenna are in parallel. To efficiently transfer AC power from the antenna to the diode the resistances of the two elements must match. Since the resistance of optical antennas is on the order of 100 Ω the resistance of the diode must be similar. The *RC* time constant for the system is then the product of the parallel resistance, approximately 50 Ω, and the diode capacitance. Providing a sufficiently low resistance in the diode requires a sufficiently large area, but the larger the diode area the larger geometric capacitance will be.

For an optimal tunneling device the largest current density that can generally obtained in a tunnel current is less than 10^7 A cm^{-2}. The smallest imaginable voltage at which such a current could be produced is at least 0.5 V. Combining these two numbers gives an absolute minimum resistance of 5×10^{-8} Ω cm^{-2}. The minimum capacitance will occur for a low dielectric constant and large insulator thickness. For an insulator with a very low relative dielectric constant of 2 and a large thickness (for a tunneling device) of 5 nm, the geometric capacitance is 4×10^{-7} F cm^{-2}. Multiplying the resistance and capacitance values gives $RC = 20$ fs. The peak of the solar spectrum is at a wavelength of approximately 2 µm, which corresponds to a frequency of $f = 0.15$ PHz $(0.15\times10^{15}$ Hz). Rectifying this requires a response time of $1/2\pi f = 1$ fs, which is a factor of 20 smaller than the lowest possible *RC* time constant. The *RC* time constant for practical diodes will be even greater than 20 fs,[74] so that it is not feasible to rectify visible-light frequencies using rectennas with parallel-plate diodes.

16.7.2 New Concepts for High Frequency

Because of the *RC* limitations in MIM diodes discussed above, several alternative diodes for optical rectennas have been instigated, as discussed below.

16.7.2.1 *Double-insulator MIIM Diodes*

Forming a double-insulator MIIM diode can provide improved *I(V)* characteristics over single-insulator MIM diodes. The application of resonant MIIM diodes for rectenna solar cells was proposed, simulated, and demonstrated at DC by Eliasson and Moddel[16,33] and demonstrated at 60 GHz[34] and infrared frequencies at Phiar Corporation. Hegyi *et al.*[87] simulated MIIM characteristics in the absence of resonance. Grover and Moddel[88] analyzed MIIM diodes in detail and compared their characteristics to single-insulator devices. Analysis of the *I(V)* characteristics of MIM and MIIM diodes requires a simulator, as use of analytical tunneling theory gives incorrect results, particularly for low barrier diodes.[88] Alimardani *et al.*[89] demonstrated double-insulator diodes and Maraghechi *et al.*[90] demonstrated high-barrier triple-insulator diodes.

The advantages of MIIM diodes arise from one of two mechanisms,[88] as shown in Figure 16.20. For the example, the two types of diodes are identical except for thickness. When the left hand insulator, which has a larger electron affinity than the right hand insulator, is sufficiently thick a resonant quantum well forms in its conduction band. When the Fermi level of the left hand

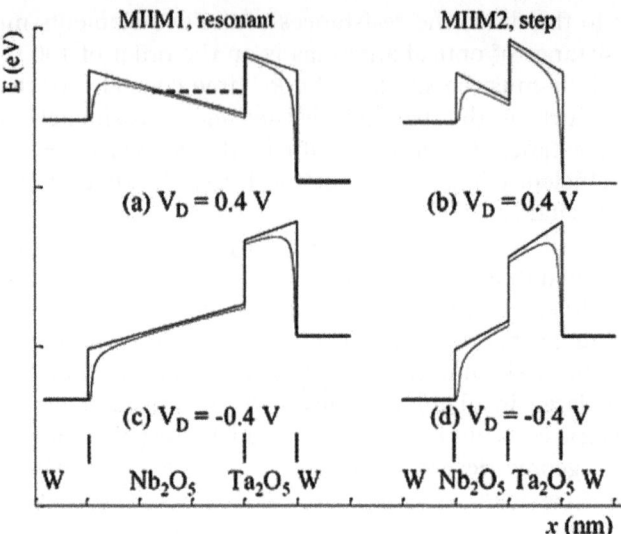

Figure 16.20 Mechanisms for enhanced nonlinearity in MIIM diodes. Energy-band profiles are shown for resonant and step MIIM diodes. Forward and reverse bias profiles are shown respectively in (a) and (c) for a resonant diode, and in (b) and (d) for a step diode. The dotted lines show the profiles with image force barrier lowering. The thickness of the Nb_2O_5 layer is the only difference between the two diodes. Reproduced from Grover and Moddel.[88]

metal rises to match the resonant level in the insulator the level provides a transport path for electrons. This produces a sharp increase in the current because the tunneling distance for electrons is reduced, and the current increases exponentially with decreasing tunneling distance. For the resonant tunneling mechanism, forward bias occurs when the Fermi level is raised for the metal adjacent to the higher electron affinity insulator. For an applied voltage of the opposite polarity, shown in Figure 16.20(c), the electrons must tunnel through all or most of the two insulators, and the current is therefore smaller than under forward bias and corresponds to reverse bias.

Alternatively, when the left hand insulator is thinner any resonant level formed would be too high to be useful. Under the polarity of bias that corresponded to forward bias for the resonant diode, shown in Figure 16.20(b), here the electrons must tunnel through both insulators. With the opposite polarity bias shown in Figure 16.20(c) electrons tunnel through only the higher-conduction-band insulator, and are then injected into the conduction band of the other insulator. For this 'step' diode the latter condition corresponds to forward bias, opposite to the resonant case. Both the resonant and the step diode mechanisms are useful. The choice depends upon the available materials and desired characteristics.

Double-insulator MIIM diodes tend to show greater nonlinearity in their $I(V)$ characteristics than single-insulator MIM diodes made using similar

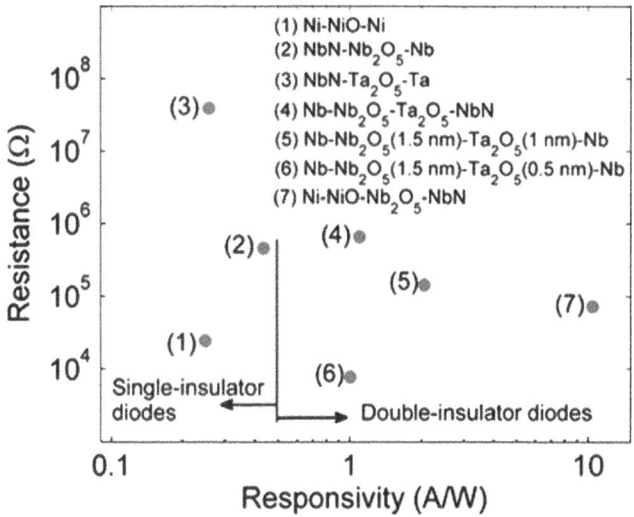

Figure 16.21 Comparison of single- and double-insulator diodes, showing resistance versus responsivity at zero bias. Rectennas require high responsivity and low resistance. The double-insulator diodes show improved performance, having smaller resistance and larger responsivity. The area for the diodes is 100×100 nm^2.
Reproduced from Grover.[52]

materials. For a rectenna two key desirable characteristics are (1) low resistance, to couple efficiently to the antenna resistance, and (2) high responsivity, which is a function of the $I(V)$ curvature as defined in Equation (16.17). In Figure 16.21 these two parameters are shown for a realistic set of materials for single and double insulator diodes. Desirable characteristics are in the lower right hand region of the plot. As can be seen from the examples shown, double-insulator diodes fall in this region and single-insulator diodes do not.

16.7.2.2 Sharp-tip Diodes

The RC time constant for planar MIM diodes is independent of area because the $R \propto 1/$area and $C \propto$ area. One way to decrease the RC is to change the shape of the diode. Miskovsky *et al.*[49] are developing a rectenna with a sharp tip because the constant RC for planar MIM devices is replaced with an RC that scales with (area)$^{1/4}$ for a spherical tip. In Figure 16.22 a schematic for a sharp tip rectenna is shown. Fabricating such a device, in which the tip is separated from the electrode by a tunneling distance of only ∼1 nm, is a challenge. The ingenious way that Miskovsky *et al.*[49] accomplish this is to initially form large metal/vacuum/metal structures and then add material to the tip using atomic layer deposition (ALD). As the spacing approaches 1 nm the reactants can no longer access the tip region, and so the growth stops in a self-limiting process.

A different type of sharp tip asymmetric tunneling diode was developed by Choi *et al.*[50] It makes use of a coplanar MIM diode formed by deposition,

Figure 16.22   Schematic of a sharp tip rectenna.
Reproduced from Miskovsky *et al.*[49]

electron-beam lithography and then strain-assisted self liftoff to form the diodes. Coupled with antennas the devices have shown asymmetric $I(V)$ characteristics and response to RF[50] and to 10.6 μm infrared radiation.[91]

16.7.2.3   Traveling-wave Diodes

Another way to avoid the diode RC time constant constraint is to using a traveling-wave configuration, so that the diode impedance is like that of a transmission line rather than a lumped element.[35,36] In such a case the characteristic impedance is largely determined by size and spacing of the structure rather than the tunneling properties. A schematic of the traveling-wave diode is shown in Figure 16.23.

Using a finite element analysis the characteristics of the traveling-wave diode were simulated.[36,92] When used as a detector with an applied bias, the traveling-wave configuration shows a responsivity for 3 μm radiation that is nearly three orders of magnitude larger than its lumped element counterpart, as shown in Figure 16.24. Further work is required to determine whether the traveling-wave diode advantages in a detector provide similar advantage for an energy harvesting device.

16.7.2.4   Geometric Diodes

The main source of capacitance in MIM diodes is between the parallel plates of the two electrodes. A totally planar device would have substantially lower

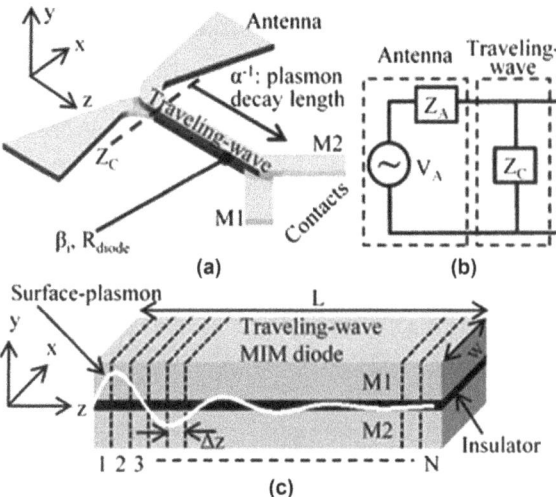

Figure 16.23 Traveling-wave diode configuration. (a) View showing the antenna arms converging into a parallel-plate waveguide with a thin insulator between the metals M1 and M2. The load is connected to the contact pads. (b) Small signal circuit model for the rectenna. The characteristic impedance of the traveling-wave diode can be readily matched to that of the antenna. (c) A 3D view of the traveling-wave MIM diode showing a surface plasmon traveling down the insulator, and decaying as it produces a rectified current.
Reproduced from Grover *et al.*[36]

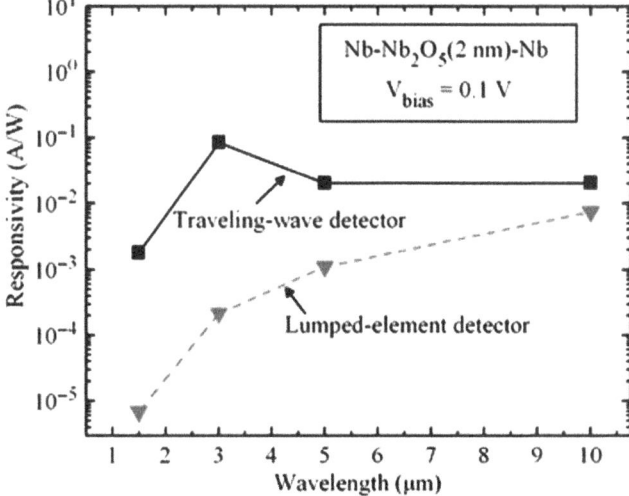

Figure 16.24 Calculated responsivity comparison of lumped-element and the traveling-wave detectors. The traveling-wave detector shows significantly better performance at smaller wavelengths.
Reproduced from Grover *et al.*[36]

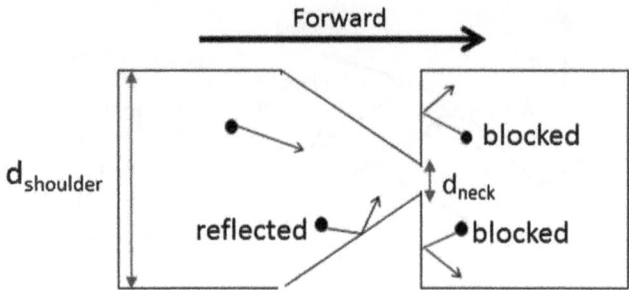

Figure 16.25   (a) Geometric diode structure. The neck width (d_{neck}) must be on the order of the carrier mean-free path length for charge carriers in the material. The charge carriers reflect at the boundaries of the device. Reproduced from Zhu et al.[37]

capacitance, and hence a potentially smaller *RC* time constant. The geometric diode configuration shown in Figure 16.25 is such a planar diode, and is formed from a thin conductor. On the left-hand side of the neck, the carriers moving to the right can either directly channel through the neck or reflect off the tapering edges and keep moving forward. On the right-hand side of the neck, the vertical edge blocks most of the carriers moving to the left. Hence the diode forward direction for carriers is left-to-right. The planar capacitance of such is device is calculated to be a few attofarads.[37]

To sense the device geometry the mean-free path length of the charge carriers must be on the order of the neck width of the diode. Because graphene has a particularly long mean-free path length, approaching 1 µm, it serves as an excellent material for such a diode. Optical response of graphene geometric diodes has been demonstrated for 10.6 µm radiation (corresponding to 28 THz).[37] Thus far the *I*(*V*) characteristics of such devices has have been insufficiently nonlinear to provide efficient rectification, and will need to be improved if the technology is to become practical.

16.8   Summary and Conclusions

Quantum rectennas for solar application accept sunlight by using optical antennas, followed by rectification process for DC power output. This concept was proposed in 1970s and has been investigated via analogs in the fields of radio astronomy, microwave transmission and biological optical detectors. Relevant fundamental problems have been discussed in this chapter. An example of this is the partial coherence of sunlight, which limits size of the antenna or array of antennas. Quantitative analysis has demonstrated a reduced coherence efficiency at a larger field point separation. 90% coherence efficiency can only be achieved at a separation of 19 µm or less. Several approaches to antenna design, such as conical spiral antennas, have been proposed to accept both polarizations plus a bandwidth much broader than that of conventional microwave antennas. In terms of practical issues, design and fabrication of nanoscale elements are essential, due to the shorter plasmonic wavelength and the ultrafast rectification speed. Rectification at optical frequencies also requires quantum analysis of the transport of charge carriers. The *RC* time constant of MIM diodes is too restricted, it may still be possible for new rectifier techniques to bypass this limit. Alternatives to MIM diodes include multi-insulator metal-insulator diodes, sharp-tip diodes, traveling-wave diodes, and geometric diodes. Element matching between the antenna, rectifier and filters put further restrictions on design of the rectenna solar cell.

The fundamental efficiency limit for the rectenna solar cell is still under investigation. In this chapter two different approaches are discussed, corresponding to the respective scenarios of incoherent and coherent sources. The first approach is based on rectification of thermally agitated noise and it is valid for purely incoherent black-body radiation and broadband absorption. In this situation the incident light and the antenna together are treated as a heat source connected to a cold rectifier. That rectenna system is essentially a heat engine. The rectification efficiency increases towards the Carnot efficiency for a larger temperature contrast, reaching 49% for 6000 K. Alternatively, to avoid cancellation of current in the diode due to absorption of incoherent radiation by the antenna, the antenna size and hence the current magnitude is limited. In quantum rectification in optical rectennas, the operating (bias) voltage plays the role that bandgap plays in semiconductor solar cells. The limited current magnitude along with quantum rectification process limit the efficiency for rectifying broadband solar radiation to 44%.

Acknowledgements

The contributions of FY and RC to this Program have been supported by the Australian Government through the Australian Renewable Energy Agency (ARENA). Responsibility for the views, information or advice expressed herein is not accepted by the Australian Government.

References

1. L. Novotny and N. van Hulst, *Nat. Photonics*, 2011, **5**, 83.
2. J. J. Greffet, *Science*, 2005, **308**, 1561.
3. R. L. Olmon and M. B. Raschke, *Nanotechnology*, 2012, **23**, 444001.
4. J. Alda, R.-G. J. M. J. López-Alonso and G. Boreman, *Nanotechnology*, 2005, **16**, S230.
5. D. C. Skigin and M. Lester, *J. Nanophotonics*, 2011, **5**, 050303.
6. R. Corkish, M. A. Green, T. Puzzer and T. Humphrey, Proceedings of 3rd World Conference on Photovoltaic Energy Conversion, 2003, **3**, 2682.
7. P. Muhlschlegel, H. J. Eisler, O. J. F. Martin, B. Hecht and D. W. Pohl, *Science*, 2005, **308**, 1607.
8. M. A. Green and S. Pillai, *Nat. Photonics*, 2012, **6**, 130.
9. V. P. Zhdanov and B. Kasemo, *Appl. Phys. Lett.*, 2004, **84**, 1748.
10. H. Tan, R. Santbergen, A. H. Smets and M. Zeman, *Nano Lett.*, 2012, **12**, 4070.

11. H. A. Atwater and A. Polman, *Nat. Mater.*, 2010, **9**, 205.
12. B. F. Burke and F. Graham-Smith, *An Introduction to Radio Astronomy*, Cambridge University Press, Cambridge, 1997.
13. R. Corkish, M. A. Green and T. Puzzer, *Sol. Energy*, 2002, **73**, 395.
14. D. Y. Goswami, S. Vijayaraghavan, S. Lu and G. Tamm, *Sol. Energy*, 2004, **76**, 33.
15. T. M. Razykov, C. S. Ferekides, D. Morel, E. Stefanakos, H. S. Ullal and H. M. Upadhyaya, *Sol. Energy*, 2011, **85**, 1580.
16. B. J. Eliasson, *Metal-insulator-metal diodes for solar energy conversion*, PhD Thesis, University of Colorado, 2001.
17. *Rectenna Solar Cells*, ed. G. Moddel and S. Grover, Springer, New York, 2013.
18. R. L. Bailey, *J. Eng. Power*, 1972, **April**, 73.
19. R. L. Bailey, P. S. Callahan and M. Zahn, Electromagnetic wave energy conversion research, Report NASA-CR-145876 (N76-13591), University of Florida, 1975.
20. USA Pat., 3,760,257, 1973.
21. USA Pat., 4,445,050, 1984.
22. USA Pat., 4,574,161, 1986.
23. USA Pat., 4,720,642, 1988.
24. USA Pat., 4,972,094, 1990.
25. A. M. Marks, Proceedings of the 26th Intersociety Energy Conversion Engineering Conference, ed. D. L Black, American Nuclear Society, La Grange Park, IL, 1991, 5, pp. 74.
26. Polarized Solar Electric Co., Lumeloid solutions, Available: http://www.polar-solar.com/lumeloi-solutions.html, 2009, [18 January 2013].
27. E. A. Farber, Antenna Solar Energy to Electricity Converter (ASETEC), Report AF C F08635-83-C-0136, Task 85-6, University of Florida, 1988.
28. J. D. Kraus, *Antennas*, McGraw-Hill, New York, 2nd edn., 1988.
29. J. D. Kraus and R. J. Marhefka, *Antennas for all Applications*, McGraw-Hill, Boston, 3rd edn., 2002.
30. G. H. Lin, R. Abdu and J. O. M. Bockris, *J. Appl. Phys.*, 1996, **80**, 565.
31. B. Berland, L. Simpson, G. Nuebel, T. Collins and B. Lanning, Proceedings of National Center for Photovoltaics, Program Review Meeting, 2001, pp. 323–324.
32. B. Berland, Photovoltaic Technologies Beyond the Horizon: Optical Rectenna Solar Cell, Report NREL/SR-520-33263, ITN. Energy Systems Inc., Littleton, Colorado, 2003.
33. B. J. Eliasson and G. Moddel, US Patent 6,534,784. 2003.
34. S. Rockwell, S. D. Lim, D. B. A. Bosco, J. H. Baker, B. Eliasson, K. Forsyth and M. Cromar, in Radio Frequency Integrated Circuits (RFIC) Symposium, 2007 *IEEE*, 171.
35. M. J. Estes, G. Moddel, US Patent 7,010,183, 2006.
36. S. Grover, O. Dmitriyeva, M. J. Estes and G. Moddel, *IEEE Trans. Nanotechnol.*, 2010, **99**, 716.
37. Z. Zhu, S. Joshi, S. Grover and G. Moddel, *J. Phys. D: Appl. Phys.*, 2013, **46**, 185101.

38. Y. Wang, K. Kempa, B. Kimball, J. B. Carlson, G. Benham, W. Z. Li, T. Kempa, J. Rybczynski, A. Herczynski and Z. F. Ren, *Appl. Phys. Lett.*, 2004, **85**, 2607.

39. M. Sarehraz, K. Buckle, T. Weller, E. Stefanakos, S. Bhansali, S. Krishnan and Y. Goswami, Proceedings of 31st IEEE Photovoltaic Specialists Conference, 2005, p. 78.

40. USA Pat., US2007096990-A1; US7362273-B2, 2008.

41. USA Pat., US2007069965-A1; US7486236-B2, 2009.

42. R. H. Dicke, *Rev. Sci. Instrum.*, 1946, **17**, 268.

43. I. M. Sokolov, *Europhys. Lett.*, 1998, **44**, 278.

44. D. K. Kotter, S. D. Novack, W. D. Slafer and P. Pinhero, ASME Conference Proceedings, 2008, 43208, 409.

45. R. Osgood III, S. Giardini, J. Carlson, G. E. Fernandes, J. H. Kim, J. Xu, M. Chin, B. Nichols, M. Dubey, P. Parilla, J. Berry, D. Ginley, P. Periasamy, H. Guthrey and R. O'Hayre, SPIE (*Int. Soc. Opt. Eng.*), Plasmonics: Metallic Nanostructures and Their Optical Properties IX, ed. M. I. Stockman, 2011, 8096, 809610.

46. J. M. Nunzi, ed. D. L. Andrews, J. M. Nunzi and A. Ostendorf, SPIE, Nanophotonics III, 2010, 7712, 771204.

47. R. M. Metzger, *Synth. Met.*, 2009, **159**, 2277.

48. H. Mashaal and J. M. Gordon, *Opt. Lett.*, 2011, **36**, 900.

49. N. M. Miskovsky, P. H. Cutler, A. Mayer, B. L. Weiss, B. Willis, T. E. Sullivan and P. B. Lerner, *J. Nanotechnol*, 2012, 512379.

50. K. Choi, *IEEE Trans. Electron Dev.*, 2011, **58**, 3519.

51. M. Gallo, L. Mescia, O. Losito, M. Bozzetti and F. Prudenzano, *Energy*, 2012, **39**, 27.

52. S. Grover, *Diodes for optical antennas*, PhD Thesis, University of Colorado, Boulder, 2011.

53. S Grover, S. Joshi and G. Moddel, *J. Phys. D: Appl. Phys.*, 2013, **46**, 135106.

54. S. Joshi and G. Moddel, *Appl. Phys. Lett.*, 2013, **102**, 083901.

55. F. V. Dwivedi, Proceedings of International Symposium on Microwave, Antenna, Propagation and EMC Technologies for Wireless Communications, 2005, 342.

56. G. A. E. Vandenbosch and M. Zhongkun, *Nano Energy*, 2012, **1**, 494.

57. J. C. Mankins, *Space Solar Power. The First International Assessment of Space Solar Power: Opportunities, Issues and Potential Ways Forward*, International Academy of Astronautics, 2011.

58. N. Komerath, V. Venkat and J. Fernandez, Space, Propulsion & Energy Sciences International Forum SPESIF-2009, Huntsville, Alabama, 2009, 149.

59. W. C. Brown, *IEEE Trans. Microwave Theory Technol.*, 1984, **MTT-32**, 1230.

60. R. Lomas, *The Man Who Invented the Twentieth Century*, London, 1999.

61. B. P. Motjolopane and R. van Zyl, *J. Eng., Design Technol.*, 2009, 7, 282.

62. B. F. Burke and F. Graham-Smith, *An Introduction to Radio Astronomy*, Cambridge University Press, Cambridge, 1997.

63. M. A. Green, *Third Generation Photovoltaics: Ultra-High Efficiency at Low Cost*, Springer, New York, 2006.

64. É. Verdet, *Leçons d'optique physique*, Imerie Impériale, 1870.

65. M. Born and E. Wolf, *Principles of Optics*, Cambridge University Press, 7th edn., 1999.
66. L. Mandel and E. Wolf, *Optical Coherence and Quantum Optics*, Cambridge University Press, 1995.
67. R. Winston, Y. Sun and R. G. Littlejohn, *Opt. Commun.*, 2002, **207**, 41.
68. G. S. Agarwal and E. Wolf, *Opt. Lett.*, 2004, **29**, 459.
69. A. M. Song, *Appl. Phys. A: Mater. Sci. Process.*, 2002, **75**, 229.
70. R. H. DuHamel and J. P. Scherer, *Antenna Engineering Handbook*, ed. R. C. Johnson McGraw-Hill, New York. 3rd edn., 1993, p. 14.
71. D. S. Filipovic and T. Cencich, *Antenna Engineering Handbook*, ed. J. L. Volakis. McGraw-Hill, 4th edn. 2007.
72. V. A. Podolskiy, A. K. Sarychev and V. M. Shalaev, *J. Nonlinear Opt. Phys. Mater.*, 2002, **11**, 65.
73. O. Y. Semchuk, M. Willander and M. Karlsteen, *Semicond. Phys., Quantum Electron. Optoelectron*, 2001, **4**, 106.
74. S. Grover and G. Moddel, *IEEE J. Photovoltaics*, 2011, **1**, 78.
75. J. O. McSpadden, L. Fan and K. Chang, *IEEE Trans. Microwave Theory Technol.*, 1998, **46**, 2053.
76. T. Razban, M. Bouthinon and A. Coumes, *IEE Proc., H: Microwaves, Opt. Antennas*, 1985, **132**, 107.
77. J. J. Nahas, *IEEE Trans. Microwave Theory Technol.*, 1975, **MTT-23**, 1030.
78. Raytheon, Reception - conversion subsystem (RXCV) for microwave power transmission System, Final Report, 1975, Report ER75-4386, NASA-CR-145917, Raytheon Co., Pasadena, 1975.
79. W. C. Brown, *J. Microwave Power*, 1970, **5**, 279.
80. E. C. Jordan and K. G. Balmain, *Electromagnetic Waves and Radiating Systems*, Prentice-Hall, Englewood Cliffs, 2nd edn. 1968.
81. P. K. Tien and J. P. Gordon, *Phys. Rev. Lett.*, 1963, **129**, 647.
82. J. R. Tucker and M. F. Millea, *Appl. Phys. Lett.*, 1978, **33**, 611.
83. G. Moddel, *Rectenna Solar Cells Rectenna Solar Cells*, ed. G. Moddel and S. Grover, Springer, New York, 2013, p. 3.
84. J. A. Hagerty, T. Zhio, R. Zane and Z. Popovic, Proceedings of The Government Microcircuit Applications and Critical Technology Conference, Las Vegas, 2005, 1.
85. W. Shockley W and H. J. Queisser, *J. Appl. Phys.*, 1961, **32**, 510.
86. D. Trivich and P. A. Flinn, in *Solar Energy Research*, ed. J. A. Duffie and F. Daniels, University of Wisconsin Press, Madison, 1955.
87. B. Hegyi, A. Csurgay and W. Porod, *J. Comput. Electron.*, 2007, **6**, 159.
88. S. Grover and G. Moddel, *Solid-State Electron.*, 2012, **67**, 94.
89. N. Alimardani and J. F. Conley, *Appl. Phys. Lett.*, 2013, **102**, 143501.
90. P. Maraghechi, A. Foroughi-Abari, K. Cadien and A. Y. Elezzabi, *Appl. Phys. Lett.*, 2012, **100**, 113503.
91. F. Yesilkoy, S. Potbhare, N. Kratzmeier, A. Akturk, N. Goldsman, M. Peckerar and M. Dagenais, in *Rectenna Solar Cells*, Springer, New York, 2013, p. 163.
92. X. Lei and V. Van, *Opt. Commun.*, 2013, **294**, 344.

Real World Efficiency Limits: the Shockley–Queisser Model as a Starting Point

PABITRA K. NAYAK AND DAVID CAHEN*

Department of Materials and Interface, Weizmann Institute of Science, Rehovot, 76100, Israel
*Email: david.cahen@weizmann.ac.il

17.1 Introduction

Alongside progress in development of established (better known and developed) solar cell types, newer ones that involve organic, molecular or polymeric and organic–inorganic hybrid materials, in some cases as nano-scale composites or with nano-structured inorganic materials, are researched intensively. The reason is that they hold the promise of drastic cost reduction of cell and module fabrication, that can make them economically attractive, *per se* or to complement the workhorse Si cells. Given the impressive improvements in the energy conversion efficiencies of nearly all cell types in the last few years, we ask if there are efficiency limits for these potentially low-cost cells, beyond the Shockley–Queisser (S-Q) limit[1] that has been approached closely for inorganic, 'classical' Si and GaAs cells. This issue is important as it can set practical goals, allow for realistic prognoses for their use and stimulate efforts to circumvent such limits.

The fundamental limit for PV conversion with a single absorber system was articulated by Shockley and Queisser,[1] based on detailed balance and calculated as a function of the minimum photon energy required for photo-

RSC Energy and Environment Series No. 11
Advanced Concepts in Photovoltaics
Edited by Arthur J Nozik, Gavin Conibeer and Matthew C Beard
© The Royal Society of Chemistry 2014
Published by the Royal Society of Chemistry, www.rsc.org

carrier generation at room temperature for an idealized model system. Recent results with GaAs[2] have shown its validity better than ever and thus confirm its use as benchmark. The S-Q limit is based purely on our understanding of the relevant physics and thermodynamics. No other material property than the optical bandgap E_g of the (single) absorber is taken into account to calculate this limit. Furthermore the device is considered to have an absorption edge that is described by a unit step function, unit quantum yield for conversion of each absorbed photon to a free electron–hole pair, infinite carrier mobilities, and radiative recombination as the only pathway of carrier recombination. The efficiency of the solar cell is determined by the detailed balance between incoming radiation and emission by radiative recombination, and the resulting photocurrent and photovoltage that are possible in the ideal case of zero non-radiative recombination.

For a given solar cell the question then is, how good can we expect it be? Although the S-Q efficiency limit is taken as a universal measure to gauge maximum thermodynamically possible efficiencies of different solar cell technologies, the fact is that for many cell types it presents only the first step in an efficiency analysis, because, as already noted, the S-Q analysis is for an ideal model system. For the real world limit, one has to take into account many factors that are simplified or ignored in the S-Q treatment. For example, the assumption of abrupt optical absorption at a certain energy value (the bandgap) is a good approximation only for high quality single crystalline absorber materials. Real materials have carrier mobilities that have finite values, they experience non-radiative recombination pathways, and devices made with them will have non-ideal diode quality factors. Thus, while all solar cells use common basic principles, the materials on which they are based are critical for how they work in detail and how well they perform.

Working with small, laboratory cells allows for the testing of the materials, cell structure and processes and to determine the approximate efficiency that is achievable with a certain combination of materials. However, there are additional issues that are important to achieve the final goal, which is the practical use of the cell for electrical power generation. Except for applications that use strongly concentrated sunlight, the cell has to be larger than some minimal size, so that practically it will be possible to build large-scale modules. It also has to be stable during prolonged operation under sunlight, even if the incident UV radiation is filtered out (or used beneficially, by spectral shifting) and if the devices are encapsulated to prevent detrimental chemical reactions with water and oxygen. Finally, cell fabrication should have an acceptable production yield (equal to the fraction of the cells and modules that are fabricated and are within the performance specifications) to make its manufacture economically viable.

Before we consider the possible additional factors that need to be taken into account to understand the gap between S-Q and actual efficiencies of most cells, we note that great attempts are being made to find practical approaches, based on what is called photon management, and on multiple exciton generation, to arrive at photovoltaic cell structures that can exceed the S-Q limits.[3,4] Even those efforts will, though, be subject to the possible existence of additional general or specific limitations, beyond those imposed by the ideal S-Q analysis. Understanding such limitations should stimulate work to determine if they can be circumvented and to what extent they present practical and basic scientific barriers; this will help focus efforts on finding the best possible routes for progress in improving the performance of different types of solar cells.

Here we look for, and identify some such possible bounds from summaries and analyses of the experimental performance data, obtained on the various single junction solar cell types.

17.2 Efficiency of Different Single-junction Cells and Performance Analysis Based on Empirical Criteria

Figure 17.1 shows the best laboratory efficiencies for various kinds of single junction solar cells operated at room temperature and the S-Q limit based on

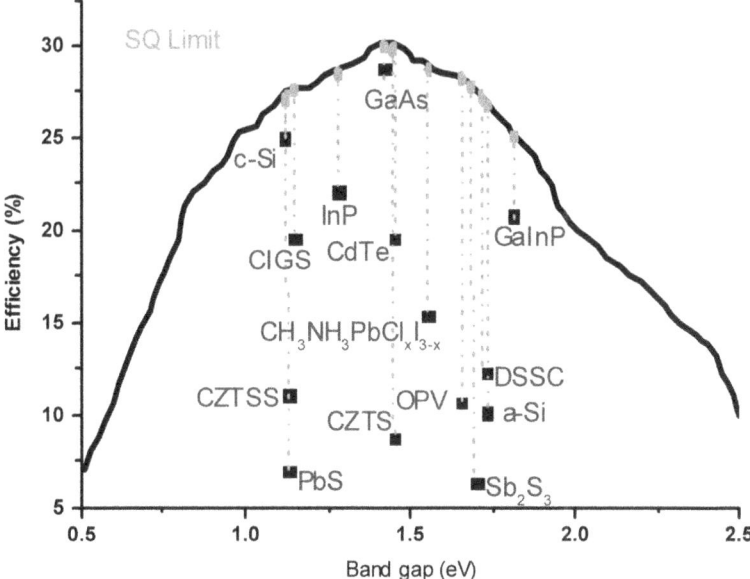

Figure 17.1 Solar to electrical power conversion efficiencies, as calculated within the Shockley–Queisser (S-Q) model (solid line) as a function of the band gap of the absorber. The pale grey dots represent the ideal S-Q efficiencies for a given bandgap and the black squares give the actual efficiencies of the best real-world single-junction cells with absorbers with that same bandgap. GaAs- and c-Si-based cells are close to the limit where as new generation cells are far from the limit. Abbreviations are CIGS, $Cu(In_{\sim 0.7}Ga_{\sim 0.3})Se$; CZTS, Cu_2ZnSnS_4; CZTSS, $Cu_2ZnSn(S,Se)_4$; OPV, organic photovoltaic (cell); DSSC, dye-sensitized solar cell.

the bandgap of the materials used as absorber. We have included cells which have efficiencies $> 5\%$ at 1 sun intensity. It can be seen from Figure 17.1 that most of the cells are far from the S-Q limit, with c-Si and GaAs as exceptions. In order to gauge the photovoltaic performance of different cells we define the following empirical performance metrics:

- *Cell-to-module loss*: How do the efficiencies of the best module that is commercially available (CM) or that was fabricated in the laboratory (LM), compare to that of the best laboratory cell (LC), η_{CM}/η_{LC} and η_{LM}/η_{LC}.
- *Photon flux to current conversion loss*: This is the ratio of the current density of the best laboratory cell at short circuit condition, to the maximal theoretically possible current density, $J_{SC,max}$. The optical absorption edge, λ_{edge}, of the absorber in the cell determines the maximal current density by the following equation (assuming 1 electron/absorbed photon:

$$J_{SC,max} = q \int_{\lambda=0}^{\lambda=\lambda_{edge}} \varphi(\lambda) \, d\lambda \qquad (17.1)$$

where q is the elementary charge, λ is the wavelength, and φ is the photon flux. Furthermore, we determine the ratio of the cell's current density at maximum power (MP), J_{MP}, to $J_{SC,max}$. Thus, these ratios are $J_{SC}/J_{SC,max}$ and $J_{MP}/J_{SC,max}$. Additionally, J_{MP}/J_{SC} reveals how close the material properties, device structure and contacts allow the cell to approach ideal diode behavior,[5] in terms of the photogenerated current.

- *Energy loss, relative to the optical bandgap energy*: We consider the ratio of the available free energy (light-induced difference in quasi-Fermi levels, *i.e.*, difference in the [quasi]-chemical potential of the electron), corresponding to the voltage at open-circuit, qV_{OC} or that at maximum power point, qV_{MP}, to the optical bandgap energy, E_g.

Before being converted into the energy stored in the electron–hole pair separation (or exciton, depending on the type of solar cell), this energy represents the minimum free energy that a photon needs to have to start the PV process. In addition to the relative bandgap energy loss of the cell, which can be expressed as qV_{OC}/E_g or qV_{MP}/E_g, this loss is expressed by $(E_g - qV_{MP})$ a quantity that we have termed [photovoltaic] over-potential, because it represents a loss, incurred to make the cell work. It is composed of two parts, a thermodynamic one (which will vanish at 0 K) and everything else, which can be viewed as kinetic ones (recombination losses, losses to allow high charge transfer rates ($=$ current) at the maximum power point. Some of those losses will even be incurred at open circuit voltage (energy level mismatches to minimize recombination). Because the S-Q limit on photovoltage for a given E_g can be estimated quite accurately,[6,7] the $E_g - qV_{MP}$ value shows the degree to which there is an energy loss beyond that dictated by the S-Q analysis Comparing the energy [= voltage ×charge] loss to the S-Q (detailed balance) loss reflects extra losses, that can be related to material properties, non-ideal processes operating in the cell, and generally to the state of technology development for the given cell type.

The total absorbed solar power, for a given absorber with bandgap E_g is $(E_g/q \cdot J_{SC,max})$. Thus, the efficiency for absorbed solar to electrical power can be written as:

$$\eta = [V_{MP} \cdot J_{MP}]/[(E_g/q) \cdot J_{SC,max}] \tag{17.2}$$

Ideally, V_{MP} is determined by V_{OC} (which depends on E_g) and the diode quality factor.[5] J_{MP} is ideally governed by $J_{SC,max}$ and the diode quality factor. Thus, comparing the ratios $J_{MP}/J_{SC,max}$ and qV_{MP}/E_g for various solar cell types (with a range of E_g) can point to underlying differences between cell types and limitations to a given cell type that can affect the prognosis for practical development.

We note here that for cells based on poorly crystalline and particularly, disordered absorbers, such as amorphous and organic polymer-based ones, E_g (or the optical absorption edge, λ_{edge}), is not well-defined due to energy tails extending into the normally forbidden bandgap region; this causes a gradual optical absorption onset. In some cases, no absorption spectra for the cell light absorbers exist in the open literature, in which case we had no choice but to use the external quantum efficiency (EQE) spectrum to determine E_g, which can lead to an E_g value, higher in energy, and a $J_{SC,max}$ value, smaller than the actual one. The wavelength onset of the absorption or the EQE spectrum was taken as E_g and was determined by fitting a straight line to the low energy edge of the optical absorption/EQE spectrum and extrapolating the line to the base line (for the absorption spectrum) and to zero current (for the EQE spectrum).

17.2.1   Possibilities for Technological Progress

η_{LM}/η_{LC} and η_{CM}/η_{LC} (see Table 17.1) reflect the state of manufacturing development of a cell type into *modules* and then into commercially mass-produced modules, respectively. In principle, the basic science underlying the PV conversion by a given cell does not change from cell to laboratory module to commercially available modules. Any deviation in efficiency should mark the lack in reproducibility of cell manufacture and/or technological difficulties in mass production. We note that the deviation has no bearing on basic scientific understanding, but is critical for reaching the ultimate goal of practical application of the cell technology. From Table 17.1 it can be seen that, for c-Si the best laboratory module efficiency is at 92% of the best laboratory cell and the best commercial modules are at 86%. These high percentages reflect the maturity of the technology in Si-based cells that has evolved over the past 6 decades.

Table 17.1 Comparison of PV conversion efficiency values of the best single-junction cells (efficiency > 10%), best laboratory modules and best commercial modules.

Cell type (area if < 1 cm²)	Best lab. cell efficiency, η_{LC} (%)	Best lab. module efficiency, η_{LM} (%) (area if < 500 cm²)	Manufacturer	η_{LM}/η_{LC}	Commercial module, η_{CM} (%)	Manufacturer	η_{CM}/η_{LC}
sc-Si	25.0	22.9	UNSW	0.92	21.5	Sunpower	0.86
GaAs	28.8	24.1	Alta Device	0.84	–	–	–
GaInP (0.25)	20.8	–	–	–	–	–	–
CIGS	19.6[a]	15.7	Miasole	0.80	14.3	TSMC	0.73
CdTe	19.6	16.1	First Solar	0.82	~11.5[c]	GE	0.59
a-Si	10.1	8.7	Univ. of Neuchatel	0.87	~7[c]	Moserbaer, Sungen, Leonics	0.69
CZTSS (0.45)	11.1	–	–	–	–	–	–
DSSC	12.3[b]	8.8 (399)	Sharp	0.72	–	–	–
OPV (thin film)	10.7[a]	8.2 (25)	Toshiba	0.76	–	–	–
Perovskite $CH_3NH_3PbI_{3-x}Cl_x$ (0.08)	15.4[b]	–	–	–	–	–	–

Efficiency data for solar cells and modules are from ref. 2 unless noted otherwise. Efficiencies for commercial modules are from manufacturers' information.
[a]Cells with < 1 cm² area have higher efficiencies as follows: CIGS (20.4)², OPV (11.1)².
[b]Efficiency values of DSSC and for the perovskites: $CH_3NH_3PbI_{3-x}Cl_x$ are from ref. 14 and ref. 36, respectively.
[c]Calculated from the total area of the module.
CIGS: $Cu(In_{\sim0.7}Ga_{\sim0.3})Se$; CZTSS: $Cu_2ZnSn(S,Se)_4$; OPV: organic photovoltaic (cell).

For GaAs cells, the laboratory module efficiency is at 84% of the champion cell. While we do not have (access to) reliable data on commercially available GaAs modules, we can expect a η_{CM}/η_{LC} value of 0.7–0.8 by looking at the case of c-Si.

Recent activities in CIGS- and CdTe-based solar cell modules show great improvement in the past few years and their modules are, in terms of performance, catching up with modules based on mc-Si cells. We can expect even more improvements for modules in the future, beyond the 2–3 % increases over the past 5 years, because of the new records in laboratory cells that have been reported in the recent past.[2]

The low η_{LM}/η_{LC} values for dye-sensitized and organic-based thin film cells and the high ones for c-Si cell types represent extreme cases. Repeating our assumption above that there are no changes in PV science of cells and modules we can expect the latter to approach the ratios of the former, assuming that a similar degree of control over manufacturing on a large scale is achieved.

The newer types cells, based on Cu_2ZnSnS_4 (CZTS), $Cu_2Zn(S,Se)_4$ (CZTSS), $CH_3NH_3PbX_3$ (X = halide) and PbS-quantum dots (PbS-QD) (not included in Table 17.1) are still in the early stage of development. Thus, they likely need to overcome many technological and economical hurdles before commercial products with these types of materials will be available.

17.2.2 Current Efficiency ($J_{SC}/J_{SC,max}$, $J_{MP}/J_{SC,max}$ and J_{MP}/J_{SC})

The $J_{SC}/J_{SC,max}$ data (Figure 17.2 and Table 17.2) show that most developed cells achieve close to theoretical limits of (short-circuit) current efficiency. For cells, made with inorganic crystalline and high quality polycrystalline electronic materials, the ratio is >90%, reaching near theoretical values (marked by a gray stripe in Figure 17.2), so that no or only minor improvements in current efficiency in these cells are expected.

For c-Si cell, this efficiency is ~99% and for GaAs ~94%. Remarkably, the best polycrystalline thin film CdTe cell yields the same 94% efficiency. While we can expect this efficiency to increase for GaAs cells, that is not so obvious for the CdTe cells, because of unavoidable scattering-related absorption in polycrystalline films. The lower value (82%) for the single crystal InP cell indicates possible room for development. While we could expect that the other leading polycrystalline thin film-based cell, CIGS, will catch up in terms of current efficiency with CdTe, we need to consider their different grain boundary (GB) chemistry and physics. In the case of CdTe the regions around the GBs are active in terms of charge separation and collection, in addition to the junction with CdS. This leads to high collection efficiencies (at the cost of the photovoltage).[8,9] In the case of CIGS cells charge separation and collection occurs primarily at the interface with the heteropartner material (mostly CdS) (*cf.* section 17.2.3).

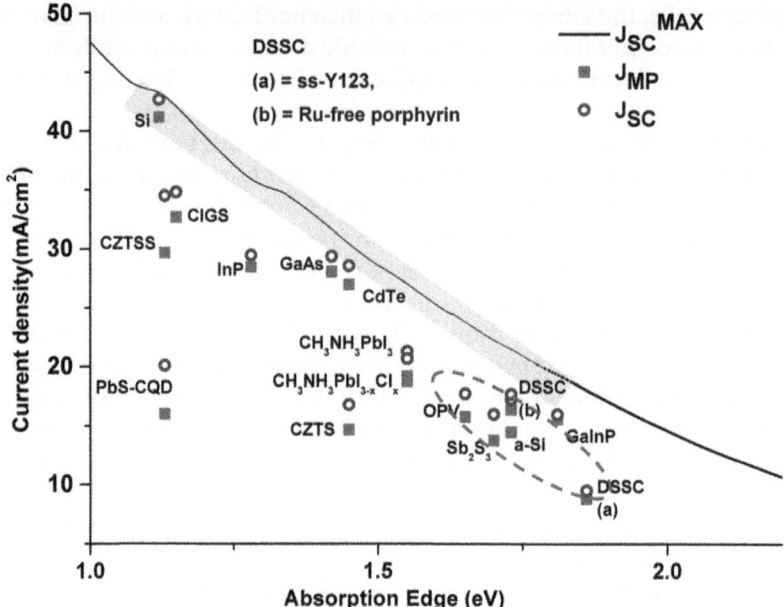

Figure 17.2 Maximal possible short circuit current density, *i.e.*, each absorbed photon generates a free electron (black line) and observed short-circuit current, J_{SC} (open circle) and current at maximum power, J_{MP} (filled square), for the best solar cells at AM 1.5 *versus* absorption edge energy ($E_g = h\lambda_{edge}/c$, with h = Planck's constant, and c the speed of light). Labels in the figure correspond to the left-most column in Table 17.2. The cells that suffer from poor mobility and parasitic absorption are within the dashed-line oval. The stippled area overlaps with cells with >90% current efficiency.
(Figure is adapted from Nayak *et al.*[7]) CQD, Colloidal quantum dot. Other abbreviations are as in Figure 17.1.

a-Si based solar cells show 80% current efficiency, lower than for inorganic polycrystalline inorganic materials. The presence of trap states and the low mobility of carriers are likely reasons for the lower current efficiency.

In the case of dye-sensitized and organic thin film-based solar cells the $J_{SC}/J_{SC,max}$ values are 60–80%. These low values can be attributed to the difference in the process of charge separation (at the interface) at organic/organic and organic/inorganic interfaces, compared to that in the all-inorganic crystalline semiconductors. While charge separation efficiency ~100% (reflected in EQE) can be achieved in OPVs at certain wavelengths,[10] this value does not hold over the complete spectral range where the materials absorb, due to the nature of (1) the *absorber* (the difference in optical absorption between molecular levels and extended solid band to band excitation; and (2) the *photogeneration* of free charge carriers.[11–13] In the case of the best-performing DSCC laboratory cell,[14] the problem due to the molecular nature of absorption is partly circumvented by using co-sensitizers (two types of dyes with complementary absorption spectra). This effort led to

$J_{SC}/J_{SC,max}$ of 82%, the highest value among solar cells involving an organic material as absorber. Due to the low mobility of charge carriers in organic semiconductors, photogenerated carriers may recombine before being collected by the contacts, even at short-circuit, thus reducing J_{SC}.

This low $J_{SC}/J_{SC,max}$ of organic-based PV cells can, in part, be attributed to lack of structural order and the presence of tail (near the band edge) or deeper trap states. It was found that in solid state organic absorber based PV systems delocalization of the exciton is crucial for efficient charge separation and that the degree of delocalization depends on structural order in the absorber.[13] In the absence of high structural order in the film, the excitons are lost as heat, due to other competing non-radiative recombination processes. Moreover, even after photo-induced charge transfer happens at the interface, the separated charge carriers can be trapped by deep states at the interface.[15] These trapped charge carriers then can recombine with oppositely charged ones, thus reducing the short circuit current. This phenomenon is analogous to SRH-type recombination in an inorganic PV system. A thorough explanation on this subject can be found in the the studies by Nayak *et al.*[13] and Street *et al.*[15,16]

J_{MP}/J_{SC} (*cf.* Table 17.2) is an indicator of the diode character of the solar cell and the closer the diode ideality factor is to 1, the better is the (dark)

Table 17.2   Comparison of reported V_{OC}, V_{MP}, J_{SC} and J_{MP} values for cells with best conversion efficiencies, and their relation to the cell's optical bandgap/absorption edge.

Absorber in the cell (area if <1 cm²)	RT abs. edge (bandgap), E_g (eV)	qV_{OC} (eV)	$E_g - qV_{OC}$ (eV)	qV_{MP} (eV)	Over-potential: $E_g - qV_{MP}$ (eV)	$J_{SC}/J_{SC,max}$ (%)	J_{MP}/J_{SC} (%)
sc-Si	1.12	0.71	0.41	0.61	0.51	99	96
GaAs	1.42	1.12	0.30	1.00	0.42	94	97
InP	1.28	0.88	0.40	0.75	0.53	82	97
GaInP	1.81	1.45	0.36	1.34	0.47	82	98
CdTe	1.45	0.86	0.59	0.73	0.72	~94	94
CIGS	~1.15^a	0.72	~0.42	0.60	~0.55	~83	94
a-Si:H	~1.75	0.88	~0.87	0.70	~1.05	~80	84
DSSC Ru-free	~1.73	0.93	0.80	0.73	~1.0	~82	93
porphyrin: DSSC ss- Y123	~1.86^a	0.99	0.87	0.81	~1.05	~53	93
OPV Mitsubishi	~1.65^a	0.87	~ 0.78	0.67	~0.98	~75	89
CZTSS (0.45)	1.13^a	0.46	0.67	0.37	0.76	~ 80	86
CZTS (0.45)	1.45^a	0.71	0.74	0.57	0.88	~55	87
CH₃NH₃PbI₃ (0.28)	1.55	0.99	0.56	0.78	0.77	~71	93
CH₃NH₃PbI₃₋ₓClₓ (0.08)		1.07	0.48	0.82	0.73	~79	88
PbS-QD (0.05)	1.13	0.60	0.53	0.44	0.69	47	80
Sb₂S₃ (0.16)	~1.7	0.59	1.11	0.45	1.25	~72	86

^{a}From EQE, as no published optical absorption data of the cell are available/were found.

diode, including the contacts of the cell. While in the case of highly crystalline materials, like c-Si, GaAs and InP, the current loss is < 3–4% of the
total current generated in the solar cell, for polycrystalline materials like
CdTe and CIGS the loss is 5–6%. This higher loss can be ascribed to loss of
carriers at grain boundaries before collection due to the increased surface
area, which implies increased surface state density per unit volume. Cells
made with polycrystalline CZTS and CZTSS have a 14–15% loss, which we
can expect to decrease to 5–6%, to become consistent with the more developed inorganic polycrystalline material-based CdTe and CIGS cells.
Similar expectations may well be applicable for $CH_3NH_3PbI_{3-x}Cl_x$-based
cells, as the crystalline quality of that polycrystalline material, is superb.

The loss in current is $\sim$7–9% in DSSC cells and 11% in OPV ones. Taking
into account that many interfaces are involved and limited (or absence of)
long-range order in several of the cell components, these values are reasonable, though higher than for the inorganic polycrystalline material-
based cells.

After decades of development a-Si shows the highest loss among the cells,
listed here $\sim$16%. As a-Si is also the most disordered of PV materials, as well
as one that has been studied and developed longer than most others, we take
this to indicate that such strong disorder implies intrinsic recombination
pathways that cannot be eliminated.

17.2.3   Photon Energy Loss: Present Status of Single-junction Solar Cells

The energy that can be converted to useful work is the free energy. The free
energy per carrier is the electrochemical potential, and characterized by the
Fermi level. Creation of excess electrons and holes, as a result of photo-
excitation, produces a separation of the quasi-Fermi levels, E_{Fn} and E_{Fp}, for
electrons and holes, respectively, in the absorber. The photovoltage is given by:

$$qV_{OC} = E_{Fn} - E_{Fp} \tag{17.3}$$

The upper limit of qV_{OC} for a given bandgap can be calculated quite accurately by taking all the fundamental loss factors into account. qV_{OC}/E_g and
qV_{MP}/E_g and particularly $(E_g - qV_{MP})$ express the limitations to many cells
and values for these are shown in Table 17.2. $(E_g - qV_{MP})$, shows photon
energy loss beyond that dictated by the S-Q analysis. Comparison of the
actual value of $E_g - qV_{MP}$ to the ideal value for S-Q reflects additional losses,
associated with material properties, non-ideality, and the level of technology
development for the given cell type.

The minimum value of $E_g - qV_{MP}$ (or fundamental S-Q limit, loss),
required for the cell to operate at maximum efficiency, is given by the
line of filled dots in Figure 17.3 (cf. Nayak et al.[7]). Absorbed photon
entropy[6,17] is independent of the bandgap and is one of the fundamental
thermodynamic losses associated with solar cells. The Carnot efficiency,

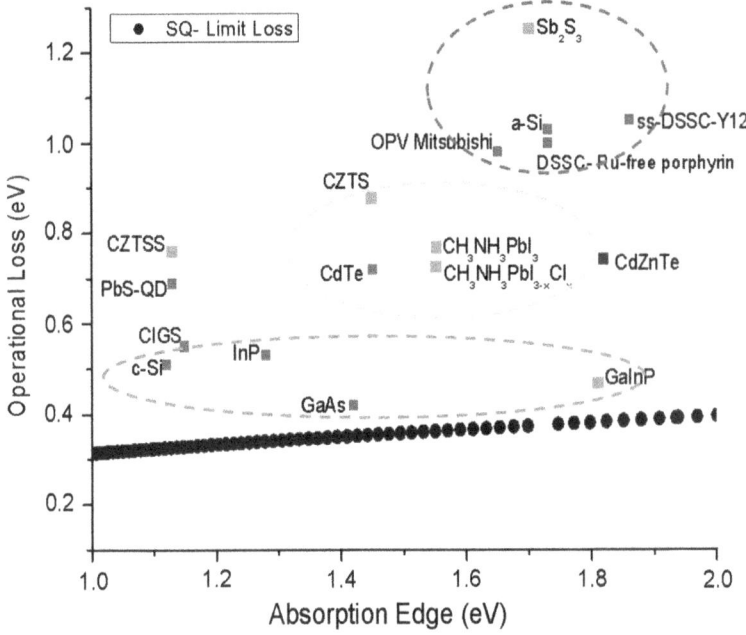

Figure 17.3   The $(E_g - V_{MP})$ loss within the Shockley–Queisser model (filled black circles) *versus* absorption edge and that loss for actual solar cells, according to the bandgap of their absorber. Labels in the figure correspond to the left-most column in Table 17.2. Data for the CdZnTe cell are from Carmody *et al.*[24] Mid-grey squares represent solar cells with area <1 cm². Single crystal, polycrystalline and amorphous/excitonic absorber-based solar cells are indicated by mid-grey (except for CIGS, a polycrystalline absorber cell), pale grey and black ovals, respectively.
(Figure is adapted from Nayak *et al.*[7]) Abbreviations used in the figure are as in Tables 17.1 and 17.2.

another fundamental loss, contributes a $\sim 0.05 E_g$ loss at room temperature. Putting all the possible fundamental losses together, materials with 1.1–1.9 eV bandgap should have $E_g - qV_{MP}$ values of 0.32–0.39 eV (see Nayak *et al.*[7]), *i.e.*, in this range the difference in $E_g - qV_{MP}$ values should be ~ 70 meV. However, the actual difference is found to be an order of magnitude larger between different types of cells. What is the reason for this? In the next few paragraphs we try to find an answer.

17.2.3.1   Case of Single Crystal-based Solar Cells

The solar cells that are based on crystalline inorganic materials are sometimes referred to as 'first generation' solar cells. The absorber thickness depends on the material that is used in the cell. In crystalline silicon solar cells, the thickness of the absorber is ~ 200 μm. This material is accompanied with selective contacts, which are a p–n junction as the electron extraction contact and a highly p-doped layer as hole extraction contact. But the thickness of these contact layers is several hundred nm, at most, and their widths are negligible compared to that of the absorber. As a result, the thermalized carriers that determine the positions of the Fermi levels are mostly in the absorber itself, and, provided that defects are well-passivated at the surface, recombination occurs in the bulk absorber material. Therefore, the crystalline silicon solar cell should be able to approach in practice the S-Q limit (*cf.* see Figure 17.3). The V_{OC} of the best cell is 0.71 V, significantly below the S-Q limit value (0.87 V). Table 17.3 shows that the solar cell parameters of the best Si cell of any type are very close to the theoretically expected ones, where the maximal V_{OC} value is the bandgap minus the photon entropy ('étendue') and the Carnot loss, and considering also the effect of carrier cooling[6,17]).

Though c-Si is a highly ordered, near-perfect crystalline material, its low (internal) photoluminescence quantum yield ($\sim 20\%$)[18] implies significant non-radiative recombination, understandable from its indirect bandgap, which decreases its PV efficiency. As shown by Ross,[19] the luminescence yield of the absorber is a factor in determining the light-induced chemical potential difference for a photochemical system. For solid state solar cells, Rau and co-workers,[20,21] Miller *et al.*[22] and Smestad[23] showed that the V_{OC} of all cells is affected by the quantum yield of the cells' external luminescence, η_{ext}. The V_{OC} of a real cell is given by:

$$qV_{OC,\text{real cell}} = qV_{OC,\text{ideal}} - kT|\ln(\eta_{ext})| \qquad (17.4)$$

where η_{ext} is the external quantum efficiency and depends on the internal luminescence quantum efficiency. The V_{OC} of best c-Si cells is slightly less than the ideal limit and because the internal photoluminescence efficiency is an intrinsic material property of Si, any significant improvement in photovoltage is expected to be slim, at least with the crystalline Si material types used today.

The V_{OC} values of the best GaAs and (GaIn)P cells are within a few tens of mV from the values, predicted from calculated fundamental losses (see Nayak and Cahen[7]), *i.e.*, GaAs and GaInP materials and cell science evolution is reaching the fundamental (voltage) limits.

CdZnTe (bandgap = 1.6 or 1.8 eV) -based single junction solar cells were reported with V_{OC} values = ~ 1.2 V or ~ 1.34,[24,25] respectively. These cells have the potential to give higher voltages if further studied and developed.

Table 17.3 The very best experimental solar cell performance parameters (in bold), as fraction of the theoretical (diode equation with $n = 1$ or S-Q model) values for single crystal Si and GaAs cells.

Cell type	V_{OC}, relative to theoretical limit (%)	J_{SC}, relative to theoretical limit (%)	Fill factor, relative to theoretical limit (%)
p–n GaAs	97	94	97
p–n Si	82	99	95

The small $E_g - qV_{OC}$ and high V_{OC}/E_g of GaAs, GaInP and CdZnTe can be attributed to a combination of high material quality and high luminescence quantum efficiency[22,26,27] (*e.g.*, for GaAs the internal luminescence quantum efficiency is 99.7%), which make them resemble the idealized S-Q model system. We note that use of back reflecting mirror (>95% efficient) at the back surface of GaAs cells makes the radiative emission to occur only from the front surface. Emission from the front-surface only is assumed in the S-Q model.

17.2.3.2 Cells Based on Inorganic Materials with Structural Disorder

The cost of solar cell production can be reduced if polycrystalline or amorphous absorbers are used, primarily because of the decreased cost of material preparation. We will first consider inorganic materials-based cells of this type.

Mostly, these cells are based on Si, in forms from amorphous (a-Si), to nano-, micro-, poly- and multi-crystalline (nc-Si, mc-Si, pc-Si, mc-Si; typical grain sizes are nm, μm, 10–100 μm and mm to cm); such cells are also based on copper indium gallium diselenide (Cu(In,Ga)Se$_2$, CIGS) or cadmium telluride (CdTe), the last two with μm grain sizes. Disorder in these materials affects the photovoltage efficiencies and it can be seen from Table 17.2 that their overall photovoltage yield is less than that of the single crystalline cells.

Figure 17.4 is a schematic band/energy level diagram of a solar cell with a single crystalline (Figure 17.4a) and a polycrystalline material (Figure 17.4b)

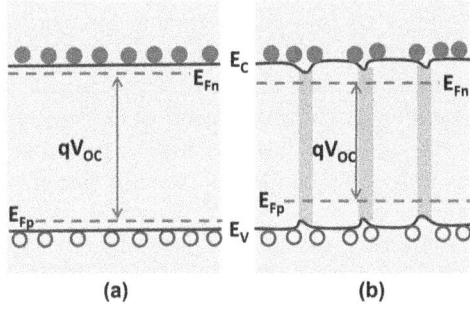

(a)   (b)

Figure 17.4 Graphic energy level/band diagram of a solar cell, formed by a single crystal material (a) and polycrystalline material (b). The material is used as the absorber, and electron *cum* hole transport material. Absorption of light generates free electrons (filled circles) and holes (open circles) that relax to the bottom of the conduction band (E_C) and top of the valence band (E_V). Splitting of the quasi-Fermi levels of electrons (E_{Fn}) and holes (E_{Fp}), due to excess carriers produced by photon absorption and shown by double arrows, determines the photo-voltage of the cell. Electrons in the conduction band (E_C) relax by recombining with holes in the valence band. In polycrystalline material grain boundaries shown (in pale grey), with slight bending of the bands (towards mid-gap), are mostly thought to be the preferred carrier recombination sites.

as absorbers. The working principle is similar for both a single crystal-based and polycrystalline cell, but the presence of grain boundaries (GB), presents energetically preferred non-radiative recombination paths, in addition to the regular band-to-band radiative recombination paths. As a result the cell's performance deviates more and more from the band-to-band radiative recombination limit that holds for the single crystal-based solar cells. The additional non-radiative recombination pathways in polycrystalline material decrease the overall photo-voltage output.

There is another effect associated with GB, though, that has received less attention, which is that of near band edge tail states. The broken periodicity of the crystal at the GBs introduces electronic states within the bandgap; this creates, for a truly amorphous semiconductor (a-Si), a continuum of states that decrease in density from the band edges to the center of the gap. States of this type are present in polycrystalline materials, although to a much smaller extent, and their density increases with the increase in structural disorder in the material and a decrease in crystallinity.

The effect of the tail states within the bandgap is to decrease the amount of absorbed photon energy that is actually available for conversion and to decrease the achievable photovoltage by limiting the separation between quasi-Fermi levels, produced upon photo-excitation.

The first effect, of sub-bandgap absorption, which removes photons that could be used in a multi-cell arrangement, is negligible. Photons, absorbed into (near-VB) tail states with energies that can excite an electron beyond the mobility edge, can contribute to the photocurrent. However, the cross-sections for excitation from the VB to near-CB tail states and from the near-VB tail states to CB are >100-fold less than that for band–band transitions; this makes these absorption processes relatively minor. We note that it is the band-to-band transition that is taken as the bandgap for these materials.

The second effect concerns thermalization of carriers, photogenerated by photons having energies greater than the bandgap E_g photons, into the tail states. As the photovoltage that can be extracted is determined by the quasi-Fermi level separation (*i.e.*, $(E_{Fn} - E_{Fp})/q$), this will decrease as tail states, rather than band (mobility) edge energies, become the ones pivotal for defining the limits of quasi-Fermi level separation. Tiedje was the first to realize this when analysing the physics of a-Si:H cells, and concluded that the amorphous structure of the absorber and the resulting tail states exacts an extra price of several 100 mV, compared to what is the case for perfect crystalline cells.[28]

Comparison of an In_2O_3/CdTe cell, based on a CdTe single crystal and made 25 years ago, with today's best polycrystalline CdTe cell, reveals that even a single crystal CdTe cell, which likely was far from optimized, had a 35 mV higher V_{OC} than today's best polycrystalline CdTe/CdS cells. However, this is not the case for CIGS-based cells, where single crystal cells perform worse than their polycrystalline counterpart. This has been explained in terms of the remarkable grain boundary chemistry and physics of the polycrystalline CIGS film, which makes it behave, for electrons (the minority carriers in p-CIGS), as a single crystal.[29] The result also points towards the inability to make sufficiently high quality single crystals.

Comparison of newer type of polycrystalline cells like CZTSS, (which are developed to provide cells made from earth-abundant materials towards a sustainable thin-film technology) with CIGS, the more developed type that it is meant to replace, can give us an idea of expected development pathways.

By virtue of similar crystal structures and bandgaps of CZTSS and CIGS, cells that are based on them should have very similar performance parameters. However, while indeed J_{SC} values are comparable (*cf.* Figure 17.2), CZTSS cells have still 260 mV lower V_{OC} than CIGS ones. As this difference can be due to deep electronic traps in the CZTSS gap, better material control may improve the situation. However, orders of magnitude higher reverse saturation currents and larger diode quality factor (1.5–2) for CZTSS cells than for CIGS cells reflect the fact that unlike the case with CIGS, in CZTSS the surfaces and grain boundaries (GBs) are not (yet) smooth in terms of their electrostatic potential landscape. Temperature dependent V_{OC} of both types of cells also indicates the difference in interfacial recombination pathways. While for CIGS cells a plot of qV_{OC} *versus* temperature yields, upon extrapolation to 0 K, a value equal to E_g, a similar extrapolation for CZTSS cells yields a value well below E_g of the absorber. Thus, further success of CZTSS cells may depend also on finding ways to prepare CZTSS material with more benign (lower loss) GBs.

Organometallic Perovskite ($CH_3NH_3PbI_3$-, $CH_3NH_3PbI_{3-x}Cl_x$- and $CH_3NH_3PbBr_3$)-based cells show remarkably low values of $(E_g - qV_{MP})$, even though these cells are at a very early stage of development. The high voltage efficiency can, esp. for the $CH_3NH_3PbI_{3-x}Cl_x$- and $CH_3NH_3PbBr_3$-based ones, be linked to the superb crystalline order, benevolent grain boundaries, and large carrier diffusion lengths. We note that much remains to be done to further improve these cells, particularly to reproducibly prepare stable large area ones with acceptable yields.

17.2.3.3   Cells Based on Strongly Disordered, Organic Absorber Materials

In recent years, there has been intense research on preparing solar cells using a combination of materials to separate the steps involved in the photovoltaic effect. The optically absorbing material is mixed with other materials that can quickly extract the carriers and transport these to the respective electrodes. This mixing is done on a scale of nanometres, so that a thick enough layer of an absorber (from 100 nm to 10 μm, depending on the extinction coefficient of the absorber) can be used for sufficient optical absorption without having too high an electrical resistance. Typical examples of this approach of combining the absorber and the electron and hole transport materials (ETM and HTM, respectively) are the dye-sensitized cell (DSC) and organic bulk heterojunction (BHJ) cells. In a BHJ, either ETM or HTM (in some cases both) play the role of absorbing layer. ETM and HTM in

BHJ are also known as Acceptor and Donor, respectively. Working principles of a BHJ solar cell and DSC can be found in elsewhere in this book.

17.2.3.4   Origin and Quantification of Tail States

The origin of the tail states in organic materials can be different for different materials. In some cases, the presence of chemical impurities dominates as a source for tail states. Tail states have been observed in organic single crystals also, because organic molecular solids are prone to structural disorder due to weak interaction among the molecules (compared to the interaction between ions in an inorganic salt, or compared to the covalent bonding in inorganic solids or in solids with no or small electronegativity difference between constituents, *i.e.*, with low ionicity). Even inert gas (N_2) exposure of an organic solid can increase the density of bandgap states.[30] In a typical organic material, the total density of states (DOS) in its HOMO or LUMO level is of the order $\sim 10^{19}$–10^{20} cm^{-3}, which is comparable to the density of states of inorganic materials like Si or GaAs. The density of mid gap states in organic materials can be of the order $\sim 10^{18}$ cm^{-3},[31] and the profile can be modelled as having a Gaussian shape, an exponentially decreasing (towards mid-gap) one or a combination of both. For example, measurement of the DOS by chemical capacitance, as for P3HT:PCBM in regular and inverted configurations, show that the DOS in the PCBM has an exponential shape that sometimes can be modelled as a Gaussian as well.

Under normal insolation the photogenerated carrier density for a material with an optical bandgap of ~ 1.85 eV is between 10^{16} and 10^{17} cm^{-3}, depending on various material parameters, especially the recombination rates.[32] As the carriers in organic semiconductors move via hopping (*i.e.*, slowly and thus susceptible to trapping), photogenerated carriers always populate these tail states. Recombination of electrons and holes occurs via these tail states, and their rate of recombination is such that, at 1 sun illumination, these tail states are not filled completely.

In the next paragraphs we analyse the effect of the midgap states on the V_{OC} of the photovoltaic cell.

The V_{OC} of a photovoltaic cell can be described as follows:[33]

$$qV_{OC} = E_g - k_B T \ln\left(\frac{N_h N_e}{n_h n_e}\right) \qquad (17.5)$$

where E_g is the bandgap, k_B is the Boltzmann's constant, n_h and n_e are photogenerated holes and electron, respectively. N_h and N_e are the total densities of states for holes and electrons. For donor/acceptor-based solar cells, the effective bandgap can be given by the difference between the donor ionization energy (IE) and the acceptor electron affinity (EA) and is denoted as $E_{g,DA}$.

In BHJ cells, one can also apply the same logic, where the bandgap states are linked to the tails in the DOS distribution profile. If a Gaussian distribution is considered for the electronic states for holes (in donor material) and electrons (in acceptor material), the photogenerated carriers thermalize in the Gaussian tail, following Boltzmann statistics in the case of low occupancy, with an average equilibrium energy $\sigma_n{}^2/k_BT$ below E_{LUMO}. Such energy signals the mean energy level of the charge carriers and is located above the concentration-dependent Fermi energy. A straightforward calculation allows the determination of the Fermi level positions, from which an expression for V_{OC} can be obtained:[34,35]

$$qV_{OC} = E_{g,DA} - \frac{\sigma_h^2 + \sigma_e^2}{k_B T} - k_B T \ln\left(\frac{N_h N_e}{n_h n_e}\right) \qquad (17.6)$$

where σ_h and σ_e are the variances of the Gaussian distributions that represent the densities of states in the donor and acceptor, respectively. The increase in the density of bandgap states in donor and/or acceptor materials results in higher values of σ_h and/or σ_e. As can be seen from Equation (17.6), an increase in bandgap states will decrease the V_{OC} of the PV cell. The second term on the right hand side of Equation (17.6) contains the term related to the electron and hole equilibration energies, which establish an upper limit to the achievable photovoltage (*i.e.*, $\sim$400 mV reduction in photovoltage for $\sigma \approx 100$ meV at room temperature).

In the case of an exponential distribution of states, the $k_B T$ term in Equation (17.5) is replaced by a decay constant E_t.[34] A higher value for E_t, as a result of an increase in bandgap states, implies a lower V_{OC}.

Thus, from the above discussion we see that whatever may be the profile of the gap state distribution, an increase in the gap states density will always result in lower V_{OC} than the value, expected for crystalline materials.

We note that in case of a DSC (with TiO_2 as n-type semiconductor/ETM) the tail states, though present, dynamically fill after photo-induced charge transfer from the dye. Hence the electron quasi-Fermi level can be close to the CB of TiO_2. The same process does not take place in the organics used for OPV, because recombination from tail states is likely to be much faster than in the case of TiO_2. This can be due to the higher degree of local disorder in the materials used in a BHJ OPV than in a DSC.

17.3 Fill Factor and Disorder

The fill factor (FF) of a PV cell can be described empirically by the expression:[5]

$$FF = \frac{v_m}{v_m + 1} \frac{v_{OC} - \ln(v_m + 1)}{v_{OC}[1 - \exp(-v_{OC})]} \qquad (17.7)$$

where $v_{OC} = V_{OC}/nk_B T$ and $v_m = v_{OC} - \ln(v_{OC} + 1 - \ln v_{OC})$.

From Equation (17.7) we can deduce that the FF depends on V_{OC} and on the diode quality factor, n, provided that the shunt and series resistance approach infinity and zero, respectively. The presence of more than one recombination path in a disordered material makes $n > 1$. For close to perfect crystalline cells like GaAs or c-Si, the FF is 80–86%, while for cells, based on disordered materials, it is 65–70%,[2] with intermediate values for polycrystalline inorganic ones, even though the values of V_{OC} for several of the cells are comparable.

17.4 Conclusion and Outlook

The S-Q limit serves as a benchmark, upper thermodynamic limit to the highest power conversion efficiency attainable in an ideal PV cell where the only loss pathway is radiative recombination; hence it helps to assess the potential for improved performance of real PV cells. These are, of course, non-ideal and the resulting additional constraints on efficiency must be understood and optimized for progress to be achieved. The best single crystal GaAs solar cell closely resembles an ideal system and in fact exhibits an efficiency (after decades of research and development) close to the S-Q limit. Material parameters that are ignored in an ideal system pose additional limits. As cells approach the S-Q limit, low luminescence quantum efficiency seems to be an important loss factor. GBs determine the current and voltage efficiency in polycrystalline material-based solar cells by providing additional pathways of recombination and/or actual electrostatic barriers that lead to voltage loss. Most low-cost preparation processes yield materials with structural disorder and this disorder affects the electronic energies for electron (or hole) transport. Tail states in the bandgap arising from dangling bonds, chemical impurities, band edge fluctuations and/or structural disorder also reduce cell performance by increasing non-radiative recombination of photogenerated carriers. We pay a price in terms of photovoltage due to energy dispersion in the bandgap (instead of sharp band edges) as it limits the magnitude of quasi-Fermi level splitting.

Realizing the real-world limits of real-world materials can help us optimize cells realistically, including new ones. Indeed, the appearance of a highly crystalline and electronic material of high quality 'out of nowhere', the organometallic perovskites, shows how far PV is from a mature science, as new surprises occur again and again.

Acknowledgements

We thank our Weizmann Institute and many other colleagues for valuable inputs. We acknowledge with thanks grants from the Leona M. and Harry B. Helmsley Charitable Trust and from Mr. Martin Kushner Schnur for partial support. DC holds the Rowland and Sylvia Schaefer Chair in Energy Research.

References

1. W. Shockley and H. J. Queisser, *J. Appl. Phys.*, 1961, **32**, 510.
2. M. A. Green, K. Emery, Y. Hishikawa, W. Warta and E. D. Dunlop, *Prog. Photovoltaics: Res. Appl.*, 2013, **21**, 827.

3. H. Shpaisman, O. Niitsoo, I. Lubomirsky and D. Cahen, *Sol. Energy Mater. Sol. Cells*, 2008, **92**, 1541.
4. J. F. Guillemoles, *Fundamentals of Materials for Energy and Environmental Sustainability*, Cambridge University Press, Cambridge, 2012.
5. M. A. Green, *Solar Cells*, 1982, **7**, 337–340.
6. L. C. Hirst and N. J. Ekins-Daukes, *Prog. Photovoltaics: Res. Appl.*, 2011, **19**, 286.
7. P. K. Nayak and D. Cahen, *Adv. Mater.*, 2013, **26**, 1622–1628.
8. I. Visoly-Fisher, S. R. Cohen, A. Ruzin and D. Cahen, *Adv. Mater.*, 2004, **16**, 879.
9. I. Visoly-Fisher, S. R. Cohen, K. Gartsman, A. Ruzin and D. Cahen, *Adv. Funct. Mater.*, 2006, **16**, 649.
10. S. H. Park, A. Roy, S. Beaupré, S. Cho, N. Coates, J. S. Moon, D. Moses, M. Leclerc, K. Lee and A. J. Heeger, *Nat. Photonics*, 2009, **3**, 297.
11. A. A. Bakulin, A. Rao, V. G. Pavelyev, P. H. M. van Loosdrecht, M. S. Pshenichnikov, D. Niedzialek, J. Cornil, D. Beljonne and R. H. Friend, *Science*, 2012, **335**, 1340.
12. S. D. Dimitrov, A. A. Bakulin, C. B. Nielsen, B. C. Schroeder, J. Du, H. Bronstein, I. McCulloch, R. H. Friend and J. R. Durrant, *J. Am. Chem. Soc.*, 2012, **134**, 18189.
13. P. K. Nayak, K. L. Narasimhan and D. Cahen, *J. Phys. Chem. Lett.*, 2013, **4**, 1707.
14. A. Yella, H.-W. Lee, H. N. Tsao, C. Yi, A. K. Chandiran, M. K. Nazeeruddin, E. W.-G. Diau, C.-Y. Yeh, S. M. Zakeeruddin and M. Grätzel, *Science*, 2011, **334**, 629.
15. R. A. Street, M. Schoendorf, A. Roy and J. H. Lee, *Phys. Rev. B*, 2010, **81**, 205307.
16. R. A. Street, A. Krakaris and S. R. Cowan, *Adv. Funct. Mater.*, 2012, **22**, 4608.
17. T. Markvart, *J. Opt. A: Pure Appl. Opt.*, 2008, **10**, 015008.
18. T. Trupke, J. Zhao, A. Wang, R. Corkish and M. A. Green, *Appl. Phys. Lett.*, 2003, **82**, 2996.
19. R. T. Ross, *J. Chem. Phys.*, 1967, **46**, 4590.
20. U. Rau, *Phys. Rev. B*, 2007, **76**, 085303.
21. T. Kirchartz, A. Helbig, W. Reetz, M. Reuter, J. H. Werner and U. Rau, *Prog. Photovoltaics: Res. Appl.*, 2009, **17**, 394.
22. O. D. Miller, E. Yablonovitch and S. R. Kurtz, *IEEE J. Photovoltaics*, 2012, **2**, 303.
23. G. Smestad, Ph.D Thesis, The Swiss Federal Institute of Technology (EPFL), 1994, 10.5075/epfl-thesis – 1263.
24. M. Carmody, S. Mallick, J. Margetis, R. Kodama, T. Biegala, D. Xu, P. Bechmann, J. W. Garland and S. Sivananthan, *Appl. Phy. Lett.*, 2010, **96**, 153502.
25. M. Carmody and A. Gilmore, 2011. Available: http://www.nrel.gov/docs/fy11osti/51380.pdf, [5 February 2014].
26. I. Schnitzer, E. Yablonovitch, C. Caneau and T. J. Gmitter, *Appl. Phys. Lett.*, 1993, **62**, 131.

27. A. Martí, J. L. Balenzategui and R. F. Reyna, *J. Appl. Phys.*, 1997, **82**, 4067.
28. T. Tiedje, *Appl. Phys. Lett.*, 1982, **40**, 627.
29. D. Azulay, O. Millo, I. Balberg, H.-W. Schock, I. Visoly-Fisher and D. Cahen, *Sol. Energy Mater. Sol. Cells*, 2007, **91**, 85–90.
30. T. Sueyoshi, H. Kakuta, M. Ono, K. Sakamoto, S. Kera and N. Ueno, *Appl. Phys. Lett.*, 2010, **96**, 093303.
31. W. L. Kalb, S. Haas, C. Krellner, T. Mathis and B. Batlogg, *Phys. Rev. B*, 2010, **81**, 155315.
32. P. K. Nayak, G. Garcia-Belmonte, A. Kahn, J. Bisquert and D. Cahen, *Energy Environ. Sci.*, 2012, **5**, 6022.
33. P. Würfel, in *Physics of Solar Cells*, Wiley-VCH Verlag GmbH, Weinheim, 2007, pp. I–XII.
34. J. C. Blakesley and D. Neher, *Phys. Rev. B*, 2011, **84**, 075210.
35. P. K. Nayak, G. Garcia-Belmonte, A. Kahn, J. Bisquert and D. Cahen, *Energy Environ. Sci.*, 2012, **5**, 6022–6039.
36. M. Liu, M. B. Johnston and H. J. Snaith, *Nature*, 2013, DOI: 10.1038/nature12509.

CHAPTER 18

Grid Parity and its Implications for Energy Policy and Regulation

MURIEL WATT*[a] AND IAIN MACGILL[b]

[a] School of Photovoltaics and Renewable Energy Engineering, University of NSW, Sydney NSW 2052, Australia; [b] School of Electrical Engineering and Telecommunications and Centre for Energy and Environmental Markets (CEEM), University of NSW, Australia
*Email: m.watt@unsw.edu.au

18.1   Introduction

18.1.1   Photovoltaics' Early Promise and Progress

The transformational potential of photovoltaics (PV) was recognized immediately upon the announcement by Bell Laboratories of the first silicon solar cell in 1954, with the *New York Times* proclaiming on its front page

> the beginning of a new era, leading eventually to the realization of one of mankind's most cherished dreams – the harnessing of the almost limitless energy of the sun for the uses of civilization.[1]

Here was a solid state device, with no moving parts, powered by nothing but sunshine and with no noise or emissions at the point of end use, that directly produced electricity. A commercial PV cell followed only a year later, albeit priced at over US$1800/W. Early applications were, therefore, highly

RSC Energy and Environment Series No. 11
Advanced Concepts in Photovoltaics
Edited by Arthur J Nozik, Gavin Conibeer and Matthew C Beard

specialized. The first solar powered satellite was launched in 1958, and a growing number of terrestrial applications emerged, such as power for navigation aids, telecommunications, monitoring and signaling devices.[2]

Many governments began to support PV research and development, as illustrated in previous chapters of this book. While there was an early focus on new materials, reliability and efficiency, the opportunity to achieve cost reductions and hence drive broader deployment was always a major motivation, particularly as the two oil crises of the 1970s focused attention on renewable energy alternatives to fossil fuels. Some 30 years after the first cell was built, the global PV market was some 20 MW per year.[3] While commercial markets were still being developed for off-grid applications, policy support from governments began to focus on PV for aid funded development projects. Many solar home system programs were deployed by governments, as well as the major development agencies, such as the World Bank. The aim of these projects was to provide power to communities with little chance of grid connection.

In the 1980s, policy support for off-grid applications in developed countries also began. Again the aim was to provide assistance to those who were unlikely to be connected to major grids in the short term. However, the policy reasoning behind such programs ranged widely, and included: reduced reliance on imported diesel, reduced pressure for grid extensions, support for local PV manufacturing and associated jobs, and local environmental aims. As these markets grew, the issues facing the PV industry focused more on cost and also on the environmental credentials of the technology, particularly energy payback times.[†] Note that while grid parity is not strictly relevant here – there is no grid after all – the growing competitiveness of PV against other options, including diesel generators and potential grid extension, were a key reasoning behind the policy support.

18.1.2   Photovoltaics Goes Mainstream

A key policy transition for PV was the decision by a number of countries, starting around two decades ago, to provide active support for development of the grid-connected PV market. In addition to the general drivers for renewable energy deployment, development of this market was seen to address other key issues: overcoming acknowledged energy market deficiencies, notably that greenhouse gas emissions, resource depletion and other benefits were not priced, that new technologies could not easily enter the established markets and that scale deployment was needed to reduce prices. The latter was a recognition that a new, modular technology like PV would

---

[†]For many years there was concern that the energy used to produce PV cells and modules exceeded the energy generated by the product in operation. While this may have been the case in very early pilot production, by the 1970's PV was already shown to be a net energy producer ([47]). Nevertheless, PV modules and systems continue to be judged by energy, and more recently emissions, payback times; a criterion not routinely applied to any other energy technology.

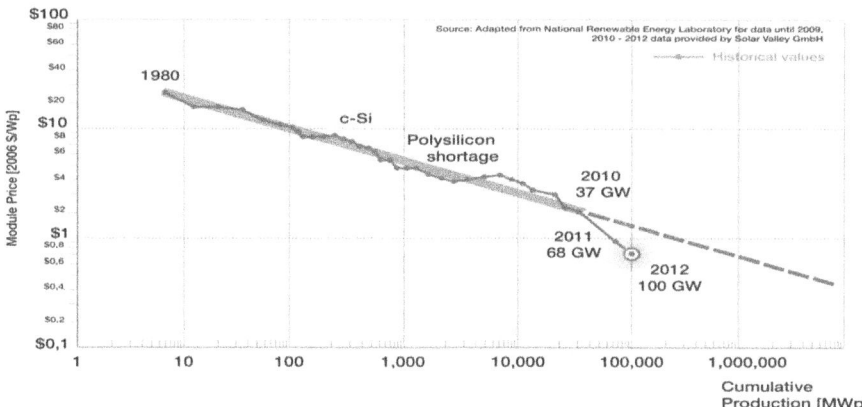

Figure 18.1   Learning curve for PV module prices 1980–2012.[4]

benefit from learning by doing and economies of scale in manufacture. The implicit, and in some cases explicit, target was 'grid parity', the point at which the cost of electricity from PV was the same as the cost of conventional grid supply from a customer perspective, at which time, many argued, government support would no longer be necessary.

PV policy support was generally financial, *via* payment towards up-front costs or electricity production. R&D support continued, and selected countries also provided manufacturing incentives. Up-front capital grant programs for grid connected PV began in Japan in the early 1990s and was followed from the 2000s by PV 'feed-in tariffs' in Germany for PV generated electricity fed back into the main grid. Both types of support led to major boosts in PV uptake, and policy replication in many other countries, as discussed in section 18.5.2. Other policy support has been provided *via* renewable energy portfolio standards, government procurement and tax incentives.

These policies have proven transformative. Underpinned by ongoing R&D progress, these large deployment programs have seen PV prices plummet, with growing global uptake (and cumulative production), as shown in Figure 18.1.

Over 100 GW is now installed world-wide, most of this over the past five years, with very significant levels of deployment in countries such as Germany (over 32 GW) and Italy (over 16 GW).‡ Even countries like Australia, with a relatively small installed capacity of around 3.2 GW by the end of 2013, has states with residential sector penetration levels over 25% and areas on the network with penetration levels exceeding 35% of owner occupied houses.[5]

---

‡PV systems are rated at their output under 1000 W of light at 25 degrees Celsius. The intensity of sunlight is not consistently at 1000 W, since it varies by location and cloud cover and increases from sunrise to a maximum at midday, and then decreases again. Hence, the capacity factor of flat plate, non-tracking PV systems is usually in the range 18–22%, meaning that they produce, on average, only 18–22% of their rated output. Over a year, the 100 GW of PV now installed would be expected to produce the equivalent of an 18–22 GW power plant operating constantly. Of course, no generating plant has a 100% capacity factor, with even coal and nuclear plant typically operating at capacity factors of less than 90%.

The associated publicity and visibility of PV have also facilitated widespread public knowledge and acceptance of the technology. As discussed in section 18.4.3, this is an important element in successful deployment of new technologies. Equally important to the direct financial incentives have been the associated development of supportive frameworks for deployment, including product standards and accreditation systems, obligations on network service providers to connect small PV systems, standardized connection procedures and net metering arrangements (by which, PV generation at a customer location offsets their electricity demand). This retail electricity market support framework has played a key, if somewhat invisible, role in facilitating uptake. Absence of any elements of these frameworks, and associated balance of system components and costs, have been identified both in the US and Europe as reasons for price discrepancy across regions or countries.[6,7]

18.1.3   Where next for Photovoltaics

The question is increasingly now one of what happens next. 'Grid parity' has been seen as a major policy goal, opening up the large grid-connected electricity market to PV. Depending on definition, grid parity is now close, or has already been reached in retail electricity markets for some customer segments in some locations. Supportive PV policies are now being wound back in numerous countries as PV system prices have fallen and PV support programs have driven levels of uptake far beyond that envisaged by the policy makers when the schemes were formulated. Increasingly, the view in some jurisdictions is that PV does not require additional financial support in key market segments, such as residential systems. Instead, the customer savings it can provide in terms of lower electricity bills from reduced 'net' consumption, and income from PV electricity exports, is sufficient to cover system capital costs. Or, in other words, grid parity has arrived. Of course, grid parity depends on the retail electricity tariffs that end users pay. In many jurisdictions, retail customer electricity tariffs may involve some fixed daily charge, but are primarily based on kWh consumption charges. This consumption may be charged at a flat rate (c/kWh), or based on the time of use (TOU). The revenue from these tariffs goes towards paying for the energy consumed but also the network infrastructure required to deliver it.

The competitiveness of PV in other market segments, such as commercial and industrial building applications, depends on the electricity tariffs available for businesses. In some countries these are less than residential electricity tariffs and also more heavily weighted to demand (kW peak) rather than energy consumption charges. Still, the falling prices of PV have opened up new markets here as well.

However, progress towards PV grid parity has raised a set of new issues and challenges for policy makers and electricity industry stakeholders. There are, of course, still market segments which are not yet competitive for PV without more direct policy support, such as, in most cases, utility scale plants. Indeed, the term grid parity is not generally used for PV, or any other technology for that matter, in the utility market. However, further falls in PV prices and changes in broader energy markets that push up conventional generation costs, such as carbon pricing, may well see PV growing more competitive over the coming decade. Programs such as the 'US Sunshot', aim specifically at reducing solar costs to US \$0.06/kWh by 2020, a level expected to be competitive in wholesale markets.[8]

PV deployment rates remain strong, even as support programs are reduced. Analysts project various installed capacity levels for the future: 330 GW by 2020[9] between 208 and 343 GW by 2016[10] and 465–665 GW by 2020.[11] European markets, which have dominated world growth over the past decade, are expected to stabilize while new markets develop in China, India, Africa, South America and the Middle East.

More generally, there is the issue of grid parity itself and what it actually means. Clearly, the highly variable and somewhat unpredictable nature of PV generation is not 'functionally equivalent' to a grid supply. Furthermore, the retail electricity tariffs faced by customers in many jurisdictions do not actually reflect the highly complex underlying economics of grid provision. Also, PV is raising significant challenges for existing utility business models and in the broader arrangements in areas as diverse as workforce training, energy consumer engagement in the electricity industry and planning frameworks for the built environment. PV is now highlighting the broader challenge of distributed resources within an electricity industry established to suit central generation and radial 'uni-directional' electricity flows.

In this chapter we explore the challenge for energy policy and regulation as PV transforms from a promising but still emerging clean energy technology whose many advantages were tempered by its high costs, to an increasingly competitive energy market competitor that brings novel and disruptive technical, economic, commercial and broader social characteristics into the electricity industry. In particular, we consider how the key issues for PV deployment have rapidly moved from PV technology and economics to those of electricity system integration and energy market structure. We therefore examine the broader policy and regulatory context within the electricity industry beyond explicit PV support itself.

The chapter is structured as follows. In the next section definitions of grid parity are explored; section 18.3 assesses past and projected PV cost trends; section 18.4 looks at the wider issues around deployment of new technologies; section 18.5 examines the broader policy context into which PV is placed; and section 18.6 looks at the implications of PV grid parity for established energy markets.

18.2 What is Photovoltaics Grid Parity?

The popular definition of PV grid parity is when the levelized cost of PV electricity is the same as, or lower than, the prevailing retail 'grid' electricity tariff being paid by a customer (and potential PV system deployer), not including the value of specific PV support mechanisms. Hence, the electricity grid is seen as a key reference point and measured by the costs to deliver electricity to the end-user; parity is judged over the longer-term costs of technology production, installation, maintenance and financing. Though not generally used for other electricity technologies, the term is widely used in the PV policy literature.[11–14] Most of the other key renewable energy technologies are deployed at large scale, and thus compete in established central energy markets where technologies connect into the transmission network, bid in \$/MWh for dispatch of energy, and networks are separately costed. PV can also operate in these large-scale markets, but the small-scale retail market, where connections are made to the distribution network, has been an important market component in many countries. It has been driven by a new category of individual investors who use grid electricity but also generate their own, sometimes referred to as 'prosumers',[15] and a new terminology was needed. Some of the key issues this terminology raises are discussed below.

Being able to explain cost effectiveness to this new group of energy investors has been critical to the marketing of PV and the concept of grid parity has therefore been useful. Traditional energy sector investors are well versed in discount rates, finance costs and the time value of money. For individual investors, comparing a high upfront, very low ongoing cost PV system against grid power, which has generally minimal upfront but significant ongoing costs, is not easy. There are also important uncertainty issues, including future electricity costs, future finance costs, whether or not finance is sourced *via* the home mortgage rate, and variable, weather dependent, PV output. Comparing the levelized cost of PV against current grid electricity prices is of course not an appropriate comparison. Strictly, the actual comparison would need to be with the levelized cost of the grid electricity displaced by the PV system over the system life, incorporating expected electricity price increases. The PV system owner also needs to be able to capture the full parity benefit; a point to which we will return.

This does mean that, in practice, calculating grid parity can be complex and requires significant assumptions including, particularly, the end-user's time value of money or implicit discount rate applied to upfront expenditure with longer-term annual returns. Other complexities can include:

- The different retail electricity tariff arrangements that energy users may be paying, whether they are on a flat kWh rate, have different rates at different times, have any demand (kW peak) component in their electricity bills, or are paying mostly on energy usage.
- What arrangements are in place for exported PV generation, whether they are under a gross feed-in tariff, and paid for all electricity generated, or whether they have net metering arrangements, where time varying PV generation is first used to meet local demand and is therefore only exported to the grid if it exceeds demand at any time; if the latter, whether they are paid the same rate as for purchased electricity, or whether a separate feed-in tariff applies.

18.2.1   Issues Around 'Grid Parity'

Although useful for consumers, the term 'grid parity' has also raised some more fundamental challenges and there is a growing appreciation of its limitations. One limitation is how it can over-simplify the actual nature of end-user decision making. For many energy users, the decision on whether or not to install PV will not be made on a strictly financial basis. For residential customers particularly, a PV system is purchased for a number of reasons that can include environmental concern, hedging against future electricity price rises, taking advantage of government incentives, meeting building energy ratings, increasing self-reliance, as well as cost effectiveness. For many customers, even if the PV salesperson can provide an indication of cost effectiveness, there are inherent uncertainties and risks attached to such analyses and the decision to buy will hinge crucially on these other factors. It is well appreciated that extremely cost-effective energy efficiency options may still not be widely adopted by energy users. Similar decision making limitations will also exist for PV. One that appears to play a key role is affordability, that is, whether or not the customer can easily raise the funds necessary to purchase, finance or lease the system. For many years, installed PV system prices in Australia stayed around AUD10 per W_p. At this price, customers needed to find AUD20 000 to fund a 2 kW_p system, which would provide say 50% of their electricity needs. This was a major household expenditure decision and relatively large incentives were usually needed to encourage investment. In Australia by 2013, a 2 kW_p PV system can be installed for less than AUD5000, even though subsidies have fallen substantially, making it much more affordable for average households. In addition, there are a growing number of PV companies offering leasing arrangements, where no up-front payments are needed, or who are selling PV electricity *via* systems installed on the customer's roof, or nearby, but where the company, not the customer, purchases the system.

Hence, PV has become a consumer appliance, widely available from home builders, electricians, homeware stores, electricity retailers and specialist PV companies, but now increasingly also readily available *via* lease or power purchase arrangements. In response, uptake levels have risen rapidly, mainly in the fixed and middle income household bracket, where rising electricity costs are causing concern and there is sufficient income to support PV purchase.

The definition of grid parity used above is of course only taken from the customer's viewpoint. As PV penetration levels have risen, more discussion is now beginning on how the grid itself should be valued if customers are relying on it for an electricity infrastructure service, but generating most of their own energy.

A fundamental problem here is that retail electricity tariffs are not economically efficient in terms of matching underlying electricity industry economics with the commercial 'price' signals being sent to market participants. Electricity industry economics are highly complex; the industry must balance supply and demand at all times and all locations on a dedicated electrical network. Moreover, many industry assets, including generation and networks, are highly capital intensive and are thus amortized over long periods. Furthermore, the electricity industry has significant environmental externalities that need to be appropriately priced into energy services in order to determine genuine societally economic prices. Economically efficient pricing would require time and location varying electricity prices, incorporation of environmental externalities, transparent treatment of social cross-subsidies and appropriate future prices to support investment.[16,17]

In reality, retail tariffs are generally fixed for some years ahead and, certainly for residential customers, often provided as a flat rate. Even the more cost reflective Time-of-Use tariffs, or network tariffs, such as peak capacity charges, are still far from economically efficient. Furthermore, these tariffs typically involve, often by design, significant cross subsidies between customer classes, and customers within these classes. This reflects the essential public good nature of electricity provision and recent industry restructuring towards more competitive arrangements have still largely failed to address these issues. The issues of environmental externalities are also still largely unaddressed, despite recent efforts in some jurisdictions, such as the introduction of carbon pricing.

Hence, energy users making decisions about PV deployment based on current and expected future retail prices and arrangements (for example, net metering and PV export payments) are not seeing the full societal costs and benefits of these decisions. For example, PV may be very attractive from the end-user/private investor viewpoint, using their private valuations and discount rates and taking into account their own view of externalities, such as climate change, social standing or aesthetics. From a societal viewpoint, the conclusions may be different, yet customers have no way of ascertaining the actual costs or benefits of their PV system to the wider market or society. For example, the benefits of universal grid access and standardized grid electricity tariffs may be threatened by individual PV customers paying less than their fair share into the common electricity pool, if their bill reductions do not actually reduce industry costs to the same extent. Alternatively, a decision not to deploy PV on the basis of low retail tariffs from an emissions intensive industry could mean that the environmental value of PV is not being appropriately factored into the decision.

A particular issue is that PV electricity may displace other energy sources but, while the PV system remains connected to the grid, it is not necessarily displacing the cost of that grid. As mentioned previously, the highly variable and somewhat unpredictable generation profile of a PV system is not functionally equivalent to a grid that meets the general expectation in developed countries of 'on demand' supply at all times and under all circumstances. This reliability relies on both significant generation reserves from a diverse mix of generation, and a robust grid. The issue of who pays for the grid is particularly problematic. Network expenditure is very much driven by peak network demands and PV systems may not reduce these peaks in many circumstances.[§]

Hence it has been argued that parity is not reached for any PV system, whether distributed or central grid connected, until PV electricity prices reach the wholesale electricity market price.[¶] This is equivalent to saying that, despite distributed PV providing power at the point of end use, its value is limited to the value of energy displaced from central generators long distances away. Grid operators also point to added costs of grid management as PV penetration rises. These may be countered by claims of network value which could be provided by PV, for instance when generation coincides with peak network loads, or where voltage support is needed.

When considering grid parity for central power generation, where PV is typically competing with coal, gas, large hydro, nuclear and more recently wind power, the parity equation may be simpler, but does depend on how the energy market is set up, and whether or not there are costs incurred to compensate for fluctuating output, seasonal variation, transmission and so on. Wholesale price parity seems likely to be reached without subsidy in some countries from 2015. However, there are large parts of the world reliant on diesel generation and mini-grids, where PV has already reached parity, with the same provisos as described for central plants. PV is already used as a fuel saver in many of these locations. For the purposes of this chapter, we will consider mini-grids to be 'off-grid' applications of PV and hence beyond the current discussion of grid parity.

The recent uptake of PV in many jurisdictions has occurred in a complex and changing context that involves other issues. In Australia, for example, high grid reliability is a strong political issue and is said to contribute significantly to grid costs.[18] Inappropriate regulation has also led to what could be over-expenditure on grid infrastructure, just when customers are embracing the new energy services offered by PV, solar water heaters, energy efficient homes and appliances and smart demand control. Even if sections of the grid end up as under-utilized stranded assets, it will need to be paid for, unless governments are willing to write off their investments. PV may not be the main contributor to reduced demand, but as utilities, regulators and governments struggle to adjust to new patterns of energy supply and use, PV is certainly in the spotlight, particular in areas where subsidies have been available. The advent of affordable storage will change some of these aspects, but will not resolve the general issue of who pays for the grid.

To conclude, grid parity has been an important policy and industry target for what has now been decades, and has played a useful role in motivating action and clarifying some of the opportunities that a distributed generation technology such as PV can provide. However, it has some significant

---

[§]Of course, it should be noted that customers causing these peaks are almost certainly also not paying appropriately for their actions given economically inefficient retail tariff arrangements.

[¶]In Australia, this can mean the difference between competing with a retail tariff of 25–30c/kWh or a wholesale price of 7–10c/kWh.

limitations that are becoming more apparent as PV succeeds in achieving mass acceptance in some market segments. As such, policy makers must, and are currently struggling to, adjust to a very new and different policy and regulatory context for PV, and distributed resources within the electricity system more generally. This context includes, critically, retail electricity tariff arrangements, but also broader retail market arrangements such access and connection.

The history of PV costs and prices, as well as projections of both PV and grid costs are discussed next, so that the likely market outlook can be examined.

18.3 Past and Projected Photovoltaics and Grid Cost Trajectories

18.3.1 Photovoltaics Costs

PV development, like other semiconductor based technologies before it, including computers and mobile phones, has been an extraordinary success, with continued price reductions opening up new markets and increasingly challenging established electricity generation technologies and distribution options. Over the last decade, and in the past 4 years in particular, PV module prices have fallen rapidly in response to technology and market developments. PV production costs have fallen, due to (1) technology advances, such as increased efficiency, new cell designs and new materials, (2) improved manufacturing processes, such as manufacturing scale-up, automation, reduced material use and reduced losses and (3) PV manufacturing and market subsidies provided by many governments in order to grow the local market or capture manufacturing bases. In response to the rapid market growth, the module market has become more competitive, with many new manufacturers and major investments in manufacturing capacity for both modules and upstream components, particularly silicon wafers.[||]

Taking a longer-term perspective, crystalline silicon PV module manufacturing costs have fallen more or less in line with a learning curve of 20–22% over the past three decades, although, as is clear from Figure 18.1, prices have been above the line for periods when demand outstripped supply and below the line when supply outstripped demand. It is not clear how long this learning rate will remain valid; as production technologies evolve it may move to a new rate. Separate learning curves are followed for thin film products, where the market is dominated by First Solar CdTe modules. New learning rates may also be established as the market moves to other technologies, including multi-junction cells and specialized PV products.

[||] Entry of Chinese companies into the PV manufacturing market has resulted in significant cost reduction, although some of this may well have occurred even if the investments had been made elsewhere, since much of the most recent price reductions have been driven by oversupply.

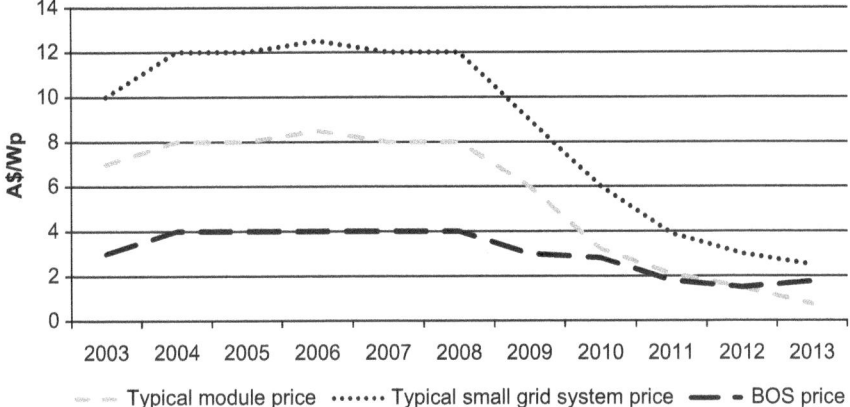

Figure 18.2 Module, balance of system and total costs for residential PV systems in Australia over the past decade (current AUD/W_p).[21]

As module costs drop, the balance of system costs are becoming a more significant component of the overall installed cost. These include the inverter, wiring, fuses and switches, racks, installation and, for large systems, land and associated costs. Over the last 3 years, inverter costs have also dropped rapidly, while market maturity and competition has seen installation costs drop in countries with sufficient market volume to encourage standardized delivery models.

Overall, balance of system (BOS) costs have also experienced a 19% to 22% learning rate[19] and installation costs have been reduced as both institutional (including permitting) and industry deployment processes have been streamlined to deliver higher volumes.[20] The latter has been especially noticeable in countries with high installed capacity. In Australia, module prices have fallen from AUD8 to AUD0.70 between 2008 and 2013, while non-module costs for small systems have fallen from $AUD4/W_p$ to $AUD1.8/W_p$ in 2013, as shown in Figure 18.2. Although installed system prices vary across the world, the trends have been similar. PV has moved from being one of the most expensive of the new energy technologies to one of the cheapest, particularly for end-use consumers. PV has therefore become one of the energy technologies of choice for consumers, and also for utilities in some regions, and thus had the highest deployment of any electricity generation technology in 2011.[10]

18.3.2 Grid Costs

The other side of the grid parity equation is of course the cost of grid supply. This varies across countries and regions, and is usually determined by a mix of technical, economic and political drivers. As discussed previously, costs are not always transparent, cross subsidies are common and the introduction of new supply-side technologies is traditionally carefully controlled. Electricity prices, and in developing regions, the promise of grid extension, are often used for political purposes. Even in developed countries, pricing rules can be changed in response to public pressure, while underlying policy drivers, including investment criteria, taxes and environmental aims, can change with a change in government. Technologies other than PV can also play significant roles in driving grid costs. For example, recent rapid retail grid electricity tariff increases in Australia have been largely driven by increased network expenditure resulting from a range of factors, but including the growing deployment and use of residential air-conditioning, which has been a major contributor to peak load growth. Nevertheless, these cost increases have been used by Australian State governments to reduce their renewable energy support programs, and consider other energy market regulations and tariffs to slow PV uptake. Hence, it can be difficult for PV proponents to know what costs they need to compete with, or what other requirements will be made before grid access is allowed. Again, it is clear that, no matter which definition is chosen, 'grid parity' can be a constantly moving target (Figure 18.3).

18.3.3   Implications for Residential Photovoltaics Systems

PV systems in residential markets connect into the distribution network and typically compete with retail electricity tariffs. In jurisdictions where customers are allowed to use PV electricity to displace their own electricity consumption, and can earn the same retail kWh tariff for exported PV power as they pay for imports from the grid (net billing) grid parity has now been reached in countries like Australia.

Reduced PV module prices alone were not sufficient to bring installed PV system prices down to the level we are now seeing. Innovation has also been required for balance of system components, design and delivery. Streamlined business models and standardized system design were first seen in Germany and are now available in Australia and elsewhere. They include 'bulk purchase' options, where, for instance, 50 or more houses in a neighborhood can sign up to access a special low rate. Standardized systems may be delivered and installed with no prior house inspection, just an internet map search. Installation times have fallen from 1 or 2 days per system to three systems per day. Of course, quality control, optimum orientation, potential for shading and other factors may not be as well considered as when individual system designs and prior site inspections were done. Also, the quality of components may have fallen in recent years. The implications of all this may not be known for some years. Good quality modules have come with 25 year warranties for some time, and many have survived much longer. So called '2nd' and '3rd' tier modules may only provide 15 year warranties and questions are already being raised about their validity now that so many manufacturers are closing down or being merged.

In general, the trend in many jurisdictions towards falling PV prices through both global and local industry scale-up and learnings, and increasing retail electricity tariffs, driven by a range of factors, suggests that PV deployment will increasingly be driven less by explicit PV policy support (which is being

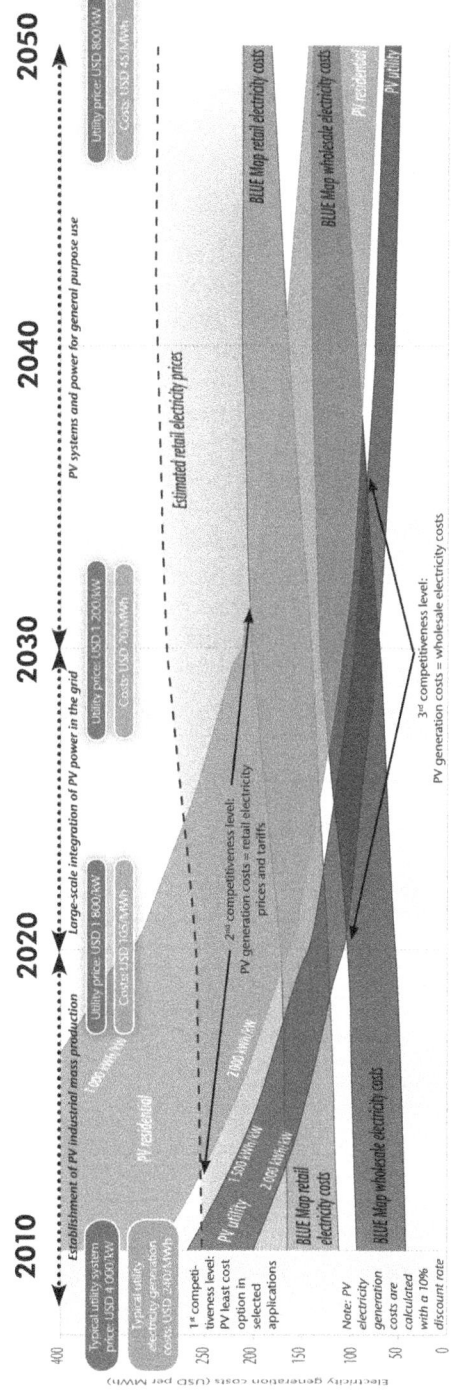

Figure 18.3 Generic trajectories of PV and electricity costs.[22]

removed or wound back in almost all jurisdictions) and more by the cost savings it offers energy-users. However, as discussed above, the very success of this market segment in some jurisdictions is driving growing discussions amongst utility regulators regarding the sustainability, or lack thereof, of current retail grid electricity tariff structures (particularly the mix of fixed, consumption and peak demand based charges) and PV connection arrangements.

18.3.4 Implications for Utility-scale Photovoltaics in Wholesale Energy Markets

For large-scale PV systems connected to transmission grids and thus competing with wholesale rather than retail electricity prices, cost competitiveness can be more difficult to achieve, even though there are economies of scale in deployment. This is illustrated for Australia in Figure 18.4. Wholesale market prices are generally more economically efficient than retail electricity tariffs, varying at least in part by time and location according to ongoing industry operation. A key limitation in many jurisdictions is the present lack of appropriate externality pricing associated with the significant environmental impacts of conventional fossil-fuel and nuclear generation as well as a range of well entrenched subsidies. The introduction of carbon pricing can assist project viability, but specific market support for PV is also likely to be required in many markets until 2020 or beyond, depending on prevailing electricity prices and solar conditions.

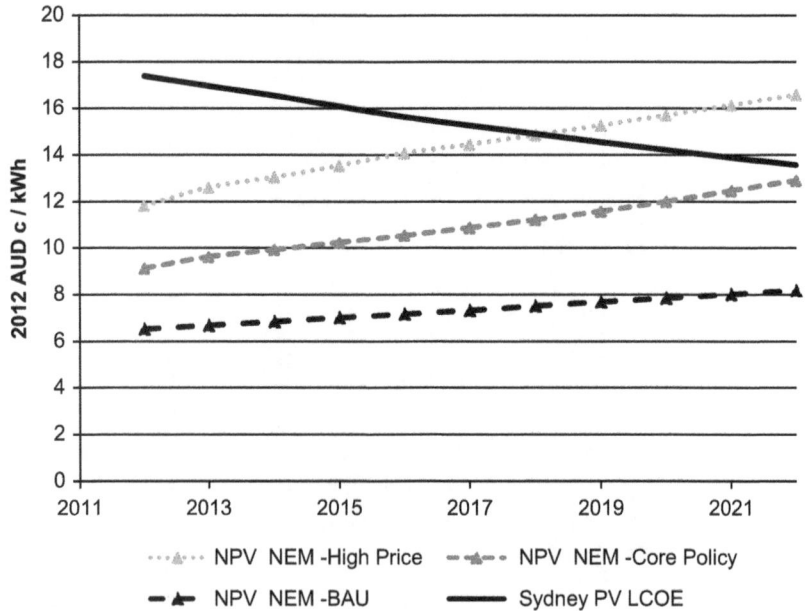

Figure 18.4 Central grid parity projections in the National Electricity Market (NEM), Australia using c-Si modules.

As in the residential market, in areas where a significant number of larger systems have been built, there have been noticeable learning effects and cost reduction associated with streamlined scheduling, site preparation and installation. Where local experience is lacking, per unit prices are noticeably higher. Companies such as First Solar have moved from CdTe module production into integrated project delivery, where they not only manufacture the module, but develop, install and operate the power station. They have developed their own streamlined approaches and could presumably transfer their knowledge to other module types in future, if they chose to enter the general market for PV system delivery.

In general, trends towards falling PV project costs and rising wholesale electricity prices are improving the cost competitiveness of PV, however, the gap is still significant in many jurisdictions. Policy makers hoping to facilitate utility scale PV therefore still need to be considering explicit support measures, as well as addressing some of the broader failings of wholesale electricity markets in terms of externalities and subsidies.

18.4 The Broader Context of Photovoltaics Deployment

Even where economic viability can be shown, the successful introduction of a new technology requires success in three other key interlinked areas: technology development, social acceptance and market access. This is illustrated in Figure 18.5. Problems with any one of these aspects will hinder deployment, as has been shown in many instances.[23,24]

Photovoltaics is arguably the most advanced of the new renewable technologies in terms of meeting all four criteria:

- PV technology is sufficiently developed to be in commercial production, although it is still in its infancy and has significant opportunities for further development
- PV is economically viable in a number of markets
- PV generally has high levels of public support
- Access has been provided to some markets through specific programs

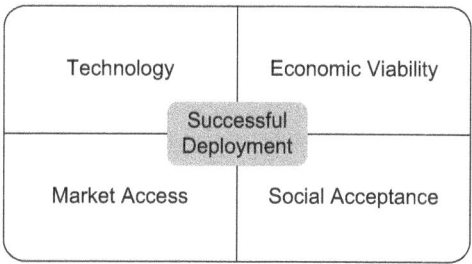

Figure 18.5 Criteria for successful deployment of a new technology, based on ref. 25.

Nevertheless, many hurdles remain; technology development and cost reduction are still required, many mainstream markets are still not accessible, while new technical and regulatory issues are now appearing as higher uptake levels are reached.

The grid parity concept addresses economic viability, keeping in mind all the complexities noted above. We now examine the issues surrounding the other three criteria.

18.4.1 Technology

PV could be considered a mature technology, having been sold commercially for over 40 years now, but it has only recently reached mainstream energy markets and so is expected to continue to develop over the next decade or so. As discussed in other chapters of this book, there are many promising new technologies, using nanotechnology, organic processes, different materials, combinations of materials, new manufacturing processes, and more. These offer continuing improvements in conversion efficiencies, as well as a range of purpose designed PV products for the future, which would add to the generic module market. The range of installation options also continues to develop, from new roof mounting structures, through various tracking and concentrating systems. Similarly, inverter technology is improving and changing, with a current trend to module inverters and so called 'ac panels' with potential for reduced mismatch and shading issues.

Although previous generations of PV modules were built to withstand rugged conditions in remote areas or in space, recent efforts to reduce costs for the grid-connect market have seen new module structures and materials in use, combined with much larger-scale manufacture and shipping. Issues of the longer-term life and performance of both these new module types and the balance of system components, remain a critical concern, as does the quality of installations, now that deployment efficiency is key to the success of PV businesses.

Of equal importance, technical issues are moving from a focus on the PV systems to their integration into the electricity system: the network itself and the electricity management system. This is a necessary but challenging change from a technical perspective, and will continue to evolve as grid penetration of PV increases.

18.4.2 Market Access

Although market access has links to economic viability, it goes beyond this to questions of interconnection and utility interface: rights and cost sharing, as well as availability of a trained and capable workforce, and responsibilities for ongoing maintenance and support. As will be discussed in section 18.6.2.2, we are starting to see instances of utilities placing hosting limits and export controls on PV. Local government planning laws, particularly those impacting solar access, heritage laws or streetscape regulations can also be an issue.[26] These can add high up-front costs to gain installation approval, or can prevent the use of PV altogether.

18.4.3   Social Acceptance

PV is now widely recognized by the public and generally viewed positively, with numerous surveys showing high levels of support for its continued development and deployment. For example, a recent survey of Australian householders showed that, despite already high levels of uptake, home owners in particular remained interested in PV, and were also potentially interested in PV-battery systems.[27] Some issues remain, and new ones will inevitably arise, as might be expected with a disruptive technology. These include concerns about electromagnetic interference from PV inverters, use of toxic materials in manufacture, roof reflections and product imports. Nevertheless, on the whole, PV as an alternative to buying utility electricity has a surprising resonance, with recent government supported deployment programs assisting in social acceptance and moving the technology rapidly along the Everett Rogers diffusion curve[28] from early adopters to the early majority.

18.5   A Changing Context for Photovoltaics Policy Support

18.5.1   The Rationale for Photovoltaics Policy Support

As noted above, while householders have a range of reasons for purchasing PV, governments have also had a range of reasons to provide incentives. These have included—depending on jurisdiction—a lack of local energy resources, greenhouse gas or general environmental policies and support for local manufacturing.[29] Depending on these reasons, the structure of the energy sector and the local political situation, different support strategies have been used. The key approaches used for PV are discussed below, although there have been variations and combinations also used. One especially useful aspect of government support programs is the associated ability to introduce industry and component standards and accreditation. This has been used in many countries as an important mechanism for industry development and remains a legacy even when subsidies are removed.

Of course, any subsidy is associated with market distortion. In some cases, PV businesses which had been established for decades, supplying off-grid and telecommunications markets, have been put out of business when generous subsidy programs have resulted in the establishment of many new businesses, taking advantage of the subsidy driven market and using new business models. There is always the danger that these new businesses will disappear when the subsidies are removed, leaving the industry worse off. Product cost is also impacted by the subsidy, with the potential for prices to remain high while demand is high. This can be countered by supply increases and associated competition, but because PV is an international commodity market, the overall world market needs to be high enough to drive this. The 2012–13 market over-supply, and resultant low prices, followed major capacity expansion coming online at the same time as key subsidy programs in Europe were reduced.

Subsidy programs are also associated with local boom and bust cycles for the industry, which can be more or less severe, depending on how well the programs are structured. Anticipation of a new program—for instance, if promised in an election campaign—can see existing markets stop suddenly, with disastrous consequences for industry. Restrictions placed on uptake, such as capacity limits, can result in very high initial uptake, sometimes accounting for the entire program in the first day it opens. This may leave entire businesses with no installations for the year. Imminent changes to the support available, whether or not built into the program up front, such as reduced rates once a particular capacity or cost has been reached, can also lead to surges in uptake, followed by market collapse for many months afterwards. In Australia, early notification of feed-in tariff program closure has resulted in significant capacity overshoot in many States, as well as associated problems such as rushed installs, poor siting and company collapses in the following market bust. These cycles inhibit the orderly industry growth which is needed for a sustainable industry, with training, infrastructure and other longer term needs.[30]

18.5.2   Photovoltaics Specific Policy Approaches to Date

Despite these problems, it is clear that government support programs have been instrumental in building the grid-connect PV market, and hence the PV industry expansion which was required to access the economies of scale and the industry innovation in both production and deployment. Despite current industry difficulties, which themselves indicate consolidation and maturation, PV is now a mainstream technology and will establish itself in diverse markets over the coming decade. Without government support programs, it would have taken many more decades for this to have occurred. A key question now is the potential role that these support programs may play in the changing PV grid parity context we face next. It is likely that support measures will become more focused and aimed at specialized products or services which better reflect the value PV can provide to the electricity sector.

18.5.2.1   Capital Subsidies

Capital subsidies have been and continue to be used for off-grid PV applications, but were first used in the 1990s in Japan to drive the grid market and to assist the local manufacturing industry. Public interest grew rapidly and subsidies were gradually reduced and eventually stopped before starting again recently. Government subsidies not only assisted with the cost, but also appeared to provide a level of public confidence in the technology and its take-up, because uptake continued steadily, even when subsidies were reduced to 10% of costs, but declined rapidly once they were removed altogether in 2005.[31]

Because PV systems, and hence the 25 years or so of electricity they will produce, are paid for up front, capital cost has been an issue until the more recent offer of solar leases or solar electricity supply contracts. Householders are used to paying for electricity as they use it, with the capital cost having been raised and amortized by governments and utilities. Hence capital subsidies were especially useful when PV was an expensive investment. They can be paid to the purchaser or to the installer, at a fixed amount or as a percentage of cost. While, ideally, degression rates and conditions should be built into the program, in practice this has only moderated, not stopped the market swings occurring around each change.

In addition to the tendency, perhaps common to all support programs, to keep prices high, capital cost subsidies have been criticized because they contain no reference to system performance. Hence, installers have little incentive to optimize PV design or orientation and sub-optimal installations may result.

In summary, there are some acknowledged failings in this approach, particularly given the widely varied quality and costs of PV equipment and system design and implementation. Falling PV capital costs also reduce the advantage of capital grants to assist in overcoming financing hurdles. In future, capital grants may usefully be directed to aspects of grid integration, including the addition of storage, ancillary services from inverters, and upgrading grids with the communications and controls systems necessary for the future introduction of a range of distributed energy options.

18.5.2.2 Feed-in Tariffs

PV feed-in tariffs (FiTs), or buy-back rates for all electricity sent back into the grid from PV systems, or just the portion in excess of a consumer's load and hence exported to the grid, have been another popular support mechanism. They have been used in Germany, in particular, as the key mechanism to support increased uptake of a range of renewables, although their origin is probably the US PURPA Act of 1978, which compelled public utilities to purchase power from independent producers, where this would be a lower cost option. In Austria, and then in Germany, enhanced feed-in tariffs, where the buy-back rate is higher than the prevailing tariff, were introduced in 2000, initially in conjunction with low interest loans, but then on their own, starting at around three times the prevailing tariff for all PV generation (*i.e.* gross metered). They have gradually been reduced as uptake rose and prices dropped. The tariff available at the time of installation is paid for 20 years. Germany also introduced manufacturing incentives, especially in East Germany, and now has the highest uptake of PV worldwide and did have one of the largest PV industries and levels of employment.

The FiT mechanism initially caused PV prices to rise due to high demand, and certainly was a key factor in keeping prices relatively high until about 2008, when substantial new capacity came online in China. With Chinese products now dominating the market and bringing prices down, political support for maintaining PV feed-in tariffs in Germany is waning, although it still has a substantial installer industry. Feed-in tariff support is now focused on maximizing own use and minimizing electricity exports in order to promote energy self-sufficient buildings, as well as to reduce network impacts.

Feed-in tariffs have been introduced in many other countries and with many variations,[32] but perhaps not as successfully as in Germany. In countries like Spain and Australia, feed-in tariffs were introduced then removed in a very short time, due to high uptake rates and concerns about the long-term payment liability. In Germany, the feed-in tariffs are paid *via* electricity bills, so that the payments are spread amongst all customers. However, there is often political concern about the impact on electricity prices, and in other countries, government budget allocations are often used to pay the feed-in tariff liabilities. As with any mechanism relying on annual budget allocations, the latter approach is also prone to abrupt changes.

Differentiated feed-in tariffs can be used to promote specific PV applications. In France, higher tariffs are paid for building integrated PV products, as a means of stimulating local industry. In Spain, initial support was also available for large utility-scale systems, and this section of the market boomed. Initial support in Germany also targeted farmers and other commercial enterprises, resulting in many mid-range (50 kW to 5 MW) systems being installed. These larger-scale installations were instrumental in ramping up market volume faster than was possible with smaller residential systems.

Feed-in tariffs are a good mechanism for encouraging installation quality and performance, because customers are focused on their system output. The tariff structure is critical to the impact on customer load patterns and hence networks. As shown in Western Australia,[33] net feed-in tariffs, where an enhanced tariff is paid for exported generation, leads to customers minimizing their own use during the times of PV generation, with the potential to increase evening peaks. In contrast, net metered arrangements where the value of own use is higher, encourages daytime use and is more likely to reduce peaks.

In summary, the flexibility of FiTs means that the impact of schemes on other electricity industry stakeholders can vary significantly depending on scheme design. Good policy designs can encourage appropriate system design and implementation, while not raising particular problems for the electricity industry due to factors such as integration with retail grid electricity tariff structures. Poor designs may, however, exacerbate these issues, as seen when high FiTs encourage socially costly system implementations. For utility scale systems, a major problem with conventional flat rate FiT arrangements is that they do not expose PV projects to wholesale market signals that reflect the time and location varying value of generation, or associated market signals such as those of ancillary services involved in securely and reliably integrating these renewables. As the PV industry matures, new FiT designs are needed which capture some of this complexity and add to the overall value of PV in the network.

18.5.2.3 Renewable Energy Targets or Portfolio Standards

Targets for renewable energy have been widely used by State and central governments as a means of encouraging diversification, cleaner generation and as a means of meeting greenhouse gas targets.[32] The mechanisms used to achieve the targets vary. They can be set as a MW target, perhaps ramping up at a set rate over time, as in Australia,[34] or as a percentage by a particular year. They can be technology neutral, have separate targets for a specified list of technologies, or provide different levels of support for each technology, depending on cost, stage of development or percentage penetration, for instance.[35] Many operate *via* a tradable renewable energy certificate scheme, which has a market independent of the main electricity market. Typically, each certificate represents 1 MWh of generation.

In Australia, small-scale PV is supported under the Renewable Energy Target *via* a deeming mechanism which allows small generators to create a set number of certificates, depending on location, for generation over 15 years. This therefore operates as an up-front capital rebate, although certificate prices are set in the market and can vary. However, it removes the necessity for small generators to bid into the central power market. Systems larger than 100 kW compete in the large-scale RET market and earn certificates after generation.

In the US, almost all states have some type of renewable-energy target or portfolio standard, with PV included in all programs. Though rules differ between states, trading of certificates is allowed between some jurisdictions.[36]

Targets often operate in conjunction with other support mechanisms, especially where certificate prices are low and would therefore not provide sufficient revenue for PV or other new technologies.

In summary, these approaches expose PV projects to wholesale or retail market signals while providing an additional production incentive for renewable generation. Depending on their structure, and that of the underlying electricity market, there are a range of potential issues associated with these designer markets in terms of the market complexities, potential exercise of market power and associated transaction costs.[37,38] Nevertheless, they remain a potentially strong driver for establishing renewable energy markets where conventional generation is entrenched and market access is otherwise difficult.

18.5.2.4 Tax Incentives

Tax incentives can operate at a number of levels, for instance, as exemptions from taxes, such as sales, payroll or import taxes; as tax deductions for individuals or businesses; or as tax credits.

Exemptions from tax have most commonly been used during the industry development phase, although more recently, taxes on imports are being used in Europe and the US to protect local industry against cheaper module imports, which result from industry support programs introduced by other governments. Tax deductions or credits are more focused on the end user and on deployment. The US has used tax credits of 30% for businesses as its key support mechanism for PV.[39] The tax credit is deducted directly from tax payable and any unused amount, if tax payable is less than the 30% credit, can be carried forward to the next tax year. For electricity utilities, this has been an important driver for large-scale PV systems, as it can be used to meet their Renewable portfolio standards.

As with any program funded from a government budget, the length the program runs is important. Longer running programs reduce the boom-bust cycles and allow more orderly industry development. Also, a tapered reduction is preferable to a complete stop, as weaning an industry off subsidies is always difficult.

18.5.3   Broader Policy Settings

Many PV related policy discussions neglect the broader context of energy policy for PV, and the key role that it can play in facilitating, or in some cases creating barriers to greater PV deployment. Some are PV specific, while others more general but relevant to PV in particular circumstances. They include policies covering some issues raised above including PV system equipment and design standards and associated accreditation processes.

18.5.3.1   *Building and Planning Codes*

Building codes can provide a useful indirect incentive to install PV. Energy rating schemes which provide credits for PV have been used in NSW, Australia, to meet the BASIX requirements for new buildings or substantial renovations. They have been particularly useful for larger homes, which otherwise struggle to meet the low energy use criteria. Commercial buildings often use PV to achieve higher energy ratings, especially as the latter can mean higher rentals.[40] Although PV obviously adds to the cost of a building, it appears also to add to resale value[41] especially in jurisdictions which require an energy rating on building sales.

Now that the PV module market is reasonably established and affordable, specialized building products are attracting more interest. Germany, France and other countries have provided higher incentives for building integrated products, including facades and windows, to encourage innovation. Dual function products are also of interest: electricity plus light, heat, shading.

Planning codes can also be useful in encouraging optimum orientation for new developments, which is critical to the opportunities then available for PV, as well as passive solar design and solar thermal devices. Solar access regulations are increasingly important to prevent future overshadowing problems as more building owners invest in solar products.

18.5.3.2   Grid Connection Arrangements

Jurisdictions can have widely varying arrangements for the connection of PV systems. Connection requirements for residential PV systems may be regulated and specify a process and charges associated with grid connection. Larger systems will often require more detailed, complex and often expensive connection arrangements that may involve network studies. This is standard practice for utility scale projects of all generation technologies. In some jurisdictions these processes are not always as transparent as they might be, and can pose significant barriers for small and medium-scale projects, such as commercial and industrial PV systems, which may be too large to fall within standardized arrangements available for residential systems.

Net and gross metering refers specifically to whether the output of a PV system is netted against the load associated with the energy user who has installed the system, or if all (gross) PV generation must be run through a separate meter. Note that a PV system installed at an energy user's premises behind the utility meter can be argued, in some regards, to represent an electrical appliance whose only impact on the grid is through reduced electricity consumption. In these systems, periods of time when PV generation exceeds the local load, of course, raise some complexities. In many jurisdictions, metering arrangements have depended on the PV policy support arrangements in place at the time of system installation.

These arrangements can, as noted earlier, have significant revenue implications, particularly in terms of PV offsetting electricity consumption which would otherwise have been charged for at retail tariffs. Payments made for PV exports can vary widely from full net metering where exports are paid at the standard retail tariff, as widely seen in the US, to tariffs which are higher than the retail tariff, as previously used in Germany, to low export tariffs that reflect estimates of wholesale electricity prices.[42] This is now becoming a critical issue for future PV prospects in many jurisdictions, particularly if rates can be changed at any time.

18.6   Implications of Photovoltaics 'Grid Parity' for Energy Markets

18.6.1   Implications of High Photovoltaics Penetration on Other Stakeholders

PV output is beginning to impact energy markets around the world, initially driven by the various support programs introduced over recent decades, but increasingly by grid parity and affordability. The technical and market impacts are similar, in some regards, to those seen with the take-up of energy efficiency, with the added complication of variable supply. In markets with significant PV uptake, impacts can now be clearly seen on both wholesale and retail energy markets.[43] The appropriate responses needed to facilitate continued uptake vary with grid architecture, market design and policy drivers.

For PV connected to transmission networks and contributing to central power supplies, there are both technical and market impacts. Wholesale energy prices may be reduced due to low market bids by PV plant operators, leading to displacement of other generators where market access is determined by merit order. This is possible because they have a zero marginal cost of generation. However, this impact will depend on market design and especially whether PV generators have access to other income streams, *e.g.*, renewable energy certificates. Low energy market prices can mean that PV displaces other generators, but will also make the PV plants less viable in the longer term.

For PV connected to distribution grids, as are most residential and many commercial systems, grid parity, followed by increasing PV uptake, can have a range of impacts and implications. These include reduced retail sales and reduced network revenue, especially where revenue streams are linked to kWh sales. If there are no tariff incentives offered to customers to change their load patterns, PV can lead to a peakier load profile, even if the overall peak is reduced, because effective daytime load is reduced while evening peaks remain.[33,44]

Because existing grids were not designed for two-way power flow, and also because many were built before electronic control systems were available, most grids are likely to need changes to grid management strategies and some level of upgrading to incorporate high levels of PV. The level of changes necessary as distributed generation penetration levels rise have been detailed in.[45] PV inverters of course can also offer some ancillary services, to offset PV impacts, or more generally. For grid operators, the reduced revenue does not reduce the need to maintain supply, with or without PV. If this is not adequately compensated for, the formulae used to calculate retail tariffs may need to change. High distributed PV penetration levels can also reduce demand from wholesale energy markets and thus reduce market prices, which impacts the income of other generators. When the energy markets operate on the basis of lowest market bids, all generators potentially receive a lower price when demand is reduced. This is of course good for the customer, if price reductions are passed through, but if not, it can result in windfall profits to retailers operating under regulated tariffs set from previous year price forecasts. However, it does also mean that the value of PV generation may reduce over time.

18.6.2   Emerging Issues and Responses

18.6.2.1   *Non-discriminatory Frameworks*

Although the rapid take-up of PV has highlighted the problems with centrally controlled electricity systems attempting to integrate distributed generation, it is important that policy responses should not be ad hoc and PV specific, but rather should aim to create new market structures suitable for the full range of possible distributed energy solutions likely to emerge in coming years.[46] This will create a future electricity industry where central and distributed generation, energy efficiency, demand management, storage and smart network controls will all be more closely integrated, providing diverse options and opportunities to facilitate more sustainable energy services delivery.

18.6.2.2 Hosting Conditions

Responses to the impacts discussed above vary with markets and regulatory arrangements. Unless connection and power purchase from new renewables is mandatory, as it is for instance under the German feed-in tariff laws, incumbent utilities may place restrictions on PV installations, either on total capacity and/or by limits to system size. Again, unless buy-back rates are stipulated, utilities may prohibit export of surplus PV electricity, not pay for it, or offer wholesale market prices only. They may also levy high up-front connection charges, change the customer's tariff structure to one with high fixed charges and low energy rates, or set technical requirements for management of PV output, including mandatory disconnect or reduction of output under set conditions.

New technical approaches may be needed to cater for increased variable supply, including staged uptake, mandatory curtailment when required and increased storage. There may also be a need for new market approaches to cater for the variable supply, including forecasting models and different market participant categories. In the Australian energy market, a new semi-scheduled category was introduced for wind generators, and this will also be available for large PV plants.

18.6.2.3 Net Metering

Net metering is widely used as a convenient and easily understood means of connecting small PV systems. PV system owners can displace their own electricity needs when the PV system is producing, call on the grid otherwise and export surplus power at the same rate as their purchasing tariff. It is mandated in most US states and widely used in other countries where PV is not specifically supported or is supported *via* capital grants rather than feed-in tariffs. In most jurisdictions, there is a capacity or percentage limit in place, beyond which new arrangements come into play. It provides a simple means of facilitating installations in the early years of market development.

As PV penetration levels increase, issues arise around the efficiency and equity of grid usage and the value of PV exports, specifically: how should the costs of the electricity network and retailing best be covered what value does distributed generation bring to the electricity industry and, in particular, what is the right value of PV electricity. Of course, with current electricity pricing typically comprising bundled energy and network charges which do not properly capture or reflect network value to the electricity user, it is difficult to allocate appropriate values and costs to PV, or any other end-user technology for that matter. Note that the current debate about PV's impact on network costs and revenue appears to show discrimination between treatment of electricity loads, such as air-conditioners, which can significantly increase network requirements, and of PV, which may either increase or decrease such network requirements depending on the particular context. Under common retail electricity tariffs, air-conditioners which create a large and highly correlated peak demand in very hot weather but only operate very occasionally create a significant cross subsidy between those households that do not have them, and those that do. This is largely a result of poor regulatory design, with economically inefficient retail tariffs and utility businesses reliant on increasing kWh sales and network expansion to maintain growing revenue. By contrast with PV, however, there has been relatively little discussion of the inefficiencies and inequities of such arrangements in many jurisdictions. Policy discussion would be better focused on increasing retail electricity tariff alignment with underlying electricity industry economics, as well as development of generic regulatory frameworks that will facilitate all distributed energy options in future, based on the overall value that they bring to the electricity industry including, of course and most importantly, its customers. Where technical and economic complexities and equity concerns preclude such tariff changes, policy makers must ensure that new technologies such as PV are not discriminated against.

18.7 Conclusion: Photovoltaics as Part of a Broader Transformation

Despite the issues arising with high PV penetration and energy markets, the major change, with as yet little known long term market implications, is the widespread availability of affordable PV appliances producing electricity at prices cheaper than grid power. A new category of 'prosumer' is developing,[15] with behavior and options evolving as new, largely electronically controlled, power supply, demand management, energy efficiency and storage options become available. PV is the leading technology in this rapid change and, like mobile phones, personal computers and other semiconductor based devices before it, will transform the sector. PV has many other advantages: it is easy to deploy in urban areas, has no moving parts, noise or local emissions and has low or even positive visual impact. Surveys around the world routinely find high levels of support for PV. This means that it will be much easier to deploy large amounts of PV than it has been for other new energy technologies, including wind, hydro and biomass. But most of all, it can be both a consumer device and a central generating technology. PV is, therefore, changing our energy future. The extent and nature of such change and the overall value that PV provides, however, will greatly depend on enlightened policy and regulatory arrangements within retail electricity markets, and more broadly, policies that facilitate this highly promising yet still emerging clean energy technology to achieve its potential.

References

1. J. Perlin, *Let It Shine: The 6000 Year Story of Solar Energy.* 2013.
2. S. Wenham, M. A. Green, M. E. Watt, R. P. Corkish and A. B. Sproul, *Applied Photovoltaics* 3rd edn. Earthscan, Abingdon, 2012.
3. P. Mints, *The photovoltaics industry: Against all odds, strong growth continues. SolarServer Magazine.* 4 November 2008.
4. I. Schwirtlich, *Strategies and Practices in the German Leading-Edge Cluster Solarvalley Mitteldeutschland: PV-Systems for the World Market.* 2013. Cleantech Made in Germany - Industry Symposium, Kuala Lumpur.
5. Australian PV Institute. Solar Map. [Online]. Australian PV Institute. Available: http://pv-map.apvi.org.au/. [16 January 2014].
6. K. Ardani, G. Barbose, R. Margolis, R. Wiser, D. Feldman and S. Ong, *Benchmarking non-hardware balance of system (soft) costs for US PV systems using a data driven analysis from PV installer survey results.* s.l. : NREL and LBNL, 2012.
7. K. Garbe, M. Latour, and P. M. Sonvilla, *Reduction of Bureaucratic Barriers for Successful PV Deployment in Europe.* s.l. : PV Legal, 2012.
8. UD Department of Energy. Vision and Goals. Sunshot. [Online] Available: http://www1.eere.energy.gov/solar/sunshot/mission_vision_goals.html. [31 January 2014.]
9. Global Data. Solar Photovoltaic (PV) 2012 - Global Market Size, Average Installation Price, Market Share, Regulations and Key Country Analyses to 2020. s.l. : Global Data, 2013.
10. EPIA, Global Market Outlook for Photovoltaics until 2016. s.l. : EPIA, 2012.
11. K. Aanesen, S. Heck, and D. Pinner, *Solar Power: Darkest Before Dawn.* s.l. : McKinsey & Co., 2012.
12. P. Lorenz, D. Pinner and T. Seitz, The economics of solar power. *McKinsey Quarterly.* June, 2008.
13. International Energy Agency, *Energy Technology Roadmaps: Report to the G8 2009 Summit.* International Energy Agency, 2009.
14. C. Breyer and A. Gerlach, *Prog. Photovoltaics,* 2013, **21**, 121–136.
15. R. Schleicher-Tappeser, *Energy Policy,* 2012, **48**, 64–75.
16. I. MacGill, *Energy Policy,* 2010, **38**, 3180–3191.
17. B. Elliston, I. MacGill and M. Diesendorf, Solar 2010: proceedings of the annual conference of the Australian Solar Energy Society.
18. Australian Productivity Commission. Electricity Network Regulatory Frameworks, Vols 1 & 2. Productivity Commission. 2013. Productivity Commission Inquiry Report, No. 62.
19. IPCC. *Renewable Energy Sources and Climate Change Mitigation.* s.l. : Intergovernmental Panel on Climate Change, 2012.

20. D. Feldman, *et al.* *PV Price Trends: Historical, Recent and Near-Term Projections*. s.l. : NREL and LBNL, 2012. Technical Report DOE/GO — 102012 — 3839.

21. M. Watt, Distributed Energy Markets - Introduction. *Distributed Energy Markets - Stakeholder Workshop*. Canberra, ACT, Australia : s.n., 19 July 2013.

22. International Energy Agency, *Technology Roadmap - Solar Photovoltaic Energy*. OECD/IEA, Paris, 2010.

23. NREL, Renewable Electricity Futures Study, Vol 2: Renewable Electricity Generation and Storage Technologies. Golden, Colarado: NREL, 2012.

24. International Energy Agency, Energy Technology Perspectives 2012 - Pathways to a Clean Energy System. s.l. International Energy Agency, 2012.

25. R. Haas, Marketing Strategies for PV Systems in the Built Environment, Task 7. s.l. IEA-PVPS, 2001.

26. Australian PV Association, *Best Practice Guidelines for Local Government Approval of Solar PV Systems*, June 2009.

27. L. Romanach, Z. Contreras and P. Ashworth, *Australian householders' interest in active participation in the distributed energy market: Survey results. Report no. EP 133598*. Pullenvale: CSIRO, 2013.

28. E. M. Rogers, *Diffusion of Innovations*. Glencoe Free Press, 1962.

29. M. Watt and H. Outhred, *Electricity Industry Sustainability: Policy Options*. Australian CRC for Renewable Energy, Murdoch, 1999.

30. New South Wales Auditor-General, *Special Report Solar Bonus Scheme*. Sydney: s.n., 2011.

31. IEA PVPS, *Trends in Photovoltaic Applications: Survey Report of selected IEA countries between 1992 and 2011. IEA-PVPS T1 — 21:2012*. s.l. : International Energy Agency PV Power Systems Programme, 2012.

32. REN21. *Renewables 2012 Global Status Report*. REN21 Secretariat, Paris, 2012.

33. B. Jones, N. Wilmot and A. Lark, *Study on the impact of PV generation on peak demand*. Western Power, Perth, 2012.

34. Australian Government. Department of Industry, [Online]. 2013. Innovation, Climate Change, Science, Research and Tertiary Education. Renewable Energy Target. Available: http://www.climatechange.gov.au/reducing-carbon/renewable-energy/renewable-energy-target. [4 September 2013.].

35. UK Department of Energy and Climate Change. [Online] 2013. The Renewable Energy Obligation. Available: https://www.gov.uk/government/policies/increasing-the-use-of-low-carbon-technologies/supporting-pages/the-renewables-obligation-ro. [4 September 2013.]

36. DSIRE. DSIRE Solar. DSIRE. [Online]. 2013. Available: http://www.dsireusa.org/solar/index.cfm?ee=0&RE=0&spf=1&st=1. [2 September 2013.]

37. I. F. MacGill, H. R. Outhred and K. Nolles, *Energy Policy*, 2006, **34**, 11–25.

38. I. F. MacGill and R. Passey, *CEEM Submission to the Senate Economics Committee Inquiry into the Renewable Energy Bill*. August 2009.

39. DSIRE, Business Energy Investment Tax Credit. DSIRE. [Online]. 2013. Available: http://www.dsireusa.org/incentives/incentive.cfm?Incentive_Code=US02F&re=1&ee=1. [27 August 2013.]
40. F. Feurst and P. McAlister, *New Evidence on the Green Building Rent and Price Premium*. Monterey, CA : s.n., 3 April 2009.
41. B. Hoen, R. Wiser, P. Cappers and M. Thayer, An analysis of the effects of residential PV energy systems on home sales prices in California. Environmental Energy Technologies Division, Lawrence Berkely National Laboratory, Berkeley, CA. 2011.
42. S. Oliva and I. MacGill, Assessing the Impact of Household PV Systems on the Profits of All Electricity Industry Participants. San Diego, Proc. IEEE PES2012. (Accepted for publication 2013).
43. P. Kind, Disruptive Challenges: Financial Implications and Strategic Responses to a Changing Retail Electric Business. Edison Electric Institute, 2013.
44. M. Watt, R. Passey, F. Barker and J. Rivier, Newington Village - An Analysis of PV Output, Residential Load and PV's ability to reduce peak demand. Centre for Energy and Environmental Markets, University of NSW, Sydney, 2006.
45. A. P. L'Abate, G. Fulli and S. D. Peteves, The Impact of Distributed Generation on European Grids. *Cogeneration and On-Site Power Production*. [Online]. 2008. Available: http://www.cospp.com/articles/print/volume-9/issue-3/features/the-impact-of-dg-on-european-electricity-grids.-printarticle.html. [17 January 2013.]
46. R. Passey, M. Watt and N. Morris, The Distributed Energy Market: Consumer and Utility Interest and the Regulatory Requirements. Australian PV Association, 2013.
47. B. Richards and M. Watt, *Renew. Sustain. Energy Rev.*, 2007, **11**, 162–172.

Subject Index

Note: Page numbers in *italics* refer to figures and tables.